Oxidation of Organic Compounds by Dioxiranes

Oxidation of Organic Compounds by Dioxiranes

WALDEMAR ADAM
CONG-GUI ZHAO
CHANTU R. SAHA-MÖLLER
KAVITHA JAKKA

JOHN WILEY & SONS, INC., PUBLICATION

Published by John Wiley & Sons, Inc., Hoboken, New Jersey

Published simultaneously in Canada.

Library of Congress Catalog Card Number: 42-20265
ISBN 978-0-470-45407-7

Printed in the United States of America

10 9 8 7 6 5 4 3 2 1

CONTENTS

FOREWORD

Chemical synthesis is an intellectually and technically challenging enterprise. Over the many decades of progress in this discipline, spectacular advances in methods have made once intimidating transformations now routine. However, as the frontier advances and the demands for ready access to greater molecular complexity increases, so does the sophistication of the chemical reactions needed to achieve these goals. With this greater sophistication (and the attendant expectation of enhanced generality, efficiency, and selectivity) comes the challenge of adapting these technologies to the specific applications needed by the practitioner. In its 66-year history, *Organic Reactions* has endeavored to meet this challenge by providing focused, scholarly, and comprehensive overviews of a given transformation.

By any yardstick, the oxidation of organic compounds is one of the most important methods for the introduction and manipulation of functional groups. No fewer than 24 of the 239 chapters in the *Organic Reactions* series are dedicated to oxidation in one if its manifold forms. Among the most modern of these transformations is the use of the fascinating family of oxidants known as dioxiranes. Although humble in size, dioxiranes can effect a remarkable array of oxidative processes by transferring one oxygen atom to organic substrates. In addition, to their virtuosity, dioxiranes can be generated catalytically and produce no toxic waste stream as part of the oxidation process. Moreover, some of the most exciting recent developments involve the enantioselective introduction of oxygen atoms via chiral dioxiranes.

The *Organic Reactions* series is fortunate to have published two chapters on dioxirane oxidations, one in Volume 61 on alkene oxidation and one in Volume 69 on oxidation of all other organic substrates. Both of these comprehensive chapters were authored by one of the internationally recognized leaders in this field, Prof. Waldemar Adam and his coworkers. Although many reviews and book chapters have been written on dioxirane oxidations, remarkably, there is no single book dedicated to this important topic. Thus, in keeping with our educational mission, the Board of Editors of *Organic Reactions* has decided to combine these two chapters to create the definitive resource for dioxirane oxidations. In addition, to keep pace with the rapid development of this field, Professors Adam and Zhao have compiled updated lists of references for each of the chapters that bring the literature coverage up to June 2008. These lists are appended at the end of each chapter and are organized by the Tabular presentation of the different classes of oxidation.

The publication of this book represents the third soft cover reproduction of *Organic Reactions* chapters. The success of the first two soft cover books ("The Stille Reaction" taken from Volume 50 and "Handbook of Nucleoside Synthesis" taken from Volume 55) has convinced us that the availability of low-cost, high-quality publications that cover broadly useful transformations is addressing an unmet need in the litarature of organic synthesis. Thus we will continue to identify appropriate candidates for the compilation of such individual volumes as opportunities present themselves.

Scott E. Denmark
Urbana, Illinois

PREFACE AND ACKNOWLEDGMENTS

Although the dioxirane structure was proposed as an intermediate in the Baeyer-Villiger oxidation already in 1899, the chemistry of dioxiranes did not commence until 1974, when Montgomery observed the ketone-catalyzed decomposition of potassium monoperoxysulfate (Oxone®). Shortly after this pivotal discovery, Edwards inferred that the plausible intermediate in this reaction should be the dioxirane molecule. A microwave structure determination confirmed the unusual three-membered-ring peroxide. In the next thirty-five years, the worldwide chemical community witnessed an exponential development of dioxirane chemistry. So far well over *one thousand* research articles have appeared that discuss the structure, properties, reactivity, selectivity, theoretical aspects, and synthetic applications of these highly reactive, yet mild oxidants. Through these efforts, dioxiranes, especially dimethyldioxirane (DMD), have become indispensable oxidants for the synthetic organic chemist.

The high popularity of these novel compounds is due to the fact that dioxiranes constitute an ideal oxidant in many respects. These reagents perform a variety of versatile oxidations, exhibit unusual reactivity, and conduct the oxygen transfer highly chemo-, regio-, and stereoselectively. They are readily prepared from a suitable ketone, such as acetone, and Oxone®, which are cheap yet environmentally benign bulk chemicals. Furthermore, since the dioxirane is reduced to the ketone precursor during the oxygen transfer, the oxidation may be run catalytically.

Dioxirane oxidations are carried out either with the isolated dioxirane (as a solution) or with the in-situ-generated dioxirane from a suitable ketone precursor. The advantage of using isolated dioxirane solutions is that the oxidations can be performed under strictly neutral and anhydrous conditions, which allows the isolation of some very elusive and sensitive oxidation products. This method is limited because the oxidation has to be carried out on a small scale due to the low dioxirane concentration in the isolated solution. Such problems are readily circumvented by the use of in-situ-generated dioxiranes, which may be scaled up. Although the in-situ mode cannot be applied to substrates and products that are sensitive to water, the oxidation can be conducted catalytically. The catalytic mode of operation has become important for the dioxirane-mediated asymmetric oxidations, since the ketone precursors are generally not readily available. Dioxiranes can be employed to oxidize heteroatoms, epoxidize π bonds, and insert into C-H and Si-H bonds. The difficulty of the oxidation increases in that order for these electrophilic oxidants.

The book contains two chapters that cover the diverse oxidations that have been reported. The first chapter comprises the dioxirane oxidations of alkenes, the most studied oxygen-transfer reaction. The second chapter covers all of the remaining types of dioxirane oxidations, namely heteroatom substrates, saturated groupings, and organometallic compounds. For each specific type of oxidation, the presentation begins with a brief discussion of the reaction mechanism, followed by the scope and limitations of the oxidation under consideration, together with a brief comparison with other reported oxidation methods. Our incentive is to give the reader a general idea of the usefulness of these oxidants in sufficient detail so that an intelligent choice may be made, especially by those who use this oxidant for the first time. In the case of the epoxidations, since the reactivity of the alkene substrates and the selectivities of the oxygen-transfer process are critical in making a proper choice, we cover the scope and limitations individually. Similarly, the scope and limitations of the oxidations of alkynes and arenes, diverse heteroatoms, alkanes and silanes, and organometallic compounds are discussed separately in view of the different nature and/or outcome of these oxygen-transfer processes.

In the sections on experimental conditions, some general guidelines on how to select the reaction parameters are given. Some representative experimental procedures are then provided that should be helpful in carrying out these oxidations. Whereas for a given oxidation the reaction conditions may still need to be optimized, these procedures should help the readers to understand the most important features of these experiments. Finally a tabular survey of the reported dioxirane-mediated oxidations is collected. Chapter 1 contains nine tables, whereas Chapter 2 comprises as many as 21 tables. Most of these tables are organized according to reaction types. For the same type of oxidation, if the substrate, selectivity, or the reaction conditions are of crucial importance to the oxidation, then these cases are presented in separate tables. Reaction conditions, such as temperature, reaction time, solvent, are specified explicitly. The product structures and their yields, as well as the diastereomeric ratios and enantiomeric excess values, if available, are presented in these tables. In each of these tables, the reactions are arranged in ascending order of carbon atoms of the substrates to facilitate browsing.

Since dioxirane chemistry has expanded quickly in recent years, an update of the most recent publications is also provided on the work that has appeared after these two chapters were published. Particular emphasis is placed on synthetic applications and asymmetric epoxidations to apprise the reader of the most recent advances. These literature references have been collected in separate sections according to the sequence of the tables in the Tabular Survey section of these two chapters. In each of the sections, the individual citations have been arranged in alphabetic order of the author names.

Although there are several excellent and extensive review articles available on dioxirane oxidations, most of these are of a specific rather than general character. The purpose of this book is to summarize all the past achievements in the field and provide the reader with a detailed account of the experimental procedures. We believe that the elaborate tabular compilations should be of help to all organic chemists engaged in oxidation chemistry, whether of mechanistic or synthetic interest.

ACKNOWLEDGMENTS

Generous financial support of the *Deutsche Forschungsgemeinschaft (Schwerpunktprogramm "Peroxidchemie: Mechanistische und Präparative Aspekte des Sauerstofftransfers"* and *Sonderforschungsbereich SFB 347 "Selektive Reaktionen Metall-aktivierter Moleküle")*, the *Fonds der Chemischen Industrie*, the *NIH-MBRS SCORE Program* (Grant No. S06 GM 08194; 1SC1GM082718-01A1), the *Welch Foundation* (Grant No. AX-1593) and the DAAD *(Deutscher Akademischer Austauschdienst)* are gratefully appreciated. We also thank Mrs. Ana-Maria Krause for her help in preparing the tabular and graphical material.

C.-G. Zhao
San Antonio, TX USA

W. Adam
Rio Piedras, PR USA

CHAPTER 1

DIOXIRANE EPOXIDATION OF ALKENES

WALDEMAR ADAM, CHANTU R. SAHA-MÖLLER, AND CONG-GUI ZHAO

Institut für Organische Chemie, Universität Würzburg, D-97074 Würzburg, Germany

CONTENTS

Oxidation of Organic Compounds by Dioxiranes, by Waldemar Adam, Cong-Gui Zhao, Chantu R. Saha-Möller, and Kavitha Jakka

ACKNOWLEDGMENTS

Generous financial support of the *Deutsche Forschungsgemeinschaft (Schwerpunktprogramm: Peroxidchemie)* and the *Fonds der Chemischen Industrie* and a doctoral fellowship for C.-G. Z. from the *Deutscher Akademischer Austauschdienst (DAAD)* are gratefully appreciated. The authors also thank Mrs. Ana-Maria Krause for her help in preparing the tabular and graphical material.

INTRODUCTION

An ideal oxidant should be highly reactive, selective, and environmentally benign. It should transform a broad range of substrates with diverse functional groups, preferably under catalytic conditions, and be readily generated from commercially available and economical starting materials. Of course, such an ideal oxidant has not yet been invented; however, the dioxiranes, which have risen to prominence during the past few decades, appear to fulfill these requirements in many respects. These three membered ring cyclic peroxides are very efficient in oxygen transfer, yet very mild toward the substrate and product. They exhibit chemo-, regio-, diastereo-, and enantioselectivities, act catalytically, and can be readily prepared from a suitable ketone (for example, acetone) and potassium monoperoxysulfate ($2KHSO_5 \cdot K_2SO_4 \cdot KHSO_4$, Caroate®, Oxone®, or Curox®), which are low-cost commercial bulk chemicals. Throughout the text we shall use $KHSO_5$ to specify this oxygen source, rather than refer to one of the commercial trade names.

Isolated dioxiranes (as solutions in the parent ketones) perform oxidation under strictly neutral conditions so that many elusive oxyfunctionalized products have been successfully prepared in this way for the first time. Epoxidations, heteroatom oxidations, and X-H insertions constitute the most investigated oxidations by dioxiranes. An overview of these transformations is displayed in the rosette of Scheme 1. These preparatively useful oxidations have been extensively reviewed during the last decade in view of their importance in synthetic chemistry.[1-12]

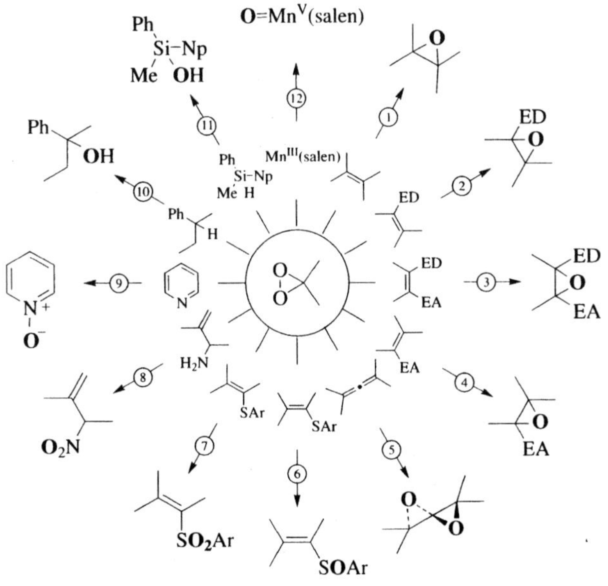

Scheme 1. An Overview of Dioxirane Oxidations (Np = 1-Naphthyl, ED = Electron Donor, EA = Electron Acceptor)

The present review deals mainly with the epoxidation of carbon-carbon double bonds [π bonds in simple alkenes and with these electron donors (ED), electron acceptors (EA), and with both ED and EA; examples 1–4 in the rosette] with either isolated or in situ generated dioxiranes (Eq. 1). In view of the vast amount of material on alkene oxidation (ca. 400 references), the epoxidation of the double bonds in cumulenes (allenes, acetylenes) and arenes is covered in a separate chapter, together with the oxidation of heteroatom functionalities (nonbonding electron pairs; examples 6–9 in the rosette), X-H insertions (σ bonds; examples 10–11 in the rosette), and transition-metal complexes (example 12 in the rosette).

$$R^1R^2C{=}CR^3R^4 \xrightarrow{\text{dioxirane } (R^5, R^6)} R^1R^2C(O)CR^3R^4 + R^5C(O)R^6 \qquad \text{(Eq. 1)}$$

MECHANISM

Dioxirane epoxidation proceeds strictly with retention of the initial alkene configuration. An impressive and mechanistically valuable example is the cis/trans pair of cyclooctenes (Eq. 2), of which the cis diastereomer leads exclusively to the cis epoxide and the trans congener to the trans epoxide in high yields.[13,14]

$$\text{cis- or trans-cyclooctene} \xrightarrow[\text{acetone, } CH_2Cl_2, \ 20°, \ 0.2 \text{ h}]{\text{dimethyldioxirane}} \text{cis-epoxide } (>99\%) \text{ / trans-epoxide } (>99\%) \qquad \text{(Eq. 2)}$$

On the basis of the relative reaction rates of cis and trans open-chain alkenes (cis alkene faster than trans alkene by a factor of ten) in the dimethyldioxirane (DMD) oxidation, a spiro transition state was suggested,[15,16] which is also supported by theoretical work.[17–23] In this oxygen transfer, the plane of the peroxide ring is oriented perpendicular to, and bisects the plane of the π system in the olefin substrate (Scheme 2, right). For this preferred geometry, DMD approaches the cis olefin from

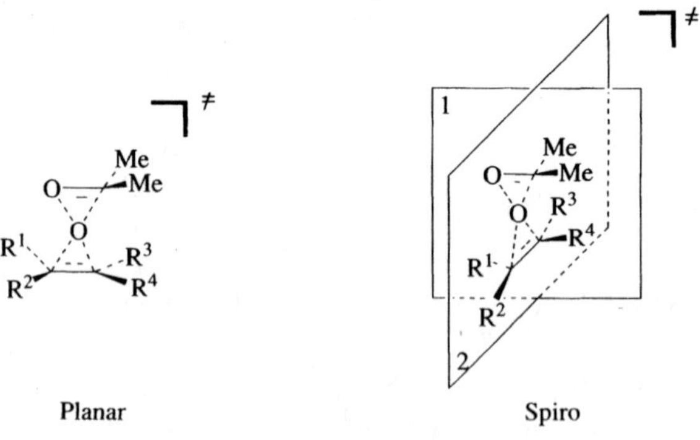

Scheme 2. Planar and Spiro Transition-State Structures for the DMD Epoxidation of Alkenes

the unsubstituted side, which avoids steric interaction between the methyl groups and the alkyl substituents of the double bond. The concerted spiro transition-state structure also explains the results of enantioselective epoxidations by optically active dioxiranes.[24,25] An exception is given in Scheme 3, for which the planar transition-state structure is favored over the spiro alternative.[26]

Ph OSiMe$_3$ Bu-t → ($KHSO_5$; CH_3CN, H_2O, $NaHCO_3$, 0°, 3 h) → (HCl; MeOH, rt) → Ph–CH(OH)–C(O)–Bu-t (27%) 52% ee

Planar Favored

Spiro Disfavored

Scheme 3. Planar and Spiro Transition-State Structures

The fact that no cis/trans isomerization is observed in the oxygen transfer speaks against the earlier proposed diradical mechanism.[5,27] Although the high stereoselectivity of DMD epoxidations may be accommodated by proposing much faster collapse of the resulting intermediary diradical than isomerization through bond rotation, it would be difficult to rationalize the established steric effects (i.e., cis olefins react faster than the trans isomers).[15,16] For the diradical mechanism one would expect that the relative rates of cis alkenes and trans alkenes should be nearly the same for such an end-on attack. Moreover, as stated above, epoxidation of trans-cyclooctene with DMD yields solely the corresponding trans epoxide (Eq 2).[13,14] In view of the high strain energy in trans-cyclooctene (23.2 kcal/mol),[28] cis/trans-isomerization would be expected, should the diradical intermediate intervene.[13] Furthermore, the lack of cyclopropylcarbinyl rearrangement in the epoxidation of 1-vinyl-2,2-diphenylcyclopropane by DMD (Scheme 4) as an ultra-fast radical clock ($k_r^{25} \approx 5 \times 10^{11}\ s^{-1}$) clearly speaks against the involvement of radical intermediates in DMD epoxidations.[13]

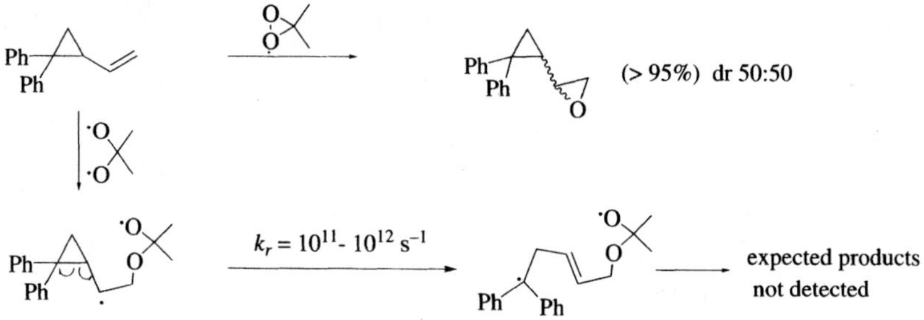

(> 95%) dr 50:50

$k_r = 10^{11}$- 10^{12} s^{-1}

expected products not detected

Scheme 4. Radical-Clock Probe for the Mechanism of DMD Epoxidation

Despite this convincing experimental and theoretical evidence in favor of the concerted oxenoid mechanism for dioxirane epoxidations, the appreciable amount of allylic oxidation of α-methylstyrene to 2-phenylpropenol and 2-phenylpropenal by DMD has been interpreted in terms of a radical mechanism.[29,30] However, when radical species are scrupulously avoided by operating in the presence of molecular oxygen (inhibitor), keeping the reaction temperature below 50° (no homolytic decomposition of the DMD), and avoiding radical-generating solvents such as $BrCCl_3$, the DMD oxidation of α-methylstyrene gives exclusively the expected epoxide.[13]

SCOPE AND LIMITATIONS

With dioxiranes, either isolated or in situ generated, different types of alkenes [i.e., unfunctionalized, electron-rich (with electron donors) and electron-poor (with electron acceptors)] have been successfully epoxidized. As a solution in acetone, DMD (isol.) is the most often used dioxirane owing to its convenient preparation and low cost. While methyl(trifluoromethyl)dioxirane (TFD) is much more reactive than DMD, applications are limited because of the high cost and volatility of trifluoroacetone. With DMD (isol.), the scale of the reaction is usually limited to no more than 100 mmol, since the DMD solution in acetone is quite dilute (ca. 0.08 M). In the case of TFD (ca. 0.6 M), the prohibitive cost of trifluoroacetone precludes scales $>$10 mmol. When a large-scale preparation is desired, the in situ mode [DMD (in situ)] is recommended, for which both biphasic[31–34] and homogeneous[35,36] media are available. Since the in situ oxidations are carried out in aqueous media, the substrate and the oxidized products must be resistant to hydrolysis and persist at temperatures above 0°. The in situ mode can also be carried out by applying a catalytic amount of ketone,[37–40] which is especially important for enantioselective epoxidations (see the Section on Enantioselectivity), since the optically active dioxiranes cannot be readily isolated.

Epoxidation of Unfunctionalized Alkenes

Some typical examples are collected in the rosette of Scheme 5. The epoxidation of unfunctionalized alkenes by dioxiranes, especially by DMD, has been mainly studied for elucidation of the reaction mechanism. The degree and pattern of

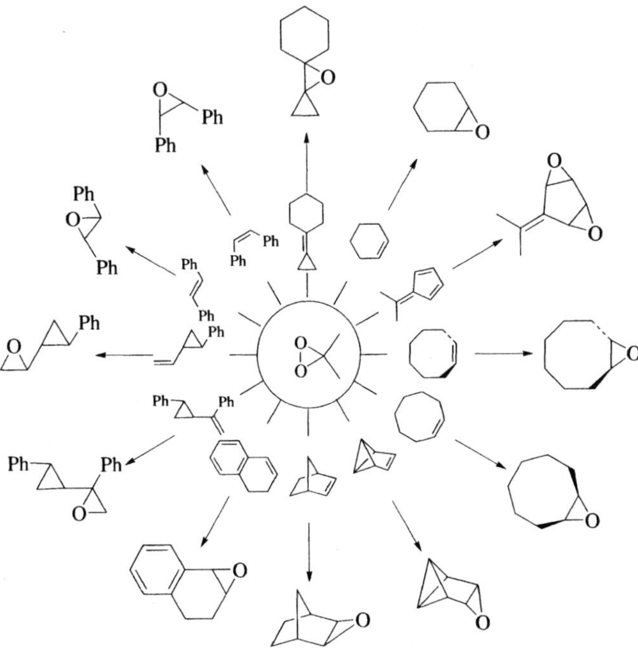

Scheme 5. Dioxirane Oxidation of Unfunctionalized Alkenes

alkyl substitution on the double bond control the reaction rate, with the more highly substituted alkenes being more reactive. The reactivity order is tetra > tri > di > mono (Table A), as is generally observed for electrophilic reagents.[15]

Steric factors also play an important role (e.g., cis alkenes are more reactive than the corresponding trans isomers).[15,16] An important exception is the cis/trans-pair of the cyclooctenes (Eq. 2), for which the trans isomer reacts about 100 times faster than the cis isomer because of ring strain.[13,14] Thus, quite generally, strained double bonds display the highest reactivity toward DMD. This fact is valuable for synthetic purposes since essentially quantitative yields of epoxides can be readily obtained with either isolated or in situ generated dioxiranes. In the latter reaction, the epoxide must resist hydrolysis. An excellent example is the epoxidation of benzvalene with DMD (isol.) at −30° (Eq. 3), which affords a rather labile epoxide in good yield.[41] This constitutes the most important advantage of DMD (isol.) since the oxidation is performed under strictly neutral and nonhydrolytic conditions, at low temperatures, with acetone as the only byproduct.

Me_2CO, Et_2O, N_2, –30°, 3 h (62%) (Eq. 3)

TABLE A. Relative Rate Constants (k_{rel}) of the Epoxidation of Alkyl-Substituted Alkenes by DMD (isol.) in Acetone at 25°.[15]

Alkene	k_{rel}
	0.5
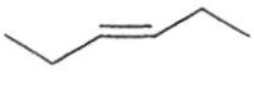	1.0[a]
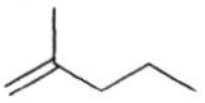	4.6
	8.5
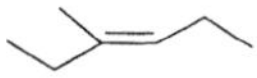	16.1
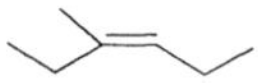	24
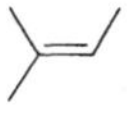	36
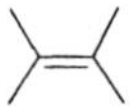	106

[a]The absolute rate constant is $0.067 \pm 0.001\ M^{-1}s$

Acid-sensitive epoxides can also be prepared by applying the in situ mode, but the pH of the medium must be carefully controlled with buffers. For example, the oxaspiropropane is obtained by epoxidation of the corresponding cyclopropylidenecyclopropane (Eq. 4).[42] Usually such highly strained and hydrolytically labile oxaspiropropanes readily rearrange to the respective cyclobutanones.[42,43]

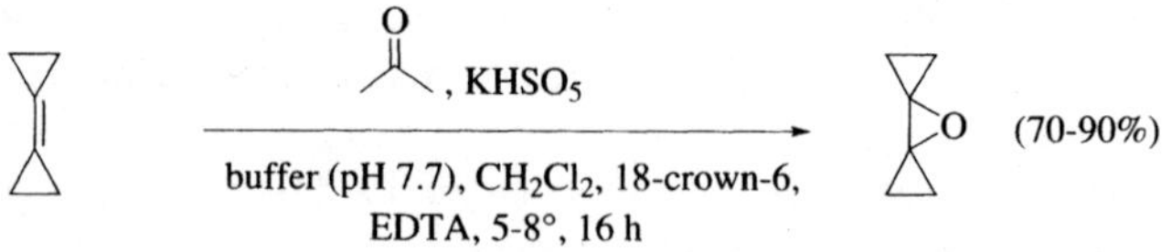

O, $KHSO_5$; buffer (pH 7.7), CH_2Cl_2, 18-crown-6, EDTA, 5-8°, 16 h → O (70-90%) (Eq. 4)

Epoxidation of Electron-Rich Alkenes

Electron-rich alkenes are usually more reactive toward oxidation than simple alkenes, and the corresponding epoxides are generally much more labile toward hydrolysis and thermolysis. With DMD (isol.), a broad variety of such epoxides has been prepared in essentially quantitative yields, and even excessively labile examples have been characterized spectroscopically (Scheme 6). For example, the

Scheme 6. Dioxirane Oxidation of Electron-Rich Alkenes

DMD epoxidation of silyl enol ethers at subambient temperature affords the corresponding epoxides in quantitative yields (Eq. 5).[44,45] These epoxides have been proposed as intermediates in the peracid oxidation of silyl enol ethers to α-hydroxy ketones (the Rubottom reaction).[46] Indeed, when the in situ mode is employed, the corresponding α-hydroxy ketones are obtained directly.[35]

acetone, CH_2Cl_2, N_2, –40°, 3 h (99%) (Eq. 5)

A labile epoxide is derived from 2,3-dimethylbenzofuran (Eq. 6), which even at $-20°$ rearranges to the *o*-quinomethide.[47–49] To characterize this epoxide by ^{1}H NMR spectroscopy, the fully deuterated DMD-d_6 (isol.), prepared in acetone-d_6, was employed for the oxidation, and the epoxide was detected at $-78°$.[47] This emphasizes the importance and convenience of the isolated dioxiranes for the

acetone-d_6, CH_2Cl_2, N_2, –20°, 20 h (97%) (Eq. 6)

synthesis of exceedingly sensitive substances. Indeed, the epoxide of 2,3-dimethylfuran, prepared by the epoxidation of the furan with DMD-d_6 (isol.), could not be detected even at $-100°$, and only the rearrangement product hex-3-ene-2,5-dione was observed.[50]

The 8,9-epoxide of aflatoxin B is a well-known carcinogen; however, because of its sensitivity toward hydrolysis, numerous efforts to prepare this epoxide failed. The synthesis is readily accomplished (no yield given) with DMD (isol.) at room temperature (Eq. 7).[51] Once this labile epoxide was available as a pure substance, the mutagenicity caused by its reaction with DNA was unequivocally confirmed.

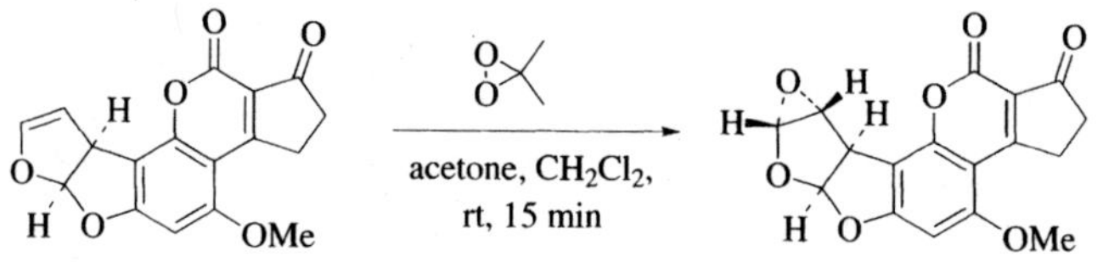

acetone, CH_2Cl_2, rt, 15 min (—) (Eq. 7)

Epoxidation of Electron-Poor Alkenes

Dioxirane oxidation of electron-poor alkenes (Scheme 7) offers general access to the corresponding epoxides, which are usually difficult to prepare with peracids, or require the use of basic conditions as in the Weitz-Scheffer reaction.[52] Since the dioxirane is an electrophilic oxidant and these alkenes are electron-poor substrates, the reaction usually needs much longer time (up to several days) compared to that of electron-rich alkenes, which are oxidized usually within minutes. For very

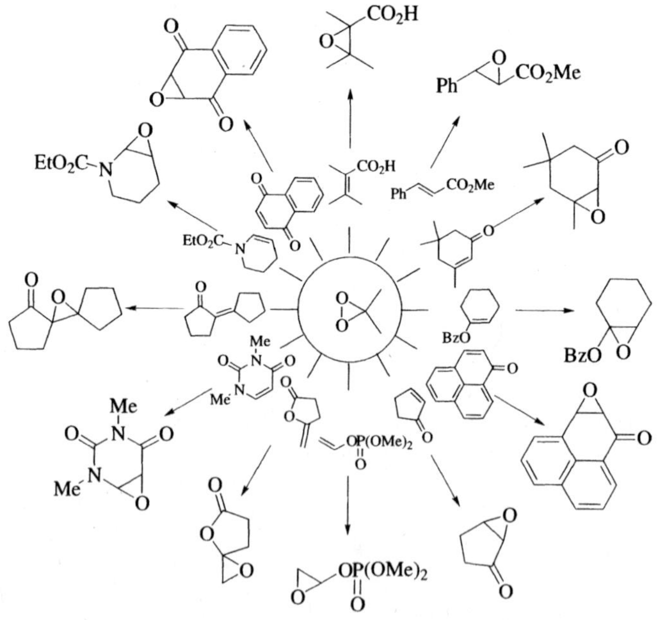

Scheme 7. Dioxirane Oxidation of Electron-Poor Alkenes

recalcitrant electron-poor substrates, a large excess of the dioxirane and elevated temperatures (heating at reflux in acetone, ca. 60°) may be needed to obtain good conversions. Nevertheless, the yields may be very high, since the product epoxides resist hydrolysis and thermolysis. For convenience, DMD (in situ) is employed; for very unreactive electron-poor alkenes, the much more reactive TFD (in situ) may have to be used.

The epoxides of a number of α,β-unsaturated aryl ketones (chalcones)[53] have been prepared in excellent yields with DMD (isol.). An illustrative example is shown in Eq. 8.[53] This example is notable because the phenolic group does not require protection in this epoxidation, which hitherto had not been possible with other oxidants.

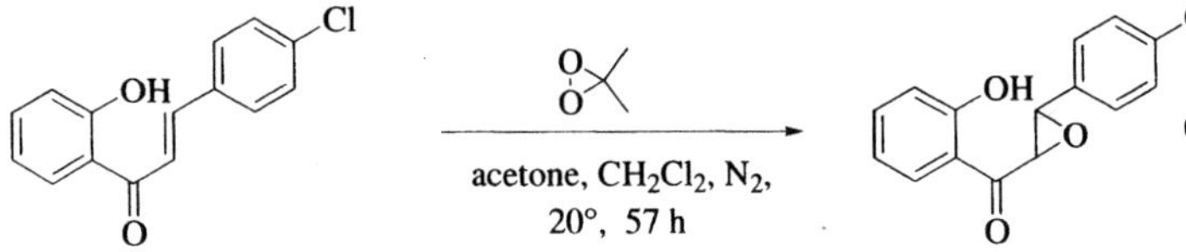

acetone, CH_2Cl_2, N_2, 20°, 57 h (100%) (Eq. 8)

Numerous examples have been documented in the literature to show the usefulness of this method. For example, the dioxirane oxidation of an α,β-unsaturated ketone in Eq. 9 is the key step in the total synthesis of verrucosan-2β-ol and homoverrucosan-5β-ol.[54]

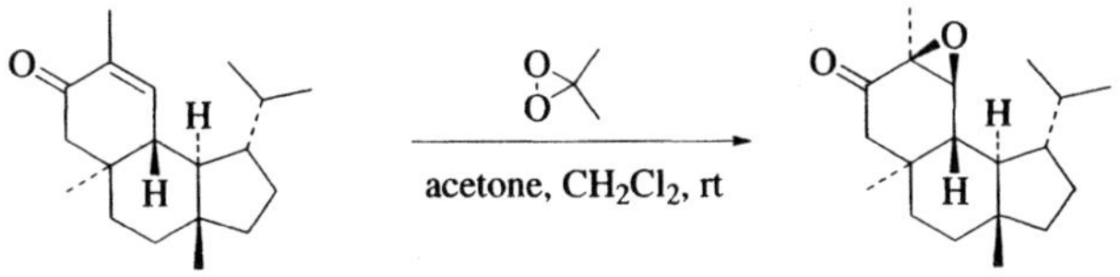

acetone, CH_2Cl_2, rt (72%) (Eq. 9)

The in situ mode is quite suitable for such electron-poor substrates, since the epoxides usually survive the aqueous conditions, longer reaction times, and higher temperatures.[55] A good example is the oxidation of *trans*-cinnamic acid by DMD (in situ), which affords the epoxide in nearly quantitative yield (Eq. 10).[31]

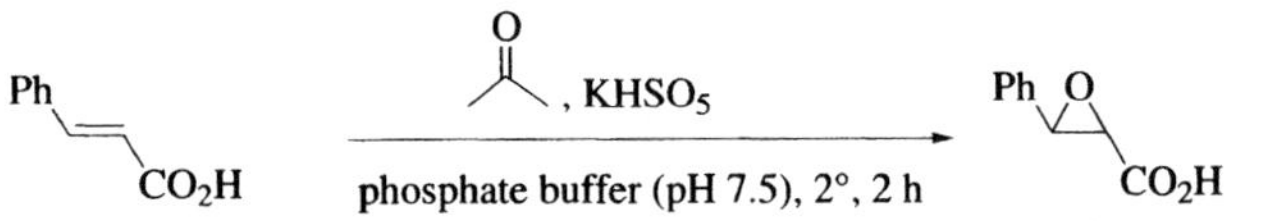

, $KHSO_5$; phosphate buffer (pH 7.5), 2°, 2 h (95%) (Eq. 10)

Epoxidation of Alkenes with both Electron Donors and Electron Acceptors

The reactivity of the substrates in Scheme 8 depends on the nature of the substituents, but usually they behave more like the electron-poor substrates, so that longer reaction times, a large excess of DMD, and elevated temperatures ($<60°$) may be required for complete conversion. β-Alkoxycyclohexenones (Eq. 11),[56] aurones,[57] flavones,[58,59] and isoflavones[57,60] are among the most studied substrates of this type. Some of these epoxides, which constitute valuable building blocks in natural product synthesis, have been made available for the first time by dioxirane

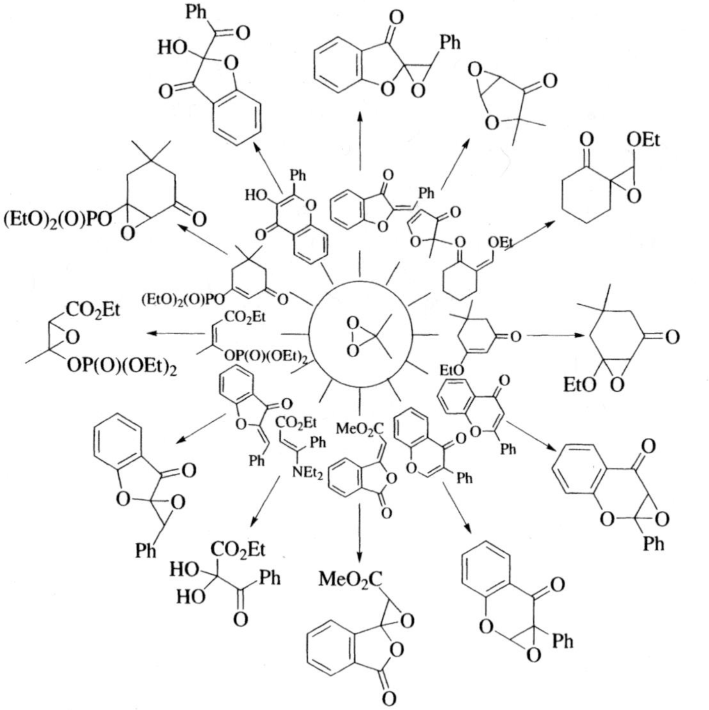

Scheme 8. Dioxirane Oxidation of Alkenes with Both Electron Donors and Acceptors

oxidation. As an example of the in situ mode of oxidation, flavone has been epoxidized by TFD (in situ) in excellent yield (Eq. 12).[35]

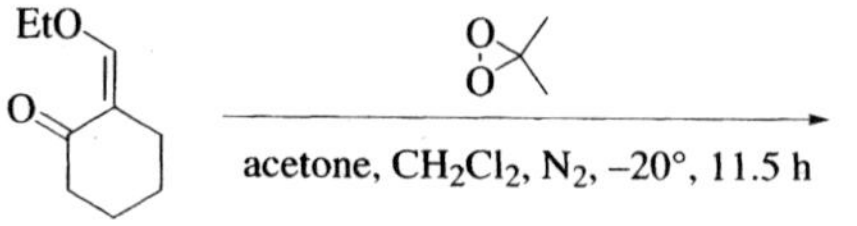

acetone, CH_2Cl_2, N_2, –20°, 11.5 h (ca. 100%) (Eq. 11)

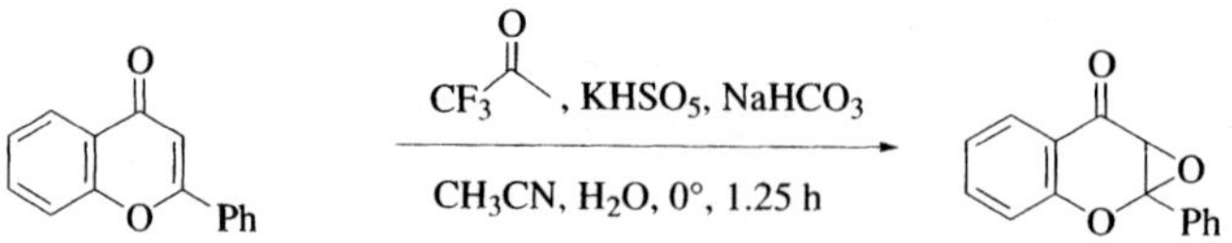

$CF_3C(O)CH_3$, $KHSO_5$, $NaHCO_3$; CH_3CN, H_2O, 0°, 1.25 h (99%) (Eq. 12)

Chemoselectivity

Chemoselectivity is a crucial feature in planning a synthetic sequence, in which a key step may utilize an epoxidation by dioxiranes on a multifunctionalized substrate. Despite the vast number of oxidations (see Tables I–V) that have already been performed with dioxiranes over the last 20 years, no deliberate efforts to test functional-group compatibility have been made, even for difunctionalized substrates. The study of such chemoselectivities would require determination of quantitative rates [e.g., as was done for the alkyl-substituted alkenes (Table A)], but such data are not available.

Based on the data in Tables I–V and our personal experience, we have compiled a chart on the tolerance of the more common functional groups in dioxirane epoxidations (Table B). These trends are based merely on the conversion of the substrate and/or the yield of the oxidation product, rather than on hard kinetic data or competition experiments. Necessarily, the classification into "*compatible, moderately compatible, and incompatible*" categories of the functional groups (Table B) is approximate. Expectedly, exceptions must be reckoned with on account of possible electronic and steric effects in a particular substrate. Nevertheless, this compilation provides a useful qualitative guide on how to avoid undesirable misplannings in a synthetic application.

TABLE B. FUNCTIONAL-GROUP TOLERANCE TO EPOXIDATION[a]

compatible	*moderately compatible*	*incompatible*
$-CO_2H$; $-CO_2R$	$>C{=}NH$; $>C{=}NR$	$-PR_2$
$-CONH_2$; $-CONR_2$	$>C{=}N{-}NR_2$	$-SH$; $-SR$
$-COR$	$>C{=}N{-}OH$; $>C{=}N{-}OR$	$-SSR$
$-C{\equiv}N$	$>{=}\bullet{=}NR$	$-NHNHR$
$-NH_3^+$; $-NR_3^+$	$>{=}\bullet{=}O$	$-NH_2$; $-NHR$; $-NR_2$
$-SO_2R$	$>{=}\bullet{=}CR_2$	enolates
$-SO_3H$; $-SO_3R$	$-N{=}C{=}S$; $-N{=}C{=}O$	$C{=}C{-}SR$
$-PO(OR)_2$	electron-rich arenes[b]	$C{=}C{-}NR_2$
$-NO_2$	polycyclic arenes[c]	$C{=}C{-}OSiR_3$
epoxide (O; R, R)	heteroarenes[d]	$C{=}C{-}OR$
$-O_2H$; $-O_2R$	$-{\equiv}{-}R$	$C{=}C{-}O_2CR$
$-X$ (F, Cl, Br)	$-CHO$	$>C{=}PR_3$
$-SiR_3$	$-CH(R)OH$	$>C{=}S$
$-SnR_3$	$-CH(R)OR'$	$>C{=}N_2$
electron-poor arenes[e]	$-CH(R)OAc$	$-N{=}O$
	N-oxides[f]	heteroarenes[g]

[a] R can be an alkyl or aryl group; the latter are usually less reactive. [b] With electron-donating substituents such as alkyl, hydroxy, and alkoxy groups. [c] Naphthalene, phenanthrene, anthracene. [d] Thiophenes. [e] With electron-withdrawing substituents such as nitro, cyano, and carboxy groups. [f] These induce dioxirane decomposition.[60a] [g] Pyridines, pyrroles, furans.

Another point that needs clarification on the compatibility trends in Table B is the fact that this chapter is restricted to epoxidations. Unquestionably, an effective discussion of functional group compatibility requires knowledge of how competitive the epoxidations are with the oxidations of heteroatom substrates, arenes and heteroarenes, acetylenes and cumulenes (heterocumulenes), alkanes and silanes, and even organometallic functionalities; however, all of these are dealt with in a separate chapter. Unfortunately, few reliable reactivity data are accessible for such dioxirane oxidations; thus, only a qualitative assessment of trends is feasible.

Provided we compare a set of substrates with similar electronic and steric features, the following reactivity order applies quite generally: heteroatom oxidation > epoxidation > CH insertion. Furthermore, among the heteroatoms, the general reactivity order P > S > N holds in their dioxirane oxidation [i.e., phosphines are more readily oxidized than sulfides (mercaptans), these in turn react faster than amines]. For epoxidations, the ease of oxidation follows the order alkenes > cumulenes > arenes > acetylenes. Arenes, particularly benzene derivatives, must be activated with hydroxy or alkoxy groups to achieve epoxidation, whereas *N*-heteroarenes undergo *N*-oxidation and thereafter are inert toward further reaction. For alkenes, the order is ED-C=C-ED>C=C-ED>C=C>EA-C=C-ED>C=C-EA>>EA-C=C-EA, in which the last substrate with two electron acceptors is unreactive (e.g., maleic and fumaric acids, esters, and nitriles are inert even toward the very reactive TFD). Finally, for CH insertions, the reactivity order is allylic > benzylic > tertiary > secondary > primary. These general directives should serve as a planning guide for synthesis; a few specific examples follow.

It should be evident from the preceding presentation of trends that certain functional group combinations are incompatible. For example, the epoxidation of a substrate with a double bond in the presence of a heteroatom functionality (sulfide or amine) is quite generally not feasible without protection. In the particular examples, the sulfide (Eq. 13)[61] is oxidized to the sulfoxide (even the sulfone is formed before the double bond is epoxidized), whereas a tertiary amine (Eq. 14), and *N*-

SPh; (1 equiv), acetone, CH_2Cl_2, –35°, 1 h → S(O)–Ph (81%); (3 equiv), acetone, CH_2Cl_2, –35°, 3 h → SO_2–Ph (91%) (Eq. 13)

acetone, 0°, 1 h → N-oxide (~100%); 1. $BF_3•Et_2O$ 2. , acetone, CH_2Cl_2, 0° 3. $KHCO_3$ → epoxide (85%) (Eq. 14)

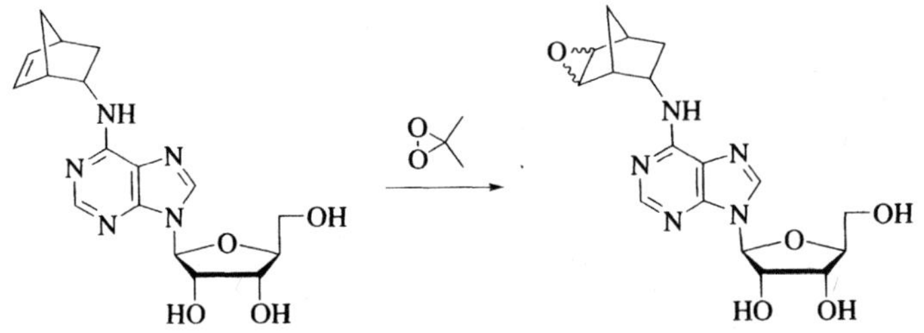

(75%) (Eq. 15)

heteroarenes, e.g., pyridines, pyrimidines, etc., are transformed into their *N*-oxides. An exception is the adenosine-substituted norbornene displayed in Eq. 15, for which the purine functionality is sufficiently less prone toward *N*-oxidation that epoxidation dominates.[61a] This chemoselectivity also holds for enamines, which are preferentially oxidized to the corresponding *N*-oxides;[62,63] exceptions are *N,N*-disilylenamides, which undergo epoxidation.[64] However, as illustrated for the pyrrolidine-substituted cyclohexene in Eq. 14, BF_3 complexation masks the amino functionality and the double bond is epoxidized without *N*-oxidation.[63] Alternatively, protonation may be employed for this purpose, since an ammonium ion also resists DMD oxidation, as demonstrated for an unsaturated amine (Eq. 16).[65] Such effective group-protected chemoselectivity still needs to be developed for the epoxidation of sulfide-functionalized alkenes, to avoid the inevitable sulfoxidation (Eq. 13).[66]

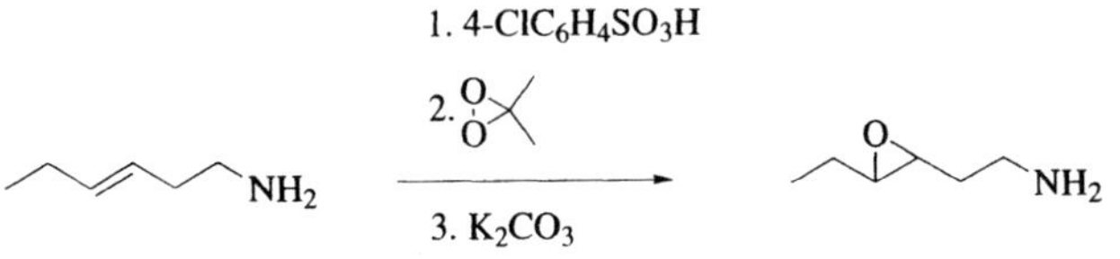

1. 4-$ClC_6H_4SO_3H$
2. DMD
3. K_2CO_3

NH_2 → NH_2 (96%) (Eq. 16)

Acylation of the amino functionality also allows epoxidation, as demonstrated for the allylic amine in Eq. 17,[67] since the amide group resists DMD as well as TFD oxidation.

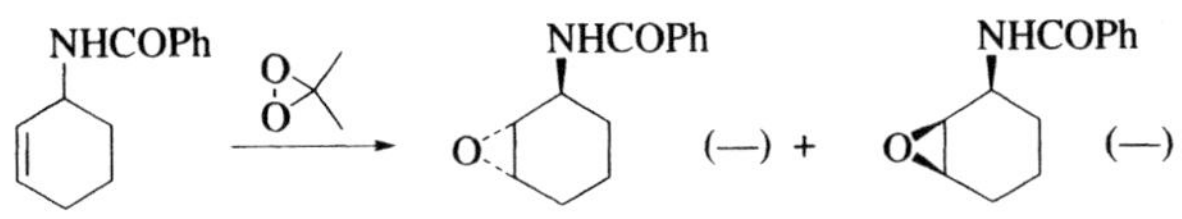

NHCOPh → NHCOPh (—) + NHCOPh (—) (Eq. 17)

That ketone functionalities are compatible, since they are not themselves oxidized by isolated dioxirane, is illustrated for the unconjugated enone in Eq. 18.[32] This is not necessarily true for in situ generated dioxiranes, because ketones that are prone to undergo the Baeyer-Villiger reaction preferentially rearrange to esters with $KHSO_5$. Since intramolecular, ketone-mediated epoxidations have been documented,[68] it is uncertain whether in this in situ oxidation the inherent keto functionality of the substrate performs the oxygen transfer or whether it is DMD (in situ), or both. A more complex example is offered in Eq. 19, which demonstrates that an allylic keto group and the remote ester functionality do not interfere in this

, $KHSO_5$, $NaHCO_3$
H_2O, pH 7-7.5,
18-crown-6, 2 to 8°, 2 h
(96%) (Eq. 18)

, $KHSO_5$, $NaHCO_3$
MeO_2C
H_2O, Bu_4NHSO_4, rt, 2 h
(Eq. 19)
MeO_2C (100%)

2-butanone-mediated, in situ epoxidation.[69] Expectedly, a cyano group is also inert toward dioxirane oxidation (DMD as well as TFD), as exemplified in Eq. 20.[67]

CN CN (—) + CN (—) (Eq. 20)

The oxidation (actually also an epoxidation) of electron-rich arenes in competition with alkene epoxidation is featured for the medium-sized ring vinylsilane substrate in Eq. 21.[70] Whereas DMD (isol.) affords a mixture of epoxidized and

MeO, MeO, Me_3Si (47%) +
MeO_2C, MeO_2C, Me_3Si (31%)
CF_3
MeO_2C, MeO_2C, Me_3Si (90%)
(Eq. 21)

arene-oxidized products, the latter are formed exclusively with TFD (isol.). Presumably, steric effects are at play, which are more pronounced for the trifluoromethyl-substituted dioxirane. However, this chemoselectivity (arene oxidation versus epoxidation) depends on the arene functionality. Thus, the DMD oxidation of the cyclic vinylsilane substrate with two methoxy groups in the benzamide ring produces only the epoxide (Eq. 22).[70]

H N O Me$_3$Si OMe OMe → H N O O Me$_3$Si OMe OMe (—) (Eq. 22)

An example of double-bond versus triple-bond chemoselectivity is shown for the conjugated enyne in Eq. 23, which also bears an alcohol functionality.[71] Not only is the latter preserved (steric effect of the *tert*-butyl group), but also the double bond is exclusively epoxidized. This reaction displays the higher reactivity of C=C vs. C≡C bonds toward epoxidation by DMD.

OH *t*-Bu → OH *t*-Bu O (—) (Eq. 23)

Competitive CH insertion vs. epoxidation is well documented for allylic alcohols. For example, whereas the *Z* isomer in Eq. 24 gives a diastereomeric mixture of threo- and erythro-epoxides with DMD (isol.),[72] equal amounts of epoxidation and allylic oxidation are formed from the *E* isomer.[73] Presumably, steric effects are again responsible since DMD is sensitive to such encumbrance.[6,74,75] A dramatic example is shown in Eq. 25, in which the introduction of an α-silyl substituent changes the exclusive epoxidation to predominant allylic oxidation; unquestionably, steric

OH → OH O (—) + OH O (—)

(Eq. 24)

OH → OH O (—) + O (—) 50:50

R OH → R OH O or R O (Eq. 25)

R = H (> 95%) R = PhMe$_2$Si (90%)

effects are at play here, because electronically, the silyl group should activate the double bond for epoxidation in view of its electron-releasing nature. In this context, an impressive example is given by the vinylsilyl-substituted cyclohexenol in Eq. 26, which yields only the corresponding enone.[74] An example of aldehyde oxidation to carboxylic acid (C-H insertion) rather than epoxidation is illustrated in Eq. 27[76] for a phosphonate-substituted enal. Here electronic effects come to bear, since the double bond is deactivated by the electron-withdrawing phosphonate ester.

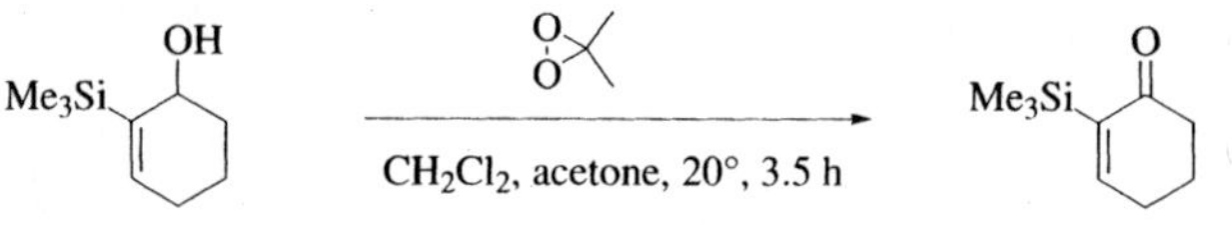

CH_2Cl_2, acetone, 20°, 3.5 h (>95%) (Eq. 26)

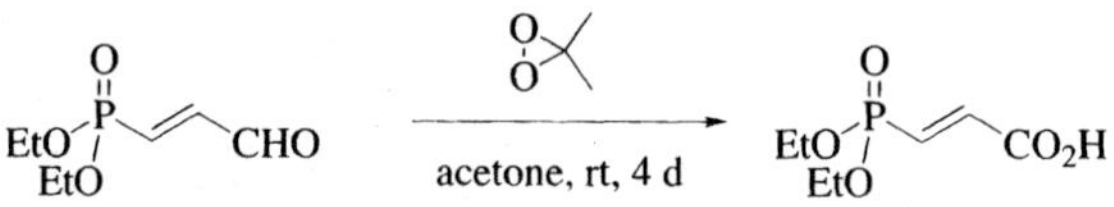

acetone, rt, 4 d (100%) (Eq. 27)

Another type of incompatibility of a hydroxy functionality is exhibited in Eq. 28.[71] The labile epoxide functionality of the dioxene is attacked intramolecularly by the nucleophilic OH group to afford the bicyclic furan product through cyclization. Although the alcohol is not oxidized (this example is not a question of oxidative compatibility), it illustrates that remote nucleophilic functionalities may not be tolerated in view of further reaction with the epoxide ring.

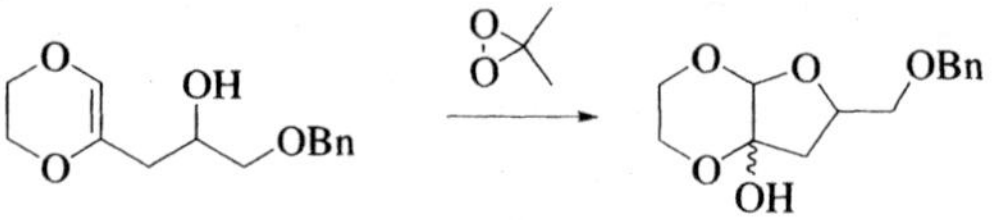

(—) (Eq. 28)

These few examples of functional-group compatibility should be helpful for orientation, but more systematic studies (especially quantitative data on the reactivity of the common functionalities (toward dioxirane oxidation) would be desirable. Nonetheless, some combinations of compatible functional groups in Table B are straightforward to deduce: Any of the functionalities in the *compatible* category may be present in a substrate molecule equipped with groups of the *incompatible* and *moderately compatible* type, as long as the groups are chemically compatible toward each other (i.e., a silyl enol ether cannot bear a carboxylic or sulfonic acid group). Not simple to decide are combinations of functional groups from the *incompatible* and *moderately compatible* categories, nor within each of these categories. The general guidelines spelled out in the beginning of this section may be helpful, but are by no means foolproof.

Regioselectivity

By regioselective epoxidations we mean the preferential oxidation of one of the double bonds in a polyunsaturated substrate. Generally, oxidations by the electrophilic DMD take place at the site of higher electron density and lower steric demand. Good regioselectivities have been observed in heteroatom oxidations[8] and C-H insertions,[76a] but only the regioselectivity in epoxidations[77] is covered in this chapter.

The regioselectivity is mainly dictated by electronic effects in DMD epoxidations; a good example is shown in Scheme 9. Whereas the selectivity of epoxidation of the electron-rich 6,7-double bond over the electron-poor 2,3-double bond is very high in the methyl-substituted compound, the selectivity drops dramatically when the methyl group is replaced by the electron-withdrawing trifluoromethyl group.[78]

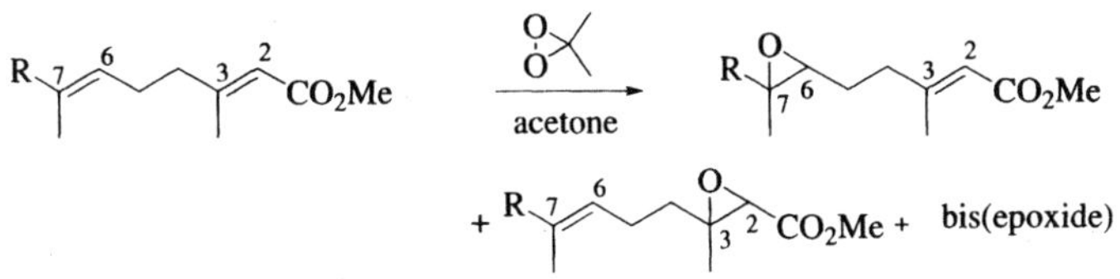

R	6,7	2,3	bis
CF_3	0	42	58
Me	100	0	0

Scheme 9. Electronic Effects on the Regioselectivity in the DMD Epoxidations of Methyl 2,6-Octadienoate Derivatives

The influence of electronic substituents may be offset by the directing effect of a hydroxy group, which operates through the assistance of intramolecular hydrogen bonding.[8,79] For instance, selectivity in the epoxidation of geraniol by DMD in protic versus aprotic solvent was shown to shift toward the 2,3-epoxide (from 88 : 12 in 9 : 1 MeOH/acetone to 51 : 49 in 9 : 1 CCl_4/acetone; Scheme 10).[72] In an aprotic medium (CCl_4/acetone), the allylic hydroxy group of geraniol stabilizes the transition-state structure through hydrogen bonding with the dioxirane and, consequently, the less nucleophilic 2,3-double bond competes more effectively in the

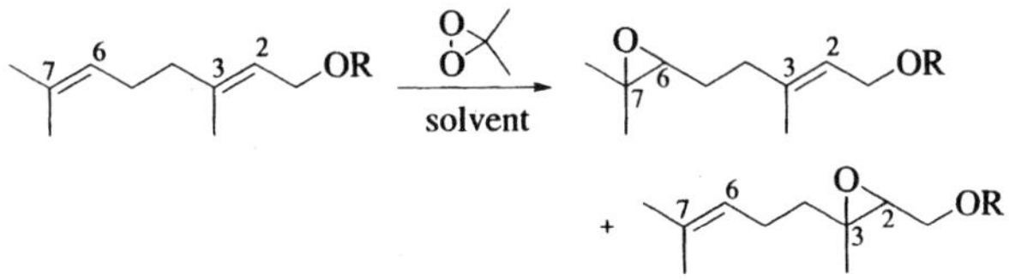

R	Solvent	6,7	2,3
H	MeOH/acetone (9:1)	88	12
H	acetone	74	26
H	CCl_4/acetone (9:1)	51	49
Me	MeOH/acetone (9:1)	88	12
Me	acetone	88	12
Me	CCl_4/acetone (9:1)	87	13

Scheme 10. Solvent Effects on the Regioselectivity in the DMD Epoxidation of Geraniol and its Methyl Ether (at 30% conversion)

epoxidation. In the protic medium (9 : 1 methanol/acetone), methanol engages in hydrogen bonding with the dioxirane and the substrate, so that the allylic hydroxy group does not assist as effectively in the transition-state structure to control the selectivity. Thus, the electronic advantage of the more nucleophilic 6,7-double bond dictates the selectivity, so that the isomer ratio represents the relative nucleophilicities of the double bonds. In contrast, the methyl ether does not display this shift in the selectivity because assistance by hydrogen bonding is lacking in the transition-state structure for oxygen transfer and only electronic factors operate.

Diastereoselectivity

Dimethyldioxirane is responsive to the presence of substituents and functionalized groups in alkenes, which often leads to highly diastereoselective epoxidations.[8,75,80,81] A comparative study on the diastereoselectivity of DMD versus *m*CPBA has revealed that DMD exhibits consistently higher diastereoselectivity than *m*CPBA as an epoxidizing reagent,[75,82] although the difference is usually small. Results of the DMD epoxidation of some 3-alkyl-substituted cyclohexenes indicate that steric interactions between the substituents and DMD control the diastereoselectivity (Eq. 29).[67,83]

R — acetone, rt → *trans* (—) + *cis* (—) (Eq. 29)

R	*trans:cis*
Me	48:52
Et	55:45
i-Pr	78:22
i-Bu	55:45
t-Bu	95: 5

Epoxidation of 2-menthene and 1,3-dimethylcyclohexene (Scheme 11) by various dioxiranes demonstrates that the size of the dioxirane substituents has a minimal effect on diastereoselectivity.[75] This is because the alkyl groups of the dioxirane cannot interact effectively with substituents at the stereogenic center in the favored transition-state structures, as shown for 2-menthene (Scheme 12).

dr (*trans/cis*)	78:22	81:19	82:18	80:20
	72:28	70:30	72:28	80:20

Scheme 11. Diastereoselective Epoxidation of 2-Menthene and 1,3-Dimethylcyclohexene by Various Dioxiranes

Scheme 12. Favored Transition-State Structure for the Dioxirane Epoxidation of 2-Menthene

Hydrogen bonding plays an important role in the diastereoselective epoxidation of chiral allylic alcohols; an instructive example is shown in Scheme 13.[75] Whereas 3-hydroxycyclohexene displays a higher cis selectivity for both DMD and *m*CPBA, the 3-hydroperoxycylcohexene favors the trans diastereomer for DMD and is unselective for *m*CPBA. Evidently, the allylic hydroxy group in cyclohexenol directs the incoming oxidant through efficient hydrogen bonding to the cis face of the double bond, whereas the hydroperoxy group obstructs such an attack because of steric and, possibly, some electrostatic effects.[77,84–86] Consequently, the trans epoxide is favored for DMD, which is more sensitive to steric repulsions with substrate substituents than *m*CPBA.

	OH	O_2H
	cis:trans	*cis:trans*
DMD	53:47	9:91
*m*CPBA	95:5	57:43

Scheme 13. Diastereoselective Epoxidation of Chiral Cyclohexenes by DMD vs. *m*CPBA

An example of the importance of medium effects on hydrogen bonding is given for the chiral 2-cyclohexen-1-ol in Eq. 30.[87] The high cis selectivity (relative to the hydroxy substituent) in the nonpolar, aprotic medium (9 : 1 CCl_4/acetone) derives from intermolecular hydrogen bonding between the hydroxy group and the attacking dioxirane. This cis selectivity remains still high in acetone, but is considerably lower in the protic medium (9 : 1 MeOH/acetone). Methanol engages in intermolecular hydrogen bonding with DMD, which interferes with the interaction between the allylic hydroxy group and DMD.

(Eq. 30)

Solvent	*cis:trans*	epoxides:enone
MeOH/acetone (9:1)	62:38	92:8
acetone	83:17	77:23
CCl_4/acetone (9:1)	96:4	92:8

Similar effects are also observed in the epoxidation of acyclic allylic alcohols.[72] In these compounds, diastereoselectivity depends on the substitution pattern of the allylic alcohol, which controls the preferred conformation via steric interactions in the ground state and presumably also in the transition state. Appreciable π-facial selectivity is expected when hydrogen bonding of DMD by the allylic hydroxy functionality operates and allylic strain helps in aligning the appropriate conformation. As shown in Scheme 14, the allylic alcohol with minimal allylic strain

R^1	R^2	Solvent	*threo:erythro*
H	H	acetone	50:50
Me	H	MeOH/acetone (9:1)	57:43
Me	H	acetone	60:40
Me	H	CCl_4/acetone (9:1)	70:30
H	Me	MeOH/acetone (9:1)	64:36
H	Me	acetone	67:33
H	Me	CCl_4/acetone (9:1)	85:5
Me	Me	MeOH/acetone (9:1)	82:18
Me	Me	acetone	87:13
Me	Me	CCl_4/acetone (9:1)	91 :9

Scheme 14. Diastereoselectivity in the DMD Epoxidation of Chiral Acyclic Allylic Alcohols with Allylic Strain

($R^1 = R^2 =$ H) displays no diastereoselectivity in dioxirane epoxidation. In the substrate with only 1,2-allylic strain ($A^{1,2}$), the diastereoselectivity is quite low, but increases steadily with the use of increasingly nonpolar solvents. Evidently, 1,2-allylic strain is not effective enough to steer the π-facial attack. In contrast, for the substrate with only 1,3-allylic strain ($A^{1,3}$; $R^1 =$ H, $R^2 =$ Me), effective control of π-facial selectivity (85:15) is observed when CCl_4 is used as cosolvent. When both 1,2- and 1,3-allylic strain ($R^1 = R^2 =$ Me) operate, expectedly, the π-facial diastereoselectivity is 91:9 in the nonprotic cosolvent CCl_4. These results provide useful mechanistic insight into the oxygen-transfer process. The two transition-state structures for the diastereomeric epoxides with both $A^{1,2}$ and $A^{1,3}$ strain are shown in Scheme 15

Scheme 15. Diastereomeric threo and erythro Transition-State Structures in the Hydroxy-Directed Epoxidation of a Chiral Allylic Substrate by DMD; $A^{1,2}$ is 1,2-Allylic and $A^{1,3}$ is 1,3-Allylic Strain for DMD. The erythro transition-state structure is disfavored because of considerable 1,3-allylic strain. In the threo transition-state structure, however, 1,3-allylic strain is not significant since the steric interaction between H and Me is ineffective and, thus, high *threo* selectivity is obtained. The best conformational alignment (minimum 1,3-allylic strain) of the allylic hydroxy group for effective hydrogen bonding occurs at a dihedral angle (α) between 120–130°, as corroborated computationally.[23,88,89]

Highly diastereoselective epoxidations by dioxiranes have been widely employed in organic synthesis. For example, glycals have been epoxidized by DMD to afford high diastereoselectivities (Eq. 31);[90] this strategy has been successfully applied in

OSiMe₃ — acetone, CH_2Cl_2, 0° — BnO ... (>99%) >95:<5

(Eq. 31)

the synthesis of biologically important oligosaccharides.[91–95] Another example of diastereoselective DMD epoxidation in natural product synthesis is shown in Eq. 32. The epoxide that is potentially useful for the synthesis of taxol was obtained in quantitative yield by DMD epoxidation.[96] High diastereoselectivity is also achieved in the DMD epoxidation of the chiral silyl enol ether and titanium enolate in Eq. 33.[97]

acetone, CH_2Cl_2, 0° (100%)

(Eq. 32)

acetone, THF, –78°; NH_4F, H_2O, rt (Eq. 33)

M	Time	Time	
$SiMe_3$	30 min	3 h	(85%) dr > 98:2
$TiCp_2Cl$	1 min	12 h	(53%) dr > 98:2

Enantioselectivity

There are several ways to obtain enantiomerically enriched epoxides with dioxiranes. One strategy is to use an achiral dioxirane, e.g., DMD, as stoichiometric oxidant and a prochiral olefin functionalized with an optically active auxiliary. The enantiomerically enriched final product is obtained after removal of the chiral auxiliary. For example, the DMD oxidation of titanium enolates with enantiomerically pure TADDOLs ($\alpha,\alpha,\alpha',\alpha'$-tetraaryl-1,3-dioxolane-4,5-dimethanols) as chiral auxiliaries has been applied in the asymmetric synthesis of enantiomerically enriched α-hydroxy ketones.[98] (*R*)-2-Hydroxy-1-phenylpropanone is obtained in up to 63% ee by DMD oxidation (Eq. 34), in which the initially formed epoxide rearranges to an α-hydroxy ketone.[98] Asymmetric epoxidation by the achiral DMD is also achieved by employing 2,2-dimethyloxazolidines as chiral auxiliaries in tiglic amide derivatives (Eq. 35).[99]

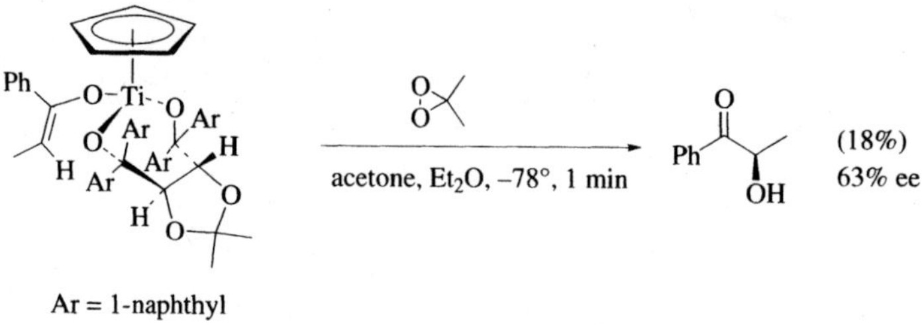

acetone, Et_2O, –78°, 1 min (18%) 63% ee (Eq. 34)

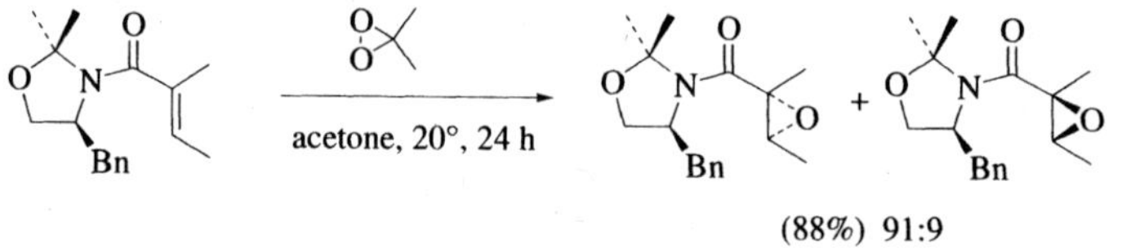

Ar = 1-naphthyl

acetone, 20°, 24 h (Eq. 35)

(88%) 91:9

An alternative approach is to employ DMD as the stoichiometric oxygen source for the preparation of an optically active oxidation catalyst. This is realized in the Jacobsen-Katsuki Mn-catalyzed[100,101] enantioselective epoxidation of prochiral chromenes, in which DMD is used as oxygen source instead of NaOCl or iodosobenzene.[102] The advantage is that the use of DMD (isol.) allows one to operate in an organic medium (acetone, CH_2Cl_2, etc.), whereas NaOCl requires a biphasic aqueous system and iodosobenzene a heterogeneous solid-liquid mixture. Since the DMD oxidation of manganese is faster than that of the olefinic substrate, the corresponding epoxides are obtained with high enantiomeric excess (Eq. 36). Similarly, isoflavone epoxides have also been obtained with good ee values by employing DMD together with the Jacobsen-Katsuki catalysts.[103,104]

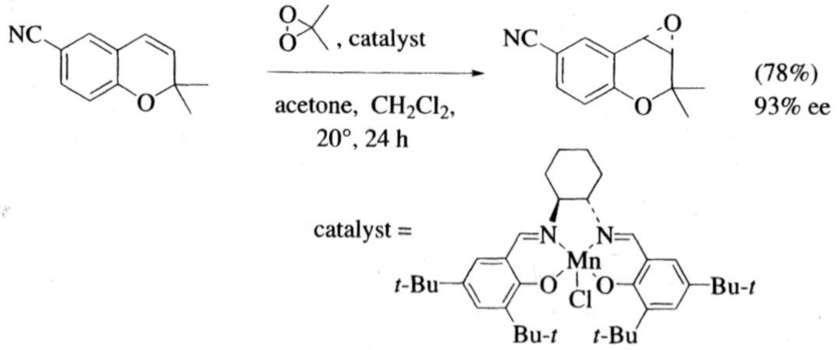

, catalyst

acetone, CH_2Cl_2, 20°, 24 h (78%) 93% ee

(Eq. 36)

catalyst =

The most convenient and direct way to effect enantioselective epoxidation is to use enantiomerically enriched dioxiranes. Such dioxiranes can be generated in situ from the corresponding enantiomerically pure ketones. Thus, asymmetric oxidation may in principle be catalytic in the ketone. Indeed, the chiral ketones in Scheme 16 were employed for the asymmetric epoxidation of olefins at the very beginning of dioxirane chemistry.[105]

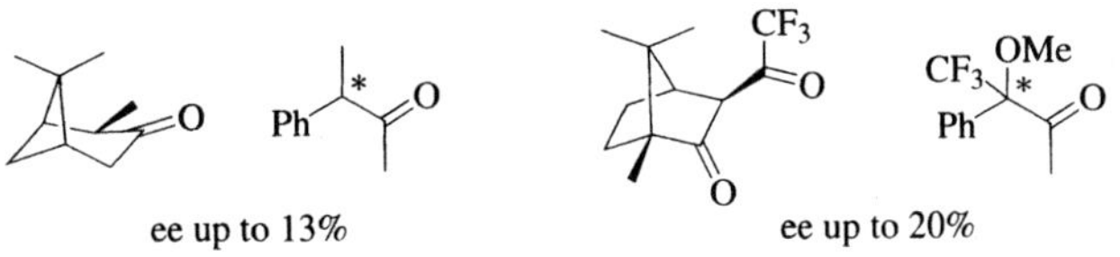

ee up to 13%   ee up to 20%

Scheme 16. Early Examples of Optically Active Ketone Catalysts for in situ Asymmetric Epoxidation of Prochiral Alkenes

Unfortunately, enantioselectivity with these ketones is quite low (≤ 13% ee),[105] with the best ee value only up to 20%.[106] However, better enantioselectivities (up to 85% ee) have been realized with catalytic amounts of the α-hetero-substituted ketones **1–3** in Eqs. 37–39.[107–109]

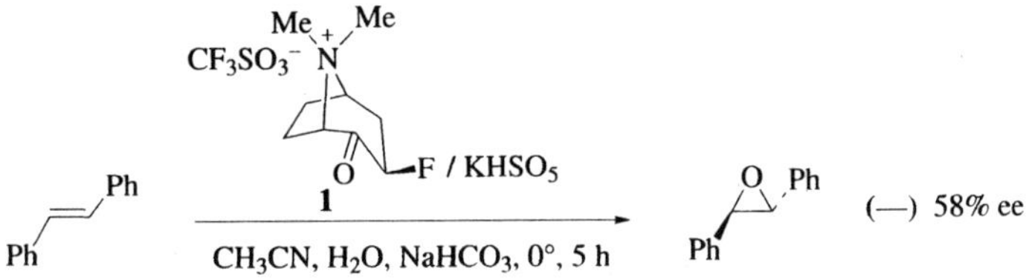

1 / $KHSO_5$, CH_3CN, H_2O, $NaHCO_3$, 0°, 5 h; (—) 58% ee (Eq. 37)

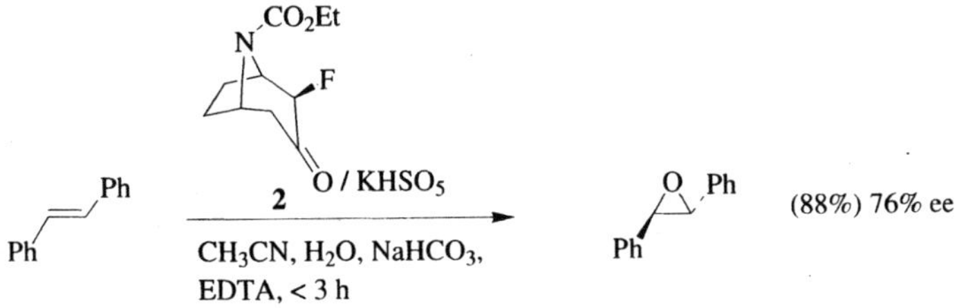

2 / $KHSO_5$, CH_3CN, H_2O, $NaHCO_3$, EDTA, < 3 h; (88%) 76% ee (Eq. 38)

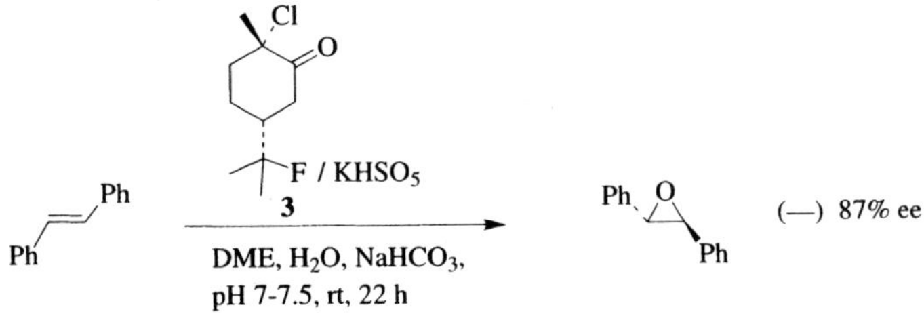

3 / $KHSO_5$, DME, H_2O, $NaHCO_3$, pH 7-7.5, rt, 22 h; (—) 87% ee (Eq. 39)

Among the most selective catalysts for enantioselective epoxidation are the C_2-symmetric ketones **4–9** in Scheme 17. The parent binaphthalene-derived catalyst **4** (R = H) epoxidizes *trans*-stilbene in 47% ee, whereas for *trans*-4,4′-

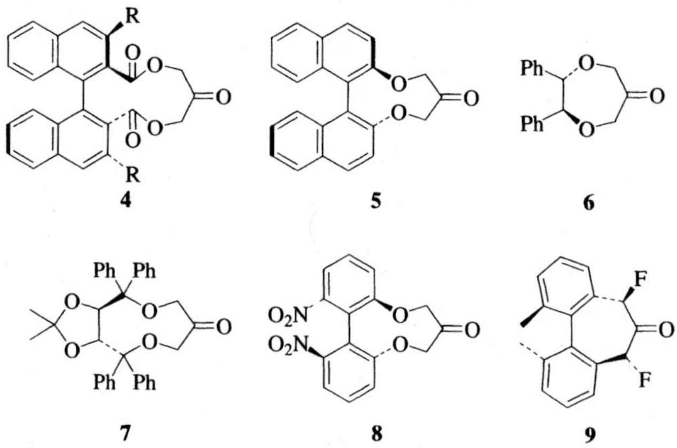

Scheme 17. Optically Active C_2-Symmetric Ketone Catalysts **4–9** for in situ Asymmetric Epoxidation

diphenylstilbene epoxide, the ee is as high as 87%.[110] Enantioselectivity is enhanced by introducing substituents at the 2 and 2′ positions of the binaphthalene moiety; thus, an ee of 84% (R = 1,3-dioxan-2-yl) is obtained even for *trans*-stilbene epoxide. However, the substrate generality for this catalyst is quite limited.[24,111]

Several other C_2-symmetric ketones 5-**9** (Scheme 17) have been reported for asymmetric epoxidation.[11,24,112,113] These ketones are readily available from the corresponding optically active diols. Although the C_2-symmetric ketone **5** contains the same chiral binaphthalene scaffold as compound **4**, ketone **5** provides much inferior enantioselectivity.[24,112,113] The TADDOL-derived ketone **7** affords the highest enantioselectivity for the epoxidation of *trans*-stilbene (65% ee) among these diol-derived ketones.[113] Whereas the C_2-symmetric dinitro-substituted biphenyl-derived ketone **8** performs poorly,[24] the difluoro-substituted ketone **9** acts catalytically under basic conditions (pH 10) and produces *trans*-stilbene epoxide in excellent enantioselectivity (94% ee); for trans alkenes without phenyl substituents, the enantioselectivity drops considerably.[11]

The chiral ketones **10** and **11** (Scheme 18), which are available from D-(−)-quinic acid, also serve as effective catalysts for asymmetric epoxidations. Whereas ketone **10** displays high enantioselectivity for trans olefins, the lengthy synthesis limits applications.[114,115] Ketone **11** is more readily prepared from quinic acid and also provides good enantioselectivity, but an excess must be used since it does not persist under the reaction conditions.[116]

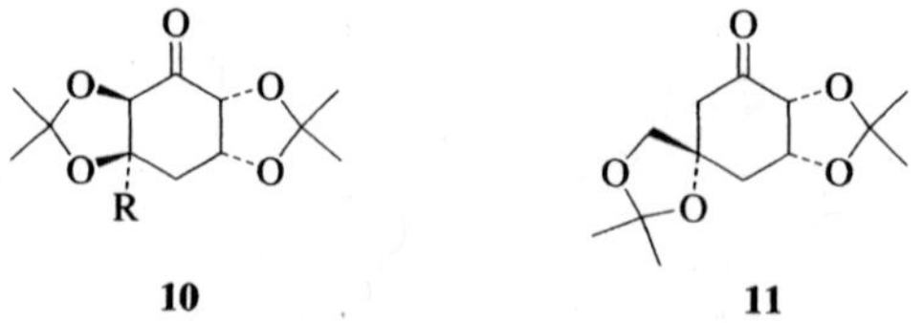

Scheme 18. Optically Active Ketone Catalysts **10** and **11** Derived from Quinic Acid for in situ Asymmetric Epoxidation

Ketone **12** (Scheme 19) represents one of the best catalysts for the asymmetric epoxidation of unfunctionalized as well as for functionalized trans-substituted and trisubstituted olefins.[25,117] Under neutral conditions (pH ca. 8, $NaHCO_3$ as buffer), a

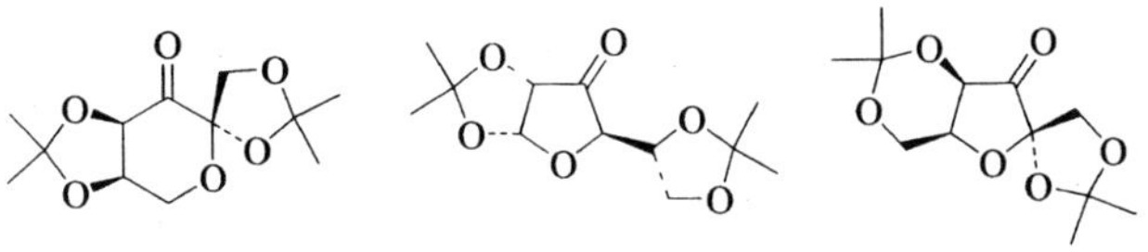

12 **13** **14**

Scheme 19. Sugar-Derived Ketone Catalysts **12–14** for in situ Asymmetric Epoxidation

stoichiometric amount of the ketone **12** is required to achieve good conversions, since this ketone undergoes oxidative degradation under the reaction conditions.[117] Nevertheless, ketone **12** can be employed in sub-stoichiometric amounts (ca. 0.3 equiv.) under basic conditions by adding K_2CO_3 (pH 10.5).[25,118] The enantioselectivity is usually improved under such basic vs. neutral conditions, especially for allylic alcohols. The ease of preparation and generality of epoxidation make ketone **12** the best choice for asymmetric epoxidation.[119]

In contrast to ketone **12**, ketone **13**, prepared from glucose, is ineffective (e.g., no reaction with *trans*-stilbene was observed).[120] Presumably, this ketone suffers Baeyer-Villiger oxidation and is destroyed faster than it epoxidizes the substrate. A similar five membered ring ketone **14**, however, converts *trans*-stilbene to its epoxide in 75% ee.[119] For comparison, the ee values for the epoxidation of *trans*-stilbene by the ketones are collected in Table C. It is evident that the best ee values are obtained with ketone **12**. The corresponding in situ generated dioxiranes have been effectively applied for asymmetric epoxidation. For example, ketone **12** has been employed for the synthesis of enantiomerically enriched α-hydroxy ketones, which are versatile building blocks for natural products, through the epoxidation of silyl enol ethers by the in situ generated dioxirane (Eq. 41).[26] This protocol also has been used to epoxidize enol esters,[121] dienes,[122] and enynes [123,124] in high enantioselectivity, as well as high chemo- and regioselectivity.

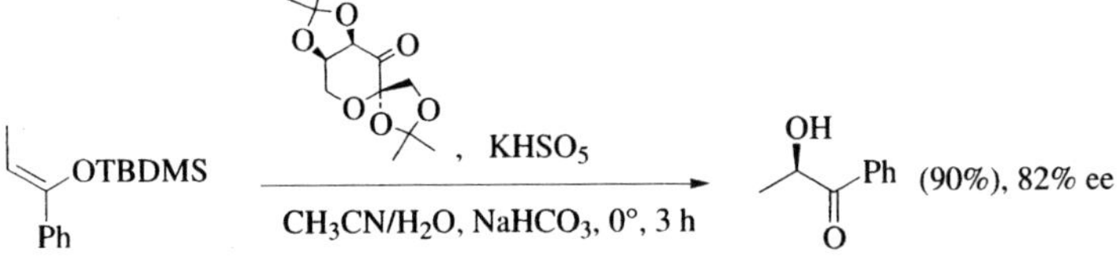

(Eq. 41)

OTBDMS, Ph → [ketone 12], $KHSO_5$; CH_3CN/H_2O, $NaHCO_3$, 0°, 3 h → OH, Ph, O (90%), 82% ee

With the in situ generated dioxirane of ketone **12** the first kinetic resolution of racemic enol acetates has also been achieved through stereoselective epoxidation.[125] An illustrative example is shown in Eq. 42. The efficiency of this method is clearly demonstrated by the high ee values of the unreacted enol acetate and the major trans epoxide product, as well as by the high diastereoselectivity (trans:cis > 95:5).

TABLE C. Comparison of Enantioselectivities in the Asymmetric Epoxidation of *trans*-Stilbene by Various in situ Generated Dioxiranes

Ph–CH=CH–Ph → (ketone, $KHSO_5$) → Ph–*CH(O)CH*–Ph (Eq. 40)

Ketone	pH of buffer	ee (%)	Configuration
1	8.0	58	*R,R*
2	8.0	76	*R,R*
3	8.0	87	*S,S*
4 (R = H)	8.0	47	*S,S*
4 (R = Br)	8.0	80	*S,S*
5	10.5	27	*R,R*
6	8.0	59	*S,S*
7	10.5	65	*R,R*
8	8.0	50	*S,S*
9	10.5	94	—[a]
10 (R = H)	10.5	93	*R,R*
10 (R = CMe_2OH)	10.5	96	*R,R*
11	8.0	85	*R,R*
12	10.5	97	*R,R*
14	10.5	75	*R,R*

[a] The absolute configuration was not given.

Bu-*t*, O_2CBu-*t* (cyclohexene substrate) → (ketone, $KHSO_5$; CH_3CN, DMM, $Na_2B_4O_7$ buffer, K_2CO_3, −10°, 2.5 h) → Bu-*t*, O_2CBu-*t* (99% ee) + (Eq. 42)

Bu-*t*, O, O_2CBu-*t* *trans* (84% ee) + Bu-*t*, O, O_2CBu-*t* *cis*

trans:cis > 95:5

A highlight of this epoxidation protocol is shown in Scheme 20.[126] The in situ generated dioxirane of ketone **12** has been used for constructing the pentaepoxide of desired configuration, so that the glabrescol analog was obtained in 31% overall yield in only two steps. The pentafuran has been used for revision of the glabrescol structure, a natural product with important biological activity.[126]

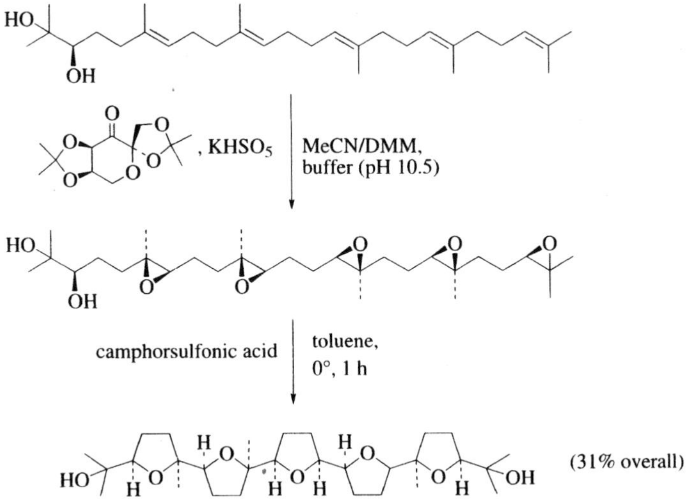

, $KHSO_5$ | MeCN/DMM, buffer (pH 10.5)

camphorsulfonic acid | toluene, 0°, 1 h

(31% overall)

Scheme 20. Synthesis of a Glabrescol Analog

COMPARISON WITH OTHER METHODS

Of the various reagents available for conducting epoxidations, none matches the versatility of dioxiranes. For the epoxidation of simple alkenes, peracids,[127,128] especially *m*CPBA, are the most widely used reagents. Although the two oxidants are mechanistically quite similar, steric effects are more pronounced for dioxiranes than peracids. Thus, when the unsaturated substrate is sterically quite encumbered, e.g., vinylsilanes, C-H oxidation may occur instead of epoxidation,[74] which is almost never observed for peracids. The disadvantage of peracids is that they cannot be applied to acid-sensitive substrates and/or products. When an aqueous buffer is employed, the alkene and the resulting epoxide must be resistant toward hydrolysis. In contrast, as already stated above, isolated dioxiranes perform epoxidations under strictly neutral (pH ca. 7) and nonhydrolytic conditions so that labile substrates and products survive. Of course, in the catalytic mode, which necessarily must be conducted in aqueous media (pH ranges between 7.5 and 10.5) for the in situ generation of the dioxirane, the problems of hydrolysis apply. Thus, the catalytic method should be avoided for the preparation of hydrolytically sensitive epoxides.

Compared to other nonmetal-mediated epoxidations[129] [e.g., perhydrates (hexa-fluoracetone/H_2O_2[130]) and $CH_3CN/H_2O_2/HO^-$ (Payne oxidation)[131]], the in situ method of dioxiranes offers advantages in terms of reactivity and selectivity. Thus, Weitz-Scheffer conditions (NaOCl, H_2O_2/KOH, *t*-BuO_2H/KOH) are suited only for electron-poor olefins,[52] and *N*-sulfonyloxaziridines are limited to electron-rich enolates and enol ethers.[132] This reagent may be activated in the form of oxaziridinium salts, which are considerably more powerful oxidants and epoxidize even simple alkenes.[133,134] In contrast, dioxiranes oxidize all types of electron-rich and electron-poor double bonds with good selectivities; moreover, the strongly basic conditions (pH $>$ 10) used in the Weitz-Scheffer[52] and Payne[131] epoxidations, as well as the acidic conditions of perhydrates,[130] are avoided with isolated dioxiranes. A comparison of the regioselective and diastereoselective epoxidations of 1-methylgeraniol by DMD (isol.) versus a few other oxidants is given in Table D.

TABLE D. Comparison of Regio- and Diastereoselectivities for the Epoxidation of 1-Methylgeraniol by DMD (isol.) vs. other Oxidants

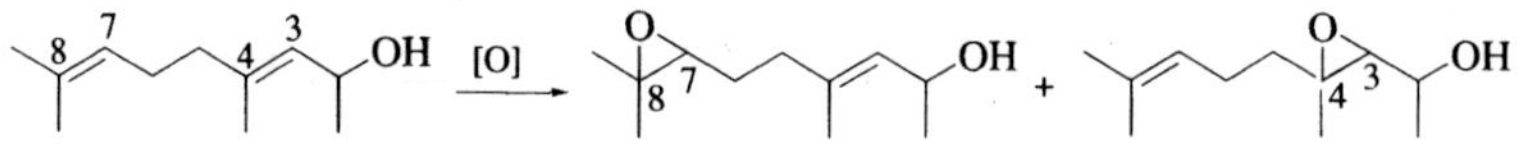

Oxidant	Solvent	Epoxide Selectivity	
		regio 7,8:3,4	diastereo[a] *threo*:*erythro*
DMD (isol.)[134a]	CCl_4	32:64	94:6
*m*CPBA[134a]	CCl_4	51:49	90:10
HFAH/H_2O_2[b,135]	CH_2Cl_2	52:48	96:4
VO(acac)$_2$/TBHP[a,c,134a]	CH_2Cl_2	< 5:95	89:11
Ti(OPr-*i*)$_4$/TBHP[a,c,134a]	CH_2Cl_2	< 5:95	98: 2
Mn(salen)PF_6/PhIO[136]	CH_2Cl_2	53:47	94:6
MTO/UHP[a,d,134a]	CCl_4	76:24	88:12

[a] Diastereoselectivity of the 3,4-epoxide regioisomer. [b] Hexafluoroacetone hydrate. [c] *tert*-Butyl hydroperoxide. [d] Methyltrioxorhenium/urea adduct of hydrogen peroxide.

Epoxides can also be prepared by treating alkenes with hydroperoxides, ClO^-, PhIO or molecular oxygen as oxygen sources in the presence of transition-metal complexes, e.g., V, Mo, Ti, Mn, Cr, or Co.[137,138] Hydrogen peroxide can be used as an oxygen donor in combination with catalysts such as tungstic acid and derivatives[139–142] as well as methyltrioxorhenium (MTO).[143–145] Representative examples are also listed in Table D. Evidently, the catalytic mode (in situ generation) of dioxiranes cannot compete in efficacy with metal-catalyzed epoxidations, but the toxicity of most transition metals must be kept in mind in view of environmental problems. Thus, in recent times, the search for nonmetal oxidants has been actively pursued, and dioxiranes can serve as an attractive alternative, especially since now a variety of dioxiranes and reaction conditions have become available. Dioxiranes are also superior to the $F_2/H_2O/MeCN$ system[146] in that dioxiranes are easier to handle and less toxic.

Furthermore, dioxiranes play a prominent role in asymmetric epoxidations. High enantioselectivities have been achieved in the asymmetric epoxidation of unfunctionalized and functionalized trans and trisubstituted olefins by in situ generated optically active dioxiranes. Nevertheless, these enantioselective oxidations hardly surpass the efficiency of environmentally benign enzymatic processes (peroxidase/H_2O_2 or RO_2H)[147] and the well-known Sharpless-Katsuki epoxidation.[148] The latter is limited to allylic alcohols, and the accessibility of large amounts of enzymes for preparative applications represents a major problem in biocatalytic reactions. Furthermore, the level of enantioselectivity in the epoxidation of cis olefins by enantiomerically enriched dioxiranes has yet to reach that of the Jacobsen-Katsuki Mn(salen)-catalyzed process.[100,101,149] Comparative examples are given in Table E for the Sharpless-Katsuki and the Jacobsen-Katsuki asymmetric epoxidations vs. that mediated by the dioxirane derived from ketone **12.**

Evidently, dioxirane chemistry is a fast-developing and growing field, and much progress is expected in the years to come. As the tabulated material (Tables 1–5) discloses, much has been achieved in the last 25 years and, unquestionably, dioxiranes have become established as important catalytic (in situ) and stoichiometric (isolated) epoxidants.

EXPERIMENTAL CONDITIONS

Oxidations with Isolated Dioxiranes

General. Although no explosions have ever been documented, the preparation and reactions of isolated dioxiranes should be carried out in a hood with good ventilation, and all safety measures should be taken. Inhalation and direct exposure to skin must be avoided because dioxiranes are strong oxidants and can potentially damage biomolecules such as DNA.[152]

Solutions of isolated dioxiranes are prepared by treating the ketone precursor with $KHSO_5$ in buffered aqueous solutions. Isolation is achieved either through distillation of readily volatile dioxiranes such as DMD (isol.)[153] and TFD (isol.),[154,155] or by salting out of the dioxirane derived from cyclohexanone[156,157] (see Experimental Procedures). In view of the much higher electrophilicity of 1,1,1-trifluoro-2-propanone compared to acetone, TFD is generated much faster than DMD (about 1 minute for TFD vs. about 20 minutes for DMD). Ketones that readily undergo Baeyer-Villiger rearrangement (cyclobutanone, cyclopentanone), that are not sufficiently electrophilic (benzophenone, acetophenone), or are sterically hindered (pinacolone, adamantanone) are not suitable precursors for the corresponding dioxiranes because either the dioxirane does not persist long enough (first category) or the dioxirane is not generated (last two categories).

The solution of the isolated dioxirane is usually dried over freshly activated molecular sieves (4 Å) or $MgSO_4$. When rigorously dry solutions are required,

TABLE E. COMPARISON OF ENANTIOSELECTIVITIES IN EPOXIDATIONS WITH KETONE **12** ($R^1 = R^2 = Me$) WITH THE SHARPLESS-KATSUKI AND THE JACOBSEN-KATSUKI ASYMMETRIC EPOXIDATIONS

Substrate	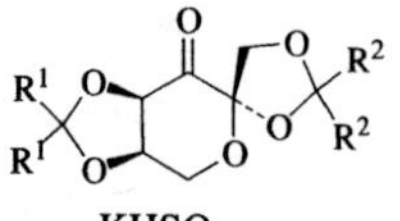 Ketone 12 (R^1, R^2), $KHSO_5$	$Ti(OPr\text{-}i)_4$, $t\text{-}BuO_2H$, tartrate	Mn(salen*) PhIO or NaOCl
 Ph, OH	(85%) 94% ee[149a]	(78%; 78% ee)[150]	—
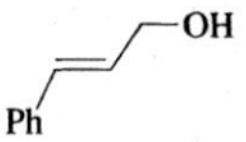 OH, Ph	(70%) 90% ee[118]	(89%; 98% ee)[150]	—
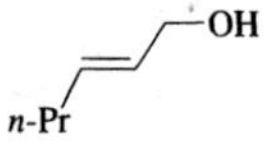 OH, *n*-Pr	(60%) 78% ee[117]	(64%; 93% ee)[150]	—
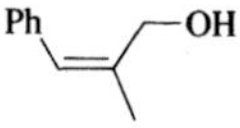 Ph, OH	(75%) 74% ee[149a]	(79%; >98% ee)[150]	—
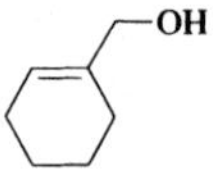 OH	(83%) 92% ee[123,124]	(77%; 93% ee)[150]	—
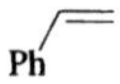 Ph	(90%) 24% ee[25]	—	(88%; 86% ee)[151]
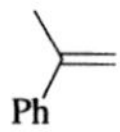 Ph	(81%) 28% ee[25]	—	(36%; 30% ee)[100]
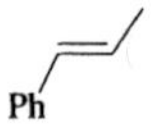 Ph	(99%) 96% ee[25]	—	(32%; 56% ee)[151]
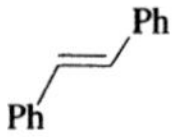 Ph, Ph	(85%) 98% ee[25]	—	(17%; 81% ee)[151]
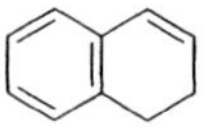	(85%) 32% ee[25]	—	(78%) 98% ee[151]
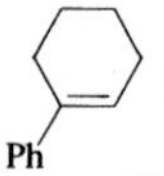 Ph	(94%) 98% ee[25]	—	(69%) 93% ee[151]
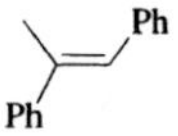 Ph, Ph	(89%) 96% ee[25]	—	(87%) 88% ee[151]

further drying over P_2O_5 and subsequently over anhydrous K_2CO_3 is recommended, with only minor loss (<5%) of the dioxirane content.[158]

Magnetic stirring is usually sufficient for oxidation with the isolated dioxirane. For most oxidations, a properly sized, stoppered Erlenmeyer flask is used as reaction vessel. An inert gas atmosphere is not required unless the substrate or the product is air-sensitive or hydrolytically labile. Pipettes fitted with a bulb filler are routinely used for the transfer of dioxirane solutions. For low-boiling dioxiranes, such as methyl(trifluoromethyl)dioxirane [TFD (isol.)], the pipette must be precooled with liquid nitrogen to minimize loss through evaporation. When an excess of dioxirane is used, it is conveniently removed during solvent evaporation, but care should be taken to condense the evaporate and thereby avoid environmental hazards.

Solvents. Most common laboratory solvents are compatible with DMD (isol.) solutions. Acetone, CH_2Cl_2, $CHCl_3$, CCl_4, and benzene are among the most frequently used cosolvents for epoxidations. With DMD (isol.), cosolvents such as 2-butanone, 1,4-dioxane, THF, *tert*-butyl methyl ether, dimethoxymethane, 1,2-dimethoxyethane, ethyl acetate, and acetonitrile can also be employed. Methanol, ethanol and 2-propanol should not be used since they react slowly with DMD (isol.) even at subambient temperature.[159] For TFD (isol.), ether solvents must be avoided because they are readily oxidized even at low temperature.[160]

Temperature. Epoxidation with DMD solutions [DMD (isol.)] is usually carried out at ambient temperature (ca. 20°). If labile substrates are used or labile products are expected, the reaction can be carried out at subambient or even at dry ice temperatures (−78°). In some cases, the reaction with slowly reacting substrates can be carried out at elevated temperature [e.g., with DMD (isol.) in boiling acetone]. Fortunately, dimethyldioxirane possesses a relatively high activation energy for decomposition (E_a ca. 24.9 kcal/mol),[161] which enables it to tolerate a temperature of 60°.

Oxidations with in situ Generated Dioxiranes

If the substrate and product are hydrolytically robust, oxidation with in situ generated dioxiranes is the most convenient mode of operation. The advantages are that a large number of ketone catalysts are available, generation of the dioxirane is more efficient, and the reaction can be carried out on a large scale under catalytic conditions and with asymmetric induction. DMD (in situ) is by far the most frequently used dioxirane, but the more reactive TFD (in situ) is the choice for less reactive substrates such as electron-poor or sterically hindered alkenes.

Oxidations under Neutral Conditions

Biphasic Media. This is the original protocol for conducting dioxirane epoxidation with DMD (in situ).[31,32] In this method, the substrate is dissolved in an organic solvent that is immiscible with water (for example, CH_2Cl_2, benzene, etc.), and the oxygen source, monoperoxysulfate, is contained in the aqueous phase, in which the dioxirane is formed from the ketone catalyst (usually acetone). Vigorous stirring with a mechanical stirrer is recommended for efficient reaction. The action of phase-transfer catalysts (18-crown-6, Bu_4NHSO_4) may be helpful. The pH of the aqueous phase is maintained at 7 to 8 by adding a suitable buffer solution or solid $NaHCO_3$; for this purpose, use of a pH-stat is convenient and advantageous. If 2-butanone is employed as the dioxirane precursor,[34] no cosolvent is necessary because the aqueous medium and 2-butanone, which has good solubilizing properties, constitute a biphasic system.

Homogeneous Media. A water-miscible organic solvent is employed to bring the substrate and/or ketone catalyst into the aqueous phase. Whereas acetonitrile is most commonly used for this purpose, dioxane, dimethoxymethane (DMM), 1,2-dimethoxyethane (DME), or mixtures thereof, have also found application. The pH of the system is readily controlled by a pH-stat, or by adding the required amount of a powdered mixture of solid $NaHCO_3$ and solid monoperoxysulfate at the beginning of the reaction. Asymmetric epoxidations have all been carried out in homogeneous media; accordingly, this mode of operation has currently gained much importance.

Oxidations under Basic Conditions in Homogeneous Media

This method is actually an extension of the above-mentioned homogeneous system in which a more basic buffer, such as 0.05 M $Na_2B_4O_7$ or K_2CO_3/AcOH buffer (pH ca. 10.5), replaces the $NaHCO_3$ buffer. The advantage of this modification is that some ketone catalysts are more effective and more persistent under basic conditions, which is crucial for conducting catalytic asymmetric epoxidations (cf. Table C). The pH value of the reaction mixture can be conveniently maintained by adding a K_2CO_3 solution.

EXPERIMENTAL PROCEDURES

Caution. Dioxiranes are usually volatile peroxides and thus should be handled with care, by observing all safety measures. The preparations and oxidations should be carried out in a hood with good ventilation. *Inhalation and direct exposure to skin must be avoided!*

Preparation of Isolated Dioxirane Solutions

$$\text{Me}_2\text{C=O} \xrightarrow[\text{H}_2\text{O, Me}_2\text{CO, 5-10°, 20 min}]{\text{KHSO}_5\text{, NaHCO}_3\text{ (buffer)}} \xrightarrow[\text{80-100 mm Hg}]{\text{distill at rt}} \text{DMD (isol.)} \sim 0.08\text{ M in Me}_2\text{CO}$$

Dimethyldioxirane [DMD (isol.)].[153] (The procedure described here is a simplified version of a protocol originally reported by Murray et al.).[162,163] A 4-L, three-necked, round-bottomed reaction flask was equipped with an efficient mechanical stirrer and an addition funnel for solids, connected by means of a U tube (i.d. 25 mm) to a 250-mL, two-necked receiving flask and the latter was cooled to −78° by means of a dry ice/ethanol bath. The reaction flask was charged with a mixture of water (254 mL), acetone (192 mL), and $NaHCO_3$ (58 g) and cooled to 5−10° with the help of an ice/water bath. With vigorous stirring and cooling, solid potassium monoperoxysulfate (120 g, 0.195 mol) was added in five portions at 3-minute intervals. Three minutes after the last addition, a moderate vacuum (80−100 mmHg) was applied, the cooling bath (5−10°) was removed from the reaction flask, and, while the mixture was stirred vigorously, the dimethyldioxirane/acetone solution was distilled (150 mL, 0.09−0.11 M, ca. 5% yield), and collected in the cooled (−78°) receiving flask. The concentration of the dimethyldioxirane solution was most conveniently determined by measuring the absorbance at λ_{max} 325 nm (ε 12.5 ± 0.5 $M^{-1}cm^{-1}$). The acetone solution of dimethyldioxirane was dried over anhydrous K_2CO_3 and stored in the freezer (−20°) over molecular sieves (4 Å).

Dimethyldioxirane-d_6 in Acetone-d_6 [DMD-d_6 (isol.)]. Occasionally it is necessary to employ the deuterated dioxirane DMD-d_6 (isol.) to facilitate direct NMR spectroscopy of the oxidation mixture, without solvent removal. The following procedure is a modification of the procedure for the small-scale preparation of DMD (isol.)[34]: A 250-mL, three-necked, round-bottomed flask, charged with water (20 mL), acetone-d_6 (10 mL), sodium bicarbonate (12 g), and a magnetic stirring bar, was equipped with an addition funnel for solids for the $KHSO_5$ (25 g, 0.041 mol), a gas-inlet tube, and a 29-cm air condenser, loosely packed with glass wool. The exit of the air condenser was connected to the top entry of a high-efficiency, double-jacketted, spiral condenser, supplied with methanol coolant (−85 to −78°) from a Colora Ultra Cryostat (Model KT 290 S). The bottom exit of the high-efficiency condenser was attached to a 25-mL receiving flask, kept at dry ice/acetone temperature (−78°). A slow stream of argon gas was passed through the reaction flask, while the solid $KHSO_5$ was added in one portion under vigorous agitation at <15° and application of water-aspirator pressure (15 mmHg). The effluent was collected as a pale yellow solution (6 mL) of DMD-d_6 in acetone (0.08 M), as determined by UV spectroscopy (cf. procedure for DMD (isol.)].

Dimethyldioxirane in Carbon Tetrachloride.[164] "Acetone-free" (actually not **all** of the acetone is removable!) solutions of DMD in the appropriate solvent may be prepared as reported.[164] Here the protocol is given for the preparation of DMD in CCl_4.

The freshly distilled DMD (isol.) solution (60 mL, 0.08 M in acetone), as prepared above, was diluted with 60 mL of cold water and extracted at 5° in a chilled separately funnel with cold CCl_4 (4 × 3 mL), to afford a total volume of 36 mL of extract as a pale yellow solution. The concentration of the DMD solution was determined by iodometry [see procedure for TFD (isol.)], and of the remaining acetone by ^{1}H-NMR spectroscopy.

To concentrate this DMD solution still further, the combined CCl_4 extracts were washed three times with 1.5 volumes of 0.01 M phosphate buffer (pH 7, prepared by dissolving 3.9 mmol of NaH_2PO_4 and 6.1 mmol of Na_2HPO_4 in 1 L of H_2O) in a cold separatory funnel at 5°. The resulting CCl_4 solution was 0.268 ± 0.082 M in DMD (by iodometry) and 0.155 ± 0.082 M in acetone (by ^{1}H-NMR spectroscopy). Further washing led only to a greater loss of DMD without higher concentrations of DMD. The recovery of DMD from the initial distillate was 41 ± 7.0%

$CF_3C(O)CH_3$ —[$KHSO_5$, $NaHCO_3$ (buffer); H_2O, TFP, 0-5°, 1 min]→ —[distill 0-5°; 716 mm Hg]→ TFD (isol.) 0.4-0.6 M in TFP

Methyl(trifluoromethyl)dioxirane in 1,1,1-Trifluoro-2-propanone [TFD (isol.)].[154,155] A 250-mL, four-necked, round-bottomed flask was equipped with an efficient mechanical stirrer, an addition funnel for low-boiling liquids [1,1,1-trifluoro-2-propanone (TFP boils at 21°; it is important to use ether-free TFP doubly distilled over P_2O_5[165])], an addition funnel for solids, and a gas-inlet tube. The exit of the gas-inlet tube was connected to the top entry of a high-efficiency, double-jacketed, spiral condenser, supplied with ethanol coolant (−80°) from a Lauda RLS 6-D cryostat (2.1 kW). The bottom exit of the high-efficiency condenser was attached to a 50-mL, pear-shaped receiving flask, cooled to −60 to −50° by an ethanol bath with liquid nitrogen. A sidearm at the condenser allowed application of a slightly reduced pressure by means of a water aspirator, which controlled the desired pressure automatically.

The 250-mL, four-necked, round-bottomed flask was charged with a slurry of $NaHCO_3$ (13.0 g, 155 mmol) in water (13 mL) while being cooled in an ice bath. Potassium monoperoxysulfate (24.0 g, 39.1 mmol) was added through the addition funnel to the vigorously stirred slurry of $NaHCO_3$, while much CO_2 gas evolved. The precooled (0°) addition funnel for liquids was quickly charged with TFP (12.0 mL, 134 mmol) after 80 seconds of CO_2 evolution, and the TFP was added to the reaction mixture within 1 minute. After a further 20 seconds the water aspirator was adjusted to a slight vacuum (650 mmHg) and the pale yellow solution of the TFD in TFP was collected in the cooled (−60 to −50°) receiving flask. After 8 minutes a new batch of potassium monoperoxysulfate (8.0 g, 13 mmol) was added, and the reaction mixture was stirred for an additional 8 minutes. The water aspirator was disconnected and the receiving flask was allowed to warm to −25° (to allow the frozen CO_2 to evolve). The flask was removed after 5 minutes from the high-efficiency condenser, tightly closed with a plastic stopper, and wrapped with aluminum foil to protect the TFD from light. IR (vapor): 1259, 1189, 971 cm^{-1};[155] ^1H NMR (200 MHz, −20°) δ 1.97 (s);[155] ^{13}C NMR (100 MHz, −20°) δ 14.5 (q), 97.3 (q, $J_{C\text{-}F}$ = 40.2 Hz), 122.2 (q, $J_{C\text{-}F}$ = 280.7 Hz);[155] ^{17}O NMR (54 MHz, −20°) δ 297 (s);[155] ^{19}F NMR (188 MHz, −20°) δ −81.5 (s);[155] UV: λ_{max} 347 nm ($\varepsilon \approx 9\ M^{-1}cm^{-1}$ at 0°).[155] The TFD yield was 2.0 ± 0.5% (relative to TFP) and was determined iodometrically (1 mL H_2O, 3 mL glacial acetic acid, 0.5 mL saturated KI solution; addition of 0.200 mL of the TFD solution at 0°; titration with a freshly standardized 0.05 N $Na_2S_2O_3$ solution). The concentration of TFD in TFP ranged from 0.4 to 0.6 M, the volume of the distillate from 4 to 6 mL (determined gravimetrically by assuming a density of that of TFP, i.e., 1.252 g/mL).

The solution can be stored for several months at −20° with only minor loss of peroxide titer (ca. 5% loss per month). A calibrated pipette, which was cooled briefly with liquid nitrogen, was used to administer conveniently amounts of the oxidant (titer ± 5% error).

Methyl(trifluoromethyl)dioxirane in Carbon Tetrachloride.[27] Ketone-free solutions of TFD in the appropriate solvent can be prepared as reported.[27] Here the protocol for TFD in CCl_4 is given.

A 50-mL, jacketed separatory funnel was charged with 4–6 mL of the yellow TFD (isol.) solution (0.5–0.6 M) in TFP, as prepared above, and 4–6 mL CCl_4. The solution was washed with 8–12 mL of cold, doubly distilled water (over $KMnO_4$) at 0°. The TFP went quickly into the aqueous phase while the "ketone-free" TFD was contained in the organic layer. The latter was dried briefly over $MgSO_4$, quickly filtered, and its TFD content was determined by iodometry [cf. procedure for TFD (isol.)]. The TFD solution in CCl_4 was stored over 4 Å molecular sieves in the freezer (−20°), and protected from exposure to light.

The TFD solution in CCl_4 was further concentrated by slow evaporation of a frozen (liquid N_2) sample under high vacuum (>0.01 mmHg). The effluent was condensed in a cold trap kept at liquid nitrogen temperature. Typically, 2 mL of an 0.85 M solution of TFD in CCl_4 was obtained from 3 mL of a 0.6 M solution.

$KHSO_5$, KOH (15%, buffer); H_2O, −10 to 0°, 5-12 min; salting out Na_2SO_4, NaH_2PO_4, Na_2HPO_4 → 0.2-0.8 M in cyclohexanone

1,2-Dioxaspiro[2.5]octane (Cyclohexanone Dioxirane).[156,157] *All materials used in the preparation, including drying agents, were cooled in an ice-salt bath before use.* Cyclohexanone (50 mL) and crushed ice (10–15 g) were stirred well with a mechanical stirrer. A cold slurry of potassium monoperoxysulfate (35 g) in water (100 mL) was added to the reaction vessel over a period of 5–10 minutes while the pH was maintained between 7.5 and 8.5 by addition of cold 15% KOH, followed by additional stirring for 1–2 minutes. At this point, the reaction mixture was dark yellow. The reaction mixture was then poured into a cold mixture of 70 g of sodium sulfate, sodium dihydrogen phosphate, and sodium monohydrogen phosphate (4:2:1). This mixture was stirred vigorously for a few seconds and then poured into a cold separatory funnel; care was taken to minimize the transfer of undissolved salts to the separatory funnel. The yellow organic layer was then separated and dried (Na_2SO_4). The separated liquid was then decanted, dried over molecular sieves, and stored in the freezer. The concentration of the dioxirane obtained in this manner was in the range 0.2–0.8 M, as determined by iodometric titration [see the TFD (isol.) procedure above.]

Epoxidations

DMD (isol.); Me_2CO, 22°, 2 h → (~100%)

Racemic *trans*-1,2-Diphenyloxirane, Method A [with DMD (isol.)].[163] To a magnetically stirred solution of *trans*-stilbene (0.724 g, 4.02 mmol) in 5 mL of acetone, contained in a 125-mL stoppered Erlenmeyer flask, was added a solution of DMD in acetone (0.062 M, 66 mL, 4.09 mmol) at room temperature (ca. 20°). The progress of the reaction was followed by GLC analysis, which indicated that *trans*-stilbene was converted into the oxide in 6 hours. Removal of the excess acetone on a rotary evaporator (20°, 15 mmHg) afforded a white crystalline solid. The solid was dissolved in CH_2Cl_2 (30 mL) and dried over anhydrous Na_2SO_4. The drying agent was removed by filtration and washed with CH_2Cl_2. The solution was concentrated on a rotary evaporator, and the remaining solvent was removed (20°, 15 mmHg) to give an analytically pure sample of the oxide (0.788 g, 100%). Crystallization from aqueous EtOH gave white plates, mp 69–70°; IR ($CHCl_3$) 3076, 3036, 2989, 1603,

1497, 1457, 870 cm^{-1}; ^{1}H NMR (300 MHz, $CDCl_3$): δ 3.86 (s, 2 H), 7.26–7.45 (m, 10 H); ^{13}C NMR (75 MHz, $CDCl_3$): δ 62.8 (d), 125.4 (d), 128.2 (d), 128.4 (d), 137.0 (s).

Method B [with DMD (in situ)].[36] To a 5-L, three-necked flask, equipped with a mechanical stirrer, was added *trans*-stilbene (18.0 g, 100 mmol), a mixture of CH_3CN and dimethoxymethane (1.17 L, 2:1 v/v) as solvent, 0.1 M aqueous K_2CO_3 (330 mL), acetone (220 mL, 3 mol), and tetrabutylammonium hydrogen sulfate (1.5 g). Potassium monoperoxysulfate (92.2 g, 150 mmol) in 330 mL of aqueous 4×10^{-4} M EDTA and K_2CO_3 (92.2 g, 667 mmol) in 330 mL of H_2O were added separately by means of addition funnels over a period of 2 hours, while the pH was adjusted to 10.5 by administering glacial acetic acid dropwise. Subsequently, the reaction mixture was extracted with hexane (3×1.5 L), the extracts were washed with brine (1×1 L), dried (Na_2SO_4), concentrated (20°, 15 mmHg), and the residue was purified by flash chromatography on silica gel (deactivated with 1% Et_3N in hexane) to yield *trans*-1,2-diphenyloxirane as a white solid (17.0 g, 87%).

12

KHSO$_5$, NaHCO$_3$ (buffer)

H$_2$O, MeCN, *n*-Bu$_4$NHSO$_4$,

Na$_2$EDTA, 0°, 2 h

(73%), 95% ee

(*R*,*R*)-1,2-Diphenyloxirane with Ketone 12, Method A (Neutral, in situ Noncatalytic Conditions).[25,117] An aqueous solution of Na_2EDTA (1×10^{-4} M, 10 mL) and a catalytic amount of tetrabutylammonium hydrogen sulfate (15 mg) were added under vigorous magnetic stirring at 0° to a solution of *trans*-stilbene (0.18 g, 1.00 mmol) in acetonitrile (15 mL), contained in a stoppered 100-mL Erlenmeyer flask. A mixture of potassium monoperoxysulfate (3.07 g, 5.00 mmol) and sodium bicarbonate (1.3 g, 15.5 mmol) was pulverized and a small portion of this solid mixture was added to the reaction mixture to bring the pH to above 7. After 5 minutes, 0.77 g (3.0 mmol) of ketone **12** was added in equal portions over a period of one hour. Simultaneously, the rest of the pulverized solid mixture of $KHSO_5$ and sodium bicarbonate was added in equal portions within 50 minutes to maintain the pH at 7. After completion of the addition, the reaction mixture was magnetically stirred for another hour at 0°, diluted with water (30 mL), and extracted with hexane (4×40 mL). The combined extracts were washed with brine, dried (Na_2SO_4), filtered, and concentrated (20°, 15 mmHg). The product was purified by flash chromatography on silica gel (deactivated with 1% Et_3N in hexane), first eluted with hexane and subsequently with a 50:1 mixture of hexane/Et_2O to afford (*R*,*R*)-1,2-diphenyloxirane as a white solid (0.149 g, 73%, 95.2% ee). The ee value was determined by ^{1}H-NMR spectroscopy with Eu(hfc)$_3$ as the shift reagent.

Method B (Basic, in situ Catalytic Conditions).[25] To a solution of *trans*-stilbene (0.181 g, 1.00 mmol) in CH_3CN/DMM mixture (15 mL, 1:2 v/v), contained in a stoppered, 100-mL Erlenmeyer flask, was added an aqueous solution (10 mL) of 0.05 M $Na_2B_4O_7 \cdot 10H_2O$, which contained 4×10^{-4} M Na_2EDTA, tetrabutylammonium hydrogen sulfate (15 mg, 0.04 mmol), and 0.0774 g (0.30 mmol) of ketone **12**. An aqueous solution (6.5 mL) of potassium monoperoxysulfate (1.00 g, 1.60 mmol) contained in 4×10^{-4} M Na_2EDTA and an aqueous solution (6.5 mL) of K_2CO_3 (0.93 g, 6.74 mmol) were added dropwise with vigorous magnetic stirring at 0° over a period of 30 minutes through two separate addition funnels. The reaction mixture was then diluted with water (30 mL) and extracted with hexane (4 × 40 mL). The combined extracts were washed with brine, dried (Na_2SO_4), filtered, and concentrated (20°, 15 mmHg). The product was purified by flash chromatography on silica gel (deactivated with 1% Et_3N in hexane), by first eluting with hexane and subsequently with a mixture of 50:1 hexane/Et_2O, to afford (*R,R*)-1,2-diphenyloxirane as a white solid (0.166 g, 85%, 97.9% ee), mp 68–70° (hexane); $[\alpha]_D^{25}$ +356.1° (*c* 0.95, benzene). The ee value was determined by ^{1}H-NMR spectroscopy with Eu(hfc)$_3$ as the shift reagent.

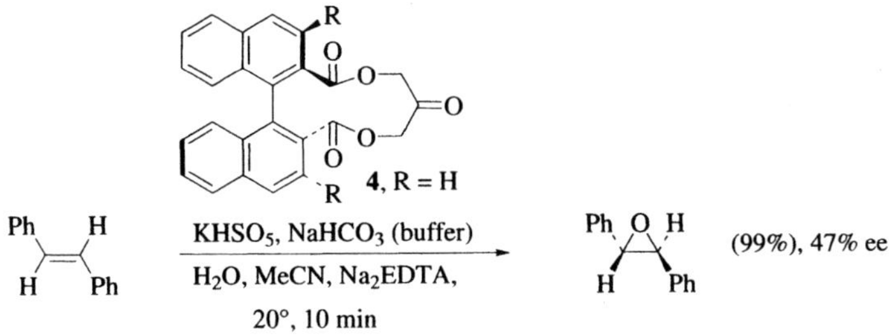

4, R = H

KHSO$_5$, NaHCO$_3$ (buffer)
H$_2$O, MeCN, Na$_2$EDTA,
20°, 10 min

(99%), 47% ee

(*S,S*)-1,2-Diphenyloxirane with Ketone 4 (in situ Catalytic Conditions).[24] To a solution of *trans*-stilbene (18.0 mg, 0.10 mmol) and ketone **4** (R = H) (29.6 mg, 0.10 mmol) in CH_3CN (1.5 mL) contained in a stoppered, 25-mL Erlenmeyer flask at room temperature (ca. 20°) was added a 4×10^{-4} M aqueous Na_2EDTA solution (1 mL). To this solution was added in portions a solid mixture of pulverized sodium bicarbonate (130 mg, 1.55 mmol) and potassium monoperoxysulfate (614 mg, 1.00 mmol), while vigorously stirring magnetically. After 7 minutes, the reaction mixture was poured into water (20 mL), extracted with CH_2Cl_2 (3 × 20 mL), and dried (Na_2SO_4). Upon removal of the solvent (20°, 15 mmHg), the residue was purified by flash chromatography on silica gel (deactivated with 2% Et_3N in hexane), by first eluting with hexane (50 mL), followed by a 1:19 mixture of EtOAc/hexane (100 mL), which gave (*S,S*)-1,2-diphenyloxirane [19.4 mg, 99%, 47% ee]. Subsequent elution with a 2:3 mixture of EtOAc/hexane afforded 27.6 mg (93% recovery) of ketone **4**. The ee was determined by ^{1}H-NMR spectroscopy with Eu(hfc)$_3$ as chiral shift reagent.

$$\text{PhCH=CHCH}_2\text{OH} \xrightarrow[\text{H}_2\text{O, MeCN, pH 7-7.5, 20°, 1.5 h}]{\text{KHSO}_5\text{, NaHCO}_3\text{ (buffer)}} \text{epoxide (95\%)}$$

Racemic 2,3-Epoxy-3-phenyl-1-propanol with 1,1-Dioxotetrahydrothiopyran-4-one as Ketone Catalyst.[38] A 50-mL stoppered Erlenmeyer flask was charged with a CH_3CN solution (9 mL) of cinnamyl alcohol (2.0 mmol), ketone catalyst (5 mol%), and an aqueous Na_2EDTA solution (6 mL, 4×10^{-4} M). At room temperature was added, with vigorous magnetic stirring, a pulverized solid mixture of $KHSO_5$ (1.84 g, 3.0 mmol) and $NaHCO_3$ (0.78 g, 9.3 mmol) within 1.5 hours. The reaction progress was monitored by TLC (silica gel) and after complete consumption of the substrate, the reaction mixture was extracted with EtOAc (2×50 mL). The combined organic layers were dried ($MgSO_4$) and concentrated (20°, 15 mmHg). The residue was purified by flash chromatography on silica gel to afford the epoxide (95% yield) and ketone catalyst (80% recovery).

$$\text{PhCH=CHCH}_2\text{OH} \xrightarrow[\text{DMM, MeCN, pH 9.3, Na}_2\text{EDTA, } n\text{-Bu}_4\text{NHSO}_4\text{, 10°, 3 h}]{\textbf{12}\text{, KHSO}_5\text{, K}_2\text{CO}_3\text{,}} \text{epoxide (85\%), 94\% ee}$$

(*R*,*R*)-2,3-Epoxy-3-phenyl-1-propanol with Ketone 12 (in situ Catalytic Conditions).[149a] A 100-mL, two-necked, round-bottomed flask, equipped with a magnetic stirrer, was charged with cinnamyl alcohol (136 mg, 1.0 mmol), ketone **12** (77.4 mg, 0.30 mmol), and tetrabutylammonium hydrogen sulfate (15.0 mg, 0.016 mmol) in 10 mL of DMM/CH_3CN (2:1 v/v). An aqueous solution (7 mL) of K_2CO_3/HOAc [prepared by mixing 100 mL of 0.1 M aqueous K_2CO_3 and 0.5 mL HOAc (pH 9.3)] was added with stirring and cooling in a NaCl/ice bath (−10°). This was followed by the dropwise addition of an aqueous solution (5 mL) of $KHSO_5$ (0.85 g, 1.38 mmol) in 4×10^{-4} M Na_2EDTA and an aqueous solution (5 mL) of K_2CO_3 (0.80 g, 5.8 mmol), separately by means of syringe pumps, over a period of 3 hours. After CH_2Cl_2 (20 mL) and water were added, the aqueous layer was extracted with CH_2Cl_2 (3×20 mL), and the combined organic phases were washed with brine, dried (Na_2SO_4), and concentrated (20°, 15 mmHg). The residue was purified by flash chromatography on silica gel [deactivated with 1% Et_3N in hexane/Et_2O (2:1 v/v)] by elution with a hexane/Et_2O mixture [first 2:1 (v/v) and subsequently 1:1 (v/v)], to afford 0.128 g (85% yield) of the epoxide (94% ee, by chiral HPLC on a Chiracel OD column).

OH O–O Me$_2$CO, 20°, 2 h OH O + OH O (97%) 76:24 *threo* *erythro*

***threo*- and *erythro*-α,3,3-Trimethyloxiranemethanol [with DMD (isol.)].**[72] A stoppered 25-mL Erlenmeyer flask was charged at 20° with 72 mg (0.72 mmol) of 4-methylpent-3-en-2-ol and a 0.069 M solution (10 mL, 0.73 mmol) of DMD (isol.). The mixture was stored at 20° in the dark until the peroxide test (KI/HOAc) was negative. The solvent was removed (20°, 20 mmHg) and 81 mg (97% yield) of a 76:24 mixture of threo/erythro epoxides (by ^{1}H-NMR analysis) were obtained as a colorless oil.

OH O–O CF_3 CF_3COMe, 0 - 10°, 15 - 60 min OH O + OH O (<95%) 85:15 *threo* *erythro*

***threo*- and *erythro*-α,2,3-Trimethyloxiranemethanol [with TFD (isol.)].**[80] A stoppered 25-mL Erlenmeyer flask was charged at 0–10° with 45 mg (0.45 mmol) of (*Z*)-3-methylpent-3-en-2-ol and a 0.05625 M solution (8 mL, 0.45 mmol) of TFD (isol.). The mixture was stored in the dark at 0–10° until the peroxide test (KI/HOAc) was negative. The solvent was removed (20°, 15 mmHg) and 49.6 mg (95% yield) of a 85:15 mixture of threo/erythro epoxides (by ^{1}H-NMR analysis) were obtained as a colorless oil.

O–O CH_2Cl_2, Et_2O, N_2, –30°, 3 h O (62%)

7-Oxatetracyclo[4.1.0.0^{2,4}.0^{3,5}]heptane (Benzvalene Epoxide) [with DMD (isol.)].[41] A 1-L, two-necked, round-bottomed flask was equipped with a magnetic stirring bar, gas-inlet and gas-outlet tubes, and charged with an anhydrous solution of benzvalene (1.97 g, 25.2 mmol) in Et_2O (60 mL). The flask was cooled to −30° under a nitrogen atmosphere and, while stirring magnetically, 400 mL of a 0.0483 M solution of DMD (19.3 mmol) was added at such a rate that the reaction temperature was kept below −20°. After complete addition, the mixture was stirred at −25° for another 3 hours and then concentrated (0°, 15 mmHg). The residue was distilled (bulb-to-bulb, 20°, 0.01 mmHg), the receiving flask being kept at −78°, to afford 1.12 g (62%) of the epoxide as a colorless liquid containing a trace of solvent.

TBDMSO / Ph —[**12**; $KHSO_5$, $NaHCO_3$ (buffer); H_2O, MeCN, Na_2EDTA, 0°, 3 h]→ [MeOH, 20°, 2 h]→ Ph(C=O)CH(OH)CH$_3$ (69%) 82% ee

(*R*)-2-Hydroxy-1-phenylpropan-1-one with Ketone 12 under in situ Conditions.[26,166] To a solution of (*Z*)-*tert*-butyldimethyl[(1-phenyl-1-propenyl)oxy]-silane (24.8 mg, 0.100 mmol) in CH_3CN (1.5 mL) was added with vigorous magnetic stirring at 0° an aqueous solution (4×10^{-4} M) of Na_2EDTA (1.0 mL). A mixture of $KHSO_5$ (307 mg, 0.500 mmol) and $NaHCO_3$ (130 mg, 1.55 mmol) as a pulverized solid was added in small portions over 50 minutes. Ketone **12** (77.5 mg, 0.300 mmol) was added in portions simultaneously. The mixture was stirred magnetically for 18 hours at 0°, and was then diluted with water and extracted with hexane. The hexane extracts were combined and concentrated (40°, 250 mmHg). Methanol (5 mL) was added to the residue and the solution was stirred for 2 hours for complete desilylation. After removal of the methanol (20°, 15 mmHg), the crude product was purified by silica gel chromatography (1:2 Et_2O/ petroleum ether as eluent) to give the title compound as a colorless oil (10.4 mg, 69%, 82% ee). The ee value was determined by HPLC analysis on a Chiralcel OD column, 9:1 hexane/ 2-propanol, flow 0.6 mL/min. IR (film): 3700–3100 (br), 3063, 2981, 2933, 1682, 1597, 1578, 1451, 1271 cm^{-1}. ^1H NMR (250 MHz, $CDCl_3$): δ 1.42 (d, J = 7.0 Hz, 3 H), 3.81 (s, 1 H, OH), 5.14 (q, J = 7.0 Hz, 1 H), 7.44–7.59 (m, 3 H), 7.87–7.95 (m, 2 H). ^{13}C NMR (63 MHz, $CDCl_3$): δ 22.5 (q), 69.6 (d), 128.9 (d), 129.1 (d), 133.6 (s), 134.2 (d), 202.6 (s).

PhC(O)NH–CH=CH–CH$_3$ —[O–O / CD_3 CD_3; $CDCl_3$, Me_2CO-d_6, N_2, −78°, 1 h]→ PhC(O)NH-epoxide (~100%)

***trans*-(2-Methyloxiranyl)benzamide [with DMD-d_6 (isol.)].**[167] A 5-mm NMR tube was charged at −78° under a N_2 atmosphere with 9.0 mg (55.8 μmol) of the enamide in 50 μL of $CDCl_3$. By means of a syringe, 750 μL (55.9 μmol) of a well-dried (over 4 Å molecular sieves) 0.074 M solution of DMD-d_6 (isol.) in acetone-d_6 was added rapidly at −78°. After 1 hour, the NMR tube was submitted to low-temperature (−50°) ^1H-NMR and ^{13}C-NMR spectroscopy, which showed that the epoxide was obtained quantitatively. At temperatures higher than −50°, the epoxy enamide deteriorated into an intractable, undefined product mixture. ^1H NMR (acetone-d_6, 200 MHz, −50°): δ 1.37 (d, J = 6.8 Hz, 3 H, CH_3), 3.36 (m, 1 H, CH), 4.82 (m, 1 H, CH), 7.49–7.59 (m, 3 H), 7.86–7.99 (m, 2 H), 8.14 (br s, 1 H; NH); ^{13}C NMR (acetone-d_6, 50 MHz, −50°): δ 16.8 (q, CH_3), 53.3 (d, CH), 61.6 (d, CH), 127.8 (2xd), 129.1 (2xd), 132.6 (d), 134.2 (s), 168.8 (s, CO).

DMD (isol.), CH_2Cl_2

N_2, –78 to –20°, 2 h

(>95%)

2,3-Epoxy-2,3-dihydro-2,3,4-trimethylbenzo[*b*]furan [with DMD (isol.)].[48] A cooled (−78°, 0.086 M) solution of dimethyldioxirane in acetone (15 mL, 1.30 mmol), dried over 4 Å molecular sieves, was rapidly added to a cooled (−78°), magnetically vigorously stirred solution of 2,3,4-trimethylbenzofuran (160 mg, 1.00 mmol) in dry CH_2Cl_2 (2 mL) under a N_2 atmosphere. Stirring was continued for 3 hours at −78 to −20°. The solvent was removed (−20°, 0.001 mmHg) to afford the benzofuran epoxide essentially quantitatively, as confirmed by ^{1}H-NMR and ^{13}C-NMR analysis at −20°. ^{1}H NMR (200 MHz, acetone-d_6, −20°): δ 1.82 (s, 3 H), 1.84 (s, 3 H), 2.46 (s, 3 H), 6.74–6.78 (m, 2 H), 7.12–7.70 (m, 1 H). ^{13}C NMR (50 MHz, acetone-d_6, −20°): δ 13.9 (q), 14.7 (q), 18.8 (q), 67.2 (s), 94.8 (s), 109.2 (d), 123.7 (d), 128.4 (s), 129.6 (d), 136.6 (s), 160.4 (s).

PvO, PvO, PvO, O

$CF_3C(O)CH_3$, $KHSO_5$, $NaHCO_3$

H_2O, MeCN, pH 7, Na_2EDTA, 0°, 1.3 h

MeOH

20°,1 h

PvO, PvO, PvO, O, OMe, OH (63%) + PvO, HO, PvO, PvO, O, OMe (21%)

Methyl 3,4,6-Tri-*O*-pivaloyl-*β*-D-glucopyranoside and Methyl 3,4,6-Tri-*O*-pivaloyl-*α*-D-mannopyranoside [with TFD (in situ)].[35] A stoppered 10-mL Erlenmeyer flask, equipped with a magnetic stirring bar, was charged with a CH_3CN solution (1 mL) of 3,4,6-tri-*O*-pivaloyl-D-glucal (40 mg, 0.1 mmol) and an aqueous solution of Na_2EDTA (0.4 mL, 4×10^{-4} M). After cooling to 8° by means of an ice bath, TFP (0.2 mL) was added by means of a precooled syringe. A powdered mixture of solid $KHSO_5$ (0.307 g, 1.00 mmol) and solid $NaHCO_3$ (0.130 g, 155 mmol) was added to the vigorously stirred mixture. The reaction was complete within 15 minutes, as assessed by TLC (silica gel) monitoring. Anhydrous Na_2SO_4 (3.0 g) was added at 20° followed by dry MeOH (10 mL). TLC (silica gel) monitoring indicated that ring-opening of the epoxide was complete within 1 hour. After the addition of water (20 mL), the reaction mixture was extracted with CH_2Cl_2 (3×20 mL), and the combined extracts were dried (Na_2SO_4) and concentrated (20°, 15 mmHg). The residue was purified by flash column chromatography on silica gel, with a 23:77 mixture of EtOAc and hexane as eluent, to afford 28.2 mg (63% yield) of glucopyranoside as a colorless solid and 9.5 mg (21% yield) of mannopyranoside as a colorless oil.

MeO OH H H O $^{-}$OTf Me N $C_{12}H_{25}$-*n* $KHSO_5$, KOH (buffer) H_2O, CH_2Cl_2, pH 7.5, 0° O OH OH H H O (81%)

17α-Hydroxy-21-hydroxypregna-4.9(11)-diene-3,20-dione with 1-Dodecyl-1-methyl-4-oxopiperidinium Trifluoromethanesulfonate as Ketone Catalyst.[168] A 250-mL, three-necked, round-bottomed flask was equipped with a Brinkmann-Heidolph overhead stirrer (Model No. 2050). The glass stirring shaft was fitted with an elliptically shaped Teflon stirring paddle (40-mm long × 18-mm high × 3-mm wide). A glass pH probe/electrode (Broadley-James Model No. C1207A-121-A03BC), connected to a Brinkmann pH-stat (Brinkmann Models: E512 pH meter, Impulsomat No. 473, and Dosimat No. E412), was inserted into the flask at one of the side necks of the flask and clamped such that the bottom of the probe was 0.5–1.0 cm from the bottom of the flask. The third neck was fitted with two separate Teflon tubes, one that delivered KOH (2 N) (the rate of addition was controlled by the pH-stat) and the other that delivered aqueous $KHSO_5$ [(0.45 M, stabilized by 0.43 mM Na_2EDTA), its rate of addition controlled by a syringe pump (Sage Instruments Model 355)]. The 250-mL flask was charged with phosphate buffer (23 mL, pH 7.8), CH_2Cl_2 (20 mL), the steroid (75 mg, 0.21 mmol), and the ketone catalyst (9 mg, 0.021 mmol). The contents were cooled to 0° by means of an ice bath, and with vigorous stirring (800–1000 rpm), the aqueous $KHSO_5$ solution (4.6 mL, 1.98 mol) was added by means of the syringe pump at such a rate that the pH of the reaction mixture was kept constant at 7.5 (± 0.1) by automated simultaneous addition of the 2 N aqueous KOH solution (ca. 80 minutes). The temperature was kept at 0° by cooling with an ice bath. The reaction progress was monitored by TLC (silica gel), and after complete consumption of the steroid, the organic layer was removed. To the aqueous layer was added brine (60 mL) and the solution was extracted with CH_2Cl_2 (4 × 10 mL). The combined organic layers were dried ($NaSO_4$) and concentrated (20°, 15 mmHg). The residue was purified by silica gel chromatography with a 1:1 mixture of EtOAc/hexane as eluent to afford 61 mg (81% yield) of the colorless product; mp 208–209° after recrystallization from CH_2Cl_2/hexane.

OH O Ph O–O CH_2Cl_2, Me_2CO, 20°, 24 h OH O O Ph (~100%)

2,3-Epoxy-1-(2-hydroxyphenyl)-3-phenyl-1-propanone [with DMD (isol.)].[34] A stoppered 100-mL Erlenmeyer flask equipped with a magnetic stirring bar was charged with chalcone (205 mg, 0.92 mmol) in CH_2Cl_2 (10 mL), and 17.5 mL of a

0.060 M (1.05 mmol) acetone solution of DMD (isol.) was added rapidly with stirring at 20°. After 12 hours of stirring, a fresh batch of 17.5 mL of a 0.060 M (1.05 mmol) acetone solution of DMD (isol.) was added rapidly and stirring was continued for another 12 hours at 20°. The reaction progress was monitored by TLC (silica gel). After complete consumption of the chalcone, the solvent was removed (20°, 15 mmHg) and 220 mg (~100%) of pure epoxide were obtained as a colorless powder, mp 73–74° (from $CHCl_3$/petroleum ether).

MeO Ph — $KHSO_5$, phosphate buffer, pH 7.3-7.5, KOH (3%), 20°, 48 h (2-butanone catalyst) → MeO Ph (94%)

1-(4-Anisyl)-2,3-epoxy-3-phenyl-1-propanone with 2-Butanone as Ketone Catalyst [in situ Catalytic Conditions].[34] To a vigorously stirred (mechanical agitation) mixture of the chalcone (3.57 g, 15.0 mmol), 2-butanone (150 mL), and phosphate buffer, prepared by dissolving 0.177 g of KH_2PO_4 and 0.648 g of Na_2HPO_4 in 150 mL of water, contained in a 1-L, 3-necked, round-bottomed flask, was slowly added (6 hours) at room temperature a saturated aqueous solution (200 mL) of $KHSO_5$ (45 g, 0.073 mol). The pH of the mixture was kept at 7.3–7.5 by continuously administering an aqueous KOH (3%) solution. After 18 hours of additional stirring, a new batch of a saturated, aqueous solution (200 mL) of $KHSO_5$ (45 g, 0.073 mol) was added slowly (6 hours) and the mixture was stirred for an additional 18 hours. The reaction progress was monitored by TLC (silica gel). After complete consumption of the chalcone, solid NaCl was added to the cloudy reaction mixture until saturation, the organic phase was separated by decantation, and the aqueous phase was extracted with CH_2Cl_2 (4 × 50 mL). The combined organic layers were dried ($MgSO_4$), filtered, and the solvent was evaporated (20°, 15 mmHg) to afford 3.60 g (94%) of the epoxide in high purity (by ^{1}H-NMR spectroscopy) as colorless plates, mp 75–76°.

Ph — DMD (isol), CH_2Cl_2, 0°, 36 h → Ph (~100%)

1a,7a-Dihydro-1a-phenyl-7*H*-oxireno[*b*][1]benzopyran-7-one (Flavone Epoxide) [with DMD (isol.)].[59] A stoppered 125-mL Erlenmeyer flask was charged with 187 mg (0.840 mmol) of flavone in 10 mL CH_2Cl_2. While stirring magnetically, a total of 61.5 mL of a 0.084 M (5.14 mmol) DMD solution in acetone was added at 20° in equal portions at three 12-hour intervals. After 36 hours, the conversion of the flavone was complete, as confirmed by TLC (silica gel), and the solvent was removed (20° at 15 mmHg) to yield 200 mg (~100%) of the epoxide as colorless needles, mp 99–100° (from $CHCl_3$/petroleum ether).

(6 equiv), *S,S* catalyst (16 mol%), CH_2Cl_2, Me_2CO, 20°, 10 d

(1a*S*,7a*R*)-epoxide (23%), 90% ee

(1a*S*,7a*R*)-1a,7a-Dihydro-4-mesyloxy-7a-(2-methoxyphenyl)-7*H*-oxireno-[*b*][1]benzopyran-7-one with Jacobsen's (*S*,*S*)-Catalyst [with DMD (isol.)].[103,104] A stoppered 250-mL Erlenmeyer flask equipped with a magnetic stirring bar was charged with the isoflavone (346 mg, 1.00 mmol) and the *S*,*S*-configured Jacobsen's Mn(III)salen complex (101 mg, 0.16 mmol, 16 mol %) in dry CH_2Cl_2 (10 mL). While stirring magnetically, 12.0 mL of a 0.05 M (0.6 mmol) acetone solution of DMD (isol.) was added at 20°. The reaction progress was monitored by TLC (silica gel), and new batches (as above) were added in 24-hour intervals until the consumption of the isoflavone halted [40% conversion after 10 days reaction time, during which a total of 120 mL (6.0 mmol) of DMD (isol.) in acetone was administered]. The solvent was evaporated (25, 15 mmHg) and the product was purified by silica gel chromatography, with toluene/EtOAc (4:1 v/v) as eluent, to afford 83.3 mg (23% yield) of the epoxide; colorless plates (methanol), mp 145–146°; 90% ee, determined by chiral HPLC analysis (Chiracel OD column), with hexane/2-propanol (9:1) as eluent (flow rate 0.6 mL/minute).

BF_3•OEt_2, Et_2O, Ar, –70°, 30 min; 1. CF_3-dioxirane, CH_2Cl_2, Ar, –20°, 30 min; 2. $KHCO_3$ (aq), 20° (82%)

***N*-Benzyl-3,4-epoxypyrrolidine [with TFD (isol.)].**[63] A 10-mL, two-necked, round-bottomed flask with gas-inlet and gas-outlet tubes and magnetic stirring bar was charged with a solution of *N*-benzyl-3-pyrroline (159 mg, 1.0 mmol) in Et_2O (2 mL). To this solution was added with stirring under argon 135 μL (1.05 mmol) of freshly distilled $BF_3 \cdot Et_2O$ at $-70°$. The mixture was stirred for 30 minutes under these conditions, and a white solid precipitated. The solvent was removed ($-20°$, 1 mmHg), to afford the BF_3 adduct of the amine quantitatively. ^{1}H-NMR analysis confirmed that all of the ether had been rigorously removed. ^{1}H NMR (300 MHz) δ 3.89 (d, J = 14 Hz, 2 H), 4.13 (s, 2 H), 4.26 (d, 2 H), 5.48 (s, 2 H), 7.30–7.60 (m, 5 H). ^{13}C NMR (75 MHz) δ 59.9 (2 × t), 60.4 (2 × t), 124.7 (2 × d), 128.6 (2 × d), 129.4 (d), 131.7 (2 × d).

The BF_3 adduct was dissolved in dry CH_2Cl_2 (3 mL) at $-70°$, and the solution was added at this temperature with stirring to 1.9 mL of a 0.60 M solution (1.1 mmol) of TFD (isol.). The reaction mixture was allowed to warm to $-20°$ and stirred for 30 minutes for complete consumption of the TFD, as confirmed by the negative peroxide test (KI/HOAc). The reaction mixture was poured into 100 mL of a 0.1 M aqueous $KHCO_3$ solution at 20°, and extracted with *tert*-butyl methyl ether (2 × 50 mL). The combined organic phases were dried ($MgSO_4$) and concentrated (20°, 15 mmHg). The residue was purified by alumina (grade V) chromatography, first with hexane as eluent and subsequently with a 1:1 mixture of hexane and *tert*-butyl methyl ether to afford the pure epoxide (143 mg, 82% yield) as a colorless oil. IR (film): 3028, 2933, 2894, 2846, 2800, 1454, 1377, 1153, 844 cm^{-1}. ^{1}H NMR (200 MHz): δ 2.67 (d, J = 12 Hz, 2 H), 3.24 (d, J = 12 Hz, 1 H), 3.64 (s, 2 H), 3.80 (s, 2 H), 7.12–7.50 (m, 5 H). ^{13}C NMR (50 MHz): δ 53.2 (2 × t), 55.6 (2 × d), 59.8 (t), 127.3 (s), 128.3 (2 × d), 129.0 (2 × d), 137.2 (s). Anal. Calcd. for $C_{11}H_{13}NO$: C, 75.40, H, 7.48, N, 7.99. Found C, 75.43, H, 7.51, N, 8.06.

TABULAR SURVEY

The epoxidation of alkenes by dioxiranes is presented in the appended tables. The literature survey was conducted by computer search of *Chemical Abstracts* (CAS-on-line), and the references cover work inclusive of the first quarter of 2000.

The tables are arranged in the order of the discussion in the section on Scope and Limitations. Thus, the data on epoxidation of simple alkenes, alkenes with electron donors, with electron acceptors, and with both electron donors and acceptors by isolated dioxiranes (mainly DMD) are presented in Tables IA–ID. Epoxidations with in situ generated achiral dioxiranes are shown in Table IE. Chemo- and regioselective epoxidations are collected in Tables II and III, while the diastereoselective epoxidations of chiral alkenes by isolated and in situ generated dioxiranes are shown in Table IV. Enantioselective reactions, for which the catalytic in situ method has been applied exclusively, are compiled in Table V.

The entries within each table are arranged in order of increasing carbon number of the alkene substrate. The carbon count is based on the total number of carbon atoms. In Table V, the structures of the chiral ketone catalysts are given in a separate column, left of the column on reaction conditions. Yields of products are given in parentheses, and a dash (—) indicates that no yield was reported in the original reference. If reported in the original reference, the data on conversion (% convn) are mentioned in the product column, preferentially in subtables. Ratios of different products or diastereomers are normalized to 100, and given without parentheses. In Table V, the enantiomeric excess (% ee) for the major isomer is listed in the product column without parentheses.

The following abbreviations are used in the tables:

Ac	acetyl
Bn	benzyl
CBZ	benzyloxycarbonyl
Cp	cyclopentadienyl
DMAP	4-(dimethylamino)pyridine
DMD	dimethyldioxirane
DMD (in situ)	in situ generated dimethyldioxirane
DMD (isol.)	isolated dimethyldioxirane in acetone
DMD-d_6 (isol.)	isolated hexadeuterated dimethyldioxirane in acetone-d_6
DME	1,2-dimethoxyethane
DMF	dimethylformamide
DMM	dimethoxymethane
EDTA	ethylenediaminetetraacetic acid
Na_2EDTA	disodium salt of ethylenediaminetetraacetic acid
ee	enantiomeric excess
HMPA	hexamethylphosphoric triamide
LDA	lithium diisopropylamide
Ms	methanesulfonyl (mesyl)
Oxone	potassium monoperoxysulfate
TBS	*tert*-butyldimethylsilyl
TDMPP	dianion of 5,10,15,20-tetrakis(2,6-dimethoxyphenyl)porphyrin
Cl8 TDMPP	dianion of 5,10,15,20-tetrakis(3,5-dichloro-2,6-dimethoxyphenyl)porphyrin
Cl16 TDMPP	dianion of 2,3,7,8,12,13,17,18-octachloro-5,10,15,20-tetrakis(3,5-dichloro-2,6-dimethoxyphenyl)porphyrin
Tf	trifluoromethylsulfonyl (triflyl)
TFD	methyl(trifluoromethyl)dioxirane
TFD (in situ)	in situ generated methyl(trifluormethyl)dioxirane
TFD (isol.)	isolated methyl(trifluoromethyl)dioxirane in TFP
TFP	1,1,1-trifluoro-2-propanone
THF	tetrahydrofuran
THP	tetrahydropyranyl
TMS	trimethylsilyl

Tr	trityl (triphenylmethyl)
Ts	*p*-toluenesulfonyl

TABLE 1A. EPOXIDATION OF UNFUNCTIONALIZED OLEFINS BY ISOLATED DIOXIRANES

Substrate	Conditions	Product(s) and Yield(s) (%)	Refs.
C_2			
$R^1CH{=}CR^2R^3$	Difluorodioxirane, CF_4, O_2, 77 K	2-R^1-3-R^2-3-R^3-oxirane (—)	169
C_{4-11}			
$CH_2{=}C(R^1)C(R^2)(R^3)O_2H$	DMD, acetone/CH_2Cl_2	2-R^1-2-[$C(R^2)(R^3)O_2H$]-oxirane	170
C_5			
$Me_2C{=}CHMe$	DMD, acetone, under N_2 or air, –20 to 0°, 30 min	2,2,3-trimethyloxirane (>98)	18
$R^1R^4C{=}CR^2R^3$	DMD, acetone, 25°	2-R^1-2-R^4-3-R^2-3-R^3-oxirane (—)	15

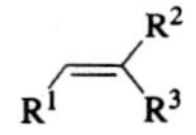

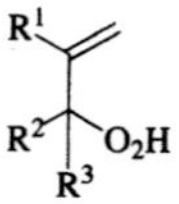

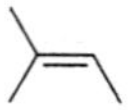

Ref. 169:

R^1	R^2	R^3
H	H	H
Me	H	H
Me	Me	H
Me	H	Me

Ref. 170:

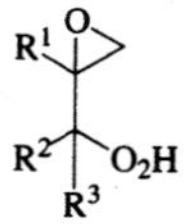

R^1	R^2	R^3	Temp	Time	
Me	H	H	0°	3 h	(>98)
H	Me	Me	20°	8 h	(82)
Me	Me	H	20°	1 h	(71)
Me	Me	Me	20°	1 h	(96)
Me	*i*-Pr	H	20°	3 h	(98)
t-Bu	H	H	20°	0.5 h	(98)
Me	*t*-Bu	H	20°	1.5 h	(>98)
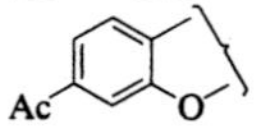 R^1–R^2 = 5-acetyl-2,3-dihydrobenzofuran-2,3-diyl		H	20°	3 h	(>98)

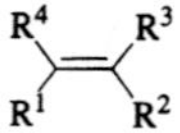

Ref. 15:

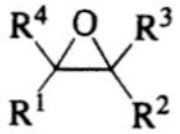

R^1	R^2	R^3	R^4	k_2 ($M^{-1}s^{-1}$)
—$(CH_2)_3$—		H	H	0.62 ± 0.01
Me	Me	Me	H	2.4 ± 0.1
Me	*n*-Pr	H	H	0.61 ± 0.01
Me	H	*n*-Pr	H	0.084 ± 0.002

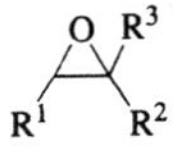

R[1]	R[2]	R[3]	R[4]	k_2 ($M^{-1}s^{-1}$)
Et	Et	H	H	0.57 ± 0.02
Et	H	Et	H	0.067 ± 0.001
Me	H	H	*n*-Pr	0.31 ± 0.01
Me	Me	Me	Me	7.1 ± 0.6
—$(CH_2)_4$—		H	Me	1.59 ± 0.06
Et	Et	Me	H	1.61 ± 0.06
Et	Me	Et	H	1.08 ± 0.04
—$(CH_2)_4$—		Me	Me	2.0 ± 0.1
n-C_8H_{17}	H	H	H	0.35 ± 0.005

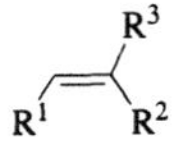

$C_{5\text{-}14}$

R[1]CH=C(R[2])R[3]

DMD, acetone, 23°

Epoxide (R[1], R[2], R[3]) (—)

R[1]	R[2]	R[3]	k_2 ($M^{-1}s^{-1}$)	
Me_2(HO)C	H	H	0.016 ± 0.002	16
—$(CH_2)_4$—		H	0.48 ± 0.02	
t-Bu	H	H	0.033 ± 0.001	
Et	Et	H	0.47 ± 0.03	
Et	H	Et	0.057 ± 0.004	
t-Bu	Me	H	0.33 ± 0.02	
t-Bu	H	Me	0.02 ± 0.01	
4-BrC_6H_4	H	H	0.089 ± 0.003	
3-$O_2NC_6H_4$	H	H	0.030 ± 0.003	
Ph	H	H	0.13 ± 0.01	
i-Pr	*i*-Pr	H	0.39 ± 0.02	
i-Pr	H	*i*-Pr	0.02 ± 0.01	
4-MeC_6H_4	H	H	0.25 ± 0.01	
Ph	Me	H	0.18 ± 0.01	
Ph	H	Me	0.29 ± 0.02	
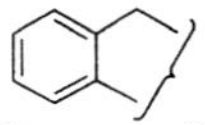 (indane-fused, R[1]–R[2])		H	0.22 ± 0.01	
t-Bu	H	*t*-Bu	0.00024 ± 0.00004	
Ph	Ph	H	0.040 ± 0.001	
Ph	H	Ph	0.043 ± 0.002	

TABLE 1A. EPOXIDATION OF UNFUNCTIONALIZED OLEFINS BY ISOLATED DIOXIRANES (*Continued*)

Substrate	Conditions	Product(s) and Yield(s) (%)	Refs.
C_6 	1. DMD, acetone/Et_2O, N_2, –30° 2. –25°, 3 h	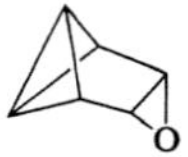 (62)	41
	1. DMD, acetone, dark, rt, 8 h 2. NaBr or LiBr, Amberlyst 15, rt, 12 h	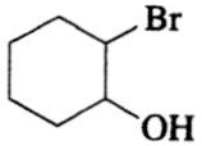 Br, OH (85)	171
	1. DMD, acetone, dark, rt, 8 h 2. NaCl or LiCl, Amberlyst 15, rt, 12 h	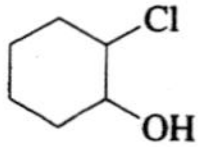 Cl, OH (76)	171
	Cyclohexanone dioxirane, cyclohexanone, –20°, 10 min	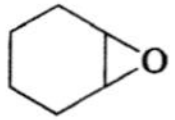 O **I** (100)	156
	DMD, $CDCl_3$, minutes	**I** (—)	172, 173
	DMD, acetone/solvent (v/v = 1:1), 25°	**I** (—)	174

Solvent	k_2 ($M^{-1}s^{-1}$)
AcOH	3.48 ± 0.39
MeOH	2.26 ± 0.19
$CHCl_3$	1.663 ± 0.025
$CDCl_3$	1.447 ± 0.026
CH_2Cl_2	1.03 ± 0.04
t-BuOH	0.942 ± 0.010
$Cl(CH_2)_2Cl$	0.920 ± 0.029
PhCl	0.641 ± 0.018

Solvent	k_2 ($M^{-1}s^{-1}$)
C_6H_6	0.488 ± 0.002
CCl_4	0.478 ± 0.002
—	0.462 ± 0.017
MeCOEt	0.416 ± 0.007
MeOAc	0.387 ± 0.016
EtOAc	0.349 ± 0.005

Substrate	Conditions	Product(s)	Refs.
+	DMD, acetone, 18°, 4 h	**I** (—)	30
	DMD, acetone, rt, 1 h	(—)	175
	DMD, acetone, rt, 2 h	(—)	175
	DMD, acetone, rt, 2 h	(—)	175
R^1 R^2 n	DMD, acetone, 15 min	O R^1 R^2 n (—)	176
C_7	DMD, acetone/CH_2Cl_2, 20°, 0.2 h	O (—) + O O (—) + CHO (—) + CHO (—)	177, 178

n	R^1	R^2
0	Et	H
0	Et	OH
1	Et	OH
2	Me	OH

TABLE 1A. EPOXIDATION OF UNFUNCTIONALIZED OLEFINS BY ISOLATED DIOXIRANES (*Continued*)

Substrate	Conditions	Product(s) and Yield(s) (%)	Refs.
	DMD, acetone/CH_2Cl_2, 20°, 0.2 h	 (>98)	14
C_{8-9} 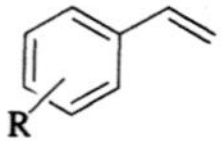 R	Phenyl(trifluoromethyl)dioxirane, CH_3CN, 20°, 30 min	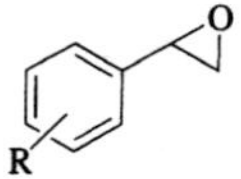 R (—) R = H, 2-Cl, 4-Me, 4-MeO	179
C_8 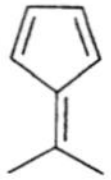	DMD (1.1 equiv), acetone, CH_2Cl_2, N_2, –10°, 3 h	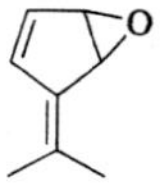 O (70) 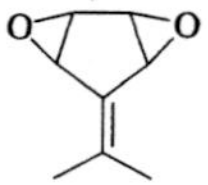 + O O **I** (30)	180
	DMD (2.5 equiv), acetone, CH_2Cl_2, N_2, 0°, 3 h	**I** (91)	180
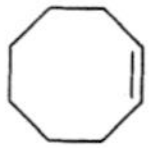	DMD, acetone/CH_2Cl_2, 20°, 0.2 h	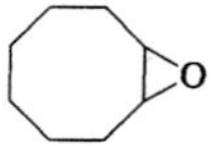 O **I** (>98)	14
	DMD, acetone, rt, 20 min	**I** (97)	181
	DMD, acetone/CH_2Cl_2, 20°, 0.2 h	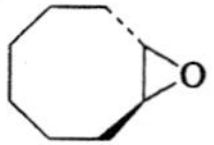 O (>98)	14, 13
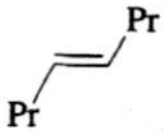 Pr Pr	Cyclohexanone dioxirane, cyclohexanone, –20°, 1 h	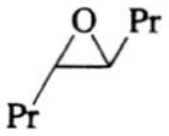 O Pr Pr **I** (100)	156
	DMD, acetone, rt, 3 h	**I** (74)	181

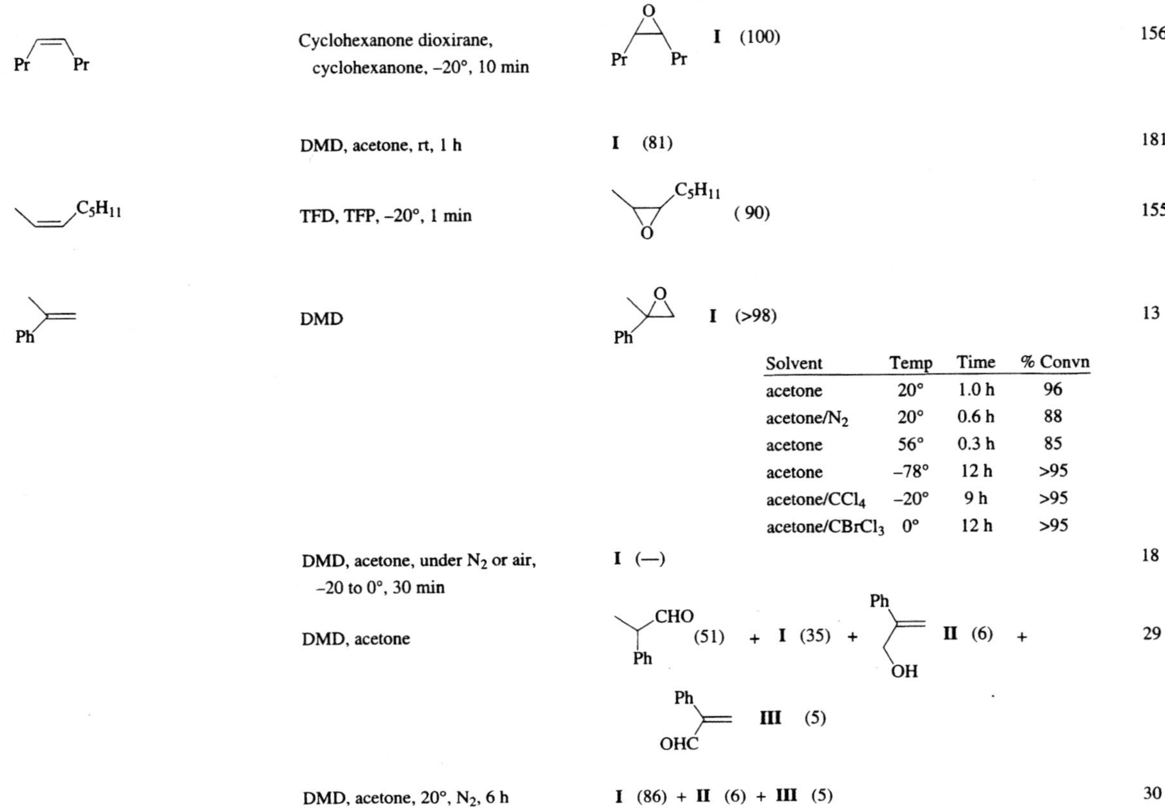

Substrate	Conditions	Product(s) and Yield(s) (%)	Refs.
(Z)-Pr–CH=CH–Pr	Cyclohexanone dioxirane, cyclohexanone, –20°, 10 min	I (100)	156
	DMD, acetone, rt, 1 h	I (81)	181
(Z)-CH$_3$–CH=CH–C_5H_{11}	TFD, TFP, –20°, 1 min	(90)	155
C_9			
Ph(CH$_3$)C=CH$_2$	DMD	I (>98)	13
	DMD, acetone, under N_2 or air, –20 to 0°, 30 min	I (—)	18
	DMD, acetone	(51) + I (35) + II (6) + III (5)	29
	DMD, acetone, 20°, N_2, 6 h	I (86) + II (6) + III (5)	30

Solvent	Temp	Time	% Convn
acetone	20°	1.0 h	96
acetone/N_2	20°	0.6 h	88
acetone	56°	0.3 h	85
acetone	–78°	12 h	>95
acetone/CCl_4	–20°	9 h	>95
acetone/$CBrCl_3$	0°	12 h	>95

TABLE 1A. EPOXIDATION OF UNFUNCTIONALIZED OLEFINS BY ISOLATED DIOXIRANES (*Continued*)

Substrate	Conditions	Product(s) and Yield(s) (%)	Refs.
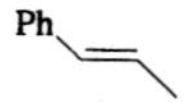 Ph	DMD, acetone, rt, 1 h	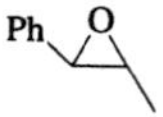 Ph O I (98)	181
	TFD, TFP, –20°, 1 min	I (90)	155
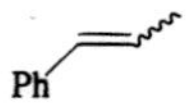 Ph	DMD, acetone, 20°, N_2, 6 h	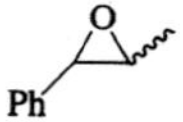 O Ph (>98)	30
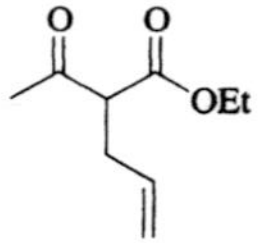 O O OEt	DMD, acetone, ~20°, 4.5 h	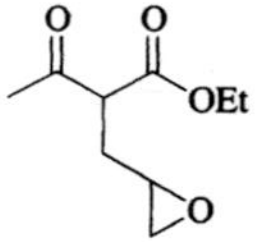 O O OEt O (>95)	182
	DMD, acetone, $Ni(acac)_2$, ~20°	mixture	182
C_{10}			
	DMD, acetone, 20°, 1 h	O (100)	183
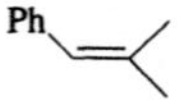 Ph	DMD, acetone, rt, 1 h	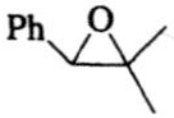 Ph O (99)	181
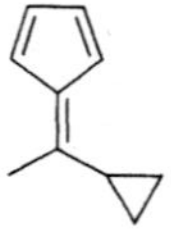	DMD (1.0 equiv), acetone, CH_2Cl_2, N_2, –10°, 3 h	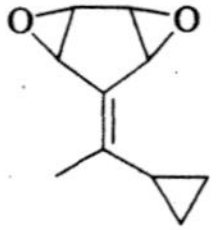 O O I (13) 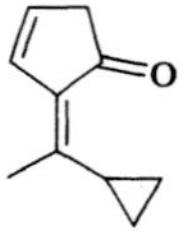 + O (72)	180
	DMD (2.5 equiv), acetone, CH_2Cl_2, N_2, 0°, 3 h	I (85)	180

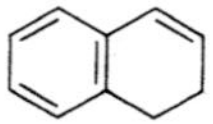

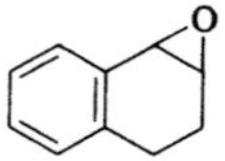

C_{10-15}

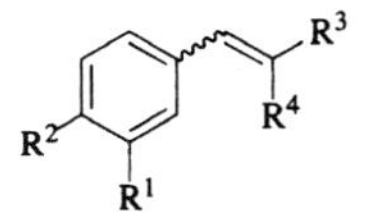

DMD, acetone, rt

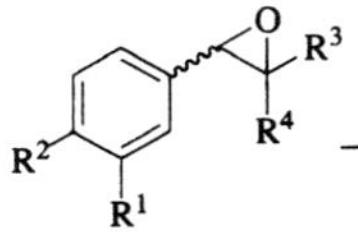

R^1	R^2	R^3	R^4	Time	
H	MeO	H	Me	40 min	(98)
MeO	MeO	H	H	40 min	(96)
MeO	MeO	H	Me	40 min	(100)
MeO	MeO	Me	Me	40 min	(100)
H	Cl	H	Ph	3 h	(87)
H	NO_2	H	Ph	24 h	(100)
H	H	H	Ph	40 min	(95)
H	MeO	H	Ph	40 min	(100)

184

C_{10-17}

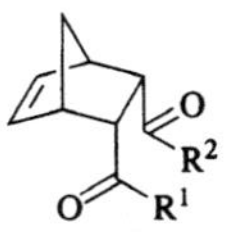

1. DMD (1 equiv), acetone, –78°
2. rt, 30 min
3. HCl (1 M) , 25°, 10 min

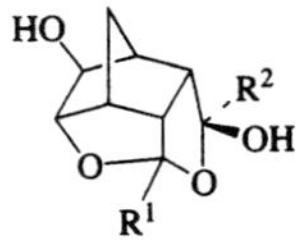

R^1	R^2	
Me	H	(75)
Me	Me	(86)
n-Bu	H	(78)
n-Bu	*n*-Bu	(84)
n-C_8H_{17}	H	(80)

185

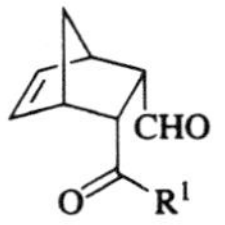

1. DMD (3 equiv), acetone, –78°
2. rt, 30 min
3. HCl (1 M), rt, 10 min

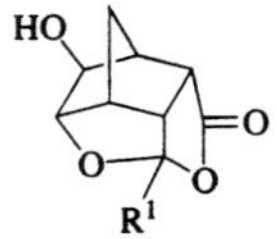

R^1	
Me	(80)
n-Bu	(84)
n-C_8H_{17}	(83)

185

C_{11}

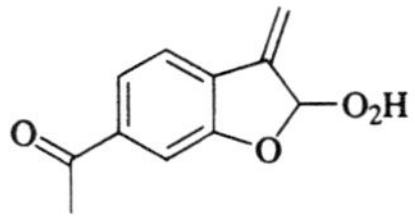

DMD, acetone/CH_2Cl_2, N_2, 20°, 3 h

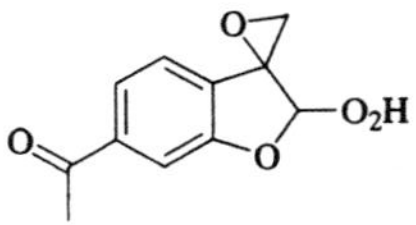

(>98)

170

TABLE 1A. EPOXIDATION OF UNFUNCTIONALIZED OLEFINS BY ISOLATED DIOXIRANES (*Continued*)

Substrate	Conditions	Product(s) and Yield(s) (%)	Refs.
C_{11-14} 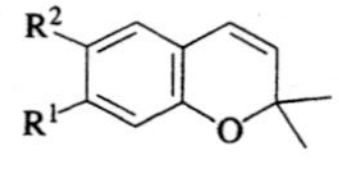 R[2], R[1]	DMD, acetone/CH_3CN	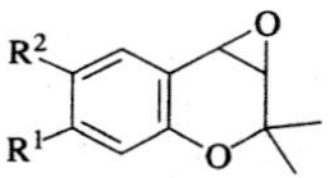 R[2], R[1], O (100) R^1 / R^2 / Temp Cl / H / 0° H / H / 0° MeO / H / –40° MeO / MeO / –40° CF_3CH_2O / MeO / –40° EtO / MeO / –40°	186
C_{11} 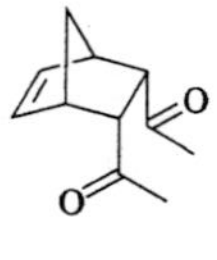 O, O	1. DMD, acetone, –78° 2. rt, 30 min	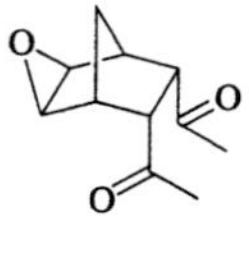 O, O, O (90)	185
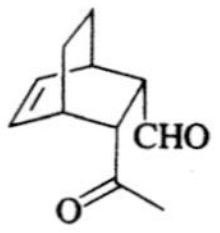 CHO, O	1. DMD (1 equiv), acetone, –78° 2. rt, 30 min 3. HCl (1 M), rt, 10 min	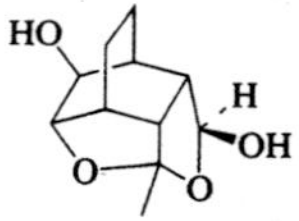 HO, H, OH, O, O (84)	185
	1. DMD (3 equiv), acetone, –78° 2. rt, 30 min 3. HCl (1 M), rt, 10 min	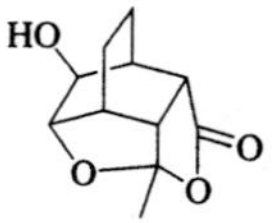 HO, O, O, O (80)	185
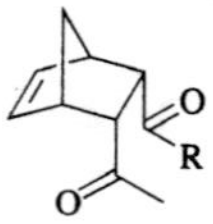 O, R, O	1. DMD, acetone, –78° 2. rt, 30 min 3. HCl (1 M), rt, 10 min	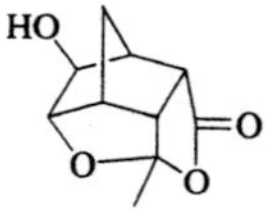 HO, O, O, O R MeO (85) MeS (85)	185

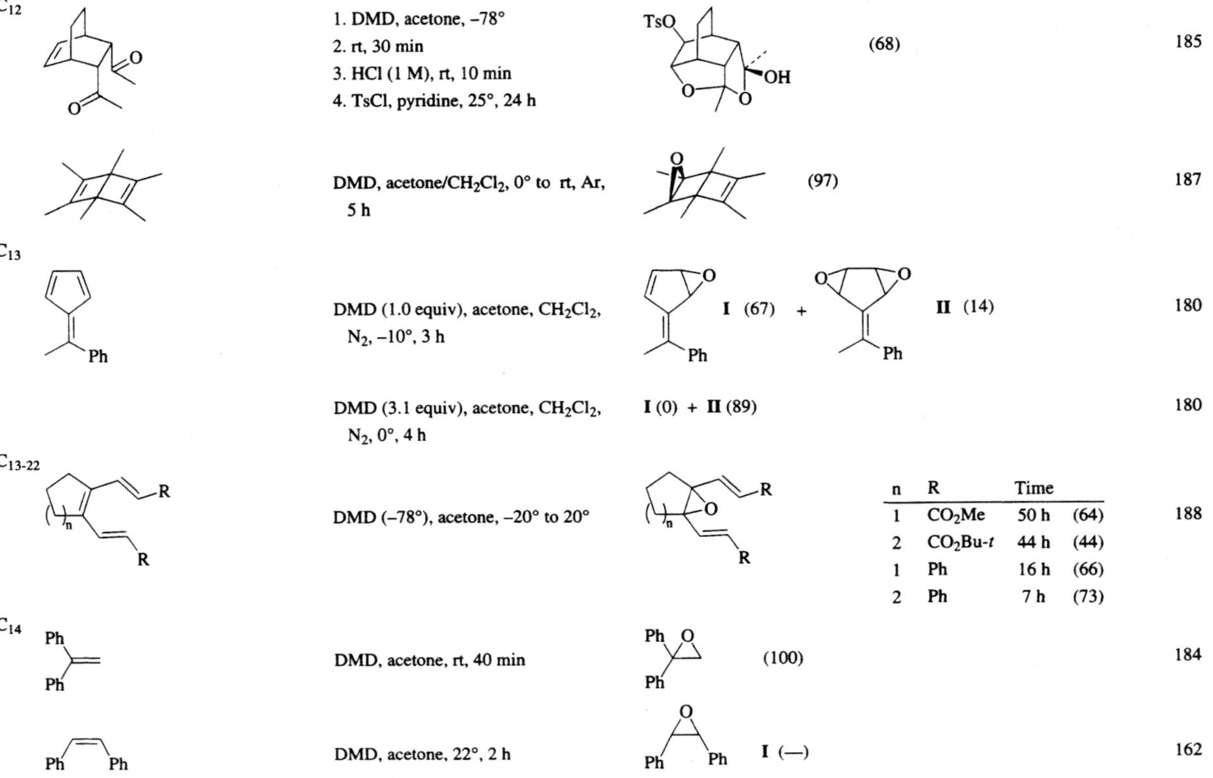

Carbon no.	Substrate	Conditions	Product(s) and Yield(s) (%)	Refs.
C_{12}		1. DMD, acetone, –78° 2. rt, 30 min 3. HCl (1 M), rt, 10 min 4. TsCl, pyridine, 25°, 24 h	TsO, OH, O (68)	185
		DMD, acetone/CH_2Cl_2, 0° to rt, Ar, 5 h	O (97)	187
C_{13}	Ph	DMD (1.0 equiv), acetone, CH_2Cl_2, N_2, –10°, 3 h	**I** (67) + **II** (14)	180
		DMD (3.1 equiv), acetone, CH_2Cl_2, N_2, 0°, 4 h	**I** (0) + **II** (89)	180
C_{13-22}	R, R, n	DMD (–78°), acetone, –20° to 20°	R, O, R, n	188
C_{14}	Ph, Ph	DMD, acetone, rt, 40 min	Ph, O, Ph (100)	184
	Ph, Ph	DMD, acetone, 22°, 2 h	O, Ph, Ph **I** (—)	162

n	R	Time	
1	CO_2Me	50 h	(64)
2	CO_2Bu-*t*	44 h	(44)
1	Ph	16 h	(66)
2	Ph	7 h	(73)

TABLE 1A. EPOXIDATION OF UNFUNCTIONALIZED OLEFINS BY ISOLATED DIOXIRANES (*Continued*)

Substrate	Conditions	Product(s) and Yield(s) (%)	Refs.
	Cyclohexanone dioxirane, cyclohexanone/CH_2Cl_2, –20°, 4.5 h	**I** (97)	156
	DMD, acetone/solvents, 2°	**I** (—)	189
	DMD, acetone, rt, 8 h	**I** (99)	181
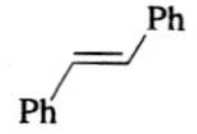 Ph–CH=CH–Ph	Ethylmethyldioxirane, 2-butanone/acetone, rt, 3 h	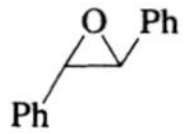 Ph–(epoxide)–Ph **I** (—)	162
	DMD, acetone, 22°	**I** (—)	162
	Cyclohexanone dioxirane, cyclohexanone/acetone, –20°, 1.5 h	**I** (100)	156
	DMD, acetone, rt, 6 h	**I** (100)	181
	1. DMD, acetone, dark, rt, 8 h 2. NaBr or LiBr, Amberlyst 15, rt, 12 h	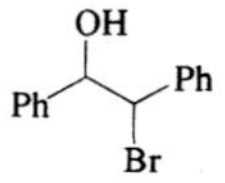 PhCH(OH)CH(Br)Ph (82)	171
C_{16} 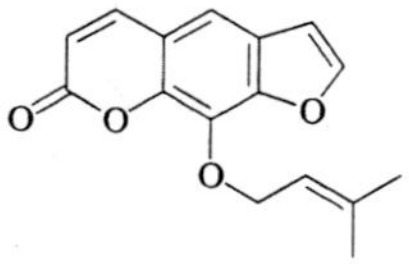 8-(3-methylbut-2-enyloxy)furocoumarin	DMD, acetone, 0°, 2 h	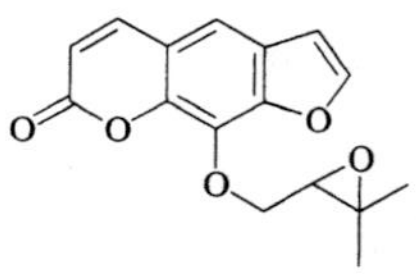 corresponding epoxide (99)	190

Substrate	Conditions	Product(s) and Yield(s) (%)	Refs.
C_{16-18}			
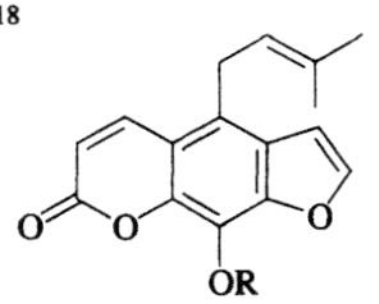 (OR structure)	DMD, acetone, 0°	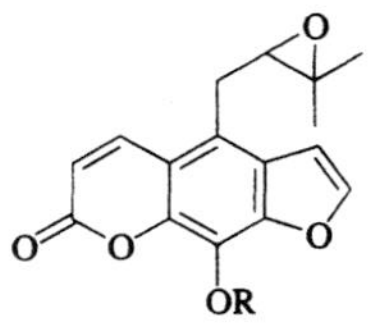 R / Time: H, 2 h (100); Me, 0.5 h (100); Ac, 0.5 h (100)	190
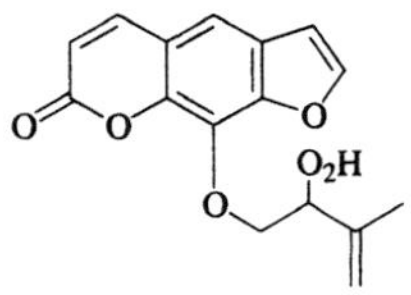 (O_2H)	DMD, acetone, 20°, 6 h	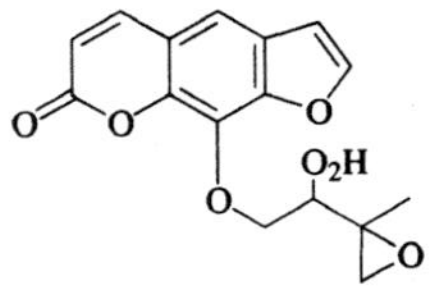 (O_2H) (96)	190
C_{17}			
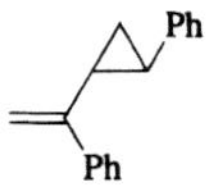 (Ph, Ph)	DMD, acetone, 20°, 3.5 h 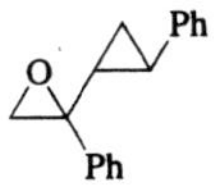	(Ph, Ph) (100)	183
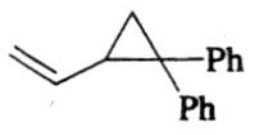 (Ph, Ph)	DMD, acetone, 20°, 1 h 	(Ph, Ph) (>95)	13
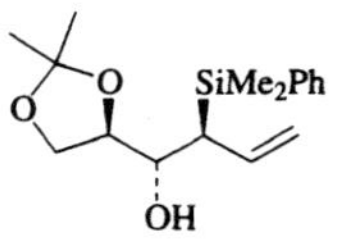 ($SiMe_2Ph$, OH)	1. DMD, acetone, 23° 2. HOAc, MeOH 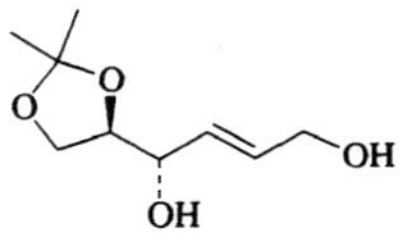	(OH, OH) (88)	191
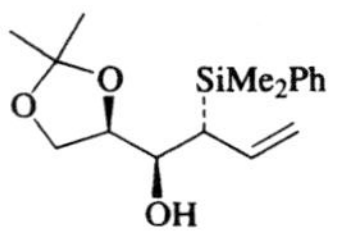 ($SiMe_2Ph$, OH)	1. DMD, acetone, 23° 2. HOAc, MeOH 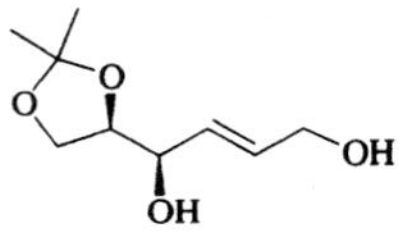	(OH, OH) (83)	191

TABLE 1A. EPOXIDATION OF UNFUNCTIONALIZED OLEFINS BY ISOLATED DIOXIRANES (*Continued*)

Substrate	Conditions	Product(s) and Yield(s) (%)	Refs.
C_{18}			
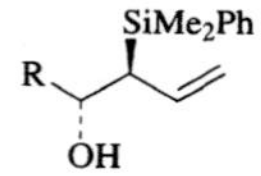 R, OH, $SiMe_2Ph$	1. DMD, acetone, 23° 2. HOAc, MeOH	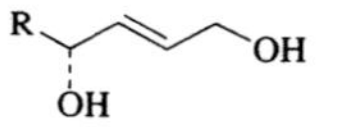 R, OH, OH; R: C_6H_{11} (95), n-C_6H_{13} (89)	191
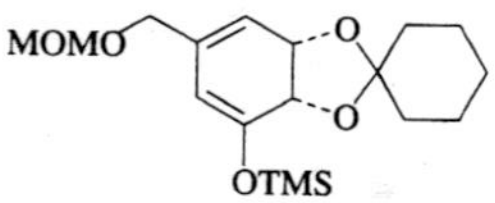 MOMO, O, O, OTMS	DMD, acetone/CH_2Cl_2, –78°, 10 min	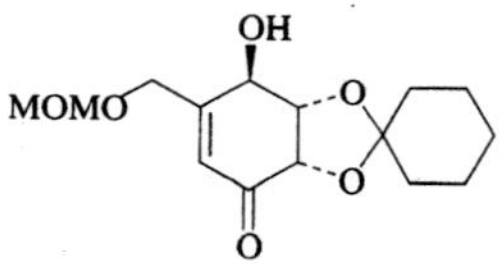 MOMO, OH, O, O, O (70)	192
C_{19}			
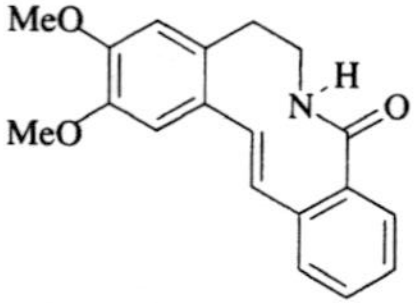 MeO, MeO, N, H, O	DMD, acetone, rt, 24 h	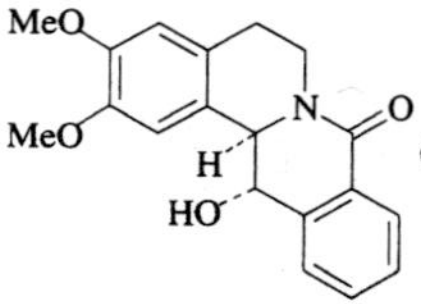 MeO, MeO, N, O, H, HO (40) 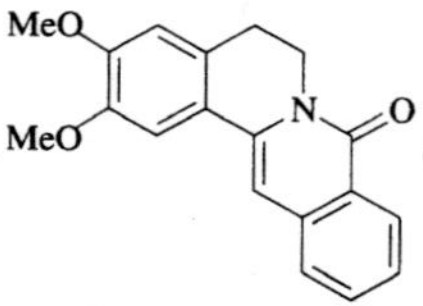 + MeO, MeO, N, O (31)	193
	DMD, acetone, rt, 2 d	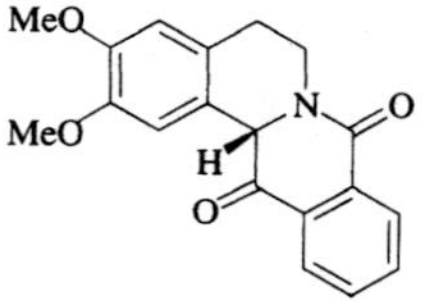 MeO, MeO, N, O, H, O (21) 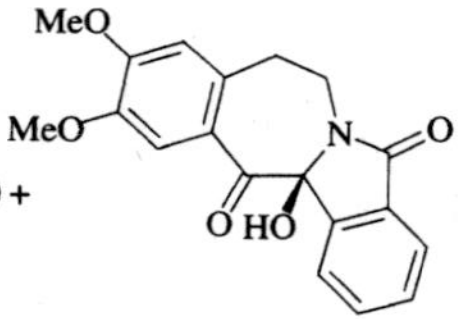 + MeO, MeO, N, O, O, HO (—)	70
C_{20}			
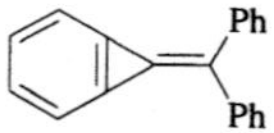 Ph, Ph	K_2CO_3, DMD (0°), acetone, N_2	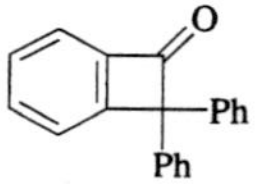 O, Ph, Ph **I**; Temp / Time: rt / — (30); 50° / 2 h (57)	194

Substrate	Conditions	Product(s) and Yield(s) (%)	Refs.
	1. K_2CO_3, DMD, acetone, N_2, –20°, 60 min 2. H_2SO_4 (2 M), –20° to rt, 60 min	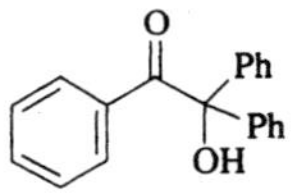 Ph, Ph, OH, O — **II** (74)	194
	1. K_2CO_3, DMD, acetone, N_2, –20°, 60 min 2. H_2O, 0°, 14 h	**I** (8) + **II** (53)	194
	1. DMD-d_6 (–90°), acetone-d_6, –30°, 50 min 2. –10°, 50 min	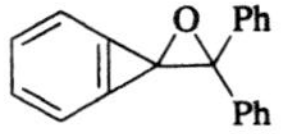 O, Ph, Ph (—)	194
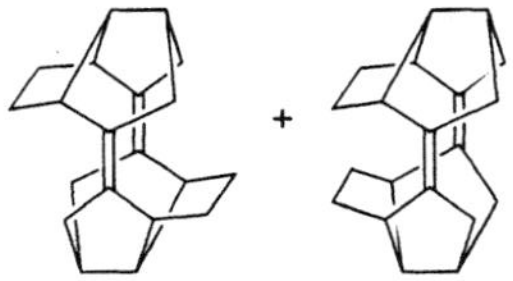 +	1. DMD, acetone/CH_2Cl_2, 0° 2. rt, 15 h	 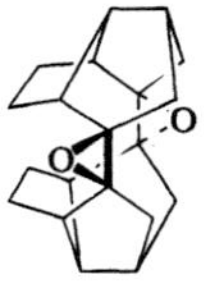 O, O **I** + O, O **II** — **I** + **II** (14)	195
C_{20-22} 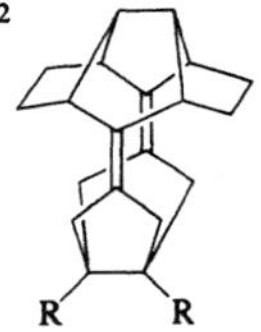 R, R	1. DMD, acetone/CH_2Cl_2, 0° 2. rt, 15 h	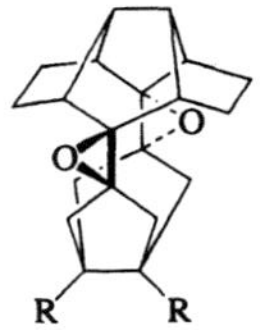 O, O, R, R — R: H (64); Me (59)	195
C_{20} 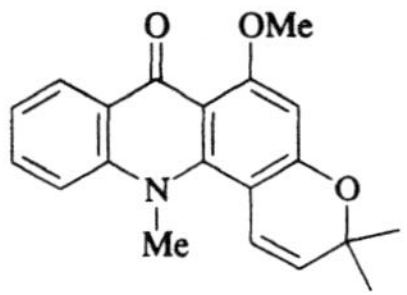 O, OMe, N–Me, O	DMD, acetone, –60°, 3 h	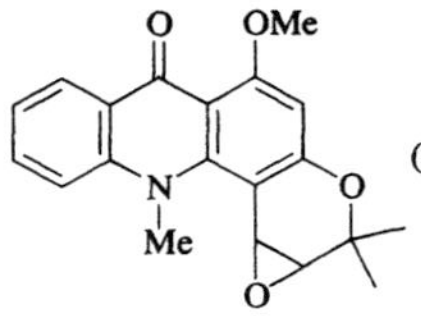 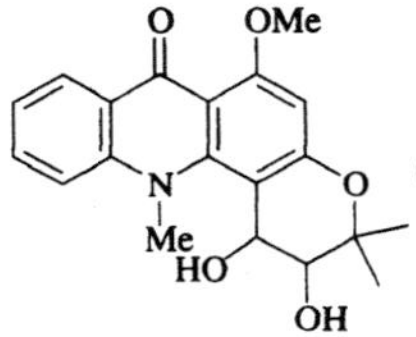 O, OMe, N–Me, O, O (—) + O, OMe, N–Me, O, HO, OH (—)	196, 197

TABLE 1A. EPOXIDATION OF UNFUNCTIONALIZED OLEFINS BY ISOLATED DIOXIRANES (*Continued*)

Substrate	Conditions	Product(s) and Yield(s) (%)	Refs.
C_{21} 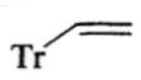 Tr	DMD, acetone	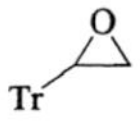 Tr (82)	198
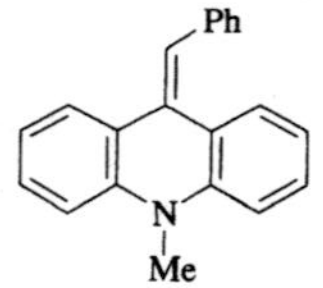 Ph, N, Me	DMD, acetone/CH_2Cl_2	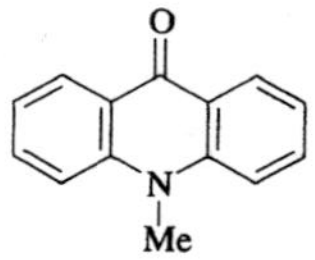 O, N, Me (—) + PhCHO (—)	197
C_{22} 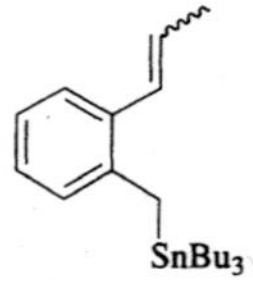 $SnBu_3$	DMD, CF_3CO_2H, dimethyl fumarate, CH_2Cl_2, rt, 2 h	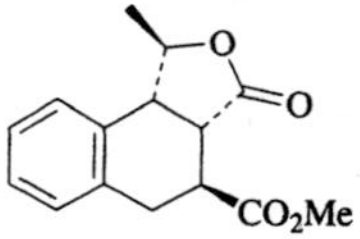 O, O, CO_2Me (78)	199
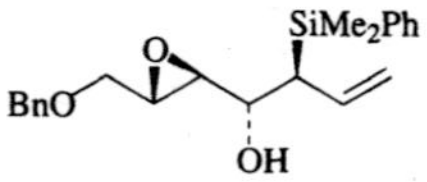 BnO, O, $SiMe_2Ph$, OH	1. DMD, acetone, 23° 2. HOAc, MeOH	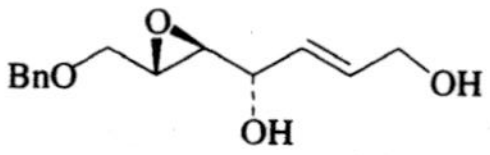 BnO, O, OH, OH (91)	191
C_{24} 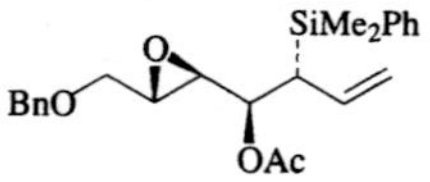 BnO, O, $SiMe_2Ph$, OAc	1. DMD, acetone, 23° 2. HOAc, MeOH	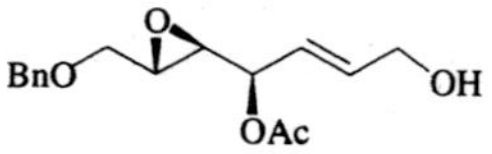 BnO, O, OAc, OH (60)	191
C_{25} 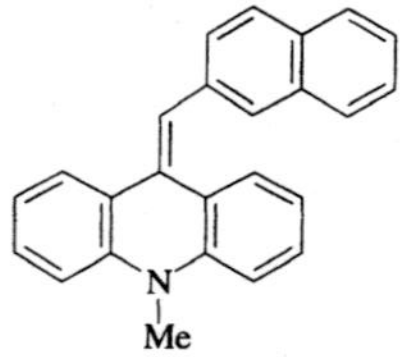 N, Me	DMD, acetone/CH_2Cl_2	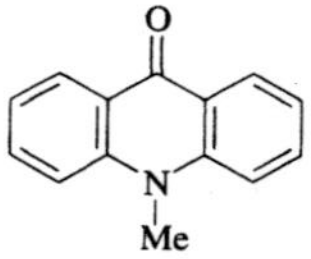 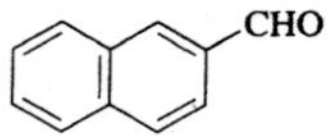 O, N, Me (—) + CHO (—)	197

C_{34}	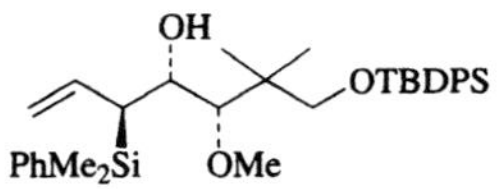 OH, OTBDPS, $PhMe_2Si$, OMe	1. DMD, acetone, K_2CO_3, 23° 2. MeOH, HOAc	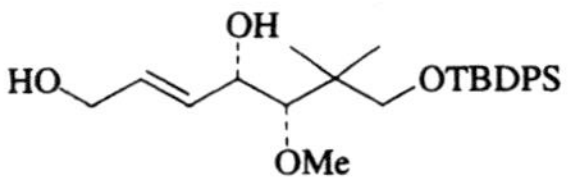 HO, OH, OTBDPS, OMe (86)	200

TABLE 1B. EPOXIDATION OF OLEFINIC SUBSTRATES WITH ELECTRON DONORS BY ISOLATED DIOXIRANES

Substrate	Conditions	Product(s) and Yield(s) (%)	Refs.
C_3			
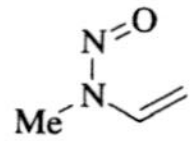	DMD, acetone/CH_2Cl_2, N_2, dark, rt, 3 h	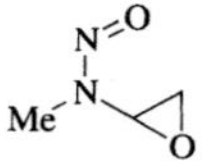 (~100)	201, 202
C_4			
	DMD, acetone/$CHCl_3$	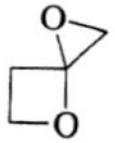 (—)	203
	DMD, acetone	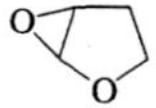 (—)	51
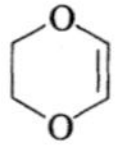	DMD	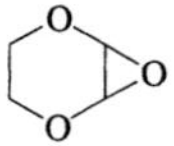 (—)	204
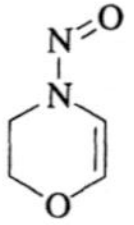	DMD, acetone/CH_2Cl_2, N_2, dark, rt, 3 h	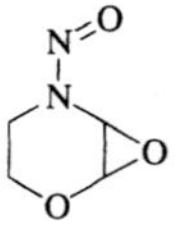 (95)	201
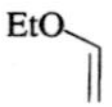	DMD	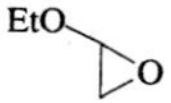 (—)	204
C_5			
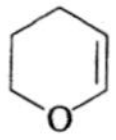	DMD, acetone	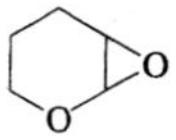 (—)	204, 51

Substrate	Conditions	Product(s) and Yield(s) (%)	Refs.
C_{5-12} 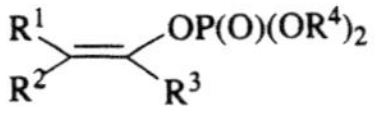 $R^1R^2C{=}C(R^3)OP(O)(OR^4)_2$	DMD, acetone/CH_2Cl_2, N_2	epoxide $R^1R^2C(O)C(R^3)OP(O)(OR^4)_2$	205, 44

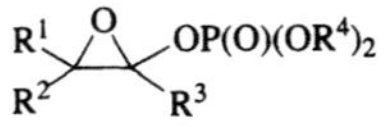

R^1	R^2	R^3	R^4	Temp	Time	
H	H	Me	Me	0°	4.5 h	(98)
H	Me	Me	Me	–10°	2.5 h	(~100)
Me	H	Me	Me	–10°	2.5 h	(~100)
H	H	Me	Et	0°	3.5 h	(99)
H	–(CH₂)₄–		Me	0°	5 h	(~100)
H	n-C_5H_{11}	H	Me	0°	6 h	(99)
H	H	Ph	Me	–10°	3.5 h	(96)
Me	–(CH₂)₄–		Et	–10°	3 h	(99)
H	H	Ph	Et	–10°	2 h	(99)

Substrate	Conditions	Product(s) and Yield(s) (%)	Refs.
C_{5-9} 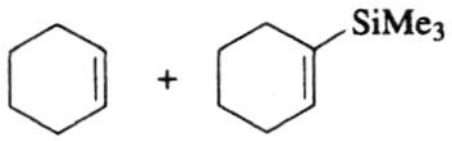 cyclohexene + 1-(trimethylsilyl)cyclohexene, 1:1	DMD (1 equiv), acetone, rt, 10 min	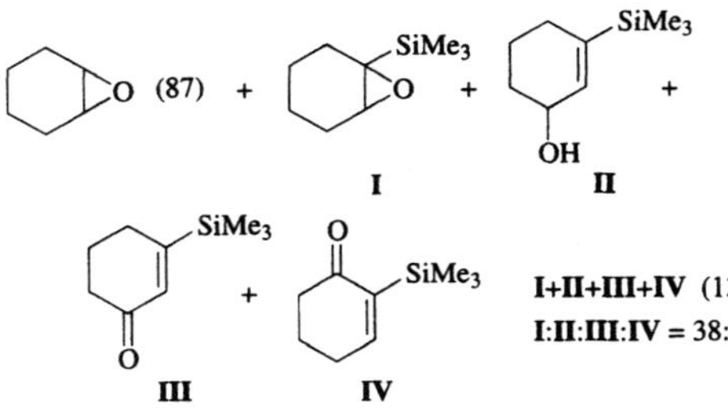 cyclohexene oxide (87) + **I** (SiMe₃ epoxide) + **II** (SiMe₃ allylic alcohol, OH) + **III** (3-SiMe₃-cyclohexenone) + **IV** (2-SiMe₃-cyclohexenone); **I+II+III+IV** (13); **I:II:III:IV** = 38:23:8:31	74
C_6 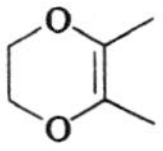 2,3-dimethyl-5,6-dihydro-1,4-dioxin	DMD-d_6, –20°, 5 min	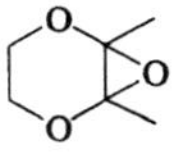 epoxide (—)	204
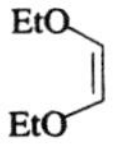 EtOCH=CHOEt	DMD-d_6, –20°, 5 min	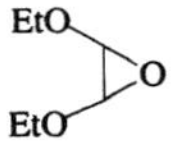 2,3-diethoxyoxirane (—)	204

TABLE 1B. EPOXIDATION OF OLEFINIC SUBSTRATES WITH ELECTRON DONORS BY ISOLATED DIOXIRANES (*Continued*)

Substrate	Conditions	Product(s) and Yield(s) (%)	Refs.
C_{7-14} 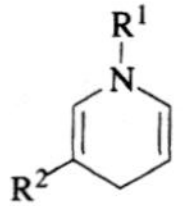	DMD, acetone, 0°, 5 min	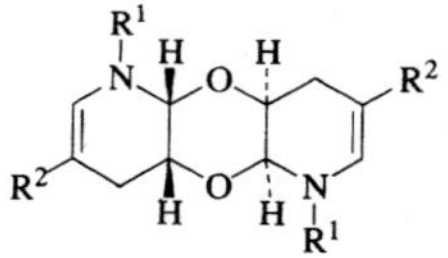 R[1] / R[2]: Me / CN (72); Me / CO_2Me (76); Bn / CN (69); Bn / CO_2Me (60)	206, 207
C_7 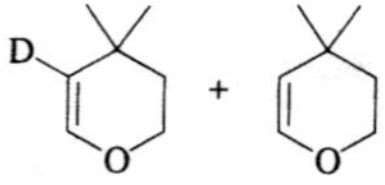	DMD, acetone, 25°	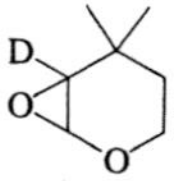 **I** (—) + 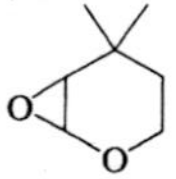 **II** (—) $k_H/k_D = 0.773 \pm 0.029$	208
	DMD, acetone, –78°	**I** (—) + **II** (—) $k_H/k_D = 0.681 \pm 0.032$	208
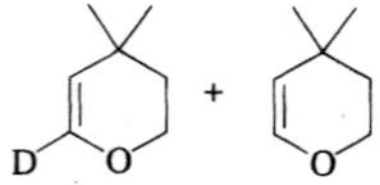	DMD, acetone, 25°	**II** (—) + 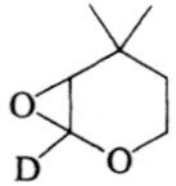 **III** (—) $k_H/k_D = 0.847 \pm 0.043$	208
	DMD, acetone, –78°	**II** (—) + **III** (—) $k_H/k_D = 0.898 \pm 0.036$	208
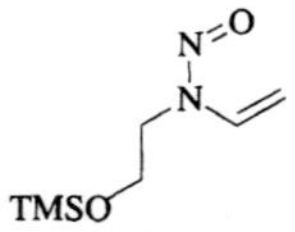 TMSO	DMD, acetone/CH_2Cl_2, N_2, dark, rt, 3 h	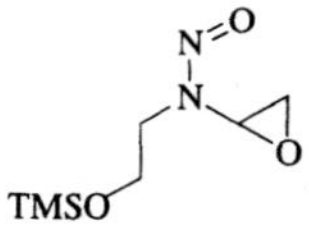 TMSO (95)	201
C_{7-13} 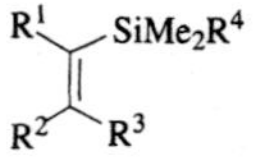	DMD, acetone, rt, 4 to 10 h	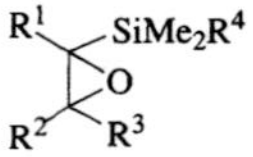	74

R^1	R^2	R^3	R^4	
Me	H	Me	Me	(>95)
H	*n*-Bu	H	Me	(>95)
Me	*n*-Bu	H	Me	(>81)
Et	*n*-Bu	H	Me	(>95)
Me	C_6H_{11}	H	Me	(>95)
Et	H	Me	Ph	(>95)
Me	Me	Me	Ph	(>95)

C_8 Ph(OLi)C=CH$_2$ | DMD, acetone/THF, Ar, –78°, 30 min | PhC(O)CH$_2$OH (77) | 209

C_{8-11} 1-OP(O)(OR)$_2$-2-R^1-cyclohexene | DMD, acetone/CH_2Cl_2, N_2, –10 to 0°, 3 to 5 h | epoxide, OP(O)(OR)$_2$ (99-100) | 44, 205

R^1	R
H	Me
Me	Et

C_{8-13} R^1R^2N–CH=CR^3R^4 | DMD, acetone, ~20°, 30 min | 2,5-bis(R^1R^2N)-3,3,6,6-(R^3,R^4)-1,4-dioxane | 211

R^1	R^2	R^3	R^4	
—$(CH_2)_4$—		Me	Me	(95)
—$(CH_2)_2O(CH_2)_2$—		Me	Me	(61)
—$(CH_2)_5$—		Me	Me	(85)
i-Pr	*i*-Pr	Me	Me	(93)
i-Bu	*i*-Bu	Me	Me	(91)
i-Pr	*i*-Pr	—$(CH_2)_5$—		(96)

C_{9-14} R(ONa)C=CHPh | DMD, acetone /THF, N_2, –78°, 2 min | RC(O)CH(OH)Ph | 210

R	
Me	(74)
MeO	(60)
Ph	(92)

TABLE 1B. EPOXIDATION OF OLEFINIC SUBSTRATES WITH ELECTRON DONORS BY ISOLATED DIOXIRANES (*Continued*)

Substrate	Conditions	Product(s) and Yield(s) (%)	Refs.
C_9 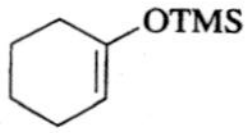 OTMS	DMD, acetone/CH_2Cl_2, N_2, –40°, 3 h	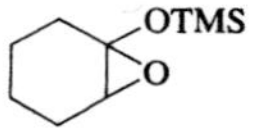 OTMS, O (99)	44, 212, 45
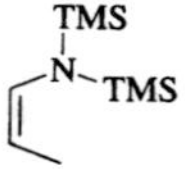 TMS, N, TMS	DMD, acetone/CH_2Cl_2, –40°, 3.5 h	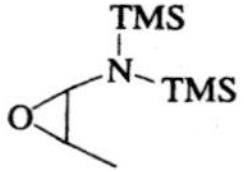 TMS, N, TMS, O (36)	64
C_{10} 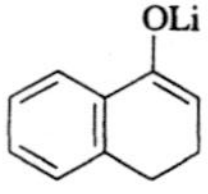 OLi	DMD, acetone/THF, Ar, –78°, 10 to 20 min	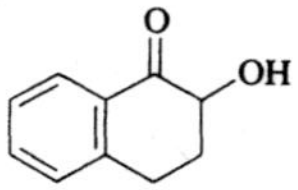 O, OH (52-82)	209
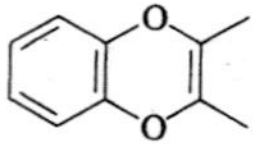 O, O	DMD, acetone/CH_2Cl_2, N_2, –70 to –30°, 0.5 to 1 h	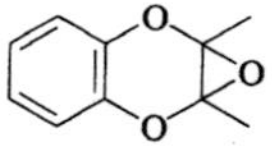 O, O, O (~100)	213, 204
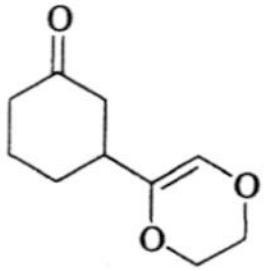 O, O, O	DMD, acetone/CH_2Cl_2, N_2, –40°	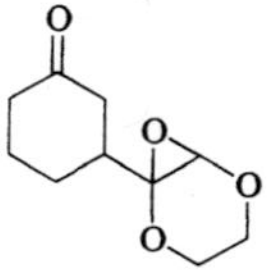 O, O, O, O (~100)	214
 LiO	DMD, acetone/THF, Ar, –78°, 15 to 30 min	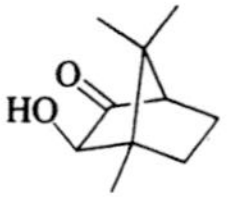 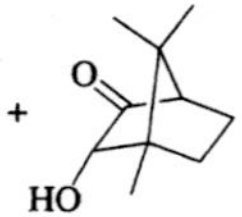 HO, O **I** + O, HO **II** (50-74) **I**:**II** = 70:30	209
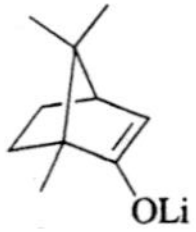 OLi	DMD, acetone/THF, N_2, –78°, 45 min	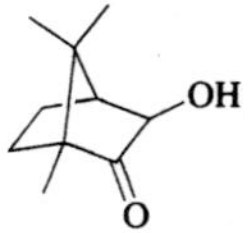 OH, O (53) 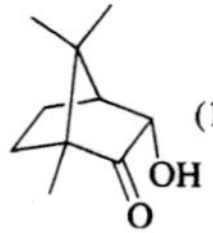 + OH, O (18)	210

Substrate	Conditions	Product(s) and Yield(s) (%)	Refs.
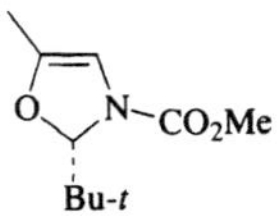 N-CO$_2$Me, Bu-*t* (oxazoline)	DMD, acetone/CH_2Cl_2, N_2, –20°, 2 h	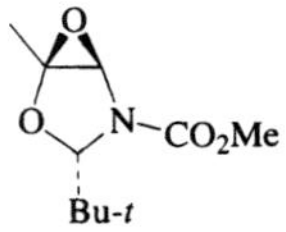 N-CO$_2$Me, Bu-*t* (epoxide)	215
C_{11} 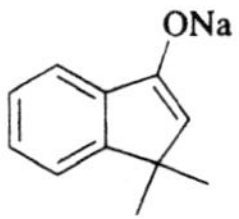 ONa	DMD, acetone /THF, N_2, –78°, 2 min	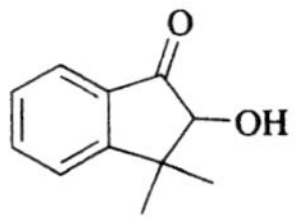 OH (80)	210
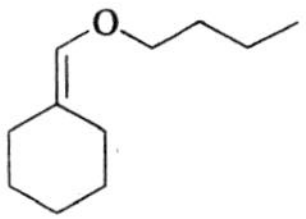 (butoxymethylene)cyclohexane	DMD, acetone, ~20°, 30 min	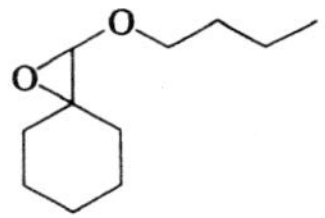 (98)	211
C_{11-16} 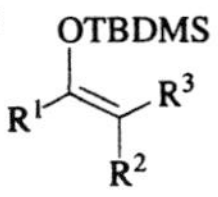 OTBDMS, R^1, R^2, R^3	DMD-d_6, acetone-d_6, –78°, 2 to 5 min	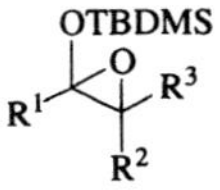 OTBDMS, R^1, R^2, R^3 (—)	216
C_{11-12} 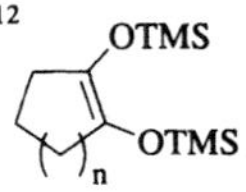 OTMS, OTMS, n	DMD, acetone/CH_2Cl_2, N_2, –30°, 3 h	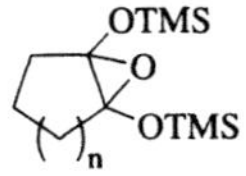 OTMS, OTMS, n	213
C_{12} 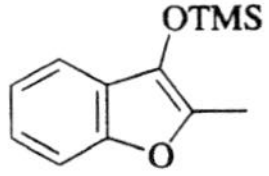 OTMS	DMD, acetone/CH_2Cl_2, N_2, –70 to –30°, 0.5 to 1 h	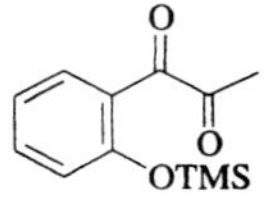 OTMS (72) 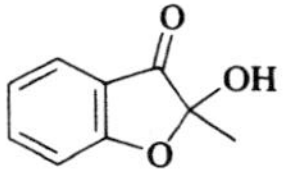 + OH (28)	213

R^1	R^2	R^3
Et	H	Me
—$(CH_2)_4$—		H
t-Bu	H	H
H	—$(CH_2)_5$—	
—$(CH_2)_4$—		Me
Ph	H	H
Ph	Me	Me

n	
1	(97)
2	(98)

TABLE 1B. EPOXIDATION OF OLEFINIC SUBSTRATES WITH ELECTRON DONORS BY ISOLATED DIOXIRANES (*Continued*)

	Substrate	Conditions	Product(s) and Yield(s) (%)	Refs.
		1. DMD, acetone/CH_2Cl_2, –70°, 10 min 2. –40°, 10 min 3. 2-Methyl-2-butene, <0°, 5 min	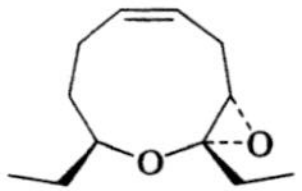 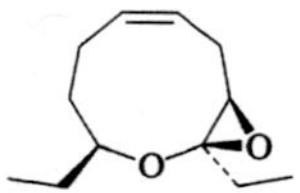 + (—)	217
	TMS N TMS	DMD, acetone/CH_2Cl_2, –50°, 1 h	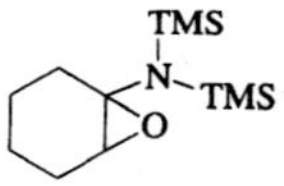 TMS N TMS (90)	64
	TMS TMS Et Et	DMD, acetone, 25°, 5 h	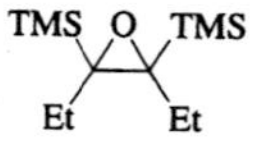 TMS O TMS Et Et (100)	218
C_{13}	OBz	DMD, acetone/CH_2Cl_2, N_2, –20°, 3 h	OBz (96)	212, 44
	Ph N O	DMD, acetone/CH_2Cl_2, –70°, 75 min	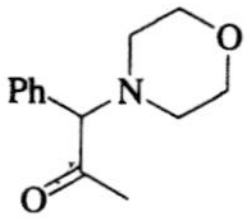 Ph N O (77)	64
	OTMS	DMD, acetone/CH_2Cl_2, N_2, –78°, 30 min	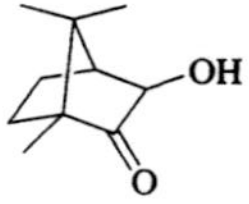 OH O (92) + OH O (7)	210
	TBDMSO OMe Bu-*t*	DMD, acetone/CH_2Cl_2, N_2, –40°, 1 h	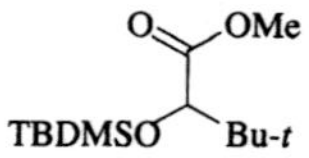 O OMe TBDMSO Bu-*t* (~100)	213

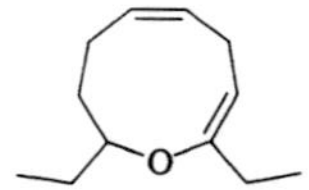

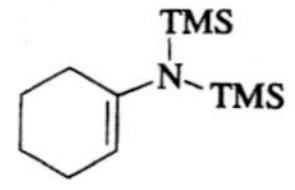

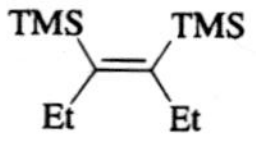

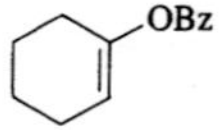

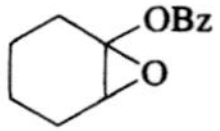

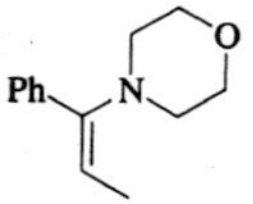

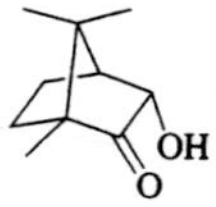

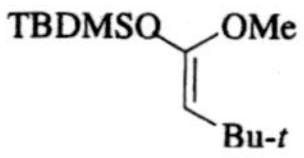

	Substrate	Conditions	Product(s) and Yield(s) (%)	Refs.
C_{14}	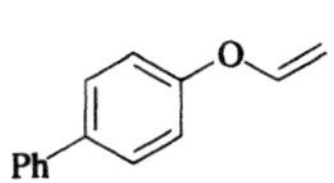 Ph, O	DMD, acetone, rt, 30 min	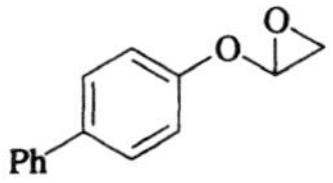 Ph, O, O (100)	219
	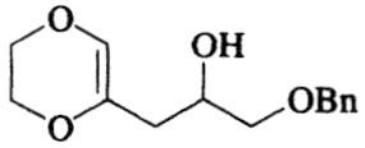 O, O, OH, OBn	DMD, acetone/CH_2Cl_2, N_2, –40°	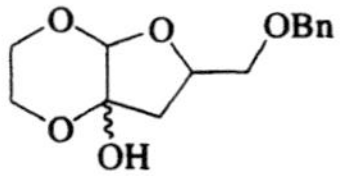 O, O, O, OBn, OH (~100)	214
	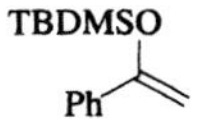 TBDMSO, Ph	DMD-d_6, acetone-d_6, –78°, 2 to 5 min	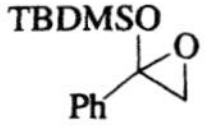 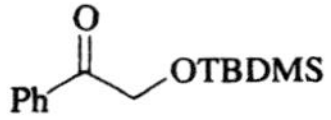 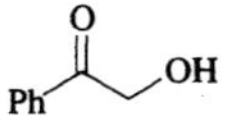 TBDMSO, O, Ph (—) + Ph, O, OTBDMS (—) + Ph, O, OH (—)	216
	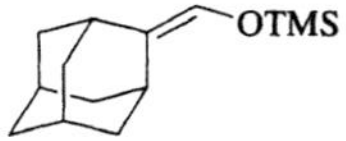 OTMS	DMD, acetone/CH_2Cl_2, N_2, –40°, 3 h	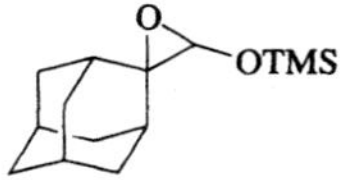 O, OTMS (98)	44, 45
C_{15}	 n-C_8H_{17}, O, C_5H_{11}	DMD, acetone/CH_2Cl_2, 0°, 7 h	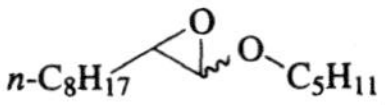 n-C_8H_{17}, O, O, C_5H_{11} (94)	220
C_{16}	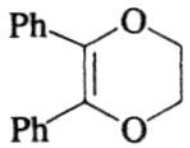 Ph, Ph, O, O	DMD, acetone, rt, < 30 sec	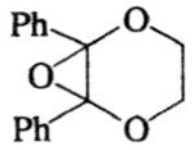 Ph, Ph, O, O, O (97)	221
	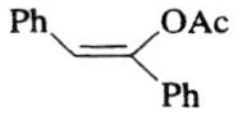 Ph, OAc, Ph	DMD, acetone/CH_2Cl_2, N_2, ~20°, 7 h	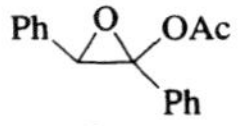 Ph, O, OAc, Ph (84)	44, 222

TABLE 1B. EPOXIDATION OF OLEFINIC SUBSTRATES WITH ELECTRON DONORS BY ISOLATED DIOXIRANES (*Continued*)

Substrate	Conditions	Product(s) and Yield(s) (%)	Refs.
MeO, OMe, Ph, Ph	MeO, O, O, Ph	MeO, O, OMe, Ph, Ph (—)	223
MeO, OMe, OMe, OMe	DMD, CH_2Cl_2/acetone, –40°, 30 min	MeO_2C, OMe, OMe, OMe (95)	224
MeO_2C, OMe, OMe, OMe	DMD, CH_2Cl_2, –30°, 1 h	MeO_2C, OMe, OMe, O, OMe (92)	224
MeO, CO_2Me, OMe, OMe	DMD, CH_2Cl_2, –30°, 1 h	MeO, CO_2Me, OMe, O, OMe (95)	224
C_{17}			
O, O, H, O, O, H, O, OMe	DMD, acetone/CH_2Cl_2, rt, 15 min	O, O, O, H, O, O, H, O, OMe (—)	225, 51

Substrate	Conditions	Product(s) and Yield(s) (%)	Refs.
t-BuO$_2$C-NH-CH(CO$_2$Me)-CH$_2$-(indol-3-yl)	DMD, acetone/CH_2Cl_2, 10°, 2 d	t-BuO$_2$C-NH-CH(CO$_2$Me)-CH$_2$-(2-oxindol-3-yl) (42) + t-BuO$_2$C-NH-CH(CO$_2$Me)-CH$_2$-(2-oxindol-3-yl) (42)	226
n-C_8H_{17}CH=CH-O-CH$_2$CH(OAc)CH$_2$OAc	DMD, acetone/CH_2Cl_2, 0°, 7 h	n-C_8H_{17}-epoxide-O-CH$_2$CH(OAc)CH$_2$OAc (97)	220
C_{18-24}			
2-adamantylidene-C(R^1)OR^2	TFD, TFP/CH_2Cl_2, –20°, 4 to 6 min	spiro-adamantane oxirane with R^1, OR^2 (92-97) R^1 / R^2: Ph / Me; 1-Naph / Me; 2-Naph / Me; Ph / Ph; Ph / Bn	227
C_{18}			
PhCH=C(Ph)-N(morpholino)	DMD (3 equiv), acetone/CH_2Cl_2, –70°	PhCOCH(OH)Ph **I** + PhCOCOPh **II** + PhCOCH(Ph)-N(morpholino) **III** (—) **I**:**II**:**III** = 19:48:33	228

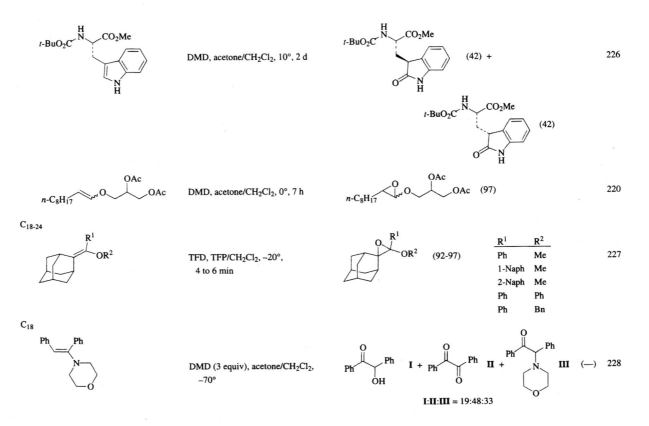

TABLE 1B. EPOXIDATION OF OLEFINIC SUBSTRATES WITH ELECTRON DONORS BY ISOLATED DIOXIRANES (*Continued*)

Substrate	Conditions	Product(s) and Yield(s) (%)	Refs.
	DMD (1.5 equiv), acetone/CH_2Cl_2, −70°	**I** + **II** + **III** (—) **I**:**II**:**III** = 21:27:52	228
C_{19} MeO_2C, OH, C_6H_{13}-*n*	DMD, acetone, rt	MeO_2C, O, OH, C_6H_{13}-*n* (24) + MeO_2C, O, O, C_6H_{13}-*n* (28) + MeO_2C, O, C_6H_{13}-*n* (24)	69
C_{20} OSiEt$_3$, Bn, O, O	DMD, acetone/CH_2Cl_2, 0°	OSiEt$_3$, Bn, O, O, O (>99)	229
O, OMe, H, O, N, S	1. DMD, K_2CO_3, CH_2Cl_2/acetone, 0° 2. $NaIO_4$, H_2O/THF	O, OMe, OHC, OHC, O, H, N, S (92)	230
P, Pr-*i*, O–Si–Pr-*i*, O, O, O, O, O	DMD, acetone/CH_2Cl_2, 0°	P, Pr-*i*, O–Si–Pr-*i*, O, O, O, O, O, O (—)	231

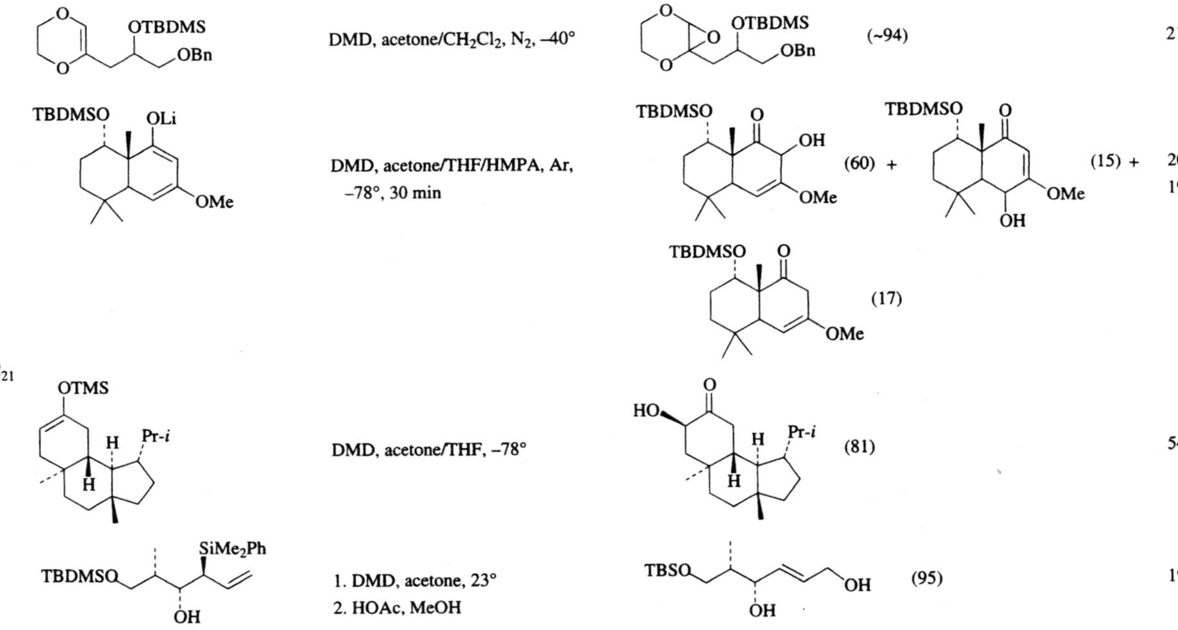

Reactant	Conditions	Product(s) and Yield(s) (%)	Refs.
	DMD, acetone/CH_2Cl_2, N_2, –40°	(~94)	214
	DMD, acetone/THF/HMPA, Ar, –78°, 30 min	(60) + (15) + (17)	209, 191
C_{21}	DMD, acetone/THF, –78°	(81)	54
	1. DMD, acetone, 23° 2. HOAc, MeOH	(95)	191

TABLE 1B. EPOXIDATION OF OLEFINIC SUBSTRATES WITH ELECTRON DONORS BY ISOLATED DIOXIRANES (*Continued*)

Substrate	Conditions	Product(s) and Yield(s) (%)	Refs.
C_{22} 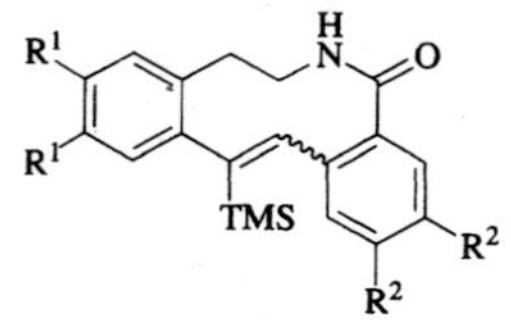	DMD (–78°), acetone, rt, 30 min	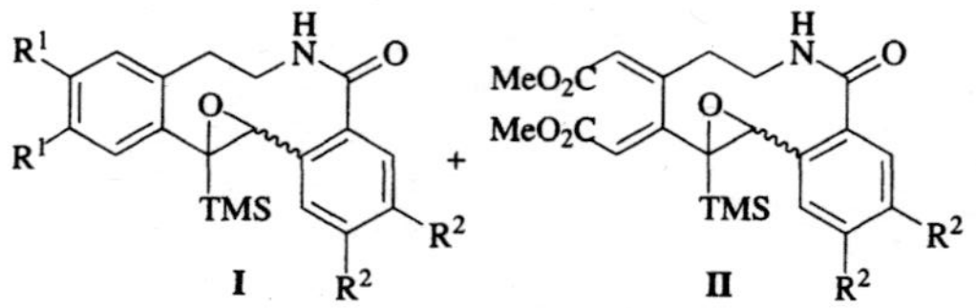 I + II	70, 193
C_{25} 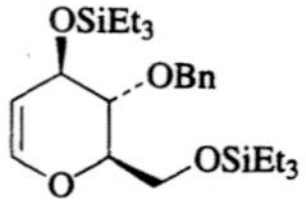	DMD, acetone/CH_2Cl_2, 0°	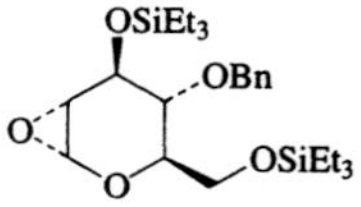 (>99)	229

R^1	R^2	Time	I	II
OMe	H	9 h	(47)	(31)
H	OMe	6 d	(96)	(0)

TABLE 1C. EPOXIDATION OF OLEFINS WITH ELECTRON ACCEPTORS BY ISOLATED DIOXIRANES

	Substrate	Conditions	Product(s) and Yield(s) (%)	Refs.
C_2	$CF_2{=}CFCl$	Difluorodioxirane, $CFCl_3$	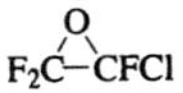 $F_2C\overset{O}{—}CFCl$ (>95)	232
	$CF_2{=}CHF$	Difluorodioxirane, $CFCl_3$	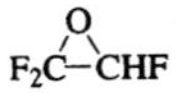 $F_2C\overset{O}{—}CHF$ (>95)	232
C_3	$CF_3CF{=}CF_2$	Difluorodioxirane, $CFCl_3$	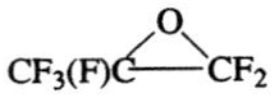 $CF_3(F)C\overset{O}{—}CF_2$ (>95)	232
$C_{5\text{-}7}$ 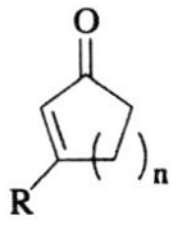	[structure: 3-R-cycloalk-2-enone, ring size n]	1. DMD, acetone, dark, rt, 8 h 2. NaBr or LiBr, Amberlyst 15, rt, 12 h	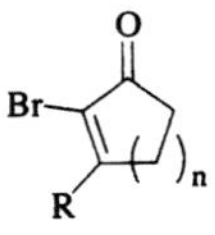 [structure: 2-Br-3-R-cycloalk-2-enone] n R 1 H (95) 1 Me (96) 2 H (95) 2 Me (96)	171
		1. DMD, acetone, dark, rt, 8 h 2. NaCl or LiCl, Amberlyst 15, rt, 12 h	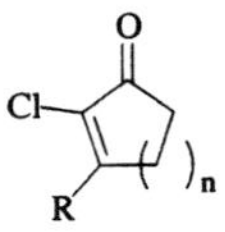 [structure: 2-Cl-3-R-cycloalk-2-enone] n R 1 H (88) 1 Me (93) 2 H (90) 2 Me (92)	171
$C_{5\text{-}8}$ 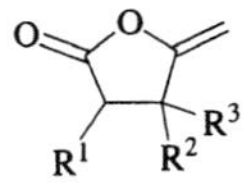	[structure: α-methylene lactone with R^1, R^2, R^3]	DMD, acetone/CH_2Cl_2, N_2, –20°	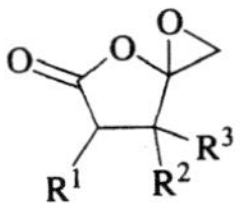 [structure: spiro-epoxide lactone with R^1, R^2, R^3] R^1 R^2 R^3 Time H H H 2 h (91) H Me H 3.5 h (95) Me H H 3.5 h (94) H Me Me 3 h (96)	44, 212

TABLE 1C. EPOXIDATION OF OLEFINS WITH ELECTRON ACCEPTORS BY ISOLATED DIOXIRANES (*Continued*)

Substrate	Conditions	Product(s) and Yield(s) (%)	Refs.
C_{5-13}	DMD, acetone/CH_2Cl_2, ~20°	R[1] / R[2] / R[3] / R[4] / Time: Me / CO_2H / H / H / 23 h (93) Me / CO_2H / Me / Ph / 20 h (96) CO_2Me / Me / Me / Ph / 20 h (99) Me / CO_2Me / Me / Ph / 16 h (98)	53
C_{6-7}	DMD, acetone/CH_2Cl_2/CH_3CN, porphyrin catalyst, imidazole, 20°	R[1] / R[2] / Catalyst: H / H / FeTDMPPCl[a] (20) H / H / MnTDMPPCl[a] (40) H / H / Mn(Cl16)TDMPPCl[b] (68) H / Me / MnTDMPPCl[a] (52) H / Me / Mn(Cl16)TDMPPCl[b] (70) Me / H / MnTDMPPCl[a] (21) Me / H / Mn(Cl16)TDMPPCl[b] (49)	233
	DMD, acetone/CH_2Cl_2, 25°	I + II + III R[1] / R[2] / I / II / III: H / H / (10) / (53) / (25) Me / H / (35) / (30) / (8) H / Me / (37) / (53) / (0)	234, 235
	DMD, acetone/CH_2Cl_2, Na_2SO_4, 25°	H / H / (50) / (—) / (—)	

C_{6-10}

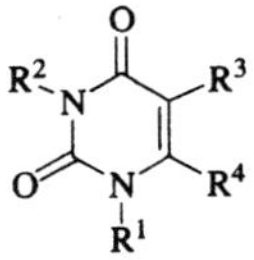

DMD, acetone/MeOH, 25°

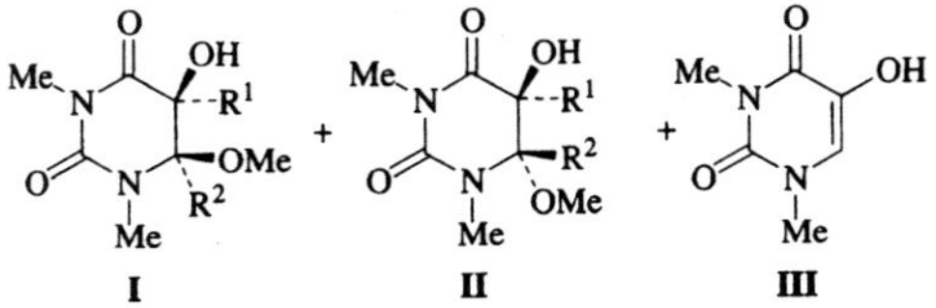

I + II + III — 234, 235

R^1	R^2	I	II	III
H	H	(38)	(35)	(11)
Me	H	(45)	(40)	(0)
H	Me	(82)	(0)	(0)

DMD, acetone/CH_2Cl_2, additive, 25°

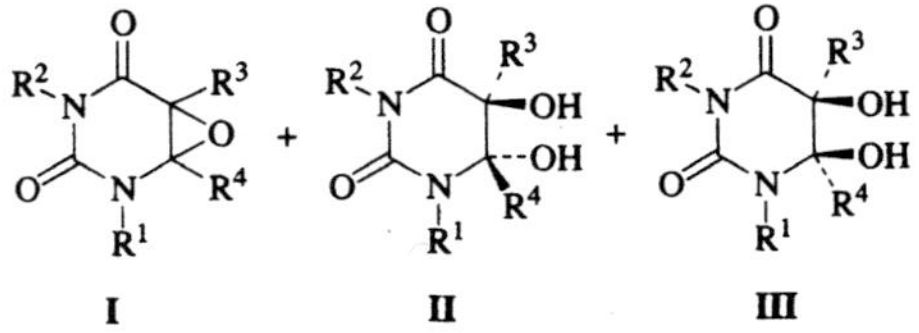

I + II + III — 235

R^1	R^2	R^3	R^4	Additive	I	II	III
Me	Me	H	H	—	(10)	(25)	(50)
Me	Me	H	H	Na_2SO_4	(50)	(7)	(15)
Me	Me	H	H	H_2O	(0)	(30)	(65)
Me	Me	Me	H	—	(30)	(57)	(0)
Me	Me	H	Me	—	(37)	(tr)	(53)
Deoxyribose	H	H	H	—	(—)	(—)	(70)
Deoxyribose	H	Me	H	—	(10)	(12)	(33)

C_6

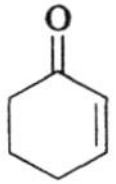

DMD, acetone, 30°

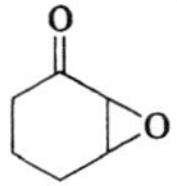

(—) $k_2 = (1.3 \pm 0.09) \times 10^{-3}\ M^{-1}s^{-1}$ — 236

TABLE 1C. EPOXIDATION OF OLEFINS WITH ELECTRON ACCEPTORS BY ISOLATED DIOXIRANES (*Continued*)

Substrate	Conditions	Product(s) and Yield(s) (%)	Refs.
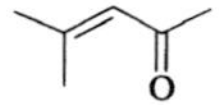 [structure] O	DMD, acetone, 30°	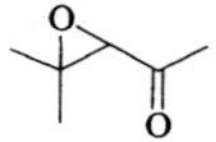 [structure] O, O (—) $k_2 = (5.7 \pm 0.9) \times 10^{-3}\ M^{-1}s^{-1}$	236
C_7 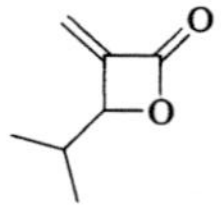 [structure] O, O	DMD, acetone, 20°, 6 d	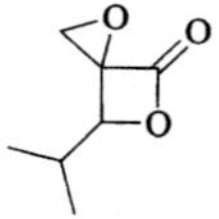 [structure] O, O, O (40)	237
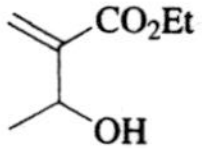 [structure] CO_2Et, OH	DMD, acetone, 20°, 3 d	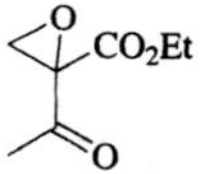 [structure] O, CO_2Et, O (37)	237
C_8 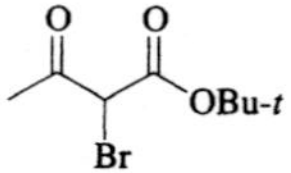 [structure] O, O, OBu-*t*, Br	DMD, acetone/CH_2Cl_2, DMAP, 0° to rt	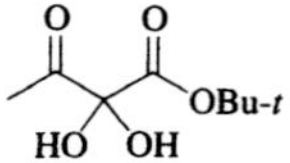 [structure] O, O, OBu-*t*, HO, OH (96)	240
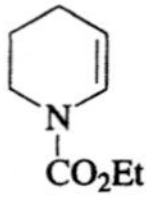 [structure] N, CO_2Et	DMD, acetone, 0° to rt, overnight	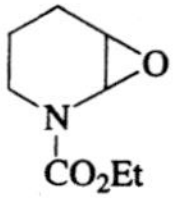 [structure] O, N, CO_2Et (97)	239
C_9 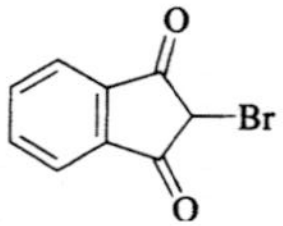 [structure] O, Br, O	DMD, acetone/CH_2Cl_2, DMAP, 0° to rt	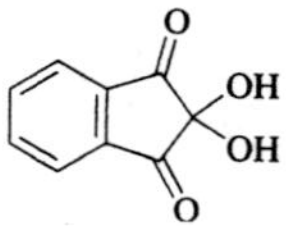 [structure] O, OH, OH, O (80)	240

C_{9-10} NHCHO, R

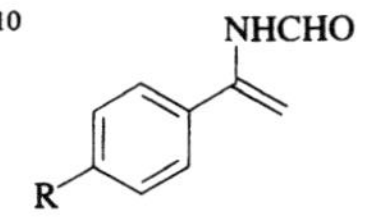

1. DMD, acetone/CH_2Cl_2, –40°
2. $(CF_3SO_2)_2O$, *i*-Pr_2NEt, –78°

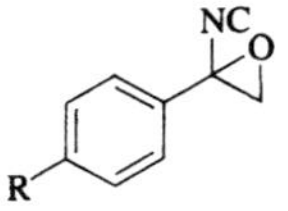

NC, O, R **I** + NC, R **II** 238

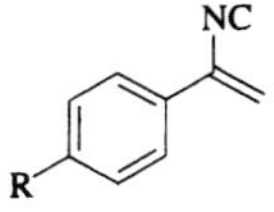

R	I	II
Br	(41)	(8)
Cl	(43)	(7)
CN	(68)	(0)

C_9 Ph, CO_2H

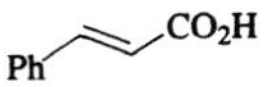

DMD, acetone, 30°

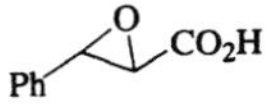

O, Ph, CO_2H (—) $k_2 = (0.66 \pm 0.07) \times 10^{-3}\ M^{-1}s^{-1}$ 236

C_{9-18} O, R^1, R^2, R^3, OH

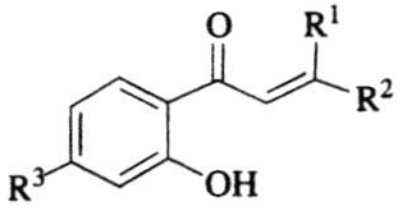

DMD, CH_2Cl_2/acetone,
0 to 5°

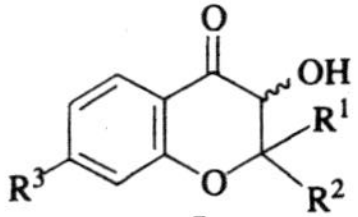

O, OH, R^1, R^2, R^3, O **I** + O, R^1, R^2, R^3, O, OH **II** + HO, O, R^1, R^2, R^3, OH **III** 241, 242

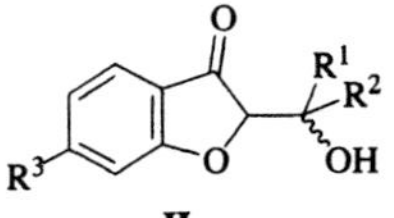

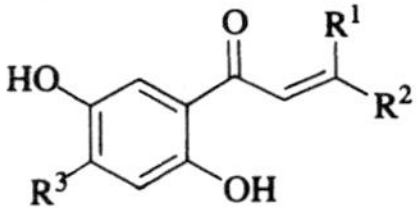

R^1	R^2	R^3	I	II	III
H	H	H	(5)	(53)	(0)
Me	Me	H	(39)	(26)	(0)
Me	H	MeO	(4)	(31)	(0)
Me	Me	MeO	(22)	(5)	(6)
Me	Me	TsO	(55)	(14)	(0)

TABLE 1C. EPOXIDATION OF OLEFINS WITH ELECTRON ACCEPTORS BY ISOLATED DIOXIRANES (*Continued*)

Substrate	Conditions	Product(s) and Yield(s) (%)	Refs.
C_{9-10}			
R[1], R[2], R[3]	DMD, acetone/H_2O, 25°	I + II + III	234
		R[1] R[2] R[3] I+II III I:II	
		H OH CH_2OH (85) (0) 83:17	
		Me H CO_2H (37) (39) —	
		Me H CH_2OH (68) (23) —	
C_9			
	Cyclohexanone dioxirane, cyclohexanone/CH_2Cl_2, −10°, 6 h	(83)	156
R[1], R[2]	DMD, acetone/CH_2Cl_2, N_2, 20°	R[1] R[2] Time	34, 53
		H Me 20 h (86)	
		Me H 24 h (90)	
OBu-*t*, OSO_2Me	DMD, acetone/CH_3CN, NaOMe, 0°	OBu-*t* (85)	240

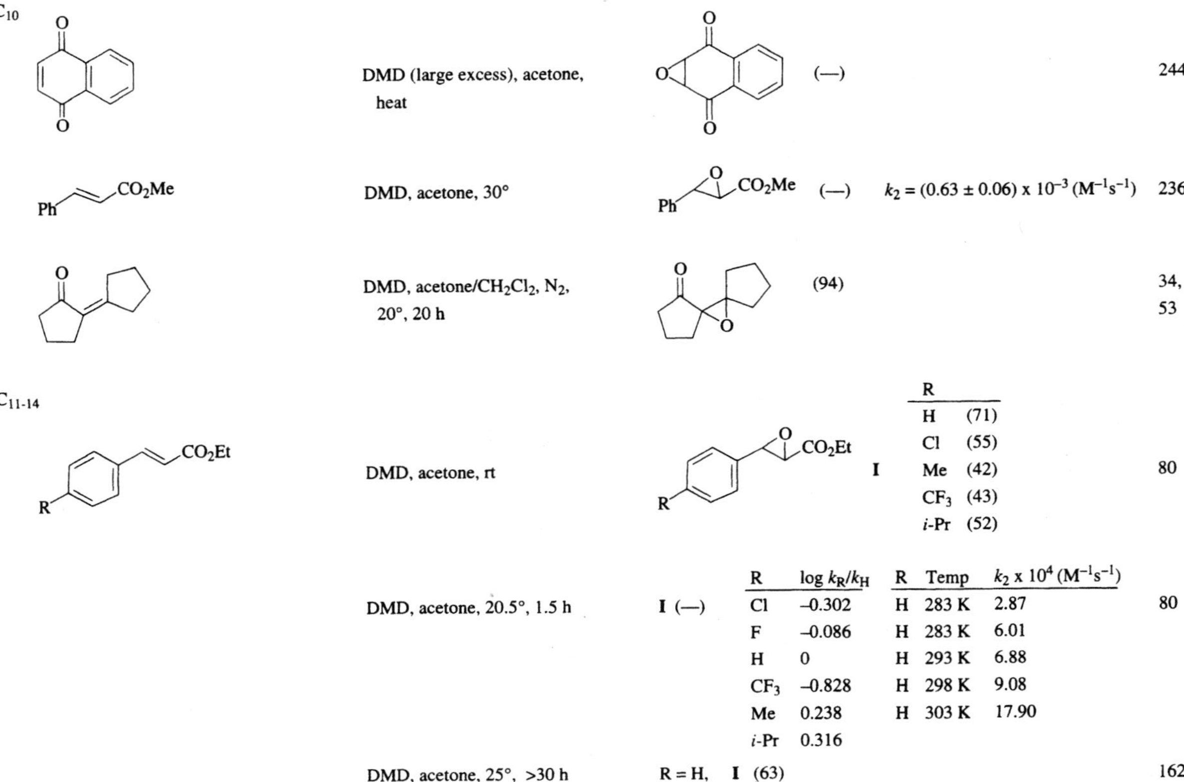

Reactant	Conditions	Product(s) and Yield(s) (%)	Refs.
C_{10}			
1,4-naphthoquinone	DMD (large excess), acetone, heat	(—)	244
Ph–CH=CH–CO_2Me	DMD, acetone, 30°	(—) $k_2 = (0.63 \pm 0.06) \times 10^{-3}\ (M^{-1}s^{-1})$	236
2-cyclopentylidenecyclopentanone	DMD, acetone/CH_2Cl_2, N_2, 20°, 20 h	(94)	34, 53
C_{11-14}			
4-R-C_6H_4–CH=CH–CO_2Et	DMD, acetone, rt	**I** R = H (71), Cl (55), Me (42), CF_3 (43), *i*-Pr (52)	80
	DMD, acetone, 20.5°, 1.5 h	**I** (—)	80
	DMD, acetone, 25°, >30 h	R = H, **I** (63)	162

R	$\log k_R/k_H$
Cl	–0.302
F	–0.086
H	0
CF_3	–0.828
Me	0.238
i-Pr	0.316

R	Temp	$k_2 \times 10^4\ (M^{-1}s^{-1})$
H	283 K	2.87
H	283 K	6.01
H	293 K	6.88
H	298 K	9.08
H	303 K	17.90

TABLE 1C. EPOXIDATION OF OLEFINS WITH ELECTRON ACCEPTORS BY ISOLATED DIOXIRANES (*Continued*)

Substrate	Conditions	Product(s) and Yield(s) (%)	Refs.
C_{11}			
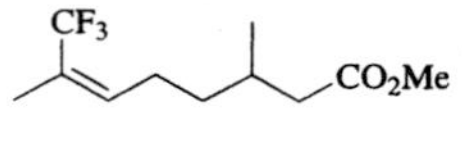 CF_3 / CO_2Me	DMD, acetone/CH_2Cl_2, 0°, 48 h	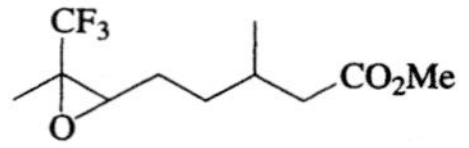 CF_3 / O / CO_2Me **I** (93)	78
	TFD, TFP/(CH_2Cl_2), 0°, 30 min	**I** (93)	78
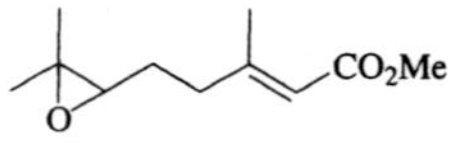 O / CO_2Me	DMD, acetone/CH_2Cl_2, 0°, 48 h	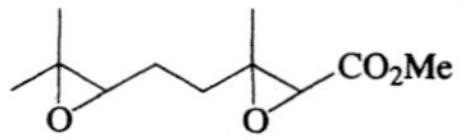 O / O / CO_2Me **I** (90)	78
	TFD, TFP/(CH_2Cl_2), –70°, 2 h	**I** (90)	78
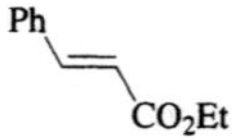 Ph / CO_2Et	DMD, acetone, rt, 24 h	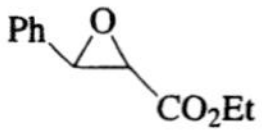 Ph / O / CO_2Et **I** (100)	181
	DMD, acetone/solvent (v/v = 1:1), 25°	**I** (—)	174

Solvent	k_2 x 10^4 ($M^{-1}s^{-1}$)
AcOH	29.66 ± 0.12
sulfolane	25.26 ± 0.26
$CHCl_3$	16.30 ± 0.30
CH_2Cl_2	14.66 ± 0.57
$Cl(CH_2)_2Cl$	12.55 ± 0.52
PhCl	19.00 ± 0.23
C_6H_6	9.41 ± 0.10
acetone	8.66 ± 0.22
2-butanone	8.10 ± 0.14
CCl_4	7.98 ± 0.02
AcOMe	7.74 ± 0.04
AcOEt	6.81 ± 0.20

Substrate	Conditions	Product(s) and Yield(s) (%)	Refs.
C_{12} 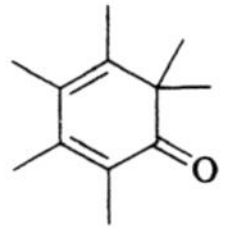	DMD (3 equiv), acetone, rt, 27 h	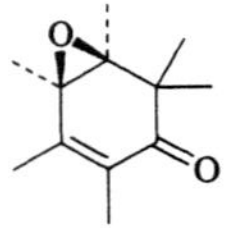 I 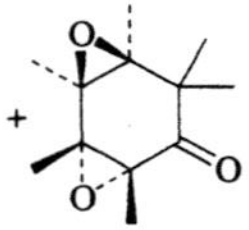 (—) + II (—) I:II = 20:80	245
	DMD (4 equiv), acetone, rt, 47 h	II (90)	245
C_{13} 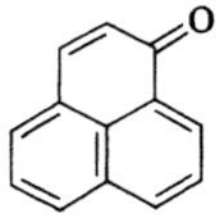	DMD, acetone, 30°	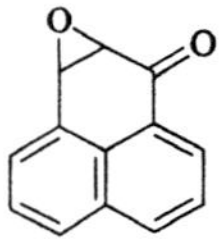 (—) $k_2 = (0.68 \pm 0.01) \times 10^{-3}$ ($M^{-1}s^{-1}$)	236
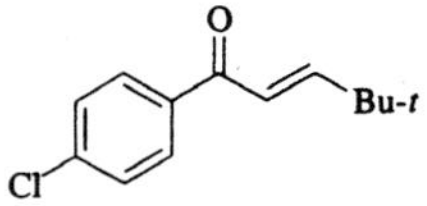	DMD, acetone, 30°	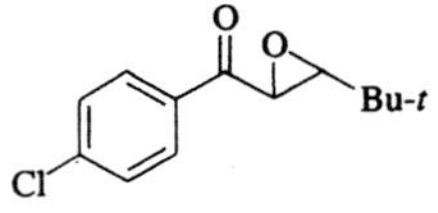 (—) $k_2 = (0.70 \pm 0.02) \times 10^{-3}$ ($M^{-1}s^{-1}$)	236
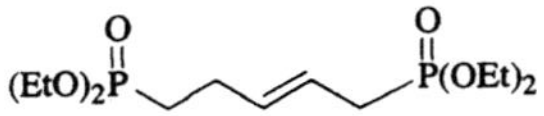	DMD, acetone, rt, 24 h	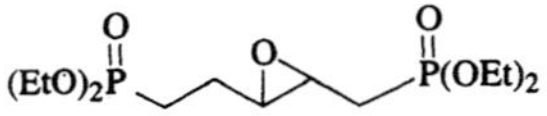 (100)	76
C_{14} 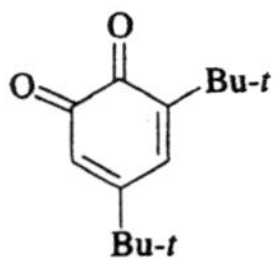	DMD, acetone	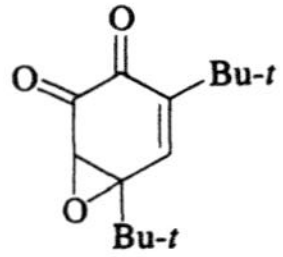 I 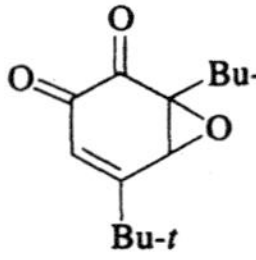 + II (52) I:II = 20:1	246
C_{15-17}	DMD, acetone/CH_2Cl_2, N_2, ~20°	see sub-table below	57

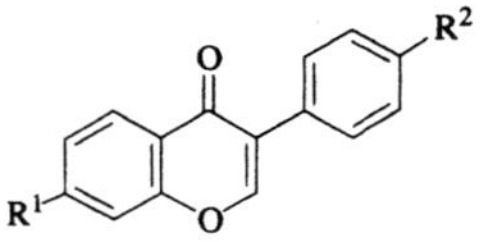

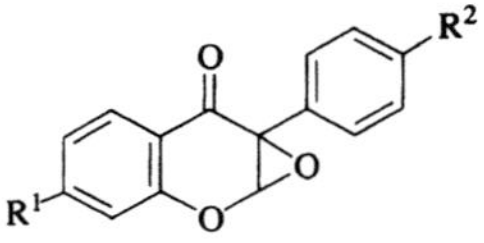

R^1	R^2	Time	
H	Cl	110 h	(100)
OH	H	128 h	(100)
MeO	H	121 h	(100)
AcO	H	135 h	(100)

TABLE 1C. EPOXIDATION OF OLEFINS WITH ELECTRON ACCEPTORS BY ISOLATED DIOXIRANES (*Continued*)

Substrate	Conditions	Product(s) and Yield(s) (%)	Refs.
C_{15-17}			
R[2], R[1], O, R[3]	DMD, acetone/CH_2Cl_2, N_2, ~20°, 20 to 24 h	R[2], R[1], O, O, R[3]	34, 53
C_{15-16}			
R[1], O, R[2]	DMD, acetone, 30°	R[1], O, O, R[2] (—)	236

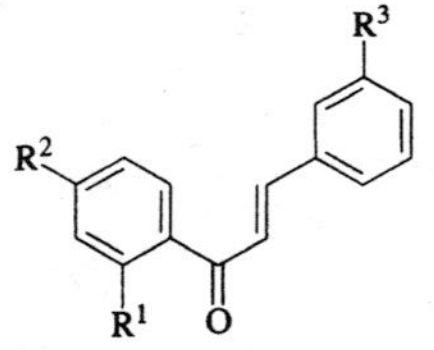

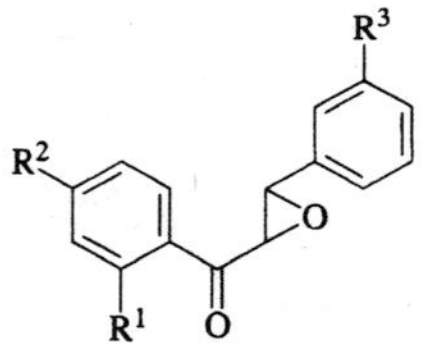

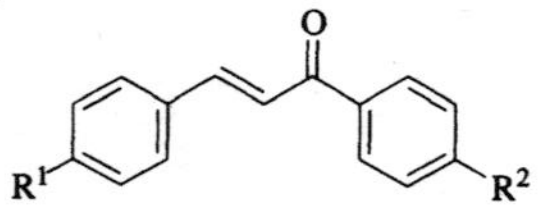

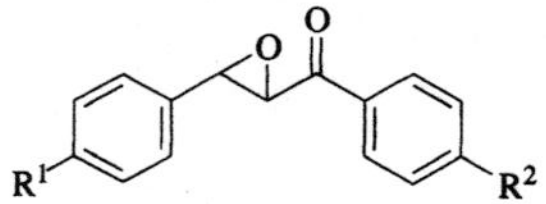

R^1	R^2	R^3	
H	Br	H	(98)
H	H	NO_2	(57)
OH	H	H	(~100)
H	H	H	(97)
H	Br	MeO	(98)
H	H	Me	(~100)
H	Me	H	(~100)
H	H	MeO	(99)
H	MeO	H	(99)
OH	H	Me	(~100)
OH	H	MeO	(~100)
H	Me	Me	(99)
H	Me	MeO	(99)
H	MeO	Me	(99)
H	MeO	MeO	(96)

C15-16

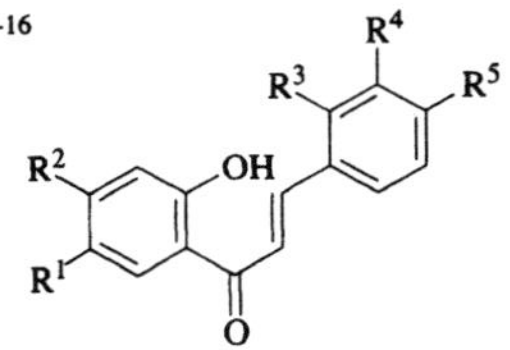

R2, R1, OH, R3, R4, R5, O

DMD, acetone/CH_2Cl_2, N_2

R2, R1, OH, O, R3, R4, R5, O

247

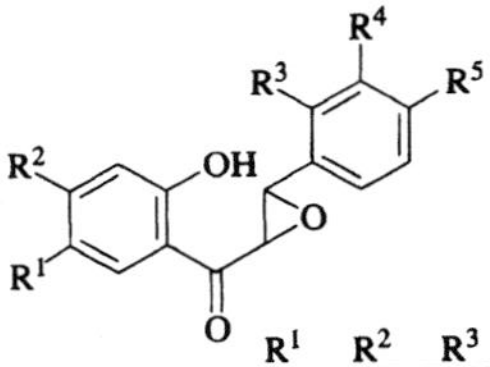

R^1	R^2	$k_2 \times 10^3$ ($M^{-1}s^{-1}$)
Br	H	1.03 ± 0.05
H	Br	1.39 ± 0.05
Cl	H	0.95 ± 0.04
H	Cl	1.47 ± 0.03
F	H	1.58 ± 0.01
H	F	1.54 ± 0.04
NO_2	H	0.25 ± 0.03
H	NO_2	1.07 ± 0.13
H	H	1.54 ± 0.13
CN	H	0.36 ± 0.03
H	CN	1.24 ± 0.05
Me	H	3.1 ± 0.2
H	Me	1.86 ± 0.04
MeO	H	11.5 ± 0.1
H	MeO	2.12 ± 0.06

R^1	R^2	R^3	R^4	R^5	Temp	Time	
Cl	H	H	H	H	~20°	59 h	(100)
H	H	H	H	Cl	~20°	57 h	(100)
Cl	H	H	H	MeO	–5°	50 h	(100)
H	Me	H	H	H	~20°	30 h	(100)
H	H	MeO	H	H	0°	62 h	(100)
H	H	H	MeO	H	0°	56 h	(100)
Me	H	H	H	MeO	0°	37 h	(99)

TABLE 1C. EPOXIDATION OF OLEFINS WITH ELECTRON ACCEPTORS BY ISOLATED DIOXIRANES (*Continued*)

Substrate	Conditions	Product(s) and Yield(s) (%)	Refs.
C_{15} 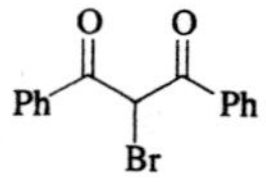 Ph, O, O, Ph, Br	DMD, acetone/CH_2Cl_2, DMAP, 0° to rt	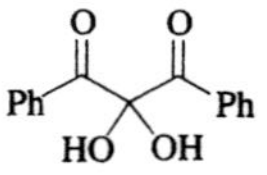 Ph, O, O, Ph, HO, OH (70)	240
C_{15-19} 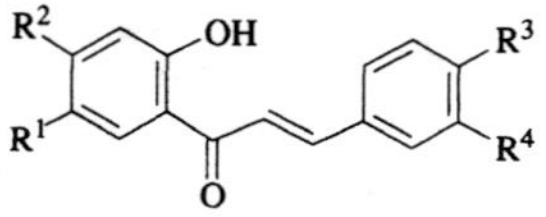 R^2, OH, R^3, R^1, O, R^4	DMD, CH_2Cl_2/acetone, 0°	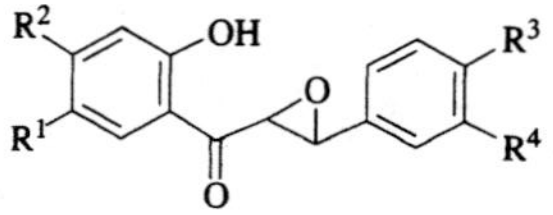 R^2, OH, R^3, R^1, O, O, R^4 (—)	241, 242

R^1	R^2	R^3	R^4
H	H	Br	H
Cl	H	H	H
H	H	F	H
H	H	H	H
H	OH	OH	OH
H	H	CN	H
H	H	Me	H
H	Me	H	H
H	H	MeO	H
H	OH	—$O(CH_2)_2O$—	
H	MOMO	—$O(CH_2)_2O$—	

Substrate	Conditions	Product(s) and Yield(s) (%)	Refs.
C_{15} 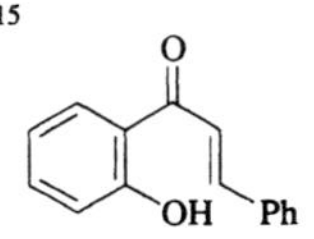	DMD, acetone/CH_2Cl_2, 0° to rt, ~81.5 h	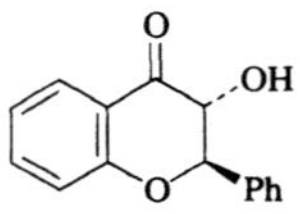 **I** (74) + 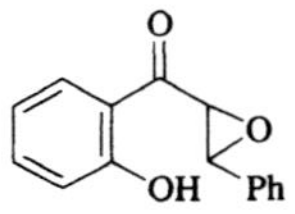 **II** (20)	248
	1. DMD, acetone/MeCN, AcOH/NaOAc, 0° 2. rt, ~25 h	**II** (60)	248
C_{15-47} 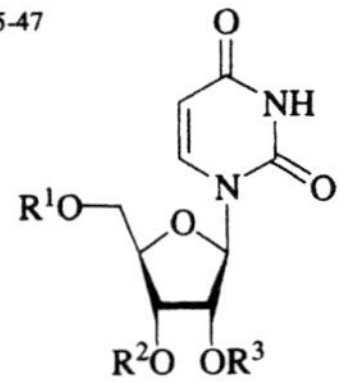	DMD, CH_2Cl_2, 25°	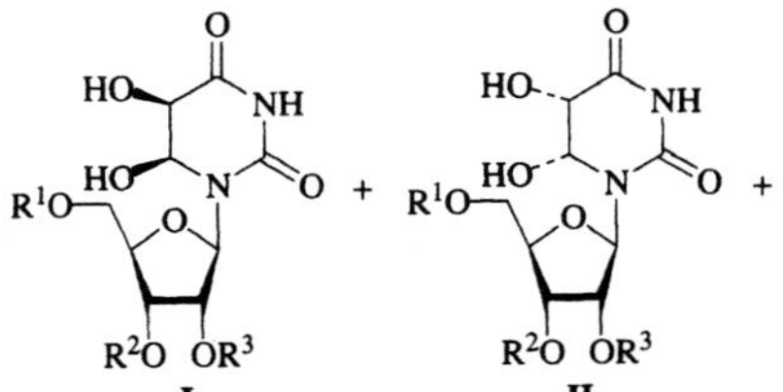 **I** + **II** + 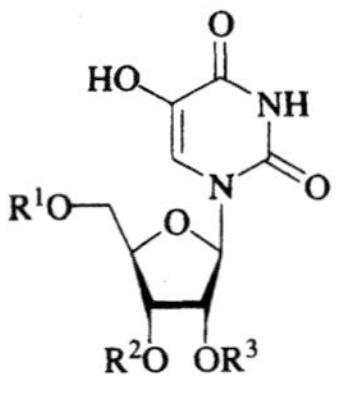 **III**	234, 235

R^1	R^2	R^3	**I+II**	**III**
Ac	Ac	Ac	(70)	(18)
Tr	H	H	(57)	(27)
PhCO	PhCO	PhCO	(65)	(0)
Tr	H	Tr	(63)	(0)
Tr	Tr	H	(69)	(0)

TABLE 1C. EPOXIDATION OF OLEFINS WITH ELECTRON ACCEPTORS BY ISOLATED DIOXIRANES (*Continued*)

Substrate	Conditions	Product(s) and Yield(s) (%)	Refs.
	DMD, acetone/MeOH, 25°	**III** 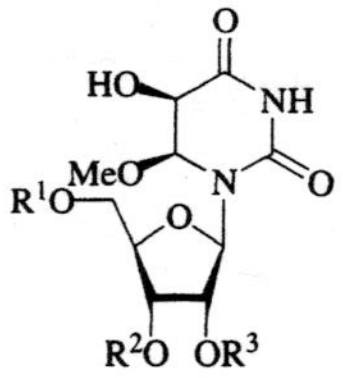 + **IV** 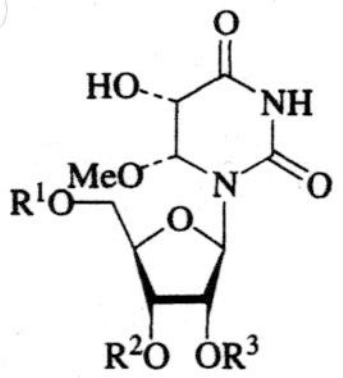 + **V** R^1 R^2 R^3 — **III** — **IV+V** Ac Ac Ac — (11) — (75)	234, 235
C_{16-17} 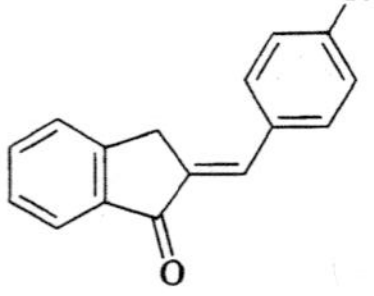	DMD, acetone/CH_2Cl_2, rt	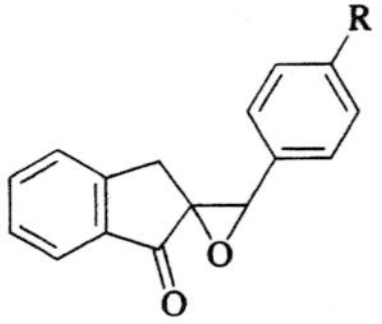 R — Time Br — 168 h — (76) Cl — 168 h — (72) F — 120 h — (83) H — 72 h — (71) CN — 168 h — (77) Me — 72 h — (94) MeO — 72 h — (87)	249
C_{16-18} 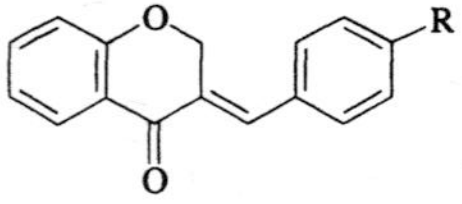	DMD, acetone/CH_2Cl_2, dark, rt	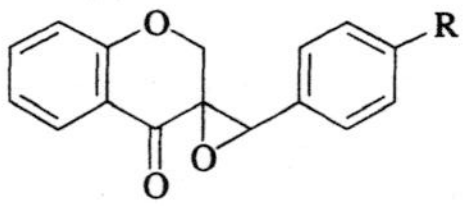 **I** + 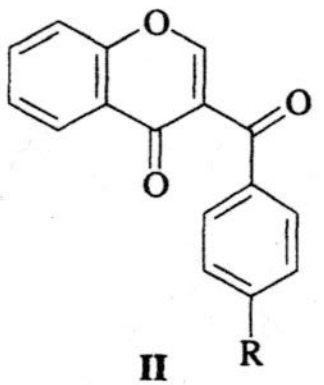 **II** +	250

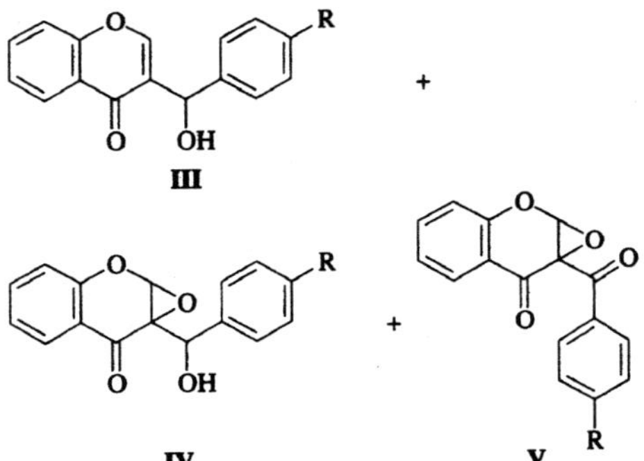

III

IV

V

R	Time	I	II	III	IV	V
Br	360 h	(48)	(29)	(0)	(0)	(0)
Cl	456 h	(47)	(30)	(0)	(0)	(0)
H	168 h	(47)	(21)	(0)	(0)	(5)
Me	144 h	(51)	(28)	(0)	(4)	(0)
MeO	144 h	(38)	(19)	(5)	(0)	(7)
EtO	168 h	(41)	(16)	(0)	(0)	(8)

TABLE 1C. EPOXIDATION OF OLEFINS WITH ELECTRON ACCEPTORS BY ISOLATED DIOXIRANES (*Continued*)

Substrate	Conditions	Product(s) and Yield(s) (%)	Refs.
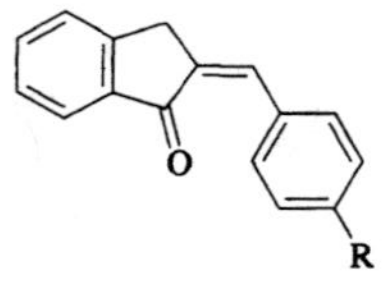	DMD, acetone/CH_2Cl_2, rt	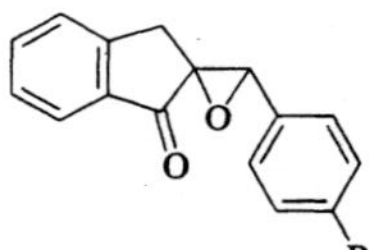 I + 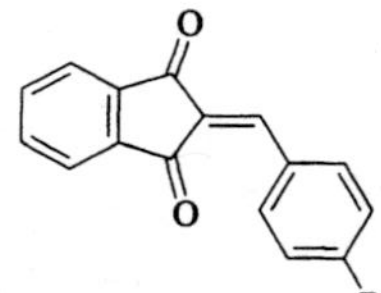 II + 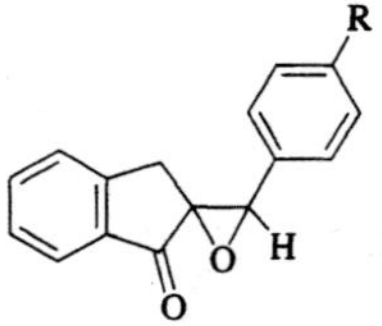 III (9-10)	251
C_{16-17} 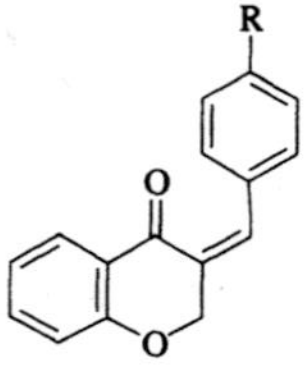	DMD, acetone/CH_2Cl_2, dark, rt	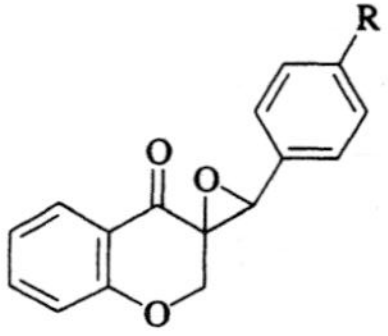 I + 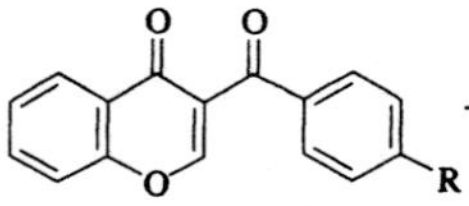 II + 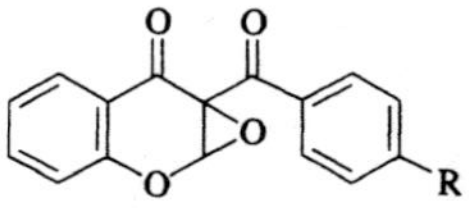 III	250

R	Time	I	II
Cl	192 h	(56)	(0)
H	120 h	(39)	(0)
Me	96 h	(35)	(5)

R	Time	I	II	III
Cl	528 h	(23)	(43)	(0)
H	288 h	(36)	(23)	(0)
Me	192 h	(42)	(21)	(0)
MeO	360 h	(28)	(19)	(4)

Substrate	Conditions	Product(s) and Yield(s) (%)	Refs.
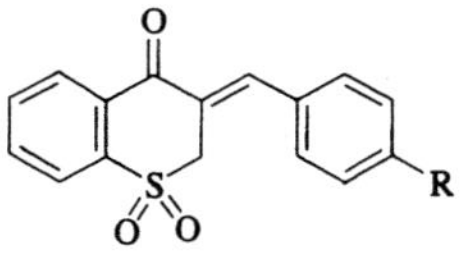	TFD, TFP/CH_2Cl_2, 20°	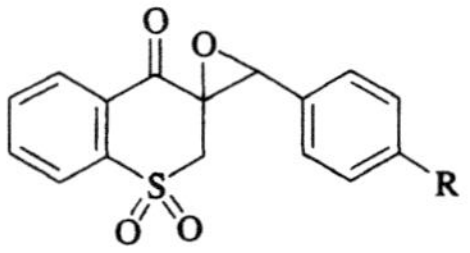 R / Time: Cl 3 d (69); H 2.5 d (31); Me 3 d (14)	66
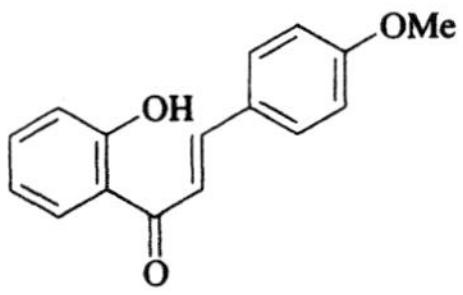	DMD, acetone/CH_2Cl_2, 0° to rt, 24 h	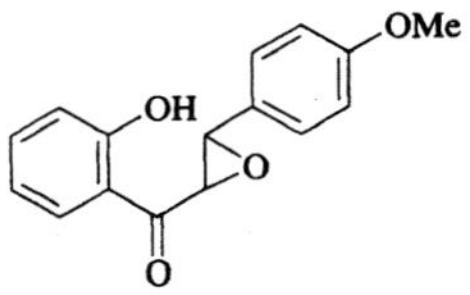 (100)	248
	DMD, acetone/MeCN, AcOH/NaOAc, 0° to rt, 20 h	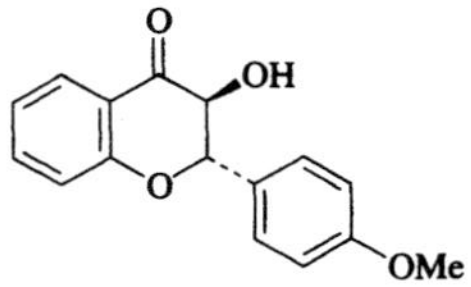 (62)	248
C_{16-17} $CF_3CO_2^-$	DMD, acetone/CH_2Cl_2, 0°, 4 h or 1. TFD, –78° 2. –20°, 30 min	$CF_3CO_2^-$ (—) + $CF_3CO_2^-$ (—) R = H, Me, MeO	60a

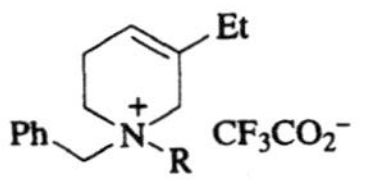

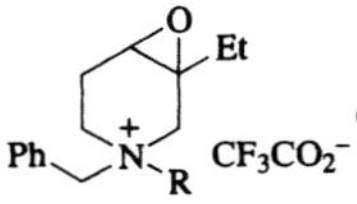

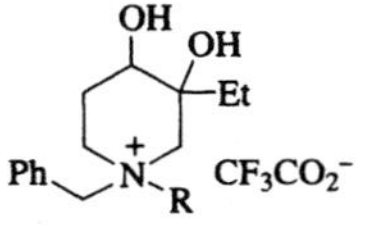

TABLE 1C. EPOXIDATION OF OLEFINS WITH ELECTRON ACCEPTORS BY ISOLATED DIOXIRANES (*Continued*)

Substrate	Conditions	Product(s) and Yield(s) (%)	Refs.
C_{16} 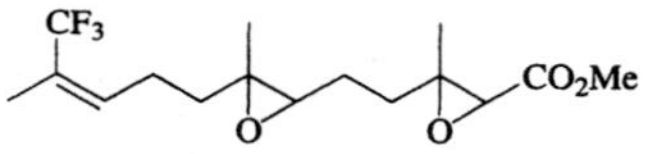 CF_3 … CO_2Me	DMD, acetone/(CH_2Cl_2), –20°, 8 d	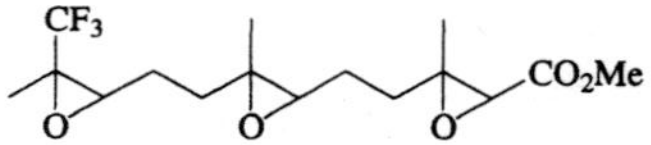 CF_3 … CO_2Me **I** (93)	78
	TFD, TFP (CH_2Cl_2), –20°, 1 h	**I** (94)	78
C_{17-18} 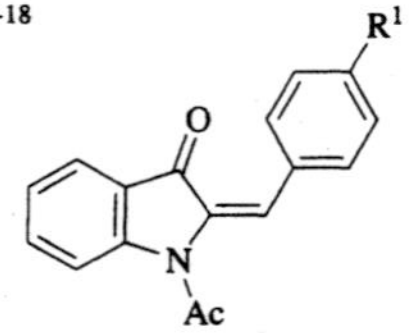 R^1 O N Ac	DMD, acetone/CH_2Cl_2, rt, dark	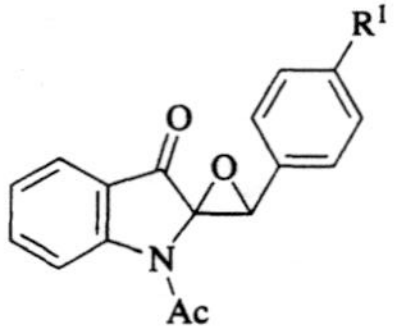 R^1 O O N Ac	252
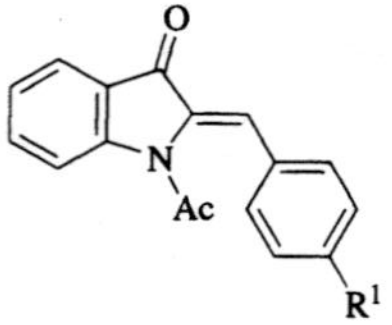 O N Ac R^1	DMD, acetone/CH_2Cl_2, rt, dark	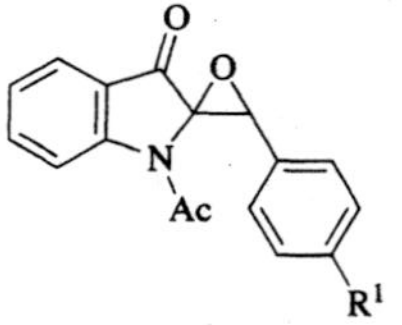 O O N Ac R^1	252

First C_{17-18} entry (ref. 252):

R^1	Time	
Br	113 h	(86)
Cl	144 h	(85)
F	109 h	(92)
H	109 h	(71)
Me	74 h	(81)
MeO	113 h	(—)

R^1	Time	
Br	121 h	(87)
Cl	144 h	(87)
F	109 h	(89)
NO_2	150 h	(84)
H	109 h	(83)
Me	89 h	(85)
MeO	—	(—)

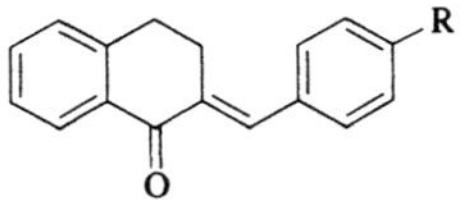

DMD, acetone/CH_2Cl_2, dark, rt

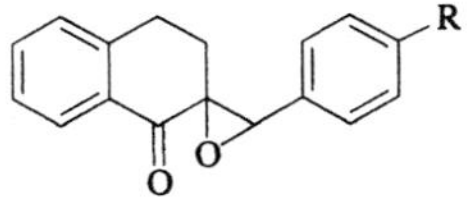

R	Time
Br	72 h (89)
Cl	72 h (81)
F	48 h (79)
H	48 h (79)
CN	72 h (83)
Me	48 h (85)
MeO	48 h (85)

253

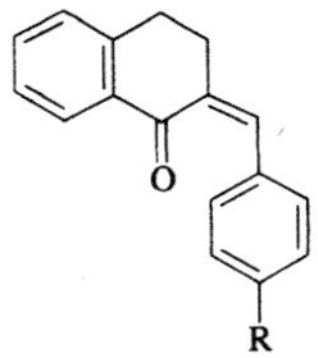

DMD, acetone/CH_2Cl_2, dark, rt

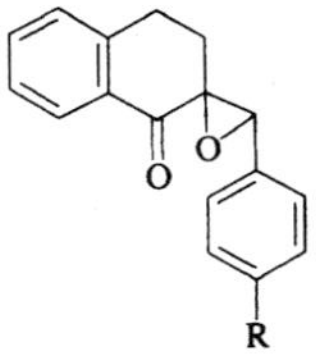

R	Time
Cl	20 h (30)
H	144 h (28)
Me	120 h (23)

253

C_{18-19}

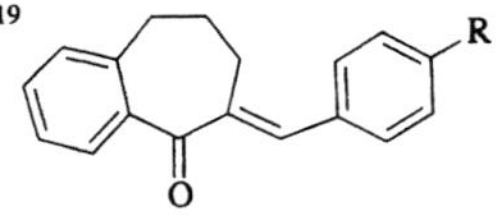

DMD, acetone/CH_2Cl_2, rt

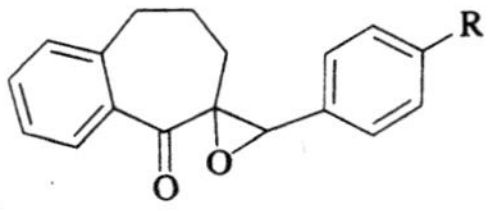

R	Time
Br	216 h (73)
Cl	192 h (79)
F	192 h (79)
H	168 h (83)
CN	216 h (96)
Me	168 h (90)
MeO	120 h (78)

251

TABLE 1C. EPOXIDATION OF OLEFINS WITH ELECTRON ACCEPTORS BY ISOLATED DIOXIRANES (*Continued*)

Substrate	Conditions	Product(s) and Yield(s) (%)	Refs.
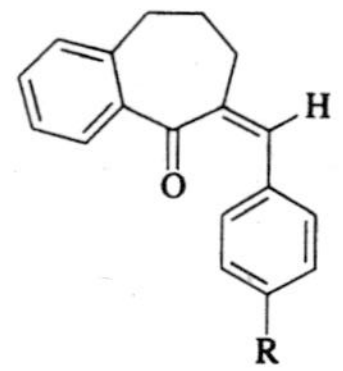 R	DMD, acetone/CH_2Cl_2, rt	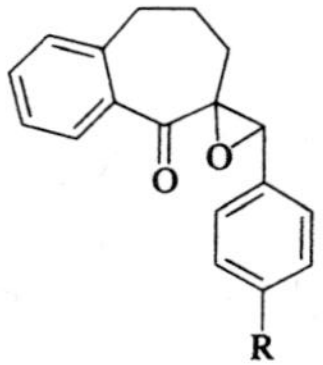 I 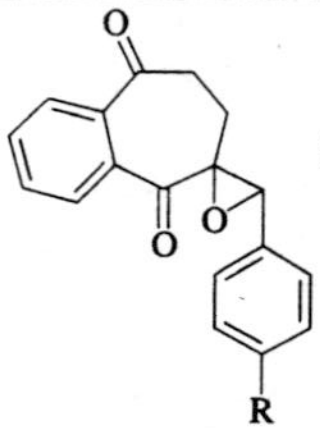 + II R Time I II Cl 336 h (61) (8) H 240 h (50) (0) Me 216 h (53) (7)	251
C_{19} 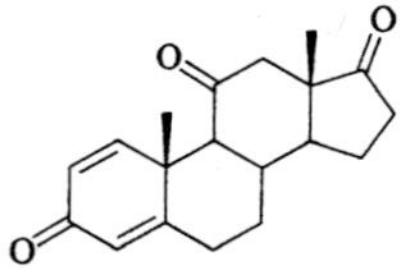	DMD, CH_2Cl_2/acetone, ~20°, 20 h	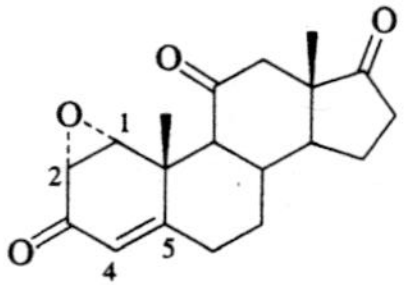 (80) + 4β,5β-epoxide (15) + 4α,5α-epoxide (5)	77
	DMD, CH_2Cl_2/acetone, ~20°, 28 h	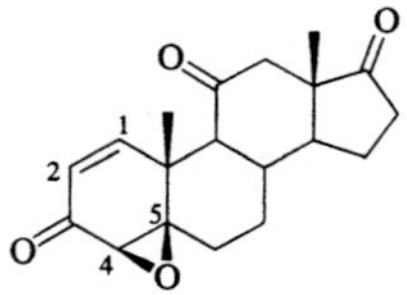 (81) + 4α,5α-epoxide (12) + 1α,2α-epoxide (6) + 1β,2β-epoxide (1)	77
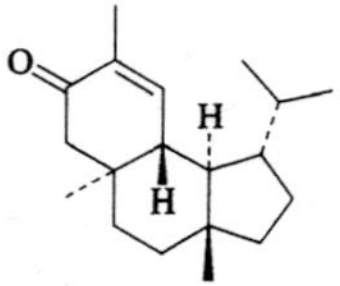	DMD, acetone/CH_2Cl_2, rt	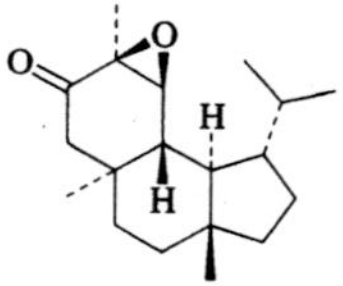 (72)	54

Substrate	Conditions	Product(s) and Yield(s) (%)	Refs.
C_{20} OMe, OMe, O, O, N, O 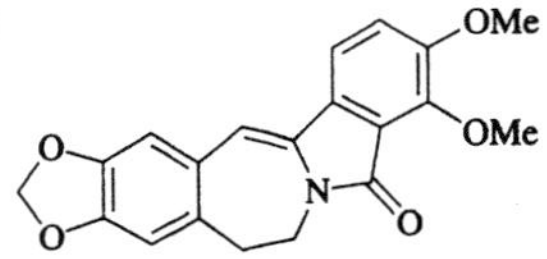	1. DMD, acetone/CH_2Cl_2, 0° 2. $NaHCO_3$	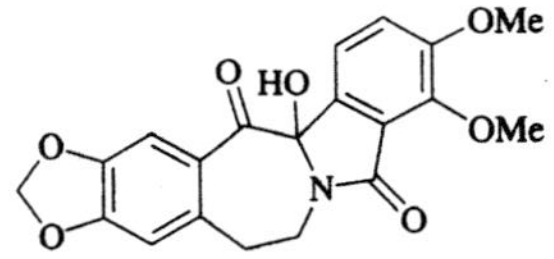 O, HO, OMe, OMe, O, O, N, O (38)	254
C_{21} 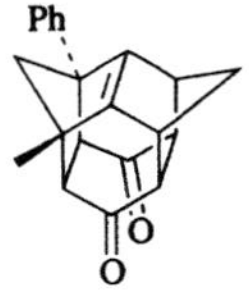 Ph, O, O	DMD, acetone, –30°, 10 min	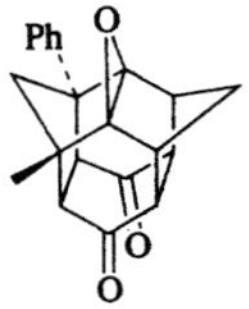 Ph, O, O, O (—)	256
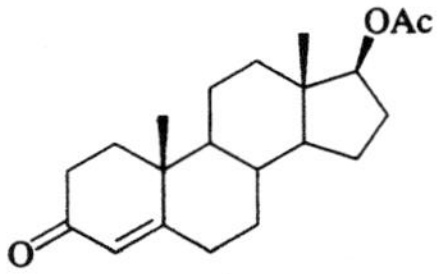 OAc, O	1. DMD, acetone, dark, rt, 8 h 2. NaBr or LiBr, Amberlyst 15, rt, 12 h	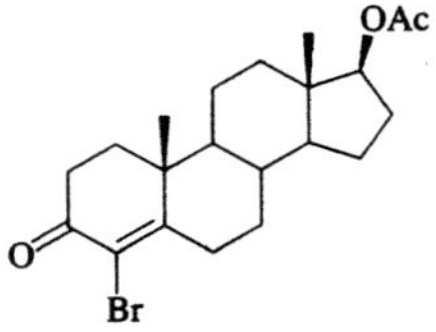 OAc, O, Br (97)	171
	1. DMD, acetone, dark, rt, 8 h 2. NaCl or LiCl, Amberlyst 15, rt, 12 h	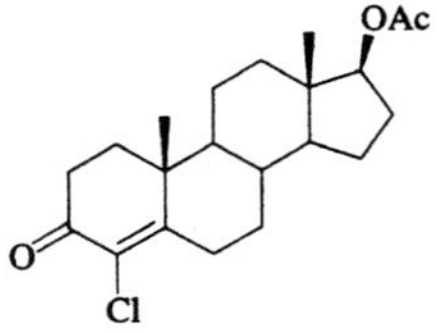 OAc, O, Cl (89)	171
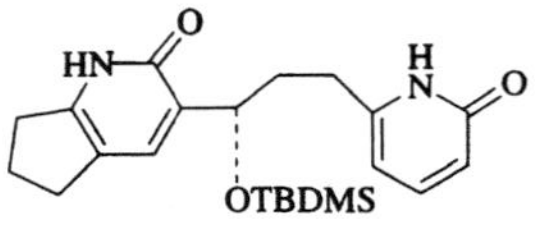 HN, O, H, N, O, OTBDMS	1. *h*ν, toluene, <5° 2. DMD, 0 to 5°	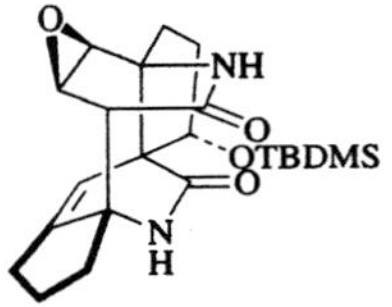 O, NH, O, OTBDMS, O, N, H (—)	257

TABLE 1C. EPOXIDATION OF OLEFINS WITH ELECTRON ACCEPTORS BY ISOLATED DIOXIRANES (*Continued*)

Substrate	Conditions	Product(s) and Yield(s) (%)	Refs.
C_{23} 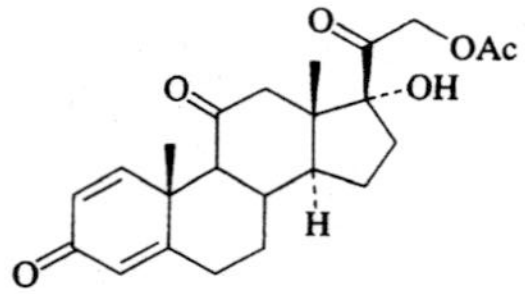 [structure: OAc, OH, O, H]	DMD, acetone, 20°, 20 h	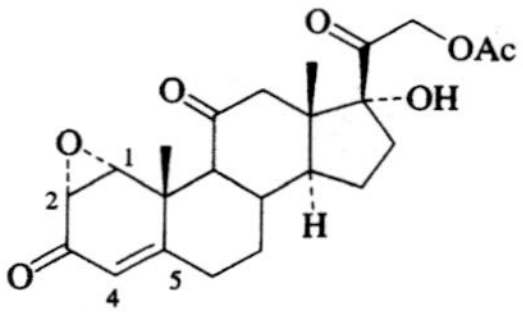 [structure: OAc, OH, O, H, 1, 2, 4, 5] **I** (80)	258
	DMD, acetone/CH_2Cl_2, ~20°, 20 h	**I** (80) + 4β,5β-epoxide (16) + 4α,5α-epoxide (4)	77
C_{27} 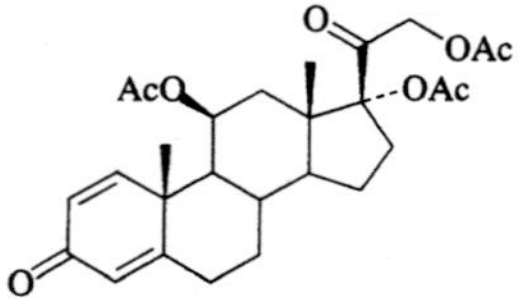 [structure: OAc, OAc, AcO, O]	DMD, acetone/CH_2Cl_2, ~20°, 24 h	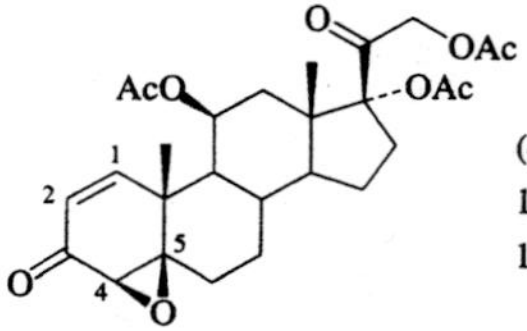 [structure: OAc, OAc, AcO, 1, 2, 4, 5, O] (49) + 4α,5α-epoxide (11) + 1α,2α-epoxide (17) + 1β,2β-epoxide (3)	77
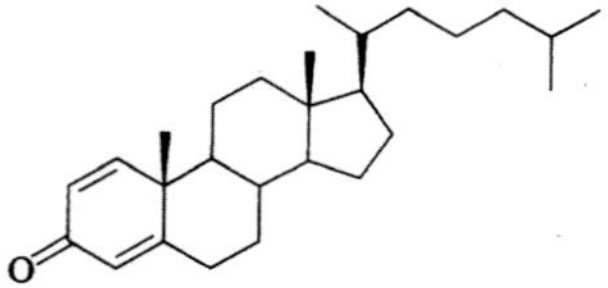 [structure]	DMD, acetone/CH_2Cl_2, ~20°, 24 h	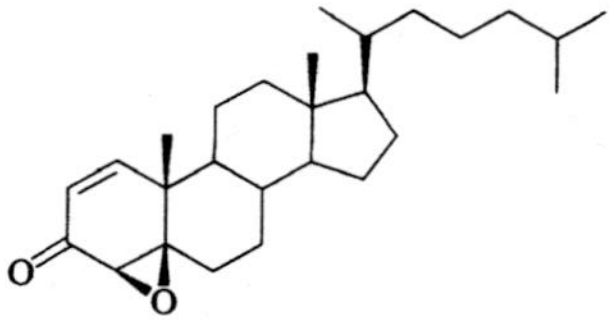 [structure] (39)	77
	DMD, acetone/CH_2Cl_2, ~20°, 72 h	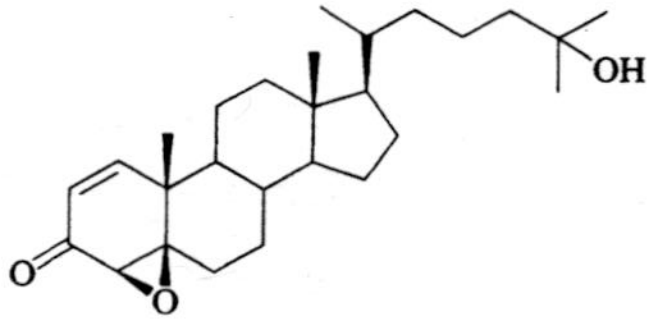 [structure: OH] (40)	77

Substrate	Conditions	Product(s) and Yield(s) (%)	Refs.
C_{29} Tetraphenylcyclopentadienone 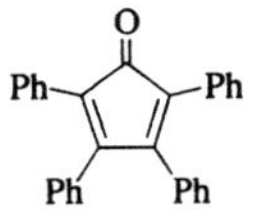	DMD (large excess), acetone, heat	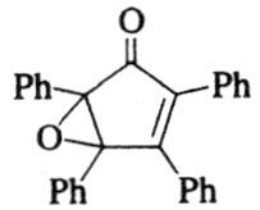 Monoepoxide (—)	5
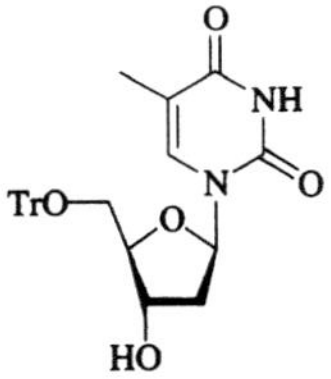 5'-O-Trityl thymidine	DMD, acetone/CH_2Cl_2, 25°	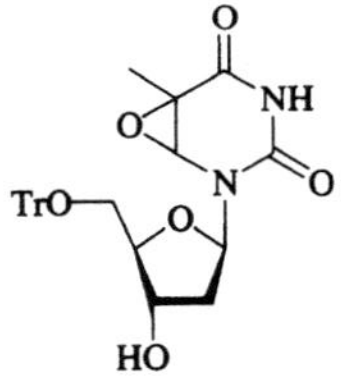 Epoxide (10) 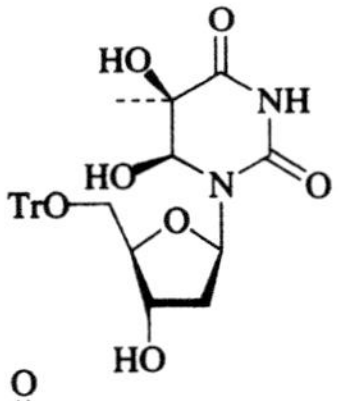 + cis-diol (43) 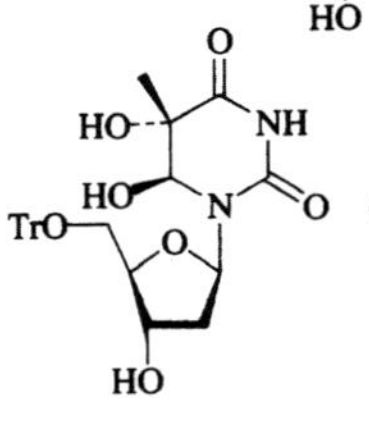 + cis-diol isomer (12)	234
C_{29-} 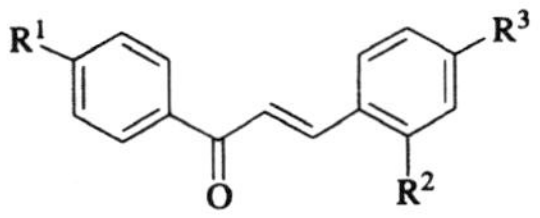 Chalcones (R^1, R^2, R^3 below)	DMD, acetone/CH_2Cl_2, 20°, 48 h	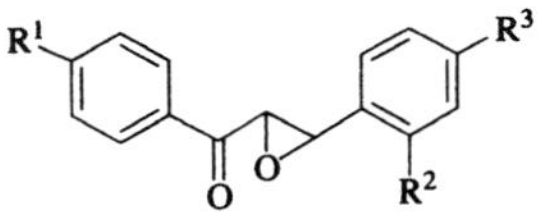 Chalcone epoxides	259

R^1	R^2	R^3	
O-β-D-glucosyl-Ac_4	H	H	(84)
H	*O*-β-D-glucosyl-Ac_4	H	(88)
H	H	*O*-β-D-glucosyl-Ac_4	(87)
O-β-cellobiosyl-Ac_7	H	H	(78)
H	*O*-β-cellobiosyl-Ac_7	H	(82)
H	H	*O*-β-cellobiosyl-Ac_7	(81)

TABLE 1C. EPOXIDATION OF OLEFINS WITH ELECTRON ACCEPTORS BY ISOLATED DIOXIRANES (*Continued*)

	Substrate	Conditions	Product(s) and Yield(s) (%)	Refs.
C_{51} 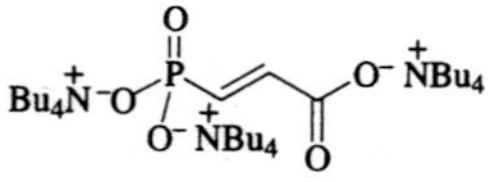	$Bu_4\overset{+}{N}\,^-O-P(=O)(O^-\,\overset{+}{N}Bu_4)-CH=CH-C(=O)O^-\,\overset{+}{N}Bu_4$	DMD, acetone, rt, 4 d	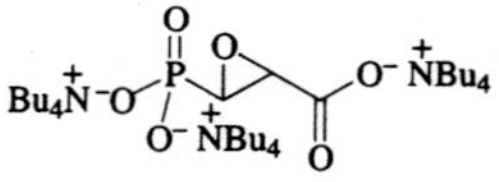 $Bu_4\overset{+}{N}\,^-O-P(=O)(O^-\,\overset{+}{N}Bu_4)$-epoxide-$C(=O)O^-\,\overset{+}{N}Bu_4$ (100)	76
C_{52} 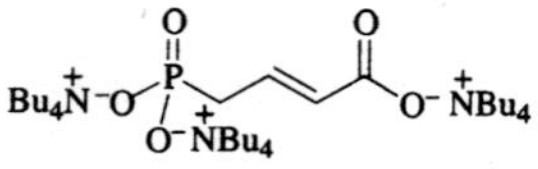	$Bu_4\overset{+}{N}\,^-O-P(=O)(O^-\,\overset{+}{N}Bu_4)-CH_2CH=CH-C(=O)O^-\,\overset{+}{N}Bu_4$	DMD, acetone, rt, 4 d	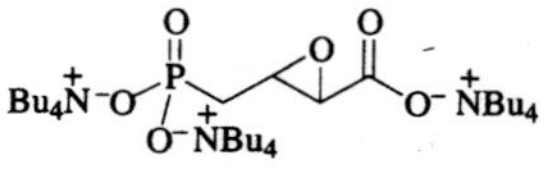 $Bu_4\overset{+}{N}\,^-O-P(=O)(O^-\,\overset{+}{N}Bu_4)-CH_2$-epoxide-$C(=O)O^-\,\overset{+}{N}Bu_4$ (100)	76

[a] TDMPP = 5,10,15,20-tetrakis(2,6-dimethoxyphenyl)porphyrin dianion

[b] (Cl16)TDMPP = 2,3,7,8,12,13,17,18-octachloro-5,10,15,20-tetrakis(2,6-dimethoxy-3,5-dichlorophenyl)porphyrin dianion

TABLE 1D. EPOXIDATION OF OLEFINS WITH ELECTRON DONORS AND ACCEPTORS BY ISOLATED DIOXIRANES

	Substrate	Conditions	Product(s) and Yield(s) (%)	Refs.
C_6	O	DMD, acetone/CH_2Cl_2, N_2, 0°, 23 h	O O (~100)	56
C_7	O O	DMD, acetone/CH_2Cl_2, N_2, –20°, 3 h	O O O (79)	44, 212
$C_{7\text{-}13}$	NHCHO R	1. DMD, acetone/CH_2Cl_2, –40° 2. $(CF_3SO_2)_2O$, *i*-Pr_2NEt, –78°	NC O R **I** + NC R **II** R / **I** / **II**: H (36) (22); Ph (33) (8)	238
C_7	N CO_2Et	DMD, acetone, 0° to rt, overnight	O N CO_2Et (97)	239
C_9	O O	DMD, acetone/CH_2Cl_2, N_2, –20°, 3 h	O O O (83)	44, 212
	O O	DMD, acetone, –78° to rt	O O O (97)	243
		1. Moist DMD, acetone, –78° to rt, 18 h 2. Acetone, $CuSO_4$, H_2SO_4, 50°, 3 d	O O O O (17)	243

TABLE 1D. EPOXIDATION OF OLEFINS WITH ELECTRON DONORS AND ACCEPTORS BY ISOLATED DIOXIRANES (*Continued*)

	Substrate	Conditions	Product(s) and Yield(s) (%)	Refs.
C_9 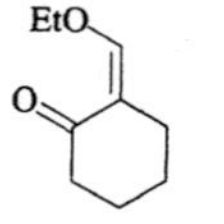	EtO, O	DMD, acetone/CH_2Cl_2, N_2, –20°, 11.5 h	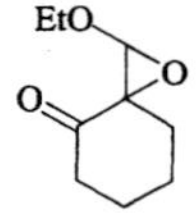 EtO, O, O (~100)	56
C_{10} 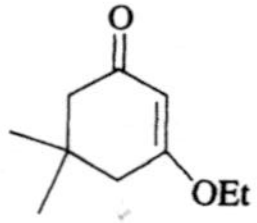	O, OEt	DMD, acetone/CH_2Cl_2, N_2, 0°, 11 h	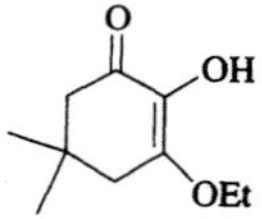 O, OH, OEt (99)	56
C_{10-14} 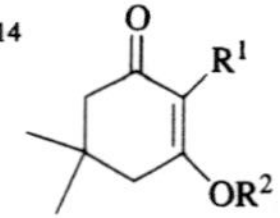	O, R^1, OR^2	DMD, acetone/CH_2Cl_2, N_2, –20°	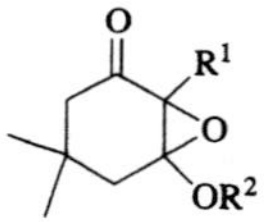 O, R^1, O, OR^2 R^1 / R^2 / Time H / Et / 26 h (98) Me / Et / 17 h (~100) H / *n*-Bu / 11 h (99) H / Ph / 24 h (97)	56
C_{10} 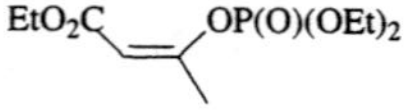	EtO_2C, $OP(O)(OEt)_2$	DMD, acetone/CH_2Cl_2, N_2, 20°, 48 h	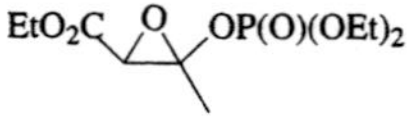 EtO_2C, O, $OP(O)(OEt)_2$ (98)	205
	Ph, H, N, O	DMD-d_6, acetone-d_6, –50°	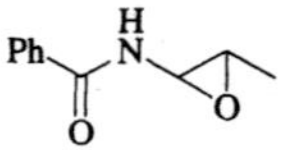 Ph, H, N, O, O (100)	167
		1. DMD, acetone/CH_2Cl_2, –78°, 3 h 2. –20°, 24 h	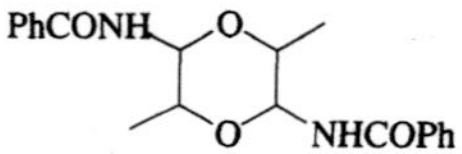 PhCONH, O, O, NHCOPh (79)	167
	Ph, H, N, O	DMD-d_6, acetone-d_6, –50°	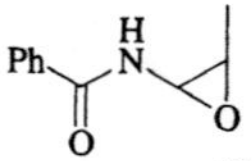 Ph, H, N, O, O (100)	167
	Ph, H, N, O	DMD, acetone/MeOH, –78°, 2 h	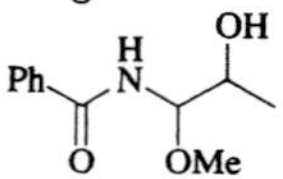 Ph, H, N, O, OH, OMe (40)	167

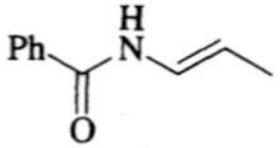

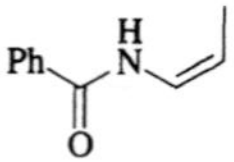

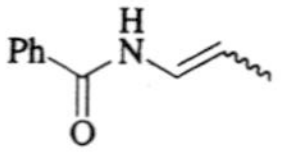

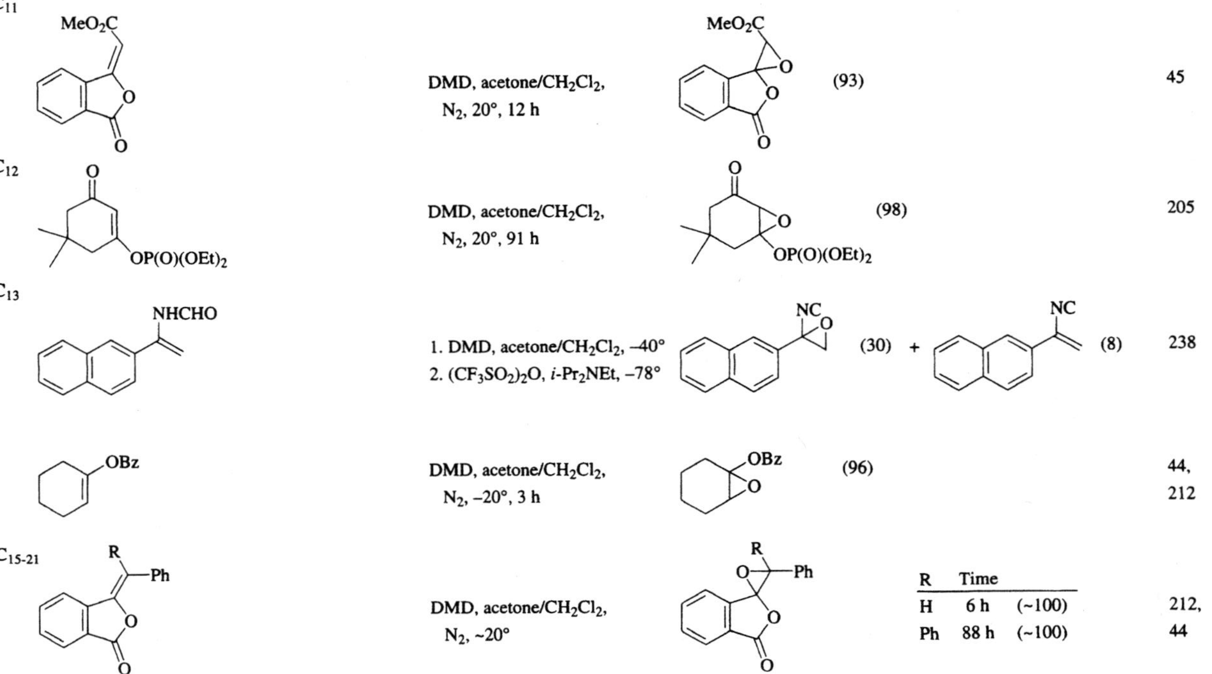

C_{11}	DMD, acetone/CH_2Cl_2, N_2, 20°, 12 h	(93)	45
C_{12}	DMD, acetone/CH_2Cl_2, N_2, 20°, 91 h	(98)	205
C_{13}	1. DMD, acetone/CH_2Cl_2, –40° 2. $(CF_3SO_2)_2O$, i-Pr_2NEt, –78°	(30) + (8)	238
	DMD, acetone/CH_2Cl_2, N_2, –20°, 3 h	(96)	44, 212
C_{15-21}	DMD, acetone/CH_2Cl_2, N_2, ~20°	R Time H 6 h (~100) Ph 88 h (~100)	212, 44

TABLE 1D. EPOXIDATION OF OLEFINS WITH ELECTRON DONORS AND ACCEPTORS BY ISOLATED DIOXIRANES (*Continued*)

Substrate	Conditions	Product(s) and Yield(s) (%)	Refs.
C_{15-17}	DMD, acetone/CH_2Cl_2, N_2		59, 58
C_{15}	1. DMD, acetone/CH_2Cl_2, Ar, 0°, 2.3 h 2. rt, ~25 h	(56)	248
	DMD, acetone/CH_2Cl_2, N_2, rt, 24 h	(80)	260

R^1	R^2	R^3	R^4	R^5	Temp	Time	
H	H	H	H	Cl	0°	61 h	(92)
H	H	H	Cl	H	~20°	28 h	(~100)
H	H	H	H	NO_2	~20°	72 h	(~100)
H	H	H	H	H	0°	36 h	(~100)
H	Me	H	H	Cl	~20°	27 h	(~100)
H	Me	H	Cl	H	~20°	37 h	(~100)
H	Me	H	H	H	0°	41 h	(~100)
H	H	H	H	Me	0°	23 h	(~100)
H	H	H	H	MeO	0°	24 h	(~100)
H	H	MeO	H	H	0°	26 h	(~100)
H	MeO	H	H	H	–10°	48 h	(~100)
MeO	H	H	H	H	–10°	46 h	(~100)
H	Me	H	H	Me	–10°	34 h	(~100)
H	Me	H	H	MeO	0°	23 h	(~100)

Substrate	Conditions	Product(s) and Yield(s) (%)	Refs.
C_{15} 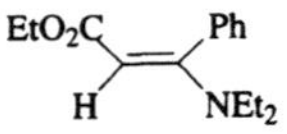	DMD, acetone/CH_2Cl_2, –70°	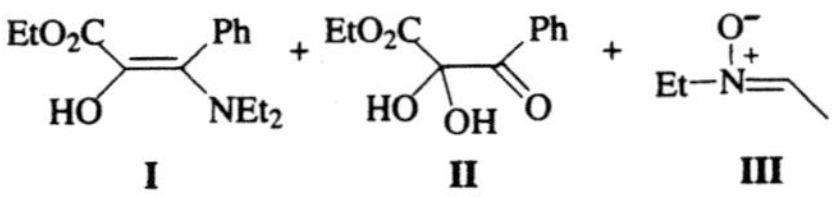 I + II + III (—) I:II:III = 27:49:24	228
	1. DMD, acetone/CH_2Cl_2, –70° 2. $^{18}O_2$	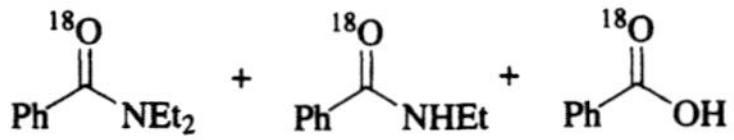 + + (—)	228
	DMD, acetone/CH_2Cl_2, 0°	II (57) + 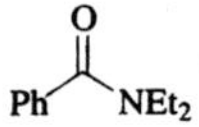 (13)	228
C_{16} 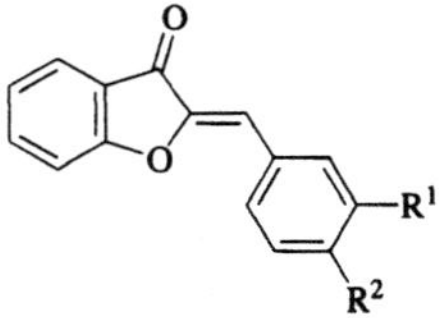	DMD, acetone/CH_2Cl_2, N_2	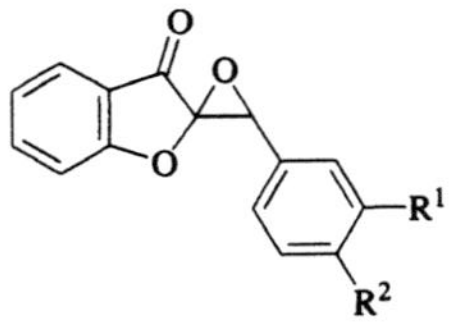 R^1 R^2 Temp Time H MeO –5° 11 h (100) MeO H 0° 22 h (100)	57
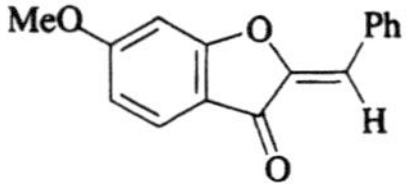	DMD, acetone/$CHCl_3$, rt, 34 h	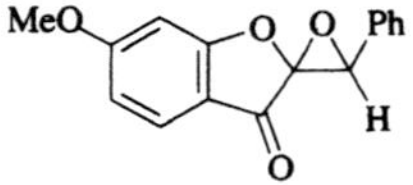 (83)	248
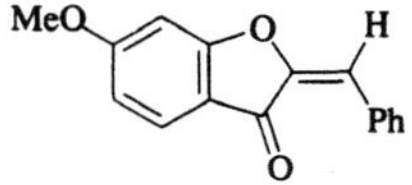	1. DMD, acetone/CH_2Cl_2, 0°, 1 h 2. rt, 36 h	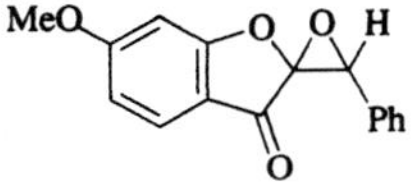 (33)	248
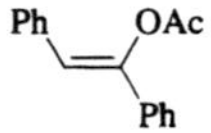	DMD, acetone/CH_2Cl_2, N_2, ~20°, 7 h	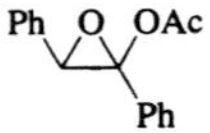 (84)	44, 45

TABLE 1D. EPOXIDATION OF OLEFINS WITH ELECTRON DONORS AND ACCEPTORS BY ISOLATED DIOXIRANES (*Continued*)

Substrate	Conditions	Product(s) and Yield(s) (%)	Refs.
C_{17-23} 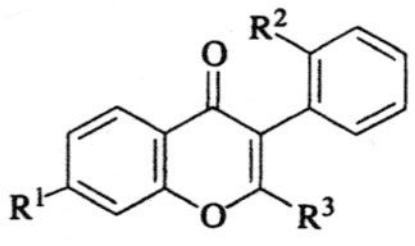 R[1], R[2], R[3]	DMD, acetone/CH_2Cl_2, ~20°, ~15 d	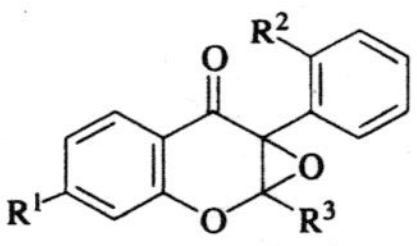 R[1] / R[2] / R[3]: MeO / H / Me (38); Ms / H / Me (23); MeO / H / Et (38); MeO / MeO / Me (14); Ms / H / Et (27); MeO / MeO / Et (17); MeO / H / Ph (16); MeO / MeO / Ph (19)	60
C_{17} 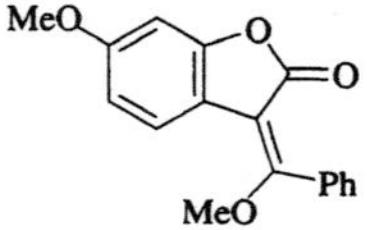 MeO, O, O, MeO, Ph	1. DMD, acetone/CH_2Cl_2, Ar, 0°, 2.3 h 2. rt, ~25 h	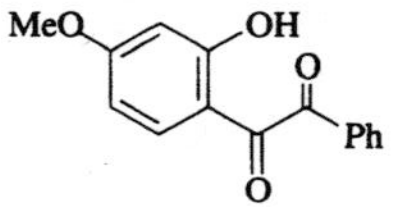 MeO, OH, O, Ph, O (33)	248
C_{20} 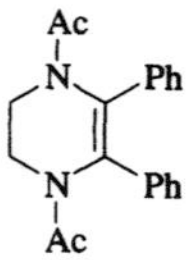 Ac, N, Ph, N, Ph, Ac	DMD, acetone, 0°, 6 h	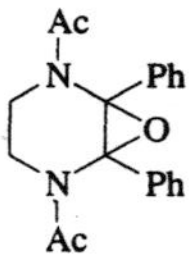 Ac, N, Ph, O, N, Ph, Ac (94)	255
C_{23} 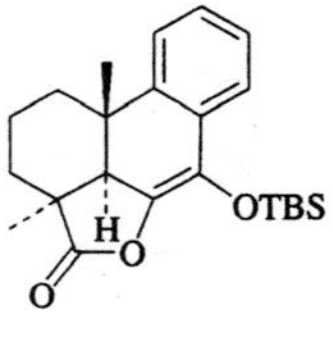 H, O, O, OTBS	DMD, acetone/CH_2Cl_2, −78°, 30 min	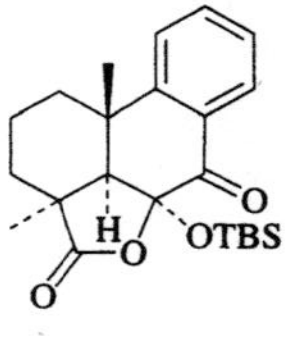 H, O, O, O, OTBS (>85%)	243

Substrate	Conditions	Product(s) and Yield(s) (%)	Refs.
C_{24}	DMD, acetone, 0°, 48 h	(—)	255
C_{25-}	DMD, acetone/CH_2Cl_2, 20°	see table below	259
C_{27-40}	1. DMD, acetone, 0°, 5 to 7 h 2. TsOH, CH_2Cl_2, 0°, 10 to 15 min	see table below	261
C_{29}	DMD, acetone/CH_2Cl_2, -40°, overnight	(>80)	262

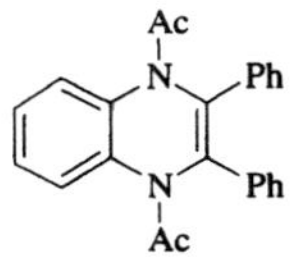

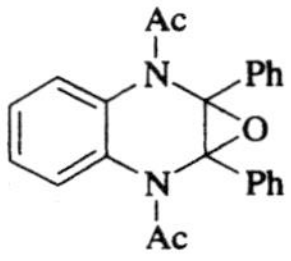

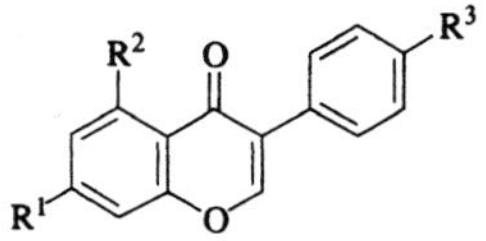

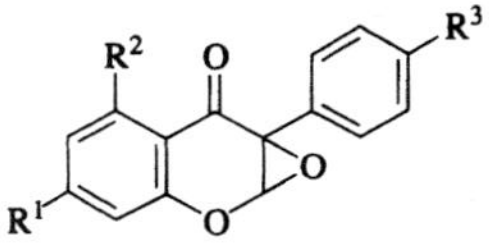

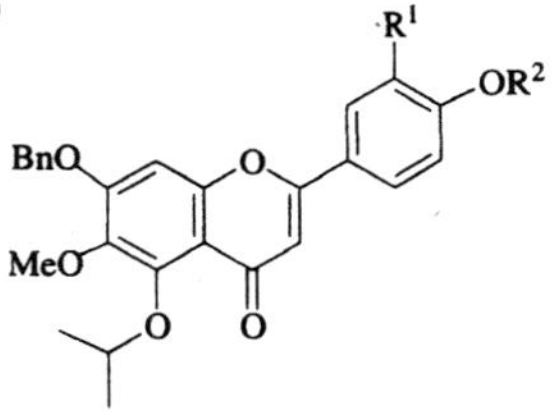

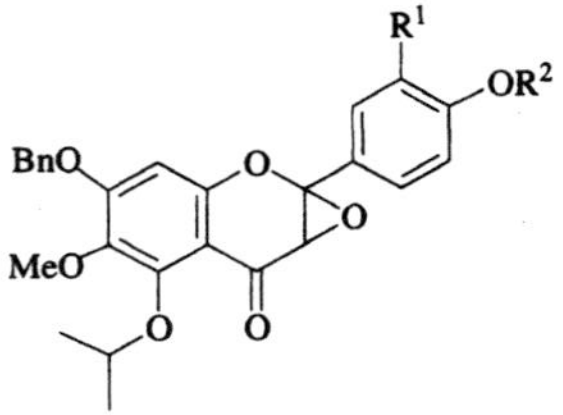

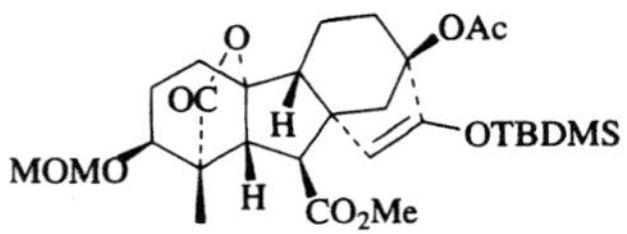

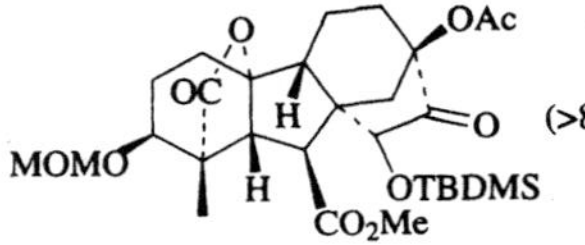

R^1	R^2	R^3	Time	
O-β-D-glucosyl-Ac_4	H	H	96 h	(76)
O-β-D-glucosyl-Ac_4	H	MeO	60 h	(88)
AcO	AcO	O-β-D-glucosyl-Ac_4	80 h	(80)
O-β-cellobiosyl-Ac_7	H	H	72 h	(83)

R^1	R^2	
H	Me	(71)
H	Bn	(67)
MeO	Bn	(56)
BnO	Me	(74)
BnO	Bn	(70)

TABLE 1D. EPOXIDATION OF OLEFINS WITH ELECTRON DONERS AND ACCEPTORS BY ISOLATED DIOXIRANES (*Continued*)

Substrate	Conditions	Product(s) and Yield(s) (%)	Refs.
C_{69} 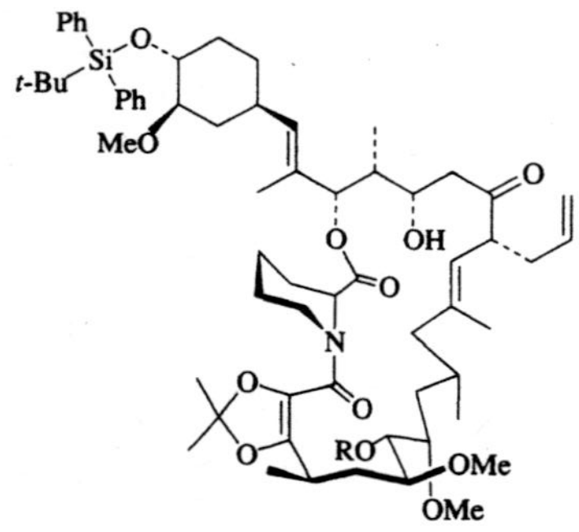 R = TBDMS	DMD, acetone, –26°, 1 h	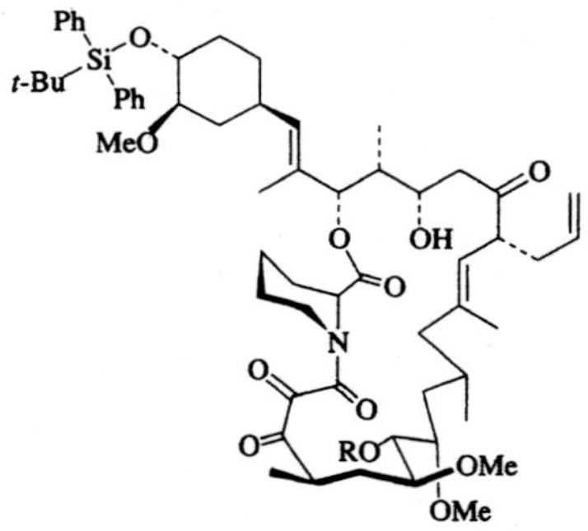 (100)	263

TABLE 1E. EPOXIDATION OF OLEFINS BY IN-SITU-GENERATED DIOXIRANES

Substrate	Conditions	Product(s) and Yield(s) (%)	Refs.
C_4 $R^1CH{=}C(R^2)CO_2H$	Acetone, Oxone®, $NaHCO_3$, H_2O, EDTA, 24° to 27°	2,3-epoxy acid (R^1, R^2, CO_2H)	264
C_{5-20} $R^1R^2C{=}CR^3R^4$	TFP, Oxone®, $NaHCO_3$, CH_3CN/H_2O, Na_2EDTA, 0 to 1°	epoxide **I**	35
	Ketone, Oxone®, $NaHCO_3$, CH_3CN/H_2O, EDTA, rt; Ketone = 2,6-diphenyl-1,1,3,5-tetramethylpiperidin-4-one-1-ium triflate (TfO^-)	**I**	38

Acetone (ref. 264):

R^1	R^2	Time	Yield (%)
Me	H	2.75 h	(62)
Ph	H	2.0 h	(92)
Ph	Me	2.5 h	(92)
Ph	Ph	2.25 h	(90)

TFP (ref. 35):

R^1	R^2	R^3	R^4	Time	Yield (%)
Me	AcO	H	H	20 min	(97)
Me	Ph	H	H	15 min	(96)
4-$CF_3C_6H_4$	H	HO_2C	H	2 h	(96)
Ph	H	MeO_2C	H	2 h	(97)
Ph	H	H	Ph	30 min	(99)
Ph	H	PhCO	H	1.3 h	(99)

Ketone (ref. 38):

R^1	R^2	R^3	R^4	Time	Yield (%)
Me	—$(CH_2)_4$—		H	0.75 h	(85)
Ph	H	H	H	7 h	(82)
n-C_6H_{13}	H	H	H	8 h	(90)
n-C_5H_{11}	H	Me	H	8 h	(87)
Ph	H	Me	H	3 h	(85)
Ph	H	CH_2OH	H	0.5 h	(96)
$(CH_2)_8CH_2OH$	H	H	H	2.5 h	(94)
Ph	H	Ph	H	1.5 h	(97)
Ph	H	H	Ph	3 h	(91)
Ph	Me	Ph	H	2.5 h	(95)
n-C_8H_{17}	H	H	$(CH_2)_7CO_2Me$	2 h	(94)
Ph	Ph	Ph	H	4.5 h	(96)

TABLE 1E. EPOXIDATION OF OLEFINS BY IN-SITU-GENERATED DIOXIRANES (*Continued*)

Substrate	Conditions	Product(s) and Yield(s) (%)	Refs.
	Ketone, Oxone®, $NaHCO_3$, CH_3CN/H_2O, EDTA, rt; Ketone = (tetrahydrothiopyran-4-one 1,1-dioxide)	I (see below)	38
	Ketone, Oxone®, CH_3CN/buffer, (pH 7.0), 0°; Ketone = (Me Me N⁺ … N⁺ Me Me ring ketone) 2 TfO⁻ =O•H_2O	I (see below)	37
C_6 (bicyclopropylidene)	Acetone, Oxone®, phosphate buffer (pH 7.65)/CH_2Cl_2, 18-crown-6, EDTA, 5 to 8°, 16 h	(dispiro epoxide) (70-90)	42

Ref. 38:

R^1	R^2	R^3	R^4	Time	
Me	—$(CH_2)_4$—		H	0.5 h	(83)
Ph	H	H	H	4.5 h	(80)
n-C_6H_{13}	H	H	H	2.5 h	(92)
n-C_5H_{11}	H	Me	H	1.5 h	(81)
Ph	H	Me	H	4.5 h	(87)
Ph	H	CH_2OH	H	1.5 h	(95)
$(CH_2)_9OH$	H	H	H	3.5 h	(95)
Ph	H	Ph	H	5 h	(95)
Ph	H	H	Ph	4 h	(95)
Ph	Me	Ph	H	4 h	(97)
n-C_8H_{17}	H	H	$(CH_2)_7CO_2Me$	3 h	(96)
Ph	Ph	Ph	H	6 h	(97)

Ref. 37:

R^1	R^2	R^3	R^4	Time	
n-C_6H_{13}	H	H	H	11 h	(74)
Ph	H	Me	H	8 h	(83)
Ph	H	H	Me	6 h	(78)
Ph	H	CH_2OH	H	11 h	(94)
Ph	—$(CH_2)_4$—		H	4 h	(93)
Me	H	$(CH_2)_3OBn$	H	22 h	(89)
Ph	H	Ph	H	22 h	(33)

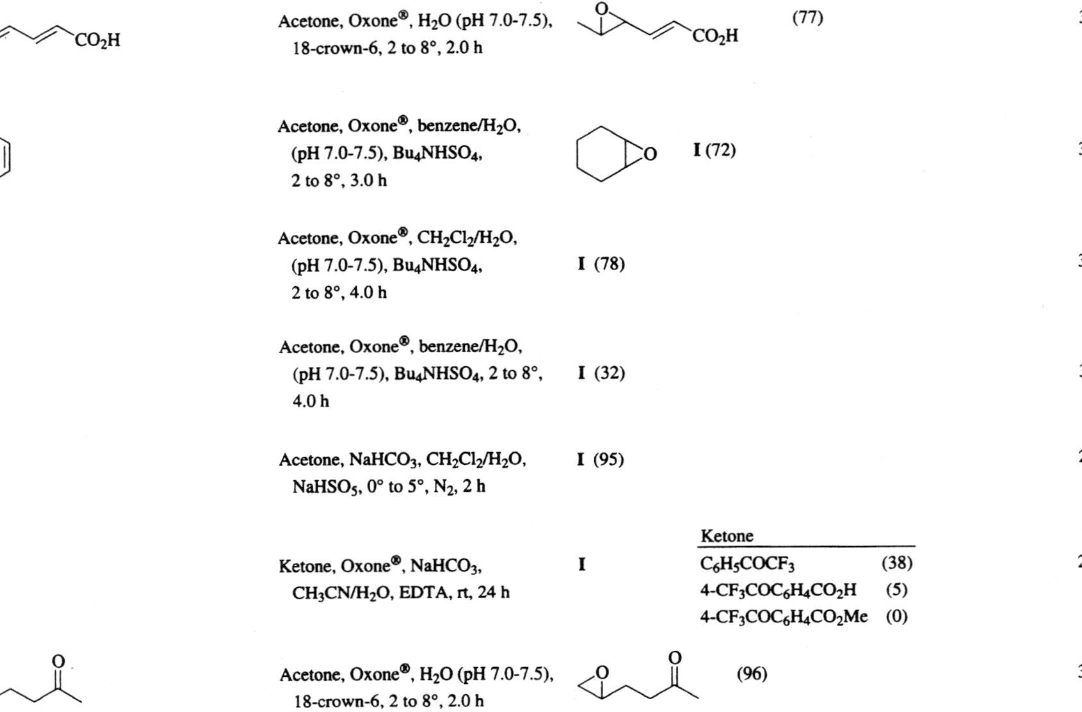

Substrate	Conditions	Product(s) and Yield(s) (%)	Refs.
sorbic acid (CO_2H)	Acetone, Oxone®, H_2O (pH 7.0-7.5), 18-crown-6, 2 to 8°, 2.0 h	epoxide, CO_2H (77)	32
cyclohexene	Acetone, Oxone®, benzene/H_2O, (pH 7.0-7.5), Bu_4NHSO_4, 2 to 8°, 3.0 h	**I** (72)	32
	Acetone, Oxone®, CH_2Cl_2/H_2O, (pH 7.0-7.5), Bu_4NHSO_4, 2 to 8°, 4.0 h	**I** (78)	32
	Acetone, Oxone®, benzene/H_2O, (pH 7.0-7.5), Bu_4NHSO_4, 2 to 8°, 4.0 h	**I** (32)	32
	Acetone, $NaHCO_3$, CH_2Cl_2/H_2O, $NaHSO_5$, 0° to 5°, N_2, 2 h	**I** (95)	265
	Ketone, Oxone®, $NaHCO_3$, CH_3CN/H_2O, EDTA, rt, 24 h	**I** Ketone $C_6H_5COCF_3$ (38) 4-$CF_3COC_6H_4CO_2H$ (5) 4-$CF_3COC_6H_4CO_2Me$ (0)	266
	Acetone, Oxone®, H_2O (pH 7.0-7.5), 18-crown-6, 2 to 8°, 2.0 h	(96)	32

TABLE 1E. EPOXIDATION OF OLEFINS BY IN-SITU-GENERATED DIOXIRANES (*Continued*)

Substrate	Conditions	Product(s) and Yield(s) (%)	Refs.
C_{6-18} $R^2R^1C{=}CR^3R^4$	Acetone, Oxone®, CH_3CN, DMM, K_2CO_3 buffer, Bu_4NHSO_4, EDTA	$R^2R^1C(O)CR^3R^4$ (epoxide)	36
C_7 norbornene	Acetone, Oxone®, H_2O (pH 7.5-8.0)/ CH_2Cl_2, 18-crown-6, 0 to 5°	exo-norbornene oxide **I** (95)	162
	2-Butanone, Oxone®, $NaHCO_3$, 20 to 30 min	**I** (68)	267

R^1	R^2	R^3	R^4	Time	(%)
n-Pr	H	CH_2OH	H	2 h	(79)
Et	H	$(CH_2)_2OH$	H	2 h	(80)
Et	H	H	$(CH_2)_2OH$	4 h	(90)
Ph	H	CH_2Cl	H	2 h	(72)
Ph	H	CH_2OH	H	2 h	(67)
Ph	Me	H	H	4 h	(91)
Ph	H	Me	H	4 h	(84)
H	—$(CH_2)_4$—		C≡CMe	1.5 h	(88)
n-C_8H_{17}	H	H	H	4 h	(67)
2-MeC_6H_4	H	Me	Me	4 h	(75)
H	—$(CH_2)_4$—		Ph	2 h	(84)
i-Pr_3SiCH_2	H	H	H	4 h	(77)
Bn	H	$(CH_2)_2CO_2Me$	H	2 h	(85)
Ph	H	Ph	H	4 h	(86)
n-C_6H_{13}	H	*n*-C_6H_{13}	H	1.5 h	(92)
Ph	Me	Ph	H	2 h	(98)
Ph	H	$CH_2OTBDMS$	H	2 h	(77)
H	Ph	*n*-$C_{10}H_{21}$	H	4 h	(72)

Substrate	Conditions	Product(s) and Yield(s) (%)	Refs.
	2-Butanone, Oxone®, iron(III) tetrakis(2,6-dichlorophenyl)-porphyrin chloride [(TDCPP)FeCl], $NaHCO_3$, 20 to 30 min	**I** (68)	267
	2-Butanone, Oxone®, $NaHCO_3$, 0°, 24 h, with prior incubation	**I** (16)	267
	2-Butanone, Oxone®, $NaHCO_3$, (TDCPP)FeCl, 0°, 24 h, with prior incubation	**I** (14)	267
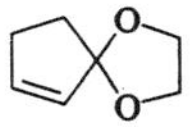	Acetone, Oxone®, K_2CO_3, CH_3CN/DMM, Bu_4NHSO_4, EDTA, rt, 4 h	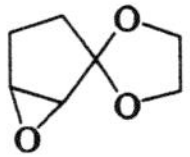 (50)	36
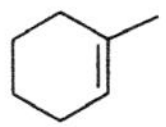	Ketone, TMSOTf, CH_2Cl_2	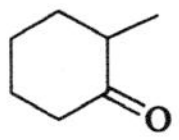	268

Ketone	Temp	Time	% Convn	
cyclohexanone	–32°	3 h	94	(97)
acetone	–32°	3 h	75	(96)
acetone	–75°	16 h	23	(95)
4-heptanone	–32°	3 h	80	(95)

TABLE 1E. EPOXIDATION OF OLEFINS BY IN-SITU-GENERATED DIOXIRANES (*Continued*)

Substrate	Conditions	Product(s) and Yield(s) (%)	Refs.
C_{8-20} 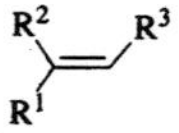 $R^2R^1C{=}CHR^3$	Ketone, Oxone®, $NaHCO_3$, CH_3CN/H_2O, EDTA, rt Ketone 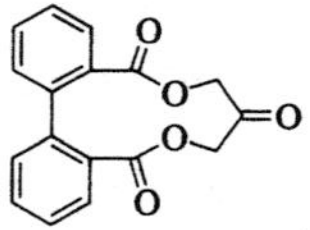 = [biphenyl-2,2'-dicarboxylate cyclic diester of 1,3-dihydroxyacetone]	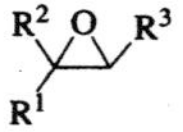 2,2,3-trisubstituted oxirane (R^2, R^1, R^3) (see subtable below)	39
C_8 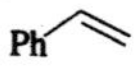 $PhCH{=}CH_2$	Acetone, $NaHCO_3$, CH_2Cl_2/H_2O, $NaHSO_5$, 0° to 5°, N_2, 2 h	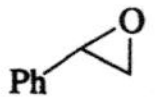 Styrene oxide (70)	265
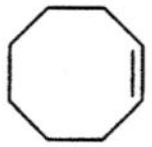 Cyclooctene	Acetone, $NaHCO_3$, CH_2Cl_2/H_2O, $NaHSO_5$, 0° to 5°, N_2, 2 h	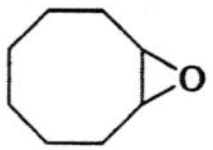 Cyclooctene oxide **I** (71)	265
	Ketone, Oxone®, $NaHCO_3$, CH_3CN/H_2O, EDTA, rt, 1.5 h Ketone 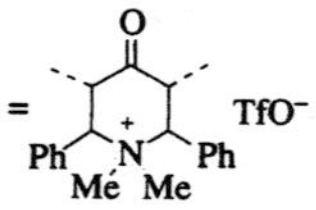 = 1,1-dimethyl-2,6-diphenyl-3,5-dimethyl-4-oxopiperidinium TfO^-	**I** (95)	38

R^1	R^2	R^3	Time	
$4\text{-}ClC_6H_4$	H	H	1 h	(80)
Me	Ph	H	1 h	(75)
Ph	H	CH_2OH	2.5 h	(92)
$4\text{-}CF_3C_6H_4$	H	CO_2H	2 h	(85)
Ph	H	CO_2Me	2 h	(96)
$(CH_2)_9OH$	H	H	5 h	(94)
Ph	—$(CH_2)_4$—		2 h	(86)
Ph	H	Ph	80 min	(94)
Ph	H	PhCO	1 h	(96)
$n\text{-}C_8H_{17}$	H	$(CH_2)_7CO_2Me$	5.5 h	(96)
Ph	Ph	Ph	160 min	(91)

Substrate	Conditions	Product(s) and Yield(s) (%)	Refs.
	Ketone, Oxone®, $NaHCO_3$, CH_3CN/H_2O, EDTA, rt, 0.5 h Ketone = 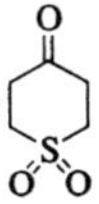	**I** (96)	38
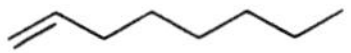 C_9	Acetone, $NaHCO_3$, CH_2Cl_2/H_2O, $NaHSO_5$, 0 to 5°, N_2, 2 h	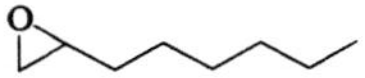 (54)	265
Ph, CO_2^-	Ketone, Oxone®, solvent, buffer	Ph, O, CO_2^- (—)	269
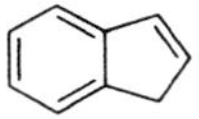	Ketone, Oxone®, CH_3CN/buffer (pH 7.0), 0°, 6 h Ketone 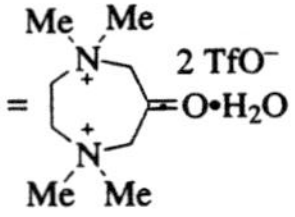 = Me, Me, N, +, 2 TfO⁻, =O•H_2O, +, N, Me, Me	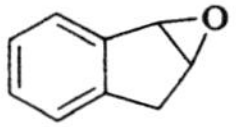 O **I** (85)	37
	Ketone, Oxone®, $NaHCO_3$, CH_3CN/H_2O, 0°, 8 h Ketone 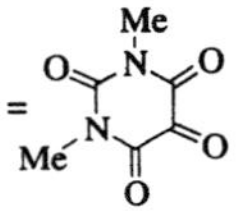 = Me, N, O, O, O, N, Me, O, O	**I** (92)	40

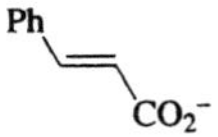

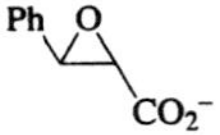

TABLE 1E. EPOXIDATION OF OLEFINS BY IN-SITU-GENERATED DIOXIRANES (*Continued*)

Substrate	Conditions	Product(s) and Yield(s) (%)	Refs.
Ph, CO_2H (trans)	Acetone, Oxone®, phosphate buffer, (pH 7.5), 2°, 2 h	Ph, O, CO_2H **I** (95)	207
	Acetone, Oxone®, H_2O (pH 7.0-7.5), 18-crown-6, 2 to 8°, 2.3 h	**I** (94)	32
	Acetone, Oxone®, benzene/H_2O (pH 7.0-7.5), Bu_4NHSO_4, 2 to 8°, 3 h	**I** (90)	32
Ph, CO_2H (cis)	Acetone, Oxone®, phosphate buffer (pH 7.0), 2°, 3.5 h	O, Ph, CO_2H **I** (90)	31
	Acetone, Oxone®, H_2O (pH 7.0-7.5), 18-crown-6, 2 to 8°, 3.5 h	**I** (>81)	32
Ph, Me	Acetophenone, Oxone®, MeOH, H_2O, 0°, 1 h	Ph, O, Me (100)	270
Ph (propenyl)	Ketone, Oxone®, $NaHCO_3$, CH_3CN/H_2O, 0°, 8 h Ketone = (Me, N, O, N, Me, O, O, O)	O, Ph (92)	40

Substrate	Conditions	Product(s) and Yield(s) (%)	Refs.
	Ketone, Oxone®, $NaHCO_3$, CH_3CN/H_2O, 0°, 8 h; Ketone = (Bn-N, N-Bn, O, O, O, O)	Ph, O (92)	40
	Ketone, $(TMS)_2SO_5$, CH_2Cl_2	Ph, O	268
		Ketone / Temp / Time / % Convn: acetone, –4°, 1.0 h, 98 (95); acetone, –32°, 3.0 h, 68 (95); cyclohexanone, –32°, 3.0 h, 88 (97)	
	p-$O_2NC_6H_4SO_2N$(imidazole, N), $[^{18}O]$acetone, H_2O_2, NaOH	O_{18} + O (38 - 43)	271
Ph, OH	Ketone, Oxone®, $NaHCO_3$, CH_3CN/H_2O, 0°, 8 h; Ketone = (Me-N, N-Me, O, O, O, O)	Ph, O, OH (90)	40
	Acetone, Oxone®, K_2CO_3, CH_3CN/DMM, Bu_4NHSO_4, EDTA, rt, 1.5 h	O (88)	36

TABLE 1E. EPOXIDATION OF OLEFINS BY IN-SITU-GENERATED DIOXIRANES (*Continued*)

Substrate	Conditions	Product(s) and Yield(s) (%)	Refs.
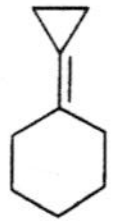	Acetone, Oxone®, phosphate buffer (pH 7.65)/CH_2Cl_2, 18-crown-6, EDTA, 5 to 8°, 16 h	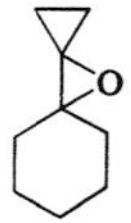 (70-90)	42
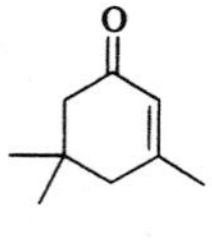	TFP, Oxone®, $NaHCO_3$, CH_3CN/H_2O, Na_2EDTA, 0 to 1°, 1.5 h	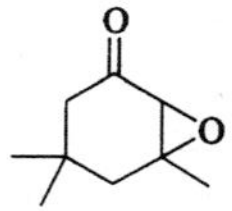 **I** (90)	35
	2-Butanone, Oxone®, phosphate buffer (pH 7.3-7.5), ~20°, 24 h	**I** (86)	34
	Ketone, Oxone®, $NaHCO_3$, CH_3CN/H_2O, EDTA, rt, 1.5 h Ketone = 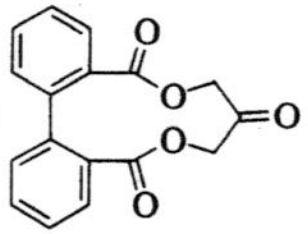	**I** (89)	39
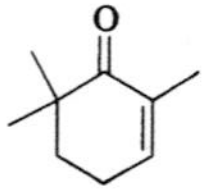	2-Butanone, Oxone®, phosphate buffer (pH 7.3-7.5), ~20°, 24 h	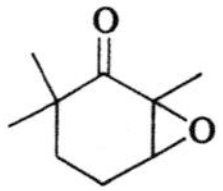 (84)	34

$C_{9\text{-}20}$

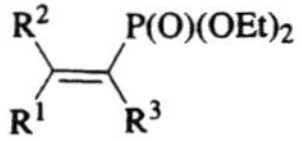

$R^2(R^1)C=C(R^3)P(O)(OEt)_2$

2-Butanone, Oxone®, CH_2Cl_2/phosphate buffer (pH 7.3-7.5), Bu_4NHSO_4, aq. KOH, 24 h

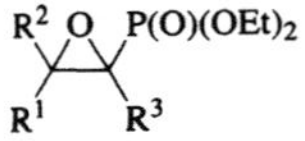

R^2 O $P(O)(OEt)_2$ / R^1 R^3 (epoxide)

R^1	R^2	R^3	
H	*n*-Pr	H	(58)
H	*n*-Bu	H	(71)
Me	*n*-Pr	H	(75)
Me	*n*-Pr	Me	(79)
H	Ph	H	(73)
Ph	*n*-Pr	H	(81)
Ph	*n*-Bu	H	(80)
n-Bu	Ph	H	(35)
n-Bu	Ph	Me	(20)
4-MeC_6H_4	Ph	H	(84)
n-C_8H_{17}	Ph	H	(19)

272

C_{10}

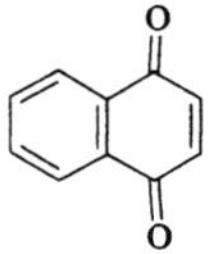

Ketone, Oxone®, $NaHCO_3$, CH_3CN/H_2O, EDTA, rt, 6 h

Ketone =

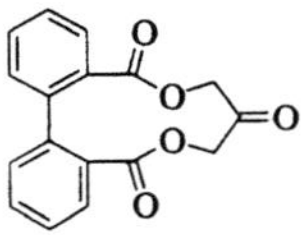

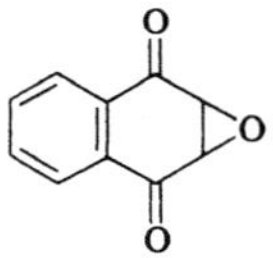

(88)

39

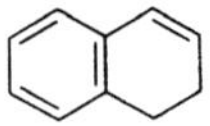

Acetone, Oxone®, K_2CO_3, CH_3CN/DMM, Bu_4NHSO_4, EDTA, rt, 4 h

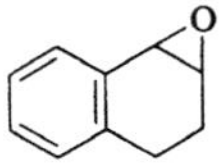

I (66)

36

TABLE IE. EPOXIDATION OF OLEFINS BY IN-SITU-GENERATED DIOXIRANES (*Continued*)

Substrate	Conditions	Product(s) and Yield(s) (%)	Refs.
	Ketone, Oxone®, $NaHCO_3$, CH_3CN/H_2O, EDTA, rt, 1.5 h; Ketone 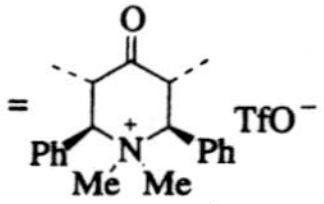 =	**I** (83)	38
	Ketone, Oxone®, $NaHCO_3$, CH_3CN/H_2O, EDTA, rt, 2.5 h; Ketone 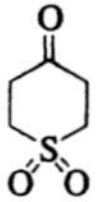 =	**I** (85)	38
	Ketone, Oxone®, $NaHCO_3$, CH_3CN/H_2O, EDTA, rt, 80 min; Ketone 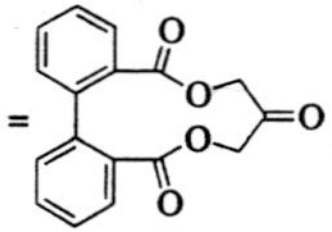 =	**I** (81)	39
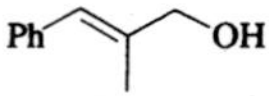	Ketone , Oxone®, $NaHCO_3$, CH_3CN/H_2O, 0°, 8 h; Ketone 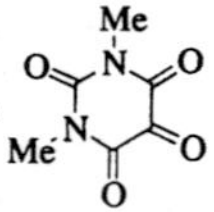 =	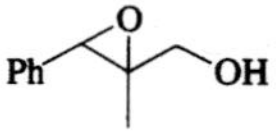 (84)	40

Substrate	Conditions	Product(s) and Yield(s) (%)	Refs.
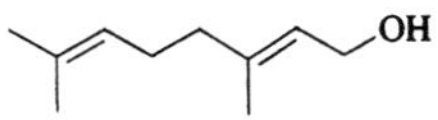 OH	Ketone, Oxone®, CH_3CN/buffer (pH 7.0), 0°, 6 h Ketone 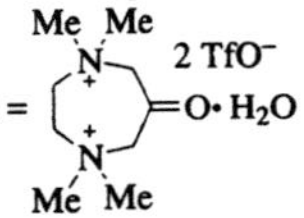 = Me Me N+ 2 TfO− =O• H_2O +N Me Me	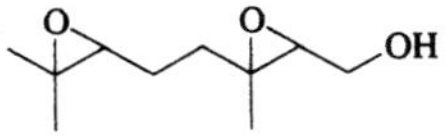 O O OH **I** (90)	37
	Ketone, Oxone®, CH_3CN/buffer, (pH 7.0), 0°, 2.4 h Ketone 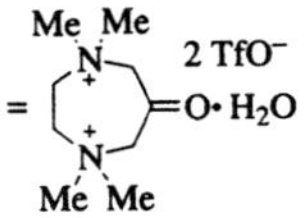 = Me Me N+ 2 TfO− =O• H_2O +N Me Me	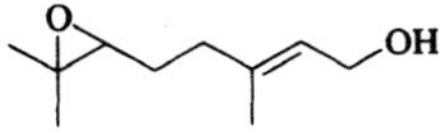 **I** + O OH **II** (89) **I**:**II** = 19:77	37
C_{12} 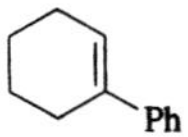 Ph	Ketone, Oxone®, $NaHCO_3$, CH_3CN/H_2O, EDTA, 0°, 3 h Ketone 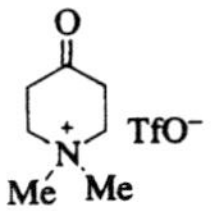 = O +N TfO− Me Me	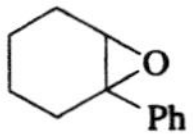 O Ph **I** (90)	273
	Ketone, Oxone®, $NaHCO_3$, CH_3CN/H_2O, 0°, 8 h Ketone 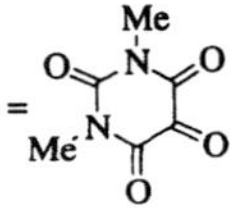 = Me N O O O N Me O O	**I** (88)	40

TABLE 1E. EPOXIDATION OF OLEFINS BY IN-SITU-GENERATED DIOXIRANES (*Continued*)

Substrate	Conditions	Product(s) and Yield(s) (%)	Refs.
	Acetone, Oxone®, K_2CO_3, CH_3CN/DMM, Bu_4NHSO_4, EDTA, rt, 2 h	**I** (84)	36
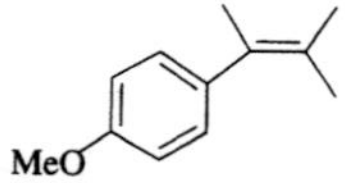 MeO	1. Acetone, Oxone®, H_2O, $NaHCO_3$, 20°, 2.5 h 2. 25°, 2 h	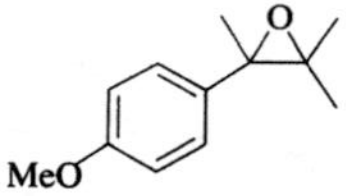 MeO, O (85)	274
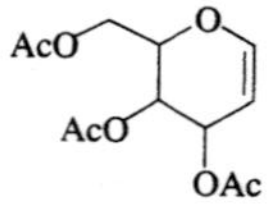 AcO, O, AcO, OAc	Ketone, Oxone®, $NaHCO_3$, CH_3CN/H_2O, EDTA, rt	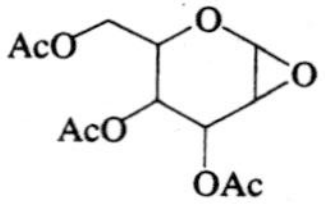 AcO, O, O, AcO, OAc Ketone / Time Tentagel S-Br 48 h (80) $C_6H_5COCF_3$ 24 h (100) 4-$CF_3COC_6H_4CO_2H$ 24 h (100) 4-$CF_3COC_6H_4CO_2Me$ 24 h (100)	266
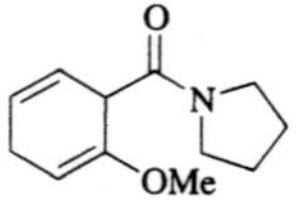 O, N, OMe	Acetone, Oxone®, phosphate buffer (pH 7.2), 0°	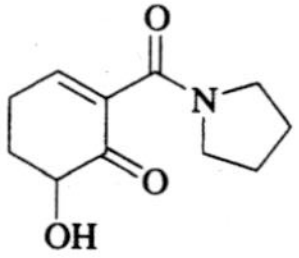 O, N, O, OH (77)	275
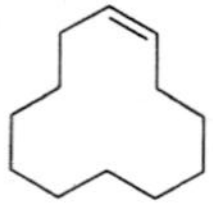	Acetone, Oxone®, benzene/H_2O, (pH 7.0-7.5), 18-crown-6, 2 to 8°, 3.0 h	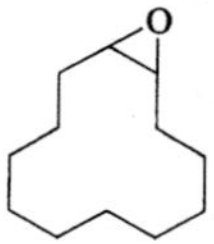 O (77)	32

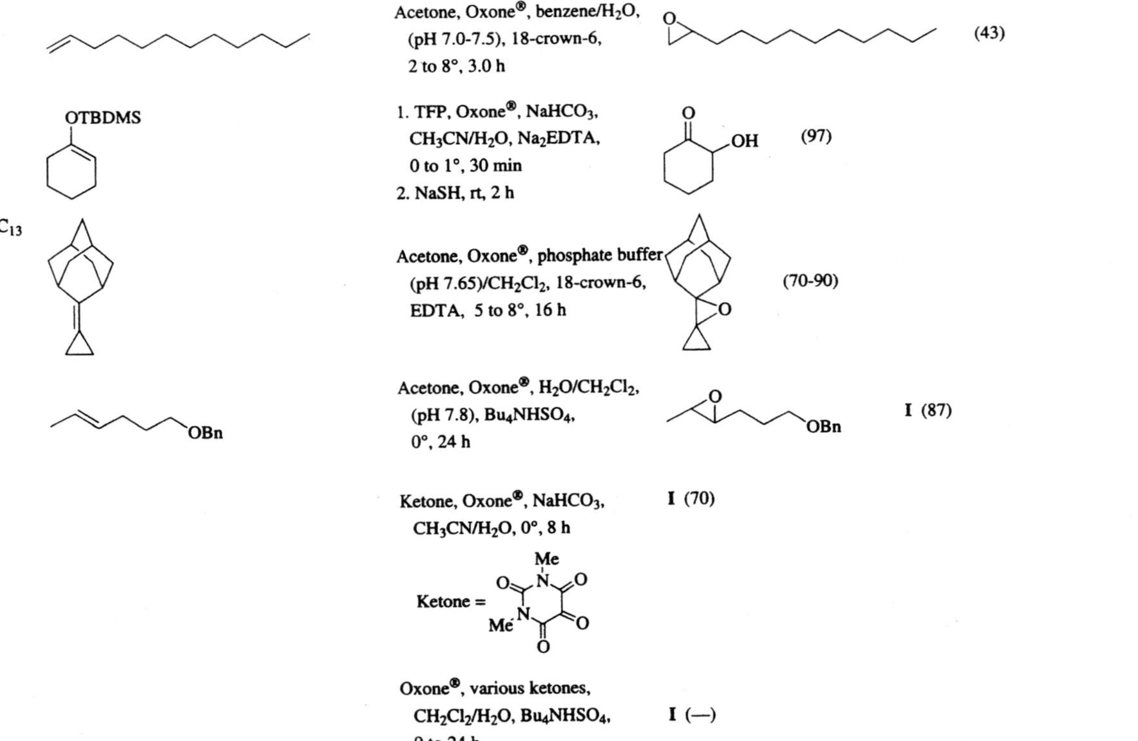

Reactant	Conditions	Product(s) and Yield(s) (%)	Refs.
	Acetone, Oxone®, benzene/H_2O, (pH 7.0-7.5), 18-crown-6, 2 to 8°, 3.0 h	(43)	32
OTBDMS	1. TFP, Oxone®, $NaHCO_3$, CH_3CN/H_2O, Na_2EDTA, 0 to 1°, 30 min 2. NaSH, rt, 2 h	OH (97)	35
C_{13}	Acetone, Oxone®, phosphate buffer (pH 7.65)/CH_2Cl_2, 18-crown-6, EDTA, 5 to 8°, 16 h	O (70-90)	42
OBn	Acetone, Oxone®, H_2O/CH_2Cl_2, (pH 7.8), Bu_4NHSO_4, 0°, 24 h	OBn **I** (87)	168
	Ketone, Oxone®, $NaHCO_3$, CH_3CN/H_2O, 0°, 8 h Ketone = (Me, N, O, Me)	**I** (70)	40
	Oxone®, various ketones, CH_2Cl_2/H_2O, Bu_4NHSO_4, 9 to 24 h	**I** (—)	107

TABLE 1E. EPOXIDATION OF OLEFINS BY IN-SITU-GENERATED DIOXIRANES (*Continued*)

Substrate	Conditions	Product(s) and Yield(s) (%)	Refs.
C_{14} 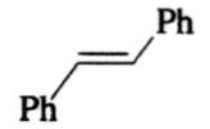 Ph–CH=CH–Ph	Ketone, Oxone®, $NaHCO_3$, CH_3CN/H_2O, 0°, 8 h Ketone 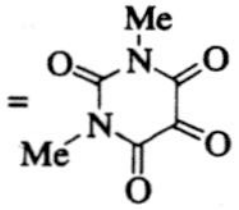 = (N,N'-dimethyl, Me on each N; ring O's)	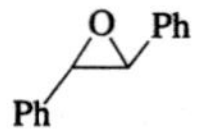 Ph–oxirane–Ph **I** (70)	40
	Ketone, Oxone®, $NaHCO_3$, CH_3CN/H_2O, 0°, 8 h Ketone 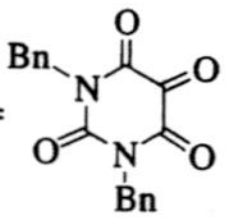 = (N,N'-dibenzyl, Bn on each N; ring O's)	**I** (70)	40
	Ketone, Oxone®, $NaHCO_3$, CH_3CN/H_2O, EDTA, rt	**I** Ketone / Time Tentagel S-Br 48 h (97) Hydroxymethyl resin 17 h (41) Hydroxymethyl resin 36 h (64) $C_6H_5COCF_3$ 24 h (70) 4-$CF_3COC_6H_4CO_2H$ 24 h (81) 4-$CF_3COC_6H_4CO_2Me$ 24 h (91)	266
	Ketone, Oxone®, $NaHCO_3$, CH_3CN/H_2O, EDTA, rt, 5 min Ketone 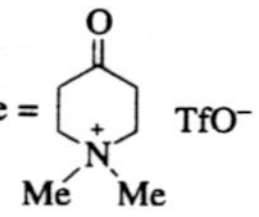 = (4-oxo-piperidinium, N^+(Me)(Me)) TfO^-	**I** (—)	38

Conditions	Product(s) and Yield(s) (%)	Refs.
Ketone, Oxone®, $NaHCO_3$, CH_3CN/H_2O, EDTA, rt, 45 min 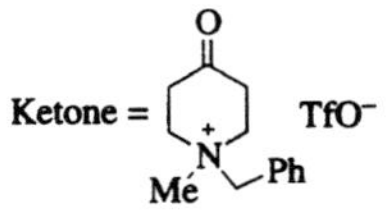 Ketone = (N-benzyl-N-methyl-4-oxopiperidinium) TfO⁻	**I** (95)	38
Ketone, Oxone®, $NaHCO_3$, CH_3CN/H_2O, EDTA, rt, 2 h 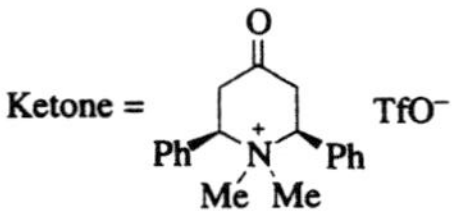 Ketone = (2,6-diphenyl-1,1-dimethyl-4-oxopiperidinium) TfO⁻	**I** (92)	38
Ketone, Oxone®, $NaHCO_3$, CH_3CN/H_2O, EDTA, rt, 2 h 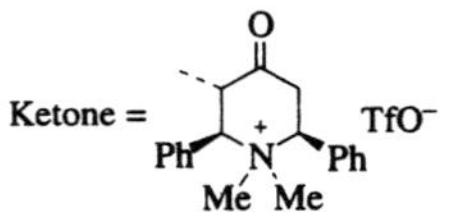 Ketone = (2,6-diphenyl-1,1,3-trimethyl-4-oxopiperidinium) TfO⁻	**I** (87)	38

TABLE 1E. EPOXIDATION OF OLEFINS BY IN-SITU-GENERATED DIOXIRANES (*Continued*)

Substrate	Conditions	Product(s) and Yield(s) (%)	Refs.
	Ketone, Oxone®, $NaHCO_3$, CH_3CN/H_2O, EDTA, rt, 15 min Ketone 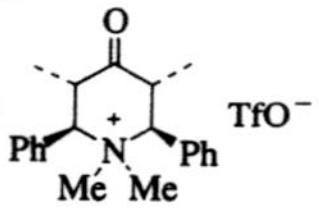 = [2,6-diphenyl-3,5-dimethyl-1,1-dimethyl-4-oxopiperidinium] TfO^-	**I** (94)	38
	Cyclohexanone, Oxone®, $NaHCO_3$, CH_3CN/H_2O, EDTA, rt, 12 h	**I** (—)	38
	Ketone, Oxone®, $NaHCO_3$, CH_3CN/H_2O, EDTA, rt, 30 min Ketone 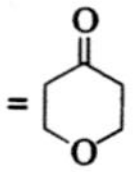 = [tetrahydro-4H-pyran-4-one]	**I** (93)	38
	Ketone, Oxone®, $NaHCO_3$, CH_3CN/H_2O, EDTA, rt, 2 to 3 min Ketone 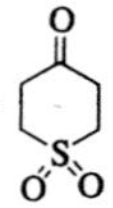 = [tetrahydrothiopyran-4-one 1,1-dioxide]	**I** (97)	38

TFP, Oxone®, $NaHCO_3$, CH_3CN/H_2O, EDTA, rt, 30 min	**I** (96)	39
Ketone, Oxone®, $NaHCO_3$, CH_3CN/H_2O, EDTA, rt, 120 min Ketone = 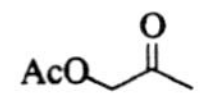	**I** (91)	39
Ketone, Oxone®, $NaHCO_3$, CH_3CN/H_2O, EDTA, rt, 50 min Ketone = 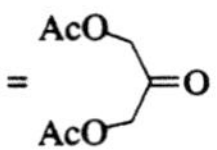	**I** (90)	39
Ketone, Oxone®, $NaHCO_3$, CH_3CN/H_2O, EDTA, rt, 35 min Ketone = 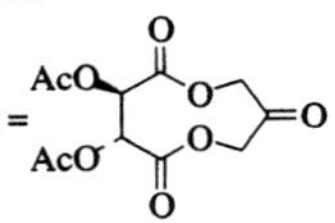	**I** (98)	39

TABLE 1E. EPOXIDATION OF OLEFINS BY IN-SITU-GENERATED DIOXIRANES (*Continued*)

Substrate	Conditions	Product(s) and Yield(s) (%)	Refs.
	Ketone, Oxone®, $NaHCO_3$, CH_3CN/H_2O, EDTA, rt, 20 min Ketone =	**I** (96)	39
	Ketone, Oxone®, $NaHCO_3$, CH_3CN/H_2O, EDTA, rt, 12 min Ketone =	**I** (97)	39
	Ketone, Oxone®, $NaHCO_3$, CH_3CN/H_2O, EDTA, rt, 10 min Ketone =	**I** (96)	39

Ketone, Oxone®, $NaHCO_3$, CH_3CN/H_2O, EDTA, rt, 10 min Ketone = 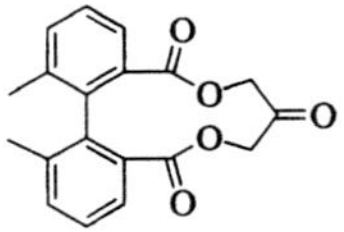	**I** (91)	39
Ketone, Oxone®, $NaHCO_3$, CH_3CN/H_2O, EDTA, rt, 6 min Ketone 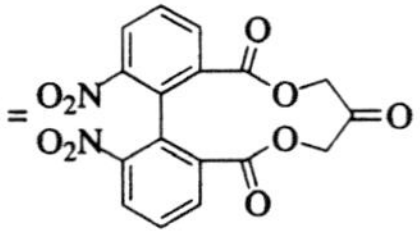 = O_2N O_2N	**I** (93)	39
Ketone, Oxone®, $NaHCO_3$, CH_3CN/H_2O, EDTA, rt, 45 min Ketone 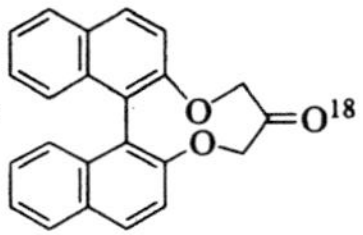 = O^{18}	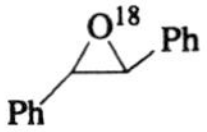 O^{18} Ph Ph (—)	112

TABLE 1E. EPOXIDATION OF OLEFINS BY IN-SITU-GENERATED DIOXIRANES (*Continued*)

Substrate	Conditions	Product(s) and Yield(s) (%)	Refs.
 Ph–CH=CH–Ph (cis)	Ketone, Oxone®, $NaHCO_3$, CH_3CN/H_2O, EDTA, rt	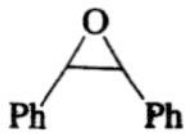 cis-2,3-diphenyloxirane Ketone / Time Tentagel S-Br 48 h (61) Hydroxymethyl resin 38 h (51) $C_6H_5COCF_3$ 24 h (100) 4-$CF_3COC_6H_4CO_2H$ 24 h (100) 4-$CF_3COC_6H_4CO_2Me$ 24 h (97)	266
C_{15} 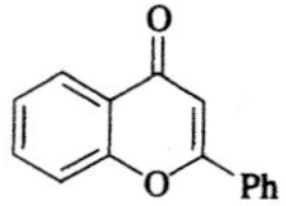 flavone (2-Ph)	TFP, Oxone®, $NaHCO_3$, CH_3CN/H_2O, Na_2EDTA, 0 to 1°, 1.25 h	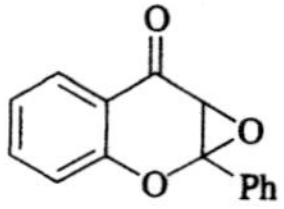 flavone 2,3-epoxide (Ph) (99)	35
C_{15-16} 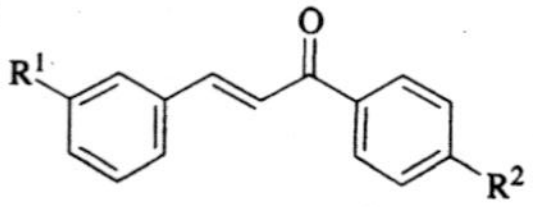 R^1-C$_6$H$_4$–CH=CH–CO–C$_6$H$_4$-R^2	2-Butanone, Oxone®, phosphate buffer (pH 7.3-7.5), ~20°, 24 h	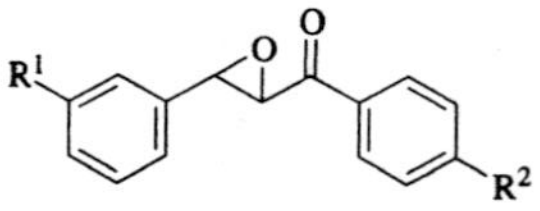 epoxide **I** R^1 / R^2 H Br (40) H H (93) Me H (98) H MeO (94) MeO H (93)	34
	2-Butanone, Oxone®, phosphate buffer (pH 7.3-7.5)/CH_2Cl_2, Bu_4NHSO_4, ~20°, 41 h	R^1 = H, R^2 = Br **I** (89)	34
C_{15} 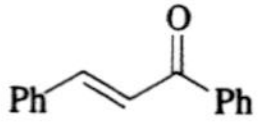 Ph–CH=CH–CO–Ph	Ketone, Oxone®, $NaHCO_3$, CH_3CN/H_2O, EDTA, rt	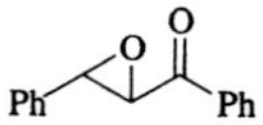 Ph-oxirane-CO-Ph Ketone / Time Tentagel S-Br 48 h (13) Hydroxymethyl resin 48 h (9) $C_6H_5COCF_3$ 24 h (11) 4-$CF_3COC_6H_4CO_2H$ 24 h (34) 4-$CF_3COC_6H_4CO_2Me$ 24 h (7)	266

Substrate	Conditions	Product(s) and Yield(s) (%)	Refs.
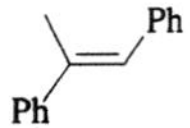	Ketone, Oxone®, $NaHCO_3$, CH_3CN/H_2O, 0°, 8 h; Ketone 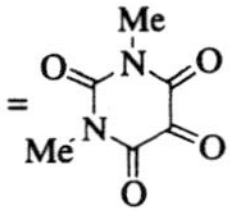 = (1,3-dimethyl-pyrimidinetetrone)	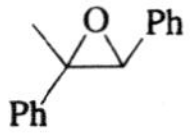 (64)	40
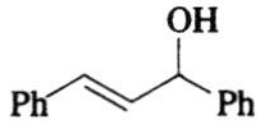	Ketone, Oxone®, $NaHCO_3$, CH_3CN/H_2O, EDTA, rt	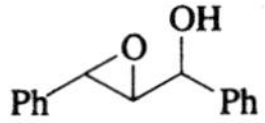 Ketone / Time: Tentagel S-Br, 48 h (90); $C_6H_5COCF_3$, 24 h (100); 4-$CF_3COC_6H_4CO_2H$, 24 h (100); 4-$CF_3COC_6H_4CO_2Me$, 24 h (100)	266
C_{16} 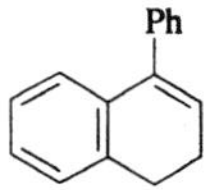	Ketone, Oxone®, $NaHCO_3$, CH_3CN/H_2O, EDTA, rt, 2 h; Ketone 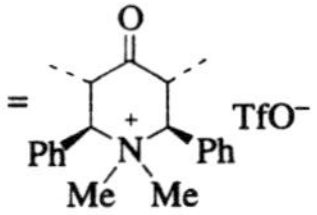 = (piperidinium ketone) TfO⁻	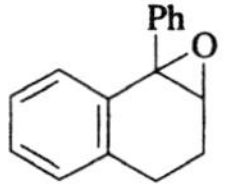 **I** (92)	38
	Ketone, Oxone®, $NaHCO_3$, CH_3CN/H_2O, EDTA, rt, 2.5 h; Ketone 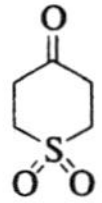 = (tetrahydrothiopyran-4-one 1,1-dioxide)	**I** (92)	38

TABLE 1E. EPOXIDATION OF OLEFINS BY IN-SITU-GENERATED DIOXIRANES (*Continued*)

Substrate	Conditions	Product(s) and Yield(s) (%)	Refs.
OBn	Acetone, Oxone®, H_2O/CH_2Cl_2, (pH 8.0), Bu_4NHSO_4, 0°, 6.5 h	O, OBn (58)	168
THPO, CO_2Et	TFP, Oxone®, $NaHCO_3$, CH_3CN/H_2O, 0°, 30 min	THPO, O, CO_2Et (72)	276
C_{19}			
MeO_2C, 7, O, C_6H_{13}	2-Butanone, Oxone®, $NaHCO_3$, H_2O, Bu_4NHSO_4, dark, rt, 2 h	MeO_2C, 7, O, O, C_6H_{13} (100)	69
MeO_2C, 7, OH, C_6H_{13}	2-Butanone, Oxone®, $NaHCO_3$, H_2O, Bu_4NHSO_4, dark, rt, 2 h	MeO_2C, 7, O, OH, C_6H_{13} (89) dr = 50:50	69
C_{22}			
MeO, OH, H, H, O	Ketone, Oxone®, buffer (pH 7.5), CH_2Cl_2, Bu_4NHSO_4, 0°, 24 h Ketone = O, $\overset{+}{N}$, TfO^-, Me, $C_{12}H_{25}$-*n*	O, OH, OH, H, H, O (81)	168

C_{27}

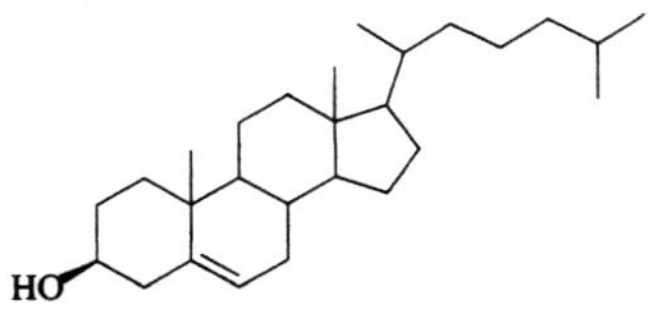

Ketone, Oxone®, $NaHCO_3$, CH_3CN/H_2O, EDTA, rt

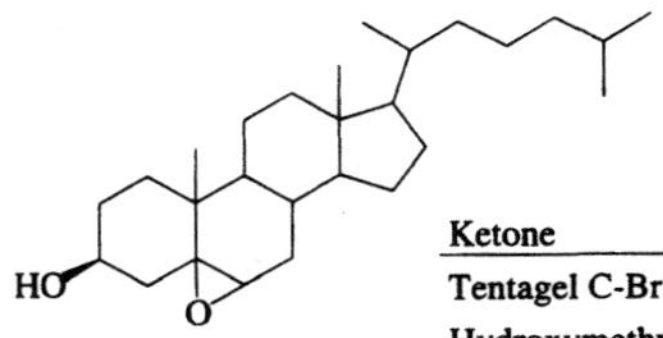

Ketone	Time	
Tentagel C-Br	48 h	(76)
Hydroxymethyl resin	48 h	(37)
$C_6H_5COCF_3$	24 h	(52)
$4\text{-}CF_3COC_6H_4CO_2H$	24 h	(100)
$4\text{-}CF_3COC_6H_4CO_2Me$	24 h	(72)

266

C_{32}

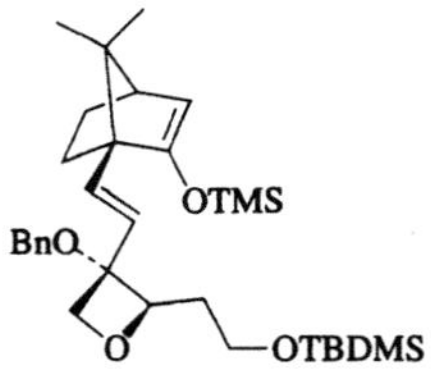

Acetone, Oxone®, $NaHCO_3$, H_2O/CH_2Cl_2, 18-crown-6, rt, > 1 h

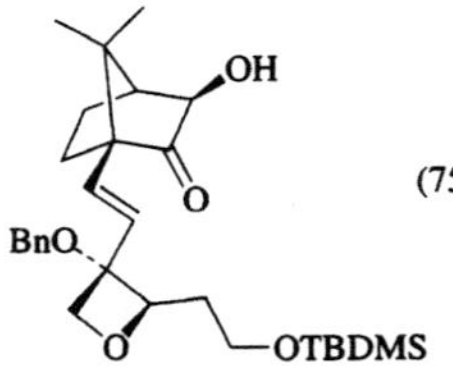

(75)

277

C_{53}

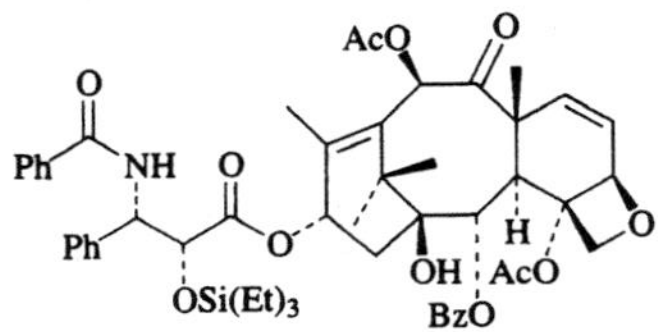

1. TFP, Oxone®, $NaHCO_3$ CH_3CN/H_2O,
2. HCl (1 N), 0°

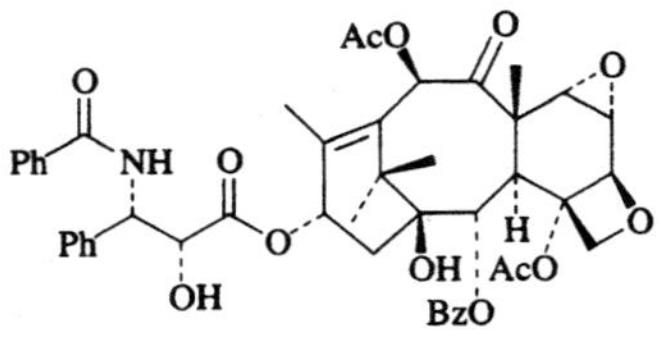

(88)

278, 278a

TABLE 2. CHEMOSELECTIVE OXIDATIONS BY ISOLATED DIOXIRANES

Substrate	Conditions	Product(s) and Yield(s) (%)	Refs.
C_3			
NH_2	1. 4-$ClC_6H_4SO_3H$, Et_2O, 0° 2. DMD, acetone/CH_3CN, 0°, 8 h 3. K_2CO_3, $MgSO_4$, CH_2Cl_2, 0°, 5 h	O NH_2 **I** (45)	65
	1. 4-$ClC_6H_4SO_3H$, Et_2O, 0° 2. TFD, CH_2Cl_2/CH_3CN, 0°, 0.1 h 3. K_2CO_3, $MgSO_4$, CH_2Cl_2, 0°, 5 h	**I** (72)	65
C_5			
N H	1. 4-$ClC_6H_4SO_3H$, Et_2O, 0° 2. DMD, acetone/CH_3CN, 0°, 8 h 3. K_2CO_3, $MgSO_4$, CH_2Cl_2, 0°, 5 h	O N H **I** (70)	65
	1. 4-$ClC_6H_4SO_3H$, Et_2O, 0° 2. TFD, CH_2Cl_2, 0°, 0.1 h 3. K_2CO_3, $MgSO_4$, CH_2Cl_2, 0°, 5 h	**I** (79)	65
C_6			
N CHO	DMD (x equiv), acetone	N^+ O^- CHO **I** + N^+ O^- $CH(OH)_2$ **II** + N CO_2H **III** + N^+ O^- CO_2H **IV**	279

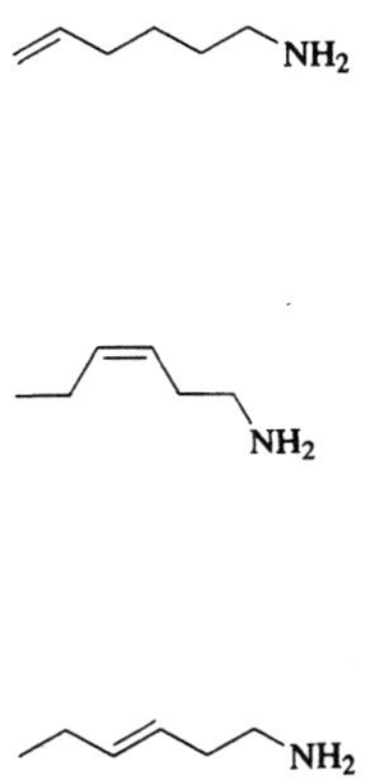

x	Temp	Time	% Convn	(**I** + **II**)	**III**	**IV**
1.0	0°	2 h	42	(42)	(—)	(—)
1.0	0°	4 h	60	(56)	(2)	(2)
1.0	0°	8 h	75	(68)	(5)	(2)
2.0	0°	2 h	72	(65)	(—)	(7)
1.0	–20°	12 h	67	(67)	(—)	(—)
1.5	–20°	12 h	79	(79)	(—)	(—)
2.0	–20°	12 h	86	(80)	(3)	(3)
1.0	20°	1 h	85	(79)	(3)	(3)

1. 4-$MeC_6H_4SO_3H$, Et_2O, 0°
2. DMD, acetone/CH_2Cl_2, 0°, 3.5 h
3. K_2CO_3, $MgSO_4$, CH_2Cl_2, 0°, 5 h

NH2 (90) 65

1. 4-$ClC_6H_4SO_3H$, Et_2O, 0°
2. DMD, acetone/CH_2Cl_2, 0°, 1.5 h
3. K_2CO_3, $MgSO_4$, CH_2Cl_2, 0°, 5 h

NH2 (81) 65

1. 4-$ClC_6H_4SO_3H$, Et_2O, 0°
2. DMD, acetone/CH_2Cl_2, 0°, 3.5 h
3. K_2CO_3, $MgSO_4$, CH_2Cl_2, 0°, 5 h

NH2 (96) 65

TABLE 2. CHEMOSELECTIVE OXIDATIONS BY ISOLATED DIOXIRANES (*Continued*)

Substrate	Conditions	Product(s) and Yield(s) (%)	Refs.
C_7			
Cl, S, Me + O, S, Me	DMD, acetone/$CHCl_3$, N_2, 0°	Cl, O, S, Me **I** (100)	280
	TFD, TFP/$CHCl_3$, N_2, 0°	**I** + Cl, O O, S, Me **II** (—) + O O, S, Me **III** (—) **I**:**II**:**III** = 61:13:26	280
Cl, O, S, Me + O, S, Me	DMD, acetone/$CHCl_3$, N_2, 0°	**II** + **III** (—) **II**:**III** = 44:56	280
	TFD, TFP/$CHCl_3$, N_2, 0°	**II** + **III** (—) **II**:**III** = 49:51	280
Cl, O, S, Me + MeO, O, S, Me	DMD, acetone/$CHCl_3$, N_2, 0°	**II** + MeO, O O, S, Me **IV** (—) **IV**:**II** = 29:71	280
	TFD, TFP/$CHCl_3$, N_2, 0°	**IV** + **II** (—) **IV**:**II** = 44:56	280
O_2N, O, S, Me + MeO, O, S, Me	DMD, acetone/$CHCl_3$, N_2, 0°	**IV** + O_2N, O O, S, Me **V** (—) **V**:**IV** = 5:95	280
	TFD, TFP/$CHCl_3$, N_2, 0°	**IV** + **V** (—) **V**:**IV** = 27:73	280

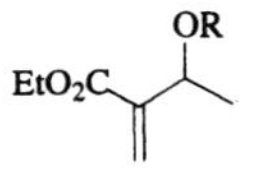

DMD, acetone, rt — I + II — 73

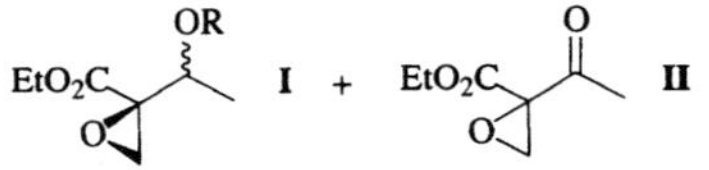

R	Time	% Convn	I	dr of I	II
H	18 h	>95	(0)	—	(100)
Ac	48 h	>95	(100)	65:35	(0)

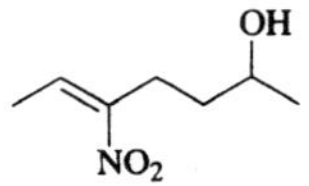

DMD, acetone, ~25°, overnight 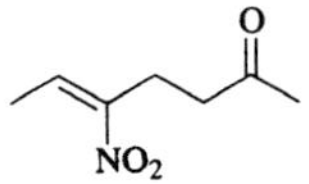 — (75) — 281

C_8

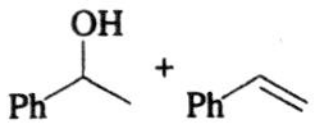

(1:1)

DMD, CH_2Cl_2/acetone, 0°, 8 h — I + II — I + II (—) I:II = 35:65 — 73

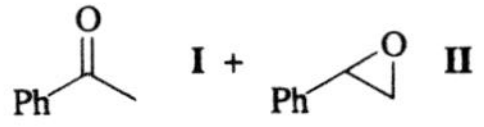

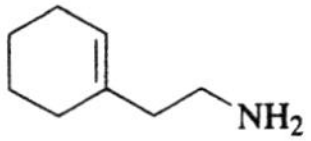

1. 4-$ClC_6H_4SO_3H$, Et_2O, 0°
2. DMD, acetone, 0°, 1 h
3. K_2CO_3, $MgSO_4$, CH_2Cl_2, 0°, 5 h

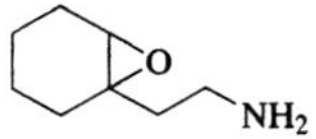

(80) — 65

$C_{8\text{-}13}$

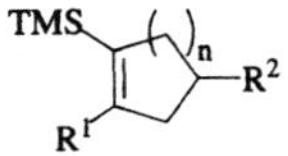

DMD, acetone, rt, 4 to 10 h — I + II + III — 74

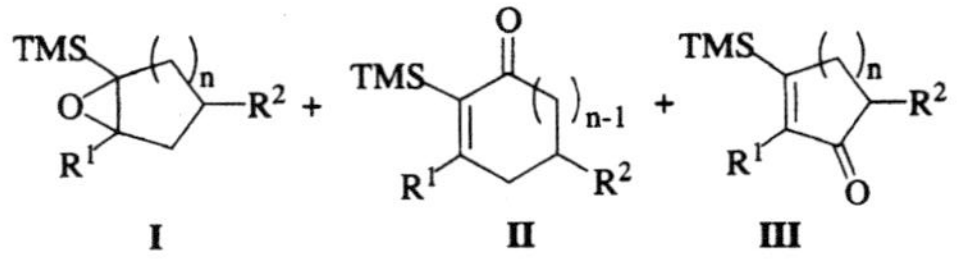

n	R^1	R^2	I	II	III
1	H	H	(90)	(4)	(6)
2	H	H	(75)	(6)	(10)
2	Me	H	(100)	(0)	(0)
4	H	H	(100)	(0)	(0)
2	H	t-Bu	(64)	(16)	(0)

TABLE 2. CHEMOSELECTIVE OXIDATIONS BY ISOLATED DIOXIRANES (*Continued*)

Substrate	Conditions	Product(s) and Yield(s) (%)	Refs.
C_{10}			
	DMD	(—) R^1 / R^2: H / *t*-Bu; Ph / Et; Ph / Ph	71
	1. 4-$ClC_6H_4SO_3H$, Et_2O, 0° 2. DMD, acetone/CH_2Cl_2, 0°, 2.5 h 3. K_2CO_3, $MgSO_4$, CH_2Cl_2, 0°, 5 h	**I** (95)	65
	1. 4-$ClC_6H_4SO_3H$, Et_2O, 0° 2. TFD, CH_2Cl_2, 0°, 0.1 h 3. K_2CO_3, $MgSO_4$, CH_2Cl_2, 0°, 5 h	**I** (95)	65
C_{11}			
	DMD, acetone, 0°, <1 h	(~100)	62
	DMD, acetone, 0°	(100)	63
	1. BF_3•Et_2O, 70° 2. TFD, CH_2Cl_2/TFP, 0° 3. $KHCO_3$	(82)	63

Substrate	Conditions	Product(s) and Yield(s) (%)	Refs.
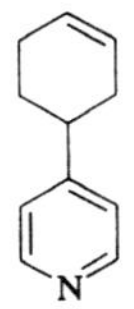	DMD, acetone, 0°	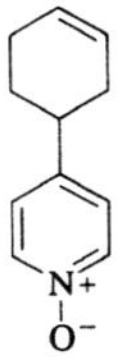 N^+ O^- (100)	63
	1. $BF_3 \cdot Et_2O$, 70° 2. DMD, CH_2Cl_2/acetone, 0° 3. $KHCO_3$	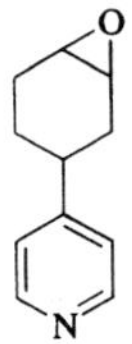 (95)	63
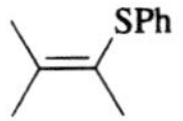 SPh	DMD (1 equiv), acetone/ CH_2Cl_2, –35°, 1 h	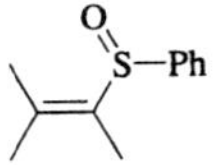 S(O)–Ph (81)	66
	DMD (3 equiv), acetone/ CH_2Cl_2, –35°, 3 h	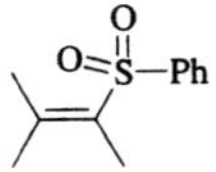 SO_2–Ph (91)	66
 C_{12} TMS TMS	DMD, acetone, 25°, 35 h	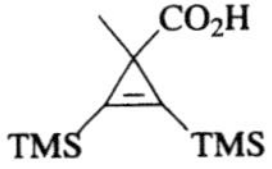 CO_2H TMS TMS (~80)	218
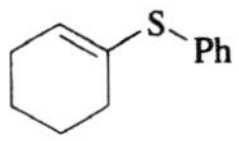 S–Ph	DMD (1 equiv), acetone/ CH_2Cl_2, –40°, 5 h	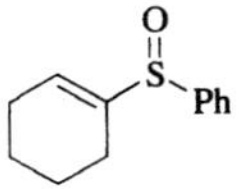 S(O)–Ph (67)	66

TABLE 2. CHEMOSELECTIVE OXIDATIONS BY ISOLATED DIOXIRANES (*Continued*)

Substrate	Conditions	Product(s) and Yield(s) (%)	Refs.
	DMD (3 equiv), acetone/ CH_2Cl_2, –40°, 3.5 h	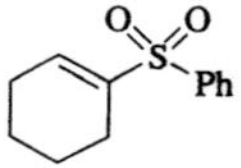 Ph (62)	66
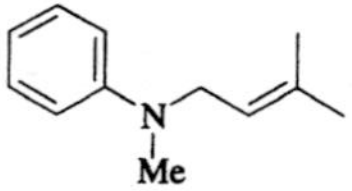 N Me	DMD, acetone, 0°, <1 h	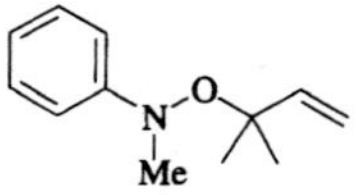 N O Me (~100) + N+ Me O⁻ (—) 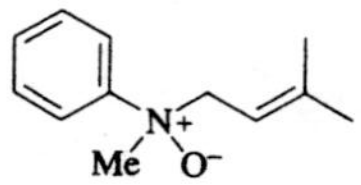	62
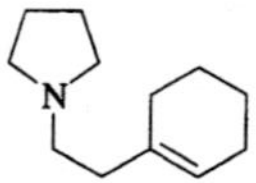 N	DMD, acetone, 0°, <1 h	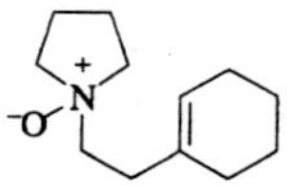 ⁻O N+ (~100)	63
	1. BF_3•Et_2O, 70° 2. DMD, acetone/CH_2Cl_2, 0° 3. $KHCO_3$	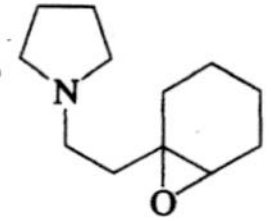 N O (85)	63
C_{13} 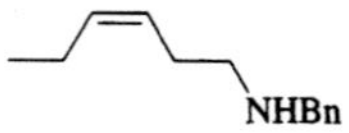 NHBn	1. 4-$ClC_6H_4SO_3H$, Et_2O, 0° 2. DMD, acetone/CH_2Cl_2, 0°, 1.5 h 3. K_2CO_3, $MgSO_4$, CH_2Cl_2, 0°, 5 h	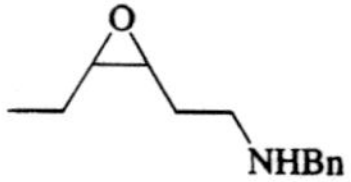 O NHBn (93)	65

Substrate	Conditions	Product(s) and Yield(s) (%)	Refs.
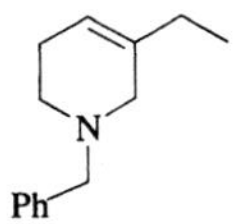 C_{14} N-benzyl (Ph)	DMD, acetone, 0°	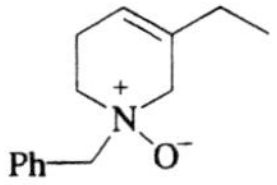 $\overset{+}{N}$–O^-, Ph (100)	63
	1. BF_3•Et_2O, 70° 2. DMD, acetone/CH_2Cl_2, 0° 3. $KHCO_3$	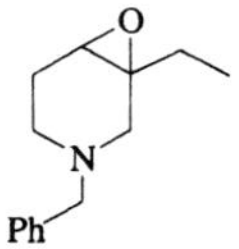 O, N, Ph (97)	63
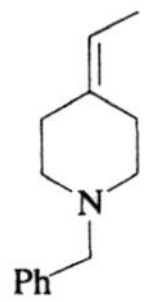 N, Ph	DMD, acetone, 0°	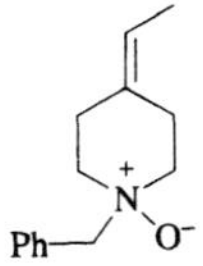 $\overset{+}{N}$–O^-, Ph (100)	63
	1. BF_3•Et_2O, 70° 2. TFD, TFP/CH_2Cl_2, 0° 3. $KHCO_3$	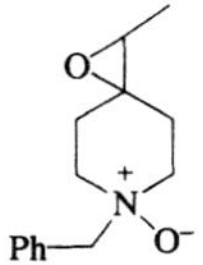 O, $\overset{+}{N}$–O^-, Ph (70)	63
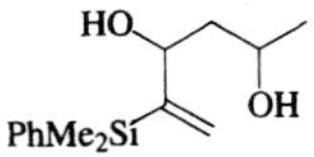 HO, OH, $PhMe_2Si$	DMD, acetone, 20°, 16 h	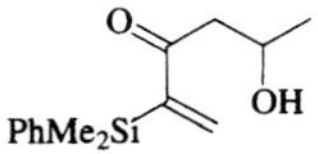 O, OH, $PhMe_2Si$ (—)	282
C_{15} OH 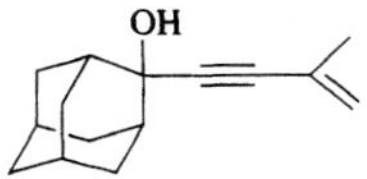 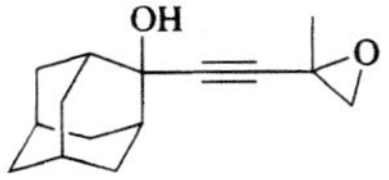	DMD	OH, O (—)	71

TABLE 2. CHEMOSELECTIVE OXIDATIONS BY ISOLATED DIOXIRANES (*Continued*)

Substrate	Conditions	Product(s) and Yield(s) (%)	Refs.
C_{15-16} 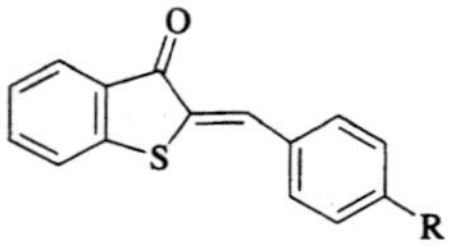	DMD (x equiv), CH_2Cl_2/acetone, 20°	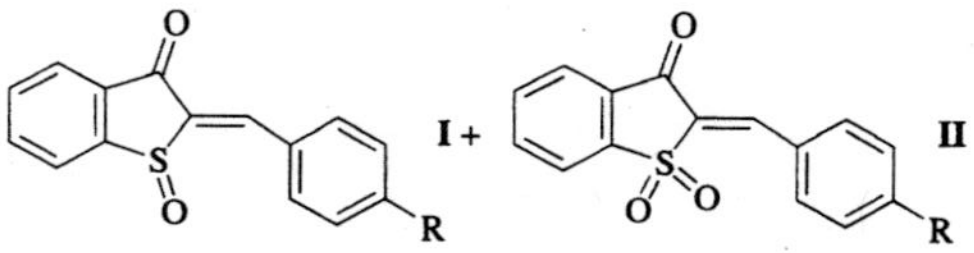 I + II	66
C_{16-23} 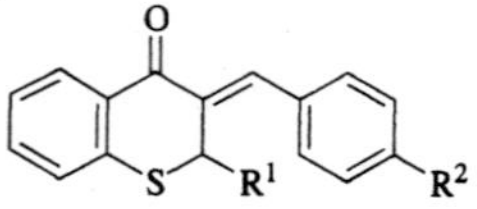	DMD (x equiv), CH_2Cl_2, acetone, 20°	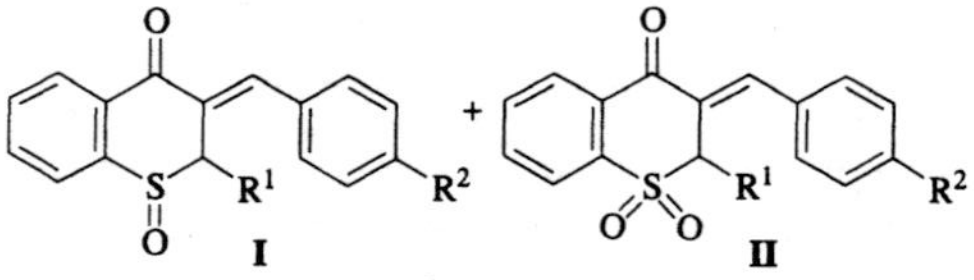 I + II	66

R	x	Time	I	II
Cl	1.4	1 h	(29)	(19)
Cl	2.2	2 h	(0)	(100)
H	1.3	1 h	(29)	(22)
H	2.2	2 h	(0)	(100)
Me	1.4	1 h	(31)	(21)
Me	2.2	2 h	(0)	(100)

R^1	R^2	x	Time	I	II
H	Br	1.3	1 h	(72)	(12)
H	Br	2.2	2 h	(0)	(100)
H	Cl	1.2	1 h	(76)	(10)
H	Cl	2.2	2 h	(0)	(100)
H	H	1.3	1 h	(75)	(14)
H	H	2.2	2 h	(0)	(100)
H	Me	1.3	1 h	(73)	(13)
H	Me	2.2	2 h	(0)	(100)
H	MeO	1.2	1 h	(78)	(10)
H	MeO	2.2	2 h	(0)	(100)

R^1	R^2	x	Time	**I**	**II**
Ph	Cl	1.3	1 h	(81)	(13)
Ph	Cl	2.2	2 h	(0)	(100)
Ph	H	1.2	1 h	(82)	(9)
Ph	H	2.2	2 h	(0)	(100)
Ph	Me	1.3	1 h	(74)	(11)
Ph	Me	2.2	2 h	(0)	(100)
Ph	MeO	1.3	1 h	(77)	(11)
Ph	MeO	2.2	2 h	(0)	(100)

Substrate	Conditions	Product(s) and Yield(s) (%)	Refs.
C_{17}	DMD	(—)	71
C_{18}	DMD	(—)	71
n-C_8H_{17} ... NH_2	1. 4-$ClC_6H_4SO_3H$, Et_2O, 0° 2. DMD, acetone/CH_2Cl_2, 0°, 2 h 3. K_2CO_3, $MgSO_4$, CH_2Cl_2, 0°, 5 h	n-C_8H_{17} ... NH_2 (98)	65

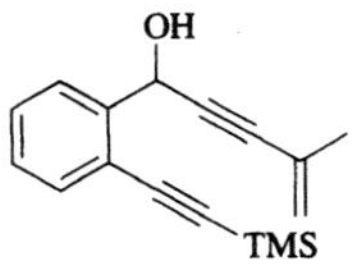

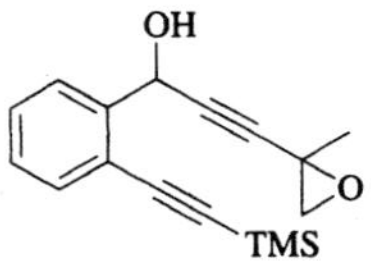

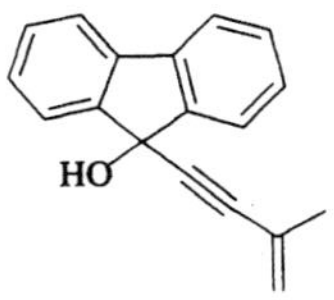

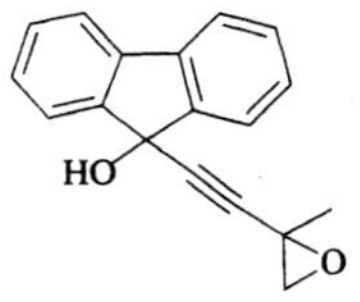

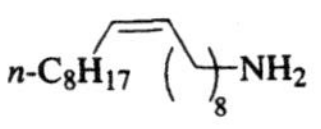

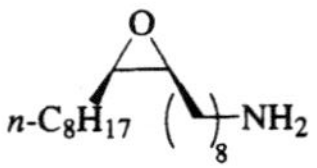

TABLE 2. CHEMOSELECTIVE OXIDATIONS BY ISOLATED DIOXIRANES (*Continued*)

Substrate	Conditions	Product(s) and Yield(s) (%)	Refs.
C_{21-28} RO (steroid)	DMD, acetone, rt	HO (epoxide) **I** + RO (epoxide) **II** + O, OH **III**	283
		R / **I** / **II** / **III**: H (—) **II**=**I** (—); Bn (30) (37) (16)	
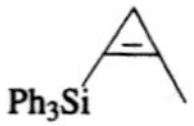 C_{22} Ph_3Si (cyclopropene)	DMD, acetone, rt, 6 h	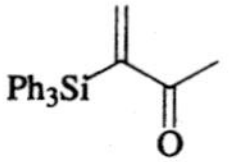 Ph_3Si, O (44) 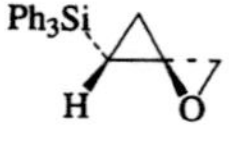 + Ph_3Si, H, O (9)	284
C_{24} 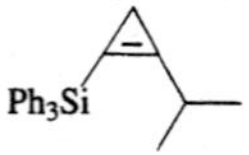 Ph_3Si (cyclopropene)	DMD, acetone, rt, 3.5 h	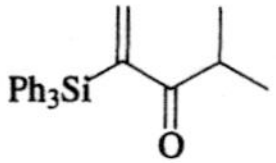 Ph_3Si, O (46) + O, Ph_3Si (22)	284

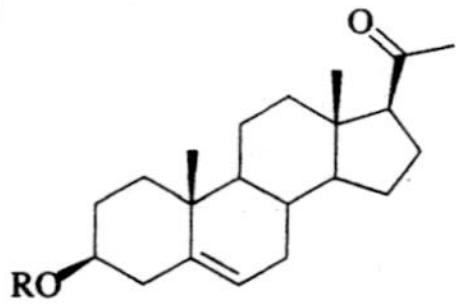

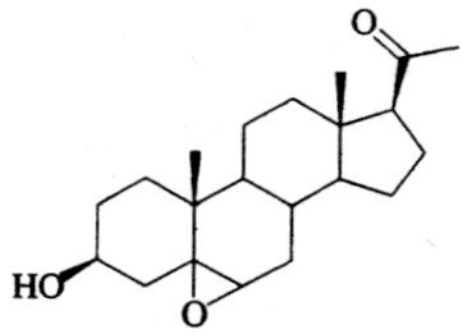

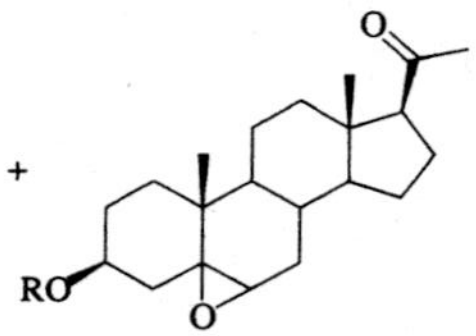

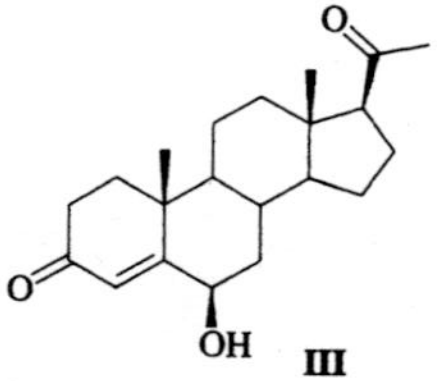

R	I	II	III
H	(—)	II=I	(—)
Bn	(30)	(37)	(16)

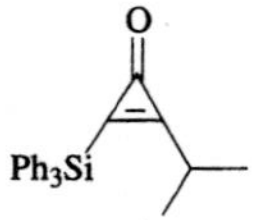

	Substrate	Conditions	Product and Yield (%)		Refs.
C_{28}	MeOCH=CHC(O)C(=PPh$_3$)CO$_2$Bu-*t*	DMD, acetone/CH_2Cl_2, –78° to 25°, 4 h	MeOCH=CHC(O)C(OH)$_2$CO$_2$Bu-*t* (70)		285
C_{29}	2-furyl/2-thienyl-C(O)C(=PPh$_3$)CO$_2$Bu-*t*	DMD, acetone/CH_2Cl_2, –78° to 25°, 4 h	2-furyl/2-thienyl-C(O)C(OH)$_2$CO$_2$Bu-*t* (97)	X = O, S	285
	Me$_2$C=CHC(O)C(=PPh$_3$)CO$_2$Bu-*t*	DMD, acetone/CH_2Cl_2, –78° to 25°, 4 h	Me$_2$C=CHC(O)C(OH)$_2$CO$_2$Bu-*t* (100)		285
C_{33}	PhCH=CHC(O)C(=PPh$_3$)CO$_2$Bu-*t*	DMD, acetone/CH_2Cl_2, –78° to 25°, 4 h	PhCH=CHC(O)C(OH)$_2$CO$_2$Bu-*t* (85)		285
C_{34}	Me$_2$C=CH(CH$_2$)$_2$CH(CH$_3$)CH$_2$C(O)C(=PPh$_3$)CO$_2$Bu-*t*	DMD, acetone/CH_2Cl_2, –78° to 25°, 4 h	Me$_2$C=CH(CH$_2$)$_2$CH(CH$_3$)CH$_2$C(O)C(OH)$_2$CO$_2$Bu-*t* (83)		285

TABLE 3. REGIOSELECTIVE EPOXIDATIONS BY DIOXIRANES

Substrate	Conditions	Product(s) and Yield(s) (%)	Refs.
C_{7-12} 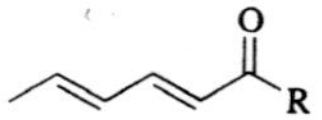	DMD, acetone/CH_2Cl_2, 0°, 3.5 h	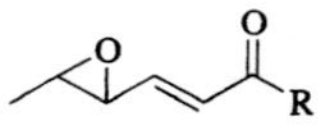 (95) R = Me, Ph	286
C_{10-15} 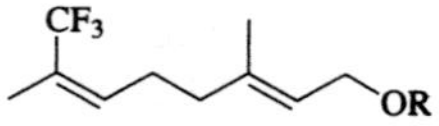	DMD, acetone or TFD, TFP/CH_2Cl_2	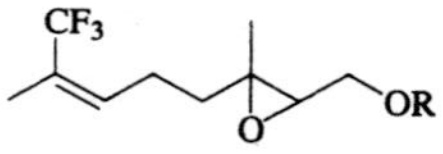 (see table below)	78
C_{10-13} 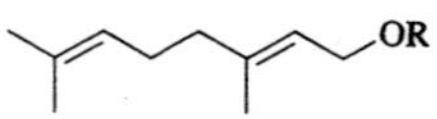	TFD, solvent, 0 to 10°, 1 to 3 h	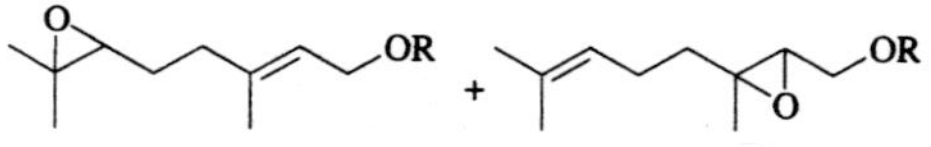 I + II (see table below)	80

R	Dioxirane	Temp	Time
H	DMD	20°	15 min (98)
H	TFD	–15°	3 min (98)
THP	DMD	20°	15 min (98)
THP	TFD	–40°	3 min (99)

R	Solvent	% Convn	I:II
H	TFP	22	68:32
H	MeOH/TFP (9:1)	21	73:27
H	CCl_4/TFP (9:1)	25	57:43
Me	TFP	32	73:27
Me	MeOH/TFP (9:1)	31	68:32
Me	CCl_4/TFP (9:1)	32	71:29
Ac	TFP	30	83:17
TMS	TFP	33	70:30

C_{10} OH

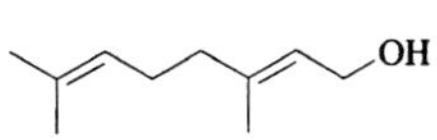

DMD (x equiv), solvent, rt, 1 to 3 h

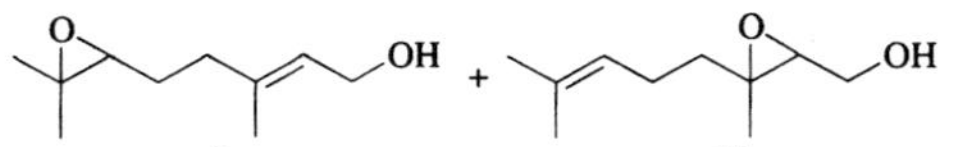

OH + OH 72

I II

+ bis(epoxide) **III**

x	Solvent	% Convn	**I**:**II**:**III**
1	acetone	93	73:17:10
0.3	acetone/MeOH (1:9)	27	88:12:0
0.3	acetone	28	74:26:0
0.3	CCl_4/acetone (9:1)	27	51:49:0
5	acetone	>95	<5:<5:95

DMD, acetone/CH_3CN, buffer, rt

I + **II** 149a

pH	% Convn	**I**:**II**
4.5	17	5.3:1
7.2	19	4.4:1
8.8	18	5.2:1
9.3	20	5.0:1
10.0	20	4.7:1
10.9	16	5.2:1

TABLE 3. REGIOSELECTIVE EPOXIDATIONS BY DIOXIRANES (*Continued*)

Substrate	Conditions	Product(s) and Yield(s) (%)	Refs.
	Fructose dioxirane, CH_3CN/H_2O, pH 7.5 Fructose dioxirane =	**I** + **II** (5) **I**:**II** = 5.7:1	149a
C_{11} CF_3 ... CO_2Me	DMD, acetone or TFD, TFP/CH_2Cl_2	CF_3 ... CO_2Me **I** (42)	78
	TFD, TFP/CH_2Cl_2, 0°, 3 min	**I** (85) + bis(epoxide) (15)	78
R^1, R^2 ... CO_2Me	DMD, acetone or TFD, TFP/CH_2Cl_2	CO_2Me **I** + CO_2Me **II**	78

R^1	R^2	Dioxirane	Temp	Time	I	II
CF_3	H	DMD (excess)	0°	12 d	(0)	(85)
CF_3	H	TFD (excess)	0°	1 h	(0)	(93)
Me	F	DMD	20°	15 min	(95)	(0)
Me	F	DMD	–40°	3 min	(95)	(0)
Me	F	DMD (excess)	0°	8 d	(0)	(90)
Me	H	DMD	20°	10 min	(96)	(0)
Me	H	TFD	–40°	3 min	(97)	(0)

Substrate	Conditions	Product(s) and Yield(s) (%)	Refs.
OH (structure)	DMD, solvent, 20°	I + II (structures)	134a
		Solvent / % Convn / I:II / dr (II) / dr (I): MeOH/acetone (9:1) 41, 95:5, 76:24, 50:50; acetone 35, 84:16, 84:16, 50:50; CCl_4 26, 32:68, 94:6, 50:50	
	TFD, CH_2Cl_2/TFP, 20°	**I** + **II** (—) **I**:**II** = 69:31 Convn = 25%; dr of **II** = 85:15	134a
	Acetone, Oxone®, CH_3CN/H_2O, 20°	**I** + **II** (—) **I**:**II** = 57:43 Convn = 18%; dr of **II** = 90:10	134a
C_{13} (structure)	DMD, acetone/CH_2Cl_2, N_2, –40°	OHC (structure) (~100)	214
O (structure)	DMD, acetone/CH_2Cl_2, N_2, –40°	OHC (structure) (~90)	214
C_{14} Bn (structure)	DMD, acetone/CH_2Cl_2, 0°	Bn I (—) + Bn II (—) **I**:**II** = 70:30	287, 287a

TABLE 3. REGIOSELECTIVE EPOXIDATIONS BY DIOXIRANES (*Continued*)

	Substrate	Conditions	Product(s) and Yield(s) (%)	Refs.
C_{16}	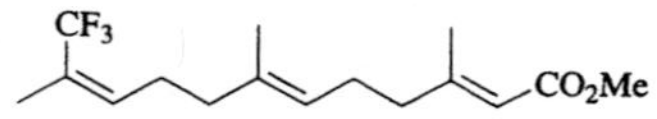 CF_3 … CO_2Me	DMD, acetone/CH_2Cl_2, 20°, 15 min	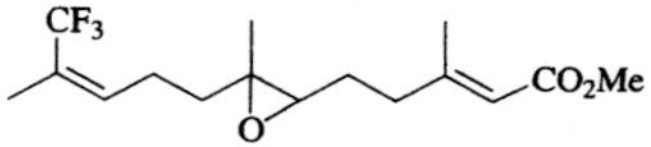 CF_3 … CO_2Me **I** (100)	78
		TFD, TFP/CH_2Cl_2, –20°, 3 min	**I** (85) 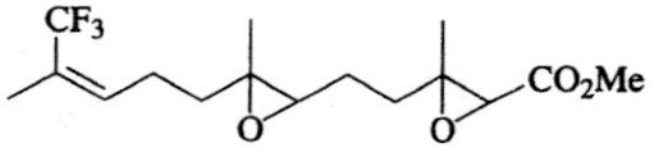 + CF_3 … CO_2Me (15)	78
	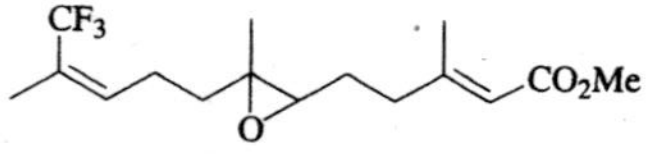 CF_3 … CO_2Me	DMD, acetone/CH_2Cl_2, 20°, 3 d	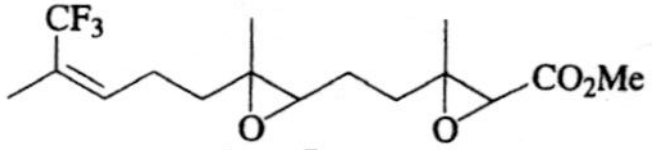 CF_3 … CO_2Me **I** (56) + triepoxide (44)	78
		TFD, TFP/CH_2Cl_2, –70°, 1 h	**I** (85) + triepoxide (8)	78
	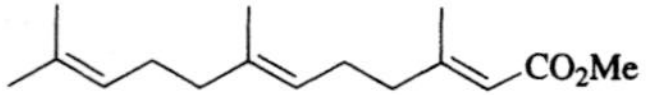 CO_2Me	DMD, acetone/CH_3CN	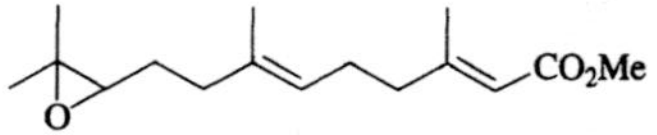 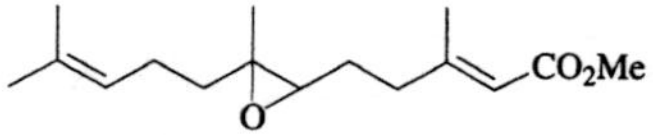 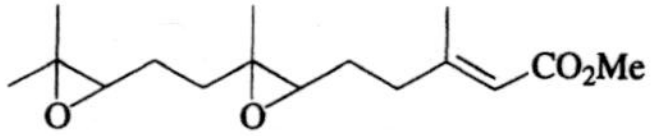 CO_2Me **I** + CO_2Me **II** + CO_2Me **III** (—)	288

Temp	% Convn	I:II:III
20°	84	52:13:22
–50°	95	55:9:28
–15°	63	42:13:7

Substrate	Conditions	Product(s) and Yield(s) (%)	Refs.
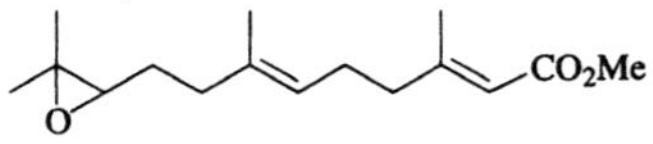 (epoxide, diene) CO_2Me	DMD (excess), acetone/ CH_2Cl_2, rt, 5 min	**III** (~100)	288
	DMD, acetone/CH_2Cl_2 rt, 5 min	**III** (>96)	288
	DMD (excess), acetone/ CH_2Cl_2, rt, 3 d	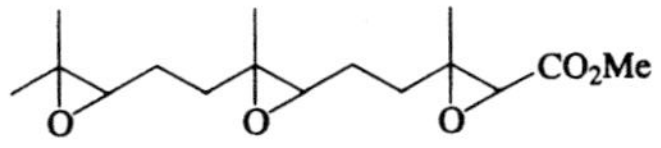 (tris-epoxide) CO_2Me (~100)	288
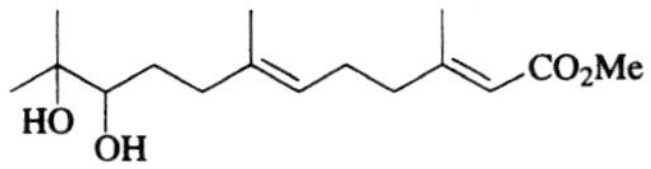 HO OH (diol, diene) CO_2Me	DMD, acetone/CH_2Cl_2, $NaHPO_4$, rt, 3 min	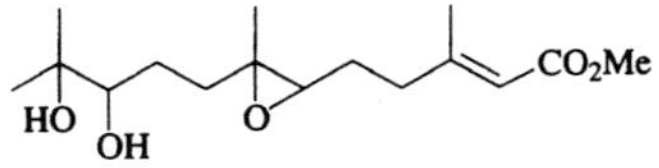 HO OH (epoxide) CO_2Me (—)	288
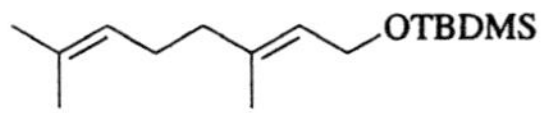 OTBDMS	Ketone **12**, Oxone®, $NaHCO_3$, CH_3CN/H_2O, EDTA, 0°	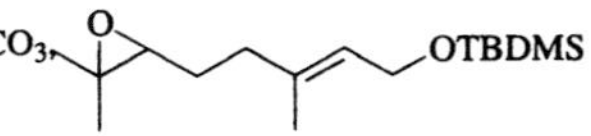 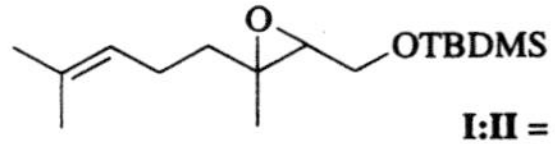 OTBDMS **I** (—) + OTBDMS **II** (—) +bisepoxide (—) **I:II** = 4.3:1	149a
C_{18} 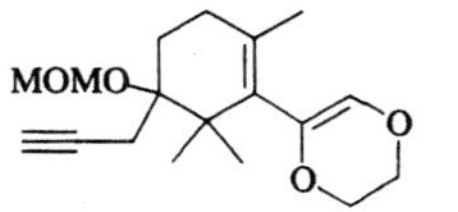 MOMO (dihydrodioxin)	DMD, acetone/CH_2Cl_2, N_2, –40°	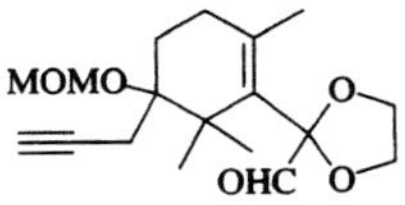 MOMO OHC (dioxolane) (~ 100)	214

TABLE 3. REGIOSELECTIVE EPOXIDATIONS BY DIOXIRANES (*Continued*)

Substrate	Conditions	Product(s) and Yield(s) (%)	Refs.
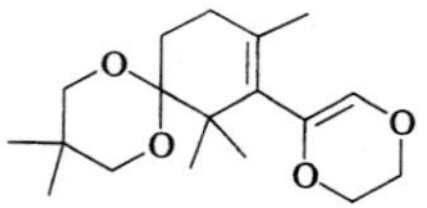 O O O O	DMD, acetone/CH_2Cl_2, N_2, –40°	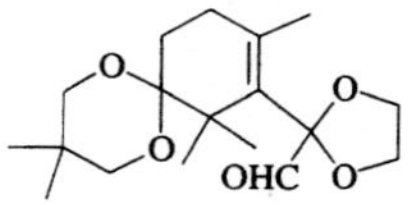 O O OHC O O (~100)	214
C_{20} 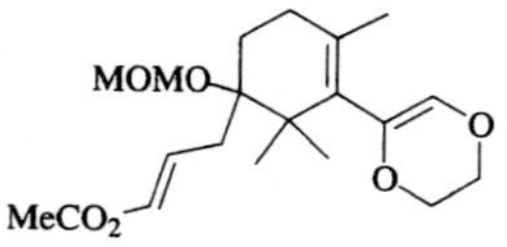 MOMO MeCO$_2$ O O	DMD, acetone/CH_2Cl_2, N_2, –40°	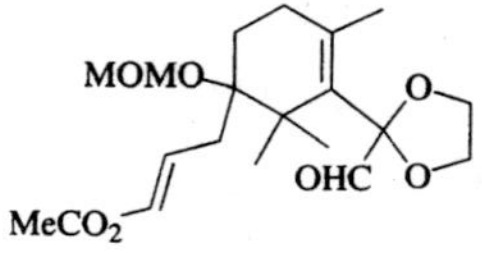 MOMO MeCO$_2$ OHC O O (84)	214
C_{23} 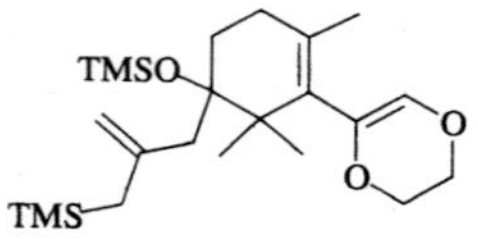 TMSO TMS O O	DMD, acetone/CH_2Cl_2, N_2, –40°	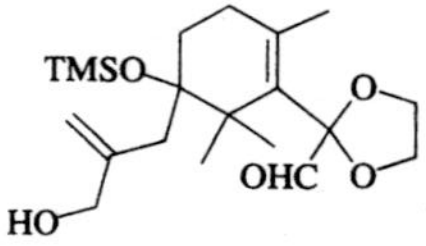 TMSO HO OHC O O (38) +	214
		TMSO TMSO OHC O O (12) +	
		TMSO OHC TMS OHC O O (24)	

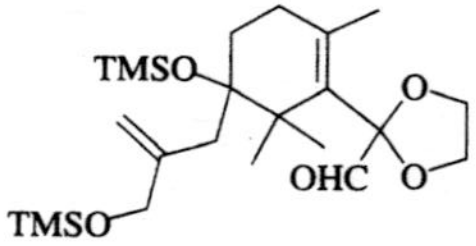

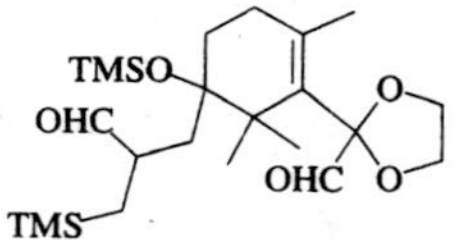

Substrate		Conditions	Product(s) and Yield(s) (%)	Refs.
C_{29}	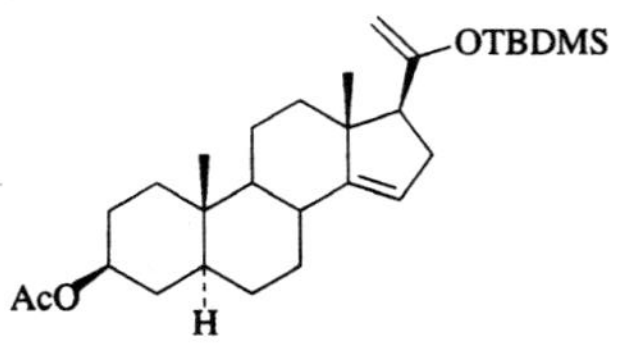 AcO, H, OTBDMS	DMD, acetone, rt, 16 h	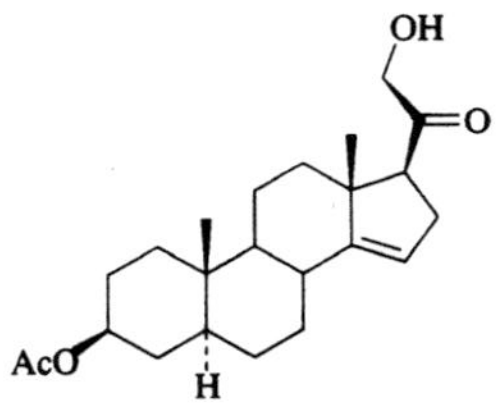 AcO, H, OH, O (85)	289
C_{30}	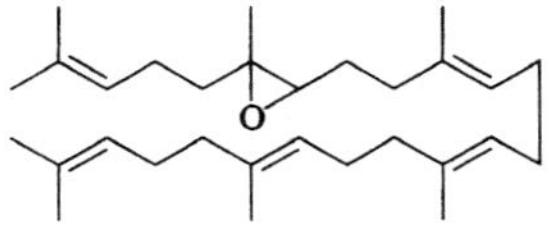 O	DMD, acetone/CH_2Cl_2, 0°, 1 h	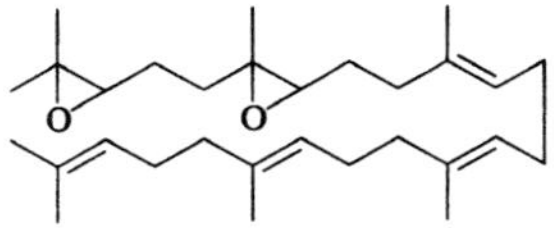 O, O (6) 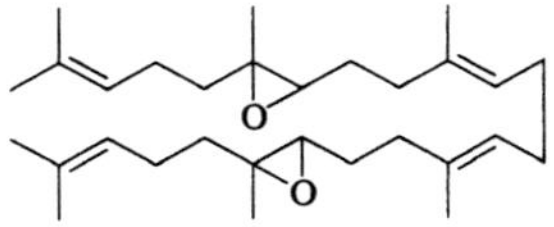 + O, O (8) 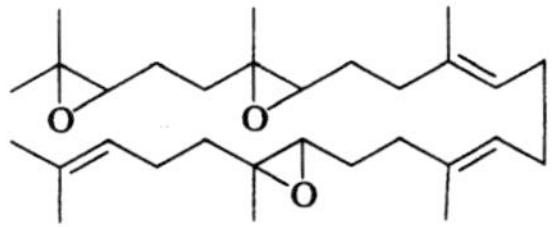 + O, O, O (9) 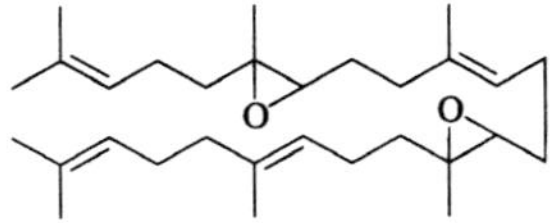 + O, O (8)	290, 291

TABLE 3. REGIOSELECTIVE EPOXIDATIONS BY DIOXIRANES (*Continued*)

	Substrate	Conditions	Product(s) and Yield(s) (%)		Refs.
C_{79}	OTBDMS; Et_3SiO; OMe; TBDMSO; OMe; O; AcO; OAc; $OSiEt_3$	DMD, acetone/CH_2Cl_2, 0°	OTBDMS; Et_3SiO; OMe; TBDMSO; OMe; O; AcO; OAc; $OSiEt_3$	(—)	90
C_{94}	OTBDMS; Et_3SiO; OMe; $OSiEt_3$; TBDMSO; OMe; i-Pr_3SiO_2C; O; AcO; OAc; $OSiEt_3$	DMD, acetone/CH_2Cl_2, 0°, 5 min	OTBDMS; Et_3SiO; OMe; $OSiEt_3$; TBDMSO; OMe; i-Pr_3SiO_2C; O; AcO; OAc; $OSiEt_3$	(—)	90

TABLE 4. DIASTEREOSELECTIVE EPOXIDATIONS OF OLEFINS BY DIOXIRANES

Substrate	Conditions	Product(s) and Yield(s) (%), Diastereomeric Ratio (dr)	Refs.
C_{4-6} (allylic alcohol: R^1, R^2, R^3, OH)	DMD, solvent, 0 to 20°, 1 to 8 h	I + II (>95)	72, 292, 293
	TFD, solvent, 0 to 10°, 15 to 60 min	I + II (>95)	80

DMD, solvent, 0 to 20°, 1 to 8 h:

R^1	R^2	R^3	Solvent	% Convn	I:II
H	H	H	acetone	89	50:50
Me	H	H	MeOH/acetone (9:1)	>95	57:43
Me	H	H	acetone	>95	60:40
Me	H	H	CCl_4/acetone (9:1)	>95	70:30
H	Me	H	acetone	78	53:47
H	Me	H	CCl_4/acetone (9:1)	>95	56:44
H	H	Me	MeOH/acetone (9:1)	86	64:36
H	H	Me	acetone	>95	67:33
H	H	Me	CCl_4/acetone (9:1)	>95	85:15
H	Me	Me	MeOH/acetone (9:1)	>95	59:41
H	Me	Me	acetone	87	76:24
H	Me	Me	CCl_4/acetone (9:1)	>95	82:18
Me	H	Me	MeOH/acetone (9:1)	>95	82:18
Me	H	Me	acetone	>95	87:13
Me	Me	H	acetone	(—)	51:49
Me	H	Me	CCl_4/acetone (9:1)	>95	91:9
Me	Me	Me	acetone	>95	>95:5

TFD, solvent, 0 to 10°, 15 to 60 min:

R^1	R^2	R^3	Solvent	% Convn	I:II
H	H	H	CCl_4/TFP (9:1)	>95	55:45
Me	H	H	TFP	>95	53:47
Me	H	H	CCl_4/TFP (9:1)	>95	62:38

TABLE 4. DIASTEREOSELECTIVE EPOXIDATIONS OF OLEFINS BY DIOXIRANES (*Continued*)

Substrate	Conditions	Product(s) and Yield(s) (%), Diastereomeric Ratio (dr)	Refs.
		(see sub-table A)	
C_{4-7}	DMD, acetone	I + II (—) (see sub-table B)	292
C_{4-12}	DMD, CH_2Cl_2/acetone, rt, 3.5 h	I + II + III (see sub-table C)	74

R^1	R^2	R^3	Solvent	% Convn	I:II
H	Me	H	TFP	95	48:52
H	Me	H	CCl_4/TFP (9:1)	76	43:57
H	H	Me	TFP	95	77:23
H	H	Me	CCl_4/TFP (9:1)	93	88:12
H	Me	Me	TFP	>95	76:24
H	Me	Me	CCl_4/TFP (9:1)	>95	82:18
Me	H	Me	TFP	>95	85:15
Me	H	Me	CCl_4/TFP (9:1)	60	90:10

R^1	R^2	R^3	I:II
H	H	H	50:50
H	Me	H	53:47
H	H	Me	67:33
Me	H	H	60:40
H	Me	Me	76:24
Me	H	Me	87:13
Me	Me	Me	95:5

R^1	R^2	R^3	I+II	III	I:II
H	Me	H	(>95)	(<5)	50:50
TMS	Me	H	(32)	(68)	60:40
TMS	–$(CH_2)_3$–		(<5)	(>95)	—
$SiMe_2Ph$	Me	H	(10)	(90)	55:45

C_5 OH

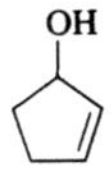

DMD, acetone, rt, 2 h

OH + OH + O

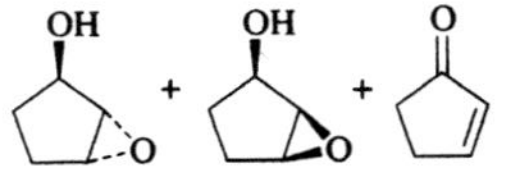

I II III

(—) **I**:**II** = 48:52
(**I** + **II**):**III** = 78:22

83

C_{5-7} OR

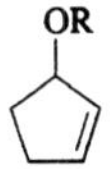

DMD or TFD, solvent, 0°, time

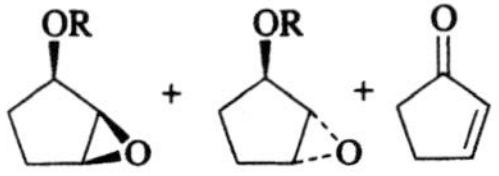

OR + OR + O

I II III

I + **II** + **III** (>95)

294

R	Dioxirane	Solvent	Time	% Convn	**I**:**II**	(**I**+**II**):**III**
H	DMD	acetone	55 min	92	53:47	70:30
H	DMD	acetone/CH_2Cl_2 (1:1)	40 min	95	61:39	90:10
H	TFD	CCl_4	10 min	95	70:30	66:34
Me	DMD	acetone	90 min	54	20:80	95:5
Me	DMD	acetone/CH_2Cl_2 (1:1)	75 min	60	30:70	95:5
Me	TFD	CCl_4	20 min	70	8:92	98:2
Ac	DMD	acetone	150 min	60	30:70	—
Ac	DMD	acetone/CH_2Cl_2 (1:1)	120 min	72	36:64	—
Ac	TFD	CCl_4	30 min	56	38:62	—

C_5 O_2H

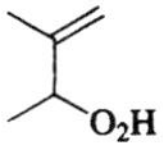

DMD, acetone/CH_2Cl_2, 20°, 1 h

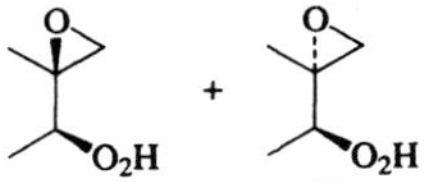

O, O_2H + O, O_2H

I II

(71) **I**:**II** = 42:58

170

TABLE 4. DIASTEREOSELECTIVE EPOXIDATIONS OF OLEFINS BY DIOXIRANES (*Continued*)

Substrate	Conditions	Product(s) and Yield(s) (%), Diastereomeric Ratio (dr)	Refs.
C_{5-6} RO-cyclopentene	DMD or TFD, solvent, 0°, time	I + II (>95)	294
C_{5-7} $R^1R^3C{=}CH\text{-}CH(OH)R^2$	DMD, acetone/CH_2Cl_2	I + II + III	73

R	Dioxirane	Solvent	Time	% Convn	I:II
H	DMD	acetone	70 min	96	54:46
H	TFD	CCl_4	18 min	85	78:22
Me	DMD	acetone	70 min	96	15:85
Me	TFD	CCl_4	18 min	85	2:98

R^1	R^2	R^3	Temp	% Convn	I:II	(I+II):III
Me	Me	H	0°	78	53:47	81:19
H	—$(CH_2)_3$—		–20°	>95	72:28	46:54
H	—$(CH_2)_3$—		0°	80	53:47	70:30
Me	Me	Me	–78°	>95	76:24	>95:5
Me	Me	Me	0°	87	77:23	>95:5
Me	*i*-Pr	H	–20°	>95	63:37	47:53
Me	*i*-Pr	H	0°	>95	65:35	90:10
H	—$(CH_2)_3$—		0°	>95	>95:5	>95:5
H	*n*-C_5H_{11}	H	0°	>95	66:34	51:49
Me	*n*-Bu	H	–20°	96	52:48	47:53
Me	*n*-Bu	H	0°	>95	52:48	71:29

C_6

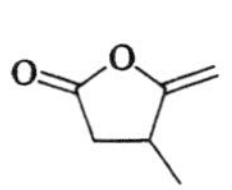

DMD, acetone/CH_2Cl_2, N_2, –20°, 3.5 h

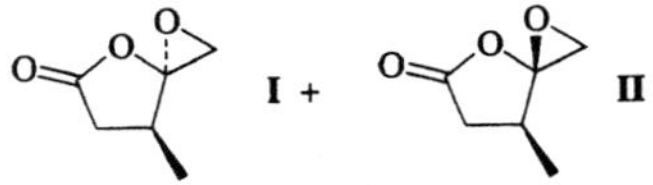

I + **II** (—) **I**:**II** = 60:40 44

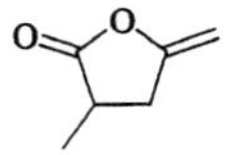

DMD, acetone/CH_2Cl_2, N_2, –20°, 3.5 h

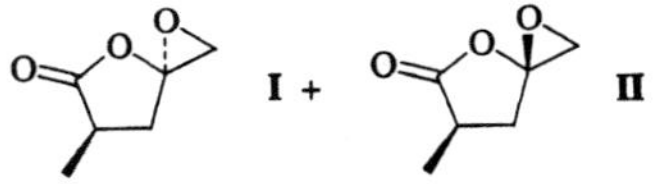

I + **II** (—) **I**:**II** = 65:35 44

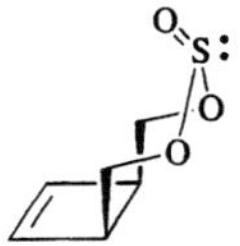

DMD, rt

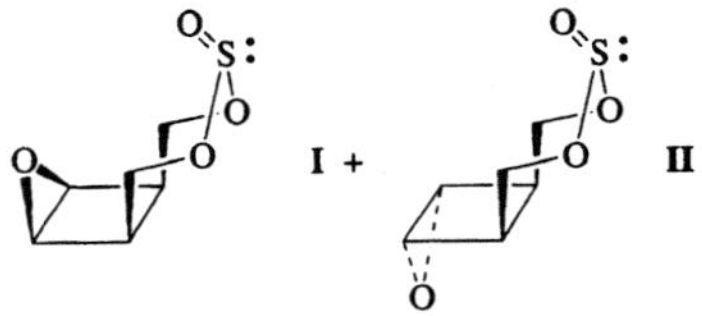

I + **II** 295

Solvent	Time	**I+II**	**I:II**
acetone	7 h	(88)	77:23
CCl_4/acetone (9:1)	12 h	(85)	70:30

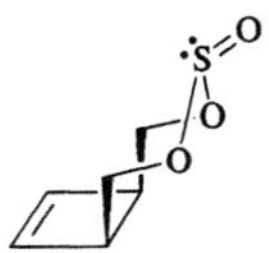

DMD, rt

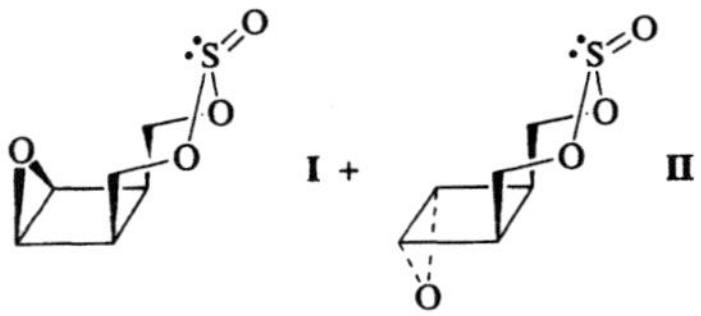

I + **II** 295

Solvent	Time	**II**	**I:II**
acetone	2 h	(95)	0:100
CCl_4/acetone (9:1)	5 h	(95)	0:100

TABLE 4. DIASTEREOSELECTIVE EPOXIDATIONS OF OLEFINS BY DIOXIRANES (*Continued*)

Substrate	Conditions	Product(s) and Yield(s) (%), Diastereomeric Ratio (dr)	Refs.
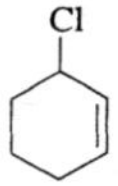 Cl	DMD, solvent, rt, 60 min	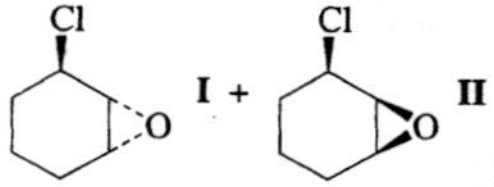 Cl O **I** + Cl O **II**	67
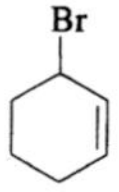 Br	DMD, acetone, rt, 60 min	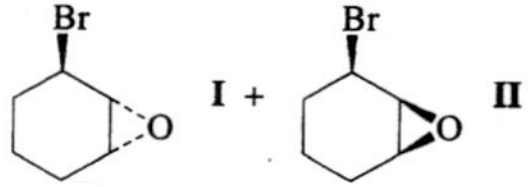 Br O **I** + Br O **II** Convn = 58%; **I**:**II** = 92:8	67
C_{6-20} 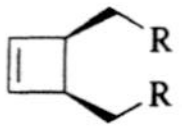 R R	DMD, rt	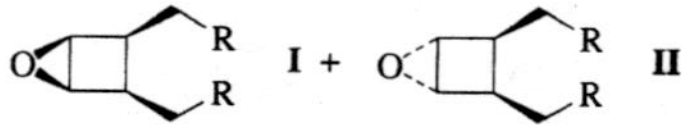 O R R **I** + O R R **II**	295

Solvent	% Convn	**I**:**II**
acetone	76	90:10
CH_2Cl_2/acetone (1:1)	74	93:7

R	Solvent	Time		**I**:**II**
H	acetone	12 min	(99)	44:56
H	CCl_4/acetone (9:1)	30 min	(99)	55:45
OH	acetone	2 h	(70)	57:43
OH	CCl_4/acetone (9:1)	1 h	(80)	67:33
OMe	acetone	7 h	(99)	49:51
OMe	CCl_4/acetone (9:1)	10 h	(99)	62:38
OMs	acetone	5 h	(85)	79:21
OMs	CCl_4/acetone (9:1)	12 h	(99)	90:10
OAc	acetone	2 h	(69)	72:28
OAc	CCl_4/acetone (9:1)	6 h	(99)	68:32
OTs	acetone	2 h	(97)	63:37
OTs	CCl_4/acetone (9:1)	8 h	(99)	82:18

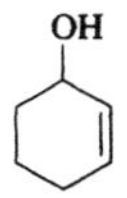

Reactant	Conditions	Product(s) and Yield(s) (%)	Refs.
OH (cyclohexenol)	DMD, acetone, 20°, 6 h	I + II + III (93) I:II = 47:53; (I + II):III = 70:30	75
	DMD, acetone/CH_2Cl_2, rt, 4 h	I + II + III I + II (80), III (10); I:II = 25:75	133
	DMD, solvent, rt, 60 min	I + II + III	296

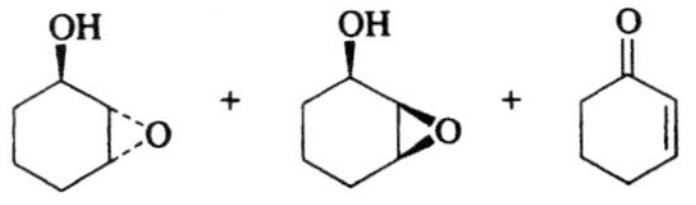

Solvent	% Convn	I:II	(I+II):III
acetone	94	54:46	46:54
CH_2Cl_2/acetone (1:1)	94	43:57	65:35
CH_2Cl_2/acetone (9:1)	89	22:78	84:16
CH_2Cl_2/acetone (97:3)	77	18:82	89:11
MeOH/acetone (9:1)	100	66:34	75:25
$CHCl_3$/acetone (9:1)	100	88:12	88:12
CCl_4/acetone (9:1)	87	15:85	52:48
CCl_4/acetone (95:5)	86	6:94	59:41

Conditions	Product(s)	Refs.
Ketone, Oxone®, CH_2Cl_2/MeOH, buffer (pH 11.0), KOH, rt	I + II	297

Ketone	Time		I:II
acetone	1.5 h	(69)	64:36
2-methylcyclohexanone	3.5 h	(63)	71:29
2,6-dimethylcyclohexanone	3.0 h	(66)	77:23

TABLE 4. DIASTEREOSELECTIVE EPOXIDATIONS OF OLEFINS BY DIOXIRANES (*Continued*)

Substrate	Conditions	Product(s) and Yield(s) (%), Diastereomeric Ratio (dr)	Refs.
	Ketone, Oxone®, $NaHCO_3$, CH_3CN/H_2O, Na_2EDTA, rt	**I** + **II** (—)	81
C_{6-10} 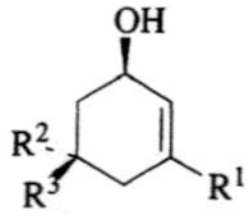	TFD, solvent, 0 to 10°, 15 to 60 min	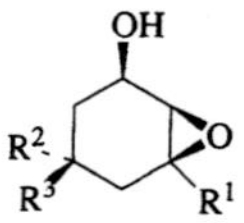 **I** + 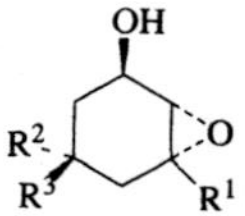 **II**	80

Ketone	I:II	Ketone	I:II
MeCOMe	1.2:1	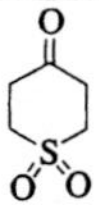	1.9:1
EtCOEt	—	TfO⁻, Ph, N, Me, Me, Ph	1.9:1
$ClCH_2COMe$	2.0:1	TfO⁻, Ph, N, Me, Me, Ph	5.9:1
$ClCH_2COCH_2Cl$	2.7:1		1.7:1
CF_3COMe	3.3:1	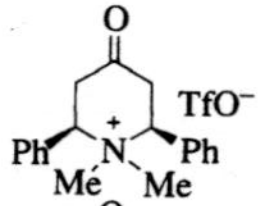	1.9:1
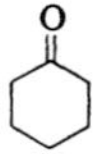	1.3:1		
	—		
Cl	2.1:1		
	1.6:1		

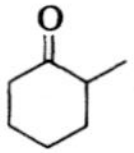

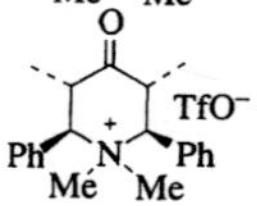

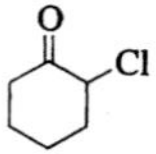

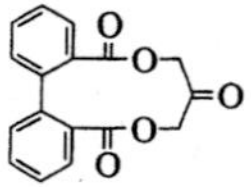

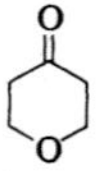

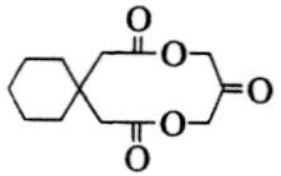

R[1]	R[2]	R[3]	Solvent	% Convn	I:II
H	H	H	TFP	70	57:43
H	H	H	CCl_4/TFP (9:1)	78	70:30
Me	Me	Me	TFP	84	88:12
Me	Me	Me	CCl_4/TFP (9:1)	70	94:6
H	H	*t*-Bu	TFP	80	73:27

C_6

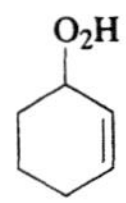

DMD, acetone, 20°, 2 to 8 h

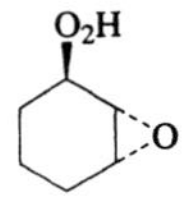 **I** + 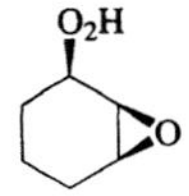 **II** (—) **I**:**II** = 91:9

75

DMD, solvent, rt, 20 h

I + **II**

Solvent	% Convn	I:II
acetone	100	84:16
CH_2Cl_2/acetone (1:1)	100	82:18

67

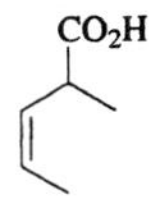

DMD, acetone, 8 h

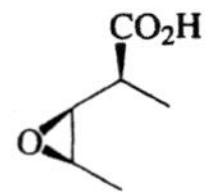 **I** + 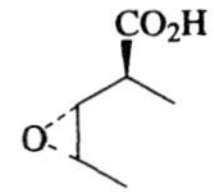 **II** (>95) **I**:**II** = 46:54

298

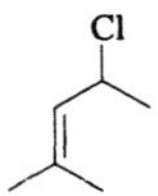

DMD, acetone, 7 h

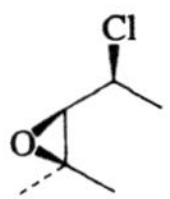 **I** + 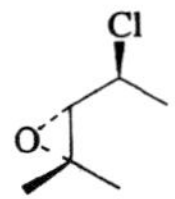 **II** (50) **I**:**II** = 90:10

298

TABLE 4. DIASTEREOSELECTIVE EPOXIDATIONS OF OLEFINS BY DIOXIRANES (*Continued*)

Substrate	Conditions	Product(s) and Yield(s) (%), Diastereomeric Ratio (dr)	Refs.
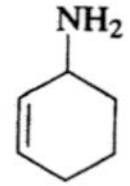 NH_2	1. 4-$ClC_6H_4SO_3H$, Et_2O, 0° 2. DMD, acetone/CH_2Cl_2, 0°, 1 h 3. K_2CO_3, $MgSO_4$, CH_2Cl_2, 0°, 5 h	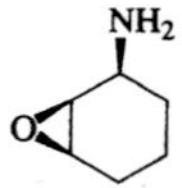 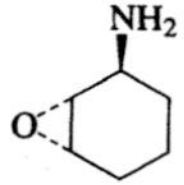 NH_2 **I** + NH_2 **II** (60) **I:II** = 90:10	65
	1. 4-$ClC_6H_4SO_3H$, Et_2O, 0° 2. TFD, CH_2Cl_2/CH_3CN, 0°, 0.1 h 3. K_2CO_3, $MgSO_4$, CH_2Cl_2, 0°, 5 h	**I** + **II** (72) **I:II** = 50:50	65
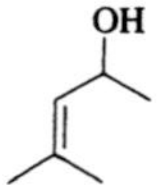 OH	DMD, solvent, ~20°, 6 h	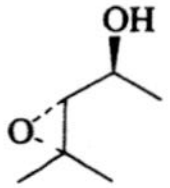 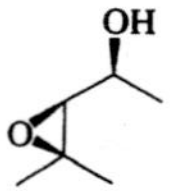 OH **I** + OH **II**	87
	TFD, acetone-d_6/$CDCl_3$, 0.5 h	**I** + **II** (>95) **I:II** =60:40	298

Solvent	**I:II**
acetone	76:24
MeOH/acetone (1:1)	60:40
AcOEt/acetone (1:1)	71:29
CCl_4/acetone (1:1)	76:24
hexane/acetone (1:1)	77:23
CCl_4/acetone (9:1)	82:18
hexane/acetone (9:1)	82:18

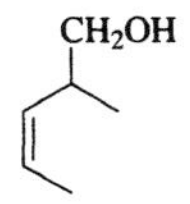

CH_2OH

TFD, CCl_4, 0.5 h — **I + II** (57) **I:II** = 92:8 — 298

DMD, acetone, 2 h — CH_2OH **I** + CH_2OH **II** (>95) **I:II** = 38:62 — 298

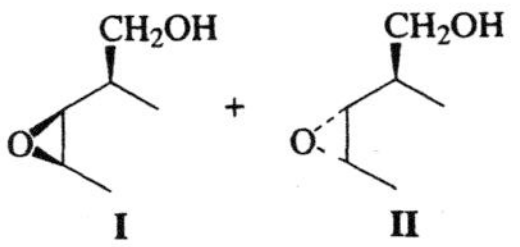

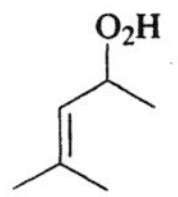

O_2H

TFD, CCl_4/TFP, 0.5 h — **I + II** (>95) **I:II** = 24:76 — 298

DMD, acetone, 5 h — O_2H **I** + O_2H **II** (>95) **I:II** = 69:31 — 298

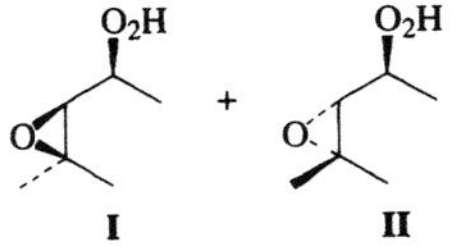

TFD, CCl_4/TFP, 0.5 h — **I + II** (>95) **I:II** = 80:20 — 298

C_7

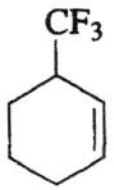

CF_3

DMD, solvent, rt, 60 min — CF_3 **I** + CF_3 **II** — 67

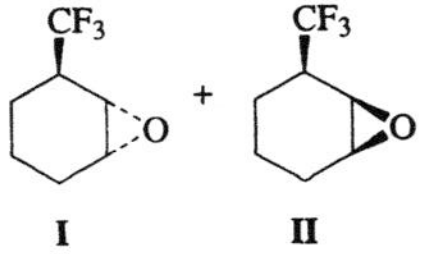

Solvent	% Convn	**I:II**
acetone	70	90:10
CH_2Cl_2/acetone (1:1)	73	94:6

TABLE 4. DIASTEREOSELECTIVE EPOXIDATIONS OF OLEFINS BY DIOXIRANES (*Continued*)

Substrate	Conditions	Product(s) and Yield(s) (%), Diastereomeric Ratio (dr)	Refs.
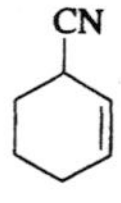 CN	DMD, solvent, rt, 60 min	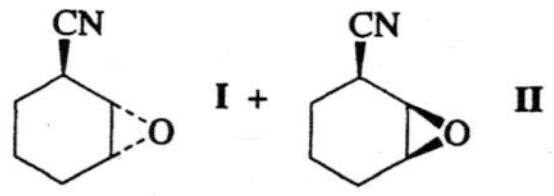 CN I + CN II	67
		Solvent / % Convn / I:II acetone 78 51:49 CH_2Cl_2/acetone (1:1) 74 57:43	
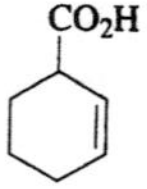 CO_2H	DMD, solvent, rt, 60 min	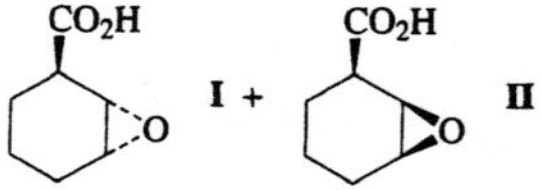 CO_2H I + CO_2H II	67
		Solvent / % Convn / I:II acetone 82 84:16 CH_2Cl_2/acetone (9:1) 100 80:20 $CHCl_3$/acetone (9:1) 54 87:13 CCl_4/acetone (9:1) 73 84:16	
C_{7-14} 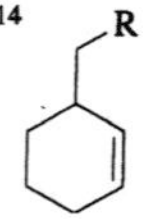 R	DMD, solvent, rt	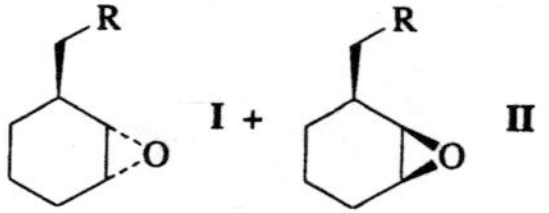 R I + R II	67

R	Solvent	Time	% Convn	**I:II**
Br	acetone	60 min	100	38:62
Br	CH_2Cl_2/acetone (1:1)	60 min	100	40:60
Br	MeOH/acetone (1:1)	90 min	100	45:55
Br	CCl_4/acetone (9:1)	5 h	100	27:73
OH	acetone	60 min	100	44:56
OH	CH_2Cl_2/acetone (9:1)	60 min	100	30:70
OH	CH_2Cl_2/acetone (95:5)	60 min	68	28:72
OH	$CHCl_3$/acetone (9:1)	60 min	100	37:63
OH	CCl_4/acetone (9:1)	60 min	92	33:67
OH	CCl_4/acetone (95:5)	60 min	84	26:74
OH	MeOH/acetone (9:1)	120 min	72	55:45
OAc	acetone	60 min	100	40:60
OAc	CH_2Cl_2/acetone (9:1)	60 min	100	40:60
CO_2Me	acetone	15 min	100	37:63
CO_2Me	CH_2Cl_2/acetone (9:1)	60 min	96	39:61
CO_2Me	CCl_4/acetone (9:1)	60 min	91	29:71
CO_2Et	acetone	15 min	100	35:65
CO_2Et	CH_2Cl_2/acetone (9:1)	60 min	87	38:62
CO_2Et	CCl_4/acetone (9:1)	60 min	89	27:73
CO_2Et	CCl_4/acetone (95:5)	60 min	80	26:74
CO_2Et	MeOH/acetone (9:1)	60 min	84	41:59
NHCOPh	acetone	15 min	100	36:64
NHCOPh	CCl_4/acetone (9:1)	60 min	100	18:82
NHCOPh	CCl_4/acetone (95:5)	60 min	100	13:87

TABLE 4. DIASTEREOSELECTIVE EPOXIDATIONS OF OLEFINS BY DIOXIRANES (*Continued*)

Substrate	Conditions	Product(s) and Yield(s) (%), Diastereomer Ratio (dr)	Refs.
C_7 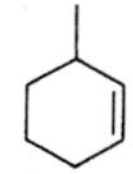	DMD, solvent, rt, 60 min	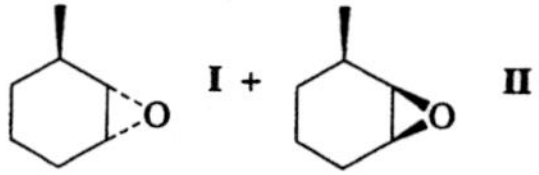 **I** + **II** Solvent / % Convn / **I**:**II** acetone / 100 / 48:52 CH_2Cl_2/acetone (9:1) / 100 / 53:47 CCl_4/acetone (9:1) / 100 / 47:53	67
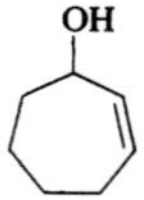 OH	DMD, acetone, rt, 2 h	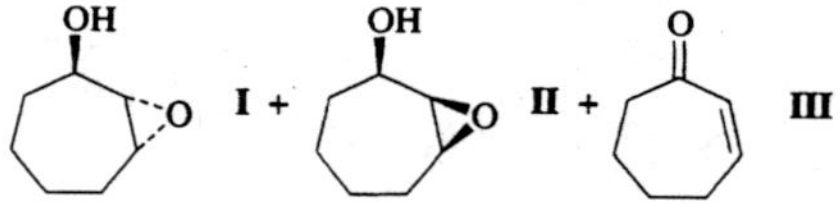 OH **I** + OH **II** + O **III** **I**:**II** = 67:33; (**I** + **II**):**III** = 62:38; **I** + **II** +**III** (—)	83
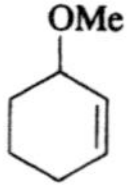 OMe	DMD, acetone, 20°, 2 h	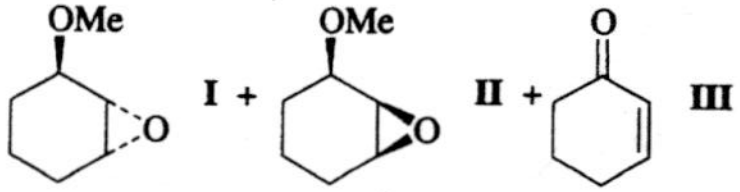 OMe **I** + OMe **II** + O **III** **I**:**II** = 75:25; (**I** + **II**):**III** = 81:19; **I** + **II** +**III** (92)	75
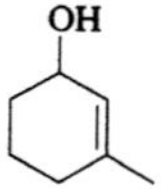 OH	DMD, acetone/CH_2Cl_2, rt, 3 h	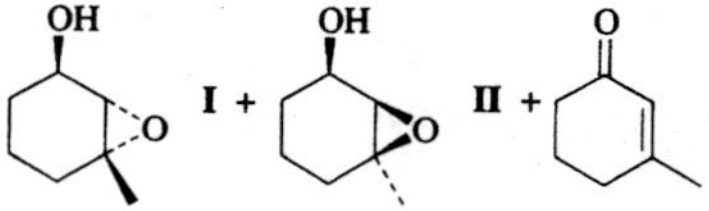 OH **I** + OH **II** + O **III** **I** + **II** (88); **III** (10); **I**:**II** = 30:70; **I** + **II** +**III** (—)	133

DMD, solvent, rt, 60 min — **I + II + III** — 67

Solvent	% Convn	**I:II**	**(I+II):III**
acetone	95	65:35	65:35
CH_2Cl_2/acetone (4:1)	93	39:61	85:15
CH_2Cl_2/acetone (9:1)	96	35:65	89:11
CH_2Cl_2/acetone (95:5)	81	13:87	94:6
CH_2Cl_2/acetone (9:1)	69	82:18	86:14
t-BuOH/acetone (9:1)	85	73:27	88:12
AcOH/acetone (9:1)	100	74:26	89:11
$CHCl_3$/acetone (9:1)	91	17:83	93:7
CCl_4/acetone (9:1)	89	21:79	76:24

Ketone, Oxone®, $NaHCO_3$, CH_3CN/H_2O, Na_2EDTA, rt — **I + II** — 81

Ketone	**I:II**	Ketone	**I:II**
MeCOMe	1.4:1		3.7:1
EtCOEt	—		
$ClCH_2COMe$	3.5:1		
$ClCH_2COCH_2Cl$	5.2:1		
CF_3COMe	9.1:1	Ph, Ph, Me, Me, TfO^-	4.4:1
	2.0:1		
	—	Ph, Ph, Me, Me, TfO^-	8.9:1
Cl	3.9:1		2.9:1
	3.3:1		3.5:1

TABLE 4. DIASTEREOSELECTIVE EPOXIDATIONS OF OLEFINS BY DIOXIRANES (*Continued*)

Substrate	Conditions	Product(s) and Yield(s) (%), Diastereomeric Ratio (dr)	Refs.
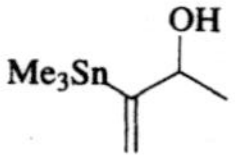 Me_3Sn … OH	DMD, CH_2Cl_2/acetone, rt, 3.5 h	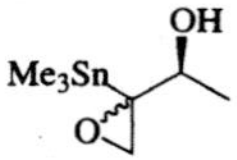 Me_3Sn … OH (23), dr 80:20 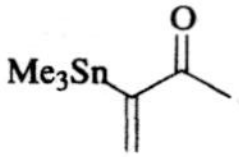 + Me_3Sn … O (77)	74
C_8			
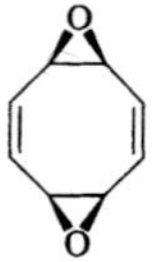 O … O	DMD, acetone, ~20°	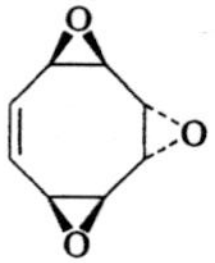 O … O … O (—)	299
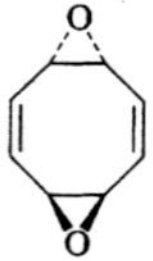 O … O	DMD, acetone, ~20°	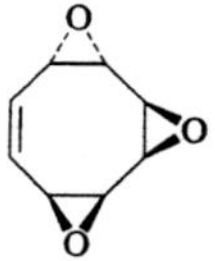 O … O … O (100)	299
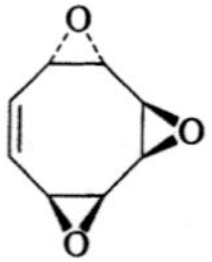 O … O … O	DMD, acetone, ~20°	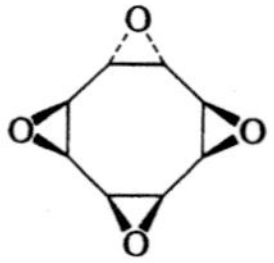 O … O … O … O (100)	299
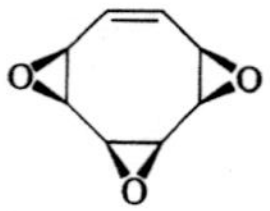 O … O … O	DMD, acetone, ~20°	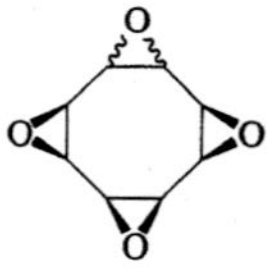 O … O … O … O (—)	299

OAc (cyclohexenyl acetate) → Ketone, Oxone®, $NaHCO_3$, CH_3CN/H_2O, Na_2EDTA, rt → OAc epoxide **I** + OAc epoxide **II** (81)

Ketone	I:II
MeCOMe	2.2:1
EtCOEt	2.7:1
$ClCH_2COMe$	2.2:1
$ClCH_2COCH_2Cl$	1.7:1
CF_3COMe	1.9:1
[cyclohexanone]	2.3:1
[2-methylcyclohexanone]	2.8:1
[2-chlorocyclohexanone] (Cl)	2.7:1
[tetrahydro-4H-pyran-4-one]	2.1:1

Ketone	I:II
[tetrahydrothiopyran-4-one 1,1-dioxide]	2.0:1
[2,6-Ph, N,N-Me piperidin-4-one] TfO⁻	1.5:1
[3,5-dimethyl-2,6-Ph, N,N-Me piperidin-4-one] TfO⁻	2.4:1
[biphenyl-based macrocyclic diester ketone]	1.6:1
[spirocyclohexane macrocyclic diester ketone]	1.4:1

TABLE 4. DIASTEREOSELECTIVE EPOXIDATIONS OF OLEFINS BY DIOXIRANES (*Continued*)

Substrate	Conditions	Product(s) and Yield(s) (%), Diastereomeric Ratio (dr)	Refs.
	DMD, acetone, 20°, 4 h	**I** + **II** (>98) **I**:**II** = 72:28	75
	DMD, acetone/CH_2Cl_2, rt, 2 h	**I** + **II** (95) **I**:**II** = 65:35	157
	DMD, solvent, rt, 60 min	**I** + **II** Solvent / % Convn / **I**:**II** acetone / 86 / 66:34 CH_2Cl_2/acetone (1:1) / 84 / 64:36 CCl_4/acetone (9:1) / 25 / 65:35	300, 67
	DMD, acetone/CH_2Cl_2, Ar, 0 to 5°, 14 h	 (—)	187
	DMD (2.5 equiv), acetone/CH_2Cl_2, Ar, 0 to 5°, 13 h	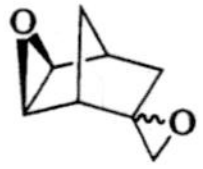 (54) dr 72:28	187

CO$_2$Me

DMD, solvent, rt, 60 min — CO$_2$Me — I + CO$_2$Me — II — 67, 300

Solvent	% Convn	**I:II**
acetone	97	55:45
CH_2Cl_2/acetone (1:1)	95	58:42
CCl_4/acetone (9:1)	56	68:32

DMD, acetone, 20°, 4 h — I + II (95) **I:II** = 72:28 — 133

Methyl(isopropyl)dioxirane, 3-methyl-2-butanone, 20°, 4 h — **I + II** (24) **I:II** = 70:30 — 75

Diethyldioxirane, 3-pentanone, 20°, 4 h — **I + II** (95) **I:II** = 72:28 — 75

Cyclohexanone dioxirane, cyclohexanone, 20°, 4 h — **I + II** (85) **I:II** = 80:20 — 75

DMD, solvent, rt, 60 min — **I + II** (—) — 67

Solvent	% Convn	**I:II**
acetone	100	73:27
CH_2Cl_2/acetone (7:3)	100	77:23

TABLE 4. DIASTEREOSELECTIVE EPOXIDATIONS OF OLEFINS BY DIOXIRANES (*Continued*)

Substrate	Conditions	Product(s) and Yield(s) (%), Diastereomeric Ratio (dr)	Refs.
	Ketone, Oxone®, CH_2Cl_2/MeOH, buffer (pH 11.0), KOH, 18-crown-6, 0 to 5°	**I** + **II** (—)	297
	Ketone, oxygen donor, CH_3CN/phosphate buffer (pH 7.4)	**I** + **II** (—)	301
	DMD, acetone, 20°, 4 h	**I** + **II** (>95) **I**:**II** = 72:28	75
	Oxone®, acetone, $NaHCO_3$, MeOH/H_2O, 20°, 4 h	**I** + **II** (73) **I**:**II** = 74:26	75
	Ketone, Oxone®, $NaHCO_3$, CH_3CN/H_2O, Na_2EDTA, rt	**I** + **II** (—)	81

Ref. 297:

Ketone	Time		I:II
acetone	3.0 h	(56)	86:14
cyclohexanone	3.5 h	(80)	91:9
2,6-dimethylcyclohexanone	3.5 h	(83)	88:12
2-phenylcyclohexanone	4.0 h	(68)	93:7
2-chlorocyclohexanone	3.0 h	(100)	96:4
camphor	18.5 h	(70)	77:23

Ref. 301:

Ketone	Oxidant	% Convn	I:II
$MeCOCO_2Me$	Oxone®	42	46:1
MeCOCOMe	Oxone®	15	100:0
$MeCOCO_2Me$	$NaNO_3$	16	4.8:1
MeCOCOMe	$NaNO_3$	23	100:0

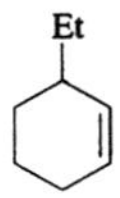

Et-cyclohexene (3-ethylcyclohexene) — DMD, acetone, rt, 60 min → I + II

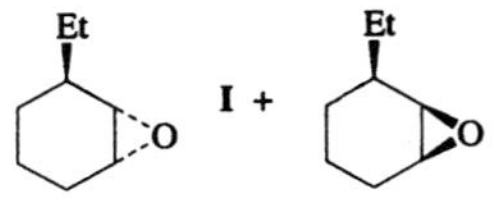

Convn = 100%
I:II = 55:45

67

Ketone	I:II	Ketone	I:II
MeCOMe	3.7:1	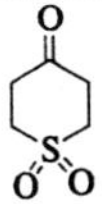 tetrahydrothiopyran-4-one 1,1-dioxide	4.7:1
EtCOEt	—	1,1-dimethyl-2,6-diphenyl-4-oxopiperidinium TfO⁻	3.8:1
$ClCH_2COMe$	4.7:1	1,1,3,5-tetramethyl-2,6-diphenyl-4-oxopiperidinium TfO⁻	18.7:1
$ClCH_2COCH_2Cl$	5.4:1	biphenyl-bridged cyclic diester ketone	4.7:1
CF_3COMe	8.4:1	spirocyclohexane cyclic diester ketone	6.3:1
cyclohexanone	4.4:1		
2-methylcyclohexanone	—		
2-chlorocyclohexanone	5.4:1		
tetrahydropyran-4-one	4.4:1		

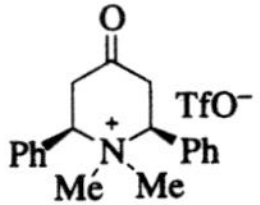

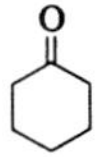

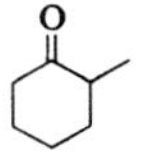

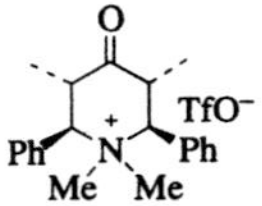

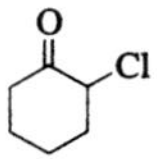

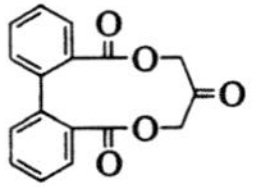

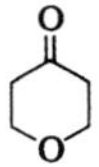

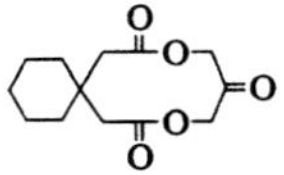

TABLE 4. DIASTEREOSELECTIVE EPOXIDATIONS OF OLEFINS BY DIOXIRANES (*Continued*)

Substrate	Conditions	Product(s) and Yield(s) (%), Diastereomeric Ratio (dr)	Refs.
OMe	DMD, solvent, rt, 60 min	OMe, O **I** + OMe, O **II** + O **III**	67
		Solvent / % Convn / **I**:**II** / (**I**+**II**):**III**	
		acetone / 100 / 95:5 / 87:13	
		CH_2Cl_2/acetone (4:1) / 75 / 95:5 / 92:8	
		CH_2Cl_2/acetone (95:5) / 92 / 95:5 / 91:9	
		CCl_4/acetone (95:5) / 86 / 5:95 / 80:20	
OH	DMD (0.9 equiv), acetone/ CH_2Cl_2, 20°, 0.2 h	OH, O **I** (90) + OH, O **II** (<2) + O, O **III** (<2) **I**:**II** = >98:<2	14
	DMD (2.2 equiv), acetone/ CH_2Cl_2, 20°, 168 h	**I** (58) + **II** (<3) + **III** (39) **I**:**II** = >95:<5	14
	Oxone®, acetone/CH_2Cl_2, phosphate buffer (pH 7.5), 18-crown-6, 6 to 8°, 4 h	**I** + **II** (91) **I**:**II** = 99:1	302

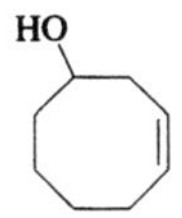

Substrate	Conditions	Product(s) and Yield(s) (%)	Refs.
HO (cyclooct-4-enol)	DMD (0.9 equiv), acetone/ CH_2Cl_2, 20°, 0.2 h	I (38) + II (43) + (6); I:II = 47:53	14
HO (cyclooct-3-enol)	DMD (0.9 equiv), acetone/ CH_2Cl_2, 20°, 0.2 h	I (48) + II (38) + (3); I:II = 56:44	14
CO_2Et	DMD, acetone, 2 h	I + II (>95) I:II = 51:49	298
C_{8-22} R, R	TFP, Oxone®, $MeCN/H_2O$, Na_2EDTA, 0°, 2 to 3 h	I + II (—)	303

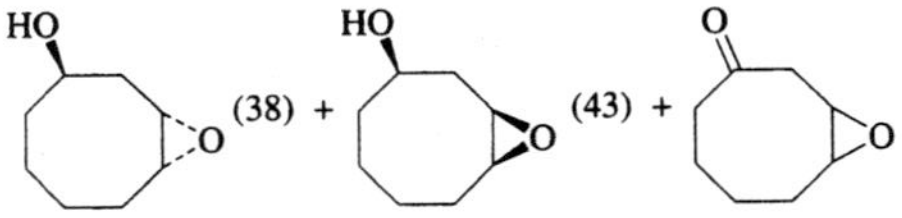

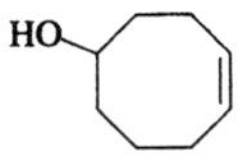

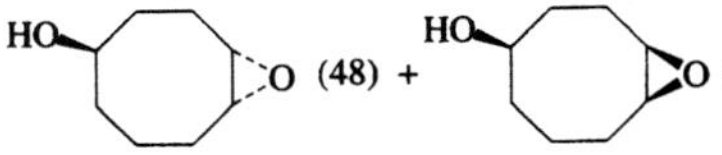

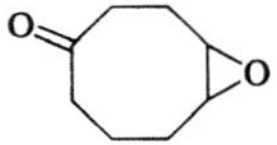

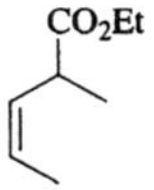

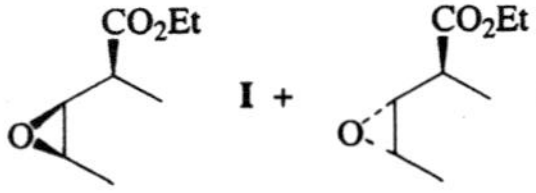

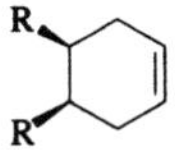

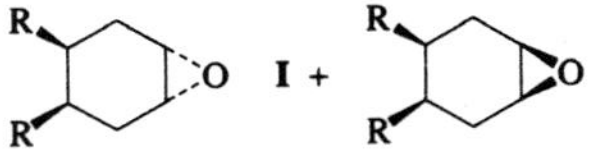

R	I:II
MeO	94:6
MeO_2C	72:28
$AcOCH_2$	90:10
Et_3SiO	98:2
TBDMSO	98:2
$PhCO_2$	81:19
$PhCO_2CH_2$	81:19

TABLE 4. DIASTEREOSELECTIVE EPOXIDATIONS OF OLEFINS BY DIOXIRANES (*Continued*)

Substrate	Conditions	Product(s) and Yield(s) (%), Diastereomeric Ratio (dr)	Refs.
C_9			
	TFP, Oxone®, $NaHCO_3$, CH_3CN/H_2O, (pH 7-7.5), 0 to 1°, 15 min	**I** (—) + **II** (—) **I**:**II** = 67:33	304
	TFP, Oxone®, $NaHCO_3$, CH_3CN/H_2O, (pH 7-7.5), 0 to 1°, 9 h	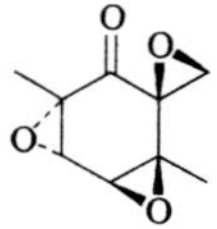 **I** (60) 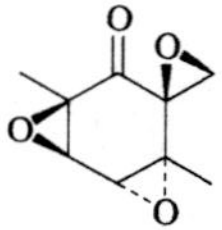 + **II** (30) **I**:**II** = 67:33	304
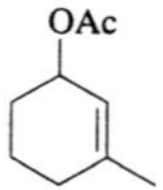 OAc	DMD, acetone/CH_2Cl_2, rt, 1.5 h	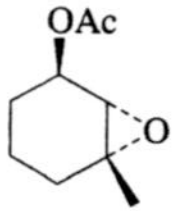 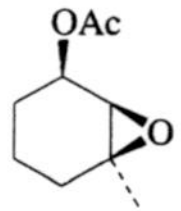 OAc **I** + OAc **II** (90) **I**:**II** = 90:10	133
	DMD, solvent, rt, 60 min	**I** + **II** (—)	67

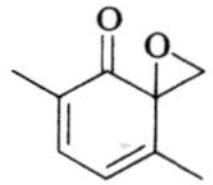

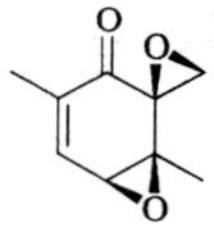

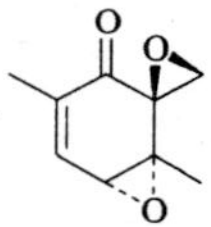

Solvent	% Convn	I:II
acetone	94	87:13
CCl_4/acetone (9:1)	63	88:12

Ketone, Oxone®, CH_3CN/ H_2O, $NaHCO_3$, Na_2EDTA, rt | I + II (—) | 81

Ketone	I:II	Ketone	I:II
MeCOMe	5.9:1		4.7:1
EtCOEt	8.5:1	Ph, Me, Me, Ph, TfO⁻	3.4:1
$ClCH_2COMe$	4.8:1	Ph, Me, Me, Ph, TfO⁻	15.1:1
$ClCH_2COCH_2Cl$	3.5:1		3.0:1
CF_3COMe	6.1:1		2.9:1
	6.6:1		
	7.8:1		
Cl	7.7:1		
	5.0:1		

DMD, acetone, rt, 16 h | (76) + (12) | 305

TABLE 4. DIASTEREOSELECTIVE EPOXIDATIONS OF OLEFINS BY DIOXIRANES (*Continued*)

Substrate	Conditions	Product(s) and Yield(s) (%), Diastereomeric Ratio (dr)	Refs.
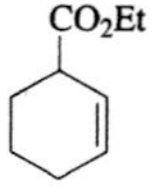 CO_2Et	DMD, solvent, rt, 60 min	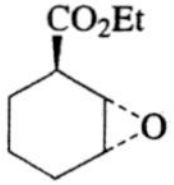 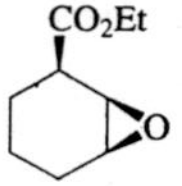 CO_2Et O **I** + CO_2Et O **II**	67
		Solvent / % Convn / **I**:**II** acetone / 82 / 66:34 CH_2Cl_2/acetone (9:1) / 41 / 62:38 CCl_4/acetone (9:1) / 22 / 62:38	
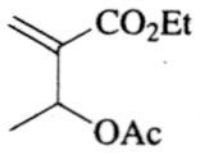 CO_2Et OAc	DMD, acetone, 20°, 3 d	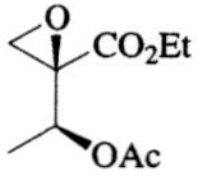 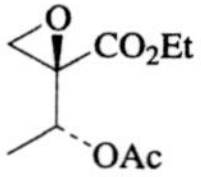 O CO_2Et OAc **I** + O CO_2Et OAc **II** (98) **I**:**II** = 35:65	237
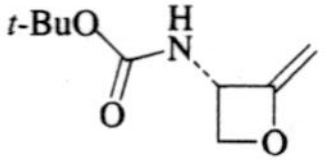 *t*-BuO H N O O	DMD, $CHCl_3$/acetone	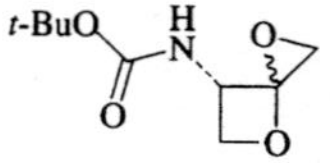 *t*-BuO H N O O O (100) dr = 86:14	203
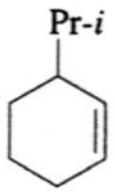 Pr-*i*	DMD, acetone, rt, 2 h	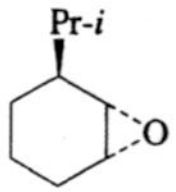 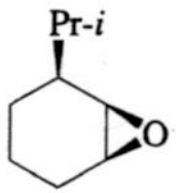 Pr-*i* O **I** + Pr-*i* O **II** (59) **I**:**II** = 78:22	83
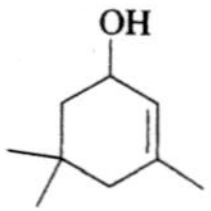 OH	DMD, solvent, ca. 20°	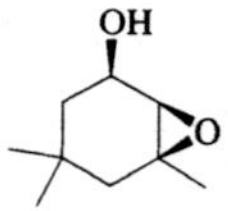 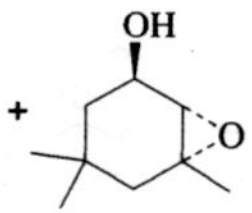 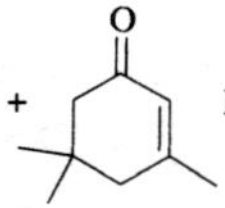 OH O **I** + OH O **II** + O **III**	87

Solvent	Time	% Convn	**I**:**II**	(**I**+**II**):**III**
MeOH/acetone (9:1)	1.5 h	66	62:38	92:8
acetone	0.5 h	>95	83:17	77:23
CCl_4/acetone (9:1)	12 h	>95	96:4	92:8

OMe MeO MeO O

DMD, acetone/CH_2Cl_2, 5°, 90 min — OMe MeO MeO O O (88) — 306

OTMS

DMD, acetone, 20°, 2 to 8 h — OTMS O **I** + OTMS O **II** (—) **I**:**II** = 88:12 — 75

DMD, acetone, 20°, 12 h — **I** + **II** + O **III** (87) **I**:**II** = 88:12 (**I** + **II**):**III** = 90:10 — 157

DMD, solvent, rt, 60 min — **I** + **II** + **III** — 67, 296

Solvent	% Convn	**I**:**II**	(**I**+**II**):**III**
acetone	100	87:13	95:5
CH_2Cl_2/acetone (1:1)	95	89:11	94:6
CCl_4/acetone (9:1)	60	99:1	90:10

TABLE 4. DIASTEREOSELECTIVE EPOXIDATIONS OF OLEFINS BY DIOXIRANES (*Continued*)

Substrate	Conditions	Product(s) and Yield(s) (%), Diastereomeric Ratio (dr)	Refs.
C_{9-18} 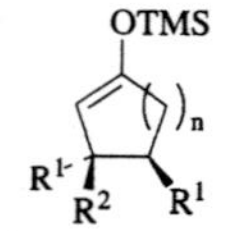 OTMS, $R^{1\prime}$, R^2, R^1, n	DMD,THF/acetone, –78°	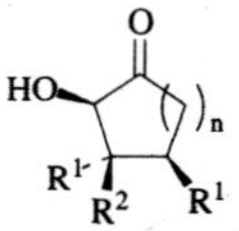 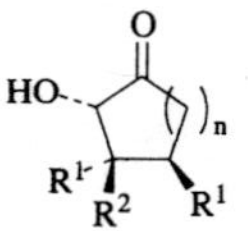 HO, O, n, $R^{1\prime}$, R^2, R^1 **I** + HO, O, n, $R^{1\prime}$, R^2, R^1 **II**	307

n	R^1	R^2	Yield (%)	I:II
1	H	Me	(92)	—
1	H	CH_2=CH	(90)	—
1	Me	Me	(79)	—
1	Me	CH_2=CH	(87)	100:0
1	Me	Ph	(—)	67:33
1	H	Bn	(92)	—
2	Me	Bn	(75)	100:0

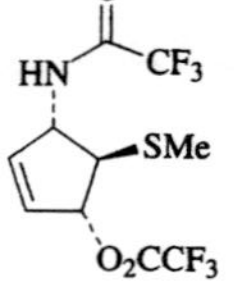

Substrate	Conditions	Product(s) and Yield(s) (%), Diastereomeric Ratio (dr)	Refs.
C_{10} O, HN, CF_3, SMe, O_2CCF_3	DMD, acetone, Ar, rt	O, HN, CF_3, SO_2Me, O, OH (91)	308
	DMD, acetone, 25°, 3 h	 (36) + 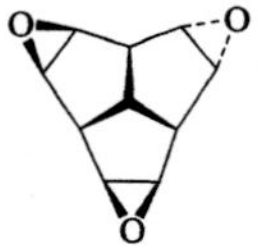 (49)	309
C_{10-13} Ph, R, O	DMD, $CHCl_3$/acetone	Ph, R, O, O	203

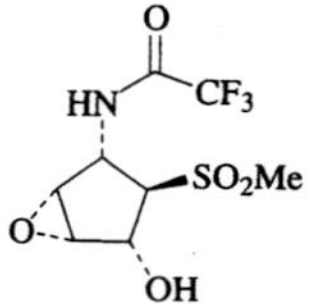

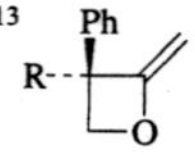

R	Yield (%)	dr
H	(99)	93:7
Me	(100)	86:14
CH_2=$CHCH_2$	(97)	75:25

C$_{10}$

Reactant	Conditions	Product(s) and Yield(s) (%)	Refs.
	DMD, acetone/CH_2Cl_2, Ar, 0°, 5 h	(35) + (40)	187
	DMD (2.5 equiv), acetone/CH_2Cl_2, Ar, 0°, 5 h	(~100)	187
	DMD, acetone/CH_2Cl_2, Ar, 0 to 5°, 14 h	(45)	187
	DMD (2.5 equiv), acetone/CH_2Cl_2, Ar, 0 to 5°, 14 h	I (51)	187
	DMD, acetone/CH_2Cl_2, Ar, 0°, 3 h	I (~100)	187
OH, Ph, R^1, R^2	DMD, acetone, rt, 15 h (95% convn)	Ph, O, R^1, R^2 I + Ph, OH, O, R^1, R^2 II + Ph, OH, O, R^1, R^2 III (—)	310

R^1	R^2	I:(II+III)	II: III
H	Me	65:35	45:55
Me	H	<5:>95	80:20

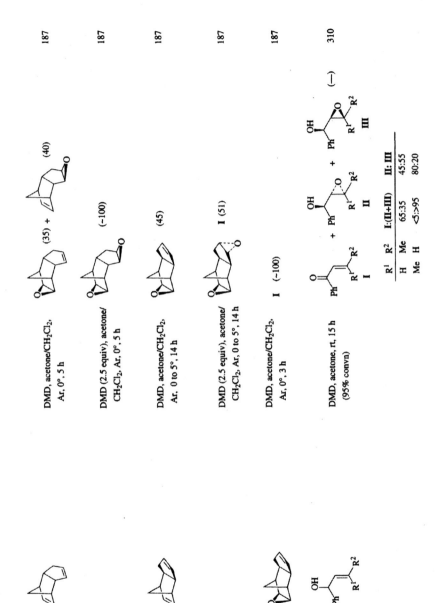

TABLE 4. DIASTEREOSELECTIVE EPOXIDATIONS OF OLEFINS BY DIOXIRANES (*Continued*)

Substrate	Conditions	Product(s) and Yield(s) (%), Diastereomeric Ratio (dr)	Refs.
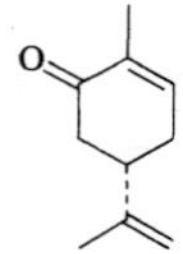	DMD, acetone, 20°, 2 to 8 h	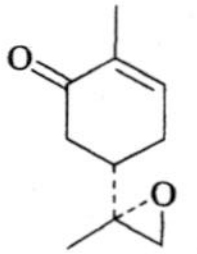 I 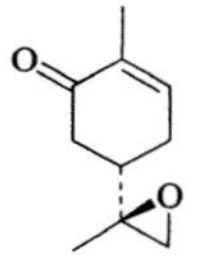 + II (>95) **I**:**II** =55:45	75
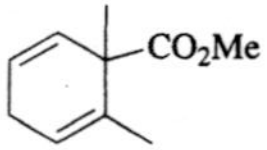 CO_2Me	Acetone, Oxone®, phosphate buffer (pH 7.2)	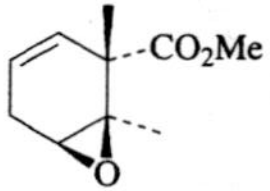 CO_2Me (50) + CO_2Me (32)	84
MeO_2C	DMD, acetone	MeO_2C (50) + MeO_2C (32) **I**:**II** = 61:39 I II	311
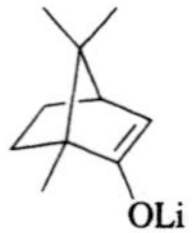 OLi	1. DMD, acetone/THF, −78°, 1.8 min 2. NH_4F, H_2O, 20°	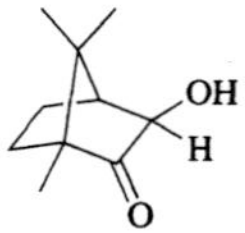 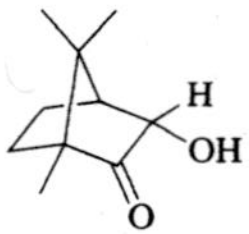 OH H + H OH (>95) **I**:**II** = 76:24 I II	312
C_{10-20} OML_3	1. DMD, THF/acetone, −78°, 1 min 2. NH_4F, H_2O, rt, 1 to 12 h	**I** + **II**	97, 210

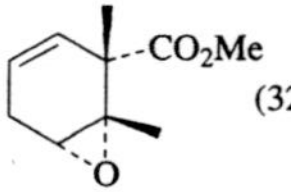

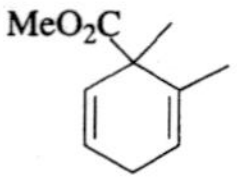

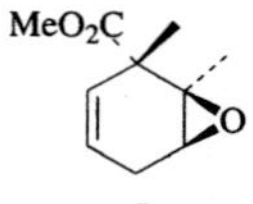

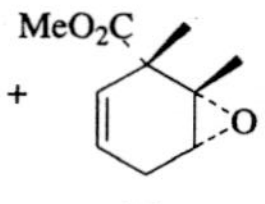

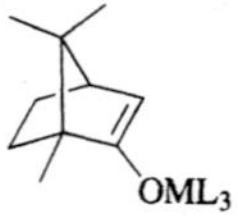

M	L_3		I:II
Na	—	(57)	67:33
Li	—	(70)	75:25
Si	Me_3	(91)	93:07
Ti	Cp_2Cl	(67)	92:08

Substrate	Conditions	Product(s) and Ratio	Refs.
C_{10} 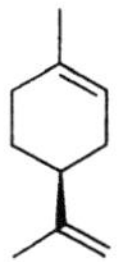	DMD, acetone, 20°, 1 to 3 h	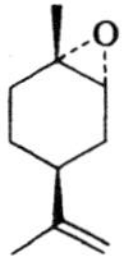 **I** + 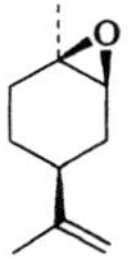 **II** + 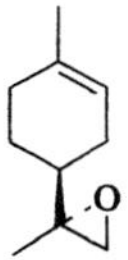 **III** + 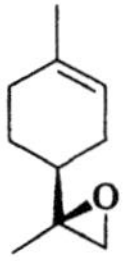 **IV** **I**:**II** = 67:33; **III**:**IV** = 50:50; (**I** + **II**):(**III** + **IV**) = 80:20	157, 75
	DMD, acetone, 20°, 1 to 3 h	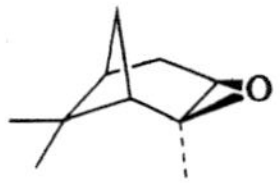 **I** + 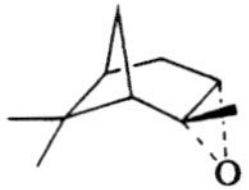 **II** **I**:**II** = >95:<5	75
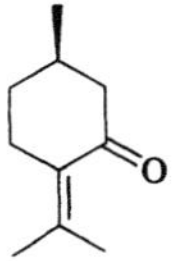	DMD, acetone, 20°, 2 to 8 h	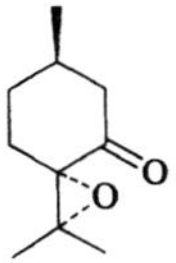 **I** + 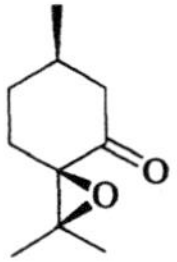 **II** **I**:**II** = 65:35	75
	DMD, acetone, 20°, 1 to 3 h	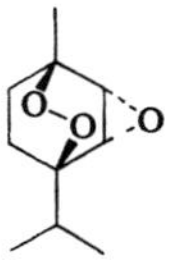 **I** + 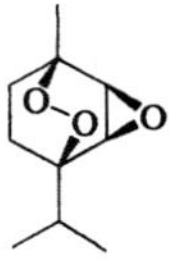 **II** (—) **I**:**II** = >95:<5	75

TABLE 4. DIASTEREOSELECTIVE EPOXIDATIONS OF OLEFINS BY DIOXIRANES (*Continued*)

Substrate	Conditions	Product(s) and Yield(s) (%), Diastereomeric Ratio (dr)	Refs.
C_{10-20} 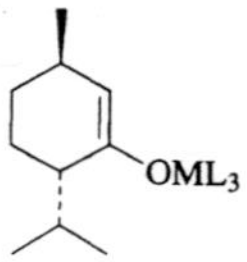 OML$_3$	1. DMD, THF/acetone, –78°, 1 min 2. NH_4F/H_2O, rt, 1 to 12 h	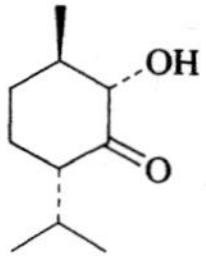 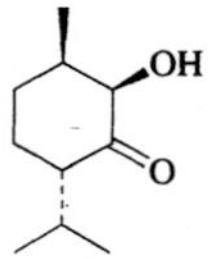 OH, O **I** + OH, O **II** M / L_3 / / I:II Na / — / (63) / 67:33 Si / Me_3 / (97) / 76:24 Ti / Cp_2Cl / (54) / 96:04	97
C_{10} 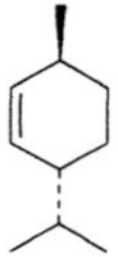	DMD, acetone, 20°, 4 h	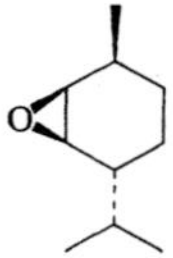 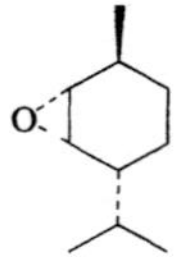 O **I** + O **II** (>95) **I**:**II** = 78:22	75
	DMD, acetone, –78°, 4 h	**I** + **II** (>95) **I**:**II** = 84:16	75
	Methyl(isopropyl)dioxirane, 3-methylbutanone, 20°, 24 h	**I** + **II** (19) **I**:**II** = 81:19	75
	Diethyldioxirane, 3-pentanone, 20°, 24 h	**I** + **II** (58) **I**:**II** = 82:18	75
	Cyclohexanone dioxirane, cyclohexanone, 20°, 24 h	**I** + **II** (14) **I**:**II** = 80:20	75

Bu-*i*

DMD, solvent, rt

Bu-*i* I + Bu-*i* II (—)

67

Solvent	Time	% Convn	I:II
acetone	60 min	100	55:45
CH_2Cl_2/acetone (9:1)	60 min	100	60:40
CCl_4/acetone (9:1)	120 min	100	54:46

Bu-*t*

DMD, solvent, rt, 60 min

Bu-*t* I + Bu-*t* II

67

Solvent	% Convn	I:II
acetone	94	95:5
CH_2Cl_2/acetone (1:1)	96	96:4

OH *t*-Bu

DMD, solvent, ~20°

OH *t*-Bu I + OH *t*-Bu II + O *t*-Bu III

87

Solvent	Time	% Convn	I:II	(I+II):III
acetone	10 h	>95	30:70	58:42
MeOH/acetone (9:1)	2 h	92	29:71	85:15
CCl_4/acetone (9:1)	8 h	>95	58:42	67:33

TABLE 4. DIASTEREOSELECTIVE EPOXIDATIONS OF OLEFINS BY DIOXIRANES (*Continued*)

Substrate	Conditions	Product(s) and Yield(s) (%), Diastereomeric Ratio (dr)	Refs.
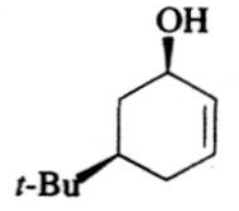 OH; t-Bu	DMD, solvent, ~20°	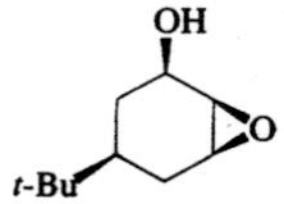 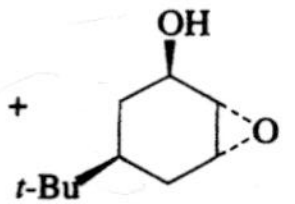 OH; t-Bu; O **IV** + OH; t-Bu; O **V** + **III**	87
		Solvent / Time / % Convn / **IV**:**V** / (**IV**+**V**):**III** acetone / 4 h / >95 / 60:40 / 43:57 MeOH/acetone (9:1) / 2 h / 60 / 38:62 / 80:20 CCl_4/acetone (9:1) / 12 h / >95 / 82:18 / 57:43	
Bu-t	DMD, acetone, 8 h	Bu-t; O **I** + Bu-t; O **II** (82) **I**:**II** = >95:5	298
OTMS	DMD, solvent, rt, 60 min	OTMS; O **I** + OTMS; O **II** + O **III**	67
		Solvent / % Convn / **I**:**II** / (**I**+**II**):**III** acetone / 100 / 95:5 / 97:3 CCl_4/acetone (9:1) / 86 / 99:1 / 96:4	
C_{11} O; O	TFP, Oxone®, $NaHCO_3$, CH_3CN/ H_2O, (pH 7-7.5), 0 to 1°, 20 min	O; O; O **I** (59) + O; O; O **II** (39) **I**:**II** = 60:40	304

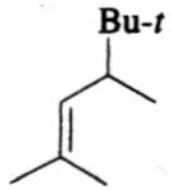

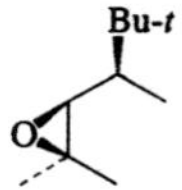

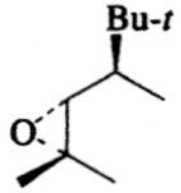

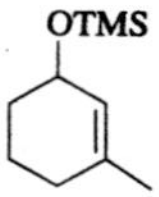

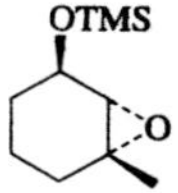

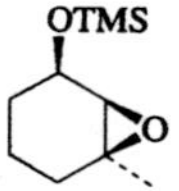

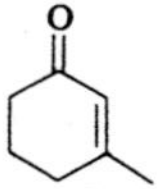

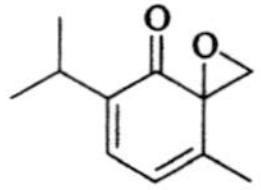

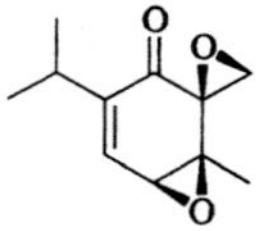

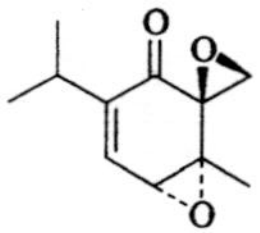

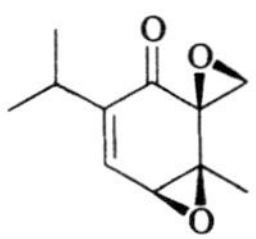

Substrate	Conditions	Product(s) and Yield(s) (%)	Refs.
	Isopropyl pyruvate, Oxone®, $NaHCO_3$, CH_3CN/H_2O (pH 7-7.5), rt, 24 h	(96)	304
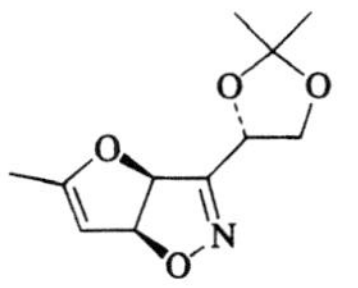	DMD, acetone/CH_2Cl_2, 0°, 3 h	(100)	313
C_{11-14}	DMD, acetone/CH_2Cl_2, 0°, 1 h	I + II (—)	82

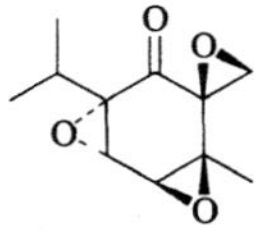

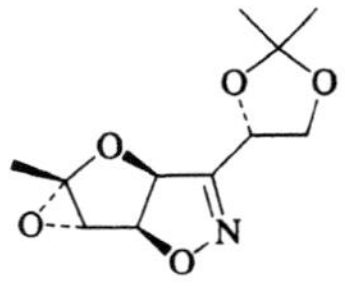

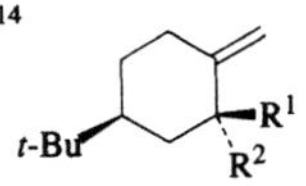

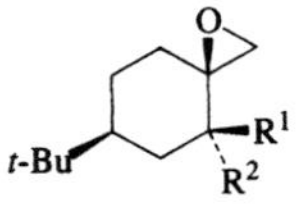

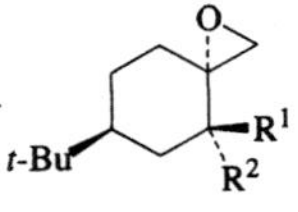

R^1	R^2	I:II
H	H	31:69
Me	H	23:77
H	Me	55:45
Me	Me	17:83
MeO	Me	31:69
Me	MeO	62:38
—$(CH_2)_3O$—		—
—$(CH_2)_3O$—		—
Me	Et	70:30
Et	Me	29:71

TABLE 4. DIASTEREOSELECTIVE EPOXIDATIONS OF OLEFINS BY DIOXIRANES (*Continued*)

	Substrate	Conditions	Product(s) and Yield(s) (%), Diastereomeric Ratio (dr)	Refs.
C_{11} 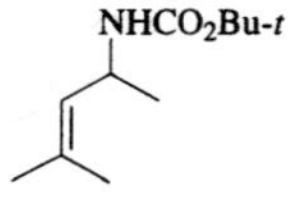	$NHCO_2Bu$-*t*	DMD, acetone, 7 h	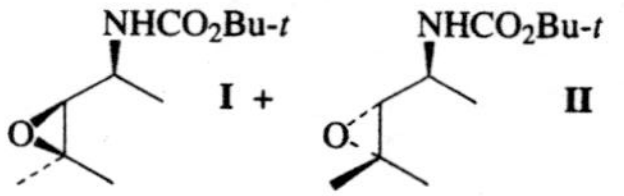 $NHCO_2Bu$-*t* **I** + $NHCO_2Bu$-*t* **II** (90) **I**:**II** = 62:38	298
C_{11-22} 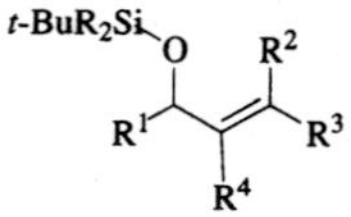	*t*-BuR_2Si O, R^1, R^2, R^3, R^4	Ketone, Oxone®, CH_2Cl_2/ MeOH, buffer (pH 11), rt, 6 h; 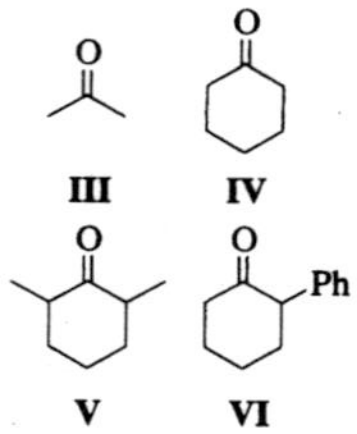 **III**, **IV**, **V**, **VI** (Ph)	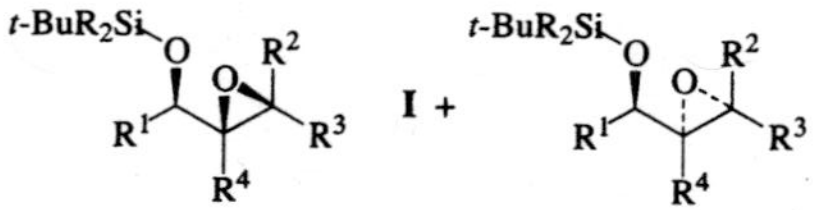 *t*-BuR_2Si O, R^1, R^2, R^3, R^4 **I** + *t*-BuR_2Si O, R^1, R^2, R^3, R^4 **II**	314
C_{12} 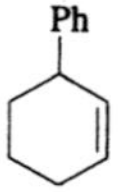	Ph	DMD, solvent, rt, 60 min	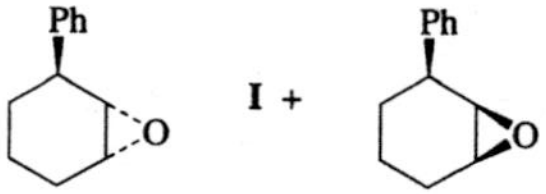 Ph **I** + Ph **II**	67

R	R^1	R^2	R^3	R^4	Ketone		I:II
Me	Me	H	Me	H	**III**	(54)	43:57
Me	Me	H	Me	H	**IV**	(84)	42:58
Me	Me	H	Me	H	**V**	(50)	29:71
Me	Me	H	Me	H	**VI**	(84)	22:78
Me	Me	H	Me	Me	**V**	(24)	58:22
Me	Me	H	Me	Me	**VI**	(13)	61:39
Me	Me	Me	Me	H	**IV**	(81)	32:68
Me	Me	Me	Me	H	**V**	(54)	30:70
Me	Me	Me	Me	H	**VI**	(99)	11:89
Me	*n*-C_5H_{11}	H	H	H	**V**	(21)	26:74
Me	*n*-C_5H_{11}	H	H	H	**VI**	(28)	25:75
Ph	Me	Me	Me	H	**V**	(28)	10:90
Ph	Me	Me	Me	H	**VI**	(94)	2:98

C_{12-22}

OML$_3$ / Bn (cyclopentenyl enolate)

1. DMD, THF/acetone, –78°, 1 min
2. NH_4F/H_2O, rt, 1 to 12 h

HO / Bn cyclopentanone **I** + HO / Bn cyclopentanone **II** — 97

Solvent	% Convn	I:II
acetone	100	82:18
CCl_4/acetone (9:1)	70	85:15
MeOH/acetone (1:1)	90	83:17

M	L_3		I:II
Na	—	(37)	53:47
Si	Me_3	(96)	35:65
Ti	Cp_2Cl	(50)	83:17

C_{12-13}

R^2, R^1, H, Ph (alkene)

DMD, acetone/CH_2Cl_2, –20°

R^2, R^1, O, H, Ph (epoxide) — 53

R^1	R^2	Time		dr
Me	CO_2H	20 h	(96)	70:30
Me	CO_2Me	16 h	(98)	70:30
CO_2Me	Me	20 h	(99)	62:38

H, N, Ts (cyclopentenyl sulfonamide)

TFP, Oxone®, $MeCN/H_2O$, Na_2EDTA, 0°, 2 to 3 h

H, N, Ts, O **I** + H, N, Ts, O **II** (—) — 303

I:**II** = 60:40

TABLE 4. DIASTEREOSELECTIVE EPOXIDATIONS OF OLEFINS BY DIOXIRANES (*Continued*)

Substrate	Conditions	Product(s) and Yield(s) (%), Diastereomeric Ratio (dr)	Refs.
C_{12-14} 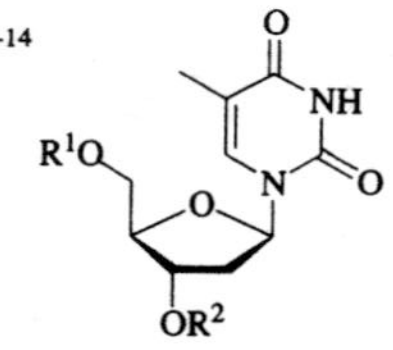	DMD, catalyst, CH_2Cl_2, 25°	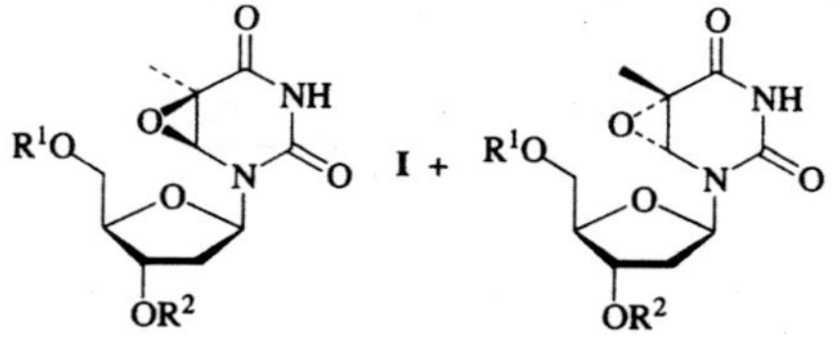 I + II	315
C_{12} 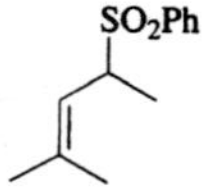	DMD, acetone, 8 h	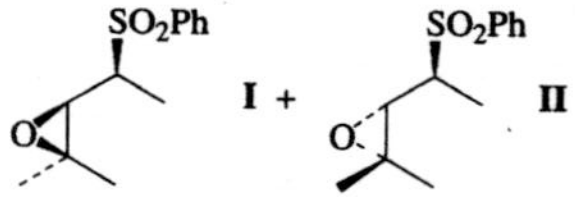 I + II (87) I:II = >95:5	298
C_{12-27} 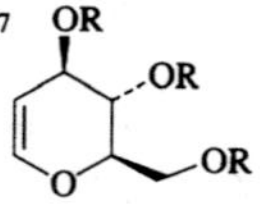	DMD, acetone/CH_2Cl_2, 0°, 1 h	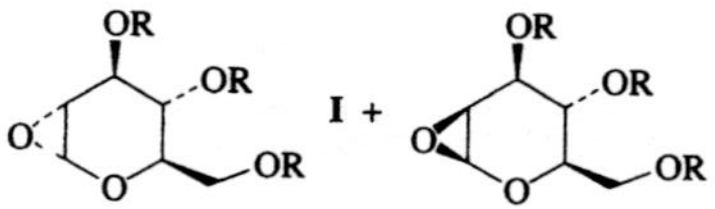 I + II	91

R^1	R^2	Catalyst		I:II
H	Ac	Mn(TDMPP)Cl[a]	(40)	50:50
H	Ac	Mn(Cl16TDMPP)Cl[b]	(57)	90:10
Ac	H	Mn(TDMPP)Cl[a]	(51)	50:50
Ac	H	Mn(Cl16TDMPP)Cl[b]	(55)	17:83
Ac	H	Mn(Cl8TDMPP)Cl[c]	(58)	67:33
Ac	Ac	Mn(TDMPP)Cl[a]	(53)	50:50
Ac	Ac	Mn(Cl16TDMPP)Cl[b]	(72)	60:40

R		I:II
Ac	(—)	mixture
Bn	(99)	~95:5
TBDMS	(100)	100:0

Reactant	Conditions	Product(s) and Yield(s) (%)	Refs.
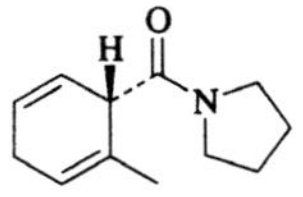	Acetone, Oxone®, phosphate buffer (pH 7.2), 0°	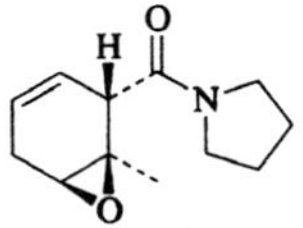 I (39) 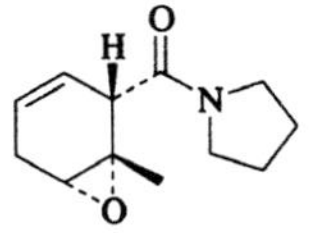 + II (27)	275
	DMD (excess), acetone/ CH_2Cl_2, Ar, 0° to rt, 5 h	 I (~100)	187
	DMD, acetone/CH_2Cl_2, Ar, 0°, 3 h	I (~100)	187
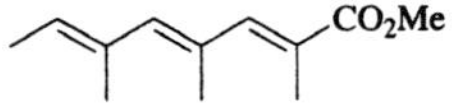 CO_2Me	DMD, acetone, 25°, 18 h	 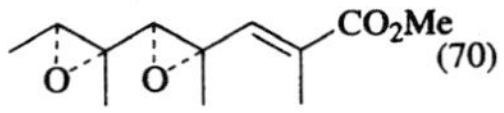 CO_2Me I + CO_2Me II (70) I:II = 9:4	86
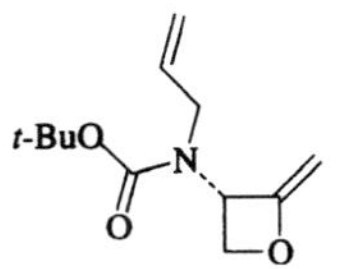 *t*-BuO, N, O	DMD, $CHCl_3$/acetone	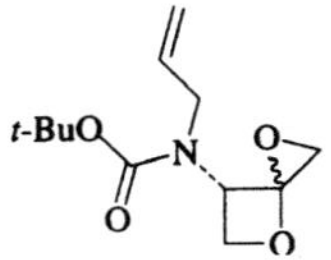 *t*-BuO, N, O (90) dr >95:<5	203

TABLE 4. DIASTEREOSELECTIVE EPOXIDATIONS OF OLEFINS BY DIOXIRANES (*Continued*)

Substrate	Conditions	Product(s) and Yield(s) (%), Diastereomeric Ratio (dr)	Refs.
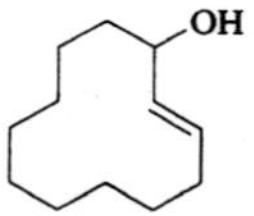 OH	Acetone, Oxone®, CH_2Cl_2, phosphate buffer (pH 7.5), 18-crown-6, 6 to 8°, 4 h	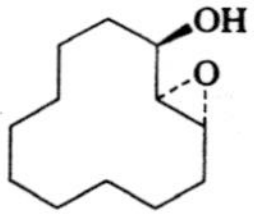 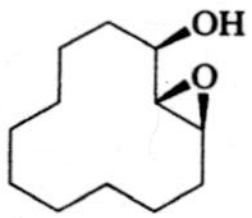 OH, O **I** + OH, O **II** (86) **I:II** = 71:29	302
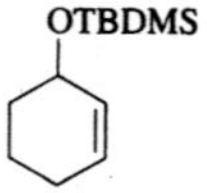 OTBDMS	Ketone, Oxone®, CH_2Cl_2/MeOH, buffer (pH 11.0), KOH, rt	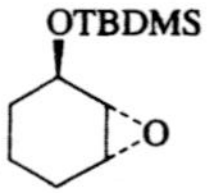 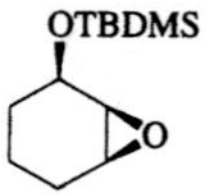 OTBDMS, O **I** + OTBDMS, O **II**	297
	Ketone, Oxone®, $NaHCO_3$, CH_3CN/H_2O, Na_2EDTA, rt	**I** + **II**	81

Ketone	Time		I:II
acetone	3.5 h	(40)	87:13
2-methylcyclohexanone	1.5 h	(77)	90:10
2,6-dimethylcyclohexan-one	3.0 h	(65)	93:7

Ketone	I:II	Ketone	I:II
MeCOMe	4.8:1	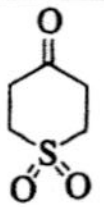 O, S, O, O	3.7:1
EtCOEt	6.3:1		
$ClCH_2COMe$	4.2:1		
$ClCH_2COCH_2Cl$	3.3:1		
CF_3COMe	5.6:1		
 O	5.6:1	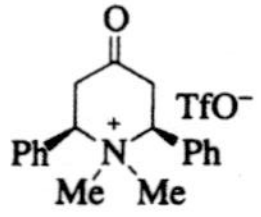 O, TfO⁻, Ph, N⁺, Ph, Me, Me	2.8:1

Ketone	I:II	Ketone	I:II
O, Me	7.0:1	O, TfO⁻, Ph, N⁺, Me, Me, Ph	4.6:1
O, Cl	6.9:1	O, O, O, O, O	2.1:1
O, O	4.6:1	O, O, O, O, O	3.1:1

C_{13} BzO, O, O — DMD, acetone — BzO, O, O, O **I** + BzO, O, O, O **II** (~100) **I:II** = 90:10 — 316

$C_{13\text{-}15}$ R, N, N, O, O, N—Ph — DMD, acetone/solvent — O, R, N, N, O, O, N—Ph **I** + O, R, N, N, O, O, N—Ph **II** — 317

R	Solvent	Time		I:II
CH_2OH	$CHCl_3$	24 h	(100)	50:50
CO_2Me	CH_2Cl_2	3 d	(96)	75:25
CH_2OAc	CH_2Cl_2	3 d	(85)	50:50

TABLE 4. DIASTEREOSELECTIVE EPOXIDATIONS OF OLEFINS BY DIOXIRANES (*Continued*)

Substrate	Conditions	Product(s) and Yield(s) (%), Diastereomeric Ratio (dr)	Refs.
	Ethylmethyldioxirane, butan-2-one, 24 h	R = CH_2OH, (66) **I**:**II** = 50:50	317
	TFP, Oxone®, $NaHCO_3$, CH_3CN/H_2O, 0°, 20 h	R = CH_2OH, (79-100) **I**:**II** = 75:25	317
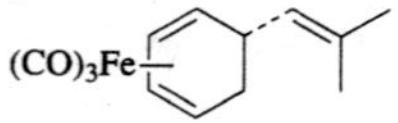 $(CO)_3Fe$	DMD, acetone, 56°	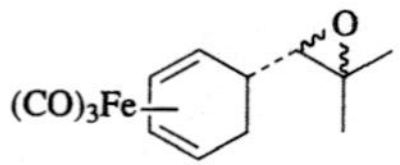 $(CO)_3Fe$ (58) dr 59:41	318
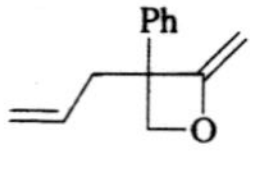 Ph	DMD, $CHCl_3$/acetone	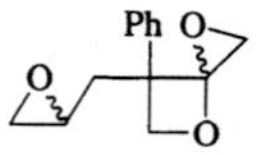 Ph (100) dr 33:33:17:17	203
NHCOPh	DMD, solvent, rt, 60 min	NHCOPh **I** + NHCOPh **II**	67
		Solvent / % Convn / **I** : **II** acetone 100 19:81 CH_2Cl_2/acetone (9:1) 100 4:96 CCl_4/acetone (9:1) 100 3:97 MeOH/acetone (9:1) 64 26:74	
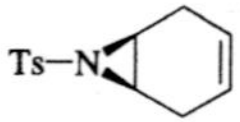 Ts—N	TFP, Oxone®, MeCN/H_2O, Na_2EDTA, 0°, 2 to 3 h	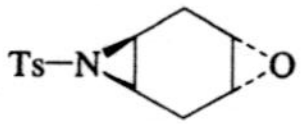  Ts—N **I** + Ts—N **II** (—) **I**:**II** = 91:9	303

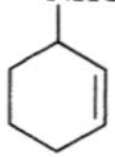

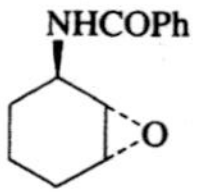

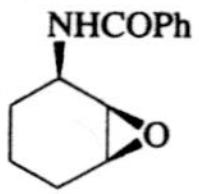

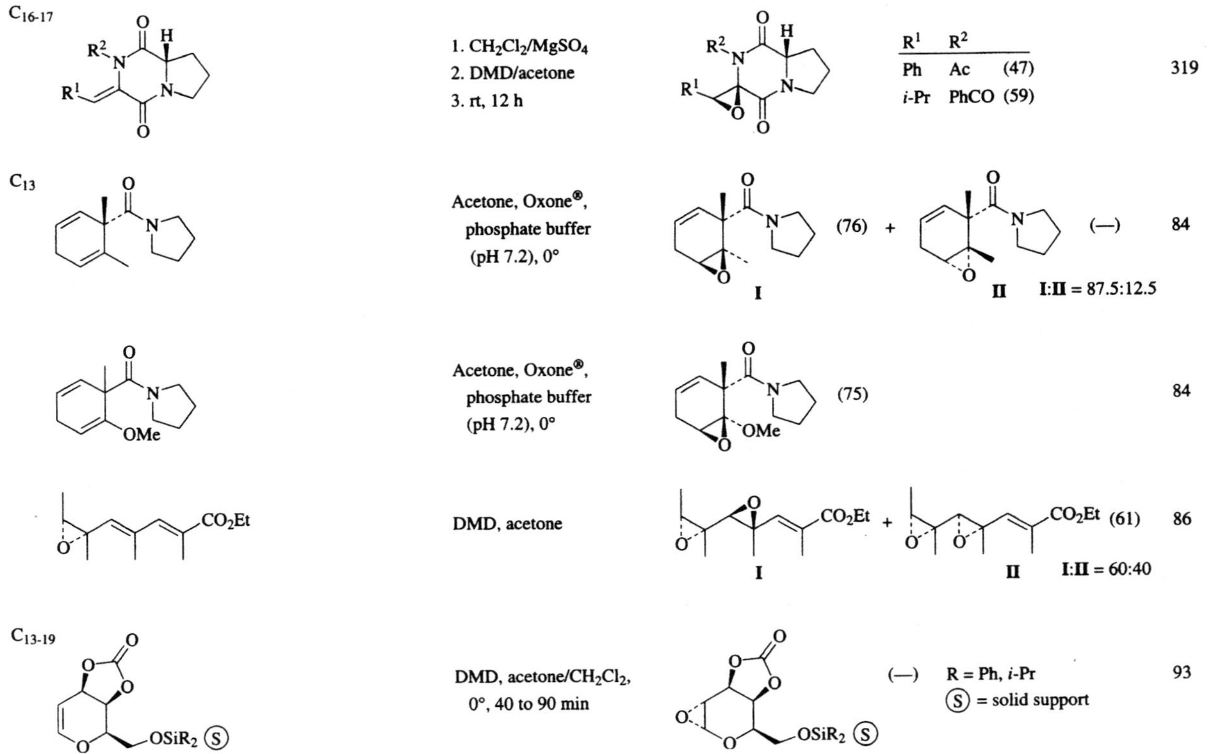

C16-17
1. CH2Cl2/MgSO4
2. DMD/acetone
3. rt, 12 h
R1 R2
Ph Ac (47)
i-Pr PhCO (59)
319
C13
Acetone, Oxone®,
phosphate buffer
(pH 7.2), 0°
(76) +
(—)
84
I
II
I:II = 87.5:12.5
OMe
Acetone, Oxone®,
phosphate buffer
(pH 7.2), 0°
(75)
84
CO2Et
DMD, acetone
CO2Et +
CO2Et (61)
86
I
II
I:II = 60:40
C13-19
OSiR2 S
DMD, acetone/CH2Cl2,
0°, 40 to 90 min
(—)
R = Ph, i-Pr
S = solid support
93

TABLE 4. DIASTEREOSELECTIVE EPOXIDATIONS OF OLEFINS BY DIOXIRANES (*Continued*)

Substrate	Conditions	Product(s) and Yield(s) (%), Diastereomeric Ratio (dr)	Refs.
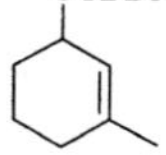 OTBDMS	Ketone, Oxone®, $NaHCO_3$, CH_3CN/H_2O, Na_2EDTA, rt	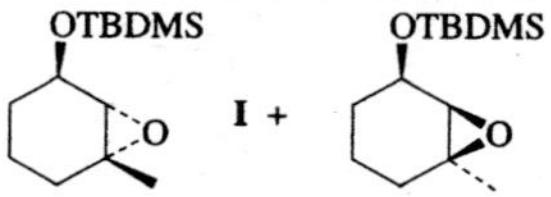 OTBDMS I + OTBDMS II	81
		Ketone — I:II: MeCOMe 13.6:1; EtCOEt 14.1:1; $ClCH_2COMe$ 7.4:1; $ClCH_2COCH_2Cl$ 4.5:1; CF_3COMe 19.3:1; 14.2:1; 16.7:1; Cl 16.6:1; 10.4:1	
		Ketone — I:II: 6.3:1; TfO⁻ Ph N⁺ Me Me Ph 4.3:1; TfO⁻ Ph N⁺ Me Me Ph 19.7:1; 4.3:1; 6.4:1	
TBDMSO	DMD, $CHCl_3$/acetone	TBDMSO (93) dr 60:40	203

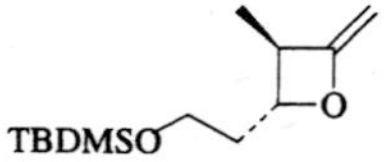

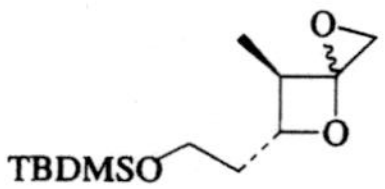

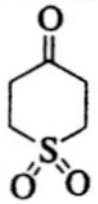

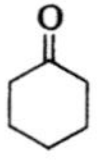

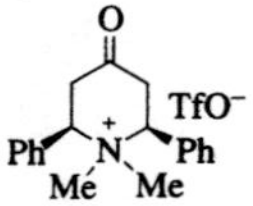

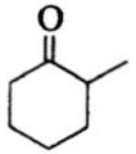

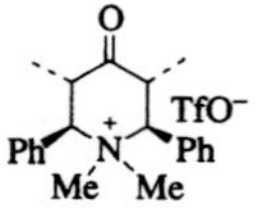

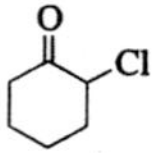

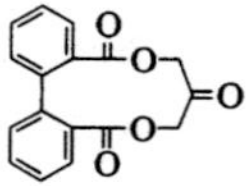

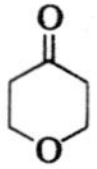

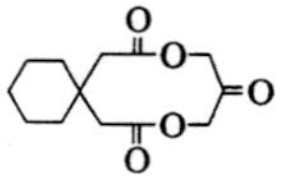

Substrate	Conditions	Product(s) and Yield(s) (%)	Refs.
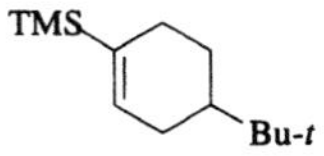 TMS, Bu-t	DMD, acetone, rt, 4 to 10 h	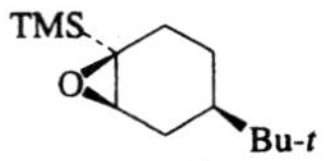 TMS, Bu-t **I** (30) 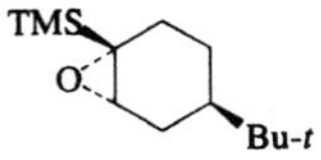 + TMS, Bu-t **II** (50) 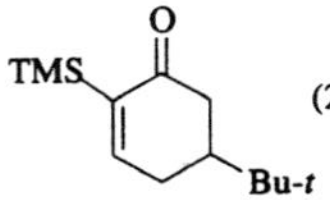 + TMS, Bu-t (20) **I**:**II** = 38:62	74
C_{14} Bn	DMD, CH_2Cl_2/acetone, rt	Bn **I** (—)	287
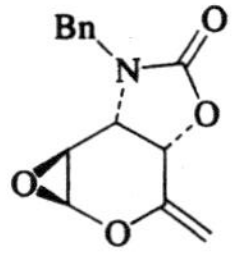 Bn	DMD, acetone/CH_2Cl_2, rt	**I** 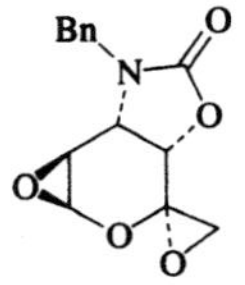 + Bn (—)	287
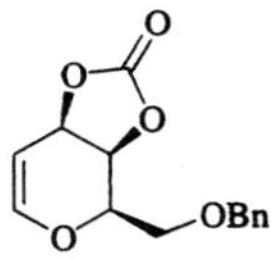 OBn	DMD, acetone/CH_2Cl_2, 0°, 1 h	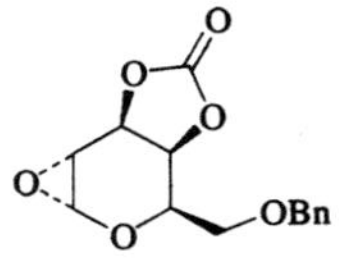 OBn (—)	95
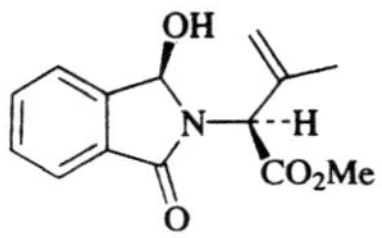 OH, H, CO_2Me	DMD, acetone/CH_2Cl_2, N_2, rt, 4 h	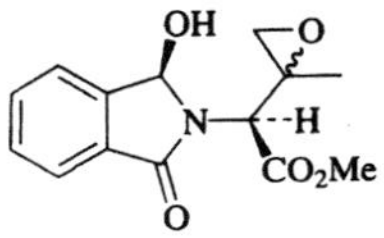 OH, H, CO_2Me (—) dr 50:50	320

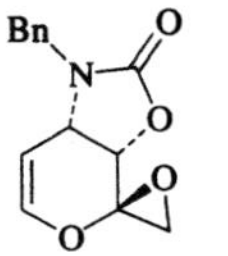

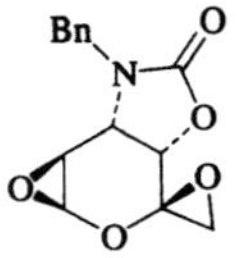

TABLE 4. DIASTEREOSELECTIVE EPOXIDATIONS OF OLEFINS BY DIOXIRANES (*Continued*)

Substrate	Conditions	Product(s) and Yield(s) (%), Diastereomeric Ratio (dr)	Refs.
Ⓟ, O, O, O, O, OH, O, O, O	DMD, CH_2Cl_2, acetone, 0°, 80 min	Ⓟ, O, O, O, O, OH, O, O, O, O (—)	94
O, HN, Ph, OH, SMe, OH	DMD, acetone, Ar, rt, >16 h	O, HN, Ph, OH, SO_2Me, O, OH (43)	308
CO_2Me	DMD, acetone	O, O, CO_2Me I + O, O, CO_2Me II (50-60) **I**:**II** = 75:25	86
	DMD, acetone, 25°, 18 h	O, O, O, CO_2Me + other diastereomers (100) dr 45:33:11:11	86
OTMS, *t*-Bu	DMD, acetone/CH_2Cl_2, 0°, 1 h	O, *t*-Bu, OH I + O, *t*-Bu, OH **II** (85) **I**:**II** = 40:60	82

Substrate	Conditions	Product(s) and Yield(s) (%)	Refs.
C_{15-25} OML_3, Ph	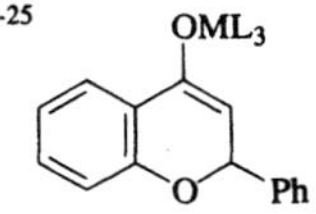 1. DMD, THF/acetone, −78°, 1 min 2. NH_4F/H_2O, rt, 1 to 12 h	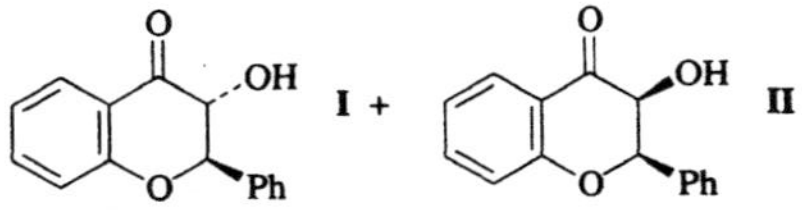 OH, Ph **I** + OH, Ph **II**	97
		M / L_3 / / **I:II** Na / — / (37) / 90:10 Si / Me_3 / (85) / >98:<2 Ti / Cp_2Cl / (53) / >98:<2	
C_{15} $(CO)_3Fe$	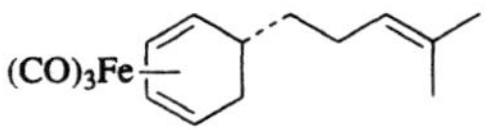 DMD, acetone, 56°	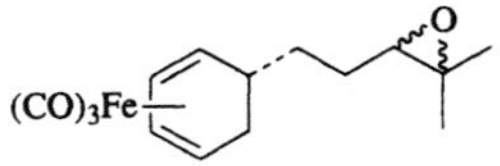 $(CO)_3Fe$ (11) dr 50:50	318
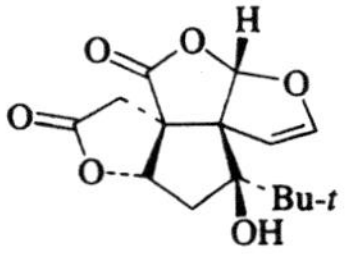 H, O, Bu-t, OH	DMD, acetone, 15 h	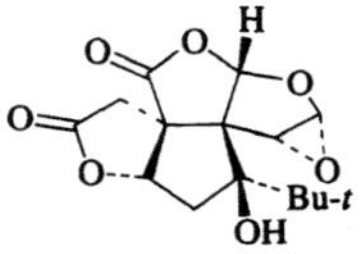 H, O, Bu-t, OH (88)	321
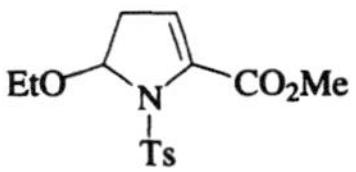 EtO, N, Ts, CO_2Me	DMD, acetone, rt, 4 h	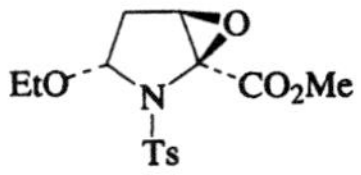 EtO, N, Ts, CO_2Me (63)	322
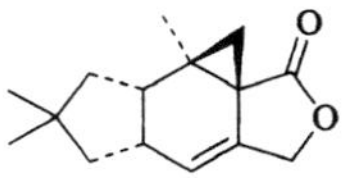 O	DMD, acetone, 45 min	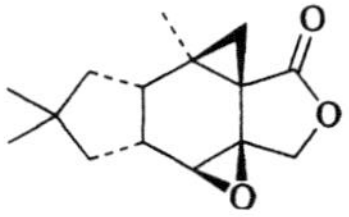 O (84)	323

TABLE 4. DIASTEREOSELECTIVE EPOXIDATIONS OF OLEFINS BY DIOXIRANES (*Continued*)

Substrate	Conditions	Product(s) and Yield(s) (%), Diastereomeric Ratio (dr)	Refs.
NMe_3^+ $ArSO_3^-$; Ar = 4-ClC_6H_4	DMD, acetone/CH_3CN, 20°, 48 h, 100% convn	NMe_3^+ $ArSO_3^-$ **I** + NMe_3^+ $ArSO_3^-$ **II** (—) **I**:**II** = 80:20	324
	DMD, acetone, 0°, 20 min	**I** (89) + **II** (—) **I**:**II** = 80:20	325
	DMD, acetone, rt, 20 h	(50) + (45)	326
C_{16} MeO, H, Bu-*t*	DMD, acetone, 18 h	MeO, H, Bu-*t* (98)	321
C_{16-19} R^2, R^2, R^1	DMD, acetone, 20°, 24 h	**I** + **II**	99

R^1	R^2	% Convn	I:II
Ph	H	>95	83:17
Bn	H	88	91:09
Bn	Me	95	90:10

Substrate	Conditions	Product(s) and Yield(s) (%)	Refs.
C_{16} 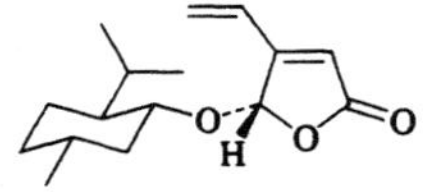	DMD, acetone, rt, overnight	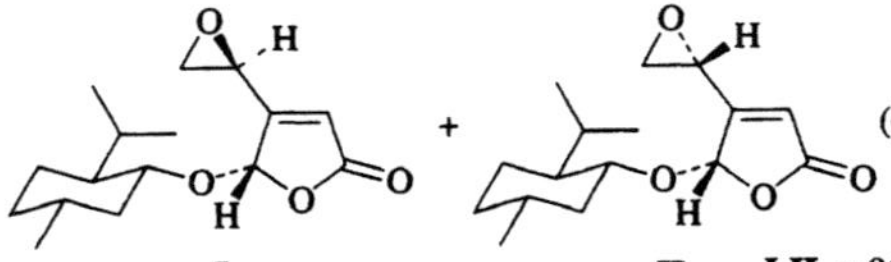 **I** + **II** (69) **I**:**II** = 86:14	327
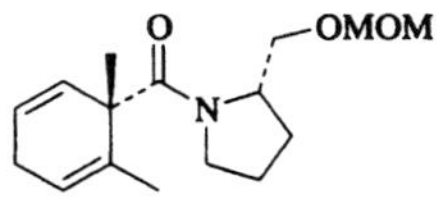 OMOM	Acetone, Oxone®, phosphate buffer (pH 7.2), 0°	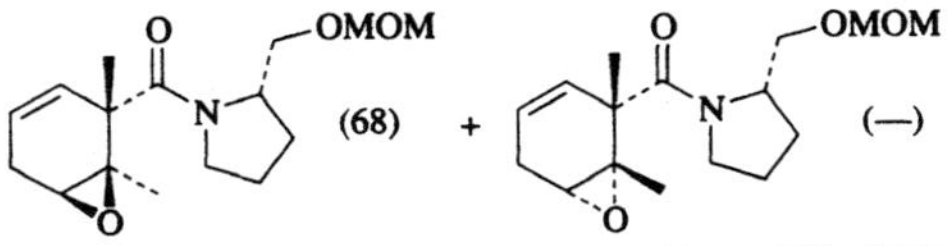 OMOM **I** (68) + OMOM **II** (—) **I**:**II** = 93:7	84
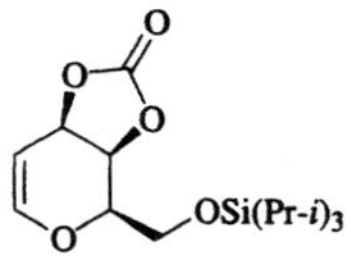 $OSi(Pr\text{-}i)_3$	DMD, acetone/CH_2Cl_2	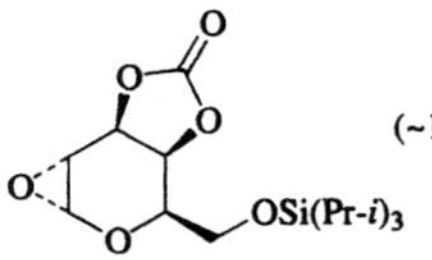 $OSi(Pr\text{-}i)_3$ (~100)	328
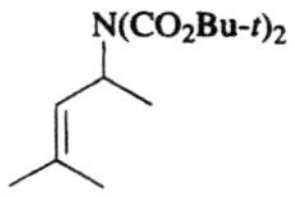 $N(CO_2Bu\text{-}t)_2$	DMD, acetone, 3 h	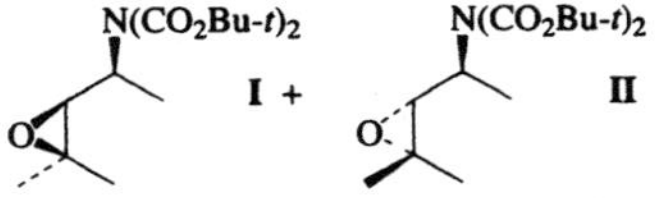 $N(CO_2Bu\text{-}t)_2$ **I** + $N(CO_2Bu\text{-}t)_2$ **II** (88) **I**:**II** = 75:25	298

TABLE 4. DIASTEREOSELECTIVE EPOXIDATIONS OF OLEFINS BY DIOXIRANES (*Continued*)

Substrate	Conditions	Product(s) and Yield(s) (%), Diastereomeric Ratio (dr)	Refs.
C_{17-18} R, Ph	DMD, acetone/CH_2Cl_2, 0°, 2.5 h	I + II (OH, OH, R, Ph)	329
NH, N, N, N, N, O, OH, OH, OH	1. DMD, acetone/MeOH, 0°, 4 h 2. rt, 2 h	O, NH, N, N, N, N, O, OH, OH, OH (75)	61a
O_2CBu-*t*, N, Ts	TFP, Oxone®, $MeCN/H_2O$, Na_2EDTA, 0°, 2 to 3 h	O_2CBu-*t*, N, Ts, O I + O_2CBu-*t*, N, Ts, O II (—) I:II = 98:2	303
TBDMSO, O, N, OTBDMS	DMD, acetone, rt, 3 d	TBDMSO, O, O, N, OTBDMS I + TBDMSO, O, O, N, OTBDMS II (93) I:II = 72:28	330

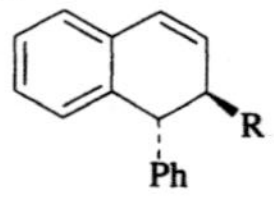

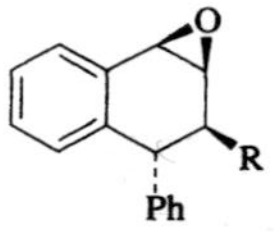

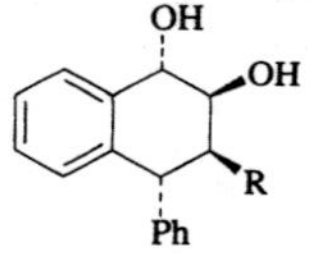

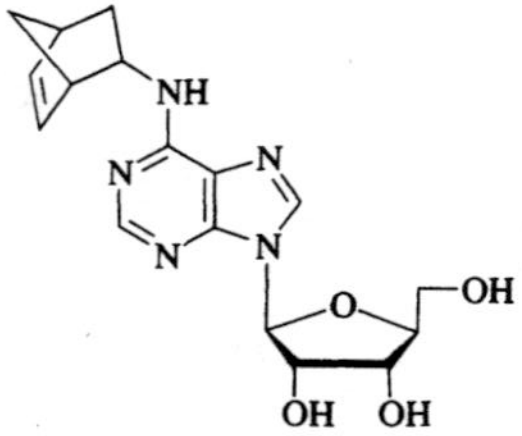

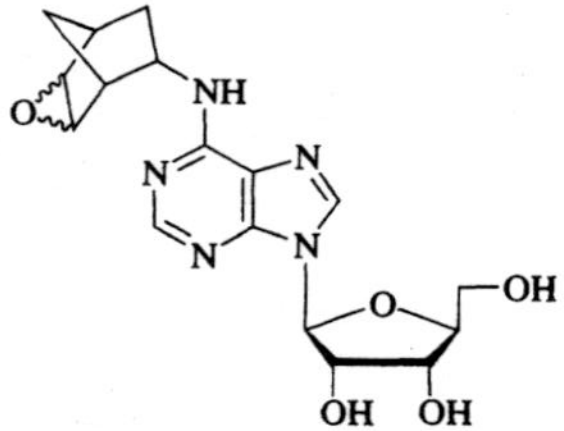

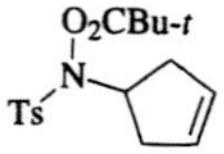

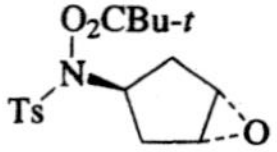

R	Additive	dr	I	II
CO_2H	—	>95:5	(89)	(0)
CH_2OH	—	>95:5	(94)	(0)
CO_2Me	—	>95:5	(98)	(0)
CO_2Me	$HClO_4/H_2O$	>95:5	(26)	(67)
CO_2Me	H_2O	—	(0)	(76)

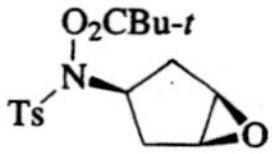

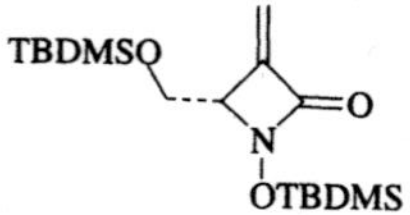

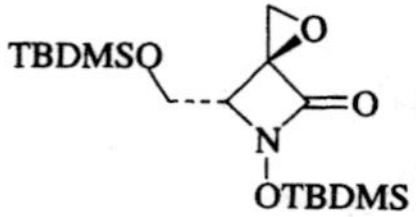

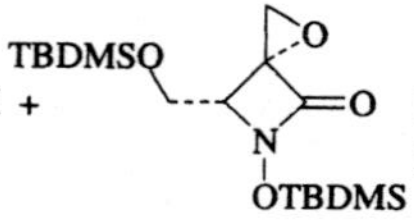

C_{18}

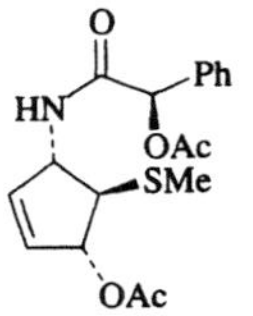

DMD, acetone, Ar, rt

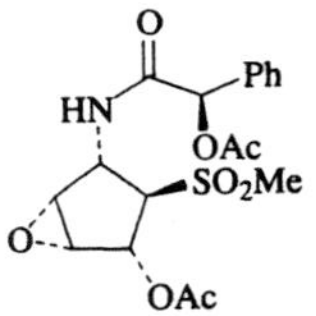

(97)

308

C_{18-20}

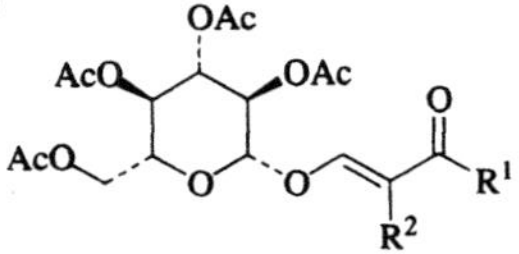

DMD, acetone

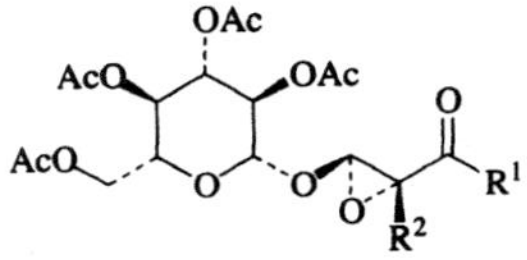

I +

331

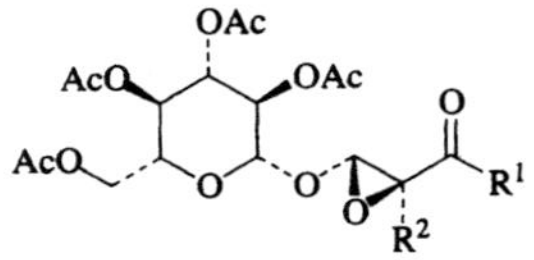

II

R^1	R^2	I	I:II
Me	H	(—)	69:31
—O(CH$_2$)$_2$—		(70)	86:14
Me	Me	(73)	90:10
EtO	H	(—)	67:33
—(CH$_2$)$_3$—		(64)	91:9
Et	Me	(52)	90:10
EtO	Me	(74)	91:9

C_{19}

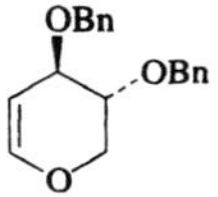

DMD, acetone/CH_2Cl_2

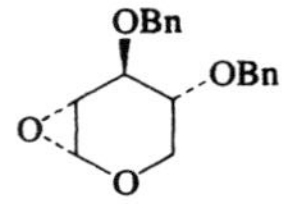

I 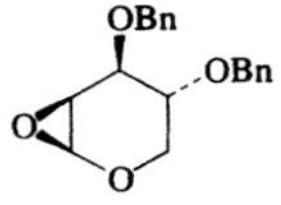 + II (—)

I:II = 80:20

328

TABLE 4. DIASTEREOSELECTIVE EPOXIDATIONS OF OLEFINS BY DIOXIRANES (*Continued*)

Substrate	Conditions	Product(s) and Yield(s) (%), Diastereomeric Ratio (dr)	Refs.
OMe; MeO_2C	DMD, acetone/CH_3CN, N_2, 0°, 36 min	OMe; MeO_2C (99)	332
OTBDMS; Ph; H	DMD, acetone/CH_2Cl_2, 0°, 1 h	OTBDMS; Ph; H **I** + OTBDMS; Ph; H **II** (90) **I**:**II** = ~50:50	91
OTBDMS; H; Ph	DMD, acetone/CH_2Cl_2, 0°, 1 h	OTBDMS; H; Ph (98)	91
C_{20-24} CBz; R	1. DMD, acetone/MeOH, rt 2. $NaBH_4$, AcOH, 25°, 3 h	OH; H; CBz; R **I** + HO; H; CBz; R **II** R / **I**:**II** Me (80) 34:66 *i*-Bu (75) 42:58	333

Substrate	Conditions	Product(s) and Yield(s) (%)	Refs.
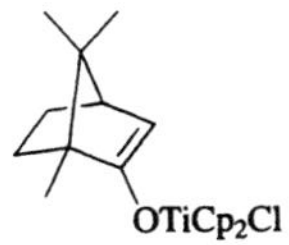 OTiCp$_2$Cl	1. DMD (1.6 equiv), acetone/THF, –78°, 108 s 2. NH_4F, H_2O, 20°, 2 to 3 h	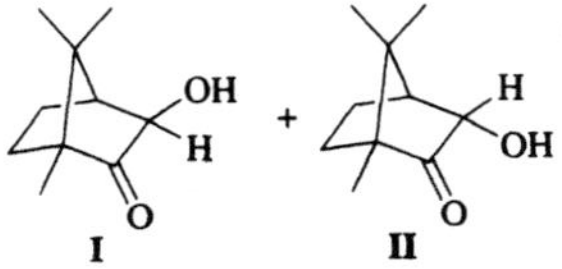 **I** (OH, H, O) + **II** (H, OH, O) (78) **I**:**II** = 94:6	311
	1. DMD (2 equiv), acetone/THF, –78°, 108 s 2. NH_4F, H_2O, 20°, 2 to 3 h	**I** + **II** (79) **I**:**II** = 96:4	311
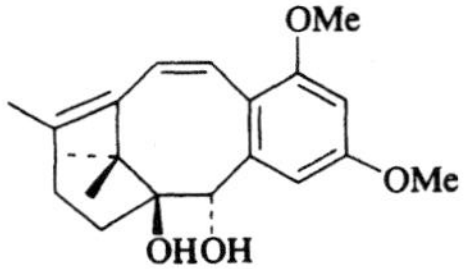 OMe, OMe, OHOH	DMD, acetone/CH_2Cl_2, 0°	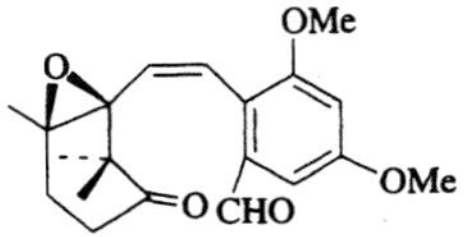 OMe, OMe, O, CHO (29)	96
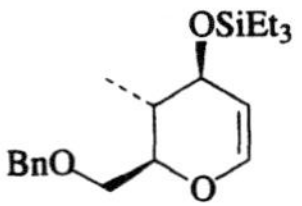 OSiEt$_3$, BnO, O	DMD, acetone/CH_2Cl_2, 0°	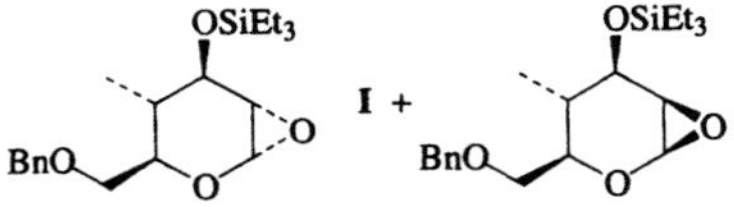 OSiEt$_3$, BnO, O, O **I** + OSiEt$_3$, BnO, O, O **II** (>99) **I**:**II** >95:5	90
C_{21} 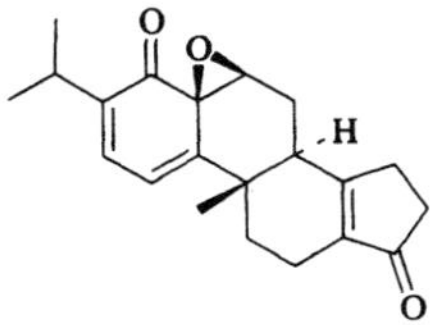 O, O, H, O	TFP, Oxone®, $NaHCO_3$, CH_3CN/H_2O, Na_2EDTA, 25°, 4 h	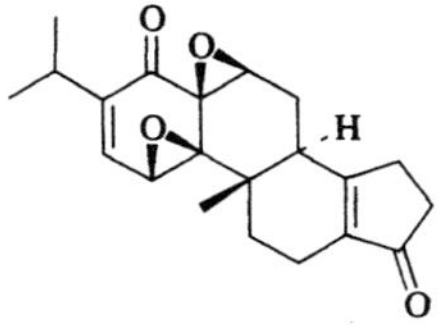 O, O, O, H, O (70)	334

TABLE 4. DIASTEREOSELECTIVE EPOXIDATIONS OF OLEFINS BY DIOXIRANES (*Continued*)

Substrate	Conditions	Product(s) and Yield(s) (%), Diastereomeric Ratio (dr)	Refs.
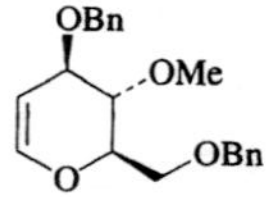 OBn, OMe, OBn	DMD, acetone/CH_2Cl_2, 0°, 30 min	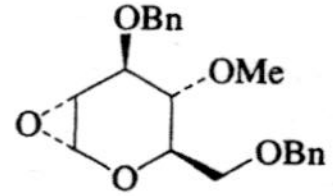 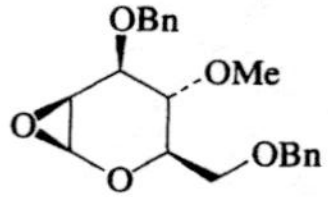 OBn, OMe, OBn, O, O **I** + OBn, OMe, OBn, O, O **II** (100) **I**:**II** = 15:1	335
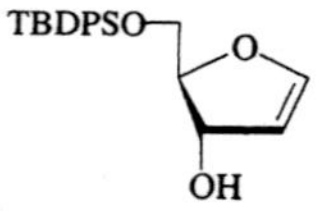 TBDPSO, O, OH	DMD, acetone/CH_2Cl_2, 0°, 40 min	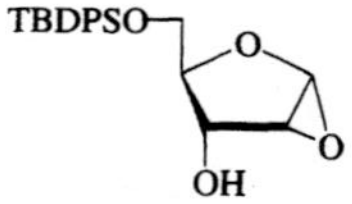 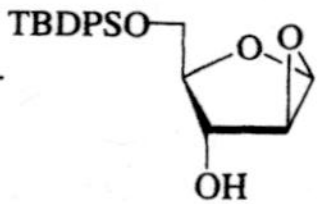 TBDPSO, O, OH, O **I** + TBDPSO, O, O, OH **II** (—) **I**:**II** = 9:1	336
C_{21-27} O, O, OSiR$_2$ (S), O, O, O, OH, O, O, O, O	DMD, acetone/CH_2Cl_2, 0°, 40 min	O, O, OSiR$_2$ (S), O, O, O, OH, O, O, O, O, O (—) R = Ph, *i*-Pr (S) = solid support	93
C_{21} O, O	DMD, acetone, CH_2Cl_2, 20°, 20 h	O, O **I** + O, O **II** (90) **I**:**II** = 80:20	258

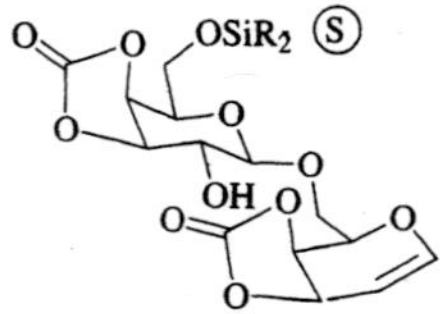

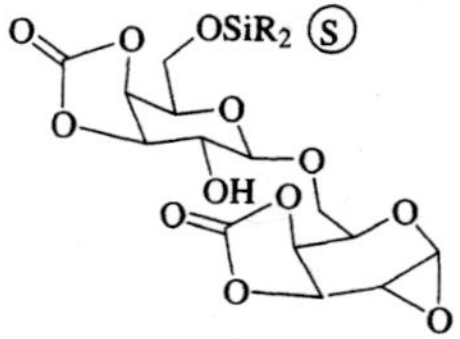

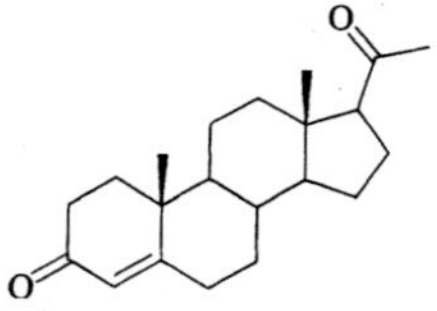

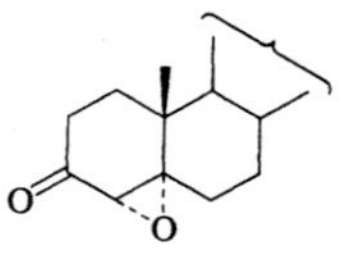

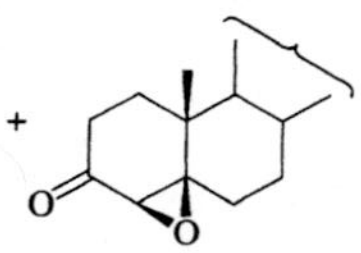

O_2CBu-*t*, O_2CBu-*t*, O_2CBu-*t*

TFP, Oxone®, $NaHCO_3$, CH_3CN/H_2O, Na_2EDTA, 8°, 15 min

I + **II** (84) **I**:**II** = 75:25 — 35

C_{22-24}

DMD, acetone/CH_2Cl_2, ~20° — 337

R^1	R^2	Time	
H	Br	408 h	(82)
H	Cl	408 h	(86)
H	H	240 h	(79)
H	CN	552 h	(80)
Me	H	312 h	(75)
H	Me	216 h	(86)
MeO	H	312 h	(73)
H	MeO	144 h	(78)
H	EtO	168 h	(79)

DMD, acetone/CH_2Cl_2, ~20°

I + **II** — 337

R	Time	**I**	**II**
Br	432 h	(19)	(0)
Cl	432 h	(15)	(0)
H	336 h	(15)	(10)
Me	228 h	(21)	(0)
EtO	336 h	(0)	(14)

TABLE 4. DIASTEREOSELECTIVE EPOXIDATIONS OF OLEFINS BY DIOXIRANES (*Continued*)

Substrate	Conditions	Product(s) and Yield(s) (%), Diastereomeric Ratio (dr)	Refs.
C_{23-23} O, O, Ph, R	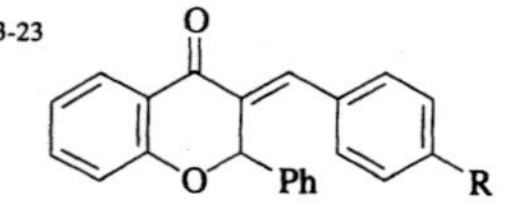 TFD, TFP/CH_2Cl_2, ~20°, 5 to 8 h	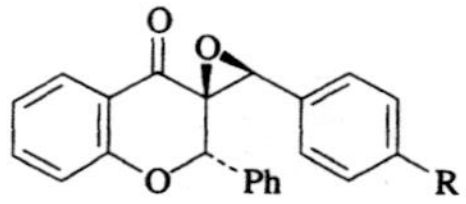 O, O, O, Ph, R (—) R = Cl, H, Me	337
C_{23} Ph, N, O, O, O, O, C_6H_4OMe-4	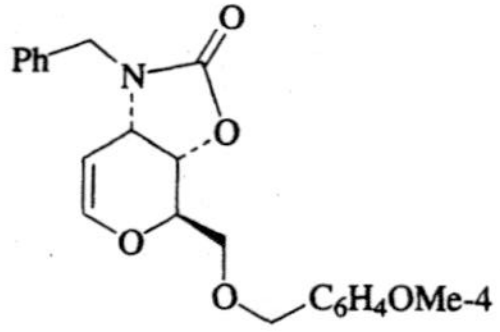 DMD, acetone/CH_2Cl_2, 0°	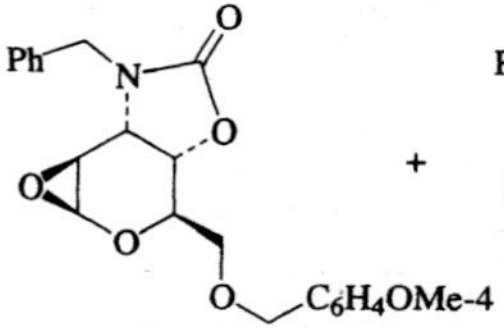 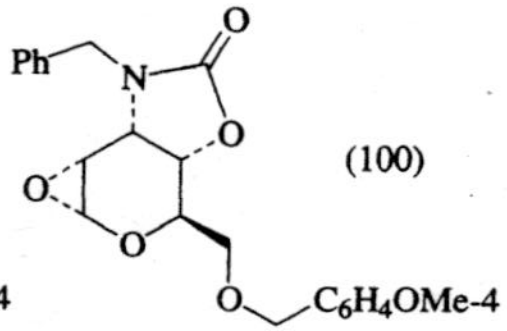 Ph, N, O, O, O, O, O, C_6H_4OMe-4 + Ph, N, O, O, O, O, O, C_6H_4OMe-4 (100)	338
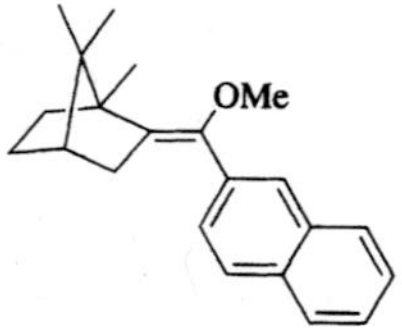 OMe	TFD, TFP/CH_2Cl_2, -20°, 5 min	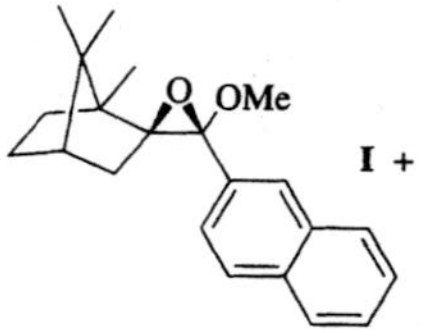 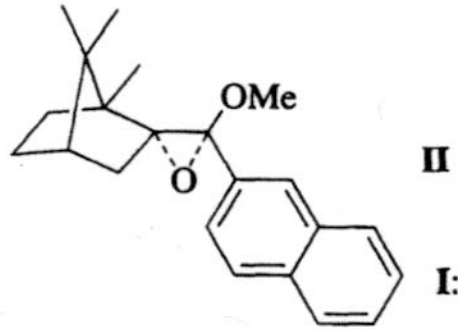 O, OMe **I** + OMe, O **II** (92) **I**:**II** = 50:50	222
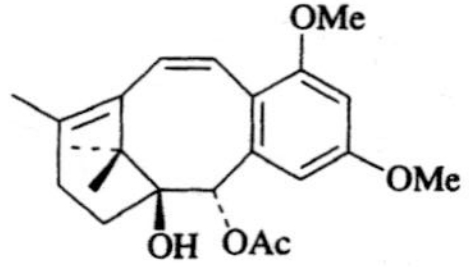 OMe, OMe, OH, OAc	DMD, acetone/CH_2Cl_2, 0°	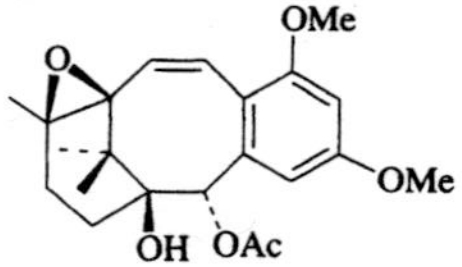 O, OMe, OMe, OH, OAc (100)	96
C_{23} O, O, O, O, OTBDPS	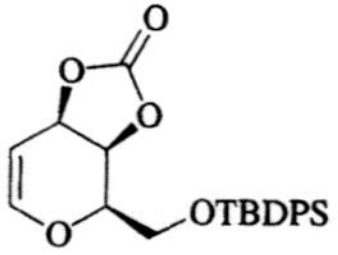 DMD, acetone/CH_2Cl_2, 0° to rt, 1 h	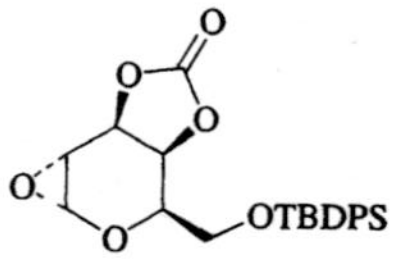 O, O, O, O, O, OTBDPS (—)	339

Substrate	Conditions	Product(s) and Yield(s) (%)	Refs.
AcO	DMD, acetone, –40°, 12 h	AcO, I + AcO, **II** (88) **I**:**II** = 60:40	258
OAc, AcO, H	DMD, acetone	OAc, O, AcO, H (90)	340
$OSi(Pr\text{-}i)_3$, O, OH	DMD, CH_2Cl_2, acetone, 0°, 30 min	$OSi(Pr\text{-}i)_3$, O, OH (>48)	94
C_{24} TBDPSO	DMD, $CHCl_3$/acetone	TBDPSO (100) dr 52:48	203
OTBDMS, OTBDMS, OTBDMS	DMD, acetone/CH_2Cl_2, 0°, 1 h	OTBDMS, OTBDMS, OTBDMS (98)	91

TABLE 4. DIASTEREOSELECTIVE EPOXIDATIONS OF OLEFINS BY DIOXIRANES (*Continued*)

Substrate	Conditions	Product(s) and Yield(s) (%), Diastereomeric Ratio (dr)	Refs.
C_{25} 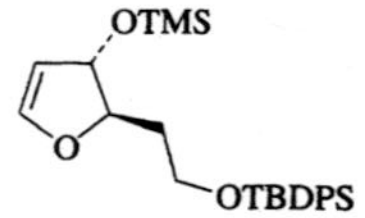 OTMS, OTBDPS	DMD, acetone/CH_2Cl_2, 0°, 30 min	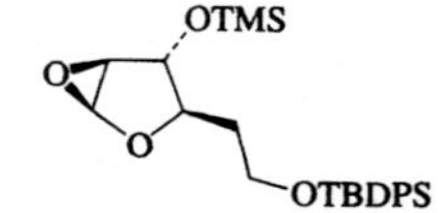 OTMS, OTBDPS (—)	336
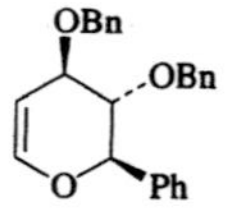 OBn, OBn, Ph	DMD, acetone	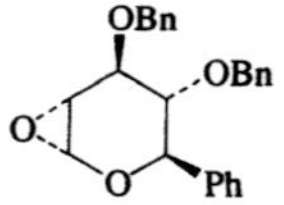 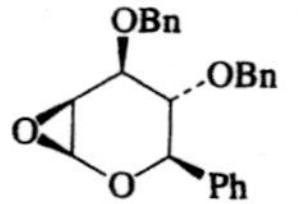 OBn, OBn, Ph I + OBn, OBn, Ph II (—) I:II = 5:1	341
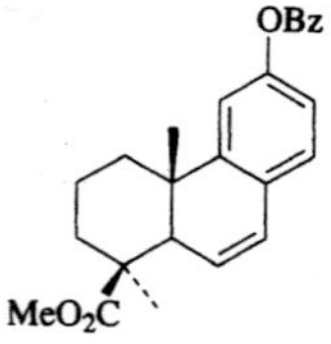 OBz, MeO_2C	DMD, acetone, N_2, rt, 30 min	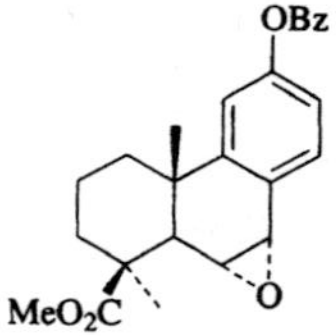 OBz, MeO_2C (65)	142
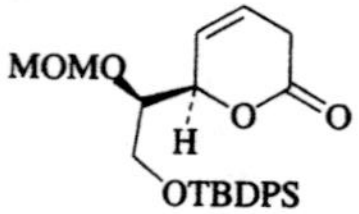 MOMO, H, OTBDPS	1. DMD, acetone, 3 h 2. Et_3N, CH_2Cl_2, 12 h	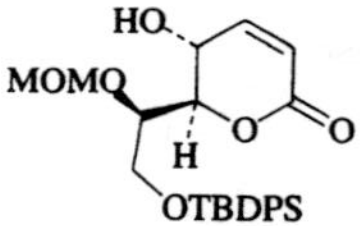 HO, MOMO, H, OTBDPS (—)	342
C_{26-33} 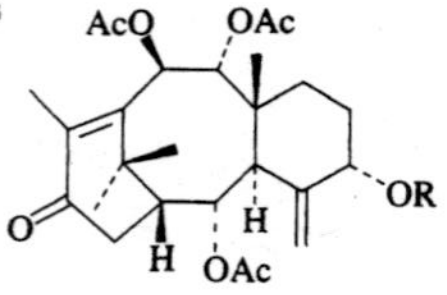 AcO, OAc, H, OAc, H, OR	DMD, acetone/CH_2Cl_2, rt, 36 h	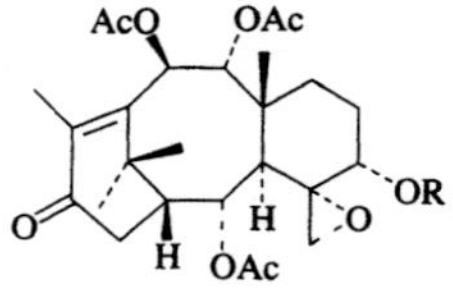 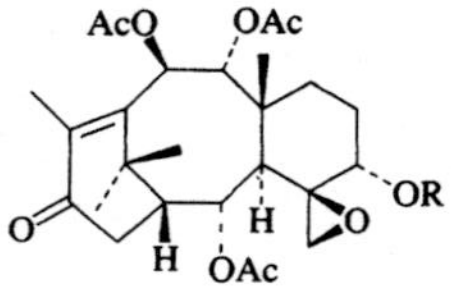 AcO, OAc, H, OAc, H, OR I + AcO, OAc, H, OAc, H, OR II	343

C_{26}

HO, O, O, O, S, N

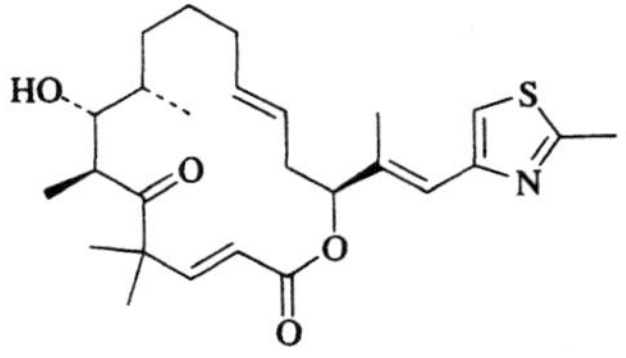

TFP, Oxone®, $NaHCO_3$,
CH_3CN/H_2O,
Na_2EDTA, 0°

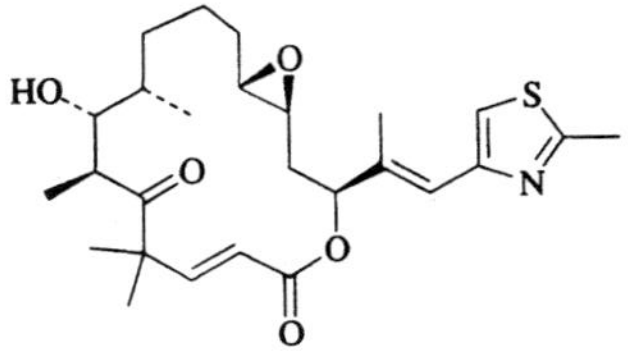

R		I:II
H	(66)	67:33
CF_3CO	(72)	50:50
Ac	(86)	67:33
Piv	(63)	33:67
Et_3Si	(86)	20:80
TBDMS	(69)	17:83
PhCO	(77)	67:33

HO, O, O, O, O, S, N

I +

HO, O, O, O, O, S, N

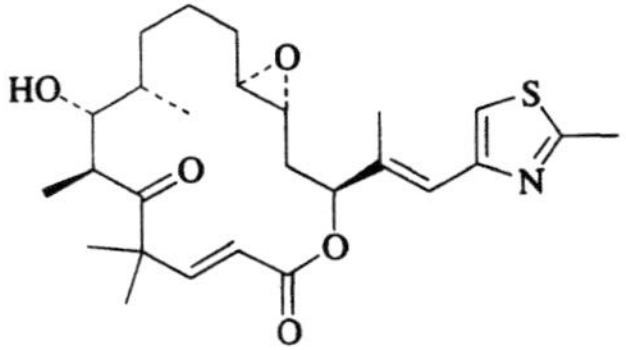

II

I(**II**) (45); **II**(**I**) (28)

344

TABLE 4. DIASTEREOSELECTIVE EPOXIDATIONS OF OLEFINS BY DIOXIRANES (*Continued*)

Substrate	Conditions	Product(s) and Yield(s) (%), Diastereomeric Ratio (dr)	Refs.
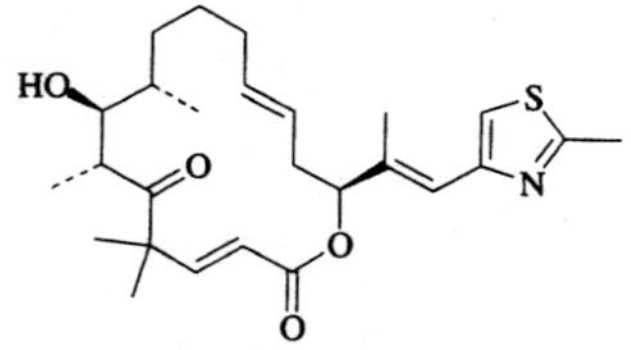	TFP, Oxone®, $NaHCO_3$, CH_3CN/H_2O, Na_2EDTA, 0°	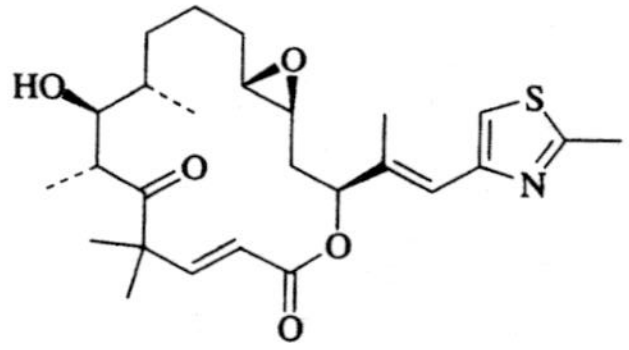 **I** + 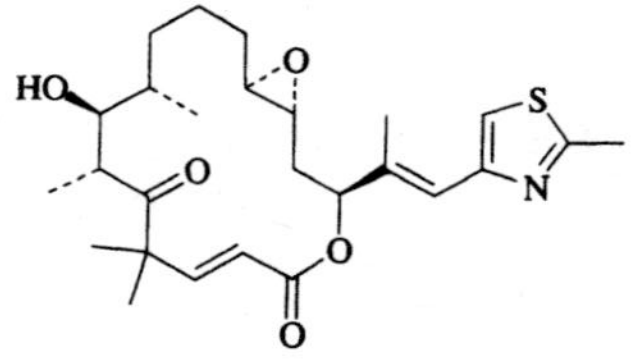 **II** **I**(**II**) (60); **II**(**I**) (15)	344
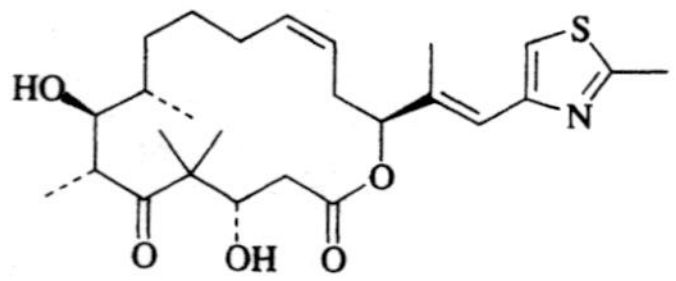	DMD, acetone/CH_2Cl_2, 0°	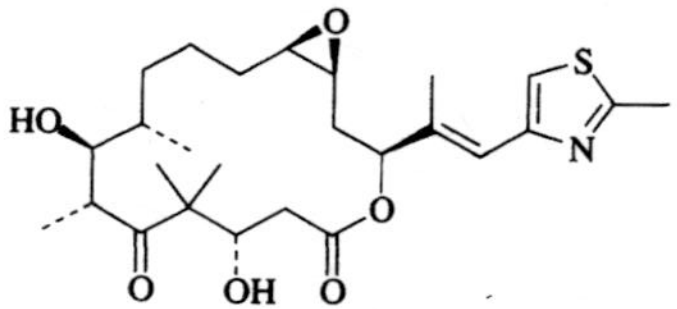 **I** (50) + 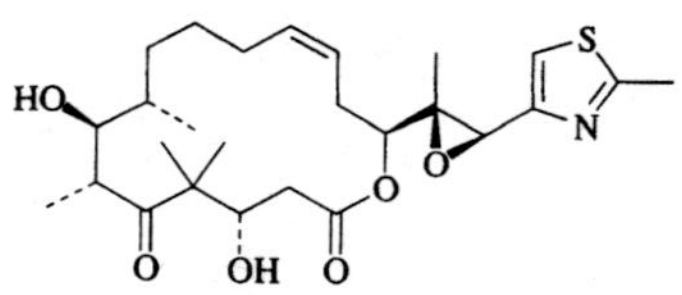 **II** (5) +	344

III (5)

TFP, Oxone®, $NaHCO_3$, CH_3CN/H_2O, Na_2EDTA, 0° — I (62) + (13) 344

TFP, Oxone®, $NaHCO_3$, CH_3CN/H_2O, Na_2EDTA, 0° — I + 344

II
I (II) (38)
II (I) (29)

TABLE 4. DIASTEREOSELECTIVE EPOXIDATIONS OF OLEFINS BY DIOXIRANES (*Continued*)

Substrate	Conditions	Product(s) and Yield(s) (%), Diastereomeric Ratio (dr)	Refs.
	DMD, acetone/CH_2Cl_2, 0°	**I** (10) + **II** (10) + **III** (40)	344
	TFP, Oxone®, $NaHCO_3$, CH_3CN/H_2O, Na_2EDTA, 0°	**I** (**II**) (45) + **II** (**I**) (35)	344

Substrate	Conditions	Product(s) (Yield %)	Refs.
	TFP, Oxone®, $NaHCO_3$, CH_3CN/H_2O, Na_2EDTA, 0°	I + II I (II) (44) II (I) (21)	344
C_{27}			
	DMD, acetone/CH_2Cl_2, 0°, 1 h	(—)	345
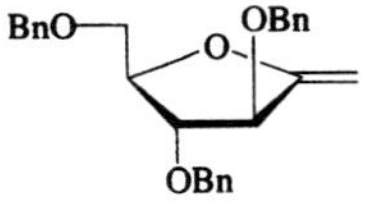	DMD, acetone/CH_2Cl_2, 0°	I + II (—) I:II = 11:1	346

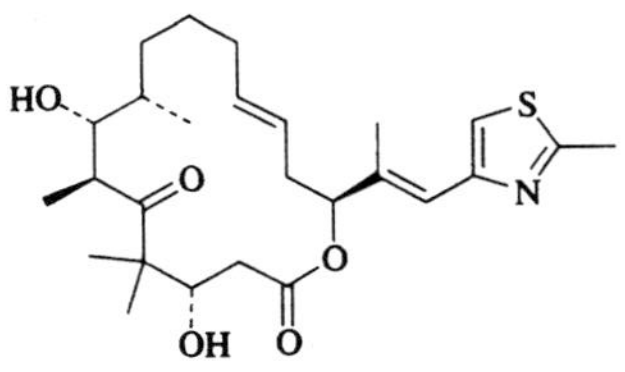

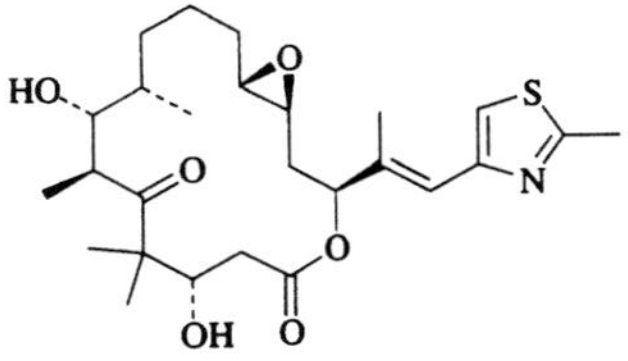

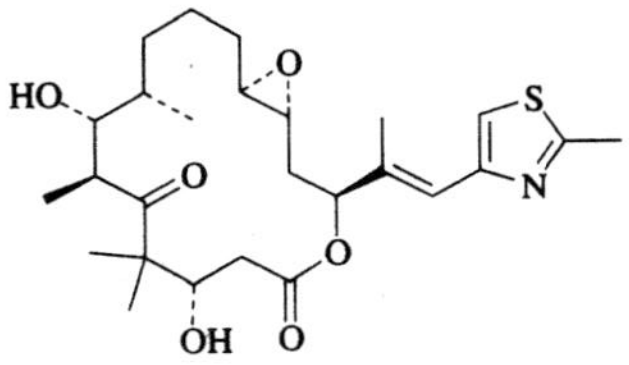

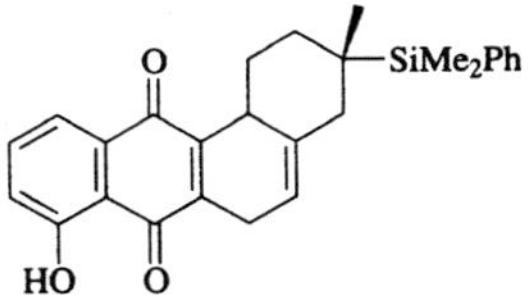

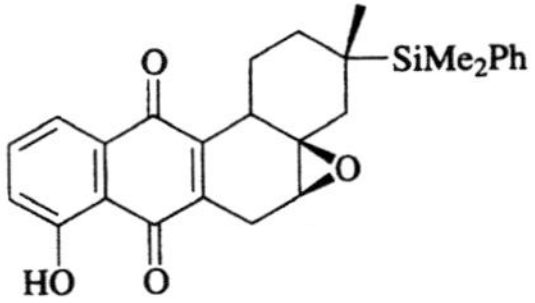

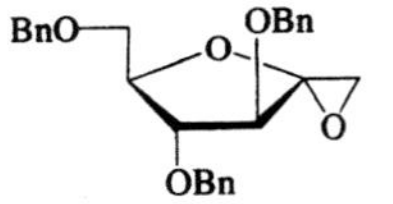

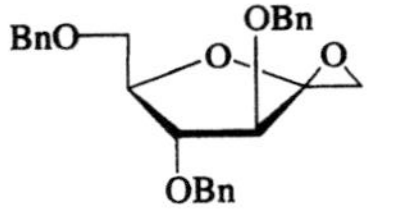

TABLE 4. DIASTEREOSELECTIVE EPOXIDATIONS OF OLEFINS BY DIOXIRANES (*Continued*)

Substrate	Conditions	Product(s) and Yield(s) (%), Diastereomeric Ratio (dr)	Refs.
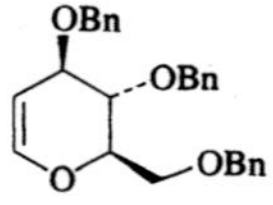 OBn, OBn, OBn	DMD, acetone/CH_2Cl_2, 0°, 6 min	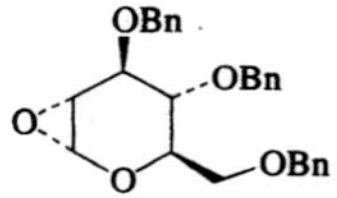 OBn, OBn, OBn, O **I** (96)	347
	DMD, acetone/CH_2Cl_2, 0°, 10 min	**I** (—)	348
	DMD, acetone/CH_2Cl_2, N_2, 0°, 15 min	**I** (—)	349
C_{27-33} 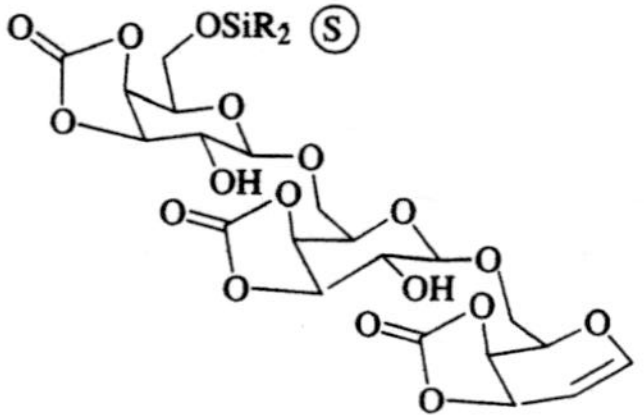 OSiR$_2$ Ⓢ	DMD, acetone/CH_2Cl_2, 0°, 40 min	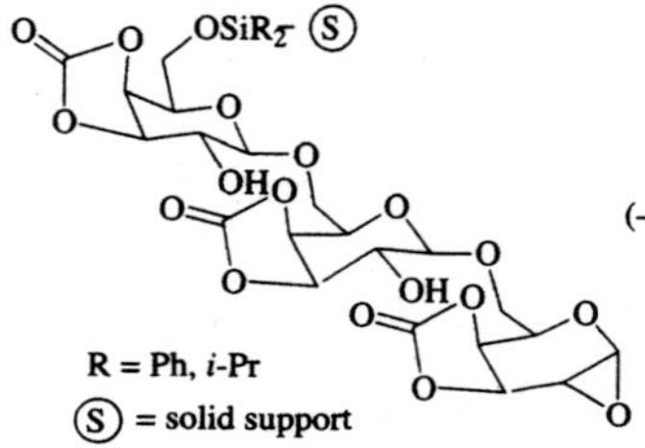 OSiR$_2$ Ⓢ (—) R = Ph, *i*-Pr; Ⓢ = solid support	93
C_{27} 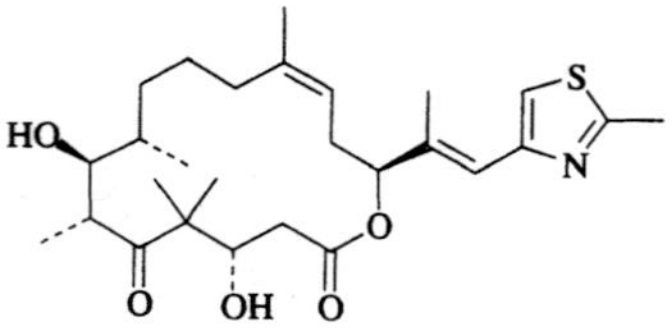 HO, OH, S, N	DMD, acetone/CH_2Cl_2, –50°	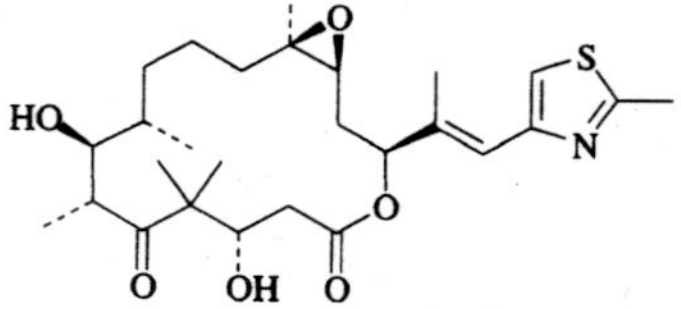 HO, OH, S, N, O **I** +	350

Substrate	Conditions	Product(s) (Yield, %)	Refs.
		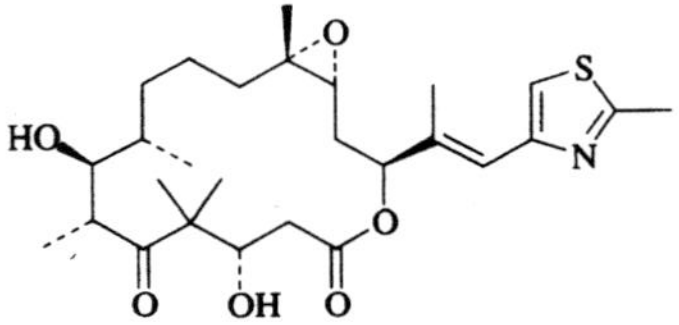 II (75) I:II = 83:17	
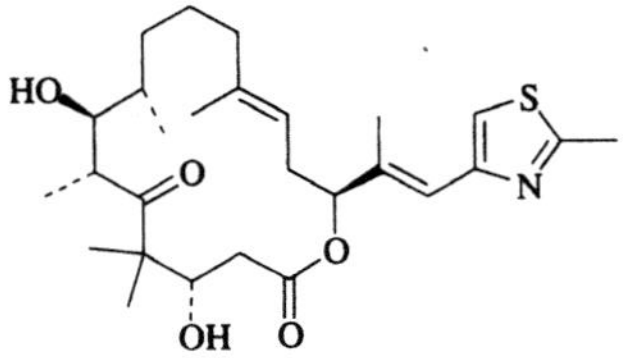	TFP, Oxone®, $NaHCO_3$, CH_3CN/ H_2O, Na_2EDTA, 0°	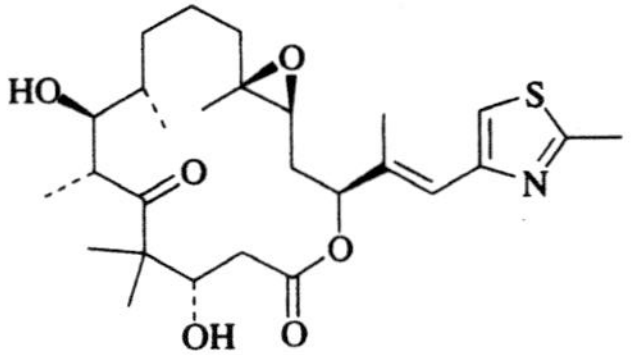 I +	350
		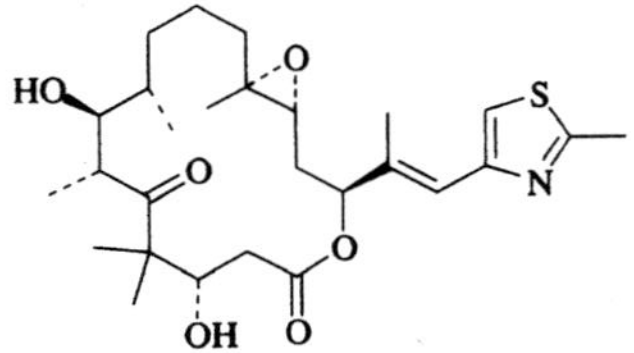 II (86) I:II = 50:50	
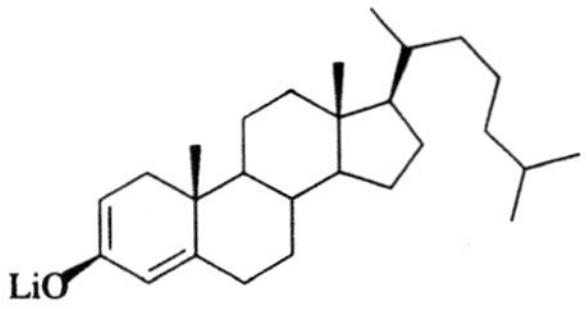 LiO	DMD, acetone/THF, Ar, –78°, 10 min	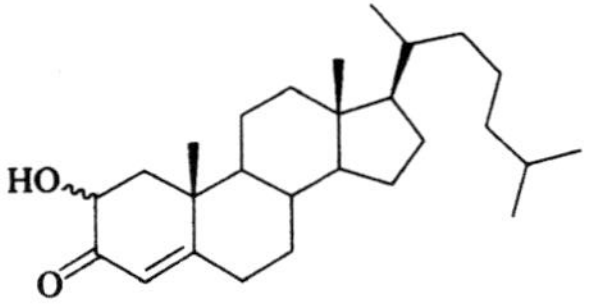 HO (60) dr 67:33	209

TABLE 4. DIASTEREOSELECTIVE EPOXIDATIONS OF OLEFINS BY DIOXIRANES (*Continued*)

Substrate	Conditions	Product(s) and Yield(s) (%), Diastereomeric Ratio (dr)	Refs.
C_{27-34}			
RO	TFD, CH_2Cl_2/TFP, –40°, 1 h	RO; R: H (72), Ac (85), 4-BrC_6H_4CO (83)	351
C_{27}			
O	DMD, acetone, 20°, 20 h	I + II (80) I:II = 75:25	258
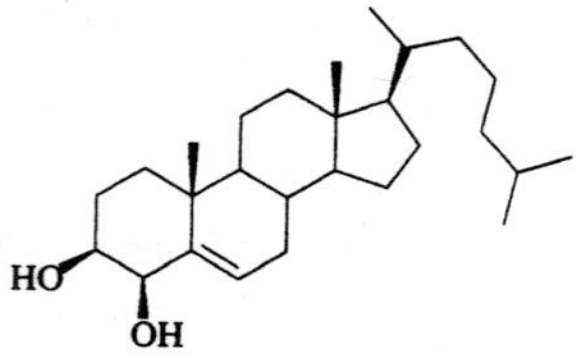 HO, OH	Oxone®, acetone/CH_2Cl_2, phosphate buffer (pH 7.5), 18-crown-6, 6 to 8°, 4 h	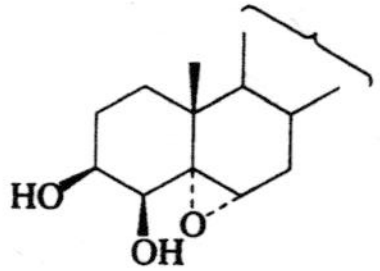 I 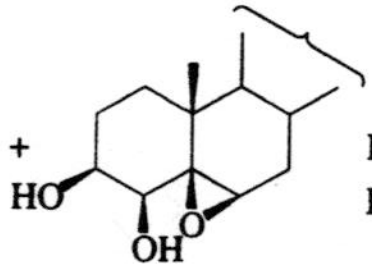 + II (50) I:II = 22:78	302
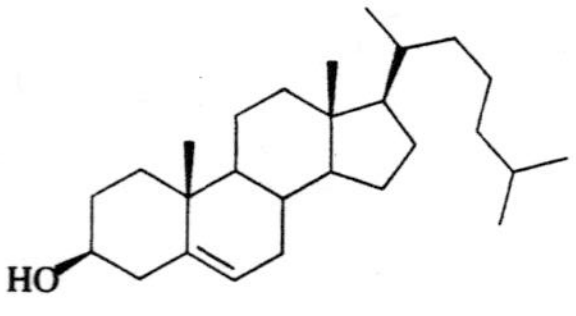 HO	Sulfonylazole, H_2O_2, acetone, THF, NaOH, 10°, ~3 h	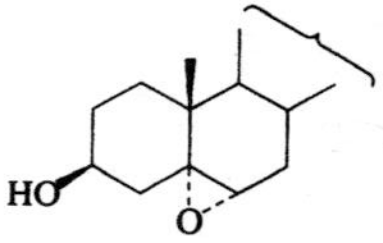 I 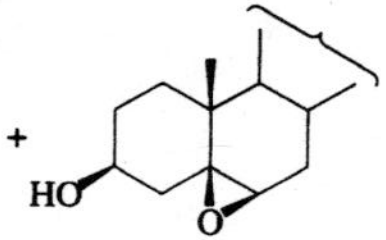 + II	271

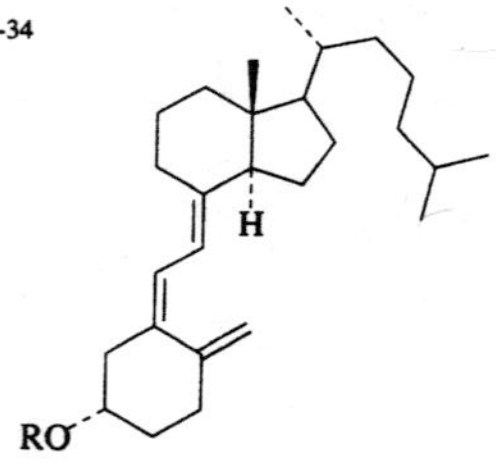

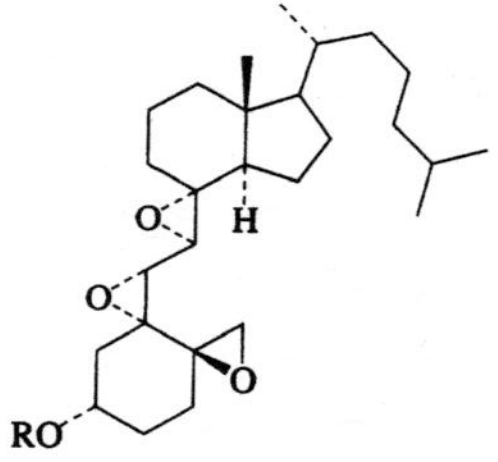

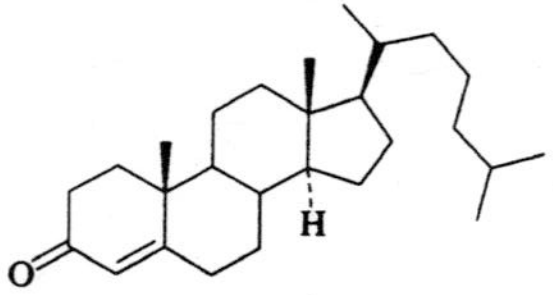

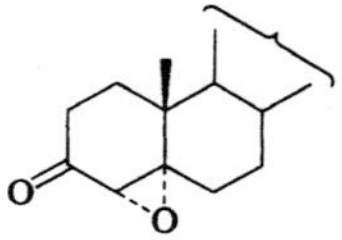

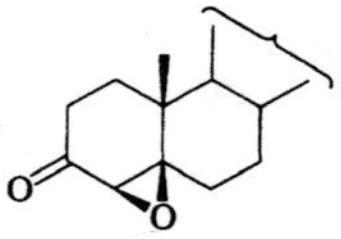

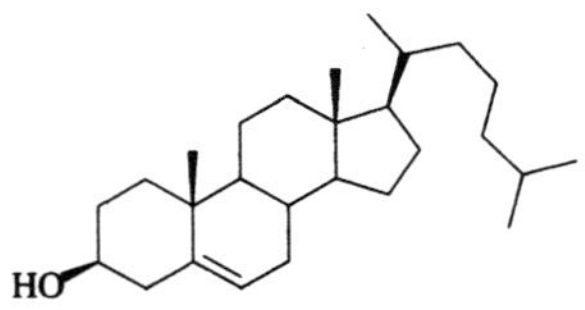

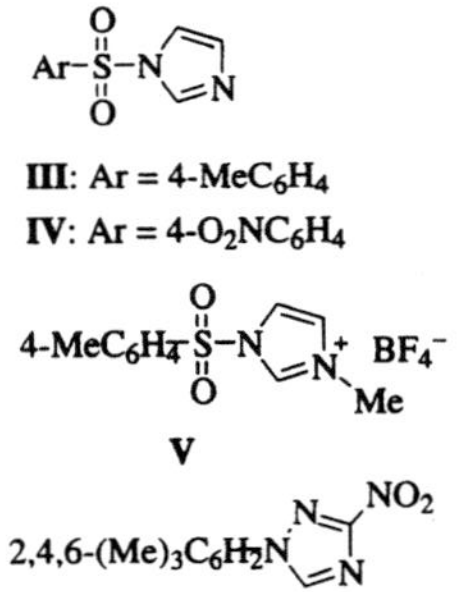

III: Ar = 4-MeC_6H_4

IV: Ar = 4-$O_2NC_6H_4$

4-MeC_6H_4-SO_2-N(imidazolium, N-Me)$^+$ BF_4^-

V

2,4,6-$(Me)_3C_6H_2$N (3-nitro-1,2,4-triazole)

VI

Sulfonylazole	Acetone (equiv)	I+II	I:II
III	1	(37)	76:24
III	4	(38)	58:42
III	10	(27)	50:50
III	20	(18)	50:50
III	40	(32)	43:57
III	60	(18)	44:56
III	80	(17)	41:59
III	100	(31)	41:59
IV	1	(54)	75:25
IV	4	(54)	56:44
IV	20	(52)	45:55
V	20	(92)	82:18
VI	4	(34)	76:24
VI	20	(18)	70:30
VI	20	(96)	91:9

Acetone, Oxone®, $NaHCO_3$, H_2O, 10°, 6 h — **I** + **II** (72) **I**:**II** = 44:56 — 271

Ketone, Oxone®, CH_2Cl_2/ phosphate buffer (pH 7.5), Bu_4NHSO_4, 0 to 5° — **I** + **II** — 352

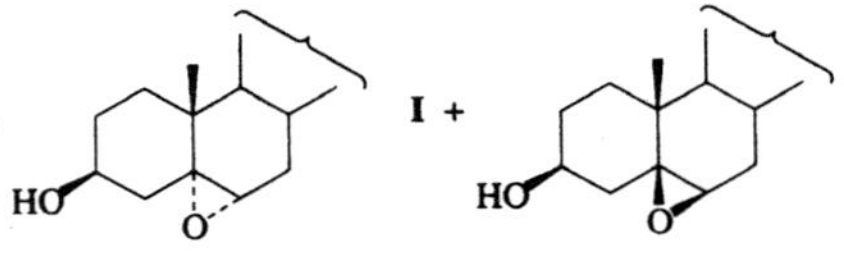

Ketone		I:II
acetone	(90)	50:50
cyclohexanone	(59)	55:45
2-methylcyclohexanone	(70)	35:65
2-hydroxy-2-methylcyclohexanone	(37)	40:60
ethyl pyruvate	(83)	64:36

TABLE 4. DIASTEREOSELECTIVE EPOXIDATIONS OF OLEFINS BY DIOXIRANES (*Continued*)

Substrate	Conditions	Product(s) and Yield(s) (%), Diastereomeric Ratio (dr)	Refs.
C_{28} OBn, O, C_6H_4OMe-4, OBn, O	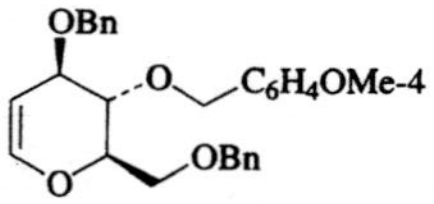 DMD, acetone	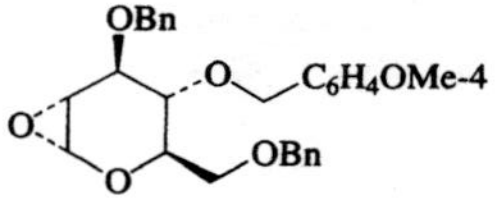 OBn, O, C_6H_4OMe-4, O, OBn, O (100)	353
C_{29} H, AcO	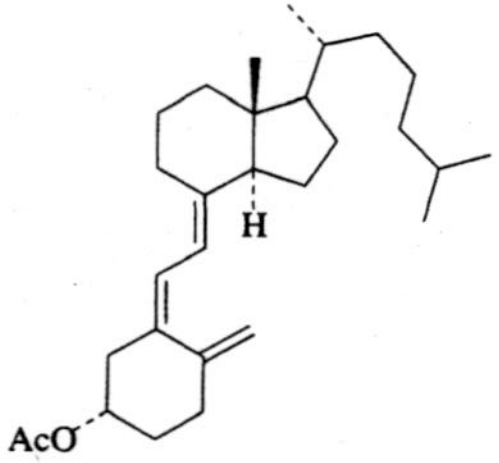 DMD, CH_2Cl_2/acetone, –40°, 2 h	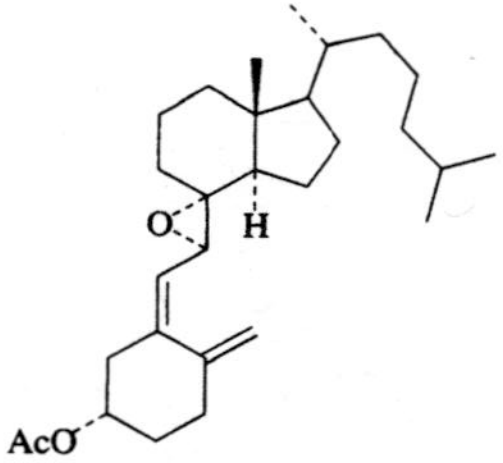 O, H, AcO (60)	351
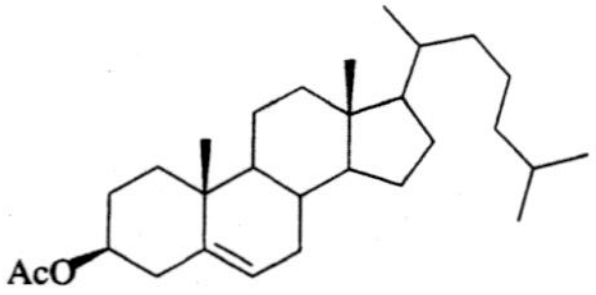 AcO	Acetone, Oxone®, CH_2Cl_2/ phosphate buffer (pH 7.5), Bu_4NHSO_4, 0 to 5°	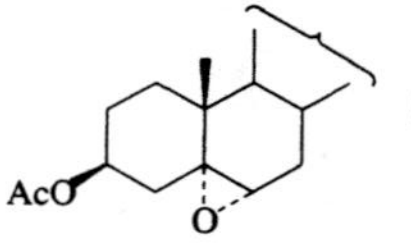 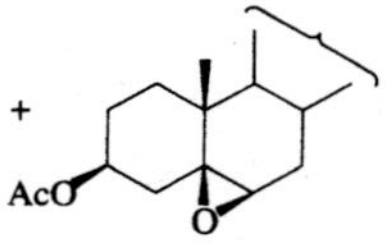 AcO, O **I** + AcO, O **II** (86) **I**:**II** = 40:60	352

Carbons	Substrate	Conditions	Product(s) and Yield(s) (%)	Refs.
C_{30}	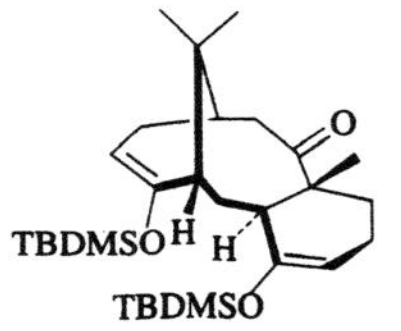 TBDMSO, H, H, TBDMSO, O	1. DMD, acetone/CH_2Cl_2, −78 to −10°, 3 h 2. $NaHCO_3$	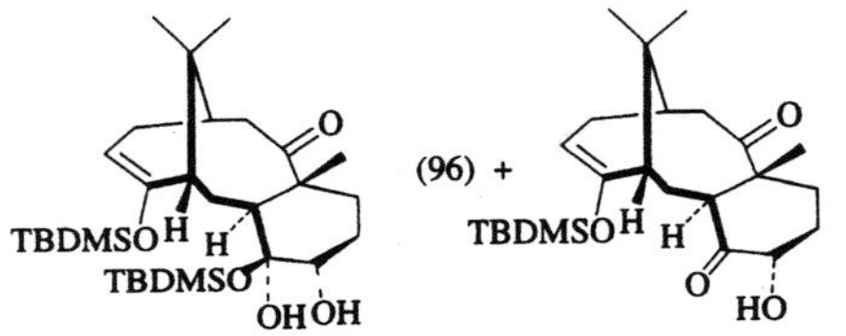 TBDMSO, H, H, TBDMSO, O, OH OH (96) + TBDMSO, H, H, O, O, HO (3)	354, 355
C_{31}	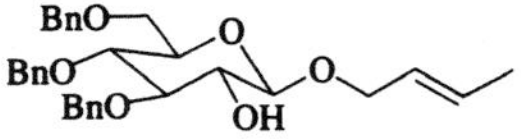 BnO, BnO, BnO, O, OH, O	DMD, acetone/CH_2Cl_2, 0°, 6 h	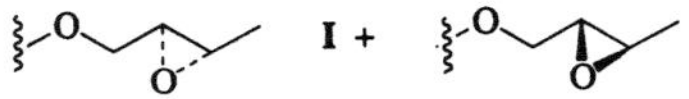 **I** + **II** (80) **I**:**II** = 50:50	356
	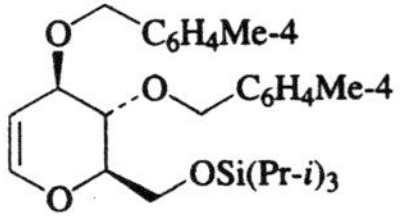 O, C_6H_4Me-4, O, C_6H_4Me-4, $OSi(Pr\text{-}i)_3$, O	1. DMD, acetone/CH_2Cl_2 2. MeOH, rt	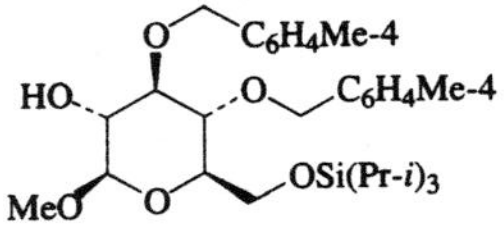 HO, MeO, O, C_6H_4Me-4, O, C_6H_4Me-4, $OSi(Pr\text{-}i)_3$ (76)	357
		DMD, acetone/CH_2Cl_2, 0°	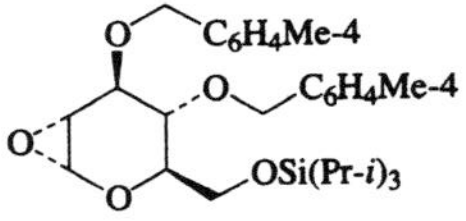 O, O, C_6H_4Me-4, O, C_6H_4Me-4, $OSi(Pr\text{-}i)_3$ (—)	358
C_{32}	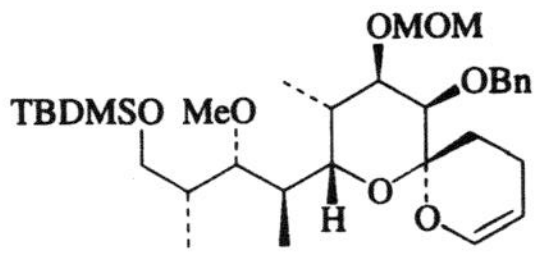 TBDMSO, MeO, OMOM, OBn, H, O, O	DMD, acetone/CH_2Cl_2, 0°	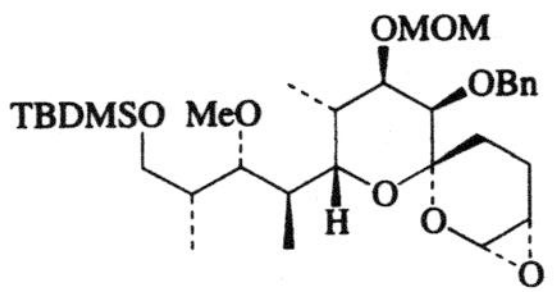 TBDMSO, MeO, OMOM, OBn, H, O, O, O (100)	359

TABLE 4. DIASTEREOSELECTIVE EPOXIDATIONS OF OLEFINS BY DIOXIRANES (*Continued*)

Substrate	Conditions	Product(s) and Yield(s) (%), Diastereomeric Ratio (dr)	Refs.
C_{34-35}			
BzO, OBz, OBz, OBz	DMD, acetone/CH_2Cl_2, N_2, –20°, 2 h	BzO, OBz, OBz, OBz (~100)	213
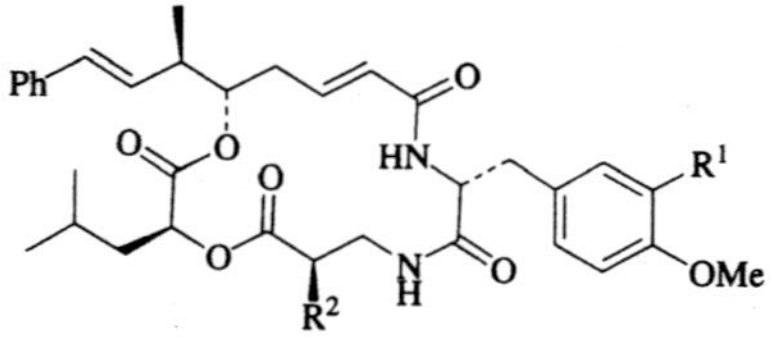 Ph, R^1, OMe, HN, N H, R^2	DMD, acetone/CH_2Cl_2, –30°, 24 h	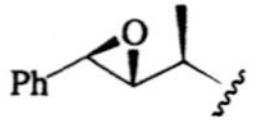 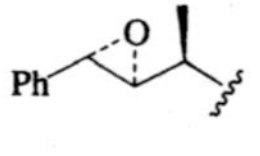 Ph I + Ph II	360
		R^1 R^2 I II H H (60) (20) Cl Me (55) (19)	
TBDMSO, TBDMSO, OTBDMS	DMD, acetone/CH_2Cl_2, N_2, –78 to –20°, 5.5 h	TBDMSO, TBDMSO, OTBDMS (79)	361

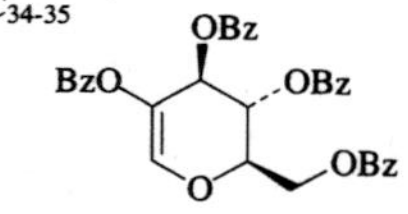

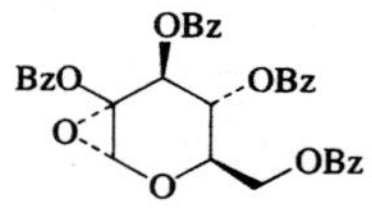

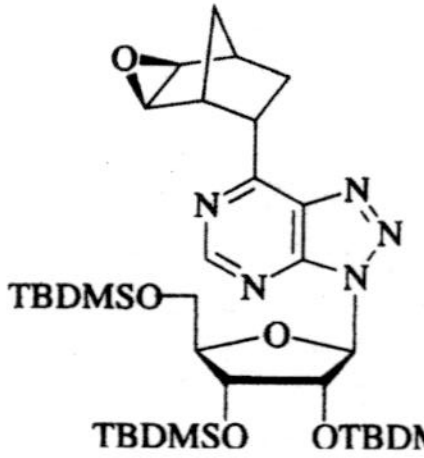

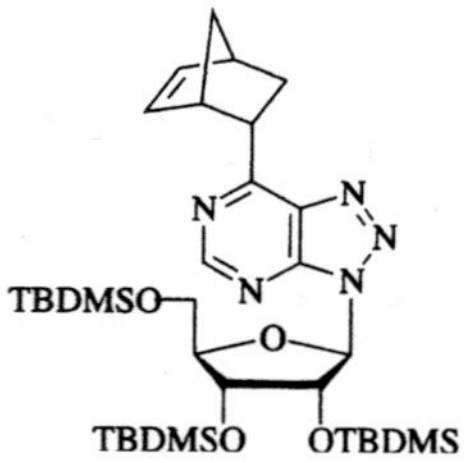

C_{35} 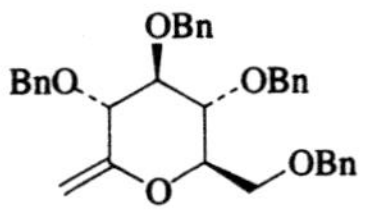	DMD	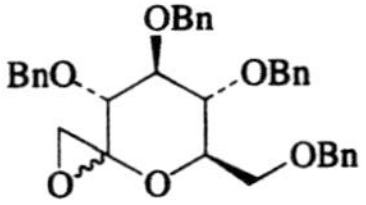 (—)	362
	DMD, acetone/CH_2Cl_2, 0°	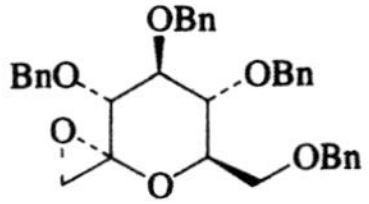 I + 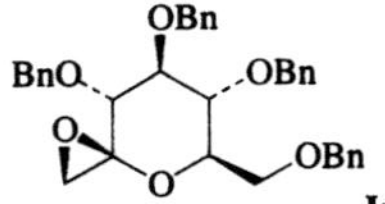 II (—) I:II = 60:40	346
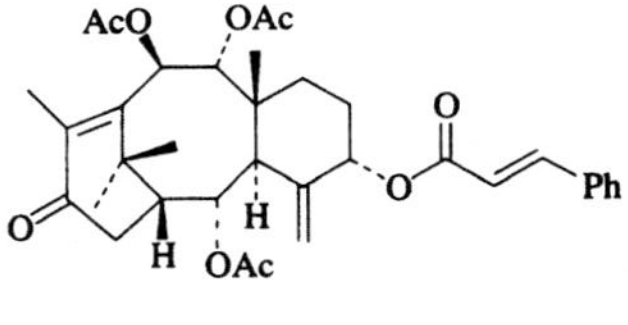	DMD, acetone/CH_2Cl_2, rt, 36 h	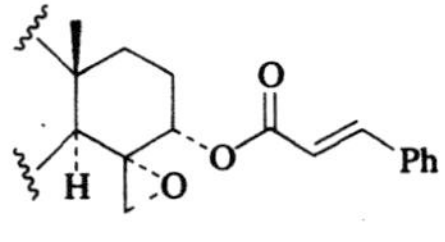 I (8) + 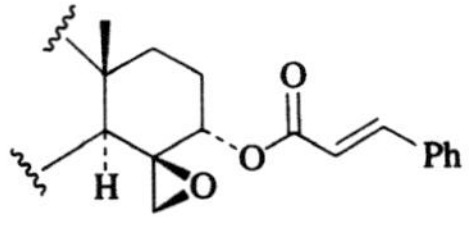 II (—) + 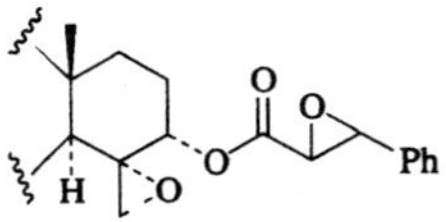 (77) I:II = 99:1	343

TABLE 4. DIASTEREOSELECTIVE EPOXIDATIONS OF OLEFINS BY DIOXIRANES (*Continued*)

Substrate	Conditions	Product(s) and Yield(s) (%), Diastereomeric Ratio (dr)	Refs.
C_{37} 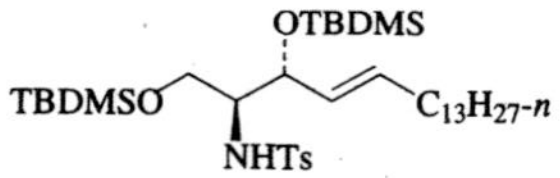 TBDMSO, OTBDMS, NHTs, $C_{13}H_{27}$-*n*	DMD, acetone/solvent	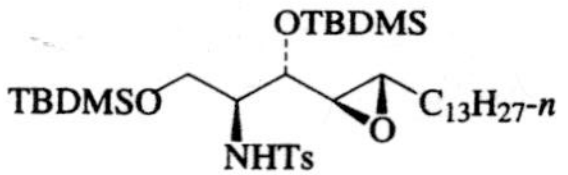 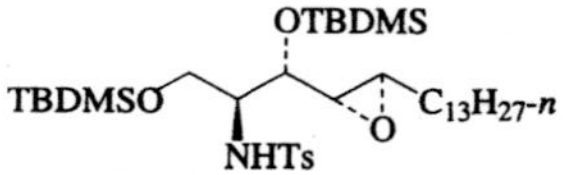 TBDMSO, OTBDMS, NHTs, O, $C_{13}H_{27}$-*n* **I** + TBDMSO, OTBDMS, NHTs, O, $C_{13}H_{27}$-*n* **II**	363
	Ethylmethyldioxirane, 2-butanone, rt, 96 h	**I** (47) + **II** (—) **I**:**II** = 82:18	363
	Ethylmethyldioxirane, 2-butanone, 4°, 96 h	**I** + **II** (tr) **I**:**II** = 72:28	363
C_{40} 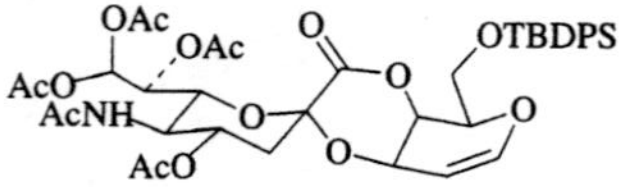 OAc, OAc, AcO, AcNH, AcO, O, OTBDPS	DMD, acetone/CH_2Cl_2, 0° to rt, 1 h	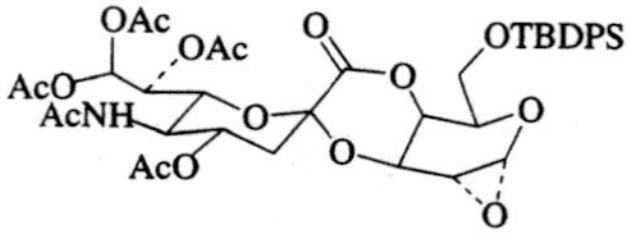 OAc, OAc, AcO, AcNH, AcO, O, OTBDPS (—)	339

Solvent	Temp	Time	I	I:II
—	rt	24 h	(56)	80:20
—	4°	72 h	(57)	84:16
—	–23°	96 h	(tr)	87:13
hexane	rt	24 h	(16)	58:42
MeOH	rt	24 h	(tr)	—
CH_2Cl_2	rt	24 h	(30)	71:29
$CHCl_3$	rt	24 h	(tr)	—
CCl_4	rt	24 h	(58)	80:20

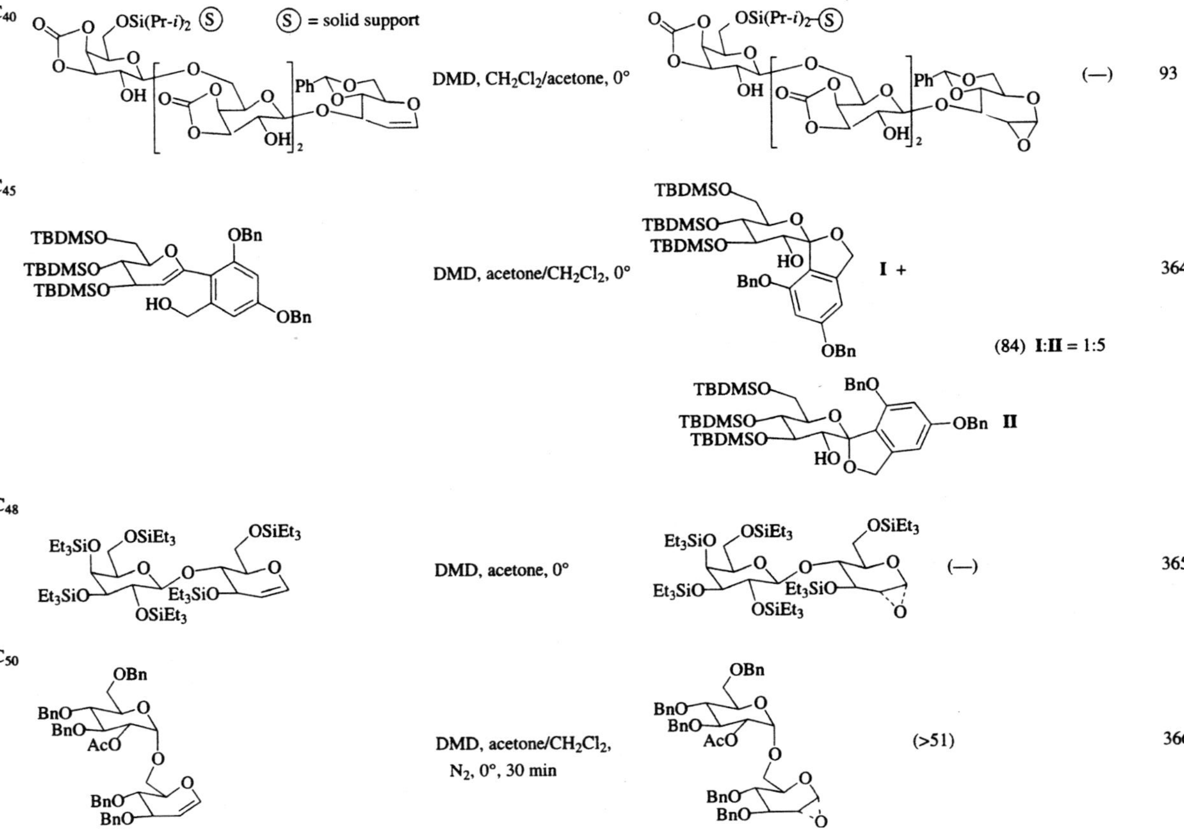

Substrate	Conditions	Product(s) and Yield(s) (%)	Refs.
C_{40} (Ⓢ = solid support)	DMD, CH_2Cl_2/acetone, 0°	(—)	93
C_{45}	DMD, acetone/CH_2Cl_2, 0°	I + II (84) I:II = 1:5	364
C_{48}	DMD, acetone, 0°	(—)	365
C_{50}	DMD, acetone/CH_2Cl_2, N_2, 0°, 30 min	(>51)	366

TABLE 4. DIASTEREOSELECTIVE EPOXIDATIONS OF OLEFINS BY DIOXIRANES (*Continued*)

Substrate	Conditions	Product(s) and Yield(s) (%), Diastereomeric Ratio (dr)	Refs.
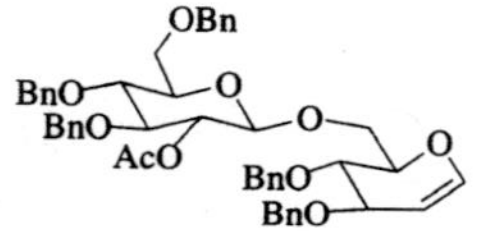 OBn, BnO, BnO, AcO, O, O, BnO, BnO, O	DMD, acetone/CH_2Cl_2, N_2, 0°, 30 min	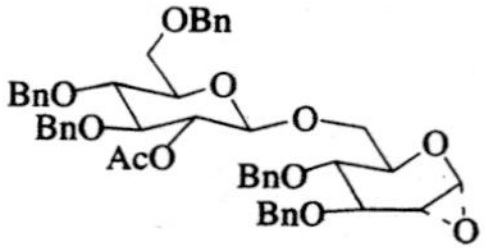 OBn, BnO, BnO, AcO, O, O, BnO, BnO, O, O (>43)	366
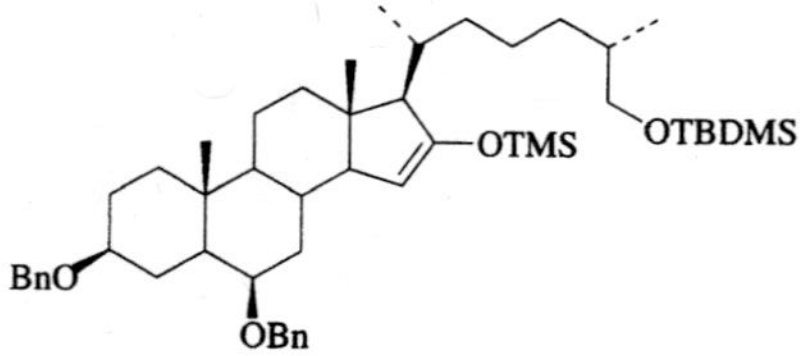 OTMS, OTBDMS, BnO, OBn	1. DMD, acetone/CH_2Cl_2, 0°, 1 h 2. Camphorsulfonic acid, acetone, 4°, 16 h	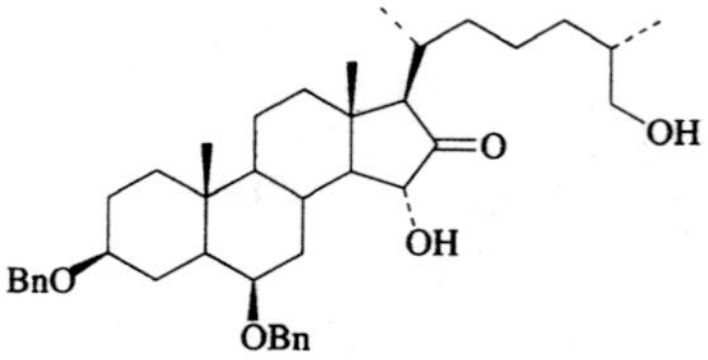 O, OH, OH, BnO, OBn (71)	367
C_{79}			
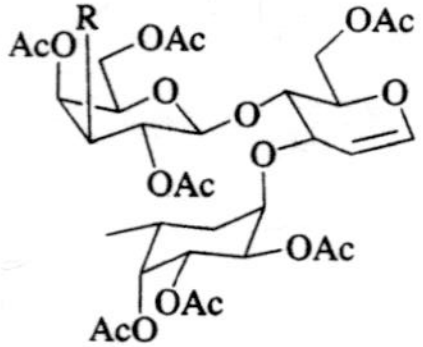 R, AcO, OAc, OAc, O, O, O, OAc, OAc, AcO, OAc	DMD, acetone/CH_2Cl_2, 0°, 45 min	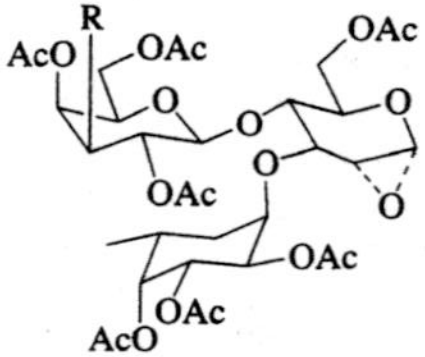 R, 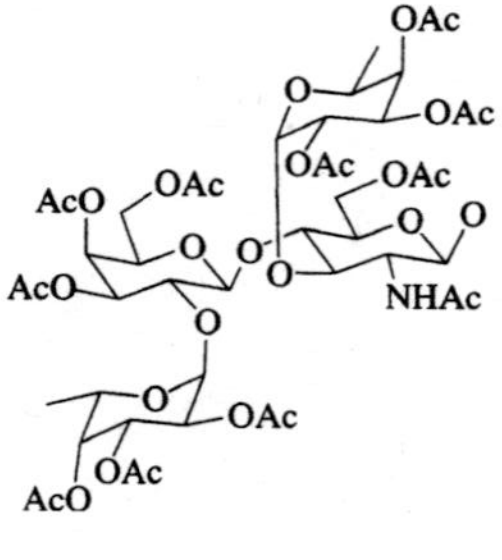 AcO, OAc, OAc, O, O, O, OAc, O, OAc, AcO, OAc (—) R = AcO, OAc, AcO, O, O, O, OAc, AcO, OAc, OAc, OAc, OAc, OAc, O, O, NHAc	95

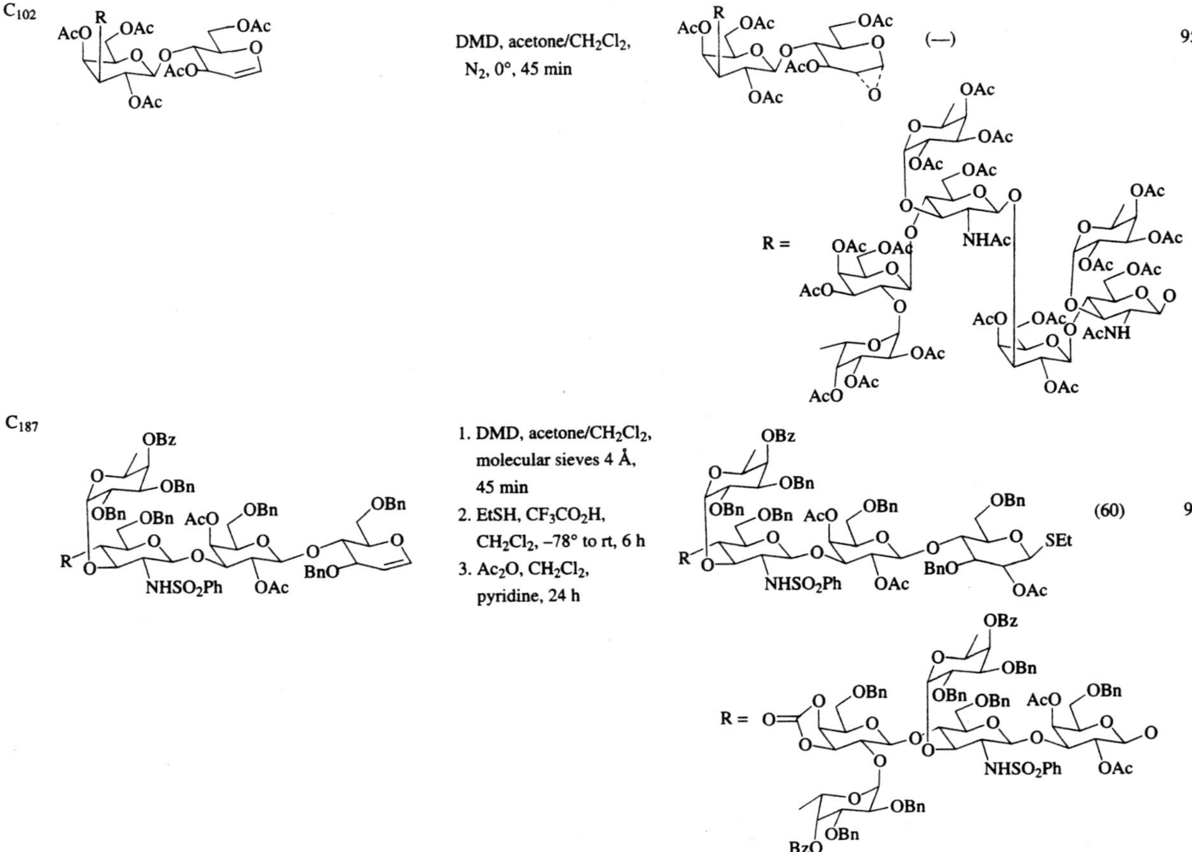

C_{102}	DMD, acetone/CH_2Cl_2, N_2, 0°, 45 min	(—)	95
C_{187}	1. DMD, acetone/CH_2Cl_2, molecular sieves 4 Å, 45 min 2. EtSH, CF_3CO_2H, CH_2Cl_2, –78° to rt, 6 h 3. Ac_2O, CH_2Cl_2, pyridine, 24 h	(60)	95

TABLE 4. DIASTEREOSELECTIVE EPOXIDATIONS OF OLEFINS BY DIOXIRANES (*Continued*)

Substrate	Conditions	Product(s) and Yield(s) (%), Diastereomeric Ratio (dr)	Refs.

[a]TDMPP is the dianion of 5,10,15,20-tetrakis(2,6-dimethoxyphenyl)porphyrin.

[b]Cl16TDMPP is the dianion of 2,3,7,8,12,13,17, 18-octachloro-5,10,15,20-tetrakis(3,5-dichloro-2,6-dimethoxyphenyl)porphyrin.

[c]Cl8TDMPP is the dianion of 5,10,15,20-tetrakis(3,5-dichloro-2,6-dimethoxyphenyl)porphyrin.

TABLE 5. ENANTIOSELECTIVE EPOXIDATION OF OLEFINS BY ENANTIOMERICALLY ENRICHED DIOXIRANES

Substrate	Ketone/Catalyst	Conditions	Product(s), Yield(s) (%) and Enantioselectivities (ee)	Refs.
C_{6-15} 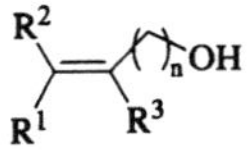	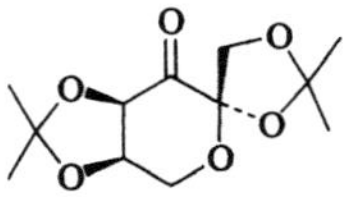	Oxone®, K_2CO_3, DMM/ CH_3CN, buffer (pH 9.3), Bu_4NHSO_4	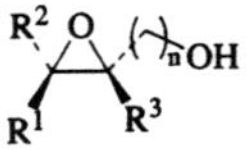 I	149a
	"	Oxone®, K_2CO_3, DME, buffer (pH 9.3), Bu_4NHSO_4, –15°	I	149a

n	R^1	R^2	R^3	Temp	Time		% ee
2	H	Et	H	–10°	3 h	(82)	90
1	H	Et	Me	–10°	3 h	(83)	91
1	H	*n*-Pr	H	–10°	3 h	(68)	91
1	H	—$(CH_2)_4$—		–10°	3 h	(92)	92
1	H	Ph	H	–10°	3 h	(85)	94
2	H	Ph	H	0°	2 h	(90)	91
1	H	Bn	H	–10°	2 h	(45)	91
2	H	*n*-C_6H_{13}	H	–10°	2 h	(86)	90
3	H	Bn	H	0°	2 h	(87)	91
1	Ph	Ph	H	0°	2 h	(87)	94

n	R^1	R^2	R^3	Time		% ee
1	H	Et	Me	5 h	(85)	92
1	H	—$(CH_2)_4$—		5 h	(93)	94
1	H	Bn	H	7 h	(43)	92
1	H	Ph	Me	5 h	(75)	74
2	H	*n*-C_6H_{13}	H	5 h	(83)	91

TABLE 5. ENANTIOSELECTIVE EPOXIDATION OF OLEFINS BY ENANTIOMERICALLY ENRICHED DIOXIRANES (*Continued*)

Substrate	Ketone/Catalyst	Conditions	Product(s), Yield(s) (%) and Enantioselectivities (ee)	Refs.
C_{6-24}				
$R^2R^3C{=}CHR^1$		Oxone®, $NaHCO_3$, CH_3CN/H_2O, EDTA, 0°, 2 h	epoxide	117

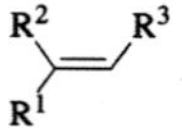

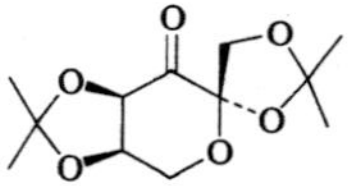

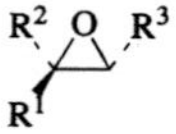

R^1	R^2	R^3		% ee
n-Pr	H	CH_2OH	(60)	78
Et	H	$(CH_2)_2OH$	(70)	70
Ph	H	CH_2Cl	(61)	93
Ph	H	CH_2OH	(81)	88
Ph	H	Me	(60)	84
Ph	—$(CH_2)_4$—		(69)	91
n-Pr	H	$CH_2OTBDMS$	(80)	93
Et	H	$(CH_2)_2OTBDMS$	(84)	87
Ph	H	Ph	(73)	95
n-C_6H_{13}	H	*n*-C_6H_{13}	(81)	90
Ph	Me	Ph	(73)	92
Ph	H	$CH_2OTBDMS$	(74)	93
Ph	Ph	*n*-$C_{10}H_{21}$	(65)	92

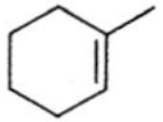

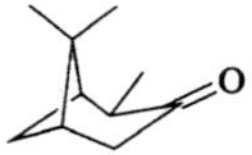

Substrate	Ketone/Catalyst	Conditions	Product(s), Yield(s) (%) and Enantioselectivities (ee)	Refs.
C_7 1-methylcyclohexene		Oxone®, CH_2Cl_2/buffer, Bu_4NHSO_4, 25°	I (87) 6% ee	105
		Oxone®, CH_2Cl_2/buffer, Bu_4NHSO_4, 5°	I (80) 12% ee	105
	Ph	Oxone®, CH_2Cl_2/buffer, Bu_4NHSO_4, 5°	(85) 5% ee	105

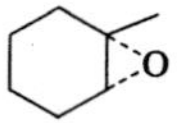

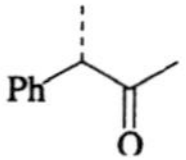

C_{7-24}

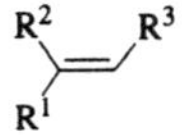

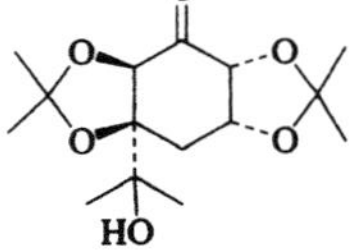

R² R³ alkene (R¹ on C bearing R²)

Oxone®, K_2CO_3, CH_3CN/ DMM, $Na_2B_4O_7$ buffer (pH 10.5)

I (epoxide) — Ref. 25

R¹	R²	R³		% ee
Me	—(CH₂)₄—		(77)	81
Me	Me	*t*-Bu	(35)	91
Ph	Me	Me	(89)	97
Me	Me	Ph	(93)	76
Ph	—(CH₂)₄—		(94)	98
$EtO_2C(CH_2)_2$	Me	*c*-C_6H_{11}	(77)	81
$EtO_2C(CH_2)_2$	Me	*n*-C_6H_{13}	(89)	94
Me	Me	*n*-$C_{10}H_{21}$	(97)	87
Ph	Me	Ph	(89)	96
Et	Et	*n*-$C_{10}H_{21}$	(94)	89
Ph	Ph	Ph	(54)	97
Ph	Ph	*n*-$C_{10}H_{21}$	(92)	97

Oxone®, K_2CO_3, DME (DMM), K_2CO_3/AcOH buffer (pH 10.5)

I 115, 114

R¹	R²	R³	Temp	Time		% ee
Me	—(CH₂)₄—		−15°	4 h	(100)	12
Ph	H	H	−15°	4 h	(79)	69
Ph	H	CH_2Cl	0°	6 h	(95)	82
Ph	H	CH_2Cl	−10°	6 h	(95)	84
Ph	H	Me	−15°	4 h	(94)	80
Ph	H	MeCO	0°	8 h	(75)	82
Ph	H	CO_2Me	0°	8 h	(35)	89
n-Bu	—(CH₂)₄—		−15°	4 h	(96)	13
Ph	—(CH₂)₄—		−10°	4 h	(94)	85
Ph	H	*i*-PrCO	0°	8 h	(70)	89
Ph	H	Ph	−10°	6 h	(91)	96
Ph	H	PhCO	0°	6 h	(85)	96
Ph	Ph	Ph	0°	6 h	(95)	92

TABLE 5. ENANTIOSELECTIVE EPOXIDATION OF OLEFINS BY ENANTIOMERICALLY ENRICHED DIOXIRANES (*Continued*)

Substrate	Ketone/Catalyst	Conditions	Product(s), Yield(s) (%) and Enantioselectivities (ee)	Refs.
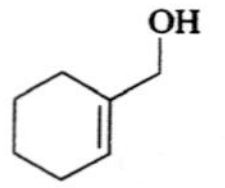 [structure]	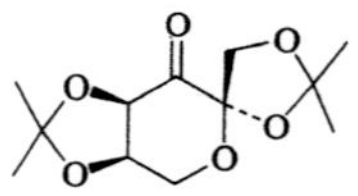 [structure]	Oxone®, CH_3CN/DMM, K_2CO_3/AcOH buffer, −10°	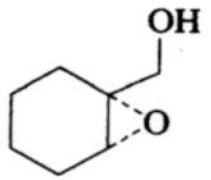 [structure] (83) 92% ee	124, 123
C_8				
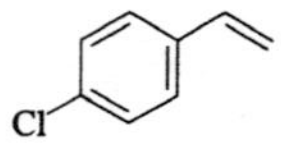 [structure]	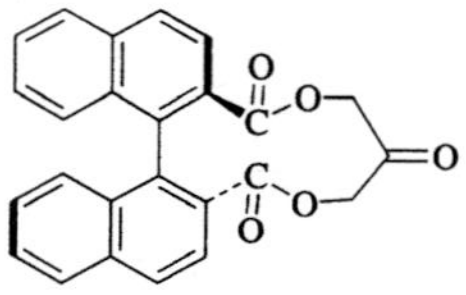 [structure]	Oxone®, $NaHCO_3$, CH_3CN/H_2O, EDTA, rt, 60 min	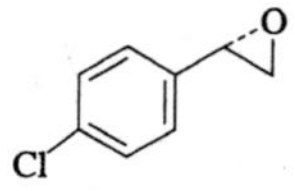 [structure] (83) 18% ee	24
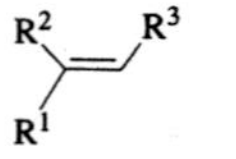 [structure]	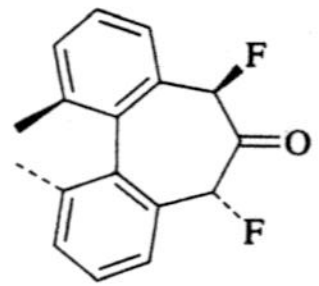 [structure]	Oxone®, K_2CO_3, CH_3CN/H_2O, 0°	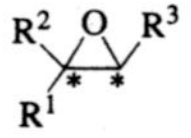 [structure]	11

R^1	R^2	R^3	Time		% ee
4-ClC_6H_4	H	H	30 h	(55)	43
Ph	H	Me	10 h	(80)	88
Ph	H	CH_2OH	7 h	(93)	89
Ph	—$(CH_2)_4$—		4 h	(78)	59
Me	H	$(CH_2)_3OBn$	11 h	(72)	68
Ph	H	Ph	72 h	(46)	94

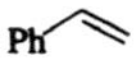

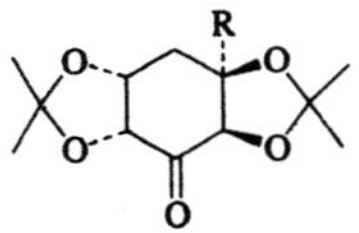

Ph

Oxone®, K_2CO_3, DME, K_2CO_3/AcOH buffer

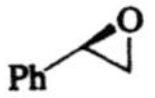

Ph I

115, 114

R	Temp	Time	% Convn	% ee
H	–15°	6 h	13	67
CH_2F	–15°	4 h	76	67
CO_2Me	–10°	4 h	90	67
CH_2OAc	–10°	4 h	96	65
CMe_2OH	–15°	4 h	79	69
CMe_2OMe	–15°	4 h	100	70
$CH_2OTBDMS$	–10°	4 h	100	66
CH_2OBz	–10°	4 h	99	65
CH_2OTs	–10°	4 h	70	67
CPh_2OH	–15°	1 h	7	59

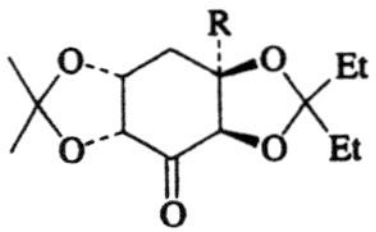

Oxone®, K_2CO_3, DME/DMM, K_2CO_3/AcOH buffer, -15°, 4 h

I

115, 114

R	% Convn	% ee
CO_2Me	94	60
CH_2OAc	66	67
CMe_2OH	31	65

TABLE 5. ENANTIOSELECTIVE EPOXIDATION OF OLEFINS BY ENANTIOMERICALLY ENRICHED DIOXIRANES (*Continued*)

Substrate	Ketone/Catalyst	Conditions	Product(s), Yield(s) (%) and Enantioselectivities (ee)	Refs.
		Oxone®, K_2CO_3, CH_3CN, $Na_2B_4O_7$ buffer, Na_2EDTA, 0°, 1.5 h	I	119
		Oxone®, K_2CO_3, CH_3CN, $Na_2B_4O_7$ buffer, Na_2EDTA, 0°, 1.5 h	I	119

First catalyst:

R^1	R^2	% Convn	% ee
Me	Me	38	38
—$(CH_2)_4$—		39	37
Et	Et	42	42
—$(CH_2)_5$—		42	36
n-Pr	*n*-Pr	29	43
Ph	Ph	30	20
Bn	Bn	6	21

Second catalyst:

R^1	R^2	R^3	R^4	R^5	% Convn	% ee
Me	Me	Me	Me	F	8	32
Me	Me	Me	Me	H	100	15
Me	Me	Et	H	H	79	10
Me	Me	*i*-Pr	H	H	60	7
Me	Me	—$(CH_2)_4$—		H	100	18
—$(CH_2)_4$—		Me	Me	H	92	27
Me	Me	Et	Et	H	92	16
Et	Et	Me	Me	H	77	12
Me	Me	—$(CH_2)_5$—		H	77	17
—$(CH_2)_5$—		Me	Me	H	93	16
—$(CH_2)_4$—		—$(CH_2)_4$—		H	100	30
Et	Et	Et	Et	H	26	12
—$(CH_2)_5$—		—$(CH_2)_5$—		H	59	14
—$(CH_2)_6$—		—$(CH_2)_6$—		H	43	17
Me	Me	Bn	Bn	H	6	3 (*S*)

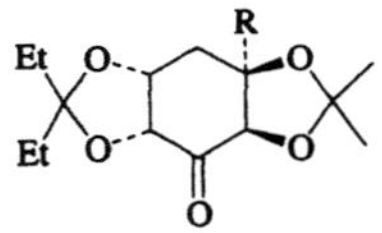

Catalyst	Conditions	Substrate	Results	Refs.
	Oxone®, K_2CO_3, DME/ DMM, K_2CO_3/AcOH buffer, -15°, 4 h	I	R / % Convn / % ee: CO_2Me 77 65 CH_2OAc 100 62 CMe_2OH 80 65	115, 114
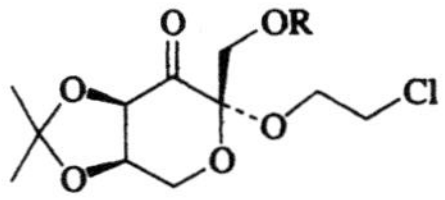	Oxone®, K_2CO_3, CH_3CN, $Na_2B_4O_7$ buffer, Na_2EDTA, 0°, 1.5 h	I	R / % Convn / % ee: H 0 — Ac 5 23 TBDMS 0 —	119
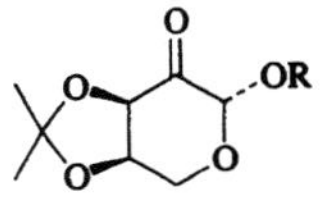	Oxone®, K_2CO_3, CH_3CN, $Na_2B_4O_7$ buffer, Na_2EDTA, 0°, 1.5 h	I	R / % Convn / % ee: Me 14 27 $(CH_2)_2Cl$ 15 29	119

TABLE 5. ENANTIOSELECTIVE EPOXIDATION OF OLEFINS BY ENANTIOMERICALLY ENRICHED DIOXIRANES (*Continued*)

Substrate	Ketone/Catalyst	Conditions	Product(s), Yield(s) (%) and Enantioselectivities (ee)	Refs.
	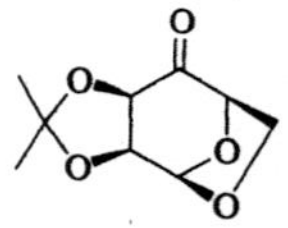	Oxone®, K_2CO_3, CH_3CN, $Na_2B_4O_7$ buffer, Na_2EDTA, 0°, 1.5 h	I (40) 10% ee	119
	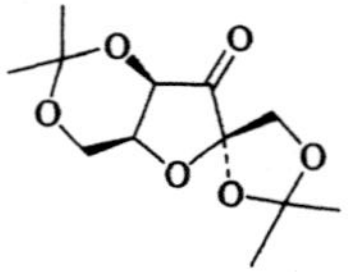	Oxone®, K_2CO_3, CH_3CN, $Na_2B_4O_7$ buffer, Na_2EDTA, 0°, 1.5 h	I (41) 15% ee	119
C_{8-10} 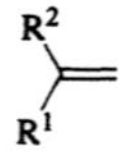 R² R¹	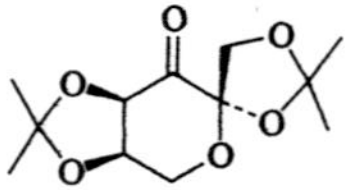	Oxone®, K_2CO_3, CH_3CN, $Na_2B_4O_7$ buffer, Na_2EDTA, −10°, 2 h	 R² O R¹ I R¹ / R² / / % ee Ph / H / (64) / 14 Ph / Me / (81) / 28 *i*-Pr_3SiCH_2 / H / (92) / 35 *n*-C_8H_{17} / H / (80) / 27	25
	"	Oxone®, K_2CO_3, CH_3CN/DMM, $Na_2B_4O_7$ buffer, Na_2EDTA, −10°, 2 h	I R¹ / R² / / % ee Ph / H / (90) / 24 Ph / Me / (95) / 20 *i*-Pr_3SiCH_2 / H / (99) / 31 *n*-C_8H_{17} / H / (92) / 17	25

C_{8-20}

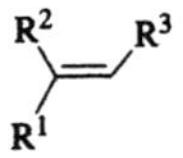

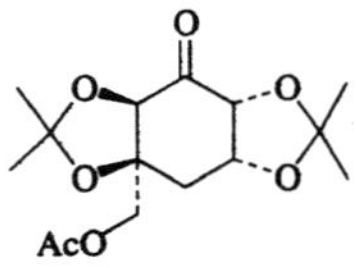

R^2, R^3, R^1 (alkene)

AcO (ketone catalyst)

Oxone®, K_2CO_3, DME (DMM), K_2CO_3/AcOH buffer

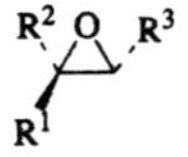

R^2 O R^3, R^1 (epoxide) I

115, 114

R^1	R^2	R^3	Temp	Time		% ee
Ph	H	H	−10°	4 h	(90)	65
Ph	H	CH_2Cl	0°	5 h	(94)	77
Ph	H	Me	−15°	4 h	(92)	75
Ph	Me	H	−10°	3 h	(92)	52
n-C_8H_{17}	H	H	−10°	4 h	(85)	15
Ph	H	CO_2Et	0°	8 h	(34)	86
2-Naphthyl	H	H	−10°	4 h	(89)	54
Ph	—$(CH_2)_4$—		−10°	4 h	(95)	68
$(CH_2)_2CO_2Me$	Me	C_6H_{11}	−10°	4 h	(96)	43
Ph	H	Ph	−10°	6 h	(95)	90
n-C_6H_{13}	H	n-C_6H_{13}	−10°	4 h	(51)	42
Ph	H	PhCO	0°	6 h	(80)	94
Ph	Ph	Ph	0°	5 h	(86)	87

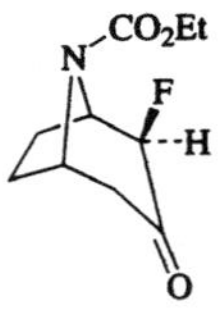

N-CO_2Et, F, H, O (ketone catalyst)

Oxone®, $NaHCO_3$, MeCN, EDTA, H_2O

I

108

R^1	R^2	R^3	Time		% ee
Ph	H	H	<2 h	(33)	29
Ph	H	CO_2Me	24 h	(33)	64
Ph	—$(CH_2)_4$—		<6 h	(97)	69
Ph	H	Ph	<3 h	(88)	76
Ph	Me	Ph	<4 h	(100)	73
Ph	Ph	Ph	<4 h	(100)	83

TABLE 5. ENANTIOSELECTIVE EPOXIDATION OF OLEFINS BY ENANTIOMERICALLY ENRICHED DIOXIRANES (*Continued*)

Substrate	Ketone/Catalyst	Conditions	Product(s), Yield(s) (%) and Enantioselectivities (ee)	Refs.
C_8 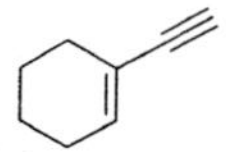	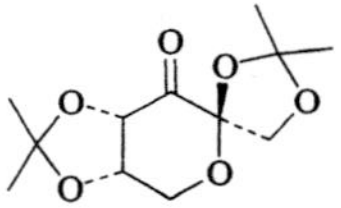	Oxone®, K_2CO_3, CH_3CN/DMM, K_2CO_3/AcOH buffer, –10°, 3 h	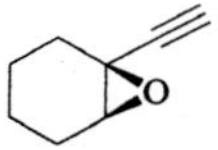 (75) 94% ee	123, 124
C_{8-11} 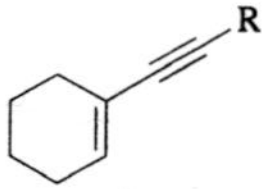	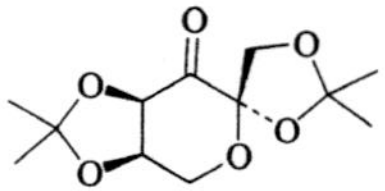	Oxone®, K_2CO_3, CH_3CN/DMM, K_2CO_3/AcOH buffer	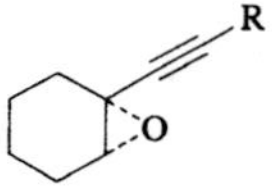 R / Temp / Time / % ee: H –10° 3 h (78) 93 Me –10° 3 h (88) 90 CO_2Et 0° 2 h (71) 93 TMS –10° 3 h (86) 93	123, 124
C_{8-14} 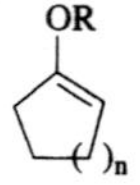	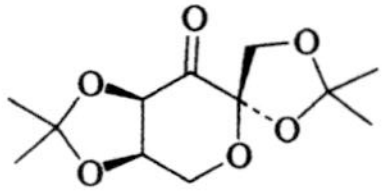	Oxone®, CH_3CN (DMM), $Na_2B_4O_7$ (or K_2CO_3/AcOH) buffer, 0°	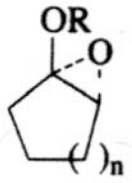 n / R / Time / % ee: 2 Ac 2.0 h (59) 74 1 PhCO 1.5 h (79) 80 2 PhCO 1.5 h (82) 93 3 PhCO 1.5 h (87) 91 4 PhCO 1.5 h (82) 95	121
C_8 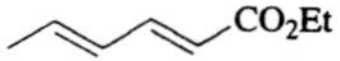	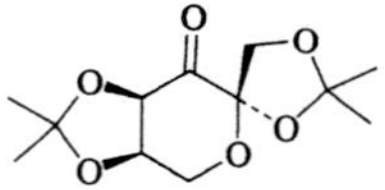	Oxone®, K_2CO_3, CH_3CN/DMM, $Na_2B_4O_7$ buffer (pH 10.5), EDTA, 0°, 4 h	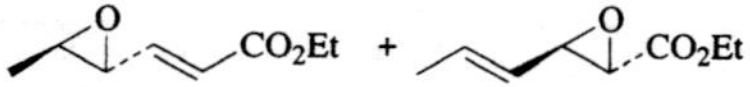 CO_2Et + CO_2Et **I** (41) 96% ee **II** **I** + **II** (—) **I**:**II** = 7:1	122

Substrate	Catalyst	Conditions	Product(s) and Yield(s) (%)	Refs.
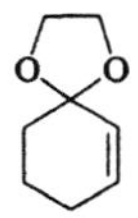	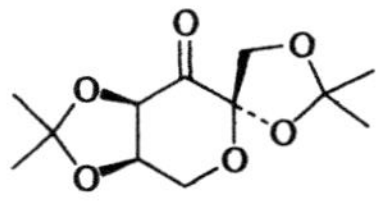	Oxone®, K_2CO_3, CH_3CN, $Na_2B_4O_7$ buffer, Na_2EDTA, −10°, 2 h	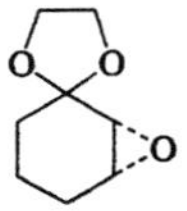 I (50) 56% ee	25
	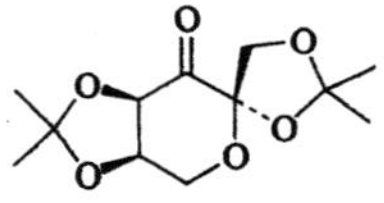	Oxone®, K_2CO_3, CH_3CN/DMM, $Na_2B_4O_7$ buffer, Na_2EDTA, −10°, 2 h	I (43) 61% ee	25
	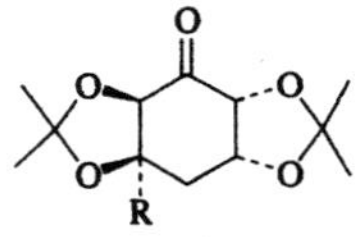	Oxone®, K_2CO_3, DME, K_2CO_3/AcOH buffer, −10°	I	115, 114
	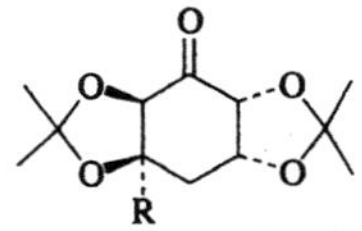 R = CH_2OAc	Oxone®, K_2CO_3, DME, K_2CO_3/AcOH buffer, −10°, 6 h	I (78) 68% ee	115, 114

R	Time	% Convn	% ee
H	6 h	9	66
CH_2F	6 h	66	71
CO_2Me	6 h	56	66
CH_2OAc	6 h	82	68
CMe_2OH	1 h	55	45
CMe_2OMe	6 h	47	40
$CH_2OTBDMS$	6 h	70	73
CH_2OBz	6 h	67	71
CH_2OTs	6 h	62	71
CPh_2OH	6 h	7	88

TABLE 5. ENANTIOSELECTIVE EPOXIDATION OF OLEFINS BY ENANTIOMERICALLY ENRICHED DIOXIRANES (*Continued*)

Substrate	Ketone/Catalyst	Conditions	Product(s), Yield(s) (%) and Enantioselectivities (ee)	Refs.
	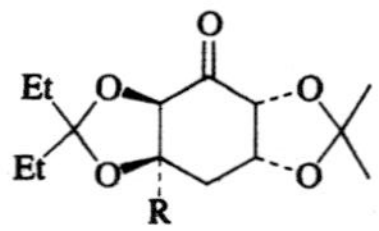	Oxone®, K_2CO_3, DME, K_2CO_3/AcOH buffer, −10°, 6 h	I; R, % Convn, % ee: CO_2Me 41 65; CH_2OAc 73 62; CMe_2OH 47 44	115, 114
	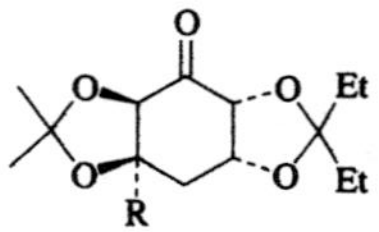	Oxone®, K_2CO_3, DME, K_2CO_3/AcOH buffer, −10°, 6 h	I; R, % Convn, % ee: CO_2Me 59 61; CH_2OAc 75 63; CMe_2OH 61 42	115, 114
	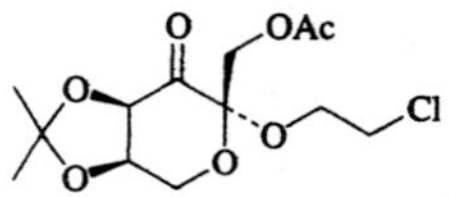	Oxone®, K_2CO_3, CH_3CN, $Na_2B_4O_7$ buffer, Na_2EDTA, 0°, 1.5 h	I (8) 44% ee	119
	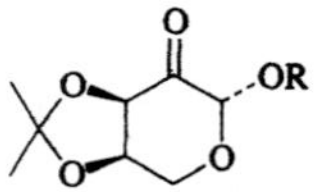	Oxone®, K_2CO_3, CH_3CN, $Na_2B_4O_7$ buffer, Na_2EDTA, 0°, 1.5 h	I; R, % Convn, % ee: Me 13 42; $(CH_2)_2Cl$ 9 41	119
	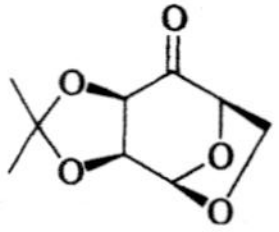	Oxone®, K_2CO_3, CH_3CN, $Na_2B_4O_7$ buffer, Na_2EDTA, 0°, 1.5 h	I (0)	119

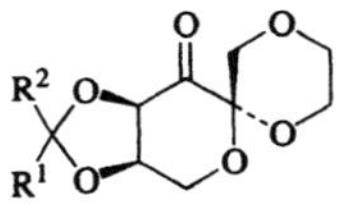

Oxone®, K_2CO_3, CH_3CN, $Na_2B_4O_7$ buffer, Na_2EDTA, 0°, 1.5 h — I — 119

R^1	R^2	% Convn	% ee
Me	Me	30	60
—$(CH_2)_4$—		29	57
Et	Et	24	49
—$(CH_2)_5$—		31	52
n-Pr	*n*-Pr	13	46
Ph	Ph	8	4
Bn	Bn	8	39

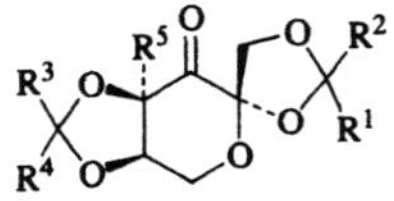

Oxone®, K_2CO_3, CH_3CN, $Na_2B_4O_7$ buffer, Na_2EDTA, 0°, 1.5 h — I — 119

R^1	R^2	R^3	R^4	R^5	% Convn	% ee
Me	Me	Me	Me	F	23	12
Me	Me	Me	Me	H	53	51
Me	Me	Et	H	H	55	42
Me	Me	*i*-Pr	H	H	55	43
Me	Me	—$(CH_2)_4$—		H	56	48
—$(CH_2)_4$—		Me	Me	H	57	49
Me	Me	Et	Et	H	19	43
Et	Et	Me	Me	H	24	55
Me	Me	—$(CH_2)_5$—		H	46	46
—$(CH_2)_5$—		Me	Me	H	50	52
—$(CH_2)_4$—		—$(CH_2)_4$—		H	73	47
Et	Et	Et	Et	H	17	47
—$(CH_2)_5$—		—$(CH_2)_5$—		H	43	44
—$(CH_2)_6$—		—$(CH_2)_6$—		H	6	51
Me	Me	Bn	Bn	H	0	—

TABLE 5. ENANTIOSELECTIVE EPOXIDATION OF OLEFINS BY ENANTIOMERICALLY ENRICHED DIOXIRANES (*Continued*)

Substrate	Ketone/Catalyst	Conditions	Product(s), Yield(s) (%) and Enantioselectivities (ee)	Refs.
	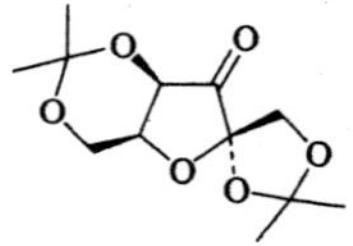	Oxone®, K_2CO_3, CH_3CN, $Na_2B_4O_7$ buffer, Na_2EDTA, 0°, 1.5 h	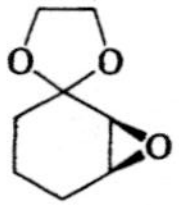 O O (46) 71% ee	119
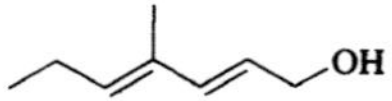 OH	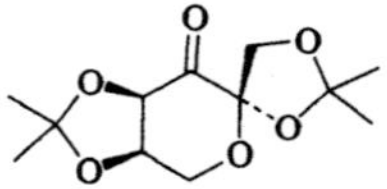 O O O O O	Oxone®, K_2CO_3, CH_3CN/DMM, $Na_2B_4O_7$ buffer (pH 10.5), EDTA, 0°, 1.5 h	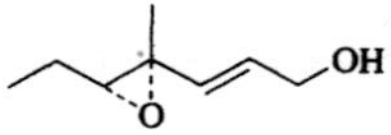 OH (68) 90% ee	122
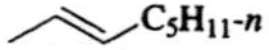 C_5H_{11}-*n*	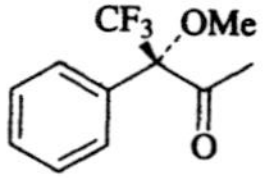 CF_3 OMe O	Oxone®, CH_2Cl_2/H_2O (pH 7.5), Bu_4NHSO_4, 2 to 5°, 48 h	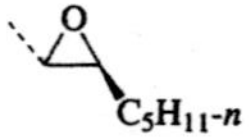 O C_5H_{11}-*n* (70) 20% ee	106
	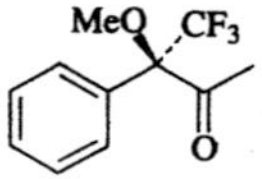 MeO CF_3 O	Oxone®, CH_2Cl_2/H_2O (pH 7.5), Bu_4NHSO_4, 2 to 5°, 48 h	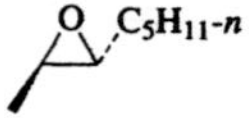 O C_5H_{11}-*n* (68) 18% ee	106
C_{9-28} 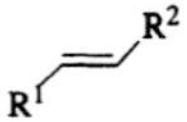 R^1 R^2	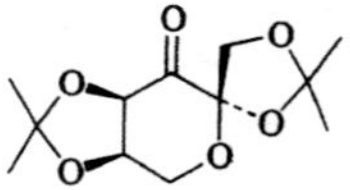 O O O O O	Oxone®, K_2CO_3, CH_3CN/DMM, $Na_2B_4O_7$ buffer, Na_2EDTA	 R^1 O R^2	25

R^1	R^2	Time	Temp		% ee
Ph	CH_2Cl	1.5 h	0°	(49)	96
Ph	Me	2 h	–10°	(94)	96
2-MeC_6H_4	Me	2 h	–10°	(91)	93
2-MeC_6H_4	*i*-Pr	2 h	–10°	(78)	96
n-Pr	$CH_2OTBDMS$	2 h	–10°	(83)	95
Et	$(CH_2)_2OTBDMS$	2 h	–10°	(85)	93
Bn	$(CH_2)_2CO_2Me$	1.5 h	0°	(68)	92
Ph	Ph	0.5 h	20°	(85)	98
Ph	Ph	1.5 h	0°	(78)	99
n-C_6H_{13}	*n*-C_6H_{13}	2 h	–10°	(89)	95
Ph	$CH_2OTBDMS$	1.5 h	0°	(71)	95
Ph	CH_2OTr	1.5 h	0°	(55)	94

C_9

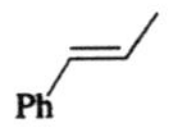

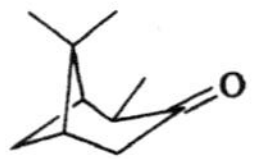

Ph

Ketone	Conditions	Product	Ref.
O	Oxone®, CH_2Cl_2, buffer, Bu_4NHSO_4, 5°	O, Ph; I (60) 12% ee	105
Ph, O	Oxone®, CH_2Cl_2, buffer, Bu_4NHSO_4, 5°	I (85) 9.5% ee	105
CF_3, O, O	Oxone®, CH_2Cl_2/H_2O (pH 7.5), Bu_4NHSO_4, 2 to 5°, 40 h	I (71) 13% ee	106
CF_3, OMe, O	Oxone®, CH_2Cl_2/H_2O (pH 7.5), Bu_4NHSO_4, 2 to 5°, 42 h	I (61) 18% ee	106

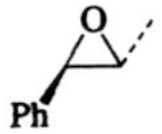

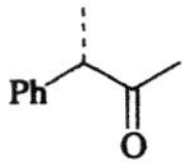

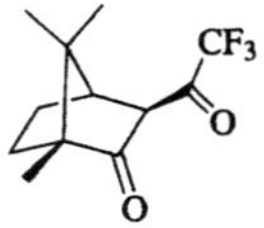

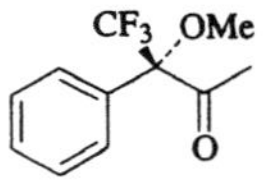

TABLE 5. ENANTIOSELECTIVE EPOXIDATION OF OLEFINS BY ENANTIOMERICALLY ENRICHED DIOXIRANES (*Continued*)

Substrate	Ketone/Catalyst	Conditions	Product(s), Yield(s) (%) and Enantioselectivities (ee)	Refs.
	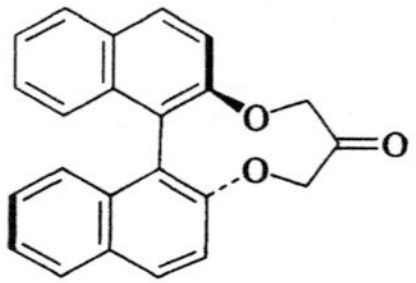	Oxone®, $NaHCO_3$, CH_3CN/H_2O, EDTA, rt, 1 h	**I** (61) 29% ee	112
		H_2O_2, CH_3CN, K_2CO_3 buffer, EDTA, 0°, 7 h	**I** (84) 92% ee	368
		Oxone®, $NaHCO_3$, CH_3CN/H_2O, EDTA, 0°	**I** Time / % ee 1 h (31) >95 2 h (39) >95 4 h (40) 89 8 h (47) 85	25
		Oxone®, K_2CO_3, CH_3CN/DMM, $Na_2B_4O_7$ buffer, Na_2EDTA	**I** Temp / % Convn / % ee –11° 99 96 –6° 97 95 –2° 98 95 2° 99 95 8° 99 94 20° 99 93 30° 97 91	25
		Oxone®, K_2CO_3, solvent, $Na_2B_4O_7$ buffer, EDTA	**I**	25

Solvent	Time	Temp	% Convn	% ee
CH_3CN	20 min	20°	100	89
CH_3CN	1.5 h	0°	96	92
EtCN	1 h	20°	11	80
DMM	1 h	20°	36	91
DME	20 min	20°	100	89
DME	1.5 h	0°	92	89
DMF	20 min	20°	95	86
Dioxane	20 min	20°	100	86
Dioxane	1.5 h	0°	96	86
THF	1 h	20°	18	74
Et_2O	1 h	20°	0	—
CH_2Cl_2	0.5 h	20°	<3	—
CH_3CN/DMM (2:1)	20 min	20°	100	90
CH_3CN/DMM (2:1)	1.5 h	0°	100	92
CH_3CN/DMM (1:1)	20 min	20°	98	91
CH_3CN/DMM (1:1)	1.5 h	0°	100	93
CH_3CN/DMM (1:2)	20 min	20°	94	92
CH_3CN/DMM (1:2)	1.5 h	0°	88	94
CH_3CN/DMM (1:4)	1.5 h	0°	77	94
DMM/DME (1:1)	25 min	20°	66	92
CH_3CN/DMM/DME (1:1:2)	20 min	20°	100	90
CH_3CN/DMM/DME (1:7:7)	20 m in	20°	89	90
CH_3CN/THF (1:1)	25 min	20°	63	82
Et_2O/CH_3CN (1:1)	25 min	20°	28	84

TABLE 5. ENANTIOSELECTIVE EPOXIDATION OF OLEFINS BY ENANTIOMERICALLY ENRICHED DIOXIRANES (*Continued*)

Substrate	Ketone/Catalyst	Conditions	Product(s), Yield(s) (%) and Enantioselectivities (ee)	Refs.
	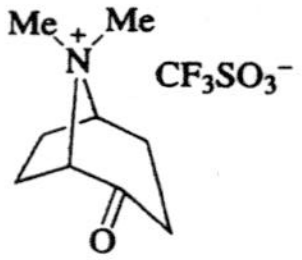 Me_2N^+ bicyclic ketone, $CF_3SO_3^-$	Oxone®, $NaHCO_3$, CH_3CN/H_2O, 0°, 5 h	I 21% convn	107
	bis(isopropylidene) ketone (R)	Oxone®, K_2CO_3, DME, K_2CO_3/AcOH buffer, −15°	I (see below)	115, 114
	Et, Et / isopropylidene ketone (R)	Oxone®, K_2CO_3, DME, K_2CO_3/AcOH buffer, −15°, 4 h	I (see below)	115, 114

R	Time	% Convn	% ee
H	4 h	29	73
CH_2F	4 h	78	73
CO_2Me	4 h	61	75
CH_2OAc	4 h	95	75
CMe_2OH	4 h	97	80
CMe_2OMe	4 h	95	80
$CH_2OTBDMS$	4 h	77	73
CH_2OBz	4 h	76	72
CH_2OTs	4 h	60	72
CPh_2OH	4 h	7	50

R	% Convn	% ee
CO_2Me	43	74
CH_2OAc	60	75
CMe_2OH	35	76

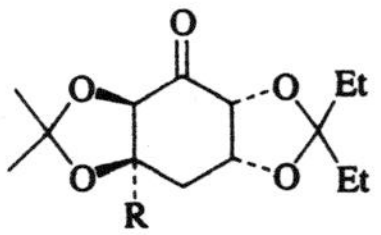

Oxone®, K_2CO_3, DME, K_2CO_3/AcOH buffer, −15°, 4 h | I | 115, 114

R	% Convn	% ee
CO_2Me	78	77
CH_2OAc	100	72
CMe_2OH	100	79

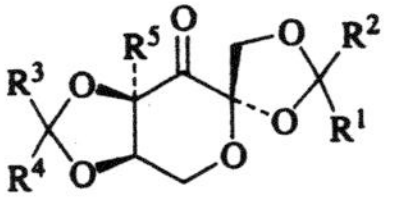

Oxone®, K_2CO_3, CH_3CN, $Na_2B_4O_7$ buffer, Na_2EDTA, 0°, 1.5 h | I | 119

R^1	R^2	R^3	R^4	R^5	% Convn	% ee
Me	Me	Me	Me	F	11	5
Me	Me	Me	Me	H	93	92
Me	Me	Et	H	H	79	91
Me	Me	*i*-Pr	H	H	64	89
Me	Me	—$(CH_2)_4$—		H	95	93
—$(CH_2)_4$—		Me	Me	H	100	93
Me	Me	Et	Et	H	91	93
Et	Et	Me	Me	H	81	90
Me	Me	—$(CH_2)_5$—		H	100	89
—$(CH_2)_5$—		Me	Me	H	100	91
—$(CH_2)_4$—		—$(CH_2)_4$—		H	89	94
Et	Et	Et	Et	H	32	86
—$(CH_2)_5$—		—$(CH_2)_5$—		H	51	87
—$(CH_2)_6$—		—$(CH_2)_6$—		H	37	91
Me	Me	Bn	Bn	H	18	66

TABLE 5. ENANTIOSELECTIVE EPOXIDATION OF OLEFINS BY ENANTIOMERICALLY ENRICHED DIOXIRANES (*Continued*)

Substrate	Ketone/Catalyst	Conditions	Product(s), Yield(s) (%) and Enantioselectivities (ee)	Refs.
	[structure: R¹,R²-acetal spiro-dioxane ketone]	Oxone®, K_2CO_3, CH_3CN, $Na_2B_4O_7$ buffer, Na_2EDTA, 0°, 1.5 h	**I** (see table A)	119
	[structure: OR / $O(CH_2)_2Cl$ ketone]	Oxone®, K_2CO_3, CH_3CN, $Na_2B_4O_7$ buffer, Na_2EDTA, 0°, 1.5 h	**I** (see table B)	119
	[structure: OR ketone]	Oxone®, K_2CO_3, CH_3CN, $Na_2B_4O_7$ buffer, Na_2EDTA, 0°, 1.5 h	**I** (see table C)	119
	[structure: bicyclic ketone]	Oxone®, K_2CO_3, CH_3CN, $Na_2B_4O_7$ buffer, Na_2EDTA, 0°, 1.5 h	**I** (5) 41% ee	119

Table A

R^1	R^2	% Convn	% ee
Me	Me	44	61
—$(CH_2)_4$—		36	61
Et	Et	33	61
—$(CH_2)_5$—		52	52
n-Pr	*n*-Pr	28	82
Ph	Ph	26	48
Bn	Bn	19	56

R	% Convn	% ee
H	8	65
Ac	5	66
TBDMS	10	40

R	% Convn	% ee
Me	15	59
$(CH_2)_2Cl$	16	67

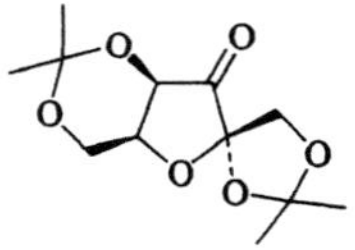

Catalyst	Conditions	Product(s) and Yield(s) (%)	Refs.
	Oxone®, K_2CO_3, CH_3CN, $Na_2B_4O_7$ buffer, Na_2EDTA, 0°, 1.5 h	**I** (41) 62% ee	119
AcO	Oxone®, K_2CO_3, Solvent, $Na_2B_4O_7$ buffer, EDTA	**I**	115, 114
MeO, CF_3	Oxone®, CH_2Cl_2/H_2O (pH 7.5), Bu_4NHSO_4, 2 to 5°, 40 h	Ph epoxide **I** (65) 17% ee	106
	Oxone®, $NaHCO_3$, CH_3CN/H_2O, EDTA, rt, 1 h	**I** (61) 20% ee	112

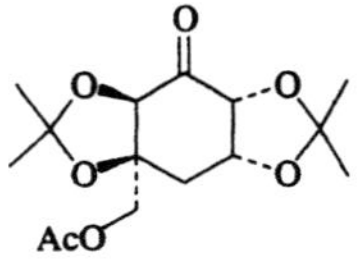

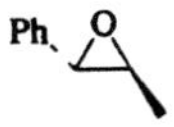

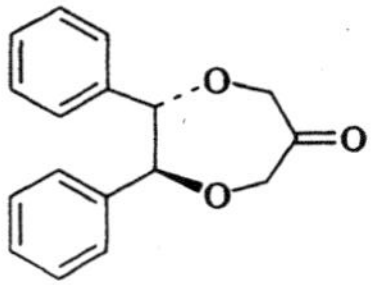

Solvent	Temp	Time	% Convn	% ee
CH_3CN	0°	4 h	58	63
DME	0°	4 h	100	70
DME	–10°	4 h	95	73
DMM	0°	4 h	43	66
Dioxane	0°	4 h	99	67
DMM/CH_3CN (2:1)	0°	4 h	91	67
DMF	0°	3 h	99	64

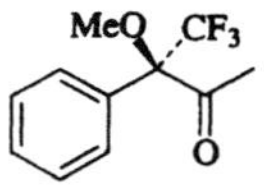

TABLE 5. ENANTIOSELECTIVE EPOXIDATION OF OLEFINS BY ENANTIOMERICALLY ENRICHED DIOXIRANES (*Continued*)

Substrate	Ketone/Catalyst	Conditions	Product(s), Yield(s) (%) and Enantioselectivities (ee)	Refs.
	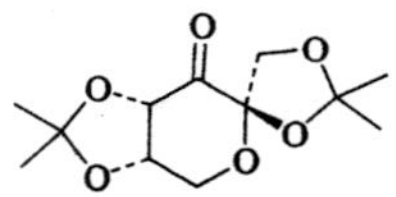	Oxone®, K_2CO_3, CH_3CN/DMM, $Na_2B_4O_7$ buffer, −10°, 2 h	**I** (94) 96% ee	25
	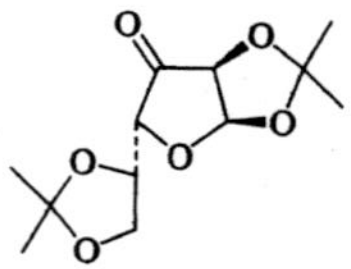	Oxone®, K_2CO_3, CH_3CN, $Na_2B_4O_7$ buffer, Na_2EDTA, 0°, 1.5 h	**I** (4) 23% ee	119
	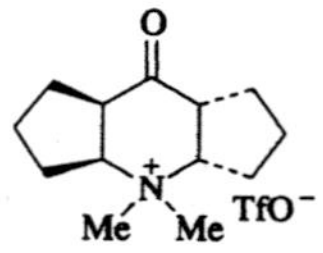 Me Me TfO⁻	—	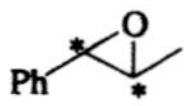 Ph	11
	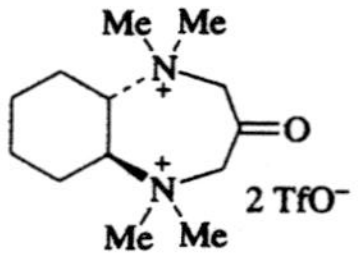 Me Me 2 TfO⁻ Me Me	Oxone®, CH_3CN, buffer, 0°, 6 h	**I** (—) 9% ee	11
	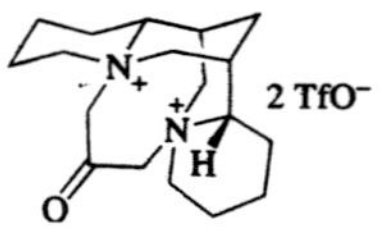 2 TfO⁻ H	Oxone®, CH_3CN, buffer, 0°, 2 h	**I** (—) 40% ee	11

Catalyst	Conditions	Product(s) and Yield(s) (%), % ee	Refs.
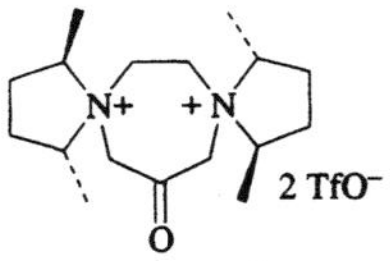 2 TfO⁻	Oxone®, CH_3CN, buffer, 0°, 6 h	I (—) 8% ee	11
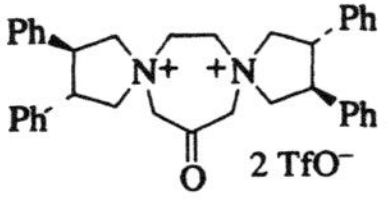 2 TfO⁻	Oxone®, CH_3CN, buffer, 0°, 3 h	I (—) 10% ee	11
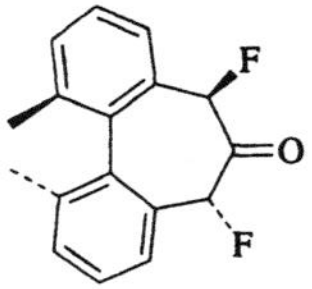	Oxone®, $NaHCO_3$, CH_3CN/H_2O, 0°, 8 h	I (—) 71% ee	11
CO_2Me, F	Oxone®, phosphate buffer, CH_2Cl_2, Bu_4NHSO_4, 0°, 4.5 h	Ph (epoxide) I (±) (—)	369
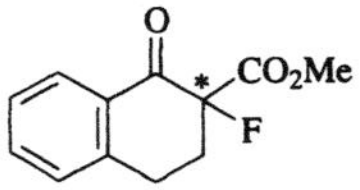 OH, F	Oxone®, phosphate buffer, CH_2Cl_2, Bu_4NHSO_4, 0°, 4.5 h	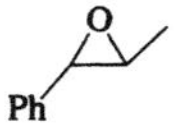 I (—)	369
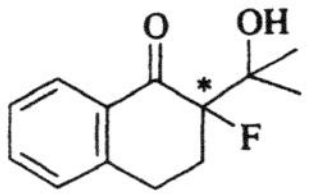 CO_2Et, F	Oxone®, phosphate buffer, CH_2Cl_2, Bu_4NHSO_4, 0°, 4.5 h	I (—)	369

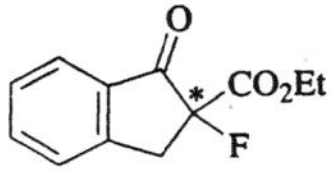

TABLE 5. ENANTIOSELECTIVE EPOXIDATION OF OLEFINS BY ENANTIOMERICALLY ENRICHED DIOXIRANES (*Continued*)

Substrate	Ketone/Catalyst	Conditions	Product(s), Yield(s) (%) and Enantioselectivities (ee)	Refs.
	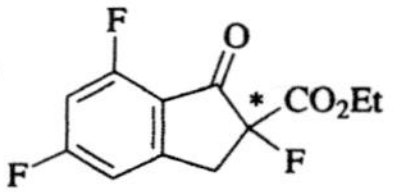 F, F, O, CO$_2$Et, F (*)	Oxone®, phosphate buffer, CH_2Cl_2, Bu_4NHSO_4, 0°, 4.5 h	**I** (—)	369
$C_{9\text{-}24}$ 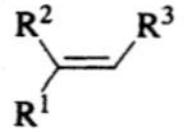 R^2, R^3, R^1	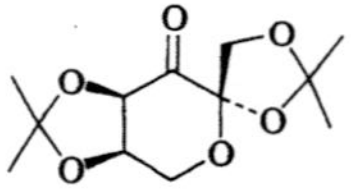 O, O, O, O, O, O	Oxone®, K_2CO_3, CH_3CN, $Na_2B_4O_7$ buffer, EDTA, 0°, 1.5 h	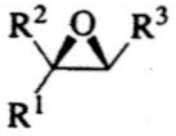 R^2, O, R^3, R^1	118

R^1	R^2	R^3		% ee
Ph	H	Me	(93)	92
Ph	H	CH_2OH	(70)	90
n-Bu	H	*n*-Bu	(70)	91
Bn	H	$(CH_2)_2CO_2Me$	(76)	91
Ph	H	Ph	(75)	97
n-C_6H_{13}	H	*n*-C_6H_{13}	(88)	93
Ph	H	$CH_2OTBDMS$	(87)	94
Ph	Ph	*n*-$C_{10}H_{21}$	(66)	94

Substrate	Ketone/Catalyst	Conditions	Product(s), Yield(s) (%) and Enantioselectivities (ee)	Refs.
$C_{9\text{-}14}$ 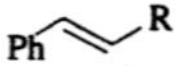 Ph, R	 Me, +, Me, N, $CF_3SO_3^-$, O, F	Oxone®, $NaHCO_3$, CH_3CN/H_2O, 0°, 5 h	 Ph, O, R	107

R		% ee
Me	(85)	35
Ph	(—)	58

Substrate	Ketone/Catalyst	Conditions	Product(s), Yield(s) (%) and Enantioselectivities (ee)	Refs.
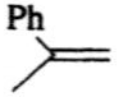 C_9 Ph	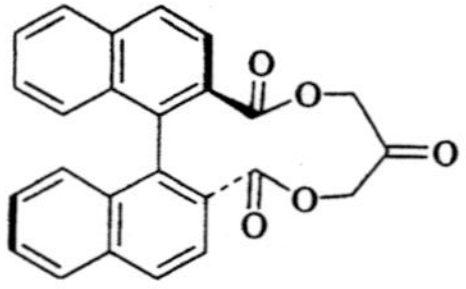 O, O, O, O, O	Oxone®, $NaHCO_3$, CH_3CN/H_2O, EDTA, rt, 210 min	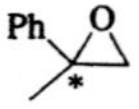 Ph, O, * (70) 18% ee	24

 Ph⁀OH	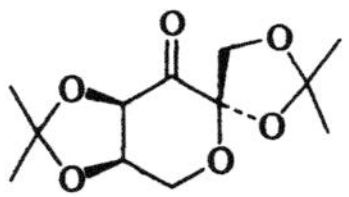	H_2O_2, CH_3CN, K_2CO_3 buffer, EDTA, 0°, 18 h	 Ph, OH (55) 89% ee	368
	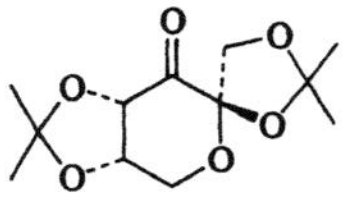	Oxone®, K_2CO_3, DMM/ CH_3CN, buffer (pH 9.3), Bu_4NHSO_4, −10°, 3 h	 Ph, OH (82) 94% ee	149a
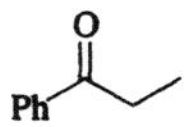 Ph	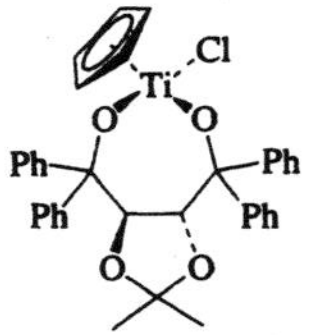 Cl, Ti, Ph, Ph, Ph, Ph	1. LDA, THF, Ar, −78°, 30 min 2. −78°, 3 h 3. DMD, acetone, −78°, 1 min	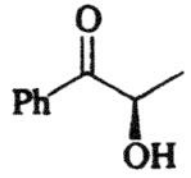 Ph, OH **I** (40) 13% ee	98
	"	1. LDA, THF, Ar, −78°, 30 min 2. −78 to 0°, 3 h 3. DMD, acetone, −78°, 1 min	**I** (64) 40% ee	98

TABLE 5. ENANTIOSELECTIVE EPOXIDATION OF OLEFINS BY ENANTIOMERICALLY ENRICHED DIOXIRANES (*Continued*)

Substrate	Ketone/Catalyst	Conditions	Product(s), Yield(s) (%) and Enantioselectivities (ee)	Refs.
	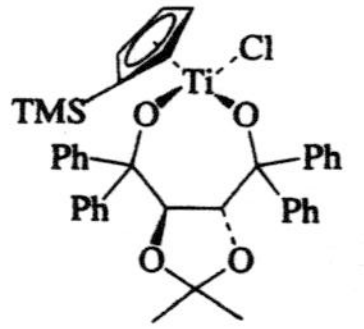 TMS, Cl, Ti, O, O, Ph, Ph, Ph, Ph, O, O	1. LDA, THF, Ar, −78°, 30 min 2. −78 to 0°, 3 h 3. DMD, acetone, −78°, 1 min	**I** (39) 38% ee	98
	Cl, Ti, O, O, Ph, Ph, Ph, Ph, O, O, Bu-*t*	1. LDA, THF, Ar, −78°, 30 min 2. −78°, 3 h 3. DMD, acetone, −78°, 1 min	**I** (36) 13% ee	98
	Cl, Ti, O, O, R, R, R, R, O, O R = (1-naphthyl)	1. LDA, THF, Ar, −78°, 30 min 2. −78 to 0°, 3 h 3. DMD, acetone/Et_2O, −78°, 1 min	**I** (18) 63% ee	98

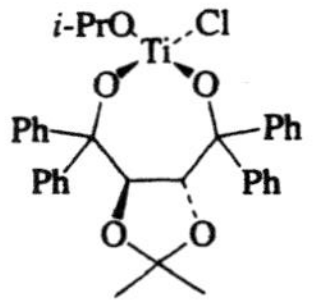

1. LDA, THF, Ar, –78°, 30 min
2. –78 to 0°, 3 h
3. DMD, acetone, –78°, 1 min

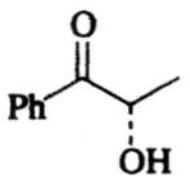

(24) 5% ee — 98

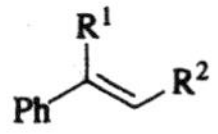

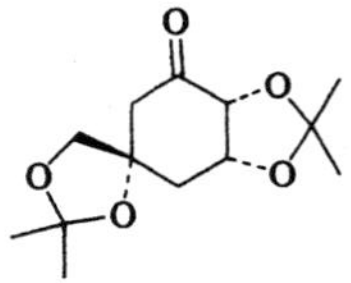

Oxone®, $NaHCO_3$, solvent, H_2O, EDTA

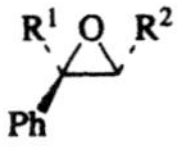

I (95) — 116

Equiv	Solvent	Temp	Time	R^1	R^2	% Convn	% ee
1	CH_3CN	0°	6.5 h	H	CH_2OH	26	18
1	CH_3CN	20°	5 h	H	Ph	26	62
1	CH_3CN	0°	5 h	H	Ph	20	78
1	CH_3CN	–10°	6 h	H	Ph	5	75
3	CH_3CN	0°	6.5 h	H	Ph	35	85
1	DME	20°	5 h	H	Ph	42	37
1	DME	–10°	6 h	H	Ph	<5	38
1	DMM	0°	5 h	H	Ph	<5	77
1	CH_2Cl_2	0°	5 h	H	Ph	0	—
1	CH_3CN	20°	5 h	H	$CH_2OTBDMS$	23	57
3	CH_3CN	0°	6.5 h	H	$CH_2OTBDMS$	29	87
3	CH_3CN	0°	5 h	Me	Ph	47	70
3	CH_3CN	0°	2.5 h	Ph	Ph	36	85
3	CH_3CN	–10°	6 h	Ph	Ph	22	84

TABLE 5. ENANTIOSELECTIVE EPOXIDATION OF OLEFINS BY ENANTIOMERICALLY ENRICHED DIOXIRANES (*Continued*)

Substrate	Ketone/Catalyst	Conditions	Product(s), Yield(s) (%) and Enantioselectivities (ee)	Refs.
	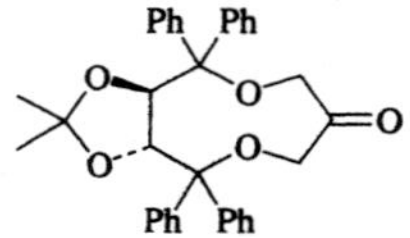 Ph Ph, Ph Ph	Oxone®, solvent, buffer (pH), Na_2EDTA, rt	I (95)	113

Solvent	pH	Time	R^1	R^2	% Convn	% ee
Dioxane	8.0	240 h	H	CH_2OH	40	14
Dioxane	10.5	5 h	H	CH_2OH	51	80
CH_3CN	8.0	72 h	H	Ph	39	1
Dioxane	8.0	168 h	H	Ph	54	33
Dioxane	10.5	5 h	H	Ph	67	65
Dioxane	10.5	5 h	H	$CH_2OTBDMS$	78	77
Dioxane	8.0	84 h	Ph	Ph	55	66
Dioxane	10.5	5 h	Ph	Ph	70	81

Substrate	Ketone/Catalyst	Conditions	Product(s), Yield(s) (%) and Enantioselectivities (ee)	Refs.
C_{9-13} 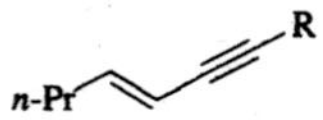 n-Pr, R	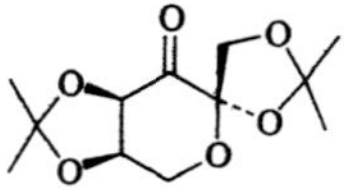	Oxone®, K_2CO_3, CH_3CN/DMM, K_2CO_3/AcOH buffer, −10°, 4 h	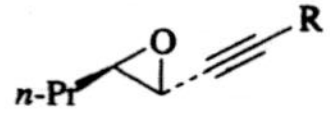 n-Pr, R	123, 124
C_9 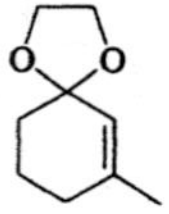		Oxone®, K_2CO_3, CH_3CN/DMM, $Na_2B_4O_7$ buffer, Bu_4NHSO_4, Na_2EDTA, −10°, 2 h	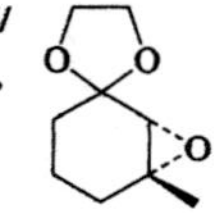 (41) 97% ee	25
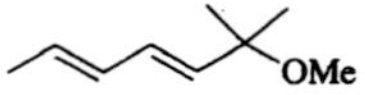 OMe		Oxone®, K_2CO_3, CH_3CN/DMM, buffer (pH 10.5), EDTA, 0°, 1.5 h	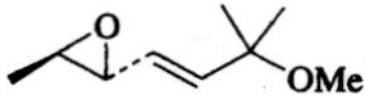 OMe (65) 89% ee	122

R		% ee
CH_2OMe	(35)	89
TMS	(71)	89
TBDMS	(69)	89

C_{10}

Substrate	Catalyst	Conditions	Product(s) and Yield(s) (%)	Refs.
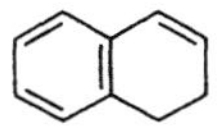	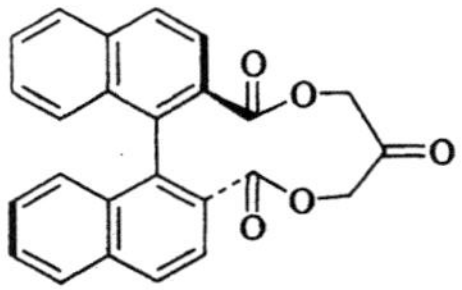	Oxone®, $NaHCO_3$, CH_3CN/H_2O, EDTA, rt, 80 min	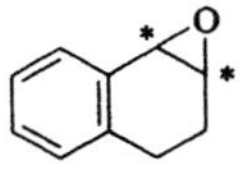 (85) <5% ee	24
	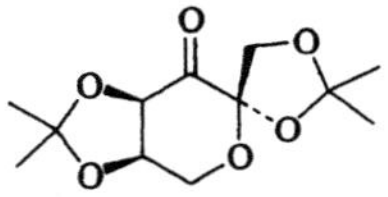	Oxone®, K_2CO_3, CH_3CN, $Na_2B_4O_7$ buffer, Na_2EDTA, –10°, 2 h	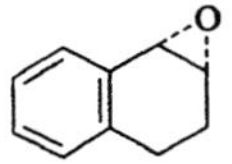 I (85) 32% ee	25
		Oxone®, K_2CO_3, CH_3CN/DMM, $Na_2B_4O_7$ buffer, Na_2EDTA, –10°, 2 h	I (92) 12% ee	25
	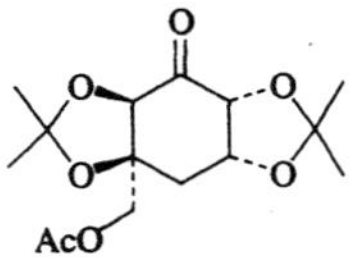 AcO	Oxone®, K_2CO_3, DME, K_2CO_3/AcOH buffer, –10°, 4 h	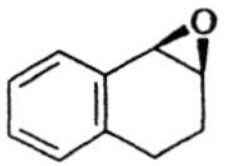 (93) 21% ee	114, 115
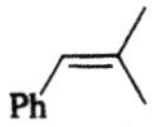 Ph	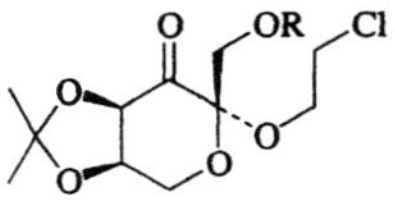 OR, Cl	Oxone®, K_2CO_3, CH_3CN, $Na_2B_4O_7$ buffer, Na_2EDTA, 0°, 1.5 h	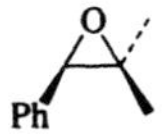 Ph I	119

R	% Convn	% ee
H	6	68
Ac	11	67
TBDMS	8	17

TABLE 5. ENANTIOSELECTIVE EPOXIDATION OF OLEFINS BY ENANTIOMERICALLY ENRICHED DIOXIRANES (*Continued*)

Substrate	Ketone/Catalyst	Conditions	Product(s), Yield(s) (%) and Enantioselectivities (ee)	Refs.
I		Oxone®, K_2CO_3, CH_3CN, $Na_2B_4O_7$ buffer, Na_2EDTA, 0°, 1.5 h	R^1 / R^2 / % Convn / % ee Me / Me / 65 / 84 –$(CH_2)_4$– / 81 / 76 Et / Et / 63 / 76 –$(CH_2)_5$– / 80 / 74 *n*-Pr / *n*-Pr / 43 / 73 Ph / Ph / 36 / 21 Bn / Bn / 17 / 29	119
I		Oxone®, K_2CO_3, CH_3CN, $Na_2B_4O_7$ buffer, Na_2EDTA, 0°, 1.5 h	R / % Convn / % ee Me / 17 / 30 $(CH_2)_2Cl$ / 18 / 29	119
I		Oxone®, K_2CO_3, CH_3CN, $Na_2B_4O_7$ buffer, Na_2EDTA, 0°, 1.5 h	(5) 35% ee	119
I		Oxone®, K_2CO_3, CH_3CN, $Na_2B_4O_7$ buffer, Na_2EDTA, 0°, 1.5 h		119

R^1	R^2	R^3	R^4	R^5	% Convn	% ee
Me	Me	Me	Me	F	23	12
Me	Me	Me	Me	H	100	72
Me	Me	Et	H	H	100	65
Me	Me	*i*-Pr	H	H	83	61
Me	Me	—$(CH_2)_4$—		H	100	69
—$(CH_2)_4$—		Me	Me	H	100	72
Me	Me	Et	Et	H	98	57
Et	Et	Me	Me	H	77	71
Me	Me	—$(CH_2)_5$—		H	100	63
—$(CH_2)_5$—		Me	Me	H	100	72
—$(CH_2)_4$—		—$(CH_2)_4$—		H	100	68
Et	Et	Et	Et	H	40	57
—$(CH_2)_5$—		—$(CH_2)_5$—		H	80	58
—$(CH_2)_6$—		—$(CH_2)_6$—		H	37	66
Me	Me	Bn	Bn	H	15	29

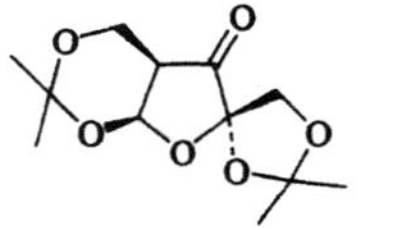

Oxone®, K_2CO_3, CH_3CN, $Na_2B_4O_7$ buffer, Na_2EDTA, 0°, 1.5 h

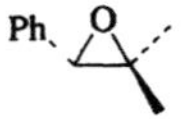

Ph (57) 20% ee 119

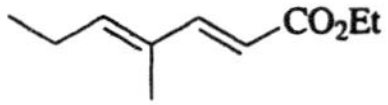

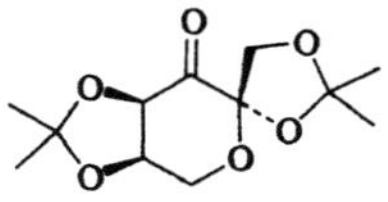

CO_2Et

Oxone®, K_2CO_3, CH_3CN/DMM, buffer (pH 10.5), EDTA, 0°, 1.5 h

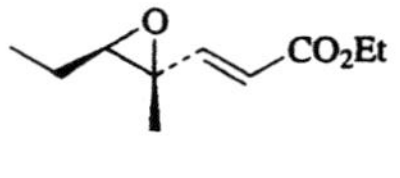

CO_2Et (82) 95% ee 122

OH

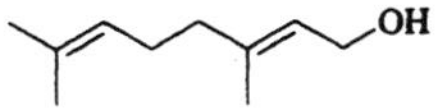

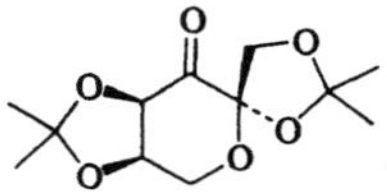

Oxone®, $NaHCO_3$, CH_3CN, EDTA, 0°

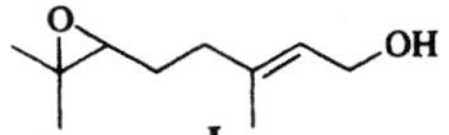

OH + 149a

I

OH

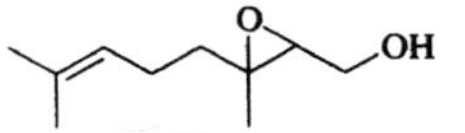

II

TABLE 5. ENANTIOSELECTIVE EPOXIDATION OF OLEFINS BY ENANTIOMERICALLY ENRICHED DIOXIRANES (*Continued*)

Substrate	Ketone/Catalyst	Conditions	Product(s), Yield(s) (%) and Enantioselectivities (ee)	Refs.
			pH / Time / % Convn / I:II 7.0-7.5 / 2 h / 12 / 2.4:1 8.0 / 1.5 h / 24 / 3.1:1 9.0 / 1.5 h / 45 / 4.6:1 10.0 / 1.5 h / 58 / 6.1:1 11.0 / 1.5 h / 58 / 5.6:1 11.5 / 1.5 h / 58 / 5.9:1	
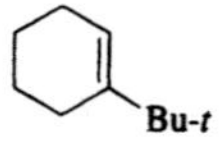 Bu-*t*	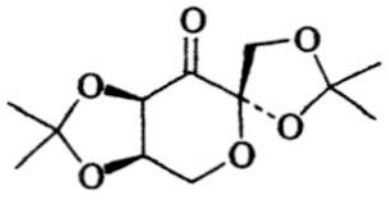	Oxone®, K_2CO_3, CH_3CN/DMM, $Na_2B_4O_7$ buffer, Na_2EDTA, 0°, 1.5 h	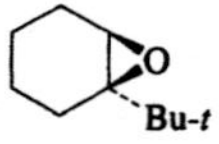 O, Bu-*t* (—) 26% ee	25
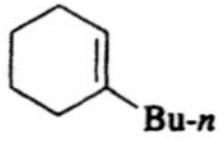 Bu-*n*	"	Oxone®, K_2CO_3, CH_3CN/DMM, $Na_2B_4O_7$ buffer, Na_2EDTA, 0°, 1.5 h	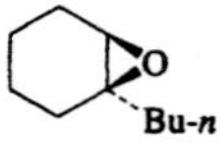 O, Bu-*n* (—) 79% ee	25
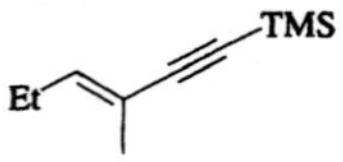 Et, TMS	"	Oxone®, K_2CO_3, CH_3CN/DMM, K_2CO_3/AcOH buffer, -10°, 3 h	 Et, O, TMS (84) 95% ee	123, 124
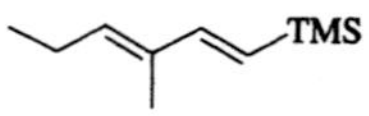 TMS	"	Oxone®, K_2CO_3, CH_3CN/DMM, buffer (pH 10.5), EDTA, 0°, 1.5 h	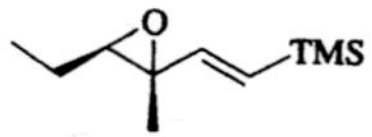 O, TMS (60) 92% ee	122
C_{10-23}				
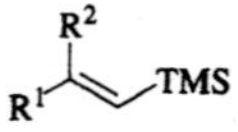 R^2, R^1, TMS	"	Oxone®, K_2CO_3, CH_3CN/DMM, $Na_2B_4O_7$ buffer, Na_2EDTA, 0°, 3 h	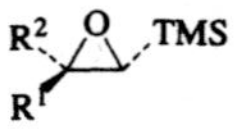 R^2, O, TMS, R^1	370

R¹	R²		% ee
$HO(CH_2)_3$	Et	(80)	90
n-Bu	Et	(51)	90
$HO(CH_2)_4$	Et	(75)	91
Ph	Me	(74)	94
Ph	Et	(82)	92
$TBDMSOCH_2$	Et	(67)	84
$PhCH_2$	*n*-Bu	(66)	93
$TBDPSO(CH_2)_2$	Et	(67)	92

C_{11}

Substrate	Catalyst	Conditions	Product and Yield(s) (%)	Refs.
Ph–CH=CH–C≡C–R	"	Oxone®, CH_3CN, DMM, K_2CO_3/AcOH buffer, K_2CO_3, 0°	R / Time / % ee: Me 3 h (35) 94; TMS 4 h (59) 96; TBDMS 4 h (60) 96	124, 123
Cl (chromene)		Oxone®, phosphate buffer, CH_2Cl_2, Bu_4NHSO_4, 0°, 4.5 h	(±) (—)	369
Ph (dioxolane)		Oxone®, $NaHCO_3$, CH_3CN/H_2O, EDTA, 0°, 2 h	(41) 93% ee	117

C_{11-16}

Substrate	Catalyst	Conditions	Product and Yield(s) (%)	Refs.
Ph–C(OAc)=CH–R	"	Oxone®, CH_3CN/DMM, $Na_2B_4O_7$ (or K_2CO_3/AcOH) buffer, 0°	R / Time / % ee: Me 2 h (66) 91; Ph 3 h (46) 91	121
	AcO	Oxone®, K_2CO_3, DME, K_2CO_3/AcOH buffer, –10°, 4 h	(83) 66% ee	115, 114

TABLE 5. ENANTIOSELECTIVE EPOXIDATION OF OLEFINS BY ENANTIOMERICALLY ENRICHED DIOXIRANES (*Continued*)

Substrate	Ketone/Catalyst	Conditions	Product(s), Yield(s) (%) and Enantioselectivities (ee)	Refs.
$C_{11\text{-}17}$ R		Oxone®, K_2CO_3, CH_3CN/DMM, K_2CO_3/AcOH buffer, −10°	R: H, Time 3 h, (35), 93 % ee; TMS, 2 h, (83), 97 % ee; TBDMS, 2 h, (93), 97 % ee	124, 123
C_{11} CO_2Et	"	Oxone®, K_2CO_3, CH_3CN/DMM, buffer (pH 10.5), EDTA, 0°, 1.5 h	CO_2Et (61) 94% ee	122
CO_2Et	"	Oxone®, K_2CO_3, CH_3CN/DMM, buffer (pH 10.5), EDTA, 0°, 1.5 h	CO_2Et (89) 94% ee	122
	"	Oxone®, K_2CO_3, CH_3CN/DMM, $Na_2B_4O_7$ buffer, Na_2EDTA, −10°, 2 h	(92) 92% ee	25
$C_{12\text{-}25}$ R^2, R^1	Mn, Cl; R = *t*-Bu	DMD, acetone/CH_2Cl_2, ~20°, 20 to 24 h	R^1, R^2	296, 102

mol %	Additive	R^1	R^2		% ee
5	—	H	CN	(67)	21
10	—	H	CN	(53)	70
10	*N*-methylimidazole	H	CN	(68)	82
15	—	H	CN	(62)	85
20	—	$4\text{-BrC}_6\text{H}_4\text{SO}_3$	$4\text{-BrC}_6\text{H}_4\text{SO}_3$	(85)	83
16	—	$PhSO_3$	$PhSO_3$	(64)	84
14	—	TsO	TsO	(68)	84

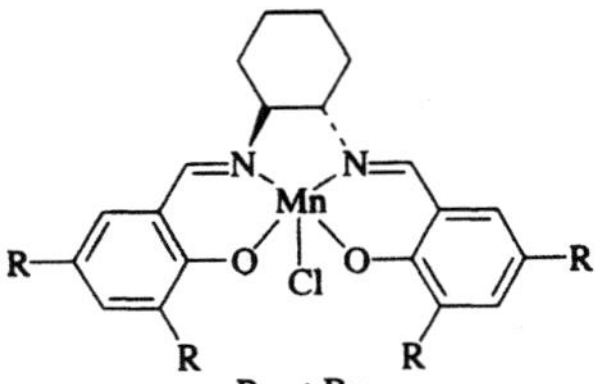

R = *t*-Bu

DMD, acetone/CH_2Cl_2, ~20°, 20 to 24 h

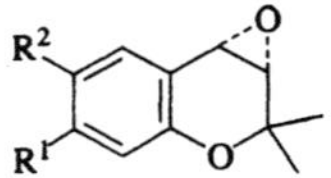

R^2 R^1 296, 102

mol %	Additive	R^1	R^2		% ee
5	—	H	CN	(55)	15
10	—	H	CN	(56)	81
10	*N*-methylimidazole	H	CN	(62)	84
15	—	H	CN	(78)	93
20	—	$4\text{-BrC}_6\text{H}_4\text{SO}_3$	$4\text{-BrC}_6\text{H}_4\text{SO}_3$	(73)	84
16	—	$PhSO_3$	$PhSO_3$	(61)	88
17	—	TsO	TsO	(69)	86

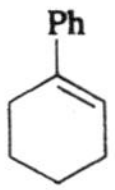

Ph

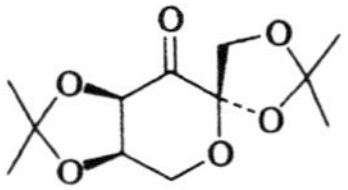

H_2O_2, CH_3CN, K_2CO_3 buffer, EDTA, 0°, 7 h

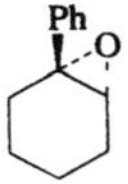

Ph (90) 95% ee 368

TABLE 5. ENANTIOSELECTIVE EPOXIDATION OF OLEFINS BY ENANTIOMERICALLY ENRICHED DIOXIRANES (*Continued*)

Substrate	Ketone/Catalyst	Conditions	Product(s), Yield(s) (%) and Enantioselectivities (ee)	Refs.
	Me Me TfO^-	—	Ph O (—) 58% ee	11
	R R	Oxone®, $NaHCO_3$, CH_3CN/H_2O, EDTA, rt, 90 min	Ph O R / Time / / % ee Br / 4 h / (81) / 64 Cl / 1.5 h / (75) / 65 H / 1.5 h / (83) / 33 CH_2OMe / 1.5 h / (80) / 67 1,3-dioxolan-2-yl / 1.3 h / (90) / 71	24, 110, 111
Ph OH		Oxone®, K_2CO_3, DMM/CH_3CN, buffer (pH 9.3), Bu_4NHSO_4, 0°, 2 h	Ph O OH (88) 91% ee	149a
		Oxone®, K_2CO_3, CH_3CN/DMM, buffer (pH 10.5), EDTA, 0°, 1.5 h	I (54) 95% ee + II (—) I:II = 12:1	122

C_{12-17}

R^1 OSiMe$_2$R^3 / R^2

Oxone®, CH_3CN, buffer (pH), Na_2EDTA, 0°

R^1 R^2 OH (O)

26

R^1	R^2	R^3	pH	Time	% Convn	% ee
Me	Ph	Me	8	1.5 h	46	54
Ph	Me	Me	8	1.5 h	35	18
Et	Ph	Me	8	1.5 h	33	67
Me	Ph	*t*-Bu	8	3.0 h	92	82
Me	Ph	*t*-Bu	10.5	1.5 h	<10	42
Ph	Ph	Me	8	1.5 h	36	61
Ph	Ph	Me	8	18 h	53	53
Ph	Ph	Me	10.5	1.5 h	38	22

C_{12}

n-Bu Pr-*n* R

Oxone®, K_2CO_3, CH_3CN, DMM, K_2CO_3/AcOH buffer, −10°

n-Bu O Pr-*n* R

R	Time		% ee
Me	3 h	(60)	93
TMS	3 h	(83)	97
TBDMS	2 h	(93)	97

124, 123

OTBDMS

Oxone®, K_2CO_3, CH_3CN/DMM, buffer (pH 10.5), EDTA, 0°, 1.5 h

O OTBDMS **I** (68) 96% ee +

122

O OTBDMS **II** (13) 91% ee

I:**II** = 4.6:1

TABLE 5. ENANTIOSELECTIVE EPOXIDATION OF OLEFINS BY ENANTIOMERICALLY ENRICHED DIOXIRANES (*Continued*)

Substrate	Ketone/Catalyst	Conditions	Product(s), Yield(s) (%) and Enantioselectivities (ee)	Refs.
$C_{13\text{-}18}$ OR, Ph		Oxone®, K_2CO_3, CH_3CN/DMM, $Na_2B_4O_7$ buffer, EDTA, –10°, 1.5 h	OR, Ph **I** + OR, O, Ph **II** + OR, O, Ph **III**	125
$C_{13\text{-}15}$ OR, Ph		Oxone®, K_2CO_3, CH_3CN/DMM, $Na_2B_4O_7$ buffer, EDTA, 2.5 h	OR, Ph **I** + OR, Ph, O **II** + OR, Ph, O **III**	125
C_{13} OTMS		Oxone®, CH_3CN, $Na_2B_4O_7$ buffer, Na_2EDTA, 0°, 1 h	O, OH (70) 83% ee	121
Ph, TMS	"	Oxone®, K_2CO_3, CH_3CN/DMM, buffer (pH 10.5), EDTA, 0°, 1.5 h	Ph, O, TMS **I** (31) 95% ee + Ph, O, TMS **II** (—) **I**:**II** = 1:1	122

R	% Convn	% ee of I	% ee of II	II:III
Me	61	95	—	6:1
TBDMS	70	99	81	4:1

R	Temp	% Convn	% ee of II	II:III
Me	–10°	65	85	6:1
COMe	0°	54	88	12:1
CO_2Et	–10°	51	97	>20:1
TMS	–10°	49	95	>20:1

Substrate	Catalyst	Conditions	Product(s) and Yield(s) (%)	Refs.
C_{14} 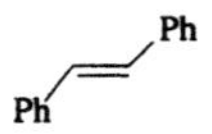 Ph–CH=CH–Ph	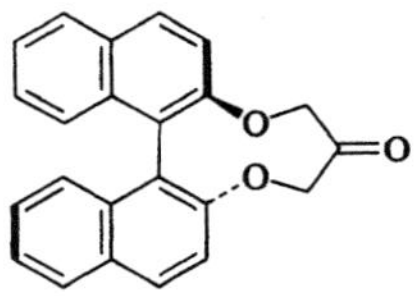	Oxone®, $NaHCO_3$, CH_3CN/H_2O, EDTA	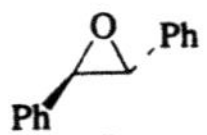 I (Ph epoxide) Temp / Time / % ee: rt, 1 h (90) 20; 0°, 5 h (79) 26	112
	"	Oxone®, $NaHCO_3$, CH_3CN/H_2O, Na_2EDTA, rt, 72 h	I (43) 27% ee	113
	"	Oxone®, K_2CO_3, CH_3CN, $Na_2B_4O_7$ buffer, rt, 1.5 h	I (55) 30% ee	113
	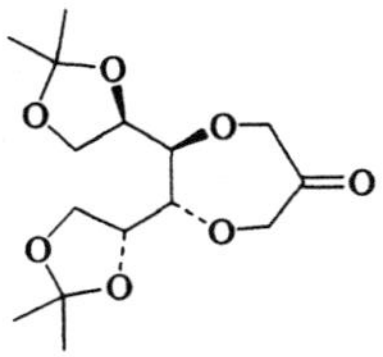	Oxone®, $NaHCO_3$, CH_3CN/H_2O, Na_2EDTA, rt, 0.4 h	I (39) 39% ee	113
	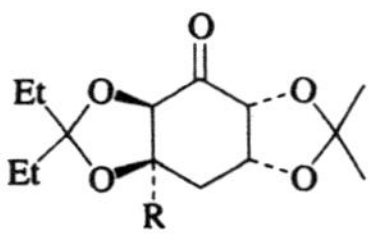 (Et, Et, R)	Oxone®, K_2CO_3, DME/DMM, K_2CO_3/AcOH buffer, –10°, 6 h	I R / % ee: CO_2Me (65) 94; CH_2OAc (77) 92; CMe_2OH (57) 95	115, 114

TABLE 5. ENANTIOSELECTIVE EPOXIDATION OF OLEFINS BY ENANTIOMERICALLY ENRICHED DIOXIRANES (*Continued*)

Substrate	Ketone/Catalyst	Conditions	Product(s), Yield(s) (%) and Enantioselectivities (ee)	Refs.
I	[structure; R]	Oxone®, K_2CO_3, DME/ DMM, K_2CO_3/AcOH buffer, –10°, 6 h	R / (yield) / % ee: H (11) 93 CH_2F (71) 89 CO_2Me (66) 95 CH_2OAc (95) 90 CMe_2OH (91) 96 CMe_2OMe (94) 96 $CH_2OTBDMS$ (77) 90 CH_2OBz (91) 90 CH_2OTs (74) 90	115, 114
I	[structure; AcO]	Oxone®, DME, buffer (pH), 0°	pH / % Convn / % ee: 8.5 15 64 9.5 32 67 10.5 44 67 11.7-12.3 50 66	115, 114
I	[structure; R, Et, Et]	Oxone®, K_2CO_3, DME/ DMM, K_2CO_3/AcOH buffer, –10°, 6 h	R / (yield) / % ee: CO_2Me (58) 93 CH_2OAc (75) 92 CMe_2OH (57) 94	115, 114

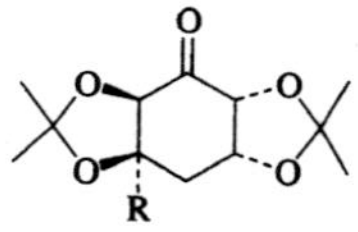

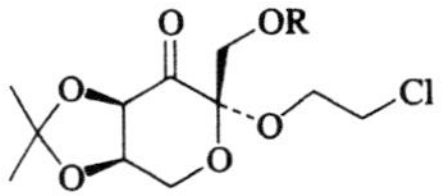

Reagent	Conditions	Product(s)	Refs.
	Oxone®, K_2CO_3, CH_3CN, $Na_2B_4O_7$ buffer, Na_2EDTA, 0°, 1.5 h	I	119

R	% Convn	% ee
H	2	—
Ac	6	96
TBDMS	0	—

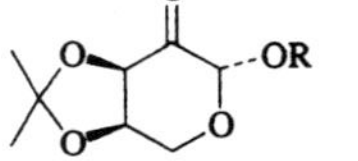

Reagent	Conditions	Product(s)	Refs.
	Oxone®, K_2CO_3, CH_3CN, $Na_2B_4O_7$ buffer, Na_2EDTA, 0°, 1.5 h	I	119

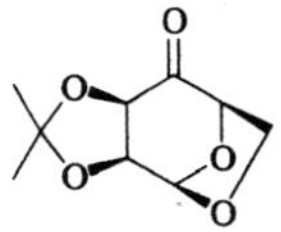

R	% Convn	% ee
Me	10	88
$(CH_2)_2Cl$	10	90

Reagent	Conditions	Product(s)	Refs.
	Oxone®, K_2CO_3, CH_3CN, $Na_2B_4O_7$ buffer, Na_2EDTA, 0°, 1.5 h	I (27) 74% ee	119
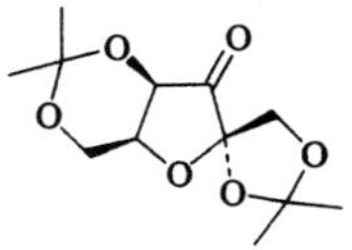	Oxone®, K_2CO_3, CH_3CN, $Na_2B_4O_7$ buffer, Na_2EDTA, 0°, 1.5 h	I (14) 75% ee	119
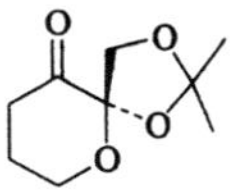	Oxone®, K_2CO_3, CH_3CN, $Na_2B_4O_7$ buffer, Na_2EDTA, 0°, 1.5 h	I (—) 78% ee	119
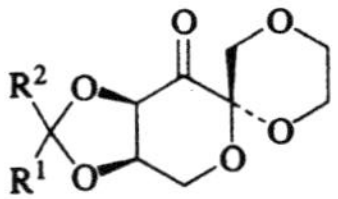	Oxone®, K_2CO_3, CH_3CN, $Na_2B_4O_7$ buffer, Na_2EDTA, 0°, 1.5 h	I	119

R^1	R^2	% Convn	% ee
Me	Me	34	90
—$(CH_2)_4$—		34	91
Et	Et	25	85
—$(CH_2)_5$—		35	78
Ph	Ph	10	67
Bn	Bn	10	82

TABLE 5. ENANTIOSELECTIVE EPOXIDATION OF OLEFINS BY ENANTIOMERICALLY ENRICHED DIOXIRANES (*Continued*)

Substrate	Ketone/Catalyst	Conditions	Product(s), Yield(s) (%) and Enantioselectivities (ee)	Refs.
	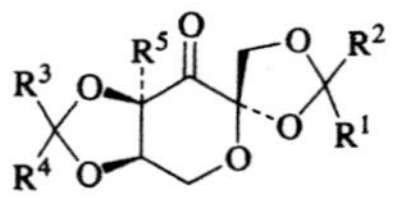	Oxone®, K_2CO_3, CH_3CN, $Na_2B_4O_7$ buffer, Na_2EDTA, 0°, 1.5 h	I	119

R^1	R^2	R^3	R^4	R^5	% Convn	% ee
Me	Me	Me	Me	F	3	11
Me	Me	Me	Me	H	75	97
Me	Me	Et	H	H	38	96
Me	Me	*i*-Pr	H	H	39	95
Me	Me	—$(CH_2)_4$—		H	52	98
—$(CH_2)_4$—		Me	Me	H	66	98
Me	Me	Et	Et	H	38	94
Et	Et	Me	Me	H	36	98
Me	Me	—$(CH_2)_3$—		H	59	92
—$(CH_2)_5$—		Me	Me	H	59	98
—$(CH_2)_4$—		—$(CH_2)_4$—		H	57	99
Et	Et	Et	Et	H	16	96
—$(CH_2)_5$—		—$(CH_2)_3$—		H	41	98
—$(CH_2)_6$—		—$(CH_2)_6$—		H	30	98
Me	Me	Bn	Bn	H	7	93

R

O
C
O
C
O
O
O
O
R

Oxone®, NaHCO$_3$,
CH$_3$CN/H$_2$O,
EDTA, rt

I

R	Time		% ee
Me	1 h	93	56
TMS	20 h	(—)	44

24

O_2N
O_2N
O
C
O
C
O
O
O

Oxone®, NaHCO$_3$,
CH$_3$CN/H$_2$O,
EDTA, rt, 35 min

Ph O Ph

I (94) 50% ee

24

R
O
C
O
C
O
O
O
O
R

Oxone®, NaHCO$_3$,
CH$_3$CN/H$_2$O,
EDTA, rt

I

R	Time		% ee
Br	3 h	(92)	75
Cl	2 h	(95)	76
I	22 h	(45)	32
H	1 h	(91)	47
CH$_2$OMe	1.8 h	(92)	66
Ph	24 h	(22)	55
	20 h	(90)	77
	20 h	(93)	84
	20 h	(91)	75

24

TABLE 5. ENANTIOSELECTIVE EPOXIDATION OF OLEFINS BY ENANTIOMERICALLY ENRICHED DIOXIRANES (*Continued*)

Substrate	Ketone/Catalyst	Conditions	Product(s), Yield(s) (%) and Enantioselectivities (ee)	Refs.
	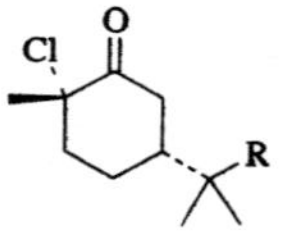 Cl	Oxone®, $NaHCO_3$, DME/H_2O, 25°, 22 h	I (—) R / % ee F 87 Cl 85 H 42 OH 81 OEt 72	109
	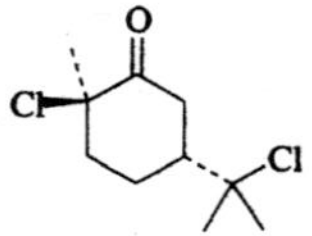 Cl, Cl	Oxone®, $NaHCO_3$, DME/H_2O, 25°	I (—) 32% ee	109
	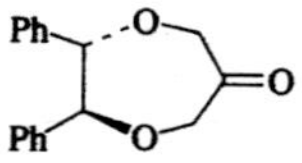 Ph, Ph	Oxone®, $NaHCO_3$, CH_3CN/H_2O, EDTA, rt, 1 h	I Temp / Time / / % ee rt 1 h (73) 30 0° 5 h (72) 59	112
	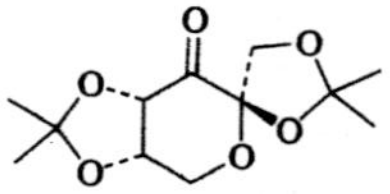	Oxone®, K_2CO_3, CH_3CN/DMM, $Na_2B_4O_7$ buffer, 0°, 1.5 h	I (81) 98% ee	25
	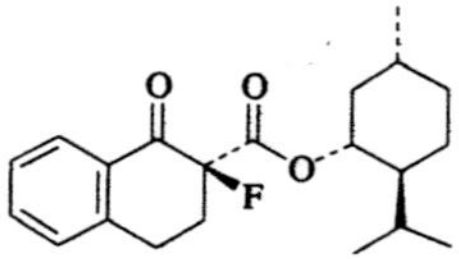 F	Oxone®, phosphate buffer, CH_2Cl_2, Bu_4NHSO_4, 0°, 4.5 h	I (±) (—)	369

		Oxone®, phosphate buffer, CH_2Cl_2, Bu_4NHSO_4, 0°, 4.5 h	**I** (±) (—)	369
C_{14-18}		Oxone®, $NaHCO_3$, DME/H_2O, 25°	(—)	109
C_{14-20}		Oxone®, $NaHCO_3$, CH_3CN/H_2O, EDTA, rt	**I**	24

R	Time	% ee
Cl	4 h	74
F	4 h	78
H	1 h	86
Me	4 h	89
MeO	0.5 h	85
AcO	1 h	74

R	Time		% ee
H	20 min	(99)	47
Me	40 min	(99)	50
Et	40 min	(96)	60
i-Pr	40 min	(98)	71
t-Bu	40 min	(95)	76
Ph	480 min	(82)	87

TABLE 5. ENANTIOSELECTIVE EPOXIDATION OF OLEFINS BY ENANTIOMERICALLY ENRICHED DIOXIRANES (*Continued*)

Substrate	Ketone/Catalyst	Conditions	Product(s), Yield(s) (%) and Enantioselectivities (ee)	Refs.
	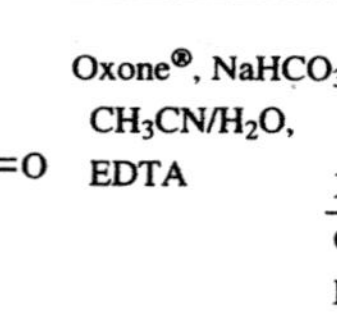 (binaphthyl diester macrocyclic ketone bearing X substituents)	Oxone®, $NaHCO_3$, CH_3CN/H_2O, EDTA	**I** (—)	24

X	R	Time	Temp	(Yield)	% ee
Cl	H	2-3 h	rt	(>90)	76
Br	H	2-3 h	rt	(>90)	75
Br	H	25 h	0-1°	(83)	80
1,3-dioxolan-2-yl	H	40 min	rt	(>90)	71
1,3-dioxolan-2-yl	H	20 h	0-1°	(>90)	84
Cl	Me	2-3 h	rt	(>90)	80
Br	Me	2-3 h	rt	(>90)	85
Br	Me	20 h	0-1°	(>90)	88
1,3-dioxolan-2-yl	Me	40 min	rt	(>90)	84
1,3-dioxolan-2-yl	Me	20 h	0-1°	(>90)	88
Cl	Et	2-3 h	rt	(>90)	85
Br	Et	2-3 h	rt	(>90)	88
Br	Et	24 h	0-1°	(51)	92
1,3-dioxolan-2-yl	Et	40 min	rt	(>90)	82
1,3-dioxolan-2-yl	Et	20 h	0-1°	(>90)	91
Cl	*i*-Pr	2-3 h	rt	(>90)	85
Br	*i*-Pr	2-3 h	rt	(>90)	90
Br	*i*-Pr	21 h	0-1°	(50)	92
1,3-dioxolan-2-yl	*i*-Pr	40 min	rt	(>90)	88
1,3-dioxolan-2-yl	*i*-Pr	20 h	0-1°	(>90)	91
Cl	*t*-Bu	2-3 h	rt	(>90)	91
Br	*t*-Bu	2-3 h	rt	(>90)	93
Br	*t*-Bu	18 h	0-1°	(22)	95
1,3-dioxolan-2-yl	*t*-Bu	40 min	rt	(>90)	90
1,3-dioxolan-2-yl	*t*-Bu	20 h	0-1°	(>90)	95

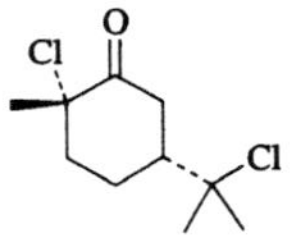

Substrate	Catalyst	Conditions	Product(s) and Yield(s) (%), % ee	Refs.
	Cl, Cl ketone	Oxone®, $NaHCO_3$, DME/H_2O, 25°	I (—)	109

R	Time	% ee
Br	4 h	75
F	4 h	79
H	1 h	86
Me	4 h	87
AcO	4 h	72
t-Bu	4 h	87

C_{14-20}

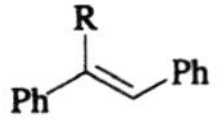

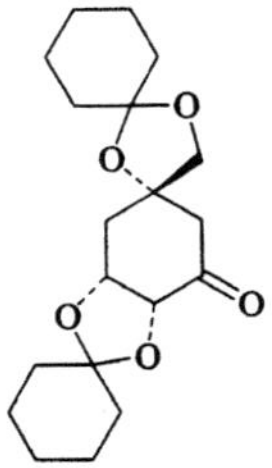

Ph(R)C=CHPh — Oxone®, $NaHCO_3$, solvent, H_2O, EDTA, 20° — 116

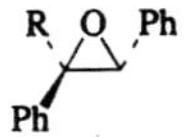

R	Solvent	Time	% Convn	% ee
H	CH_3CN	24 h	12	32
H	dioxane	3 h	31	25
H	CH_3CN/DMM (2:1)	24 h	<5	45
Ph	dioxane	3 h	32	25

C_{14}

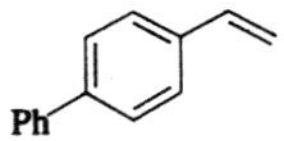

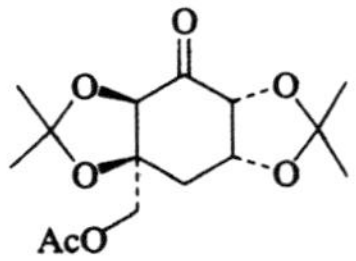

4-Ph-styrene — Oxone®, K_2CO_3, DME, K_2CO_3/AcOH buffer, −10°, 4 h 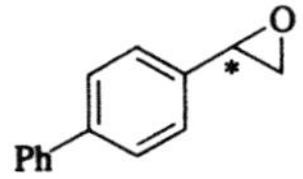 — (54) 65% ee — 115, 114

TABLE 5. ENANTIOSELECTIVE EPOXIDATION OF OLEFINS BY ENANTIOMERICALLY ENRICHED DIOXIRANES (*Continued*)

Substrate	Ketone/Catalyst	Conditions	Product(s), Yield(s) (%) and Enantioselectivities (ee)	Refs.
Ph, TMS		Oxone®, K_2CO_3, CH_3CN/ DMM, K_2CO_3/AcOH buffer, 0°, 2 h	Ph, TMS (64) 94% ee	124, 123
Ph, TMS	"	Oxone®, K_2CO_3, CH_3CN/ DMM, buffer (pH 10.5), EDTA, 0°, 1.5 h	Ph, TMS **I** (77) 94% ee + Ph, TMS **II** (—) **I**:**II** = 14:1	122
C_{14-21} n, OTBDMS	"	Oxone®, K_2CO_3, CH_3CN/ DMM, K_2CO_3/AcOH buffer, −10°, 3 h	n, OTBDMS; n 1 (97) 77 % ee; 4 (98) 96; 8 (99) 86	124, 123
C_{14} CO_2Et	"	Oxone®, K_2CO_3, CH_3CN/ DMM, buffer (pH 10.5), EDTA, 0°, 1.5 h	CO_2Et (68) 95% ee	122
C_{14-15} R, O_2CBu-*t*	"	Oxone®, K_2CO_3, CH_3CN/ DMM, $Na_2B_4O_7$ buffer, EDTA, −10°, 2.5 h	R, O_2CBu-*t* **I** + R, O_2CBu-*t* **II** + R, O_2CBu-*t* **III**	125

R	% Convn	% ee of **I**	% ee of **II**	**II**:**III**
i-Pr	59	93	85	8:1
t-Bu	54	99	84	>20:1
OTMS	61	91	76	4:1

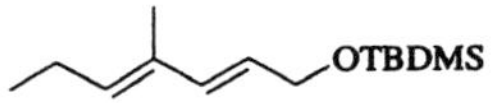

OTBDMS

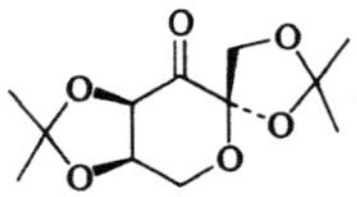

Oxone®, K_2CO_3, CH_3CN/DMM, buffer (pH 10.5), EDTA, 0°, 1.5 h

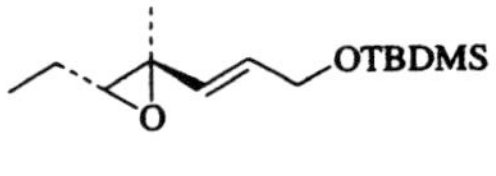

OTBDMS (81) 96% ee 122

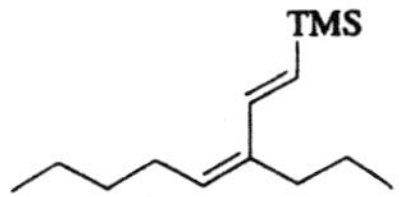

TMS

"

Oxone®, K_2CO_3, CH_3CN/DMM, buffer (pH 10.5), EDTA, 0°, 1.5 h

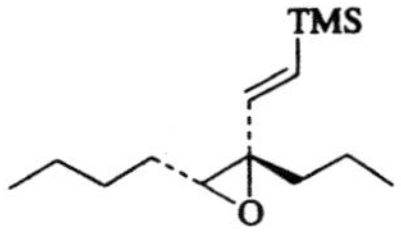

TMS

(79) 95% ee 122

C_{15-17}

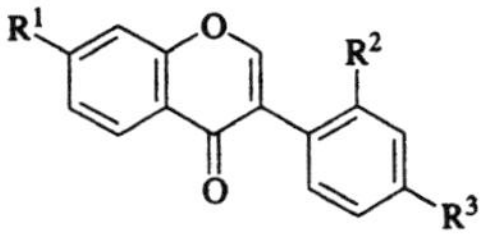

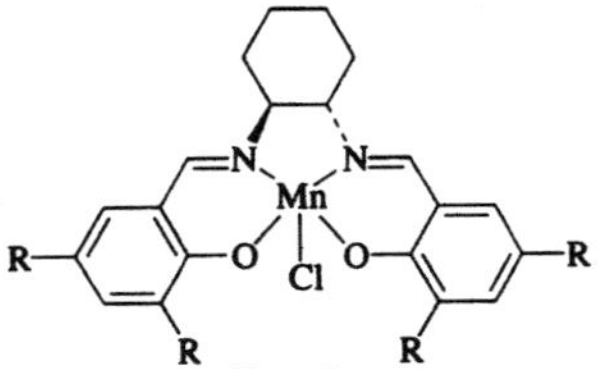

R[1] O R[2] O R[3]

R Mn N N O O Cl R R R

R = *t*-Bu

DMD, acetone/CH_2Cl_2, ~20°, 10 d

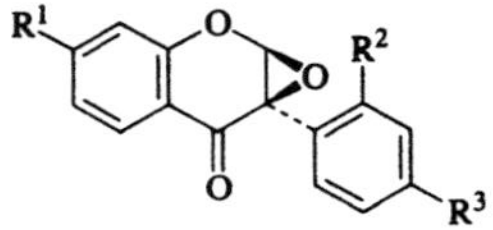

R[1] O O R[2] O R[3]

R[1]	R[2]	R[3]		% ee
H	H	H	(34)	56
MeO	H	H	(36)	39
Ms	H	H	(27)	48
MeO	MeO	H	(32)	86
MeO	H	MeO	(29)	22
Ms	MeO	H	(23)	90

103, 104

TABLE 5. ENANTIOSELECTIVE EPOXIDATION OF OLEFINS BY ENANTIOMERICALLY ENRICHED DIOXIRANES (*Continued*)

Substrate	Ketone/Catalyst	Conditions	Product(s), Yield(s) (%) and Enantioselectivities (ee)	Refs.
	R = t-Bu	DMD, acetone/CH_2Cl_2, ~20°, 10 d	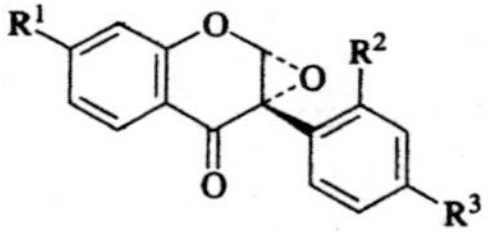 R^1 R^2 R^3 % ee H H H (31) 52 MeO H H (27) 37 Ms H H (22) 21 MeO MeO H (31) 82 MeO H MeO (39) 52 Ms MeO H (25) 72	103, 104
C_{15}				
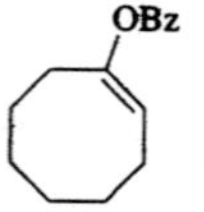 OBz	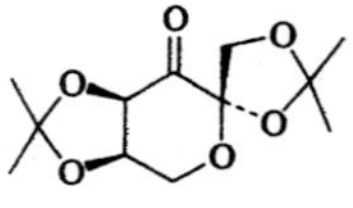	H_2O_2, CH_3CN, K_2CO_3 buffer, EDTA, 0°, 7 h	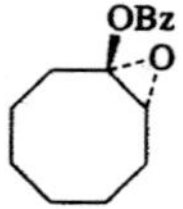 OBz (75) 93% ee	368
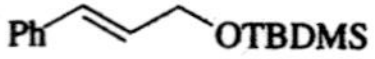 Ph OTBDMS	"	H_2O_2, CH_3CN, K_2CO_3 buffer, EDTA, 0°, 15 h	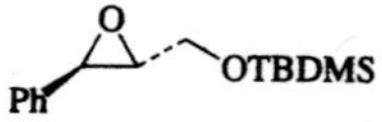 Ph OTBDMS (74) 93% ee	368
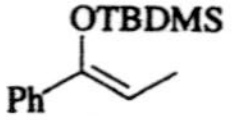 OTBDMS Ph	"	Oxone®, CH_3CN, $Na_2B_4O_7$ buffer, Na_2EDTA, 0°, 2 h	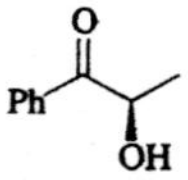 Ph OH (80) 90% ee	121

Substrate	Ketone	Conditions	Product(s) and Yield(s) (%)	Refs.
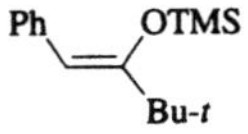 Ph, OTMS, Bu-*t*	"	Oxone®, $NaHCO_3$, CH_3CN/H_2O, Na_2EDTA, 0°, 3.0 h	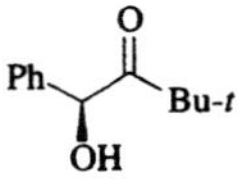 Ph, O, Bu-*t*, OH **I** (—) 52% ee	26
	"	Oxone®, K_2CO_3, CH_3CN, $Na_2B_4O_7$ buffer, Na_2EDTA, 0°, 1.5 h	**I** (—) 43% ee	26
C_{16}				
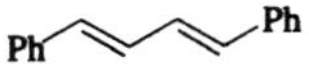 Ph, Ph	"	Oxone®, K_2CO_3, CH_3CN/DMM, buffer (pH 10.5), EDTA, 0°, 1.5 h	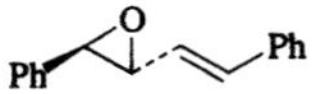 **I** (77) 97%ee 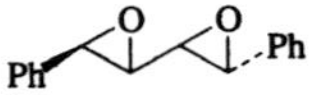 + **II** (—) **I**:**II** = 22:1	122
	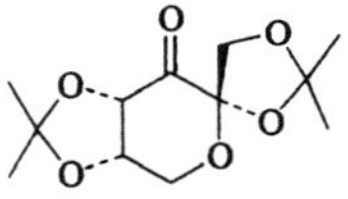 (ketone structure)	Oxone®, K_2CO_3, CH_3CN/DMM, buffer (pH 10.5), EDTA, 0°, 1.5 h	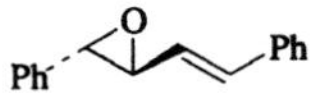 **I** (65) 97% ee 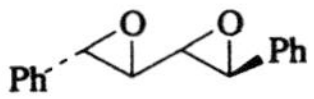 + **II** (—) **I**:**II** = 26:1	122
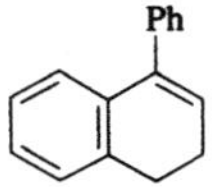 Ph	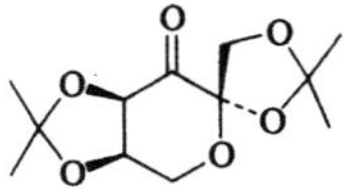 (ketone structure)	Oxone®, $NaHCO_3$, CH_3CN/H_2O, EDTA, 0°, 2 h	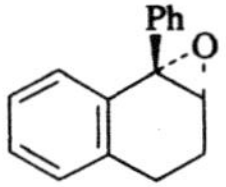 **I** (74) 95% ee	117
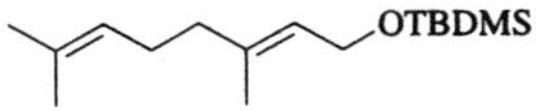 OTBDMS	"	Oxone®, $NaHCO_3$, CH_3CN/H_2O, EDTA, 0°	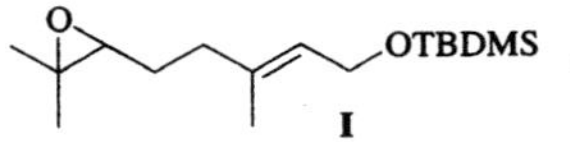 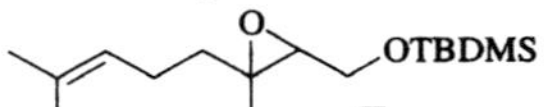 OTBDMS **I** + OTBDMS **II**	149a

pH	Time	% Convn	I:II
7.0-7.5	2 h	5	4.3:1
8.0	1.5 h	10	4.4:1
10.6	1.5 h	41	4.5;1

TABLE 5. ENANTIOSELECTIVE EPOXIDATION OF OLEFINS BY ENANTIOMERICALLY ENRICHED DIOXIRANES (*Continued*)

	Substrate	Ketone/Catalyst	Conditions	Product(s), Yield(s) (%) and Enantioselectivities (ee)	Refs.
		O, O, O, O, O	Oxone®, K_2CO_3, CH_3CN/DMM, $Na_2B_4O_7$ buffer, Bu_4NHSO_4, Na_2EDTA, –10°, 2 h	**I** (98) 95% ee	25
	Ph, TMS	"	Oxone®, K_2CO_3, CH_3CN/DMM, buffer (pH 10.5), EDTA, 0°, 1.5 h	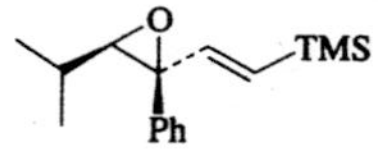 O, Ph, TMS (81) 95% ee	122
C_{17} 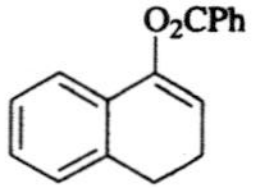	O_2CPh	"	Oxone®, CH_3CN, DMM, $Na_2B_4O_7$ buffer, EDTA, 0°, 1.5 h	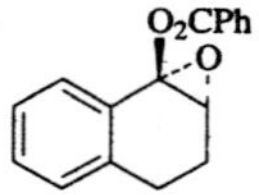 O_2CPh, O (92) 88% ee	121
	$CH(CO_2Me)_2$, Ph	"	Oxone®, K_2CO_3, CH_3CN, DMM, $Na_2B_4O_7$ buffer, EDTA, 0°, 2.5 h	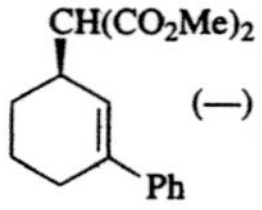 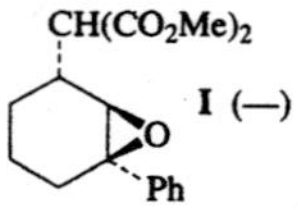 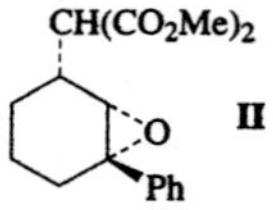 $CH(CO_2Me)_2$, Ph (—) 81% ee + $CH(CO_2Me)_2$, O, Ph **I** (—) + $CH(CO_2Me)_2$, O, Ph **II** (—) **I**:**II** = 1.7:1	125
	OTBDMS, TMS	"	Oxone®, K_2CO_3, CH_3CN/DMM, $Na_2B_4O_7$ buffer, EDTA, 2.5 h	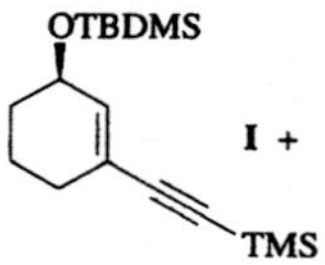 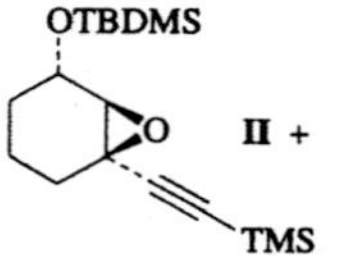 OTBDMS, TMS **I** + OTBDMS, O, TMS **II** +	125

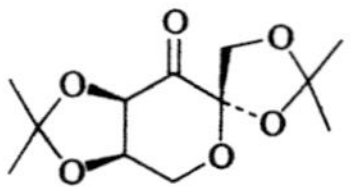

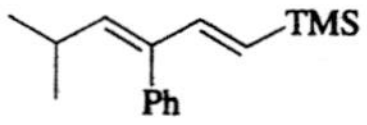

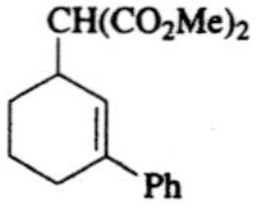

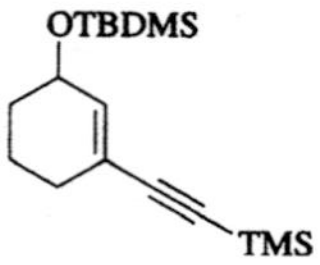

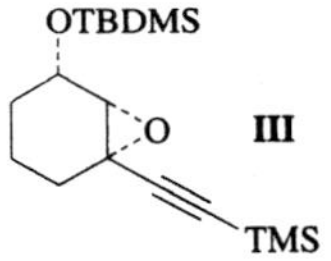

OTBDMS

O **III**

TMS

Temp	% Convn	% ee of **I**	**II**:**III**
−10°	49	75	13:1
20°	66	96	8:1

C_{18}

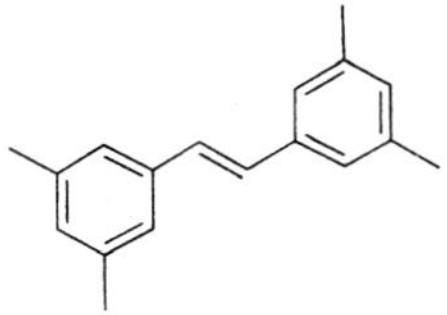

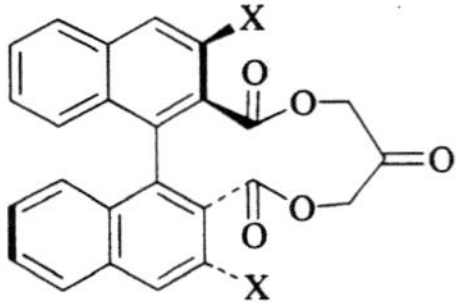

X O O O O O X

Oxone®, $NaHCO_3$, CH_3CN/H_2O, EDTA, rt

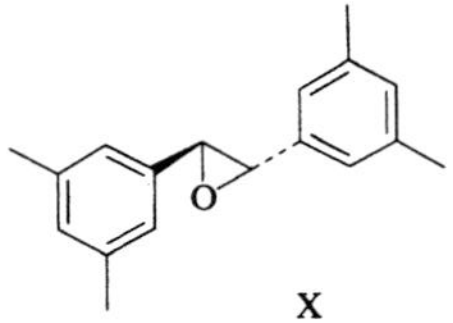

O (>90) 24

X	Time	% ee
Br	2-3 h	74
1,3-dioxolan-2-yl	40 min	73

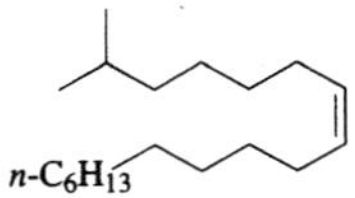

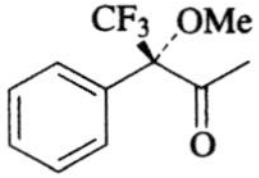

C_{19}

n-C_6H_{13}

CF_3 OMe O

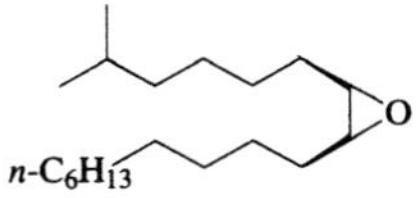

Oxone®, CH_2Cl_2/H_2O (pH 7.5), Bu_4NHSO_4, 2 to 5°, 17 h

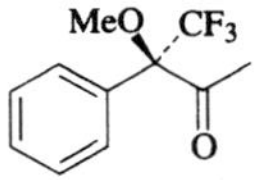

n-C_6H_{13} O (68) 16% ee 106

MeO CF_3 O

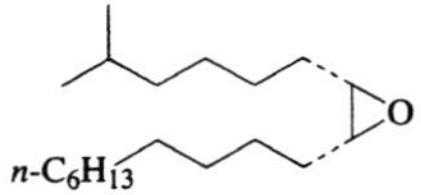

Oxone®, CH_2Cl_2/H_2O (pH 7.5), Bu_4NHSO_4, 2 to 5°, 17 h

n-C_6H_{13} O (68) 16% ee 106

TABLE 5. ENANTIOSELECTIVE EPOXIDATION OF OLEFINS BY ENANTIOMERICALLY ENRICHED DIOXIRANES (*Continued*)

Substrate	Ketone/Catalyst	Conditions	Product(s), Yield(s) (%) and Enantioselectivities (ee)	Refs.
C_{20}				
Ph, Ph, Ph (triphenylethylene)	X	Oxone®, $NaHCO_3$, CH_3CN/H_2O, EDTA, rt	Ph, O, Ph, Ph **I** X / Time / % ee Br / 24 h / (82) 81 Cl / 24 h / (96) 76 H / 75 min / (97) 48 CH_2OMe / 2.3 h / (95) 67 1,3-dioxolan-2-yl / 3 h / (90) 73	24, 110, 111
	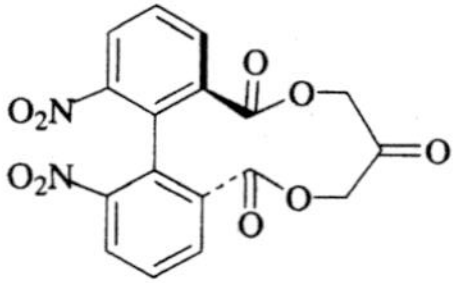 O_2N, O_2N	Oxone®, $NaHCO_3$, CH_3CN/H_2O, EDTA, rt, 60 min	**I** (82) 49% ee	24
	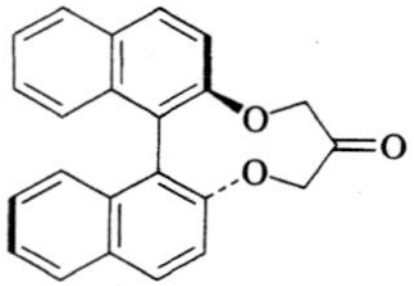	Oxone®, K_2CO_3, CH_3CN, $Na_2B_4O_7$ buffer, rt, 1.5 h	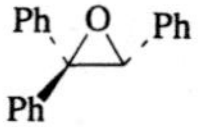 Ph, O, Ph, Ph (—) 24% ee)	113
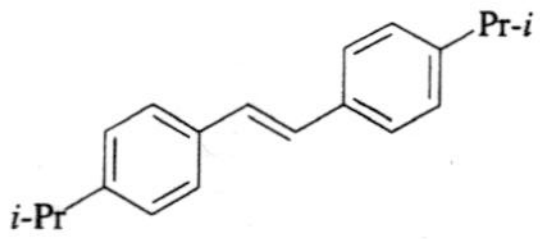 *i*-Pr, Pr-*i*	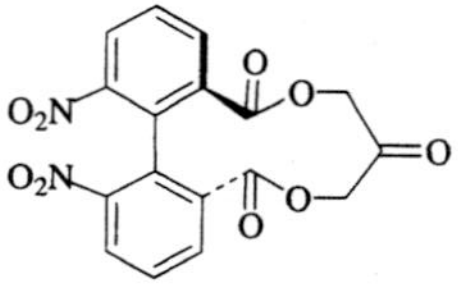 O_2N, O_2N	Oxone®, $NaHCO_3$, CH_3CN/H_2O, EDTA, rt, 35 min	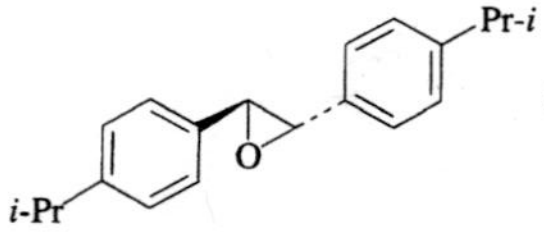 *i*-Pr, O, Pr-*i* (94) 66% ee	24

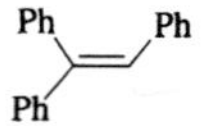

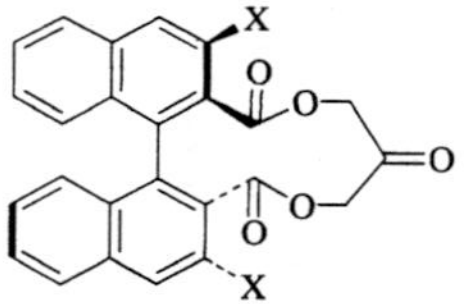

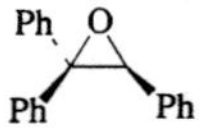

C_{22}

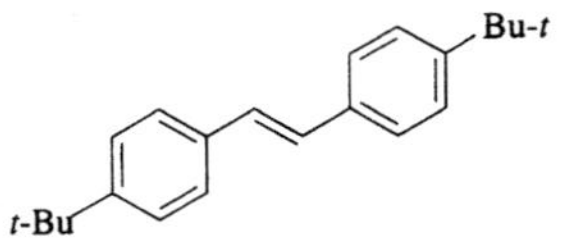

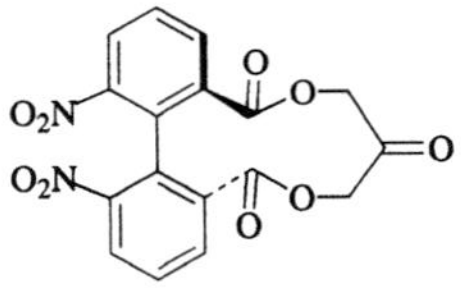

t-Bu, Bu-t

O_2N, O_2N, O, O, O, O, O

Oxone®, $NaHCO_3$, CH_3CN/H_2O, EDTA, rt, 35 min

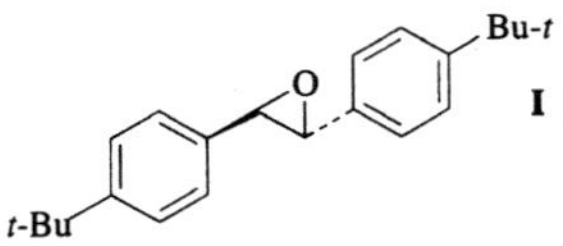

t-Bu, O, Bu-t

I (91) 77% ee 24

O, O, O, O, O

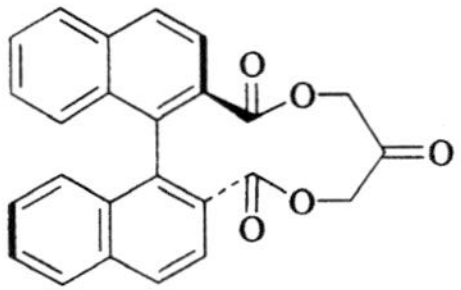

Oxone®, $NaHCO_3$, solvent, EDTA

I

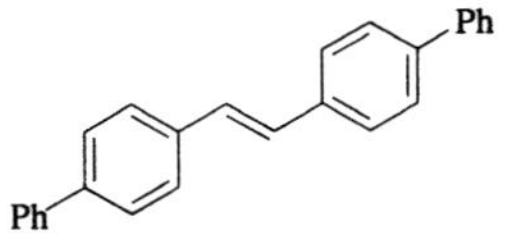

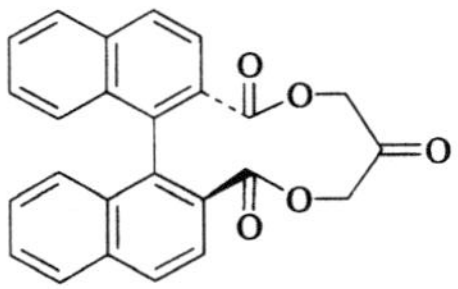

Solvent	Temp	Time		% ee
DME-H_2O	rt	0.5 h	(92)	77
CH_3CN-H_2O	rt	0.7 h	(95)	76
Dioxane-H_2O	rt	24 h	(52)	76
THF-H_2O	rt	24 h	(41)	75
DME-H_2O	0°	20 h	(91)	83
DME-H_2O	–20°	20 h	(2)	84

24

C_{26}

Ph, Ph

O, O, O, O, O

Oxone®, $NaHCO_3$, CH_3CN/H_2O, EDTA, rt, 8 h

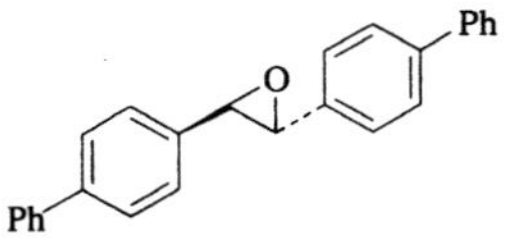

Ph, O, Ph

(83) 18% ee 24

TABLE 5. ENANTIOSELECTIVE EPOXIDATION OF OLEFINS BY ENANTIOMERICALLY ENRICHED DIOXIRANES (*Continued*)

Substrate	Ketone/Catalyst	Conditions	Product(s), Yield(s) (%) and Enantioselectivities (ee)	Refs.
C_{30}				
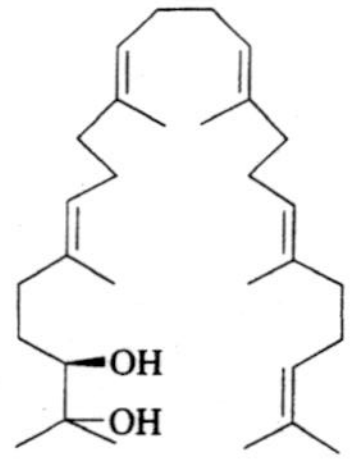 OH, OH	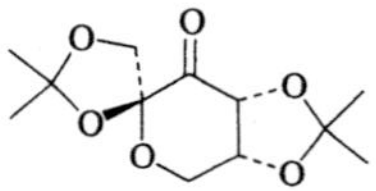	1. Oxone®, MeCN, DMM, H_2O, (pH 10.5), 0°, 1.5 h 2. CSA, toluene, 0°, 1 h,	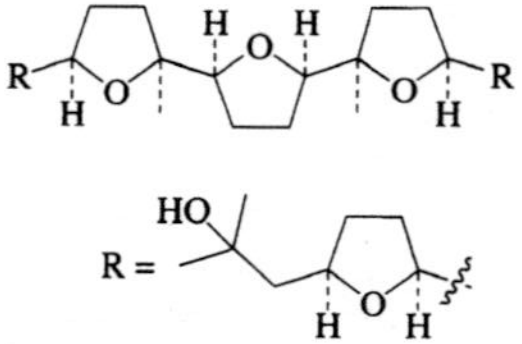 R, H, O, H, H, O, H, O, H, R (31) R = HO, H, O, H	126
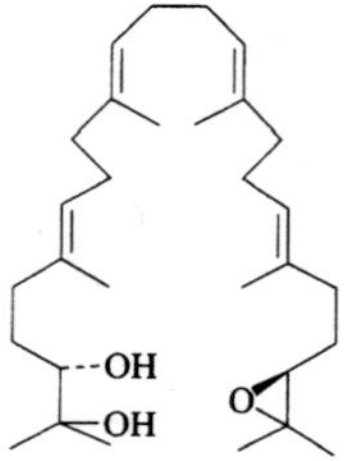 OH, OH, O	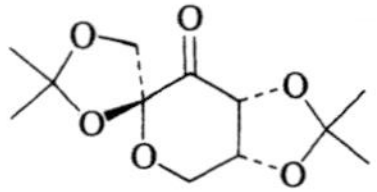	Oxone®, MeCN, DMM, H_2O, (pH 10.5), 0°, 1.5 h	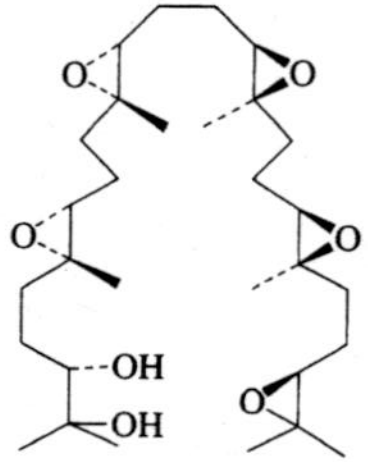 O, O, O, O, OH, OH, O (—)	126
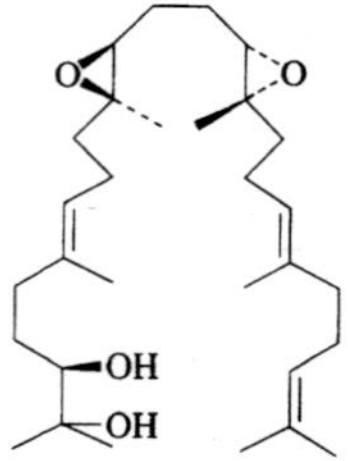 O, O, OH, OH	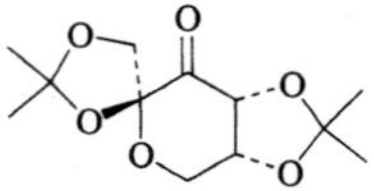	Oxone®, MeCN, DMM, H_2O, (pH 10.5), 0°, 1.5 h	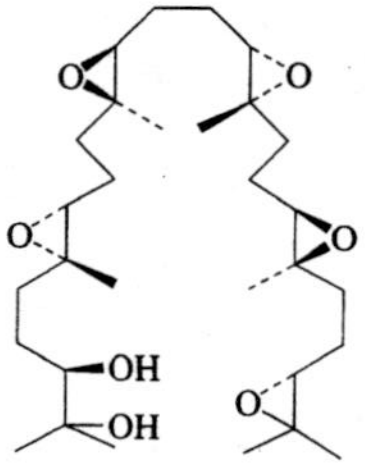 O, O, O, O, OH, OH, O (—)	126

OH, OH, O, O	O, O, O, O, O, O	Oxone®, MeCN, DMM, H_2O, (pH 10.5), 0°, 1.5 h	O, O, O, O, OH, OH, O (—)	126

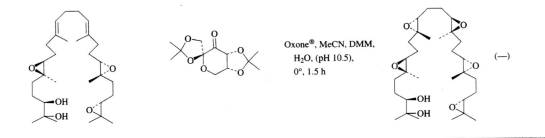

REFERENCES

1. Adam, W.; Curci, R.; Edwards, J. O. *Acc. Chem. Res.* **1989**, *22*, 205.
2. Murray, R. W. *Chem. Rev.* **1989**, *89*, 1187.
3. Curci, R. In *Advances in Oxygenated Processes*; Baumstark, A. L., Ed.; JAI: Greenwich CT, 1990, Vol. 2, Chapter I; pp. 1–59.
4. Adam, W.; Hadjiarapoglou, L. P.; Curci, R.; Mello, R. In *Organic Peroxides*; Ando, W., Ed.; Wiley: New York, 1992, Chapter 4; pp. 195–219.
5. Adam, W.; Hadjiarapoglou, L. *Top. Curr. Chem.* **1993**, *164*, 45.
6. Curci, R.; Dinoi, A.; Rubino, M. F. *Pure Appl. Chem.* **1995**, *67*, 811.
7. Lévai, A.; Adam, W.; Halász, J.; Nemes, C.; Patonay, T.; Tóth, G. *Khim.Geterotsikl. Soedin. (Engl. Trans.)* **1995**, *10*, 1345; *Chem. Abstr.* **1991**, *125*, 1049z.
8. Adam, W.; Smerz, A. K. *Bull. Soc. Chim. Belg.* **1996**, *105*, 581.
9. Murray, R. W.; Singh, M. In *Comprehensive Heterocyclic Chemistry II*; Padwa, A., Ed.; Elsevier: Oxford, 1996, Vol. 1A; pp. 429–456.
10. Adam, W.; Smerz, A. K.; Zhao, C.-G. *J. Prakt. Chem.* **1997**, *339*, 298.
11. Denmark, S. E.; Wu, Z. *Synlett* **1999**, 847.
12. Adam, W.; Degen, H.-G.; Pastor, A.; Saha-Möller, C. R.; Schambony, S. B.; Zhao, C.-G. In *Peroxide Chemistry: Mechanistic and Preparative Aspects of Oxygen Transfer*; Adam, W., Ed.; Wiley-VCH: Weinheim, 2000, pp. 78–112.
13. Adam, W.; Curci, R.; D'Accolti, L.; Dinoi, A.; Fusco, C.; Gasparrini, F.; Kluge, R.; Paredes, R.; Schulz, M.; Smerz, A. K.; Veloza, L. A.; Weinkötz, S.; Winde, R. *Chem. Eur. J.* **1997**, *3*, 105.
14. Adam, W.; Fröhling, B.; Peters, K.; Weinkötz, S. *J. Am. Chem. Soc.* **1998**, *120*, 8914.
15. Baumstark, A. L.; McCloskey, C. J. *Tetrahedron Lett.* **1987**, *28*, 3311.
16. Baumstark, A. L.; Vasquez, P. C. *J. Org. Chem.* **1988**, *53*, 3437.
17. Houk, K. N.; Liu, J.; DeMello, N. C.; Condroski, K. R. *J. Am. Chem. Soc.* **1997**, *119*, 10147.
18. Liu, J.; Houk, K. N.; Dinoi, A.; Fusco, C.; Curci, R. *J. Org. Chem.* **1998**, *63*, 8565.
19. Jenson, C.; Liu, J.; Houk, K. N.; Jorgensen, W. L. *J. Am. Chem. Soc.* **1997**, *119*, 12982.
20. Bach, R. D.; Canepa, C.; Winter, J. E.; Blanchette, P. E. *J. Org. Chem.* **1997**, *62*, 5191.
21. Bach, R. D.; Glukhovtsev, M. N.; Gonzalez, C.; Marquez, M.; Estévez, C. M.; Baboul, A. G.; Schlegel, H. B. *J. Phys. Chem. A* **1997**, *101*, 6092.
22. Miaskiewicz, K; Smith, D. A. *J. Am. Chem. Soc.* **1998**, *120*, 1872.
23. Freccero, M.; Gandolfi, R.; Sarzi-Amadè, M.; Rastelli, A. *Tetrahedron* **1998**, *54*, 12323.
24. Yang, D.; Wong, M.-K.; Yip, Y.-C.; Wang, X.-C.; Tang, M.-W.; Zheng, J.-H.; Cheung, K.-K. *J. Am. Chem. Soc.* **1998**, *120*, 5943.
25. Wang, Z.-X.; Tu, Y.; Frohn, M.; Zhang, J.-R.; Shi, Y. *J. Am. Chem. Soc.* **1997**, *119*, 11224.
26. Adam, W.; Fell, R. T.; Saha-Möller, C. R.; Zhao, C.-G. *Tetrahedron: Asymmetry* **1998**, *9*, 397.
27. Adam, W.; Curci, R.; González Nuñez, M. E.; Mello, R. *J. Am. Chem. Soc.* **1991**, *113*, 7654.
28. Shea, K. J.; Kim, J.-S. *J. Am. Chem. Soc.* **1992**, *114*, 3044.
29. Bravo, A.; Fontana, F.; Fronza, G.; Minisci, F.; Serri, A. *Tetrahedron Lett.* **1995**, *36*, 6945.
30. Bravo, A.; Fontana, F.; Fronza, G.; Minisci, F.; Zhao, L. *J. Org. Chem.* **1998**, *63*, 254.
31. Edwards, J. O.; Pater, R. H.; Curci, R.; Di Furia, F. *Photochem. Photobiol.* **1979**, *30*, 63.
32. Curci, R.; Fiorentino, M.; Troisi, L.; Edwards, J. O.; Pater, R. H. *J. Org. Chem.* **1980**, *45*, 4758.
33. Jeyaraman, R.; Murray, R. W. *J. Am. Chem. Soc.* **1984**, *106*, 2462.
34. Adam, W.; Hadjiarapoglou, L.; Smerz, A. *Chem. Ber.* **1991**, *124*, 227.
35. Yang, D.; Wong, M.-K.; Yip, Y.-C. *J. Org. Chem.* **1995**, *60*, 3887.
36. Frohn, M.; Wang, Z.-X.; Shi, Y. *J. Org. Chem.* **1998**, *63*, 6425.
37. Denmark, S. E.; Wu, Z. *J. Org. Chem.* **1998**, *63*, 2810.
38. Yang, D.; Yip, Y.-C.; Jiao, G.-S.; Wong, M.-K. *J. Org. Chem.* **1998**, *63*, 8952.
39. Yang, D.; Yip, Y.-C.; Tang, M.-W.; Wong, M.-K; Cheung, K.-K. *J. Org. Chem.* **1998**, *63*, 9888.
40. Carnell, A. J.; Johnstone, R. A. W.; Parsy, C. C.; Sanderson, W. R. *Tetrahedron Lett.* **1999**, *40*, 8029.
41. Bentley, T. W.; Norman, S. J.; Gerstner, E.; Kemmer, R.; Christl, M. *Chem. Ber.* **1993**, *126*, 1749.
42. Hofland, A.; Steinberg, H.; de Boer, T. J. *Recl. Trav. Chim. Pays-Bas* **1985**, *104*, 350.
43. Trost , B. M.; Bogdanowicz, M. J. *J. Am. Chem. Soc.* **1973**, *95*, 5321.

[44] Adam, W.; Hadjiarapoglou, L.; Jäger, V.; Klicić, J.; Seidel, B.; Wang, X. *Chem. Ber.* **1991**, *124*, 2361.

[45] Adam, W.; Hadjiarapoglou, L.; Wang, X. *Tetrahedron Lett.* **1989**, *30*, 6497.

[46] Rubottom, G. M.; Vazquez, M. A.; Pelegrina, D. R. *Tetrahedron Lett.* **1974**, *15*, 4319.

[47] Adam, W.; Hadjiarapoglou, L.; Mosandl, T.; Saha-Möller, C. R.; Wild, D. *Angew. Chem., Int. Ed. Engl.* **1991**, *30*, 200.

[48] Adam, W.; Bialas, J.; Hadjiarapoglou, L.; Sauter, M. *Chem. Ber.* **1992**, *125*, 231.

[49] Sauter, M.; Adam, W. *Acc. Chem. Res.* **1995**, *28*, 289.

[50] Adam, W.; Sauter, M. unpublished results.

[51] Baertschi, S. W.; Raney, K. D.; Stone, M. P.; Harris, T. M. . *J. Am. Chem. Soc.* **1988**, *110*, 7929.

[52] Patai, S.; Rappoport, Z. In *The Chemistry of Alkenes*; Patai, S., Ed.; Wiley: New York, 1964, Vol. 1, pp. 512–517.

[53] Adam, W.; Hadjiarapoglou, L.; Nestler, B. *Tetrahedron Lett.* **1990**, *31*, 331.

[54] Piers, E.; Boulet, S. L. *Tetrahedron Lett.* **1997**, *38*, 8815.

[55] Porter, M. J.; Skidmore, J. *J. Chem. Soc., Chem. Commun.* **2000**, 1215.

[56] Adam, W.; Hadjiarapoglou, L. *Chem. Ber.* **1990**, *123*, 2077.

[57] Adam, W.; Hadjiarapoglou, L.; Levai, A. *Synthesis* **1992**, 436.

[58] Adam, W.; Golsch, D.; Hadjiarapoglou, L.; Patonay, T. *Tetrahedron Lett.* **1991**, *32*, 1041.

[59] Adam, W.; Golsch, D.; Hadjiarapoglou, L.; Patonay, T. *J. Org. Chem.* **1991**, *56*, 7292.

[60] Lévai, A.; Adam, W.; Jekö, J.; Patonay, T.; Székely, A.; Vass, E. B. *J. Heterocycl. Chem.* **2000**, *37*, 1065.

[60a] Ferrer, M.; Sánchez-Baeza, F.; Messeguer, A.; Adam, W.; Golsch, D.; Görth, F.; Kiefer, W.; Nagel, V. *Eur. J. Org. Chem.* **1998**, 2527.

[61] Murray, R. W.; Jeyaraman, R.; Pillay, M. K. *J. Org. Chem.* **1987**, *52*, 746.

[61a] Scammells, P. J.; Baker, S. P.; Bellardinelli, L.; Olsson, R. A.; Russell, R. A.; Knevitt, S. A. *Bioorg. Med. Chem. Lett.* **1996**, *6*, 811.

[62] Ferrer, M.; Sánchez-Baeza, F.; Messeguer, A. *Tetrahedron* **1997**, *53*, 15877.

[63] Ferrer, M.; Sánchez-Baeza, F.; Messeguer, A.; Diez, A.; Rubiralta, M. *J. Chem. Soc., Chem. Commun.* **1995**, 293.

[64] Adam, W.; Ahrweiler, M.; Paulini, K.; Reißig, H.-U.; Voerckel, V. *Chem. Ber.* **1992**, *125*, 2719.

[65] Asensio, G.; Mello, R.; Boix-Bernardini, C.; González-Núñez, M. E.; Castellano, G. *J. Org. Chem.* **1995**, *60*, 3692.

[66] Adam, W.; Golsch, D.; Hadjiarapoglou, L.; Lévai, A.; Nemes, C.; Patonay, T. *Tetrahedron* **1994**, *50*, 13113.

[67] Murray, R. W.; Singh, M.; Williams, B. L.; Moncrieff, H. M. *J. Org. Chem.* **1996**, *61*, 1830.

[68] Armstrong, A.; Barsanti, P. A.; Clarke, P. A.; Wood, A. *J. Chem. Soc., Perkin Trans. 1* **1996**, 1373.

[69] Foglia, T. A.; Sonnet, P. E.; Nunez, A.; Dudley, R. L. *J. Am. Oil Chem. Soc.* **1998**, *75*, 601.

[70] Rodríguez, G.; Castedo, L.; Domínguez, D.; Saá, C.; Adam, W.; Saha-Möller, C. R. *J. Org. Chem.* **1999**, *64*, 877.

[71] Wang, X.; Ramos, B.; Rodriguez, A. *Tetrahedron Lett.* **1994**, *35*, 6977.

[72] Adam, W.; Smerz, A. K. *J. Org. Chem.* **1996**, *61*, 3506.

[73] Adam, W.; Prechtl, F.; Richter, M. J.; Smerz, A. K. *Tetrahedron Lett.* **1993**, *34*, 8427.

[74] Adam, W.; Prechtl, F.; Richter, M. J.; Smerz, A. K. *Tetrahedron Lett.* **1995**, *36*, 4991.

[75] Adam, W.; Paredes, R.; Smerz, A. K.; Veloza, L. A. *Liebigs Ann./Recl.* **1997**, 547.

[76] de Macedo Puyau, P.; Perie, J. J. *Synth. Commun.* **1998**, *28*, 2679.

[76a] Mello, R.; Fiorentino, M.; Fusco, C.; Curci, R. *J. Am. Chem. Soc.* **1989**, *111*, 6749.

[77] Bovicelli, P.; Lupattelli, P.; Mincione, E. *J. Org. Chem.* **1994**, *59*, 4304.

[78] Messeguer, A.; Fusco, C.; Curci, R. *Tetrahedron* **1993**, *49*, 6299.

[79] Adam, W.; Wirth, T. *Acc. Chem. Res.* **1999**, *32*, 703.

[80] Adam, W.; Paredes, R.; Smerz, A. K.; Veloza, L. A. *Eur. J. Org. Chem.* **1998**, 349.

[81] Yang, D.; Jiao, G.-S.; Yip, Y.-C.; Wong, M.-K. *J. Org. Chem.* **1999**, *64*, 1635.

[82] Vedejs, E.; Dent, W. H., III; Kendall, J. T.; Oliver, P. A. *J. Am. Chem. Soc.* **1996**, *118*, 3556.

[83] Adam, W.; Mitchell, C. M.; Saha-Möller, C. R. *Eur. J. Org. Chem.*, **1999**, 785.

[84] Mello, R.; Ciminale, F.; Fiorentino, M.; Fusco, C.; Prencipe, T.; Curci, R. *Tetrahedron Lett.* **1990**, *31*, 6097.

[85] Bovicelli, P.; Lupattelli, P.; Mincione, E.; Prencipe, T.; Curci, R. *J. Org. Chem.* **1992**, *57*, 5052.
[86] Ebenezer, W.; Pattenden, G. *Tetrahedron Lett.* **1992**, *33*, 4053.
[87] Adam, W.; Smerz, A. K. *Tetrahedron* **1995**, *51*, 13039.
[88] Freccero, M.; Gandolfi, R.; Sarzi-Amede, M; Rastelli, A. *J. Org. Chem.* **2000**, *65*, 2030.
[89] Adam, W.; Bach, R. D.; Dmitrenko, O.; Saha-Möller, C. R. *J. Org. Chem.* **2000**, *65*, 6715.
[90] Evans, D. A.; Trotter, B. W.; Coleman, P. J.; Côté, B.; Dias, L. C.; Rajapakse, H. A.; Tyler, A. N. *Tetrahedron* **1999**, *55*, 8671.
[91] Halcomb, R. L.; Danishefsky, S. J. *J. Am. Chem. Soc.* **1989**, *111*, 6661.
[92] Liu, K. K.-C.; Danishefsky, S. J. *J. Org. Chem.* **1994**, *59*, 1892.
[93] Randolph, J. T.; McClure, K. F.; Danishefsky, S. J. *J. Am. Chem. Soc.* **1995**, *117*, 5712.
[94] Roberge, J. Y.; Beebe, X.; Danishefsky, S. J. *J. Am. Chem. Soc.* **1998**, *120*, 3915.
[95] Deshpande, P. P.; Kim, H. M.; Zatorski, A.; Park, T.-K.; Ragupathi, G.; Livingston, P. O.; Live, D.; Danishefsky, S. J. *J. Am. Chem. Soc.* **1998**, *120*, 1600.
[96] Swindell, C. S.; Chander, M. C. *Tetrahedron Lett.* **1994**, *35*, 6001.
[97] Adam, W.; Müller, M.; Prechtl, F. *J. Org. Chem.* **1994**, *59*, 2358.
[98] Adam, W.; Prechtl, F. *Chem. Ber.* **1994**, *127*, 667.
[99] Adam, W.; Pastor, A.; Peters, K.; Peters, E.-M. *Org. Lett.* **2000**, *2*, 1019.
[100] Jacobsen, E. N. In *Catalytic Asymmetric Synthesis*; Ojima, I., Ed.; VCH: New York, 1993, pp. 159–202.
[101] Katsuki, T. *Coord. Chem. Rev.* **1995**, *140*, 189.
[102] Adam, W.; Jekö, J.; Lévai, A.; Nemes, C.; Patonay, T.; Sebök, P. *Tetrahedron Lett.* **1995**, *36*, 3669.
[103] Lévai, A.; Adam, W.; Fell, R. T.; Gessner, R.; Patonay, T.; Simon, A.; Tóth, G. *Tetrahedron* **1998**, *54*, 13105.
[104] Adam, W.; Fell, R. T.; Lévai, A.; Patonay, T.; Peters, K.; Simon, A.; Tóth, G. *Tetrahedron: Asymmetry* **1998**, *9*, 1121.
[105] Curci, R.; Fiorentino, M.; Serio, M. R. *J. Chem. Soc., Chem. Commun.* **1984**, 155.
[106] Curci, R.; D´Accolti, L.; Fiorentino, M.; Rosa, A. *Tetrahedron Lett.* **1995**, *36*, 5831.
[107] Denmark, S. E.; Wu, Z.; Crudden, C. M.; Matsuhashi, H. *J. Org. Chem.* **1997**, *62*, 8288.
[108] Armstrong, A.; Hayter, B. R. *J. Chem. Soc., Chem. Commun.* **1998**, 621.
[109] Yang, D.; Yip, Y.-C.; Chen, J.; Cheung, K.-K. *J. Am. Chem. Soc.* **1998**, *120*, 7659.
[110] Yang, D.; Yip, Y.-C.; Tang, M.-W.; Wong, M.-K.; Zheng, J.-H.; Cheung, K.-K. *J. Am. Chem. Soc.* **1996**, *118*, 491.
[111] Yang, D.; Wang, X.-C.; Wong, M.-K.; Yip, Y.-C.; Tang, M.-W. *J. Am. Chem. Soc.* **1996**, *118*, 11311.
[112] Song, C. E.; Kim, Y. H.; Lee, K. C.; Lee, S.-g.; Jin, B. W. *Tetrahedron: Asymmetry* **1997**, *8*, 2921.
[113] Adam, W.; Zhao, C.-G. *Tetrahedron: Asymmetry* **1997**, *8*, 3995.
[114] Wang, Z.-X.; Shi, Y. *J. Org. Chem.* **1997**, *62*, 8622.
[115] Wang, Z.-X.; Miller, S. M.; Anderson, O. P.; Shi, Y. *J. Org. Chem.* **1999**, *64*, 6443.
[116] Adam, W.; Saha-Möller, C. R.; Zhao, C.-G. *Tetrahedron: Asymmetry* **1999**, *10*, 2749.
[117] Tu, Y.; Wang, Z.-X.; Shi, Y. *J. Am. Chem. Soc.* **1996**, *118*, 9806.
[118] Wang, Z.-X.; Tu, Y.; Frohn, M.; Shi, Y. *J. Org. Chem.* **1997**, *62*, 2328.
[119] Tu, Y.; Wang, Z.-X.; Frohn, M.; He, M.; Yu, H.; Tang, Y.; Shi, Y. *J. Org. Chem.* **1998**, *63*, 8475.
[120] Adam, W.; Staab, E. Institut für Organische Chemie, Universität Würzburg, Würzburg, Germany, unpublished results.
[121] Zhu, Y.; Tu, Y.; Yu, H.; Shi, Y. *Tetrahedron Lett.* **1998**, *39*, 7819.
[122] Frohn, M.; Dalkiewicz, M.; Tu, Y.; Wang, Z.-X.; Shi, Y. *J. Org. Chem.* **1998**, *63*, 2948.
[123] Cao, G.-A.; Wang, Z.-X.; Tu, Y.; Shi, Y. *Tetrahedron Lett.* **1998**, *39*, 4425.
[124] Wang, Z.-X.; Cao, G.-A.; Shi, Y. *J. Org. Chem.* **1999**, *64*, 7646.
[125] Frohn, M.; Zhou, X.; Zhang, J.-R.; Tang, Y.; Shi, Y. *J. Am. Chem. Soc.* **1999**, *121*, 7718.
[126] Xiong, Z.; Corey, E. J. *J. Am. Chem. Soc.* **2000**, *122*, 4831.
[127] Swern, D. *Org. React.* **1953**, *7*, 378.
[128] Dryuk, V. G. *Russ. Chem. Rev.* **1985**, *54*, 986; *Chem. Abstr.* **1986**, *104*, 109449.
[129] Adam, W.; Saha-Möller, C. R.; Ganeshpure, P. *Chem. Rev.* **2001**, *101*, 3499.
[130] Ganeshpure, P. A.; Adam, W. *Synthesis* **1996**, 179.
[131] Plesničar, B. In *The Chemistry of Peroxides*; Patai, S., Ed.; Wiley: Chichester, 1983, pp. 567–584.

132 Davis, F. A.; Sheppard, A. C.; Chen, B.-C.; Haque, M. S. *J. Am. Chem. Soc.* **1990**, *112*, 6679.
133 Lusinchi, X.; Hanquet, G. *Tetrahedron* **1997**, *53*, 13727.
134 Minakata, S.; Takemiya, A.; Nakamura, K.; Ryu, I.; Komatsu, M. *Synlett* **2000**, 1810.
134a Adam, W.; Mitchell, C. M.; Paredes, R.; Smerz, A. K.; Veloza, L. A. *Liebigs Ann./Recl.* **1997**, 1365.
135 Adam, W.; Degen, H.-G.; Saha-Möller, C. R. *J. Org. Chem.* **1999**, *64*, 1274.
136 Adam, W.; Stegmann, V. R.; Saha-Möller, C. R. *J. Am. Chem. Soc.* **1999**, *121*, 1879.
137 Jørgensen, K. A. *Chem. Rev.* **1989**, *89*, 431.
138 Mimoun, H. In *The Chemistry of Peroxides*; Patai, S., Ed.; Wiley: Chichester, 1983, pp. 463–482.
139 Bortolini, O.; Di Furia, F.; Modena, G.; Seraglia R. *J. Org. Chem.* **1985**, *50*, 2688.
140 Prat, D.; Lett, R. *Tetrahedron Lett.* **1986**, *27*, 707.
141 Prandi, J.; Kagan, H. B.; Mimoun, H. *Tetrahedron Lett.* **1986**, *27*, 2617.
142 Venturello, C.; D'Aloisio, R. *J. Org. Chem.* **1988**, *53*, 1553.
143 Adam, W.; Mitchell, C. M. *Angew. Chem., Int. Ed. Engl.* **1996**, *35*, 533.
144 Adam, W.; Mitchell, C. M.; Saha-Möller, C. R.; Weichold, O. In *Structure and Bonding, Metal-Oxo and Metal-Peroxo Species in Catalytic Oxidations*; Meunier, B., Ed.; Springer Verlag: Berlin Heidelberg, 2000, Vol. 97, pp. 237–285.
145 Herrmann, W. A.; Fischer, R. W.; Marz, D. W. *Angew. Chem., Int. Ed. Engl.* **1991**, *30*, 1638.
146 Rozen, S.; Kol, M. *J. Org. Chem.* **1990**, *55*, 5155.
147 Adam, W.; Lazarus, M.; Saha-Möller, C. R.; Weichold, O.; Hoch, U.; Häring, D.; Schreier P. In *Advances in Biochemical Engineering/Biotechnology*; Faber, K., Ed.; Springer Verlag: Heidelberg, 1999, Vol. 63, pp. 73–108.
148 Katsuki, T.; Martin, V. S. *Org. React.* **1996**, *48*, 1.
149 Jacobsen, E. N. In *Comprehensive Organometallic Chemistry II*; Abel, E. W.; Stone, F. G. A.; Wilkinson, G.; Hegedus, L. S., Eds.; Pergamon: New York, 1995, Vol. 12, Chapter 11.1, pp. 1097–1135.
149a Wang, Z.-X.; Shi, Y. *J. Org. Chem.* **1998**, *63*, 3099.
150 Johnson, R. A.; Sharpless, K. B. In *Catalytic Asymmetric Synthesis*; Ojima, I., Ed.; VCH: New York, 1993, pp. 103–158.
151 Katsuki, T. *J. Mol. Cat. A* **1966**, *113*, 87.
152 Davies, R. J. H.; Boyd, D. R.; Kumar, S.; Sharma, N. D.; Stevenson, C. *Biochem. Biophys. Res. Commun.* **1990**, *169*, 87.
153 Adam, W.; Bialas, J.; Hadjiarapoglou, L. *Chem. Ber.* **1991**, *124*, 2377.
154 Adam, W.; Golsch, D.; Zhao, C.-G. Institut für Organische Chemie, Universität Würzburg, Würzburg, Germany, unpublished result, which was adapted from refs. 76a and 155.
155 Mello, R.; Fiorentino, M.; Sciacovelli, O.; Curci, R. *J. Org. Chem.* **1988**, *53*, 3890.
156 Murray, R. W.; Singh, M.; Jeyaraman, R. *J. Am. Chem. Soc.* **1992**, *114*, 1346.
157 Smerz, A. K. Ph. D. Dissertation, University of Würzburg, 1996.
158 Adam, W.; Makosza, M.; Stalinski, K.; Zhao, C.-G. *J. Org. Chem.* **1998**, *63*, 4390.
159 Adam, W.; Pastor, A.; Zhao, C.-G. Institut für Organische Chemie, Universität Würzburg, Würzburg, Germany, unpublished results.
160 Curci, R.; D'Accolti, L.; Fiorentino, M.; Fusco, C.; Adam, W.; González-Nuñez, M. E.; Mello, R. *Tetrahedron Lett.* **1992**, *33*, 4225.
161 Hull, L. A.; Badai, L. *Tetrahedron Lett.* **1993**, *34*, 5039.
162 Murray, R. W.; Jeyaraman, R. *J. Org. Chem.* **1985**, *50*, 2847.
163 Murray, R. W.; Singh, M. *Org. Synth. Coll. Vol. 9* **1998**, 288.
164 Gibert, M.; Ferrer, M.; Sánchez-Baeza, F.; Messeguer, A. *Tetrahedron* **1997**, *53*, 8643.
165 Ferrer, M.; Sánchez-Baeza, F.; Casas J.; Messeguer, A. *Tetrahedron Lett.* **1994**, *35*, 2981.
166 Zhao, C.-G. Ph. D. Dissertation, University of Würzburg, 1999.
167 Adam, W.; Reinhardt, D.; Reißig, H.-U.; Paulini, K. *Tetrahedron* **1995**, *51*, 12257.
168 Denmark, S. E.; Forbes, D. C.; Hays, D. S.; DePue, J. S.; Wilde, R. G. *J. Org. Chem.* **1995**, *60*, 1391.
169 Rahman, M.; McKee, M. L.; Shevlin, P. B.; Sztyrbicka, R. *J. Am. Chem. Soc.* **1988**, *110*, 4002.
170 Abou-Elzahab, M.; Adam, W.; Saha-Möller, C. R. *Liebigs Ann. Chem.* **1991**, 445.
171 Righi, G.; Bovicelli, P.; Sperandio, A. *Tetrahedron Lett.* **1999**, *40*, 5889.

[172] Sander, W.; Schroeder, K.; Muthusamy, S.; Kirschfeld, A.; Kappert, W.; Boese, R.; Kraka, E.; Sosa, C.; Cremer, D. *J. Am. Chem. Soc.* **1997**, *119*, 7265.
[173] Kirschfeld, A.; Muthusamy, S.; Sander, W. *Angew. Chem., Int. Ed. Engl.* **1994**, *33*, 2212.
[174] Murray, R. W.; Gu, D. *J. Chem. Soc., Perkin Trans. 2* **1993**, 2203.
[175] Murray, R. W.; Kong, W.; Rajadhyaksha, S. N. *J. Org. Chem.* **1993**, *58*, 315.
[176] Murray, R. W.; Gu, H. *J. Phys. Org. Chem.* **1996**, *9*, 751.
[177] Murray, R. W.; Pillay, M. K.; Jeyaraman, R. *J. Org. Chem.* **1988**, *53*, 3007.
[178] Murray, R. W.; Pillay, M. K. *Tetrahedron Lett.* **1988**, *29*, 15.
[179] Nojima, T.; Hirano, Y.; Ishiguro, K.; Sawaki, Y. *J. Org. Chem.* **1997**, *62*, 2387.
[180] Adam, W.; Hadjiarapoglou, L. P.; Meffert, A. *Tetrahedron Lett.* **1991**, *32*, 6697.
[181] Murray, R. W.; Singh, M. *Org. Synth.* **1997**, *74*, 91.
[182] Adam, W.; Smerz, A. K. *Tetrahedron* **1996**, *52*, 5799.
[183] Mock-Knoblauch, C. Ph.D. Dissertation, University of Würzburg, 2000.
[184] Bolli, M. H.; Ley, S. V. *J. Chem. Soc., Perkin Trans. 1* **1998**, 2243.
[185] Lin, H.-C.; Wu, H.-J. *Tetrahedron* **2000**, *56*, 341.
[186] Bujons, J.; Camps, F.; Messeguer, A. *Tetrahedron Lett.* **1990**, *31*, 5235.
[187] Asouti, A.; Hadjiarapoglou, L. P. *Tetrahedron Lett.* **2000**, *41*, 539.
[188] von Zezschwitz, P.; Voigt, K.; Lansky, A.; Noltemeyer, M., de Meijere, A. *J. Org. Chem.* **1999**, *64*, 3806.
[189] Gibert, M.; Ferrer, M.; Sánchez-Baeza, F.; Messeguer, A. *Tetrahedron* **1997**, *53*, 8643.
[190] Abou-Elzahab, M. M.; Adam, W.; Saha-Möller, C. R. *Liebigs Ann. Chem.* **1992**, 731.
[191] Roush, W. R.; Grover, P. T. *Tetrahedron* **1992**, *48*, 1981.
[192] Piguel, S.; Ulibarri, G.; Grierson, D. S. *Tetrahedron Lett.* **1999**, *40*, 295.
[193] Rodríguez, G.; Castedo, L.; Domínguez, D.; Saá, C.; Adam, W. *J. Org. Chem.* **1999**, *64*, 4830.
[194] Bickers, P. T.; Halton, B.; Kay, A. J.; Northcote, P. T. *Aust. J. Chem.* **1999**, *52*, 647.
[195] Camps, P.; Font-Bardia, M.; Méndez, N.; Pérez, F.; Pujol, X.; Solans, X.; Vázquez, S.; Vilalta, M. *Tetrahedron* **1998**, *54*, 4679.
[196] Reisch, J.; Top, M. *Pharmazie* **1991**, *46*, 745; *Chem. Abstr.* **1992**, *116*, 128716u.
[197] Sakanishi, K.; Kato, Y.; Mizukoshi, E.; Shimizu, K. *Tetrahedron Lett.* **1994**, *35*, 4789.
[198] Ikeno, T.; Harada, J.; Tomoda, S. *Chem. Lett.* **1999**, 409.
[199] Sano, H.; Kawata, K.; Kosugi, M. *Synlett* **1993**, 831.
[200] Roush, W. R.; Marron, T. G. *Tetrahedron Lett.* **1993**, *34*, 5421.
[201] Park, M.; Gu, F.; Loeppky, R. N. *Tetrahedron Lett.* **1998**, *39*, 1287.
[202] Okazaki, O.; Persmark, M.; Guengerich, F. P. *Chem. Res. Toxicol.* **1993**, *6*, 168; *Chem. Abstr.* **1993**, *118*, 11872v.
[203] Ndakala, A. J.; Howell, A. R. *J. Org. Chem.* **1998**, *63*, 6098.
[204] Adam, W.; Blancafort, L. *J. Org. Chem.* **1996**, *61*, 8432.
[205] Adam, W.; Hadjiarapoglou, L.; Klicic, J. *Tetrahedron Lett.* **1990**, *31*, 6517.
[206] Lavilla, R.; Gullón, F.; Barón, X.; Bosch, J. *J. Chem. Soc., Chem. Commun.* **1997**, 213.
[207] Lavilla, R.; Barón, X.; Coll, O.; Gullón, F.; Masdeu, C.; Bosch, J. *J. Org. Chem.* **1998**, *63*, 10001.
[208] Murray, R. W.; Shiang, D. L.; Singh, M. *J. Org. Chem.* **1991**, *56*, 3677.
[209] Guertin, K. R.; Chan, T.-H. *Tetrahedron Lett.* **1991**, *32*, 715.
[210] Adam, W.; Prechtl, F. *Chem. Ber.* **1991**, *124*, 2369.
[211] Adam, W.; Peters, E.-M.; Peters, K.; von Schnering, H. G.; Voerckel, V. *Chem. Ber.* **1992**, *125*, 1263.
[212] Adam, W.; Hadjiarapoglou, L.; Jäger, V.; Seidel, B. *Tetrahedron Lett.* **1989**, *30*, 4223.
[213] Adam, W.; Hadjiarapoglou, L.; Wang, X. *Tetrahedron Lett.* **1991**, *32*, 1295.
[214] Baylon, C.; Hanna, I. *Tetrahedron Lett.* **1995**, *36*, 6475.
[215] Adam, W.; Hadjiarapoglou, L.; Seebach, D. unpublished results.
[216] Chenault, H. K.; Danishefsky, S. J. *J. Org. Chem.* **1989**, *54*, 4249.
[217] Fujiwara, K.; Tsunashima, M.; Awakura, D.; Murai, A. *Chem. Lett.* **1997**, 665.
[218] Maynard, G. D.; Paquette, L. A. *J. Org. Chem.* **1991**, *56*, 5480.
[219] Sayer, J. M.; Servé, P. M.; Jerina, D. M. *J. Org. Chem.* **1994**, *59*, 977.
[220] Meyer, C.; Spiteller, G. *Liebigs Ann. Chem.* **1993**, 17.
[221] Silverman, S. K.; Foote, C. S. *J. Am. Chem. Soc.* **1991**, *113*, 7672.

222 Troisi, L.; Cassidei, L.; Lopez, L.; Mello, R.; Curci, R. *Tetrahedron Lett.* **1989**, *30*, 257.
223 Kopecky, K. R.; Xie, Y.; Molina, J. *Can. J. Chem.* **1993**, *71*, 272.
224 Adam, W.; Ahrweiler, M.; Balci, M.; Çakmak, O.; Saha-Möller, C. R. *Tetrahedron Lett.* **1995**, *36*, 1429.
225 Gopalakrishnan, S.; Stone, M. P.; Harris, T. M. *J. Am. Chem. Soc.* **1989**, *111*, 7232.
226 Saladino, R.; Mezzetti, M.; Mincione, E.; Torrini, I.; Paradisi, M. P.; Mastropietro, G. *J. Org. Chem.* **1999**, *64*, 8468.
227 Altamura, A.; Fusco, C.; D'Accolti, L.; Mello, R.; Prencipe, T.; Curci, R. *Tetrahedron Lett.* **1991**, *32*, 5445.
228 Lluch, A.-M.; Gibert, M.; Sánchez-Baeza, F.; Messeguer, A. *Tetrahedron* **1996**, *52*, 3973.
229 Evans, D. A.; Trotter, B. W.; Côté, B. *Tetrahedron Lett.* **1998**, *39*, 1709.
230 Meng, D.; Sorensen, E. J.; Bertinato, P.; Danishefsky, S. J. *J. Org. Chem.* **1996**, *61*, 7998.
231 Savin, K. A.; Woo, J. C. G.; Danishefsky, S. J. *J. Org. Chem.* **1999**, *64*, 4183.
232 Russo, A.; DesMarteau, D. D. *Angew. Chem., Int. Ed. Engl.* **1993**, *32*, 905.
233 Saladino, R.; Bernini, R.; Mincione, E.; Tagliatesta, P.; Boschi, T. *Tetrahedron Lett.* **1996**, *37*, 2647.
234 Saladino, R.; Bernini, R.; Crestini, C.; Mincione, E.; Bergamini, A.; Marini, S.; Palamara, A. T. *Tetrahedron* **1995**, *51*, 7561.
235 Lupattelli, P.; Saladino, R.; Mincione, E. *Tetrahedron Lett.* **1993**, *34*, 6313.
236 Baumstark, A. L.; Harden, D. B., Jr. *J. Org. Chem.* **1993**, *58*, 7615.
237 Prechtl, F. Diplomarbeit, University of Würzburg, 1990.
238 Baldwin, J. E.; O'Neil, I. A. *Tetrahedron Lett.* **1990**, *31*, 2047.
239 Burgess, L. E.; Gross, E. K. M.; Jurka, J. *Tetrahedron Lett.* **1996**, *37*, 3255.
240 Coats, S. J.; Wasserman, H. H. *Tetrahedron Lett.* **1995**, *36*, 7735.
241 Patonay, T.; Lévai, A.; Nemes, C.; Timár, T.; Tóth, G.; Adam, W. *J. Org. Chem.* **1996**, *61*, 5375.
242 Patonay, T.; Tóth, G.; Adam, W. *Tetrahedron Lett.* **1993**, *34*, 5055.
243 Zander, N.; Langschwager, W.; Hoffmann, H. M. R. *Synth. Commun.* **1996**, *26*, 4577.
244 Adam, W.; Schönberger, A. *Tetrahedron Lett.* **1992**, *33*, 53.
245 Murray, R. W.; Singh, M.; Rath, N. *J. Org. Chem.* **1997**, *62*, 8794.
246 Crandall, J. K.; Zucco, M.; Kirsch, R. S.; Coppert, D. M. *Tetrahedron Lett.* **1991**, *32*, 5441.
247 Adam, W.; Bialas, J.; Hadjiarapoglou, L.; Patonay, T. *Synthesis* **1992**, 49.
248 Burke, A. J.; O'Sullivan, W. I. *Tetrahedron* **1997**, *53*, 8491.
249 Halász, J.; Tóth, G.; Lévai, A.; Nemes, C.; Jámbor, Z. *J. Chem. Res. (S)* **1994**, 326.
250 Adam, W.; Halász, J.; Lévai, A.; Nemes, C.; Patonay, T.; Tóth, G. *Liebigs Ann. Chem.* **1994**, 795.
251 Adam, W.; Halász, J.; Jámbor, Z.; Lévai, A.; Nemes, C.; Patonay, T.; Tóth, G. *Monatsh. Chem.* **1996**, *127*, 683.
252 Adam, W.; Lévai, A.; Mérour, J.-Y.; Nemes, C.; Patonay, T. *Synthesis* **1997**, 268.
253 Adam, W.; Halász, J.; Jámbor, Z.; Lévai, A.; Nemes, C.; Patonay, T.; Tóth, G. *J. Chem. Soc., Perkin Trans. 1* **1996**, 395.
254 Fang, F. G.; Danishefsky, S. J. *Tetrahedron Lett.* **1989**, *30*, 2747.
255 Adam, W.; Ahrweiler, M.; Vlček, P. *J. Am. Chem. Soc.* **1995**, *117*, 9690.
256 Griesbeck, A. G.; Deufel, T.; Hohlneicher, G.; Rebentisch, R.; Steinwascher, J. *Eur. J. Org. Chem.* **1998**, 1759.
257 Sieburth, S. M.; McGee, K. F., Jr.; Al-Tel, T. H. *Tetrahedron Lett.* **1999**, *40*, 4007.
258 Bovicelli, P.; Lupattelli, P.; Mincione, E.; Prencipe, T.; Curci, R. *J. Org. Chem.* **1992**, *57*, 2182.
259 Adam, W.; Jekõ, J.; Lévai, A.; Nemes, C.; Patonay, T. *Liebigs Ann. Chem.* **1995**, 1547.
260 Lacroix, I.; Biton, J.; Azerad, R. *Bioorg. Med. Chem.* **1997**, *5*, 1369.
261 Horie, T.; Shibata, K.; Yamashita, K.; Kawamura, Y.; Tsukayama, M. *Chem. Pharm. Bull.* **1997**, *45*, 446.
262 Liu, J.; Mander, L. N.; Willis, A. C. *Tetrahedron* **1998**, *54*, 11637.
263 Ireland, R. E.; Liu, L.; Roper, T. D.; Gleason, J. L. *Tetrahedron* **1997**, *53*, 13257.
264 Corey, P. F.; Ward, F. E. *J. Org. Chem.* **1986**, *51*, 1925.
265 Jones, C. W.; Sankey, J. P.; Sanderson, W. R.; Rocca, M. C.; Wilson, S. L. *J. Chem. Res. (S)* **1994**, 114.
266 Boehlow, T. R.; Buxton, P. C.; Grocock, E. L.; Marples, B. A.; Waddington, V. L. *Tetrahedron Lett.* **1998**, *39*, 1839.

[267] Traylor, T. G.; Kim, C.; Richards, J. L.; Xu, F.; Perrin, C. L. *J. Am. Chem. Soc.* **1995**, *117*, 3468.
[268] Camporeale, M.; Fiorani, T.; Troisi, L.; Adam, W.; Curci, R.; Edwards, J. O. *J. Org. Chem.* **1990**, *55*, 93.
[269] Gallopo, A. R.; Edwards, J. O. *J. Org. Chem.* **1981**, *46*, 1684.
[270] Nakamura, N.; Nojima, M.; Kusabayashi, S. *J. Am. Chem. Soc.* **1987**, *109*, 4969.
[271] Schulz, M.; Liebsch, S.; Kluge, R.; Adam, W. *J. Org. Chem.* **1997**, *62*, 188.
[272] Cristau, H.-J.; Mbianda, X. Y.; Geze, A.; Beziat, Y.; Gasc, M.-B. *J. Organomet. Chem.* **1998**, *571*, 189.
[273] Denmark, S. E.; Wu, Z. *J. Org. Chem.* **1997**, *62*, 8964.
[274] Suprun, W. Y.; Schulze, D. *J. Prakt. Chem.* **1997**, *339*, 71; *Chem. Abstr.* **1997**, *126*, 143946u.
[275] Schultz, A. G.; Harrington, R. E.; Tham, F. S. *Tetrahedron Lett.* **1992**, *33*, 6097.
[276] Suh, Y.-G.; Min, K.-H.; Baek, S.-Y.; Chai, J.-H. *Heterocycles* **1998**, *48*, 1527.
[277] Tsui, H.-C.; Paquette, L. A. *J. Org. Chem.* **1998**, *63*, 9968.
[278a] Liang, X.; Kingston, D. G. I.; Lin, C. M.; Hamel, E. *Tetrahedron Lett.* **1995**, *36*, 2901.
[278] Altstadt, T. J.; Gao, Q.; Wittman, M. D.; Kadow, J. F.; Vyas, D. M. *Tetrahedron Lett.* **1998**, *39*, 4965.
[279] Dyker, G.; Hölzer, B. *Tetrahedron* **1999**, *55*, 12557.
[280] Ballistreri, F. P.; Tomaselli, G. A.; Toscano, R. M.; Bonchio, M.; Conte, V.; Di Furia, F. *Tetrahedron Lett.* **1994**, *35*, 8041.
[281] Ballini, R.; Papa, F.; Bovicelli, P. *Tetrahedron Lett.* **1996**, *37*, 3507.
[282] Adam, W.; Saha-Möller, C. R.; Schmid, K. S. *J. Org. Chem.* **2001**, *66*, 7365.
[283] van Heerden, F. R.; Dixon, J. T.; Holzapfel, C. W. *Tetrahedron Lett.* **1992**, *33*, 7399.
[284] Branan, B. M.; Wang, X.; Jankowski, P.; Wicha, J.; Paquette, L. A. *J. Org. Chem.* **1994**, *59*, 6874.
[285] Wasserman, H. H.; Baldino, C. M.; Coats, S. J. *J. Org. Chem.* **1995**, *60*, 8231.
[286] Ley, S. V.; Meek, G.; Metten, K.-H.; Pique, C. *J. Chem. Soc., Chem. Commun.* **1994**, 1931.
[287a] Link, J. T.; Danishefsky, S. J.; Schulte, G. *Tetrahedron Lett.* **1994**, *35*, 9131.
[287] Link, J. T.; Danishefsky, S. J.; Schulte, G. *Tetrahedron Lett.* **1995**, *36*, 3584.
[288] Messeguer, A.; Sánchez-Baeza, F.; Casas, J.; Hammock, B. D. *Tetrahedron* **1991**, *47*, 1291.
[289] Simbi, L.; van Heerden, F. R. *J. Chem. Soc., Perkin Trans. 1* **1997**, 269.
[290] Abad, J.-L.; Casas, J.; Sánchez-Baeza, F.; Messeguer, A. *J. Org. Chem.* **1993**, *58*, 3991.
[291] Abad, J.-L.; Casas, J.; Sánchez-Baeza, F.; Messeguer, A. *Bioorg. Med. Chem. Lett.* **1992**, *2*, 1239.
[292] Adam, W.; Corma, A.; Martínez, A.; Mitchell, C. M.; Reddy, T. I.; Renz, M.; Smerz, A. K. *J. Mol. Catal. A.* **1997**, *117*, 357.
[293] Adam, W.; Mitchell, C. M.; Saha-Möller, C. R. *J. Org. Chem.* **1999**, *64*, 3699.
[294] D´Accolti, L.; Fiorentino, M.; Fusco, C.; Rosa, A. M.; Curci, R. *Tetrahedron Lett.* **1999**, *40*, 8023.
[295] Freccero, M.; Gandolfi, R.; Sarzi-Amadè, M. *Tetrahedron* **1999**, *55*, 11309.
[296] Murray, R. W.; Singh, M.; Williams, B. L.; Moncrieff, H. M. *Tetrahedron Lett.* **1995**, *36*, 2437.
[297] Kurihara, M.; Ito, S.; Tsutsumi, N.; Miyata, N. *Tetrahedron Lett.* **1994**, *35*, 1577.
[298] Adam, W.; Brünker, H.-G.; Kumar, A. S.; Peters, E.-M.; Peters, K.; Schneider, U.; von Schnering, H. G. *J. Am. Chem. Soc.* **1996**, *118*, 1899.
[299] Murray, R. W.; Singh, M.; Rath, N. P. *Tetrahedron* **1999**, *55*, 4539.
[300] Adam, W.; Jekö, J.; Lévai, A.; Majer, Z.; Nemes, C.; Patonay, T.; Párkányi, L.; Sebök, P. *Tetrahedron: Asymmetry* **1996**, *7*, 2437.
[301] Yang, D.; Tang, Y.-C.; Chen, J.; Wang, X.-C.; Bartberger, M. D.; Houk, K. N.; Olson, L. *J. Am. Chem. Soc.* **1999**, *121*, 11976.
[302] Cicala, G.; Curci, R.; Fiorentino, M.; Laricchiuta, O. *J. Org. Chem.* **1982**, *47*, 2670.
[303] de Sousa, S. E.; O'Brien, P.; Pilgram, C. D.; Roder, D.; Towers, T. D. *Tetrahedron Lett.* **1999**, *40*, 391.
[304] Yang, D.; Wong, M.-K.; Cheung, K.-K.; Chan, E. W. C.; Xie, Y. *Tetrahedron Lett.* **1997**, *38*, 6865.
[305] Brimble, M. A.; Johnston, A. D.; Furneaux, R. H. *J. Org. Chem.* **1998**, *63*, 471.
[306] Marzabadi, C. H.; Spilling, C. D. *J. Org. Chem.* **1993**, *58*, 3761.
[307] Benbow, J. W.; Katoch, R.; Martinez, B. L.; Shetzline, S. B. *Tetrahedron Lett.* **1997**, *38*, 4017.
[308] King, S. B.; Ganem, B. *J. Am. Chem. Soc.* **1994**, *116*, 562.
[309] Haag, R.; Zuber, R.; Donon, S.; Lee, C.-H.; Noltemeyer, M.; Johnsen, K.; de Meijere, A. *J. Org. Chem.* **1998**, *63*, 2544.
[310] Adam, W.; Roschmann, K. unpublished results.
[311] Linker, T.; Fröhlich, L. *J. Am. Chem. Soc.* **1995**, *117*, 2694.

312 Adam, W.; Korb, M. N. *Tetrahedron* **1996**, *52*, 5487.

313 Zimmermann, P. J.; Blanarikova, I.; Jäger, V. *Angew. Chem., Int. Ed. Engl.* **2000**, *39*, 910.

314 Kurihara, M.; Ishii, K.; Kasahara, Y.; Kameda, M.; Pathak, A. K.; Miyata, N. *Chem. Lett.* **1997**, 1015.

315 Tagliatesta, P.; Bernini, R.; Crestini, C.; Monti, D.; Boschi, T.; Mincione, E.; Saladino, R. *J. Org. Chem.* **1999**, *64*, 5361.

316 Hambalek, R.; Just, G. *Tetrahedron Lett.* **1990**, *31*, 4693.

317 Bols, M.; Hazell, R. G.; Thomsen, I. B. *Chem. Eur. J.* **1997**, *3*, 940.

318 Adam, W.; Schuhmann, R. M. *J. Org. Chem.* **1996**, *61*, 874.

319 Bartels, A.; Jones, P. G.; Liebscher J. *Tetrahedron Lett.* **1995**, *36*, 3673.

320 Griesbeck, A. G.; Mauder, H.; Müller, I. *Chem. Ber.* **1992**, *125*, 2467.

321 Crimmins, M. T.; Jung, D. K.; Gray, J. L. *J. Am. Chem. Soc.* **1993**, *115*, 3146.

322 Luker, T.; Hiemstra, H.; Speckamp, W. N. *Tetrahedron Lett.* **1996**, *37*, 8257.

323 Thompson, S. K.; Heathcock, C. H. *J. Org. Chem.* **1992**, *57*, 5979.

324 Asensio, G.; Biox-Bernardini, C.; Andreu, C.; González-Núñez, M. E.; Mello, R.; Edwards, J. O.; Carpenter, G. B. *J. Org. Chem.* **1999**, *64*, 4705.

325 Nicolaou, K. C.; Prasad, C. V. C.; Ogilvie, W. W. *J. Am. Chem. Soc.* **1990**, *112*, 4988.

326 Nan, F.; Chen, X.; Xiong, Z.; Li, T.; Li, Y. *Synth. Commun.* **1994**, *24*, 2319.

327 Gerlach, K.; Hoffmann, H. M. R.; Wartchow, R. *J. Chem. Soc, Perkin Trans. 1* **1998**, 3867.

328 Randolph, J. T.; Danishefsky, S. J. *J. Am. Chem. Soc.* **1993**, *115*, 8473.

329 Linker, T.; Peters, K.; Peters, E.-M.; Rebien, F. *Angew. Chem., Int. Ed. Engl.* **1996**, *35*, 2487.

330 Anklam, S.; Liebscher, J. *Tetrahedron* **1998**, *54*, 6369.

331 Bhatia, G. S.; Lowe, R. F.; Pritchard, R. G.; Stoodley, R. J. *J. Chem. Soc., Chem. Commun.* **1997**, 1981.

332 Cambie, R. C.; Grimsdale, A. C.; Rutledge, P. S.; Walker, M. F.; Woodgate, P. D. *Aust. J. Chem.* **1991**, *44*, 1553.

333 He, F.; Foxman, B. M.; Snider, B. B. *J. Am. Chem. Soc.* **1998**, *120*, 6417.

334 Yang, D.; Ye, X.-Y.; Xu, M. *J. Org. Chem.* **2000**, *65*, 2208.

335 Gallant, M.; Link, J. T.; Danishefsky, S. J. *J. Org. Chem.* **1993**, *58*, 343.

336 Chow, K.; Danishefsky, S. J. *J. Org. Chem.* **1990**, *55*, 4211.

337 Nemes, C.; Lévai, A.; Patonay, T.; Tóth, G.; Boros, S.; Halász, J.; Adam, W.; Golsch, D. *J. Org. Chem.* **1994**, *59*, 900.

338 Link, J. T.; Gallant, M.; Danishefsky, S. J.; Huber, S. *J. Am. Chem. Soc.* **1993**, *115*, 3782.

339 Gervay, J.; Peterson, J. M.; Oriyama, T.; Danishefsky, S. J. *J. Org. Chem.* **1993**, *58*, 5465.

340 Smith, S. C.; Heathcock, C. H. *J. Org. Chem.* **1992**, *57*, 6379.

341 Berkowitz, D. B.; Danishefsky, S. J.; Schulte, G. K. *J. Am. Chem. Soc.* **1992**, *114*, 4518.

342 Grieco, P. A.; Henry, K. J.; Nunes, J. J.; Matt, J. E., Jr. *J. Chem. Soc., Chem. Commun.* **1992**, 368.

343 Hosoyama, H.; Shigemori, H.; In, Y.; Ishida, T.; Kobayashi, J. *Tetrahedron* **1998**, *54*, 2521.

344 Nicolaou, K. C.; He, Y.; Vourloumis, D.; Vallberg, H.; Roschangar, F.; Sarabia, F.; Ninkovic, S.; Yang, Z.; Trujillo, J. I. *J. Am. Chem. Soc* **1997**, *119*, 7960.

345 Krohn, K.; Micheel, J. *Tetrahedron* **1998**, *54*, 4827.

346 Nicotra, F.; Panza, L.; Russo, G. *Tetrahedron Lett.* **1991**, *32*, 4035.

347 Timmers, C. M.; van der Marel, G. A.; van Boom, J. H. *Recl. Trav. Chim. Pays-Bas* **1993**, *112*, 609.

348 Dushin, R. G.; Danishefsky, S. J. *J. Am. Chem. Soc.* **1992**, *114*, 3471.

349 Gervay, J.; Danishefsky, S. J. *J. Org. Chem.* **1991**, *56*, 5448.

350 Nicolaou, K. C.; Ninkovic, S.; Sarabia, F.; Vourloumis, D.; He, Y.; Vallberg, H.; Finlay, M. R. V.; Yang, Z. *J. Am. Chem. Soc.* **1997**, *119*, 7974.

351 Curci, R.; Detomaso, A.; Prencipe, T.; Carpenter, G. B. *J. Am. Chem. Soc.* **1994**, *116*, 8112.

352 Marples, B. A.; Muxworthy, J. P.; Baggaley, K. H. *Tetrahedron Lett.* **1991**, *32*, 533.

353 Park, T. K.; Peterson, J. M.; Danishefsky, S. J. *Tetrahedron Lett.* **1994**, *35*, 2671.

354 Paquette, L. A.; Zhao, M. *J. Am. Chem. Soc.* **1998**, *120*, 5203.

355 Paquette, L. A.; Zhao, M. *J. Am. Chem. Soc.* **1993**, *115*, 354.

356 Charette, A. B.; Côté, B. *Tetrahedron: Asymmetry* **1993**, *4*, 2283.

357 Park, T. K.; Danishefsky, S. J. *Tetrahedron Lett.* **1994**, *35*, 2667.

358 Gordon, D. M.; Danishefsky, S. J. *J. Am. Chem. Soc.* **1992**, *114*, 659.

[359] Ireland, R. E.; Armstrong, J. D.; Lebreton, J.; Meissner, R. S.; Rizzacasa, M. A. *J. Am. Chem. Soc.* **1993**, *115*, 7152.
[360] White, J. D.; Hong, J.; Robarge, L. A. *J. Org. Chem.* **1999**, *64*, 6206.
[361] Overas, A. T.; Gundersen, L.-L.; Rise, F. *Tetrahedron* **1997**, *53*, 1777.
[362] Lay, L.; Nicotra, F.; Panza, L.; Russo, G. *Synlett* **1995**, 167.
[363] Takikawa, H.; Muto, S.-e.; Mori, K. *Tetrahedron* **1998**, *54*, 3141.
[364] Friesen, R. W.; Sturino, C. F. *J. Org. Chem.* **1990**, *55*, 5808.
[365] Liu, K. K.-C.; Danishefsky, S. J. *J. Am. Chem. Soc.* **1993**, 4933.
[366] Liu, K. K.-C.; Danishefsky, S. J. *J. Org. Chem.* **1994**, *59*, 1895.
[367] Izzo, I.; De Riccardis, F.; Sodano, G. *J. Org. Chem.* **1998**, *63*, 4438.
[368] Shu, L.; Shi, Y. *Tetrahedron Lett.* **1999**, *40*, 8721.
[369] Brown, D. S.; Marples, B. A.; Smith, P.; Walton, L. *Tetrahedron* **1995**, *51*, 3587.
[370] Warren, J. D.; Shi, Y. *J. Org. Chem.* **1999**, *64*, 7675.

Supplemental References of Recent Reviews

[371] Adam, W.; Malisch, W.; Roschmann, K. J.; Saha-Möller, C. R.; Schenk, W. A. *J. Organometal. Chem.* **2002**, *661*, 3.
[372] Adam, W.; Zhao, C.-G. In *Chemistry of Peroxides*; Rappoport, Z. Ed.; John Wiley: Chichester, **2006**; Vol. 2(Pt. 2), pp. 1129–1169.
[373] Adam, W.; Zhao, C.-G. In *Handbook of C-H Transformations*; Dyker, G. Ed.; Wiley-VCH: Weinheim, **2005**, Vol. 2, pp. 507–516.
[374] Arterburn, J. B. *Tetrahedron* **2001**, *57*, 9765.
[375] Bach, R. D. In *Chemistry of Peroxides*; Rappoport, Z. Ed.; John Wiley: Chichester, **2006**; Vol. 2(Pt. 1), pp. 1–91.
[376] Barbaro, P.; Bianchini, C. *Chemtracts* **2001**, *14*, 274.
[377] Block, K.; Kappert, W.; Kirschfeld, A.; Muthusamy, S.; Schroeder, K.; Sander, W.; Kraka, E.; Sosa, C.; Cremer, D. In *Peroxide Chemistry*; Adam, W. Ed.; Wiley-VCH: Weinheim, **2000**; pp 139–156.
[378] Bortolini, O.; Fantin, G.; Fogagnolo, M. *Chim. Ind.* **2006**, *88*, 40.
[379] Curci, R.; D'Accolti, L.; Fusco, C. *Acc. Chem. Res.* **2006**, *39*, 1.
[380] Furutani, T. *Yuki Gosei Kagaku Kyokaishi* **2001**, *59*, 510.
[381] Gao, Y.-J.; Ma, J.-J.; Wang, C.; Zhang, Y.-Q.; Cui, P.-L.; Zang, X.-H.; Zhou, Y.-X.; Tang, R.-X.; Li, Y.-M. *Youji Huaxue* **2005**, *25*, 745.
[382] Ge, H.-Q. *Synlett* **2004**, 2046.
[383] Goeddel, D.; Shi, Y. In *Catalysts for Fine Chemical Synthesis*; Roberts, S. M.; Whittall, J. Eds; John Wiley: Chichester, **2007**; Vol. 5, pp. 215–224.
[384] Gong, B.; Meng, Q.-W.; Gao, Z.-X. *Youji Huaxue* **2008**, *28*, 588.
[385] Lange, A.; Brauer, H.-D. In *Peroxide Chemistry*; Adam, W. Ed.; Wiley-VCH: Weinheim, **2000**; pp 157–176.
[386] Li, D.-Y.; Li, R.-J.; Hong, G.-F.; Zhang, H.-Y.; Liu, H.-M. *Youji Huaxue* **2005**, *25*, 386.
[387] Parish, E. J.; Qiu, Z. *Lipids* **2004**, *39*, 805.
[388] Qin, W.; Cui, G.; Liang, W.; Yu, H. *Xianweisu Kexue Yu Jishu* **2000**, *8*, 58.
[389] Saladino, R. *Targets Heterocycl. Syst.* **2000**, *4*, 357.
[390] Sawwan, N.; Greer, A. *Chem. Rev.* **2007**, *107*, 3247.
[391] Seki, M. *Yuki Gosei Kagaku Kyokaishi* **2003**, *61*, 236.
[392] Seki, M. *Synlett* **2008**, 164.
[393] Seki, M.; Kawase, Y. *Shokubai* **2007**, *49*, 195.
[394] Shi, Y. *Yuki Gosei Kagaku Kyokaishi* **2002**, *60*, 342.
[395] Shi, Y. In *Modern Oxidation Methods*; Bäckvall, J.-E. Ed.; Wiley-VCH: Weinheim, **2004**; pp. 51–82.
[396] Shi, Y. In *Handbook of Chiral Chemicals (2nd Ed)*; Ager, D. Ed.; CRC Press: Boca Raton, **2006**; pp. 147–163.
[397] Shi, Y. *Acc. Chem. Res.* **2004**, *37*, 488.

[398] Srivastava, V. P. *Synlett* **2008**, 626.
[399] Tao, X.-Y. *Synlett* **2007**, 3226.
[400] Wang, C.; Gao, Y.; Zhang, Y.; Wang, Z.; Ma, J. *Huaxue Jinzhan* **2006**, *18*, 761.
[401] Wang, C.; Wu, Q.-H.; Yang, L.-H.; Tang, R.-X.; Yang, X.-Z.; Guo, X.-M.; Feng, S.; Li, G.-S. *Youji Huaxue* **2004**, *24*, 380.
[402] Wong, O. A.; Shi, Y. *Chem. Rev.* **2008**, *108*, 3958.
[403] Yang, D. *Acc. Chem. Res.* **2004**, *37*, 497.
[404] Zhang, Z.-G.; Wang, X.-Y.; Sun, C.; Shi, H.-C. *Youji Huaxue* **2004**, *24*, 7.
[405] Zheng, Y.; Tian, Z.; Jiang, A. *Huaxue Tongbao* **2002**, *65*, 261.

Supplemental References for Table 1A. Epoxidation of Unfunctionalized Alkenes by Isolated Dioxiranes

[406] Adam, W.; Peters, K.; Peters, E.-M.; Schambony, S. B. *J. Am. Chem. Soc.* **2001**, *123*, 7228.
[407] Adam, W.; Roschmann, K. J.; Saha-Möller, C. R.; Seebach, D. *J. Am. Chem. Soc.* **2002**, *124*, 5068.
[408] Albrecht, C.; Barnes, S.; Boeckemeier, H.; Davies, D.; Dennis, M.; Evans, D. M.; Fletcher, M. D.; Jones, I.; Leitmann, V.; Murphy, P.J.; Rowles, R.; Nash, R.; Stephenson, R. A.; Horton, P.N.; Hursthouse, M. B. *Tetrahedron Lett.* **2008**, *49*, 185.
[409] Annese, C.; D'Accolti, L.; Dinoi, A.; Fusco, C.; Gandolfi, R.; Curci, R. *J. Am. Chem. Soc.* **2008**, *130*, 1197.
[410] Antonioletti, R.; Malancona, S.; Bovicelli, P. *Tetrahedron* **2002**, *58*, 8825.
[411] Antonioletti, R.; Righi, G.; Oliveri, L.; Bovicelli, P. *Tetrahedron Lett.* **2000**, *41*, 10127.
[412] Appendino, G.; Cravotto, G.; Jarevang, T.; Sterner, O. *Eur. J. Org. Chem.* **2000**, 2933.
[413] Ashavina, O. Y.; Kabal'nova, N. N.; Flekhter, O. B.; Spirikhin, L. V.; Galin, F. Z.; Baltina, L. A.; Starikova, Z. A.; Antipin, M. Y.; Tolstikov, G. A. *Mendeleev Commun.* **2004**, 221.
[414] Asouti, A.; Hadjiarapoglou, L. P. *Synlett* **2001**, 1847.
[415] Bach, R. D.; Dmitrenko, O.; Adam, W.; Schambony, S. *J. Am. Chem. Soc.* **2003**, *125*, 924.
[416] Baumstark, A. L.; Franklin, P. J.; Vasquez, P. C.; Crow, B. S. *Molecules* **2004**, *9*, 117.
[417] Bentley, T. W.; Engels, B.; Hupp, T.; Bogdan, E.; Christl, M. *J. Org. Chem.* **2006**, *71*, 1018.
[418] Boyer, F.-D.; Descoins, C. L.; Thanh, G. V.; Descoins, C.; Prange, T.; Ducrot, P.-H. *Eur. J. Org. Chem.* **2003**, 1172.
[419] Brimble, M. A.; Furkert, D. P. *Org. Biomol. Chem.* **2004**, *2*, 3573.
[420] Camps, P.; Colet, G.; Delgado, S.; Munoz, M. R.; Pericas, M. A.; Sola, L.; Vazquez, S. *Tetrahedron* **2007**, *63*, 4669.
[421] Crehuet, R.; Anglada, J. M.; Cremer, D.; Bofill, J. M. *J. Phys. Chem. A* **2002**, *106*, 3917.
[422] Crich, D.; Grant, D. *J. Org. Chem.* **2005**, *70*, 2384.
[423] Crow, B. S.; Winkeljohn, W. R.; Navarro-Eisenstein, A.; Michelena-Baez, E.; Franklin, P. J.; Vasquez, P. C.; Baumstark, A. *Eur. J. Org. Chem.* **2006**, 4642.
[424] D'Accolti, L.; Annese, C.; Fusco, C. *Tetrahedron Lett.* **2005**, *46*, 8459.
[425] Deubel, D. V.; Frenking, G.; Sundermeyer, J.; Senn, H. M. *Chem. Commun.* **2000**, 2469.
[426] Duefert, A.; Werz, D. B. *J. Org. Chem.* **2008**, *73*, 5514.
[427] Dmitrenko, O.; Bach, R. D. *J. Phys. Chem. A* **2004**, *108*, 6886.
[428] Dyker, G.; Kerl, T.; Korning, J.; Bubenitschek, P.; Jones, P. G. *Tetrahedron* **2000**, *56*, 8665.
[429] Eggen, M.; Nair, S. K.; Georg, G. I. *Org. Lett.* **2001**, *3*, 1813.
[430] Enders, D.; Lenzen, A.; Backes, M.; Janeck, C.; Catlin, K.; Lannou, M.-I.; Runsink, J.; Raabe, G. *J. Org. Chem.* **2005**, *70*, 10538.
[431] Ermolenko, M. S.; Potier, P. *Tetrahedron Lett.* **2002**, *43*, 2895.
[432] Frank, D.; Kozhushkov, S. I.; Labahn, T.; de Meijere, A. *Tetrahedron* **2002**, *58*, 7001.
[433] Fyvie, W. S.; Peczuh, M. W. *J. Org. Chem.* **2008**, *73*, 3626.
[434] Gataullin, R. R.; Nasyrov, M. F.; Ivanova, E. V.; Kabal'nova, N. N.; Abdrakhmanov, I. B. *Russ. J. Org. Chem.* **2002**, *38*, 763.
[435] Gisdakis, P.; Rosch, N. *J. Phys. Org. Chem.* **2001**, *14*, 328.
[436] Gisdakis, P.; Rosch, N. *Eur. J. Org. Chem.* **2001**, 719.

[437] Grabovskiy, S. A.; Kabal'nova, N. N.; Chatgilialoglu, C.; Ferreri, C. *Helvetica Chim. Acta* **2006**, *89*, 2243.

[438] Grubbs, R. B.; Broz, M. E.; Dean, J. M.; Bates, F. S. *Macromolecules* **2000**, *33*, 2308.

[439] Gruijters, B. W. T.; Van Veldhuizen, A.; Weijers, C. A. G. M.; Wijnberg, J. B. P. A. *J. Nat. Prod.* **2002**, *65*, 558.

[440] Guiard, S.; Giorgi, M.; Santelli, M.; Parrain, J.-L. *J. Org. Chem.* **2003**, *68*, 3319.

[441] Halim, R.; Brimble, M. A.; Merten, J. *Org. Biomol. Chem.* **2006**, *4*, 1387.

[442] Han, Y.-K.; Pearce, E. M.; Kwei, T. K. *Macromolecules* **2000**, *33*, 1321.

[443] Hansen, S. U.; Bols, M. *J. Chem. Soc., Perkin Trans. 1*, **2000**, 911.

[444] Horiguchi, T.; Kiyota, H.; Cheng, Q.; Oritani, T. *Tennen Yuki Kagobutsu Toronkai Koen Yoshishu* **2000**, 745.

[445] Horiguchi, T.; Nagura, M.; Cheng, Q.; Oritani, T.i; Kudo, T. *Heterocycles* **2000**, *53*, 2629.

[446] Inoue, M.; Lee, N.; Kasuya, S.; Sato, T.; Hirama, M.; Moriyama, M.; Fukuyama, Y. *J. Org. Chem.* **2007**, *72*, 3065.

[447] Jie, M. S. F.; Lie K; Lam, C. N. W.; Ho, J. C. M.; Lau, M. M. L. *Eur. J. Lipid Sci.Tech.* **2003**, *105*, 391.

[448] Keki, S.; Nagy, M.; Deak, G.; Lévai, A.; Zsuga, M. *J. Polymer Sci. A: Polymer Chem.* **2002**, *40*, 3974.

[449] Knoell, J.; Knoelker, H.-J. *Tetrahedron Lett.* **2006**, *47*, 6079.

[450] Krohn, K.; Micheel, J.; Zukowski, M. *Tetrahedron* **2000**, *56*, 4753.

[451] Michelena-Baez, E.; Navarro-Eisenstein, A. M.; Banks, H. D.; Vasquez, P. C.; Baumstark, A. L. *Heterocycl. Commun.* **2000**, *6*, 119.

[452] Minisci, F.; Gambarotti, C.; Pierini, M.; Porta, O.; Punta, C.; Recupero, F.; Lucarini, M.; Mugnaini, V. *Tetrahedron Lett.* **2006**, *47*, 1421.

[453] Molander, G. A; Ribagorda, M. *J. Am. Chem. Soc.* **2003**, *125*, 11148.

[454] Mollenberg, A.; Spiteller, G. *Chem. Phys. Lipids* **2001**, *109*, 225.

[455] Murakami, N.; Tamura, S.; Wang, W.; Takagi, T.; Kobayashi, M. *Tetrahedron* **2001**, *57*, 4323.

[456] Murakami, N.; Wang, W.; Ohyabu, N.; Ito, T.; Tamura, S.; Aoki, S.; Kobayashi, M.; Kitagawa, I. *Tetrahedron* **2000**, *56*, 9121.

[457] Nikje, M. A. Mozaffari, Z.; Rfiee, A. *Designed Monomers and Polymers* **2007**, *10*, 119.

[458] Nikje, M. A.; Mozaffari, Z.; Rfiee, A. *Designed Monomers and Polymers* **2007**, *10*, 119.

[459] Ogawa, S.; Hosoi, K.; Ikeda, N.; Makino, M.; Fujimoto, Y.; Iida, T. *Chem. Pharm. Bull.* **2007**, *55*, 247.

[460] Organ, M. G.; Dixon, C. E.; Mayhew, D.; Parks, D. J.; Arvanitis, E. A. *Combin. Chem. High Throughput Screen.* **2002**, *5*, 211.

[461] Parrish, J. D.; Little, R. D. *Org. Lett.* **2002**, *4*, 1439.

[462] Posner, G. H.; Paik, I.-H.; Sur, S.; McRiner, A. J.; Borstnik, K.; Xie, S.; Shapiro, T. A. *J. Med. Chem.* **2003**, *46*, 1060.

[463] Sabila, P. S.; Liang, Y.; Howell, A. R. *Tetrahedron Lett.* **2007**, *48*, 8356.

[464] Schank, K.; Weiter, M.; Keasalar, R. *Helvetica Chim. Acta* **2002**, *85*, 2105.

[465] Schobert, R.; Siegfried, S.; Weingaertner, J.; Nieuwenhuyzen, M. *J. Chem. Soc., Perkin Trans.1* **2001**, 2009.

[466] Song, C. E.; Lim, J. S.; Kim, S. C.; Lee, K.-J.; Chi, D. Y. *Chem. Commun.* **2000**, 2415.

[467] Spiteller, P.; Kern, W.; Reiner, J.; Spiteller, G. *Biochim. Biophys. Acta: Mol. Cell Biol. Lipids* **2001**, *1531*, 188.

[468] Stachel, S. J.; Danishefsky, S. J. *Tetrahedron Lett.* **2001**, *42*, 6785.

[469] Stoianova, D. S.; Whitehead, A.; Hanson, P. R. *J. Org. Chem.* **2005**, *70*, 5880.

[470] Tomooka, K.; Komine, N.; Fujiki, D.; Nakai, T.; Yanagitsuru, S. *J. Am. Chem. Soc.* **2005**, *127*, 12182.

[471] Wang, X.; Lee, Y. R. *Tetrahedron Lett.* **2007**, *48*, 6275.

[472] Yu, F.-L.; Bender, W.; Fang, Q.; Ludeke, A.; Welch, B. *Cancer Detect. Prev.* **2003**, *27*, 370.

[473] Yu, F.-L.; Gapor, A.; Bender, W. *Cancer Detect. Prev.* **2005**, *29*, 383.

[474] Yu, F.-L.; Greenlaw, R.; Fang, Q.; Bender, W.; Yamaguchi, K.; Xue, B. H.; Yu, C.-C. *Eur. J. Cancer Prev.: Off. J. Eur. Cancer Prev. Org.* **2004**, *13*, 239.

[475] Zareba, M.; Legiec, M.; Sanecka, B.; Sobczak, J.; Hojniak, M.; Wolowiec, S. *J. Mol. Catal. A: Chem.* **2006**, *248*, 144.

Supplemental References for Table 1B. Epoxidation of Alkenes with Electron Donors by Isolated Dioxiranes

[476] Adam, W.; Bosio, S. G.; Turro, N. J.; Wolff, B. T. *J. Org. Chem.* **2004**, *69*, 1704.
[477] Adam, W.; Bosio, S. G.; Wolff, B. T. *Org. Lett.* **2003**, *5*, 819.
[478] Adam, W.; Schambony, S. B. *Org. Lett.* **2001**, *3*, 79.
[479] Armstrong, A.; Chung, H. *Tetrahedron Lett.* **2006**, *47*, 1617.
[480] Baumstark, A. L.; Chen, H.; Singh, S. N.; Vasquez, P. C. *Heterocycl. Commun.* **2000**, *6*, 501.
[481] Boulineau, F. P.; Liew, S.-T.; Shi, Q.; Wenthold, P. G.; Wei, A. *Org. Lett.* **2006**, *8*, 4545.
[482] Boulineau, F. P.; Wei, A. *Org. Lett.* **2002**, *4*, 2281.
[483] Caravano, A.; Baillieul, D.; Ansiaux, C.; Pan, W.; Kovensky, J.; Sinay, P.; Vincent, S. P. *Tetrahedron* **2007**, *63*, 2070.
[484] Cuzzupe, A. N.; Hutton, C. A.; Lilly, M. J.; Mann, R. K.; McRae, K. J.; Zammit, S. C.; Rizzacasa, M. A. *J. Org. Chem.* **2001**, *66*, 2382.
[485] Dixon, J. T.; van Heerden, F. R.; Holzapfel, C. W. *Tetrahedron: Asymmetry* **2005**, *16*, 393.
[486] Djung, J. F.; Hart, David J.; Young, E. R. R. *J. Org. Chem.* **2000**, *65*, 5668.
[487] Dransfield, P. J.; Dilley, A. S.; Wang, S.; Romo, D. *Tetrahedron* **2006**, *62*, 5223.
[488] Dransfield, P. J.; Wang, S.; Dilley, A.; Romo, D. *Org. Lett.* **2005**, *7*, 1679.
[489] Duffy, R. J.; Morris, K. A.; Romo, D. *J. Am. Chem. Soc.* **2005**, *127*, 16754.
[490] Ernst, C.; Klaffke, W. *J. Org. Chem.* **2003**, *68*, 5780.
[491] Ghisalberti, E. L.; Hargreaves, J. R.; Skelton, B. W.; White, A. H. *Austral. J. Chem.***2000**, *53*, 995.
[492] Haraguchi K.; Sumino M.; Tanaka H. *Nucleosides Nucleotides Nucl. Acids* **2007**, *26*, 835.
[493] Haraguchi, K.; Kubota, Y.; Tanaka, H. *J. Org. Chem.* **2004**, *69*, 1831.
[494] Haraguchi, K.; Takeda, S.; Sumino, M.; Tanaka, H.; Dutschman, G. E.; Cheng, Y.-C.; Nitanda, T.; Baba, M. *Nucleosides Nucleotides Nucl. Acids* **2005**, *24*, 343.
[495] Haraguchi, K.; Takeda, S.; Tanaka, H. *Org. Lett.* **2003**, *5*, 1399.
[496] Hayashi, Y.; Shoji, M.; Yamaguchi, S.; Mukaiyama, T.; Yamaguchi, J.; Kakeya, H.; Osada, H. *Org. Lett.* **2003**, *5*, 2287.
[497] Hodgson, D. M.; Stent, M. A. H.; Wilson, F. X. *Synthesis* **2002**, 1445.
[498] Hunt, D. K.; Seeberger, P. H. *Org. Lett.* **2002**, *4*, 2751.
[499] Jastrzebska, I.; Katrynski, K. S.; Morzycki, J. W. *ARKIVOC* **2002**, (9), 46.
[500] Konno, F.; Ishikawa, T.; Kawahata, M.; Yamaguchi, K. *J. Org. Chem.* **2006**, *71*, 9818.
[501] Koseki, Y.; Kusano, S.; Ichi, D.; Yoshida, K.; Nagasaka, T. *Tetrahedron* **2000**, *56*, 8855.
[502] Krause, N.; Laux, M.; Hoffmann-Roder, A. *Tetrahedron Lett.* **2000**, *41*, 9613.
[503] Kubota, Y.; Haraguchi, K.; Kunikata, M.; Hayashi, M.; Ohkawa, M.; Tanaka, H. *J. Org. Chem.* **2006**, *71*, 1099.
[504] Larsen, D. S.; Lins, R. J.; Stoodley, R. J.; Trotter, N. S. *J. Chem. Soc., Perkin Trans. 1* **2001**, 2204.
[505] Lee, C.-S.; Audelo, M. Q.; Reibenpies, J.; Sulikowski, G. A. *Tetrahedron* **2002**, *58*, 4403.
[506] Lee, J. S.; Fuchs, P. L. *Org. Lett.* **2003**, *5*, 2247.
[507] Leggett-Robinson, P. M.; Vasquez, P.; Baumstark, A. L. *Heterocycl. Commun.* **2003**, *9*, 433.
[508] Loeppky, R. N.; Sukhtankar, S.; Gu, F.; Park, M. *Chem. Res. Toxicol.* **2005**, *18*, 1955.
[509] Lohman, G. J. S.; Seeberger, P. H. *J. Org. Chem.* **2003**, *68*, 7541.
[510] Majumder, U.; Cox, J. M.; Rainier, J. D. *Org. Lett.* **2003**, *5*, 913.
[511] Ndakala, A. J.; Hashemzadeh, M.; So, R. C.; Howell, A. R. *Org. Lett.* **2002**, *4*, 1719.
[512] Parrish, J. D.; Little, R. D. *Org. Lett.* **2002**, *4*, 1439.
[513] Patonay, T.; Jeko, J.; Kiss-Szikszai, A.; Lévai, A. *Monatsh. Chem.* **2004**, *135*, 743.
[514] Peczuh, M. W.; Snyder, N. L.; Fyvie, W. S. *Carbohydr. Res.* **2004**, *339*, 1163.
[515] Rainier, J. D.; Cox, J. M. *Org. Lett.* **2000**, *2*, 2707.
[516] Schaumann, E.; Tries, F. *Synthesis* **2002**, 191.
[517] Scholl, P. F.; Groopman, J. D. *J. Label. Compd. Radiopharm.* **2004**, *47*, 807.

[518] Sieck, O.; Ehwald, M.; Liebscher, J. *Eur. J. Org. Chem.* **2005**, 663.
[519] Skelton, B. W.; Stick, R. V.; Stubbs, K. A.; Watts, A. G.; White, A. H. *Austral. J. Chem.* **2004**, *57*, 345.
[520] Smith, D. B.; Martin, J. A.; Klumpp, K.; Baker, S. J.; Blomgren, P. A.; Devos, R.; Granycome, C.; Hang, J.; Hobbs, C. J.; Jiang, W.-R.; Laxton, C.; Le Pogam, S.; Leveque, V.; Ma, H.; Maile, G.; Merrett, J. H.; Pichota, A.; Sarma, K.; Smith, M.; Swallow, S.; Symons, J.; Vesey, D.; Najera, I.; Cammack, N. *Bioorg. Med. Chem. Lett.* **2007**, *17*, 2570.
[521] Upreti, M.; Ruhela, D.; Vishwakarma, R. A. *Tetrahedron* **2000**, *56*, 6577.
[522] Vedejs, E.; Duncan, S. M. *J. Org. Chem.* **2000**, *65*, 6073.
[523] Wang, X.; Lee, Y. R. *Tetrahedron Lett.* **2007**, *48*, 6275.
[524] Zimmermann, P. J.; Lee, J. Y.; Hlobilova, I.; Endermann, R.; Haebich, D.; Jäger, V. *Eur. J. Org. Chem.* **2005**, 3450.

Supplemental References for Table 1C. Epoxidation of Alkenes with Electron Acceptors by Isolated Dioxiranes

[525] Adam, W.; Zhang, A. *Eur. J. Org. Chem.* **2004**, 147.
[526] Annese, C.; D'Accolti, L.; Dinoi, A.; Fusco, C.; Gandolfi, R.; Curci, R. *J. Am. Chem. Soc.* **2008**, *130*, 1197.
[527] Baumstark, A. L.; Chen, H.; Singh, S. N.; Vasquez, P. C. *Heterocycl. Commun.* **2000**, *6*, 501.
[528] Biscoe, M. R.; Breslow, R. *J. Am. Chem. Soc.* **2005**, *127*, 10812.
[529] Cermola, F.; Iesce, M. R. *Tetrahedron* **2006**, *62*, 10694.
[530] D'Accolti, L.; Fusco, C.; Rella, M. R.; Curci, R. *Synth. Commun.* **2003**, *33*, 3009.
[531] Davies, S. G.; Key, M.-S.; Rodriguez-Solla, H.; Sanganee, H. J.; Savory, E. D.; Smith, A. D. *Synlett* **2003**, 1659.
[532] Fan, C. L.; Lee, W.-D.; Teng, N.-W.; Sun, Y.-C.; Chen, K. *J. Org. Chem.* **2003**, *68*, 9816.
[533] Guiney, D.; Gibson, C. L.; Suckling, C. J. *Org. Biomol. Chem.* **2003**, *1*, 664.
[534] Ishimi, K.; Makino, M.; Fujimoto, Y.; Yabuta, R.; Iida, T.; Takagi, Y. *Kenkyu Kiyo - Nihon Daigaku Bunrigakubu Shizen Kagaku Kenkyusho* **2004**, *39*, 387.
[535] Katritzky, A. R.; Maimait, R.; Denisenko, A.; Denisenko, S. N. *ARKIVOC* **2001**, (5), 68.
[536] Klomklao, T.; Pyne, S. G.; Baramee, A.; Skelton, B. W.; White, A. H. *Tetrahedron: Asymmetry* **2003**, *14*, 3885.
[537] Lévai, A. *J. Heterocycl. Chem.* **2003**, *40*, 395.
[538] Lévai, A.; Jekő, J.; Brahmbhatt, D. I. *J. Heterocycl. Chem.* **2004**, *41*, 707.
[539] Lévai, A.; Silva, A. M. S.; Cavaleiro, J. A. S.; Patonay, T.; Silva, Vera L. M. *Eur. J. Org. Chem.* **2001**, 3213.
[540] Lévai, A; Jekő, J. *J. Heterocycl. Chem.* **2004**, *41*, 439.
[541] Mello, R.; González-Núñez, M. E.; Asensio, G. *Synlett* **2007**, 47.
[542] Molander, G. A; Ribagorda, M. *J. Am. Chem. Soc.* **2003**, *125*, 11148.
[543] Navarro-Eisenstein P., A. M.; Vasquez, C.; Franklin, P. J.; Baumstark, A. L. *Heterocycl. Commun.* **2003**, *9*, 575.
[544] Navarro-Eisenstein, A. M.; Vasquez, P. C.; Franklin, P. J.; Baumstark, A. L. *Heterocycl. Commun.* **2003**, *9*, 449.
[545] Otto, A.; Liebscher, J. *Synthesis* **2003**, 1209.
[546] Pastor, A.; Adam, W.; Wirth, T.; Tóth, G. *Eur. J. Org. Chem.* **2005**, 3075.
[547] Phutdhawong, W.; Pyne, S. G.; Baramee, A.; Buddhasukh, D.; Skelton, B. W.; White, A. H. *Tetrahedron Lett.* **2002**, *43*, 6047.
[548] Yamaguchi, T.; Nakamori, R.; Iida, T.; Nambara, T. *Synth. Commun.* **2001**, *31*, 1213.

Supplemental References for Table 1D. Epoxidation of Alkenes with Electron Donors and Acceptors by Isolated Dioxiranes

[549] Bartels, A.; J., Peter G.; Liebscher, J. *Synthesis* **2003**, 67.
[550] Boyer, F.-D.; Descoins, C. L.; Thanh, G. V.; Descoins, C.; Prange, T.; Ducrot, P.-H. *Eur. J. Org. Chem.* **2003**, 1172.

[551] Chu, H.-W.; Wu, H.-T.; Lee, Y.-J. *Tetrahedron* **2004**, *60*, 2647.
[552] Haraguchi, K.; Kubota, Y.; Tanaka, H. *Nucl. Acids Res. Suppl.* **2002**, *2*, 17.
[553] Haraguchi, K.; Takeda, S.; Tanaka, H. *Nucl. Acids Res. Suppl.* **2002**, *2*, 133.
[554] Lee, Y.-J.; Wu, T.-D. *J. Chin. Chem. Soc.* **2001**, *48*, 201.
[555] Lévai, A.; Kočevar, M.; Tóth, G.; Simon, A.; Vraničar, L.; Adam, W. *Eur. J. Org. Chem.* **2002**, 1830.
[556] Mougin, C.; Boyer, F.-D.; Caminade, E.; Rama, R. *J. Agric. Food Chem.* **2000**, *48*, 4529.
[557] Patonay, T.; Kiss-Szikszai, A.; Silva, V. M. L.; Silva, A. M. S.; Pinto, D. C. G. A.; Cavaleiro, J. A. S.; Jeko, J. *Eur. J. Org. Chem.* **2008**, 1937.
[558] Schobert, R.; Siegfried, S.; Weingaertner, J.; Nieuwenhuyzen, M. *J. Chem. Soc., Perkin Trans.1* **2001**, 2009.
[559] Watts, A. G.; Withers, S. G. *Can. J. Chem.* **2004**, *82*, 1581.
[560] Ye, W.; Sangaiah, R.; Degen, D. E.; Gold, A.; Jayaraj, K.; Koshlap, K. M.; Boysen, G.; Williams, J.; Tomer, K. B.; Ball, L. M. *Chem. Res. Toxicol.* **2006**, *19*, 506.

Supplemental References for Table 1E. Epoxidation of Alkenes by In Situ Generated Dioxiranes

[561] Alavi Nikje, M. M.; Mozaffari, Z. *Polimery* **2007**, *52*, 820.
[562] Antonioletti, R.; Righi, G.; Oliveri, L.; Bovicelli, P. *Tetrahedron Lett.* **2000**, *41*, 10127.
[563] Armstrong, A.; Draffan, A. G. *J. Chem. Soc., Perkin Trans. 1* **2001**, 2861.
[564] Bhoga, U. *Tetrahedron Lett.* **2005**, *46*, 5239.
[565] Broshears, W. C.; Esteb, J. J.; Richter, J.; Wilson, A. M. *J. Chem. Ed.* **2004**, *81*, 1018.
[566] Chen, J.; Soucek, M. D.; Simonsick, W. J., Jr.; Celikay, R. W. *Macromol. Chem. Phys.* **2002**, *203*, 2042.
[567] Cheshev, P.; Marra, A.; Dondoni, A. *Carbohydr. Res.* **2006** *341*, 2714.
[568] Cristau, H.-J.; Pirat, J.-L.; Drag, M.; Kafarski, P. *Tetrahedron Lett.* **2000**, *41*, 9781.
[569] Fawcett, J.; Griffith, G. A.; Percy, J. M.; Uneyama, E. *Org. Lett.* **2004**, *6*, 1277.
[570] Ferraz, H. M. C.; Longo, L. S., Jr. *J. Org. Chem.* **2007**, *72*, 2945.
[571] Ferraz, H. M. C.; Muzzi, R. M.; Vieira, T. de O.; Viertler, H. *Tetrahedron Lett.* **2000**, *41*, 5021.
[572] Grocock, E. L.; Marples, B. A.; Toon, R. C. *Tetrahedron* **2000**, *56*, 989.
[573] Hashimoto, N; Kanda, A. *Org. Process Res. Dev.* **2002**, *6*, 405.
[574] Jie, M. S. F.; Lie K; Lam, C. N. W.; Ho, J. C. M.; Lau, M. M. L. *Eur. J. Lipid Sci.Tech.* **2003**, *105*, 391.
[575] Kachasakul, P.; Assabumrungrat, S.; Praserthdam, P.; Pancharoen, U. *Chem. Eng. J.* **2003**, *92*, 131.
[576] Kan, J. T. W.; Toy, P. *Tetrahedron Lett.* **2004**, *45*, 6357.
[577] Lee, W.-D.; Chiu, C.-C.; Hsu, H.-L.; Chen, K. *Tetrahedron* **2004**, *60*, 6657.
[578] Legros, J.; Crousse, B.; Bonnet-Delpon, D.; Begue, J.-P. *Tetrahedron* **2002**, *58*, 3993.
[579] Legros, J.; Crousse, B.; Bourdon, J.; Bonnet-Delpon, D.; Begue, J.-P. *Tetrahedron Lett.* **2001**, *42*, 4463.
[580] Li, W.; Fuchs, P. L. *Org. Lett.* **2003**, *5*, 2853.
[581] Nikje, M. M. A.; Rafiee, A.; Haghshenas, M. *Designed Monomers and Polymers* **2006**, *9*, 293.
[582] Pasc-Banu, A.; Petrov, O.; Perez, E.; Rico-Lattes, I.; Lattes, A.; Pozzi, G.; Quici, S. *Synth. Commun.* **2003**, *33*, 4321.
[583] Rajabi, F. H.; Nikje, M. M. A; Farahani, B. V.; Saboury, N. *Designed Monomers and Polymers* **2006**, *9*, 383.
[584] Rousseau, C.; Christensen, B.; Petersen, T. E.; Bols, M. *Org. Biomol.Chem.* **2004**, *2*, 3476.
[585] Sanami, R. K. *J. Macromol. Sci. A* **2006**, *43*, 1205.
[586] Sartori, G.; Armstrong, A.; Maggi, R.; Mazzacani, Alessandro; S., Raffaella; Bigi, F.; Dominguez-Fernandez, B. *J. Org. Chem.* **2003**, *68*, 3232.
[587] Sawwan, N.; Greer, A. *J. Org. Chem.* **2006**, *71*, 5796.
[588] Yang, D.; Jiao, G.-S. *Chem. Eur. J.* **2000**, *6*, 3517.

Supplemental References for Table 2. Chemoselective Oxidations by Isolated Dioxiranes

[589] Albrecht, C.; Barnes, S.; Boeckemeier, H.; Davies, D.; Dennis, M.; Evans, D. M.; Fletcher, M. D.; Jones, I.; Leitmann, V.; Murphy, P.J.; Rowles, R.; Nash, R.; Stephenson, R. A.; Horton, P.N.; Hursthouse, M. B. *Tetrahedron Lett.* **2008**, *49*, 185.
[590] Ashavina, O. Y.; Kabal'nova, N. N.; Flekhter, O. B.; Spirikhin, L. V.; Galin, F. Z.; Baltina, L. A.; Starikova, Z. A.; Antipin, M. Y.; Tolstikov, G. A. *Mendeleev Commun.* **2004**, 221.
[591] Cermola, F.; Iesce, M. R. *J. Org. Chem.* **2002**, *67*, 4937.
[592] D'Accolti, L.; Fiorentino, M.; Fusco, C.; Crupi, P.; Curci, R. *Tetrahedron Lett.* **2004**, *45*, 8575.
[593] Djung, J. F.; Hart, David J.; Young, E. R. R. *J. Org. Chem.* **2000**, *65*, 5668.
[594] Dransfield, P. J.; Dilley, A. S.; Wang, S.; Romo, D. *Tetrahedron* **2006**, *62*, 5223.
[595] Enders, D.; Lenzen, A.; Backes, M.; Janeck, C.; Catlin, K.; Lannou, M.-I.; Runsink, J.; Raabe, G. *J. Org. Chem.* **2005**, *70*, 10538.
[596] Fawcett, J.; Griffith, G. A.; Percy, J. M.; Uneyama, E. *Org. Lett.* **2004**, *6*, 1277.
[597] Florio, S.; Makosza, M.; Lorusso, P.; Troisi, L. *ARKIVOC* **2006**, (6), 59.
[598] Gataullin, R. R.; Nasyrov, M. F.; Ivanova, E. V.; Kabal'nova, N. N.; Abdrakhmanov, I. B. *Russ. J. Org. Chem.* **2002**, *38*, 763.
[599] Ghisalberti, E. L.; Hargreaves, J. R.; Skelton, B. W.; White, A. H. *Austral. J. Chem.***2000**, *53*, 995.
[600] Guiney, D.; Gibson, C. L.; Suckling, C. J. *Org. Biomol. Chem.* **2003**, *1*, 664.
[601] Halim, R.; Brimble, M. A.; Merten, J. *Org. Biomol. Chem.* **2006**, *4*, 1387.
[602] Halim, R.; Brimble, M. A.; Merten, J. *Org. Lett.* **2005**, *7*, 2659.
[603] Ishii, A.; Kashiura, S.; Oshida, H.; Nakayama, J. *Org. Lett.* **2004**, *6*, 2623.
[604] Katritzky, A. R.; Maimait, R.; Denisenko, A.; Denisenko, S. N. *ARKIVOC* **2001**, (5), 68.
[605] Kiss-Szikszai, A.; Patonay, T.; Jekő, J. *ARKIVOC* **2001**, (3), 40.
[606] Knoell, J.; Knoelker, H.-J. *Tetrahedron Lett.* **2006**, *47*, 6079.
[607] Krause, N.; Laux, M.; Hoffmann-Roder, A. *Tetrahedron Lett.* **2000**, *41*, 9613.
[608] Kubota, Y.; Haraguchi, K.; Kunikata, M.; Hayashi, M.; Ohkawa, M.; Tanaka, H. *J. Org. Chem.* **2006**, *71*, 1099.
[609] Lee, J. S.; Fuchs, P. L. *Org. Lett.* **2003**, *5*, 2247.
[610] Lévai, A.; Jekő, J. *ARKIVOC* **2003**, (5), 19.
[611] Li, L.; Teng, G.-F.; Li, Z.-H. *Gaodeng Xuexiao Huaxue Xuebao* **2007**, *28*, 2179.
[612] Nakayama, J.; Aoki, S.; Takayama, J.; Sakamoto, A.; Sugihara, Y.; Ishii, A. *J. Am. Chem. Soc.* **2004**, *126*, 9085.
[613] Nikje, M. M. A.; Mozaffari, Z. *Designed Monomers and Polymers* **2007**, *10*, 67.
[614] Ogawa, S.; Hosoi, K.; Ikeda, N.; Makino, M.; Fujimoto, Y.; Iida, T. *Chem. Pharm. Bull.* **2007**, *55*, 247.
[615] Perales, J. B.; Makino, N. F.; Van Vranken, D. L. *J. Org. Chem.* **2002**, *67*, 6711.
[616] Rainier, J. D.; Cox, J. M. *Org. Lett.* **2000**, *2*, 2707.
[617] Rella, M. R.; Williard, P. G. *J. Org. Chem.* **2007**, *72*, 525.
[618] Richardson, R. D.; Desaize, M.; Wirth, T. *Chem. Eur. J.* **2007**, *13*, 6745.
[619] Sasaki, T.; Nakamori, R.; Yamaguchi, T.; Kasuga, Y.; Iida, T.; Nambara, T. *Chem. Phys. Lipids* **2001**, *109*, 135.
[620] Spiteller, P.; Kern, W.; Reiner, J.; Spiteller, G. *Biochim. Biophys. Acta: Mol. Cell Biol. Lipids* **2001**, *1531*, 188.
[621] Tardif, S. L.; Harpp, D. N. *Sulfur Lett.* **2000**, *23*, 169.
[622] Wen, X.; Norling, H.; Hegedus, L. S. *J. Org. Chem.* **2000**, *65*, 2096.
[623] Wender, P. A.; Hilinski, M. K.; Mayweg, A. V. W. *Org. Lett.* **2005**, *7*, 79.
[624] Ye, W.; Sangaiah, R.; Degen, D. E.; Gold, A.; Jayaraj, K.; Koshlap, K. M.; Boysen, G.; Williams, J.; Tomer, K. B.; Ball, L. M. *Chem. Res. Toxicol.* **2006**, *19*, 506.
[625] Yu, F.-L.; Bender, W.; Fang, Q.; Ludeke, A.; Welch, B. *Cancer Detect. Prev.* **2003**, *27*, 370.

Supplemental References for Table 3. Regioselective Epoxidations by Isolated Dioxiranes

[626] Appendino, G.; Cravotto, G.; Jarevang, T.; Sterner, O. *Eur. J. Org. Chem.* **2000**, 2933.

[627] Asouti, A.; Hadjiarapoglou, L. P. *Synlett* **2001**, 1847.
[628] Baumstark, A. L.; Franklin, P. J.; Vasquez, P. C.; Crow, B. S. *Molecules* **2004**, *9*, 117.
[629] Burke, C. P.; Shi, Y. *Angew. Chem., Int. Ed.* **2006**, *45*, 4475.
[630] Chu, H.-W.; Wu, H.-T.; Lee, Y.-J. *Tetrahedron* **2004**, *60*, 2647.
[631] D'Accolti, L.; Annese, C.; Fusco, C. *Tetrahedron Lett.* **2005**, *46*, 8459.
[632] Ghisalberti, E. L.; Hargreaves, J. R.; Skelton, B. W.; White, A. H. *Austral. J. Chem.***2000**, *53*, 995.
[633] Grabovskiy, S. A.; Kabal'nova, N. N.; Chatgilialoglu, C.; Ferreri, C. *Helvetica Chim. Acta* **2006**, *89*, 2243.
[634] Grocock, E. L.; Marples, B. A.; Toon, R. C. *Tetrahedron* **2000**, *56*, 989.
[635] Grubbs, R. B.; Broz, M. E.; Dean, J. M.; Bates, F. S. *Macromolecules* **2000**, *33*, 2308.
[636] Horiguchi, T.; Nagura, M.; Cheng, Q.; Oritani, T.; Kudo, T. *Heterocycles* **2000**, *53*, 2629.
[637] Jie, M. S. F. Lie K; Lam, C. N. W.; Ho, J. C. M.; Lau, M. M. L. *Eur. J. Lipid Sci.Tech.* **2003**, *105*, 391.
[638] Lee, C.-S.; Audelo, M. Q.; Reibenpies, J.; Sulikowski, G. A. *Tetrahedron* **2002**, *58*, 4403.
[639] Lee, J. S.; Fuchs, P. L. *Org. Lett.* **2003**, *5*, 2247.
[640] Lévai, A.; Jekő, J.; Brahmbhatt, D. I. *J. Heterocycl. Chem.* **2004**, *41*, 707.
[641] Lévai, A.; Silva, A. M. S.; Cavaleiro, J. A. S.; Patonay, T.; Silva, Vera L. M. *Eur. J. Org. Chem.* **2001**, 3213.
[642] Lévai, A; Jekő, J. *J. Heterocycl. Chem.* **2004**, *41*, 439.
[643] Navarro-Eisenstein P., A. M.; Vasquez, C.; Franklin, P. J.; Baumstark, A. L. *Heterocycl. Commun.* **2003**, *9*, 575.
[644] Phutdhawong, W.; Pyne, S. G.; Baramee, A.; Buddhasukh, D.; Skelton, B. W.; White, A. H. *Tetrahedron Lett.* **2002**, *43*, 6047.
[645] Rainier, J. D.; Cox, J. M. *Org. Lett.* **2000**, *2*, 2707.
[646] Schobert, R.; Siegfried, S.; Weingaertner, J.; Nieuwenhuyzen, M. *J. Chem.l Soc., Perkin Trans.1* **2001**, 2009.
[647] Spiteller, P.; Kern, W.; Reiner, J.; Spiteller, G. *Biochim. Biophys. Acta: Mol. Cell Biol. Lipids* **2001**, *1531*, 188.
[648] Wang, X.; Lee, Y. R. *Tetrahedron Lett.* **2007**, *48*, 6275.

Supplemental References for Table 4. Diasteroselective Epoxidations of Alkenes by Dioxiranes

[649] Adam, W.; Bosio, S. G.; Turro, N. J.; Wolff, B. T. *J. Org. Chem.* **2004**, *69*, 1704.
[650] Adam, W.; Bosio, S. G.; Wolff, B. T. *Org. Lett.* **2003**, *5*, 819.
[651] Adam, W.; Peters, K.; Peters, E.-M.; Schambony, S. B. *J. Am. Chem. Soc.* **2001**, *123*, 7228.
[652] Adam, W.; Schambony, S. B. *Org. Lett.* **2001**, *3*, 79.
[653] Adam, W.; Zhang, A. *Eur. J. Org. Chem.* **2004**, 147.
[654] Antonioletti, R.; Malancona, S.; Bovicelli, P. *Tetrahedron* **2002**, *58*, 8825.
[655] Appendino, G.; Cravotto, G.; Jarevang, T.; Sterner, O. *Eur. J. Org. Chem.* **2000**, 2933.
[656] Asouti, A.; Hadjiarapoglou, L. P. *Synlett* **2001**, 1847.
[657] Bartels, A.; J., Peter G.; Liebscher, J. *Synthesis* **2003**, 67.
[658] Bentley, T. W.; Engels, B.; Hupp, T.; Bogdan, E.; Christl, M. *J. Org. Chem.* **2006**, *71*, 1018.
[659] Boulineau, F. P.; Liew, S.-T.; Shi, Q.; Wenthold, P. G.; Wei, A. *Org. Lett.* **2006**, *8*, 4545.
[660] Boulineau, F. P.; Wei, A. *Org. Lett.* **2002**, *4*, 2281.
[661] Boyer, F.-D.; Descoins, C. L.; Thanh, G. V.; Descoins, C.; Prange, T.; Ducrot, P.-H. *Eur. J. Org. Chem.* **2003**, 1172.
[662] Brimble, M. A.; Furkert, D. P. *Org. Biomol. Chem.* **2004**, *2*, 3573.
[663] Caravano, A.; Baillieul, D.; Ansiaux, C.; Pan, W.; Kovensky, J.; Sinay, P.; Vincent, S. P. *Tetrahedron* **2007**, *63*, 2070.
[664] Cheshev, P.; Marra, A.; Dondoni, A. *Carbohydr. Res.* **2006** *341*, 2714.
[665] Cuzzupe, A. N.; Hutton, C. A.; Lilly, M. J.; Mann, R. K.; McRae, K. J.; Zammit, S. C.; Rizzacasa, M. A. *J. Org. Chem.* **2001**, *66*, 2382.
[666] D'Accolti, L.; Annese, C.; Fusco, C. *Tetrahedron Lett.* **2005**, *46*, 8459.

[667] Davies, S. G.; Key, M.-S.; Rodriguez-Solla, H.; Sanganee, H. J.; Savory, E. D.; Smith, A. D. *Synlett* **2003**, 1659.

[668] Duffy, R. J.; Morris, K. A.; Romo, D. *J. Am. Chem. Soc.* **2005**, *127*, 16754.

[669] Eggen, M.; Nair, S. K.; Georg, G. I. *Org. Lett.* **2001**, *3*, 1813.

[670] Enders, D.; Lenzen, A.; Backes, M.; Janeck, C.; Catlin, K.; Lannou, M.-I.; Runsink, J.; Raabe, G. *J. Org. Chem.* **2005**, *70*, 10538.

[671] Ermolenko, M. S.; Potier, P. *Tetrahedron Lett.* **2002**, *43*, 2895.

[672] Fan, C. L.; Lee, W.-D.; Teng, N.-W.; Sun, Y.-C.; Chen, K. *J. Org. Chem.* **2003**, *68*, 9816.

[673] Fawcett, J.; Griffith, G. A.; Percy, J. M.; Uneyama, E. *Org. Lett.* **2004**, *6*, 1277.

[674] Ferraz, H. M. C.; Longo, L. S., Jr. *J. Org. Chem.* **2007**, *72*, 2945.

[675] Fyvie, W. S.; Peczuh, M. W. *J. Org. Chem.* **2008**, *73*, 3626.

[676] Gataullin, R. R.; Ishberdina, R. R.; Antipin, A. V.; Suponitskii, K. Y.; Kabal'nova, N. N.; Shitikova, O. V.; Spirikhin, L. V.; Antipin, M. Yu.; Abdrakhmanov, I. B. *Chem. Heterocycl. Comp.* **2006**, *42*, 1130.

[677] Grabovskiy, S. A.; Kabal'nova, N. N.; Chatgilialoglu, C.; Ferreri, C. *Helvetica Chim. Acta* **2006**, *89*, 2243.

[678] Guiard, S.; Giorgi, M.; Santelli, M.; Parrain, J.-L. *J. Org. Chem.* **2003**, *68*, 3319.

[679] Halim, R.; Brimble, M. A.; Merten, J. *Org. Biomol. Chem.* **2006**, *4*, 1387.

[680] Halim, R.; Brimble, M. A.; Merten, J. *Org. Lett.* **2005**, *7*, 2659.

[681] Haraguchi, K.; Kubota, Y.; Tanaka, H. *J. Org. Chem.* **2004**, *69*, 1831.

[682] Haraguchi, K.; Kubota, Y.; Tanaka, H. *Nucl. Acids Res. Suppl.* **2002**, *2*, 17.

[683] Haraguchi, K.; Takeda, S.; Sumino, M.; Tanaka, H.; Dutschman, G. E.; Cheng, Y.-C.; Nitanda, T.; Baba, M. *Nucleosides Nucleotides Nucl. Acids* **2005**, *24*, 343.

[684] Haraguchi, K.; Takeda, S.; Tanaka, H. *Org. Lett.* **2003**, *5*, 1399.

[685] Haraguchi, K.; Takeda, S.; Tanaka, H. *Nucl. Acids Res. Suppl.* **2002**, *2*, 133.

[686] Horiguchi, T.; Nagura, M.; Cheng, Q.; Oritani, T.i; Kudo, T. *Heterocycles* **2000**, *53*, 2629.

[687] Jastrzebska, I.; Katrynski, K. S.; Morzycki, J. W. *ARKIVOC* **2002**, (9), 46.

[688] Jie, M. S. F. Lie K; Lam, C. N. W.; Ho, J. C. M.; Lau, M. M. L. *Eur. J. Lipid Sci.Tech.* **2003**, *105*, 391.

[689] Krause, N.; Laux, M.; Hoffmann-Roder, A. *Tetrahedron Lett.* **2000**, *41*, 9613.

[690] Krohn, K.; Micheel, J.; Zukowski, M. *Tetrahedron* **2000**, *56*, 4753.

[691] Kubota, Y.; Haraguchi, K.; Kunikata, M.; Hayashi, M.; Ohkawa, M.; Tanaka, H. *J. Org. Chem.* **2006**, *71*, 1099.

[692] Lee, C.-S.; Audelo, M. Q.; Reibenpies, J.; Sulikowski, G. A. *Tetrahedron* **2002**, *58*, 4403.

[693] Li, W.; Fuchs, P. L. *Org. Lett.* **2003**, *5*, 2853.

[694] Lohman, G. J. S.; Seeberger, P. H. *J. Org. Chem.* **2003**, *68*, 7541.

[695] Majumder, U.; Cox, J. M.; Rainier, J. D. *Org. Lett.* **2003**, *5*, 913.

[696] Murakami, N.; Tamura, S.; Wang, W.; Takagi, T.; Kobayashi, M. *Tetrahedron* **2001**, *57*, 4323.

[697] Murakami, N.; Wang, W.; Ohyabu, N.; Ito, T.; Tamura, S.; Aoki, S.; Kobayashi, M.; Kitagawa, I. *Tetrahedron* **2000**, *56*, 9121.

[698] Navarro-Eisenstein, A. M.; Vasquez, P. C.; Franklin, P. J.; Baumstark, A. L. *Heterocycl. Commun.* **2003**, *9*, 449.

[699] Parrish, J. D.; Little, R. D. *Org. Lett.* **2002**, *4*, 1439.

[700] Pastor, A.; Adam, W.; Wirth, T.; Toth, G. *Eur. J. Org. Chem.* **2005**, 3075.

[701] Patonay, T.; Kiss-Szikszai, A.; Silva, V. M. L.; Silva, A. M. S.; Pinto, D. C. G. A.; Cavaleiro, J. A. S.; Jekő, J. *Eur. J. Org. Chem.* **2008**, 1937.

[702] Peczuh, M. W.; Snyder, N. L.; Fyvie, W. S. *Carbohydr. Res.* **2004**, *339*, 1163.

[703] Phutdhawong, W.; Pyne, S. G.; Baramee, A.; Buddhasukh, D.; Skelton, B. W.; White, A. H. *Tetrahedron Lett.* **2002**, *43*, 6047.

[704] Rainier, J. D.; Cox, J. M. *Org. Lett.* **2000**, *2*, 2707.

[705] Skelton, B. W.; Stick, R. V.; Stubbs, K. A.; Watts, A. G.; White, A. H. *Austral. J. Chem.* **2004**, *57*, 345.

[706] Stachel, S. J.; Danishefsky, S. J. *Tetrahedron Lett.* **2001**, *42*, 6785.

[707] Stoianova, D. S.; Whitehead, A.; Hanson, P. R. *J. Org. Chem.* **2005**, *70*, 5880.

708 Tomooka, K.; Komine, N.; Fujiki, D.; Nakai, T.; Yanagitsuru, S. *J. Am. Chem. Soc.* **2005**, *127*, 12182.
709 Upreti, M.; Ruhela, D.; Vishwakarma, R. A. *Tetrahedron* **2000**, *56*, 6577.
710 Vedejs, E.; Duncan, S. M. *J. Org. Chem.* **2000**, *65*, 6073.
711 Watts, A. G.; Withers, S. G. *Can. J. Chem.* **2004**, *82*, 1581.
712 Zimmermann, P. J.; Lee, J. Y.; Hlobilova, I.; Endermann, R.; Häbich, D.; Jäger, V. *Eur. J. Org. Chem.* **2005**, 3450.

Supplemental References for Table 5. Enantioselective Epoxidations of Alkenes Using Dioxiranes

713 Ager, D. J.; Anderson, K. Oblinger, E.; Shi, Y.; Van der Roest, J. *Org. Process Res. Dev.* **2007**, *11*, 44.
714 Altmann, K.-H.; Bold, G.; Caravatti, G.; Denni, D.; Florsheimer, A.; Schmidt, A.; Rihs, G.; Wartmann, M. *Helvetica Chim. Acta* **2002**, *85*, 4086.
715 Armstrong, A.; Ahmed, G.; Dominguez-Fernandez, B.; Hayter, B. R.; Wailes, J. S. *J. Org. Chem.* **2002**, *67*, 8610.
716 Armstrong, A.; Dominguez-Fernandez, B.; Tsuchiya, T. *Tetrahedron* **2006**, *62*, 6614.
717 Armstrong, A.; Draffan, A. G. *J. Chem. Soc., Perkin Trans. 1* **2001**, 2861. (check)
718 Armstrong, A.; Hayter, B. R.; Moss, W. O.; Reeves, J. R.; Wailes, J. S. *Tetrahedron: Asymmetry* **2000**, *11*, 2057.
719 Armstrong, A.; Moss, W. O.; Reeves, J. R. *Tetrahedron: Asymmetry* **2001**, *12*, 2779.
720 Armstrong, A.; Tsuchiya, T. *Tetrahedron* **2006**, *62*, 257.
721 Armstrong, A.; Washington, I.; Houk, K. N. *J. Am. Chem. Soc.* **2000**, *122*, 6297.
722 Bez, G.; Zhao, C.-G. *Tetrahedron Lett.* **2003**, *44*, 7403.
723 Bian, J.; Van Wingerden, M.; Ready, J. M. *J. Am. Chem. Soc.* **2006** *128*, 7428.
724 Bortolini, O.; Fantin, G.; Fogagnolo, M.; Forlani, R.; Maietti, S.; Pedrini, P. *J. Org. Chem.* **2002**, *67*, 5802.
725 Bortolini, O.; Fantin, G.; Fogagnolo, M.; Mari, L. *Tetrahedron* **2006**, *62*, 4482.
726 Bortolini, O.; Fantin, G.; Fogagnolo, M.; Mari, L. *Tetrahedron: Asymmetry* **2004**, *15*, 3831.
727 Bortolini, O.; Fogagnolo, M.; Fantin, G.; Maietti, S.; Medici, A. *Tetrahedron: Asymmetry* **2001**, *12*, 1113.
728 Burke, A.; Dillon, P.; Martin, K.; Hanks, T. W. *J. Chem. Edu.* **2000**, *77*, 271.
729 Burke, C. P.; Shi, Y. *J. Org. Chem.* **2007**, *72*, 4093.
730 Burke, C. P.; Shi, Y. *Angew. Chem., Int. Ed.* **2006**, *45*, 4475.
731 Burke, C. P.; Shi, Y. *J. Org. Chem.* **2007**, *72*, 4093.
732 Burke, C. P.; Shu, L.; Shi, Y. *J. Org. Chem.* **2007**, *72*, 6320.
733 Cachoux, F.; Isarno, T.; Wartmann, M.; Altmann, K.-H. *ChemBioChem* **2006**, *7*, 54.
734 Cachoux, F.; Isarno, T.; Wartmann, M.; Altmann, K.-H. *Angew. Chem. Int. Ed.* **2005**, *44*, 7469.
735 Chan, W.-K.; Yu, W.-Y.; Che, C.-M.; Wong, M.-K. *J. Org. Chem.* **2003**, *68*, 6576.
736 Crane, Z.; Goeddel, D.; Gan, Y.; Shi, Y. *Tetrahedron* **2005**, *61*, 6409.
737 Cubillos, J.; Holderich, W. *Revista Facult. Ingenieria, Univ. Antioquia* **2007**, *41*, 31.
738 Curran, D. P.; Zhang, Q.; Richard, C.; Lu, H.; Gudipati, V.; Wilcox, C. S. *J. Am. Chem. Soc.* **2006**, *128*, 9561.
739 Deng, F.; Lan, Z.; Yin, D.; Xiao, Z. *Huaxue Tongbao* **2006**, *69*, 362.
740 Deng, L.; Mi, A.-Q.; Cui, X.; Jiang, Y.-Z. *Hecheng Huaxue* **2001**, *9*, 93.
741 Denmark, S. E.; Matsuhashi, H. *J. Org. Chem.* **2002**, *67*, 3479.
742 Devlin, F. J.; Stephens, P. J.; Bortolini, O. *Tetrahedron: Asymmetry* **2005**, *16*, 2653.
743 Freedman, T. B.; Cao, X.; Nafie, L. A.; Solladie-Cavallo, A.; Jierry, L.; Bouerat, L. *Chirality* **2004**, *16*, 467.
744 Fuerstner, A.; Kattnig, E.; Lepage, O. *J. Am. Chem. Soc.* **2006**, *128*, 9194.
745 Furutani, T.; Imashiro, R.; Hatsuda, M.; Seki, M. *J. Org. Chem.* **2002**, *67*, 4599.
746 Fyvie, W. S.; Peczuh, M. W. *J. Org. Chem.* **2008**, *73*, 3626.
747 Goeddel, D.; Shu, L.; Yuan, Y.; Wong, O. A.; Wang, B.; Shi, Y. *J. Org.Chem.* **2006**, *71*, 1715.
748 Halim, R.; Brimble, M. A.; Merten, J. *Org. Biomol. Chem.* **2006**, *4*, 1387.

[749] Halim, R.; Brimble, M. A.; Merten, J. *Org. Lett.* **2005**, *7*, 2659.
[750] Heffron, T. P.; Jamison, T. F. *Synlett* **2006**, 2329.
[751] Heffron, T. P.; Jamison, T. F. *Org. Lett.* **2003**, *5*, 2339.
[752] Hickey, M.; Goeddel, D.; Crane, Z.; Shi, Y. *Proc. Nat. Acad. Sci. USA* **2004**, *101*, 5794.
[753] Hioki, H.; Kanehara, C.; Ohnishi, Y.; Umemori, Y.; Sakai, H.; Yoshio, S.; Matsushita, M.; Kodama, M. *Angew. Chem. Int. Ed.* **2000**, *39*, 2552.
[754] Hoard, D. W.; Moher, E. D.; Martinelli, M. J.; Norman, B. H. *Org. Lett.* **2002**, *4*, 1813.
[755] Imashiro, R.; Seki, M. *J. Org. Chem.* **2004**, *69*, 4216.
[756] Iwasaki, J.; Ito, H.; Nakamura, M.; Iguchi, K. *Tetrahedron Lett.* **2006**, *47*, 1483.
[757] Klein, S.; Roberts, S. M. *J. Chem. Soc., Perkin Trans. 1* **2002**, 2686.
[758] Knoell, J.; Knoelker, H.-J. *Tetrahedron Lett.* **2006**, *47*, 6079.
[759] Kumar, V. S.; Wan, S.; Aubele, D. L.; Floreancig, P. E. *Tetrahedron: Asymmetry* **2005**, *16*, 3570.
[760] Lorenz, J. C.; Frohn, M.; Zhou, X.; Zhang, J.-R.; Tang, Y.; Burke, C.; Shi, Y. *J. Org. Chem.* **2005**, *70*, 2904.
[761] Madhushaw, R. J.; Li, C.-L.; Shen, K.-H.; Hu, C.-C.; Liu, R.-S. *J. Am. Chem. Soc.* **2001**, *123*, 7427.
[762] Mandel, A. L.; Jones, B. D.; La Clair, J. J.; Burkart, M. D. *Bioorg. Med. Chem. Lett.* **2007**, *17*, 5159.
[763] Matsumoto, K.; Tomioka, K. *Tetrahedron Lett.* **2002**, *43*, 631.
[764] Matsumoto, K.; Tomioka, K. *Chem. Pharm. Bull.* **2001**, *49*, 1653.
[765] Matsumoto, K.; Tomioka, K. *Heterocycles* **2001**, *54*, 615.
[766] McDonald, F. E.; Bravo, F.; Wang, X.; Wei, X.; Toganoh, M.; Rodriguez, J. R.; Do, B.; Neiwert, W. A.; Hardcastle, K. I. *J. Org. Chem.* **2002**, *67*, 2515.
[767] Morimoto, Y.; Iwai, T.; Kinoshita, T. *Tetrahedron Lett.* **2001**, *42*, 6307.
[768] Morimoto, Y.; Nishikawa, Y.; Takaishi, M. *J. Am. Chem. Soc.* **2005**, *127*, 5806.
[769] Morimoto, Y.; Okita, T.; Takaishi, M.; Tanaka, T. *Angew. Chem. Int. Ed.* **2007**, *46*, 1132.
[770] Morimoto, Y.; Takaishi, M.; Adachi, N.; Okita, T.; Yata, H. *Org. Biomol. Chem.* **2006**, *4*, 3220.
[771] Morishita, Y.; Iwai, T.; Muragaki, K.; Kinoshita, T.; Morimoto, Y. *Tennen Yuki Kagobutsu Toronkai Koen Yoshishu* **2001**, 229.
[772] Neighbors, J. D.; Mente, N. R.; Boss, K. D.; Zehnder, D. W.; Wiemer, D. F. *Tetrahedron Lett.* **2008**, *49*, 516.
[773] Nieto, N.; Molas, P.; Benet-Buchholz, J.; Vidal-Ferran, A. *J. Org. Chem.* **2005**, *70*, 10143.
[774] Nieto, N.; Munslow, I. J.; Barr, J.; Benet-Buchholz, J.; Vidal-Ferran, A. *Org. Biomol. Chem.* **2008**, *6*, 2276.
[775] O'Brien, K. C.; Colby, E. A.; Jamison, T. F. *Tetrahedron* **2005**, *61*, 6243.
[776] Olofsson, B.; Somfai, P. *J. Org. Chem.* **2003**, *68*, 2514.
[777] Patonay, T.; Jeko, J.; Kiss-Szikszai, A.; Lévai, A. *Monatsh. Chem.* **2004**, *135*, 743.
[778] Rousseau, C.; Christensen, B.; Bols, M. *Eur. J. Org. Chem.* **2005**, 2734.
[779] Rousseau, C.; Christensen, B.; Petersen, T. E.; Bols, M. *Org. Biomol. Chem.* **2004**, *2*, 3476.
[780] Sartori, G.; Armstrong, A.; Maggi, R.; Mazzacani, Alessandro; S., Raffaella; Bigi, F.; Dominguez-Fernandez, B. *J. Org. Chem.* **2003**, *68*, 3232.
[781] Seki, M.; Furutani, T.; Imashiro, R.; Kuroda, T.; Yamanaka, T.; Harada, N.; Arakawa, H.; Kusama, M.; Hashiyama, T. *Tetrahedron Lett.* **2001**, *42*, 8201.
[782] Shen, Y.-M.; Wang, B.; Shi, Y. *Tetrahedron Lett.* **2006**, *47*, 5455.
[783] Shen, Y.-M.; Wang, B.; Shi, Y. *Angew. Chem. Int. Ed.* **2006**, *45*, 1429.
[784] Shing, T. K. M.; Leung, G. Y. C.; Luk, T. *J. Org. Chem.* **2005**, *70*, 7279.
[785] Shing, T. K. M.; Leung, G.Y. C.; Yeung, K. W. *Tetrahedron Lett.* **2003**, *44*, 9225.
[786] Shing, T. K. M.; Leung, Y. C.; Y., Kwan W. *Tetrahedron* **2003**, *59*, 2159.
[787] Shing, T. K. M.; Luk, T.; Lee, C. M. *Tetrahedron* **2006**, *62*, 6621.
[788] Shu, L.; Shi, Y. *Tetrahedron Lett.* **2004**, *45*, 8115.
[789] Shu, L.; Shi, Y. *Tetrahedron* **2001**, *57*, 5213.
[790] Shu, L.; Wang, P.; Gan, Y.; Shi, Y. *Org. Lett.* **2003**, *5*, 293.
[791] Singleton, D. A.; Wang, Z. *J. Am. Chem. Soc.* **2005**, *127*, 6679.
[792] Smith, A. B., III; Fox, R. J. *Org. Lett.* **2004**, *6*, 1477.

[793] Solladie-Cavallo, A.; Azyat, K.; Jierry, L.; Cahard, D. *J. Flu. Chem.* **2006**, *127*, 1510.
[794] Solladie-Cavallo, A.; Bouerat, L. *Org. Lett.* **2000**, *2*, 3531.
[795] Solladie-Cavallo, A.; Bouerat, L.; Jierry, L. *Eur. J. Org. Chem.* **2001**, 4557.
[796] Solladie-Cavallo, A.; Jierry, L.; Klein, A. *Compt. Rend. Chim.* **2003**, *6*, 603.
[797] Solladie-Cavallo, A.; Jierry, L.; Klein, A.; Schmitt, M.; Welter, R. *Tetrahedron: Asymmetry* **2004**, *15*, 3891.
[798] Solladie-Cavallo, A.; Jierry, L.; Lupattelli, P.; Bovicelli, P.; Antonioletti, R. *Tetrahedron* **2004**, *60*, 11375.
[799] Solladie-Cavallo, A.; Jierry, L.; Norouzi-Arasi, H.; Tahmassebi, D. *J. Flu. Chem.* **2004**, *125*, 1371.
[800] Solladie-Cavallo, A.; Lupattelli, P.; Jierry, L.; Bovicelli, P.; Angeli, F.; Antonioletti, R.; Klein, A. *Tetrahedron Lett.* **2003**, *44*, 6523.
[801] Stearman, C. J.; Behar, V. *Tetrahedron Lett.* **2002**, *43*, 1943.
[802] Tian, H.; She, X.; Shi, Y. *Org. Lett.* **2001**, *3*, 715.
[803] Tian, H.; She, X.; Shu, L.; Yu, H.; Shi, Y. *J. Am. Chem. Soc.* **2000**, *122*, 11551.
[804] Tian, H.; She, X.; Xu, J.; Shi, Y. *Org. Lett.* **2001**, *3*, 1929.
[805] Tian, H.; She, X.; Yu, H.; Shu, L.; Shi, Y. *J. Org. Chem.* **2002**, *67*, 2435.
[806] Tong, R.; Valentine, J. C.; McDonald, F. E.; Cao, R.; Fang, X.; Hardcastle, K. I. *J. Am. Chem. Soc.* **2007**, *129*, 1050.
[807] Villar, H.; Guibe, F.; Aroulanda, C.; Lesot, P. *Tetrahedron: Asymmetry* **2002**, *13*, 1465.
[808] Wan, S.; Gunaydin, H.; Houk, K. N.; Floreancig, P. E. *J. Am. Chem. Soc.* **2007**, *129*, 7915.
[809] Wang, B.; Shen, Y.-M.; Shi, Y. *J. Org. Chem.* **2006**, *71*, 9519.
[810] Wang, P.-D.; Yang, N.-F.; Ling, Y.; Li, J.-C.; Cao, J. *Youji Huaxue* **2007**, *27*, 885.
[811] Wang, Z.-X. Shu, L.; Frohn, M.; Tu, Y.; Shi, Y. *Org. Synth.* **2003**, *80*, 9.
[812] Wang, Z.-X.; Miller, S. M.; Anderson, O. P.; Shi, Y. *J. Org. Chem.* **2001**, *66*, 521.
[813] Wang, Z.-X.; Shu, L.; Frohn, M.; Tu, Y.; Shi, Y. *Org. Synth.* **2003**, *80*, 9.
[814] Wiseman, J. M.; McDonald, F. E.; Liotta, D. C. *Org. Lett.* **2005**, *7*, 3155.
[815] Wong, O. A.; Shi, Y. *J. Org. Chem.* **2006**, *71*, 3973.
[816] Wu, X.-Y.; She, X.; Shi, Y. *J. Am. Chem. Soc.* **2002**, *124*, 8792.
[817] Xiao, Z.; Lan, Z.; Yin, D.; Liu, F.; Li, C. *Cuihua Xuebao* **2007**, *28*, 469.
[818] Xie, B.; Cai, X. *Huaxue Tongbao* **2006**, *69*, 552.
[819] Xiong, Z.; Corey, E. J. *J. Am. Chem. Soc.* **2000**, *122*, 9328.
[820] Yang, D.; Jiao, G.-S.; Yip, Y.-C.; Lai, T.-H.; Wong, M.-K. *J. Org. Chem.* **2001**, *66*, 4619.
[821] Yoshida, M.; Abdel-Hamid Ismail, M.; Nemoto, H.; Ihara, M. *J. Chem. Soc. Perkin 1* **2000**, 2629.
[822] Yoshida, M.; Hayashi, M.; Shishido, K. *Org. Lett.* **2007**, *9*, 1643.
[823] Zhang, Z.; Tang, J.; Wang, X.; Shi, H. *J. Mol. Catal. A: Chem.* **2008**, *285*, 68.
[824] Zheng, Y.-S.; Xiao, Q. *Gaodeng Xuexiao Huaxue Xuebao* **2001**, *22*, 1155.
[825] Zhu, Y.; Shu, L.; Tu, Y.; Shi, Y. *J. Org. Chem.* **2001**, *66*, 1818.

CHAPTER 2

DIOXIRANE OXIDATIONS OF COMPOUNDS OTHER THAN ALKENES

WALDEMAR ADAM

Institute for Organic Chemistry, Würzburg University, D-97074 Würzburg, Germany
and
Department of Chemistry, University of Puerto Rico, Rio Piedras, Puerto Rico 00931

CONG-GUI ZHAO and KAVITHA JAKKA

Department of Chemistry, The University of Texas at San Antonio, San Antonio, Texas 78249

CONTENTS

Oxidation of Organic Compounds by Dioxiranes, by Waldemar Adam, Cong-Gui Zhao, Chantu R. Saha-Möller, and Kavitha Jakka

ACKNOWLEDGMENTS

Generous financial support from the *Deutsche Forschungsgemeinschaft (Schwerpunktprogramm "Peroxidchemie: Mechanistische und Präparative Aspekte des Sauerstofftransfers"* and *Sonderforschungsbereich SFB 347 "Selektive Reaktionen Metall-aktivierter Moleküle")*, the *Fonds der Chemischen Industrie*, the *NIH-MBRS Program* (Grant S06 GM 08194), the Welch Foundation (Grant AX-1593), and the DAAD *(Deutscher Akademischer Austauschdienst)* is gratefully appreciated. The authors also thank Dr. Chantu R. Saha-Möller and Mrs. Ana-Maria Krause for their help in preparing the tabular and graphical material.

INTRODUCTION

Epoxidations, heteroatom oxidations, and Y–H insertions constitute the best investigated oxidations by dioxiranes. An overview of these transformations is displayed in the rosette of Scheme 1. These preparatively useful oxidations have been extensively reviewed during the last decade.[1–14] In a previous chapter,[15] we presented the epoxidation of double bonds [π bonds in simple alkenes and those functionalized with electron donors (ED), electron acceptors (EA), and with both ED and EA substituents; case 1 in the rosette] with either isolated or in situ generated dioxiranes. The recent developments in the dioxirane-mediated asymmetric epoxidation have also been extensively covered there.[15] The present chapter concerns the remaining oxidations in the rosette of Scheme 1, that is, epoxidation of the double bonds in the cumulenes, such as allenes (transformation 2), acetylenes (transformation 3), and arenes (transformation 4); the oxidation of heteroatom functionalities, mainly lone pairs on sulfur (transformation 5), on nitrogen (transformations 6 and 7), and on oxygen as the deoxygenation of *N*-oxides (transformation 8); the oxidation of C=Y functionalities (e.g., transformation 9), Y–H insertions (σ bonds) such as C–H in alkanes (transformation 10) and Si–H in silanes (transformation 11); and the oxidation of organometallic substrates including metal (transformation 12) and ligand-sphere oxidation.

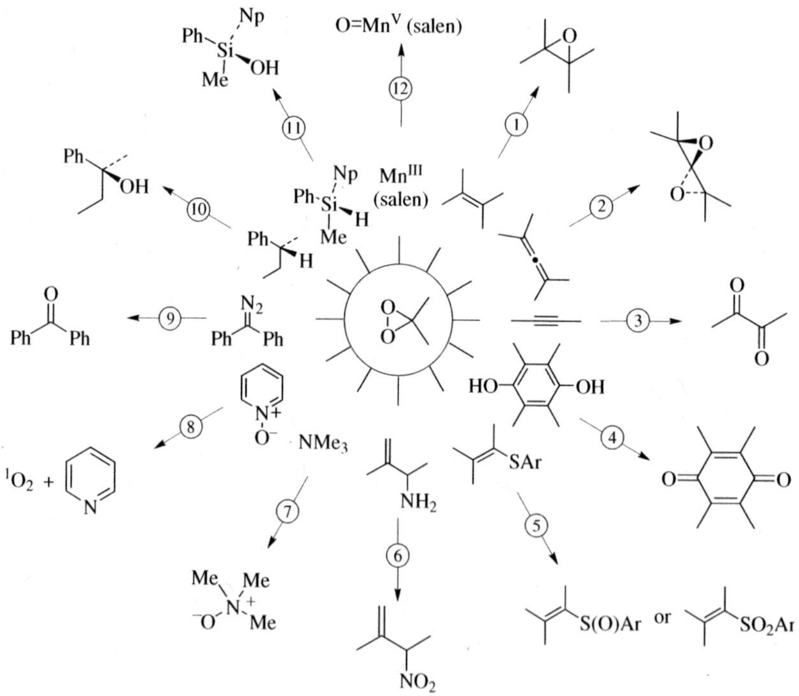

Scheme 1. An overview of dioxirane oxidations (Np = 1-naphthyl).

MECHANISM

Allenes, Alkynes, and Arenes

Although the products of the dioxirane oxidation of allenes, alkynes, and arenes are usually more complex than those of the epoxidation of simple C=C double bonds, the initial step of the oxidation is usually epoxidation. Therefore, the same mechanism that has been extensively discussed in the previous chapter[15] also applies in these reactions. The oxygen transfer proceeds with complete retention of the initial olefin configuration through the concerted spiro transition state.[15] An example is shown in Eq. 1, in which the oxidation of the chiral allene proceeds in nearly quantitative yield (95%) with preservation of the starting allene configuration in the spiro-bisepoxide.[16]

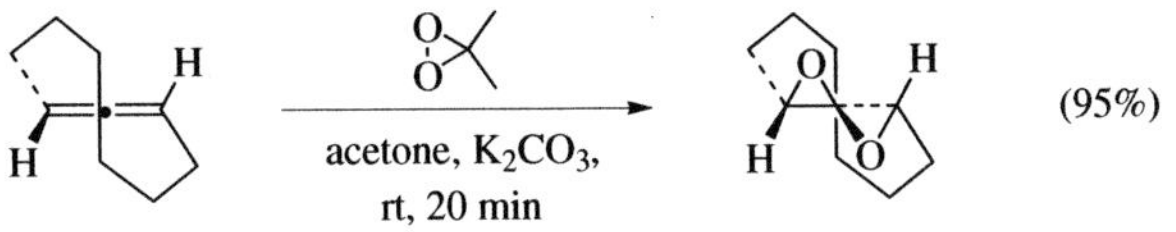

acetone, K_2CO_3, rt, 20 min (95%) (Eq. 1)

Since the initial epoxidation products of the allenes, alkynes, and arenes are usually labile substances, they may undergo subsequent reactions, which include further oxidation by dioxirane other than epoxidation. For example, in the dimethyldioxirane (DMD) oxidation of the phenanthrene derivative in Scheme 2,[17] the second oxidation by DMD involves C–H insertion instead of epoxidation.

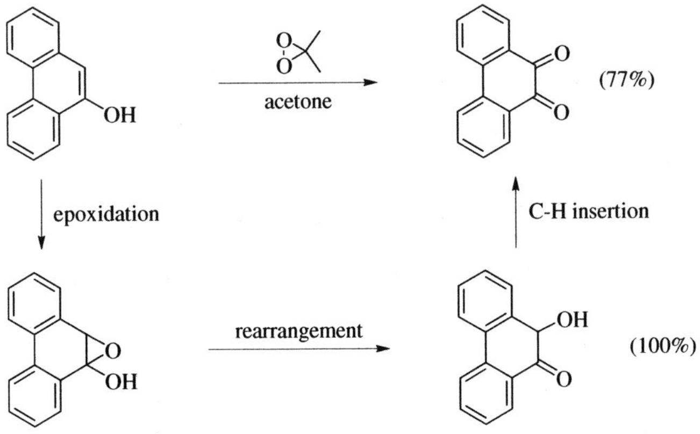

OH; acetone; O; O; (77%); epoxidation; C-H insertion; O; OH; rearrangement; OH; O; (100%)

Scheme 2. DMD oxidation of 9-hydroxyphenanthrene.

Heteroatom Substrates

Through a detailed study of the competitive oxidation of the sulfide versus sulfoxide functionalities in thianthrene 5-oxide (SSO),[18] a pronounced electrophilic character has been demonstrated for DMD and methyl(trifluoromethyl)dioxirane (TFD).[19,20] Thus, dioxiranes prefer to oxidize the sulfide over the sulfoxide functionality, a typical behavior of an electrophilic oxidant (Scheme 3). Also, the

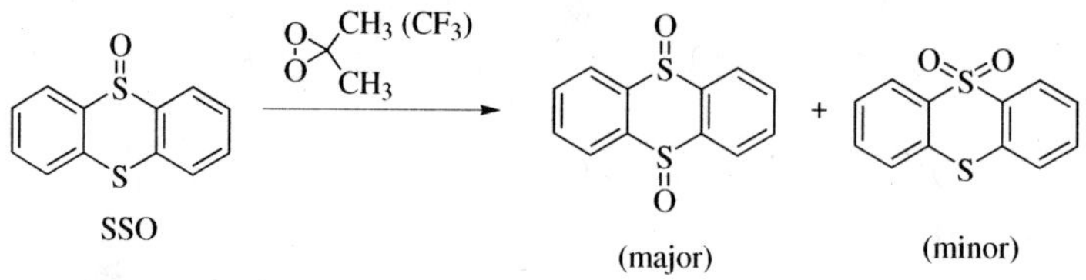

Scheme 3. Competitive oxidation of the sulfide vs. sulfoxide functionalities in thianthrene-5-oxide (SSO) by the dioxiranes DMD and TFD.

kinetic data[6] for the oxidation of sulfides and sulfoxides have revealed the electrophilic character of dioxiranes. Thus, the heteroatom oxidations by dioxirane are generally explained in terms of a S_N2-type attack of the heteroatom lone pair on the dioxirane peroxide σ^*-orbital.[21,22]

A possible single-electron-transfer (SET) mechanism in N-oxidations[23,24] has been discounted[21] on the basis of kinetic experiments by comparing the relative rates of oxygen transfer by DMD with those of alkylation by methyl iodide. For the latter, an S_N2 mechanism unequivocally applies. Similar reactivities (linear correlation of rates) for N-oxidation also establish the S_N2 pathway for dioxirane oxidations. This conclusion is supported by a kinetic study of the DMD oxidation of substituted *N,N*-dimethylanilines.[25]

The heterolytic mechanism is presumably also valid for a variety of oxygen-type nucleophiles, e.g., amine *N*-oxides, ClO^-, HO^-, HOO^-, RO^-, ROO^-, $RC(O)OO^-$, and $^-OS(O)_2OO^-$, which all catalyze the decomposition of dioxiranes with the evolution of molecular oxygen.[26,27] A typical case is illustrated with 4-dimethylaminopyridine *N*-oxide in Scheme 4.[26] The chemiluminescence emitted by the generated singlet oxygen confirms the heterolytic nature of the dioxirane decomposition.[26] Further support for this mechanism has been provided by theoretical work, from which it was concluded that the oxidation of primary amines by DMD does not proceed by a radical process.[28]

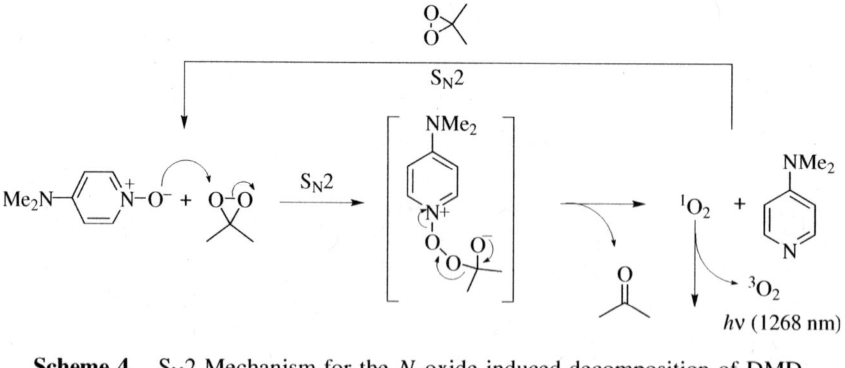

Scheme 4. S_N2 Mechanism for the *N*-oxide-induced decomposition of DMD.

Alkanes and Silanes

Two mechanisms have been suggested for the insertion of an oxygen atom into the Y–H bond of alkanes and silanes. Abundant evidence, which includes kinetics,[29] kinetic isotope effects,[30] and stereoselectivity,[31] all unequivocally support a concerted oxenoid-type mechanism (Figure 1).

Nonetheless, radical reactivity has been observed recently and interpreted in terms of the dioxirane diradical as the active oxidant, in particular, the so-called "molecule-induced homolysis."[32–35] It has also been proposed[36] that alkane hydroxylation may proceed by a rate-determining oxygen insertion into the alkane C–H bond to generate a caged radical pair, followed by very fast collapse (oxygen rebound) to hydroxylated products (Scheme 5).

That hydroxylation of (*R*)-2-phenylbutane proceeds with 100% retention to furnish (*S*)-2-phenylbutan-2-ol for both DMD[37] and TFD[31] sheds serious doubt on the involvement of out-of-cage radical intermediates in such C–H oxidations (Eq. 2).

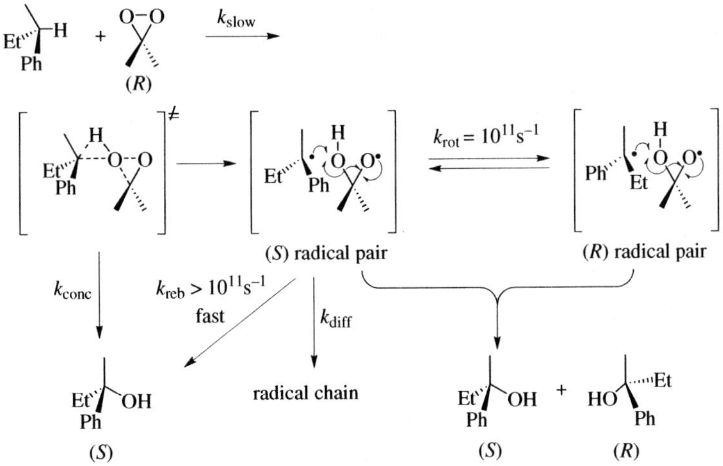

k_{slow}; (*R*); $k_{rot} = 10^{11} s^{-1}$; (*S*) radical pair; (*R*) radical pair; k_{conc}; $k_{reb} > 10^{11} s^{-1}$ fast; k_{diff}; radical chain; (*S*); (*S*) + (*R*)

Scheme 5. Concerted oxenoid-type (k_{conc}) vs. oxygen-rebound (k_{reb}) mechanisms for C–H insertion by DMD.

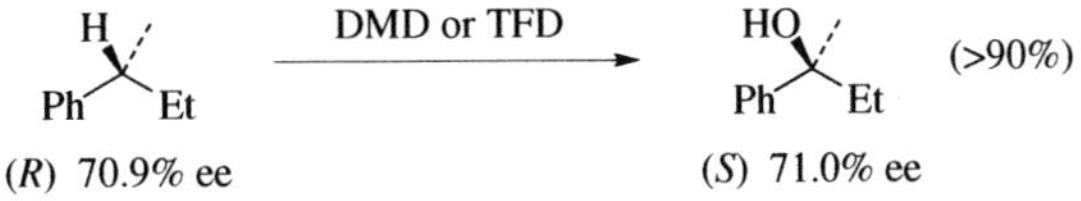

(*R*) 70.9% ee —DMD or TFD→ (*S*) 71.0% ee (>90%) (Eq. 2)

δ^+, $CH_3(CF_3)$, δ^-

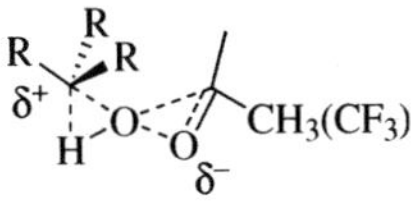

Figure 1. Concerted oxenoid-type transition state for C–H insertion.

The tertiary benzyl radical derived from this optically active substrate is one of the fastest radical clocks (the configurational persistence of this radical is estimated to be about 10^{-11} seconds)[38] and serves as a definitive probe for the intervention of radical intermediates. Thus, as shown in Scheme 5,[37] if a caged radical pair is formed, collapse with configurational conservation by oxygen rebound (k_{reb}) must be faster than diffusion out of the cage (k_{diff}), as well as in-cage isomerization (k_{rot}), since such competitive processes would lead to racemization.

As in the C–H oxidation of (*R*)-2-phenylbutane (Eq. 2), the hydroxylation of the (+)-(*S*)-(α-Np)PhMeSiH silane enantiomer by both dioxiranes DMD and TFD proceeds with complete retention of configuration to afford (+)-(*R*)-(α-Np)PhMeSiOH (Eq. 3).[39,40] Therefore, a similar mechanism would appear to apply for the oxidation of C–H and Si–H bonds.

Np(Ph)(Me)Si–H (*S*) 96.5% ee —DMD or TFD→ Np(Ph)(Me)Si–OH (*R*) 97.0% ee (>98%) (Eq. 3)

Most recent theoretical work on oxygen transfer for C–H insertion supports the concerted spiro oxenoid-type mechanism, in which the transition structure has considerable dipolar and also some diradical character.[41–43] Under typical preparative conditions, for example, in the presence of molecular oxygen, it was concluded that a concerted mechanism applies for the C–H insertion.

SCOPE AND LIMITATIONS

The oxidation of double bonds (π bonds) in cumulenes (allenes, acetylenes) and arenes, of heteroatom functionalities (lone-pair electrons), of transition-metal complexes, and Y–H insertions (σ bonds) has been successfully performed, either with isolated or with in situ generated dioxiranes. Thus, a broad spectrum of substrates has been oxidized by dioxiranes. The pertinent examples are listed in Tables 1–7 (see Tabular Survey). An isolated (distilled) acetone solution [DMD (isol.)] is the most often used dioxirane owing to its convenient preparation and relatively low cost. Although methyl(trifluoromethyl)dioxirane (TFD) is considerably more reactive than DMD, its application is limited because of its high cost and the high volatility of trifluoroacetone. With DMD (isol.), the scale of the reaction is usually limited to 100 mmol because DMD (isol.) is quite dilute (ca. 0.08 M). In the case of TFD (ca. 0.6 M), the prohibitive cost of trifluoroacetone obliges small-scale (ca. 10 mmol) applications. When a large-scale preparation is desired, the in situ mode [DMD (in situ)] is recommended, for which both biphasic[44–47] and homogeneous[48,49] media are available. It should be kept in mind that when one operates in aqueous solution, both the substrate and the oxidized products should resist hydrolysis and persist at temperatures above 0°. An advantage of the in situ mode is that it may be carried out with less than stoichiometric amounts (<0.5 equiv.) of ketone, which is important for enantioselective oxidations.[50–54]

Allenes, Alkynes, and Arenes

Representative examples of oxidations of allenes, alkynes, and arenes are collected in the rosette of Scheme 6.

The products of dioxirane oxidation of allenes depend on the reaction conditions and the substrate structure. Unfunctionalized allenes give the corresponding spiro-bisepoxides usually in good yields[16,54] at subambient temperatures when dry dioxirane solution is employed (Eq. 1).[16] If the allene is unsymmetrically substituted, a mixture of regioisomers is obtained, and the selectivity is highly dependent on the allene structure.[16,55] Since these spiro-bisepoxides are labile toward hydrolysis, the in situ oxidation mode is not recommended. If the allene substrate contains a hydroxy functionality, the latter will react with the spiro-bisepoxide intermediate to form ring-opened and/or rearranged products.[56–58] The final products may be cyclic or acyclic, depending on the reaction conditions, the chain-length of the substituent that contains the hydroxy functionality, and the other substituents on the allene. For example, when the hydroxyallene

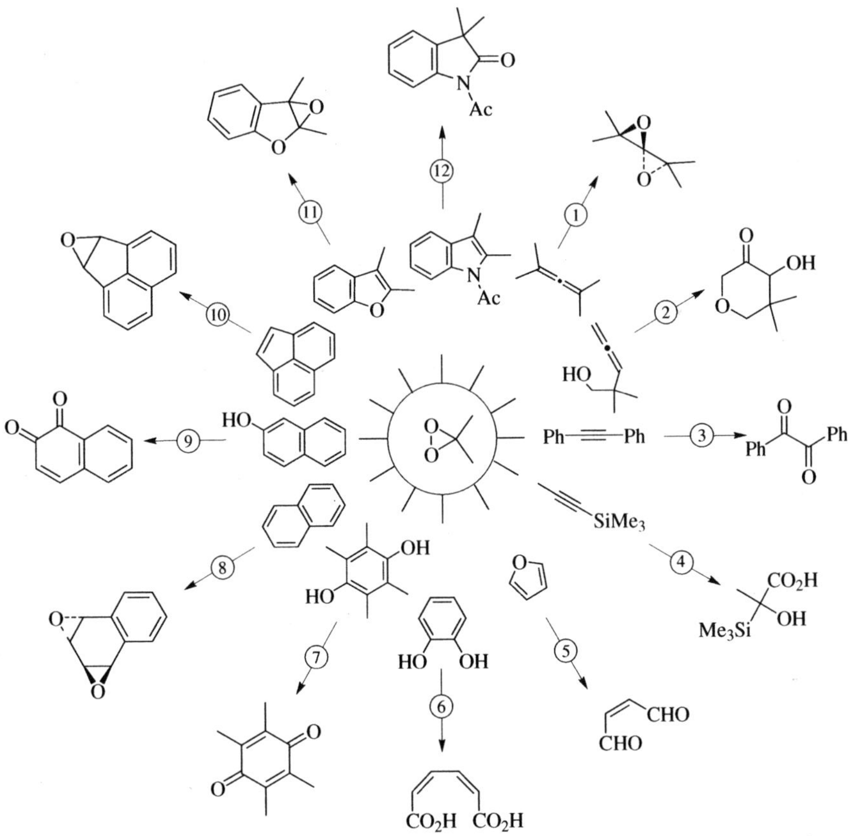

Scheme 6. An overview of dioxirane oxidations of allenes, alkynes, and arenes.

in Eq. 4 is oxidized with an acetone solution of DMD,[57] the hydroxyfuranone is obtained as the major product (upper route), together with minor amounts of open-chain material. In the presence of catalytic amounts of *p*-TsOH (lower route), however, the above hydroxy-substituted heterocycle is a minor product. On protection of the hydroxy functionality as a silyl ether, these complications are avoided, and the spiro-bisepoxide is obtained (Eq. 5).[57]

acetone, rt, 5 min: (55%) + (10%)

acetone/CH_2Cl_2, *p*-TsOH, rt, 10 min: (5%) + (80%) + (10%)

(Eq. 4)

acetone, rt: (100%) (Eq. 5)

Other reactive functionalities in the allene, such as amine,[58] amide,[58] aldehyde,[59] carboxylic acid,[60] oxime,[58] and even ketone[59] groups, will open the spiro-bisepoxide intermediate and lead to substrate-specific products. These multifunctionalized heterocyclic and acyclic products should be of potential use in organic synthesis.

The oxidation of (bis)allenes and higher cumulenes has been much less studied. Nevertheless, one example of the DMD oxidation of a bisallene yields a cyclopentenone and an exocyclic epoxide (Eq. 6).[61] The epoxide presumably arises from further oxidation of the exomethylenic double bond. A higher-order cumulene has also been oxidized with DMD to give an unusual cyclopropanone in 38% yield (Eq. 7).[62]

(1 eq), acetone, rt: + (—) (Eq. 6)

(1 eq), acetone, rt: (38%) (Eq. 7)

The oxidation of alkynes appears to be little studied, most likely because of the complexity of the product composition obtained in this oxidation. The oxyfunctionalized intermediates, presumably oxirenes, are much more labile than allene oxides and have so far not been detected. This oxidation is usually not useful for synthetic purposes since extensively rearranged products are obtained in poor yields, especially with open-chain alkynes (Eq. 8).[63] The cyclic alkyne in Eq. 9, however, gives well-defined bicyclic rearrangement products in good yields when oxidized with TFD at 0° (Eq. 9).[64]

t-Bu—≡—Bu-t → (acetone, rt, 140 h) (22%) + (8%) + (6%)

(Eq. 8)

→ ($1,1,1$-trifluoroacetone/CH_2Cl_2, 0°, 3 min) (83%) + (12%) (Eq. 9)

In contrast, dioxirane oxidation of arenes is a useful reaction and has been thoroughly studied. Among the arenes and heteroarenes, benzene is the most difficult to oxidize. It is inert toward DMD oxidation and, thus, it has been employed as solvent in the biphasic oxidation mode with in situ generated DMD.[47] Nonetheless, benzene has been oxidized with the more reactive TFD in a fluorinated solvent, affording two isomeric dialdehydes in low yield (Eq. 10).[65]

→ (1,1,1-trifluoroacetone/F113, 0°, 6 h) CHO / CHO + H ... H (4%) (Eq. 10)

40 : 60

Electron-rich substituted benzenes are more reactive. For example, phenols and naphthols have been oxidized by DMD to the corresponding quinones.[17,66] Oxidation of the arene substrate in Eq. 11 leads to the tris(epoxide) in high yield.[67] This transformation demonstrates the usefulness of dioxiranes in the oxidation of arenes, since such a tris(epoxide) would be difficult to make by any other route.

→ (acetone) (84%) (Eq. 11)

Recently, methoxy-substituted benzenes have been hydroxylated to phenols with isolated DMD under acidic conditions at subambient temperatures (Eq. 12).[68]

The overall reaction at first appears to be a direct CH insertion, but actually epoxidation takes place followed by an acid-catalyzed rearrangement of the intermediary epoxide.

acetone/2 N HCl, –30° (100%) (Eq. 12)

Indene, naphthalene, and polycyclic arenes are more reactive and thus susceptible to both DMD and TFD oxidation. For example, the tetracyclic arene in Eq. 13 is oxidized by DMD to furnish the corresponding epoxide.[69] Such arene epoxides are of special interest since they are biologically active metabolites of carcinogenic polycyclic aromatic hydrocarbons.[69,70]

acetone, 0° to 5°, 12 h (—) (Eq. 13)

A highlight of arene oxidation by dioxirane reagents is that of the fullerene C_{60}. DMD leads mainly to the monoxide,[71] but the more reactive TFD yields dioxides and even some trioxides of C_{60} (Eq. 14).[72]

(4 eq) 1,1,1-trifluoroacetone, 1,2-dichlorobenzene, 0°, 5 to 6 min (20%) +

(Eq. 14)

(21%) + isomeric $C_{60}O_2$ + isomeric $C_{60}O_3$

Furan-type heteroarenes are usually more reactive towards dioxirane oxidation than arenes. When subjected to oxidation by DMD (isol.), furan and its 2,5-disubstituted derivatives[73,74] lead to the ring-opened enediones shown in Eq. 15,[73] which are useful building blocks in synthesis.[73,75] The intermediary mono-epoxide of 2,3-dimethylfuran is presumably involved in the epoxidation with d_6-DMD (prepared in d_6-acetone), but could not be detected by NMR spectroscopy even at −100°; only the rearrangement product hex-3-ene-2,5-dione was observed.[76]

acetone, rt (>95%) R^1, R^2= H, alkyl (Eq. 15)

Related benzofurans also form labile epoxides when epoxidized by dry dioxirane solutions at low temperature under an inert atmosphere; they are sufficiently persistent to be detected at low temperatures.[77−86] Depending on the substituents of the heterocycle, the epoxide may undergo opening to dicarbonyl products. Thus, the labile epoxide derived from 2,3-dimethylbenzofuran (Eq. 16) rearranges at −20° to the *o*-quinomethide.[79,87,88] To characterize this epoxide by NMR spectroscopy, fully deuterated DMD was employed for the oxidation; the epoxide was directly detected in situ at −78° without work-up.[79] This example further emphasizes the importance and convenience of isolated dioxiranes for the synthesis of exceedingly labile oxy-functionalized substances.

acetone-d_6/CH_2Cl_2, N_2, −20°, ca. 20 h (97%) (Eq. 16)

In view of the high reactivity of the benzofuran oxides, ring-opened products are formed in the presence of nucleophiles. For example, when the epoxidation of the structurally related 8-methoxypsoralen was carried out in the presence of MeOH, the hydroxy ether was obtained in good yield as ring-opened product (Eq. 17).[85] Elevated temperatures lead to rearrangement products as in the oxidation of benzofuran. Thus, when the reaction is carried out at 0°, the rearranged product in Eq. 18 is obtained from the intermediary epoxide by migration of the hydroxymethyl group.[89]

acetone/MeOH, rt, 19 h (83%) (Eq. 17)

acetone/CH_2Cl_2, N_2, 0°, 3 h (89%) (Eq. 18)

Dioxirane oxidation of the aromatic ring in nitrogen-containing heteroarenes may be even more complex. Since amine nitrogen atoms are more nucleophilic than an arene C=C double bond, N-oxidation usually precedes epoxidation. For example, the oxidation of pyridines takes place at the nitrogen atom to give the corresponding *N*-oxides as the sole products (see the section on Heteroatom

Substrates). Nonetheless, acetylation of the nitrogen functionality may sufficiently suppress N-oxidation, as illustrated for the *N*-acetylated indole in Eq. 19.[90,91]

acetone-d_6, −78° to −40°, 10 to 30 min (95%)

acetone, 0°, 0.5 h (89%)

(Eq. 19)

Analogous to the benzofuran example above, the corresponding labile epoxide is produced (upper pathway), when the indole is oxidized with DMD at low temperature (−78°).[90] In contrast, the rearranged product[91] is obtained when the oxidation is run at subambient temperature (lower pathway). The latter rearrangement has been used for the synthesis of spiro lactams (Eq. 20).[92] Variation of the substitution pattern of indoles may permit other rearrangements to take place. For example, the oxidation of the indole shown in Eq. 21 yields the respective benzopyrroldihydroindole instead of a lactam.[92] The minor product presumably results from overoxidation of the pyrrolidone enol tautomer.

acetone/CH_2Cl_2, 20° (99%) (Eq. 20)

acetone/CH_2Cl_2, −78° (>98%) 91:9 (Eq. 21)

Currently little is known about dioxirane oxidations of unprotected indoles. One example is given in Eq. 22, for which the diastereomeric lactams are obtained.[93] Examples of dioxirane oxidations of heteroarenes with more than one nitrogen atom are also scarce. An example is the DMD oxidation of a substituted imidazole (Eq. 23), which furnishes an imidazolone in moderate yield.[93] In contrast, the DMD oxidation of 1,2,4-triazole results in a complex mixture of unidentified products.[94]

acetone/CH_2Cl_2, 10°, 2 d (42%) (42%) (Eq. 22)

Boc-NH-CH(CO₂Me)-CH₂-(imidazolyl) → [dimethyldioxirane, acetone/CH_2Cl_2, 10°, 2 d] → (55%) (Eq. 23)

The oxidation of sulfur-containing heteroarenes such as thiophenes takes place only on sulfur, as will be discussed in the next section.

Heteroatom Substrates

Electrophilic dioxiranes are particularly reactive towards heteroatom substrates, for which the electron lone-pair serves as the nucleophile in the oxygen transfer. Because of the importance of their oxidation products, substrates that contain nitrogen, sulfur, and C=Y functionalities are among the best studied. Some typical examples are collected in the rosette of Scheme 7. These oxidations will be discussed separately according to the type of heteroatom that is oxidized, mainly nitrogen and sulfur.

Scheme 7. An overview of dioxirane oxidations of heteroatom substrates.

Nitrogen. Irrespective of whether the nitrogen atom in the substrate is sp^3 or sp^2 hybridized, it is readily oxidized by either isolated or in situ generated dioxiranes. The outcome of the dioxirane oxidation of an sp^3-hybridized nitrogen depends on the structure of the amine. Tertiary amines give cleanly the *N*-oxides, as illustrated in Eq. 24. Particularly noteworthy is the selective oxidation of the nitrogen atom, without epoxidation of the double bond.[95] Oxidation of secondary amines is more complex. If the secondary amine does not bear α-hydrogen atoms, the product is usually the expected hydroxylamine; however, the latter may be further oxidized by excess DMD to the corresponding nitroxyl radical. An illustrative example is shown in Scheme 8;[96,97] it should be noted that the alcohol functionality survives, which demonstrates the greater reactivity of the amino group towards DMD oxidation.

acetone, 0°, < 1 h (ca. 100%) (Eq. 24)

If the secondary amine bears α-hydrogen atoms, the product can be either a nitrone or a hydroxylamine, depending on the reaction conditions. For example, *N*,*N*-dibenzylamine is oxidized to the hydroxylamine with DMD at 0° (upper pathway),[96] whereas the nitrone is obtained upon treatment with two equivalents of cyclohexanone dioxirane at −20° (lower pathway) (Scheme 9).[98] Other products may also be obtained, as illustrated in the DMD oxidation of piperidine; here the hydroxamic acid is obtained in good yield (Eq. 25).[99] Primary amines usually give a complex mixture of products, which may contain hydroxylamine, oxime, nitroso, nitro, and sometimes even azoxy compounds. However, reaction conditions may be chosen to favor one of these products. Thus, a preparatively valuable method is the DMD oxidation of aliphatic and aromatic amines to the corresponding nitro compounds. For example, the optically active amine in Scheme 10 is oxidized with excess DMD to the respective nitroalkane in quantitative yield

(1.2 eq) acetone, 0°, 2 h (99%)

(2 eq) acetone, 0°, 30 min (100%)

Scheme 8. DMD oxidation of a secondary amine.

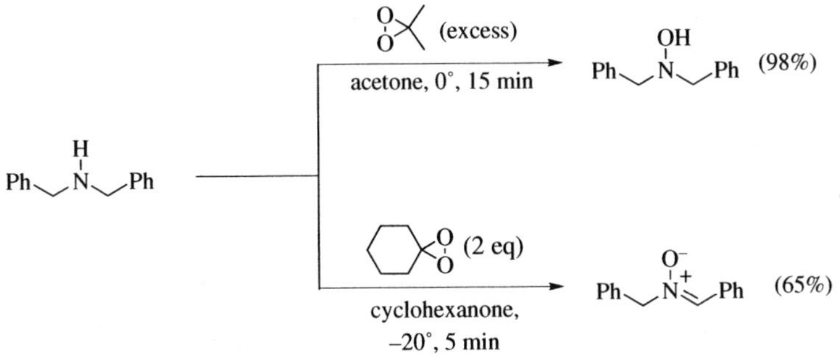

(excess)
acetone, 0°, 15 min
(98%)
(2 eq)
cyclohexanone, –20°, 5 min
(65%)

Scheme 9. DMD oxidation of *N*,*N*-dibenzylamine.

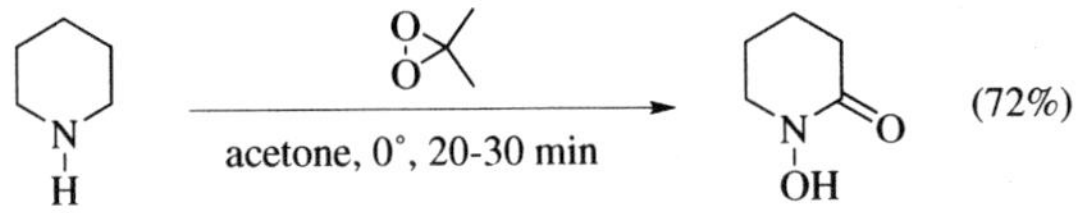

acetone, 0°, 20-30 min (72%) (Eq. 25)

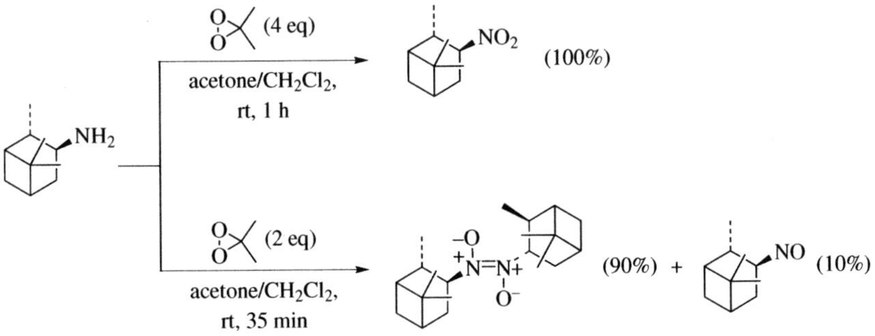

(4 eq)
acetone/CH_2Cl_2, rt, 1 h
(100%)
(2 eq)
acetone/CH_2Cl_2, rt, 35 min
(90%) + (10%)

Scheme 10. DMD oxidation of a primary amine.

(upper pathway).[22] If an insufficient amount (2 equiv.) of DMD is used, the nitroso and the azoxy products are obtained instead (lower pathway).[22]

At subambient temperature, primary amines are converted either into hydroxylamines or into oximes by DMD. This transformation can be synthetically useful, as, for example, in the oxidation of the amino sugar derivative in Eq. 26 to the corresponding hydroxylamine in good yield. However, the related amino sugar in Eq. 27 is oxidized to the oxime under identical reaction conditions.[100] Clearly, structural features of the amino sugar strongly influence the course of the oxidation.

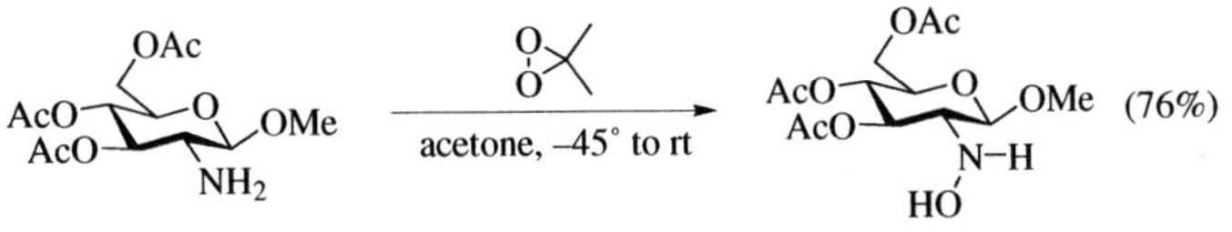

acetone, –45° to rt (76%) (Eq. 26)

acetone, –45° to rt (—) (Eq. 27)

Although amines are readily oxidized by dioxiranes, the corresponding amides persist and are, therefore, often used to protect amines. Protection of the amine by protonation may also be employed, but since ammonium salts dissociate into the free amine, it is essential to conduct the oxidation under strongly acidic conditions; otherwise the oxidation will still take place slowly (Eq. 28).[101]

acetone/H_2O, rt, 23 h (82%) (Eq. 28)

The oxidation of sp^2-hybridized nitrogen generally falls into two categories: namely, heteroarenes and isolated C=N bonds. The nitrogen-containing arenes usually afford *N*-oxides. Thus, pyridines are generally oxidized to pyridine *N*-oxides. An interesting example is given in Eq. 29 in which the pyridine nitrogen atom is selectively oxidized rather than the dimethylamino group.[21] Since the resulting *N*-oxide may further react with DMD to give *N*,*N*-dimethylaminopyridine and singlet oxygen, as illustrated in Scheme 4, it is not possible to achieve full conversion of the substrate in such situations.[21]

acetone (—) (Eq. 29)

Oxidation of the nitrogen atom in isolated C=N bonds is complicated by the fact that the initial oxidation product may be further oxidized by dioxirane with cleavage of the C=N bond. For example, N-alkylated imines are oxidized by DMD to a mixture of nitrones and ketones (Eq. 30)[102]. When imines without substituents on the nitrogen atom are treated with DMD, the main products are oximes,[102] which in turn may be further oxidized by excess dioxirane to the corresponding ketones or aldehydes. The method thus serves as a useful deprotection method for oximes. Similarly, diazoalkanes yield cleavage products (ketones, aldehydes, or their hydrates),[103–106] when oxidized by DMD (Eq. 31).[103] This oxidation has been employed to prepare tricarbonyl compounds (Eq. 32),[105] and as such comprises a convenient method for the preparation of these reactive substances.

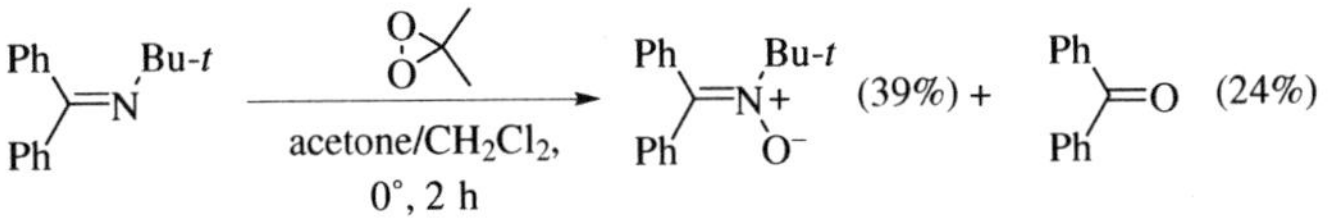

(Eq. 30)

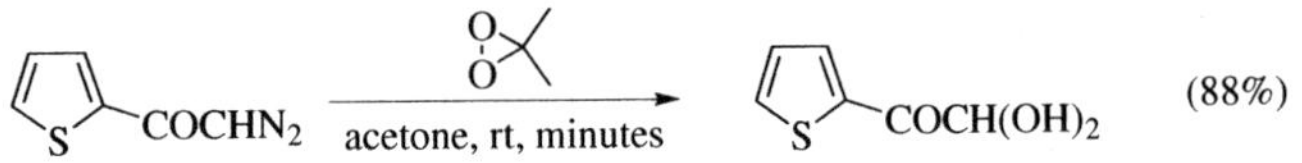

(Eq. 31)

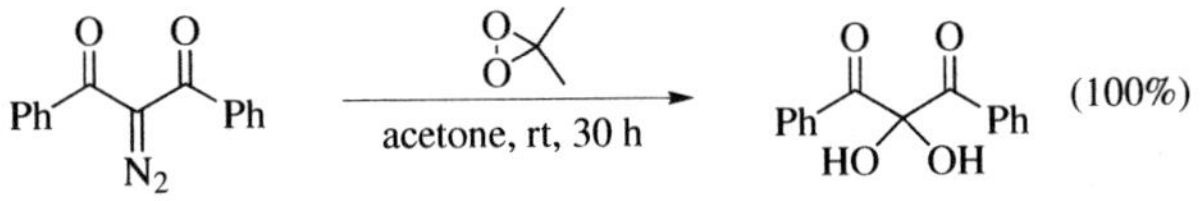

(Eq. 32)

The DMD oxidation of nitronate anions, generated in situ from nitroalkanes, also affords carbonyl compounds through cleavage of the carbon-nitrogen bond (Eq. 33),[107] and in effect constitutes an oxidative Nef reaction.[108] This efficient new method has been successfully employed in the synthesis of the AB ring system of norzoanthamine (Eq. 34).[109] In a similar manner, the σ^H adducts generated in situ from nitroarenes by the addition of a carbanion are efficiently oxidized by DMD to the corresponding phenols (Eq. 35).[110,111] This transformation comprises the first method for the direct oxidation of nitroarenes to phenols.

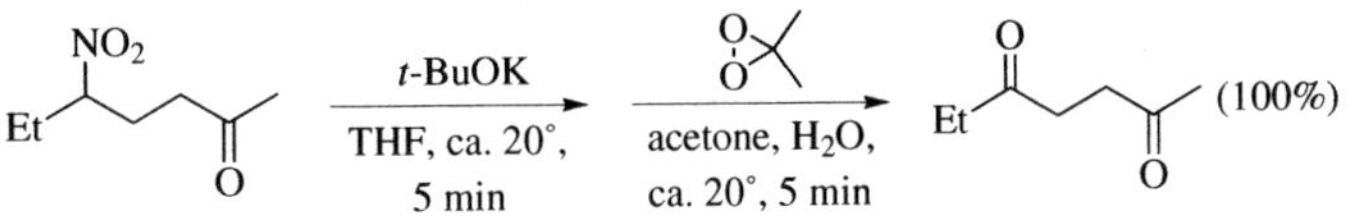

(Eq. 33)

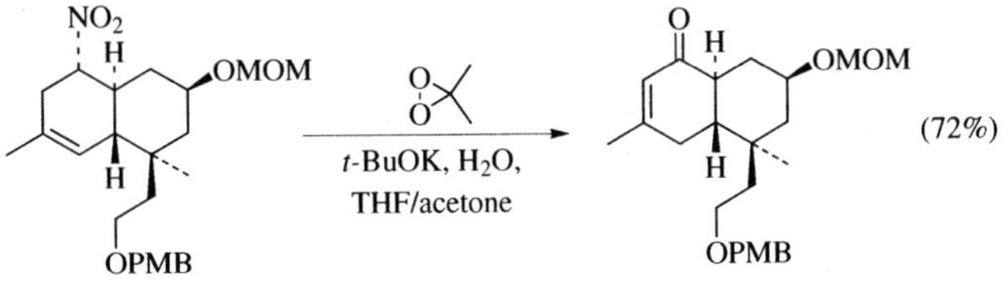

(Eq. 34)

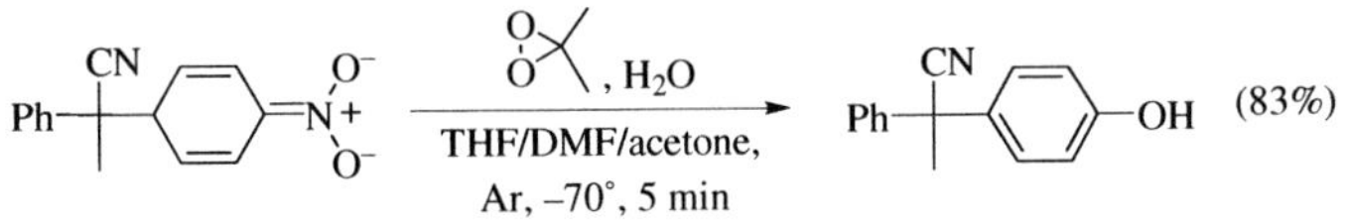

(Eq. 35)

Similar oxidative cleavage reactions have been reported for several α-amino acids. Thus, when arginine is oxidized by DMD under in situ conditions, 4-guanidinobutanoic acid is obtained in moderate yield (Eq. 36).[112]

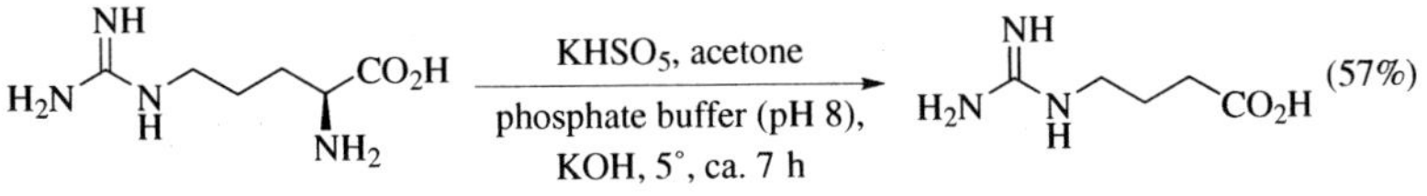

(Eq. 36)

The oxidation of N,N-dimethylhydrazone by DMD at 0° produces the corresponding nitriles (Eq. 37)[113]. What is remarkable about this useful oxidation is that no racemization takes place at the stereogenic center, which again emphasizes the mild reaction conditions.

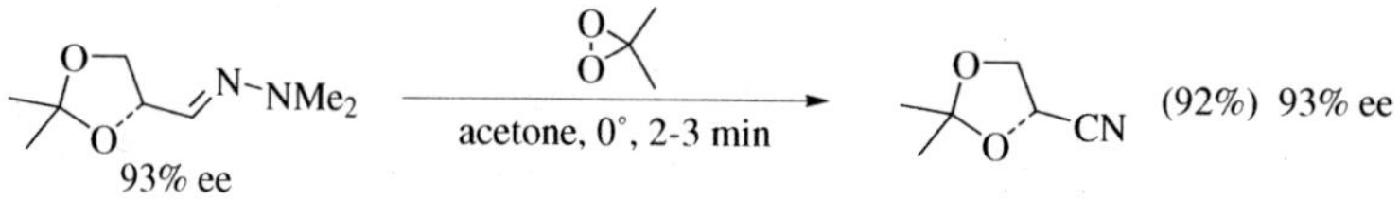

(Eq. 37)

Dioxirane-mediated epoxidation of the C=N bond of imines to oxaziridines is rare, but examples are known for the in situ method (Eq. 38).[114]

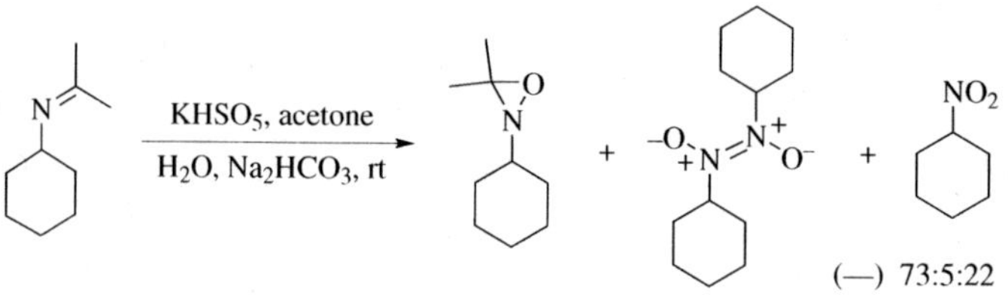

(Eq. 38)

Sulfur and Selenium. In general, sulfur-containing substrates are more reactive toward dioxirane than nitrogen compounds. Thus, the oxidation of aliphatic thiols by DMD (isol.) at low temperature leads to sulfinic acids in good yield (Eq. 39).[115] However, under the same reaction conditions, benzyl mercaptan affords a complex mixture of benzylsulfinic acid, benzylsulfonic acid, dibenzyl disulfide, dibenzyl thiosulfonate, and benzaldehyde.[115] Similarly, DMD (isol.) oxidation of *p*-methylthiophenol displays this complexity (Eq. 40).[115] The latter oxidations are, thus, not synthetically useful, given the multiple products formed; however, the oxidation of 9*H*-purine-6-thiols in the presence of an amine nucleophile produces ribonucleoside analogs in useful yields.[116–120] An example of such a mercaptan oxidation with DMD (isol.) in the presence of methylamine is illustrated in Eq. 41. This reaction confirms that thiols are more reactive toward DMD oxidation than primary amines, as would be expected from the nucleophilicity of mercaptans compared with amines.[15]

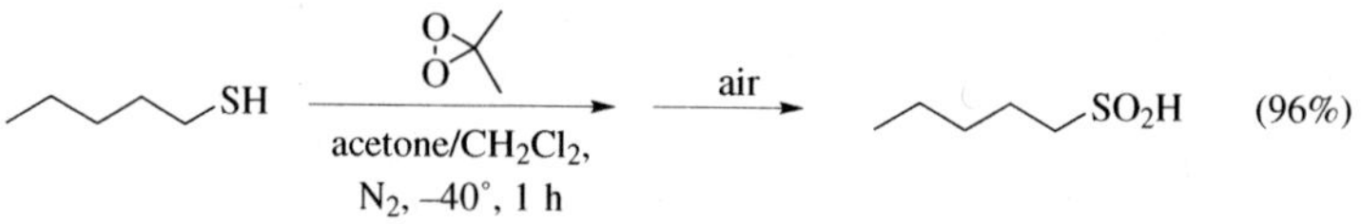

(Eq. 39)

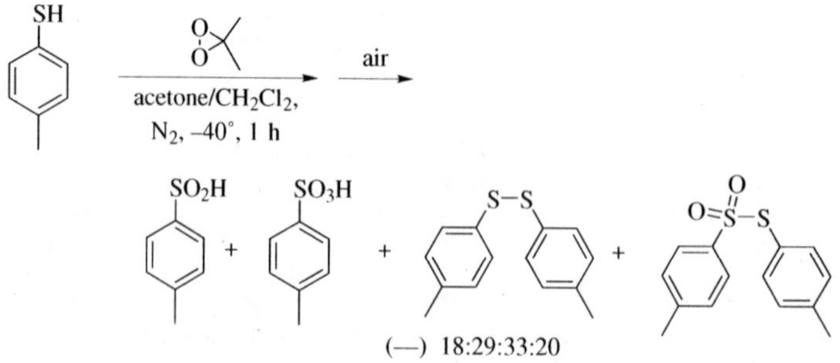

(Eq. 40)

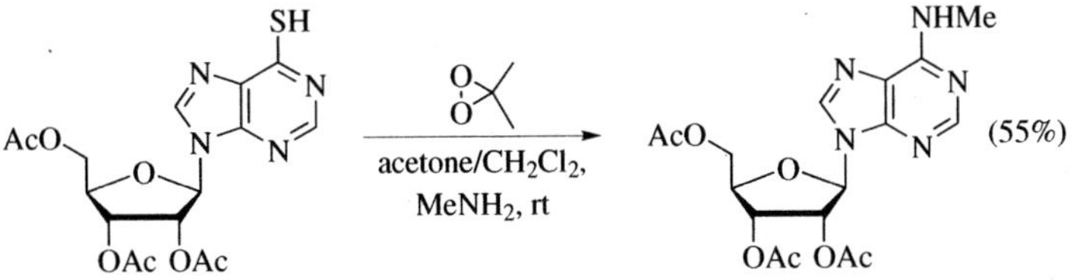

acetone/CH_2Cl_2, $MeNH_2$, rt (55%) (Eq. 41)

A sulfide may be oxidized by dioxirane to either the sulfoxide or sulfone, depending on the number of equivalents of the oxidant employed (Scheme 3). As pointed out already, sulfides are more readily oxidized than sulfoxides and, therefore, the sulfoxide product may be selectively obtained by employing only one equivalent of DMD (Eq. 42).[121] Since methyl phenyl sulfide is prochiral, DMD oxidation affords the racemic sulfoxide. Similarly, if a chiral sulfide is employed, diastereomeric sulfoxides are expected. For example, DMD (in situ) oxidation of the tetrahydrothiophene derivative shown in Eq. 43 furnishes mainly one diastereomeric *S*-oxide with excellent stereocontrol (94 : 6).[122] Such high diastereoselective oxidations are not general, as illustrated for the two similar substrates shown in Scheme 11.[123,124] For the five-membered-ring sulfide (upper equation), the cis diastereomer is formed exclusively, whereas for the six-membered cyclic sulfide (lower equation) both diastereomeric sulfoxides are obtained in about equal amounts.

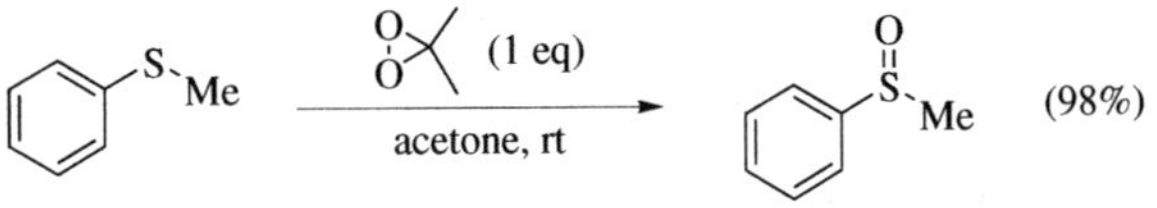

(1 eq), acetone, rt (98%) (Eq. 42)

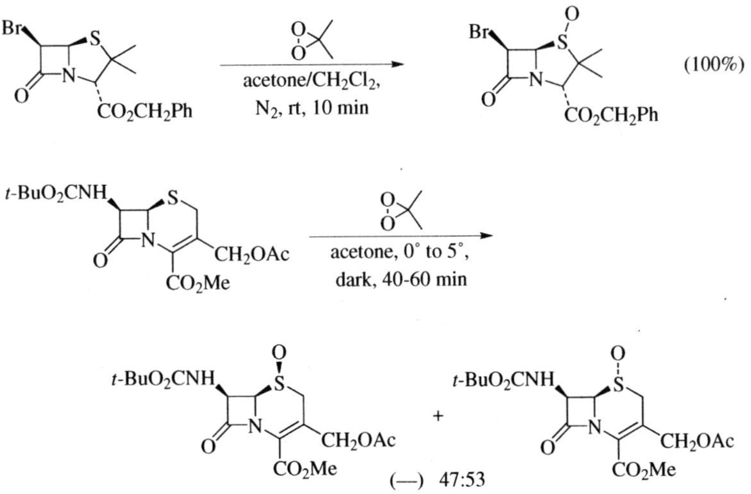

acetone/CH_2Cl_2, N_2, rt, 10 min (100%)

acetone, 0° to 5°, dark, 40-60 min

(—) 47:53

Scheme 11. Sulfoxidation of cyclic sulfides by DMD (isol.).

TsO ... $\xrightarrow[\text{acetone, } H_2O, \text{ 0°, 40 min}]{KHSO_5}$ ... (90%) 94:6 (Eq. 43)

There is only one report on the diastereoselective oxidation of a chiral acyclic sulfide by DMD. In this case, the exocyclic sulfide is oxidized with low diastereoselectivity to the corresponding sulfoxide.[125]

To achieve better diastereomeric control, prochiral sulfides have been coordinated to chiral organometallic complexes and then oxidized with DMD (isol.).[126–129] As is evident from Eq. 44,[126] the diastereoselectivity depends highly on the structure of the sulfide, and as such, the outcome is difficult to predict. After decomplexation, enantiomerically enriched sulfoxides are obtained. The overall process qualifies, therefore, as an indirect enantioselective oxidation.

Ph_2P---Ru–S(Me)(R) PF_6^- $\xrightarrow[\text{45 min}]{\text{acetone, 0°,}}$ I + II (Eq. 44)

R		I:II
i-Pr	(95%)	7:93
Ph	(90%)	73:27
$PhCH_2$	(95%)	99:1

Enantioselective oxidation of sulfides by achiral dioxirane may be performed with bovine serum albumin (BSA). In the presence of this protein, prochiral sulfides have been oxidized by TFD (in situ) to enantioenriched sulfoxides in moderate to good enantioselectivities (up to 89% ee).[130,131] A typical example is shown in Eq. 45.[130,131]

PhS-Pr-*i* $\xrightarrow[NaHCO_3, H_2O, 4°, 5 \text{ min}]{BSA, KHSO_5, CH_3COCF_3}$ PhS(O)-Pr-*i* (67%) 89% ee (Eq. 45)

Enantioselective oxidation of a prochiral sulfide with an optically active dioxirane has not yet been accomplished. An attempt with methyl phenyl sulfide as substrate and in situ generated fructose-derived dioxirane, which has been successfully employed in asymmetric epoxidation,[15,132] resulted in an enantiomeric excess of less than 5%.[94]

If two or more equivalents of dioxirane are used, the sulfone is the main product. An illustrative example with DMD (isol.) is shown in Eq. 46.[124] It is noteworthy that both alkenyl[124] and alkynyl[133] sulfides are oxidized by dioxirane to the corresponding sulfones without epoxidation of the C–C multiple bonds (Eq. 47).[124] As expected, the dioxirane oxidation of a sulfoxide affords the corresponding sulfone.[134]

PhCH$_2$CONH ... CO$_2$CH$_2$OAc —[dimethyldioxirane; acetone, dark, 0-5°, 40 to 60 min]→ sulfone, CO$_2$CH$_2$OAc (100%) (Eq. 46)

PhOCH$_2$CONH ... CO$_2$Me —[dimethyldioxirane; acetone, 0-5° 60 min]→ sulfone, CO$_2$Me (95%) (Eq. 47)

Oxidation of disulfides, trisulfides, and polysulfides by DMD usually leads to a mixture of multiple products,[135,136] and is not of synthetic importance. Useful selectivities have been observed with DMD (isol.) only when one of the sulfur atoms in the disulfide is substituted by an electron-withdrawing group (Eq. 48).[137]

TsHN ... —[dimethyldioxirane; acetone/CH$_2$Cl$_2$, –78°]→ TsHN ... + TsHN ... (—) 94:6 (Eq. 48)

Dioxirane oxidation of sulfur-containing heteroarenes, such as substituted thiophenes, takes place exclusively on the sulfur atom.[138–141] At subambient temperature, the oxidation products are usually the corresponding thiophene 1,1-dioxides. Recently, the oxidation of thiophene with DMD (isol.) below −40° afforded the elusive parent thiophene 1,1-dioxide, which was isolated (Eq. 49).[140] When the oxidation was carried out at higher temperature, the thiophene 1,1-dioxide decomposed to other products.[140,141]

thiophene —[dimethyldioxirane; acetone, < –40°]→ thiophene 1,1-dioxide (20-30%) (Eq. 49)

Dioxirane oxidation of the C=S bond in the thiourea functionality of some heterocycles leads to desulfurization products. A typical example is shown for a cyclic thiourea with DMD (isol.) in Eq. 50.[142] This oxidation is, however, complex since disulfides and other products may be formed (Scheme 12).[142–144]

cyclic thiourea —[dimethyldioxirane; acetone/MeOH, rt]→ imidazoline (53%) + imidazolidinone (15%) (Eq. 50)

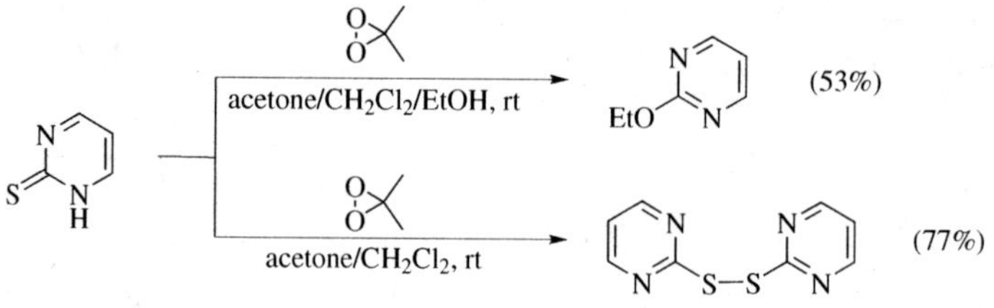

acetone/CH_2Cl_2/EtOH, rt (53%)

acetone/CH_2Cl_2, rt (77%)

Scheme 12. Oxidation of 1*H*-pyrimidine-2-thione by DMD (isol.).

The dioxirane oxidation of the C=S functionality in thioketones to thioketone *S*-oxides is quite rare. Examples are collected in Scheme 13.[145,146] The yields of the *S*-oxides from bis(*tert*-butyl)thione and adamantane-2-thione by DMD (in situ) are quite low because of further oxidation. Thiobenzaldehyde derivatives with bulky substituents, however, are oxidized by DMD (isol.) to the corresponding *S*-oxides both in good yield and with high diastereoselectivity;[147] further oxidation of the syn diastereomer by DMD (isol.) gives two unusual products in low yields (Scheme 14).[147]

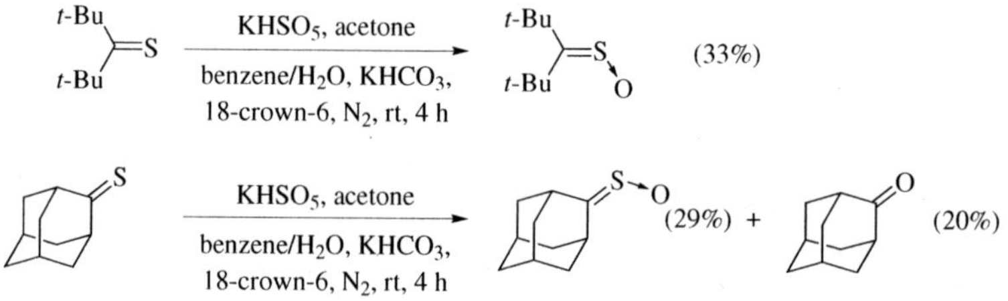

$KHSO_5$, acetone; benzene/H_2O, $KHCO_3$, 18-crown-6, N_2, rt, 4 h (33%)

$KHSO_5$, acetone; benzene/H_2O, $KHCO_3$, 18-crown-6, N_2, rt, 4 h (29%) + (20%)

Scheme 13. Oxidation of thioketones by DMD (in situ).

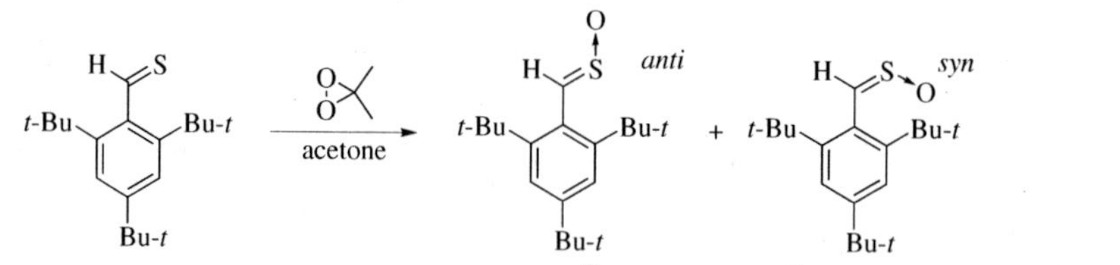

acetone

Temp		anti:syn
rt	(100%)	95:1
–78°	(100%)	100:0

acetone, –78° to rt, overnight (29%) + (9%)

Scheme 14. DMD (isol.) oxidation of a thioaldehyde and its *syn*-*S*-oxide.

Oxidation of N-tosylated or N-acylated sulfilimines with DMD (isol.) takes place selectively on sulfur to give sulfoximines in good yields (Eq. 51).[148] The oxidation by DMD (isol.) is stereoselective so that a chiral sulfoximine may be obtained from an optically active sulfilimine with complete preservation of the initial enantiomeric purity (Eq. 52).[148]

acetone, rt, 4 h (86%) (Eq. 51)

acetone, 0° to rt (97%) 80% ee (Eq. 52)

p-MeC$_6$H$_4$ 80% ee

With the more reactive TFD (isol.), some sulfoxides and sulfones are also formed via cleavage of the S=N bond.[148] When the N atom of the sulfilimine bears a heterocyclic ring instead of an acyl or tosyl group, DMD (isol.) oxidation leads to a mixture of S- and N-oxidations, as evidenced by the presence of the nitro product (Eq. 53).[149] With wet DMD (isol.), N-oxidation represents the major process (63% *vs.* 37%).[149] These results suggest that the chemoselectivity of S versus N oxidation by DMD (isol.) of sulfilimine depends on the electron density of the heteroatoms.

acetone, rt, 16 h (63%) + (37%) (Eq. 53)

In the oxidation of phosphine sulfides by DMD (isol.), desulfurization affords the corresponding phosphine oxides in nearly quantitative yields (Eq. 54).[150] Similarly, a thiophosphonate is converted into the phosphonate by oxidation with DMD (isol.), as shown in Eq. 55.[151]

Ph$_3$P=S → acetone/CH$_2$Cl$_2$, rt, 5 min → Ph$_3$P=O (ca. 100%) (Eq. 54)

acetone, rt, 3 h (97%) (Eq. 55)

Selenides react with dioxirane like their sulfur analogs to give selenoxides, but the selenium atom is more reactive than the sulfur atom. Since selenoxides are much more labile than sulfoxides, good yields are usually only obtained acetone/CH_2Cl_2, 0° to rt, 30 min (100%) (Eq. 56)

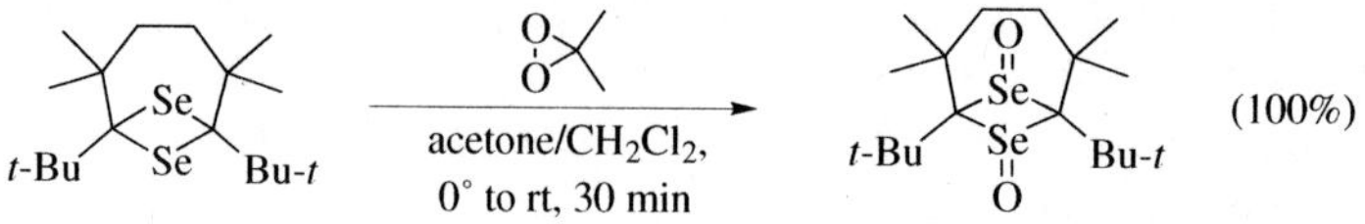

acetone, –78° (—)

1. DMD, acetone, –78°
2. rt (70%)

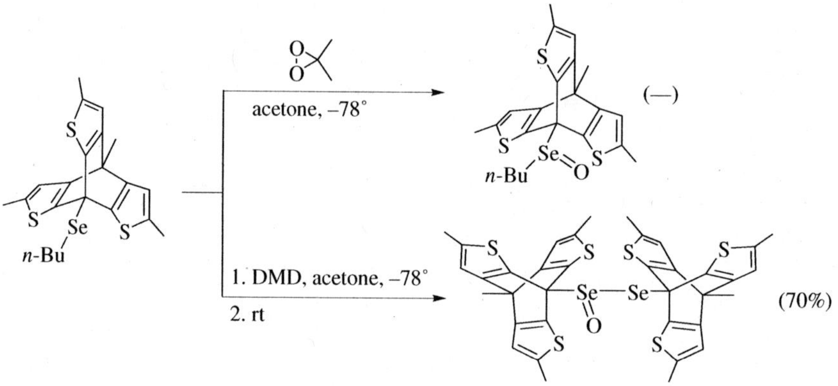

Scheme 15. Oxidation of a selenide by DMD (isol.).

for selenoxides with bulky substituents (Eq. 56).[134] In some cases more complex products may be obtained because of the labile nature of the selenoxides (Scheme 15).[152]

Oxidation of selenophenes by DMD (isol.)[153–155] at subambient temperatures gives selenophene 1-oxides or 1,1-dioxides in good yields. The amount of DMD used determines which product predominates (Scheme 16).[153] These results are comparable to those obtained with their sulfur counterparts.[138–141]

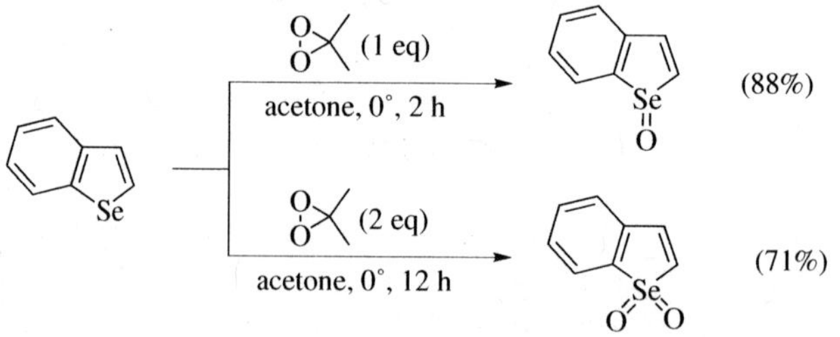

(1 eq) acetone, 0°, 2 h (88%)

(2 eq) acetone, 0°, 12 h (71%)

Scheme 16. Oxidation of benzoselenophene by DMD (isol.).

Phosphorus. Trivalent phosphorus compounds are readily oxidized by various oxidants; however, the oxidation of such substrates by dioxiranes has been sparsely studied. Since only a handful of examples are available in the literature, little may be said about general trends in reactivity and/or selectivity. Clearly, more detailed studies are needed to define the scope and limitations of this oxidation. Nontheless, the phosphorus atom is readily oxidized by other reagents, such that it is questionable whether dioxiranes need to be used.

Triphenylphosphine is a favorite substrate for testing the oxidation of trivalent phosphorus. DMD (isol.) leads to triphosphine oxide quantitatively under a variety of conditions.[121,156] DMD (in situ) has also been used, although the product yield was not specified.[121,157]

The selective oxidation of the phosphite functionality in nucleoside derivatives bearing a modified sugar on the trivalent phosphorus atom produces the corresponding nucleotides in nearly quantitative yields (Eq. 57).[158] In this reaction, the phosphorus atom is, as expected, selectively oxidized by the dioxirane, without epoxidation of the allylic double bond. Furthermore, this phosphorus oxidation may offer an expedient way of synthesizing unusual nucleotides.

NHAc N N O O–P–O O O AcHN O BzO CO_2Me OAc OAc → acetone/CH_2Cl_2, 0°, 10 min → NHAc N N O O O–P–O O O AcHN O BzO CO_2Me OAc OAc (100%) (Eq. 57)

A case of pentavalent phosphorus atom oxidation is documented for α,α-dicarbonylphosphoranes. When oxidized by DMD (isol.), the corresponding tricarbonyl compounds (as their hydrates) are obtained in excellent yields (Eq. 58).[159] These results are comparable with the DMD oxidation of similar diazo compounds[104] discussed above (cf. Eq. 32).

O O Ph OMe PPh_3 → acetone/CH_2Cl_2, rt, 1 h → O O Ph OMe HO OH (100%) (Eq. 58)

Oxygen. Dioxiranes also oxidize several types of oxygen functionalities, which include peroxides, N-oxides, and N-oxyl radicals. In most of these oxidations, molecular oxygen is produced. Thus, $KHSO_5$, which is used as reagent for the generation of dioxiranes, reacts with DMD (isol.) to generate oxygen gas, $KHSO_4$, and acetone as products (Eq. 59).[160] This reaction also takes place under in situ conditions and is responsible for the acetone-catalyzed decomposition of $KHSO_5$, the process that actually led to the discovery of dimethyldioxirane.[161] The molecular oxygen that is released in this oxidation is the electronically excited singlet oxygen, as confirmed by the characteristic chemiluminescence emission.[162] Similarly, the catalytic decomposition of peroxynitrite by ketones,

$$HO-S(O_2)-O-O^- K^+ \xrightarrow[\text{acetone/H}_2\text{O, NaHCO}_3\text{, 20°}]{\text{DMD} \rightarrow \text{acetone}} {}^1O_2 \ (—) + HO-S(O_2)-O^- K^+ \ (—) \quad \text{(Eq. 59)}$$

ONOO⁻ NO₂⁻ OMe NO₂⁻ + ¹O₂ O–O OMe ONOO⁻

Scheme 17. Catalytic decomposition of peroxynitrite by methyl pyruvate.

such as methyl pyruvate, is rationalized in terms of peroxynitrite oxidation by the in situ generated dioxirane (Scheme 17).[163]

Potassium superoxide (KO_2) decomposes DMD in acetone solution, releasing singlet oxygen as detected by chemiluminescence.[94] Furthermore, a catalytic amount of n-Bu_4NI efficiently converts two molecules of TFD into singlet oxygen and trifluoroacetone (Eq. 60).[164]

$$2\ \text{TFD} \xrightarrow[\text{CH}_2\text{Cl}_2\text{, 0°}]{n\text{-Bu}_4\text{N}^+ \text{I}^- \text{(cat.)}} CF_3C(O)CH_3 \ (—) + {}^1O_2 \ (—) \quad \text{(Eq. 60)}$$

Aliphatic and aromatic tertiary amine oxides also react with dioxiranes to generate free amines and singlet oxygen (Eq. 61).[26,165] Since the N-oxide is prepared by DMD oxidation of the tertiary amine, treatment of the latter with excess DMD causes decomposition of the DMD by the in situ formed N-oxide, with concomitant release of O_2.[26] The mechanism of this oxidation is presented in Scheme 4.

$$\text{pyridine } N\text{-oxide} \xrightarrow[\text{CDCl}_3\text{, 20°}]{\text{DMD}} {}^1O_2 \ (—) + \text{pyridine} \ (—) \quad \text{(Eq. 61)}$$

The dioxirane oxidation of an N-oxyl radical is illustrated in Eq. 62.[23,166] The reaction follows a complex radical mechanism to give two O-alkylated products.

$$\text{TEMPO (N–O•)} \xrightarrow[\text{acetone, air or N}_2\text{, 20°, 2-4 h}]{\text{DMD}} \text{N–OMe} \ (98\%) + \text{N–OCH}_2\text{COMe} \ (\leq 1\%) \quad \text{(Eq. 62)}$$

Halogens. There are only a few reports on the dioxirane oxidation of halogen-containing compounds. The oxidation of the chloride ion to the hypochlorite ion by DMD (in situ) has been known since the very beginning of dioxirane chemistry. In fact, this reaction constitutes the first example of a dioxirane oxidation.[161] Iodometry,[121,167] which utilizes the oxidation of iodide anion under acidic conditions, serves as the method for quantitative determination of dioxirane concentration.

Organoiodides are also prone to dioxirane oxidation, as illustrated by the DMD oxidation of iodobenzene to a mixture of iodosobenzene and iodylbenzene (Eq. 63).[168] In contrast, alkyl iodides afford labile primary oxidation products, which eliminate the oxidized iodine functionality resulting in alkenes (Eq. 64).[169] In such oxidations, the alkene product may be converted to an epoxide, as illustrated when the cyclic iodide in Eq. 65 is oxidized by DMD (isol.).[169] The oxidation of iodocyclohexane by DMD (isol.) under nitrogen leads to the iodohydrin and diol as unexpected products (Eq. 66).[168] The formation of iodohydrin, the major product, clearly reveals that hypoiodous acid (HOI) is generated in situ, which in turn adds to the liberated cyclohexene. Indeed, when methyl iodide is oxidized by moist DMD (isol.) at subambient temperature in the presence of cyclohexene, the corresponding iodohydrin is obtained (Eq. 67).[170] When an allene is used as substrate for this reaction, an allylic alcohol with a vinyl iodo functional group is obtained in high yield (Eq. 68).[171]

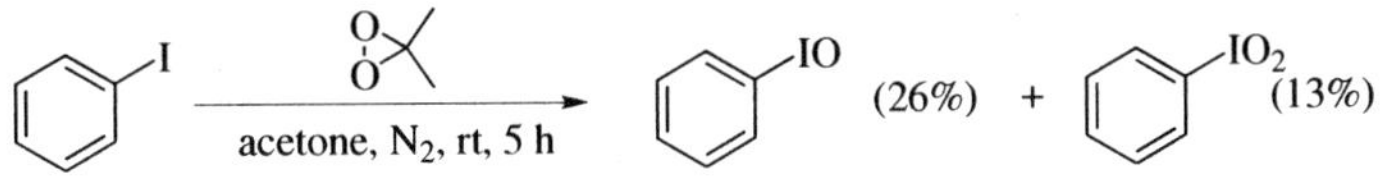

acetone, N_2, rt, 5 h; IO (26%) + IO_2 (13%) (Eq. 63)

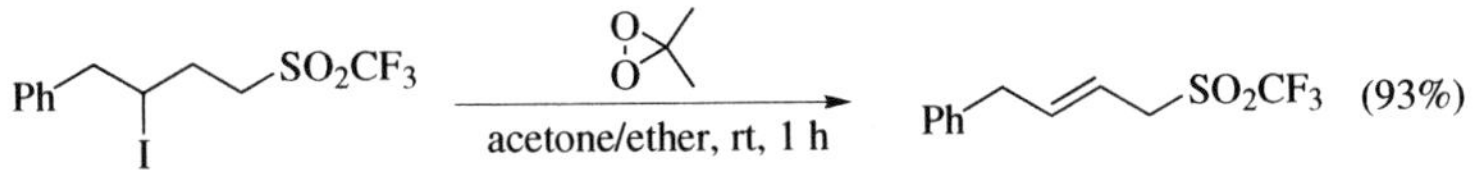

Ph, I, SO_2CF_3; acetone/ether, rt, 1 h; Ph, SO_2CF_3 (93%) (Eq. 64)

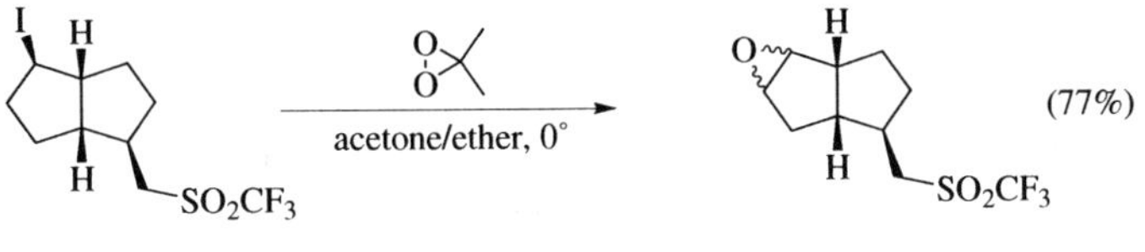

I, H, H, SO_2CF_3; acetone/ether, 0°; O, H, H, SO_2CF_3 (77%) (Eq. 65)

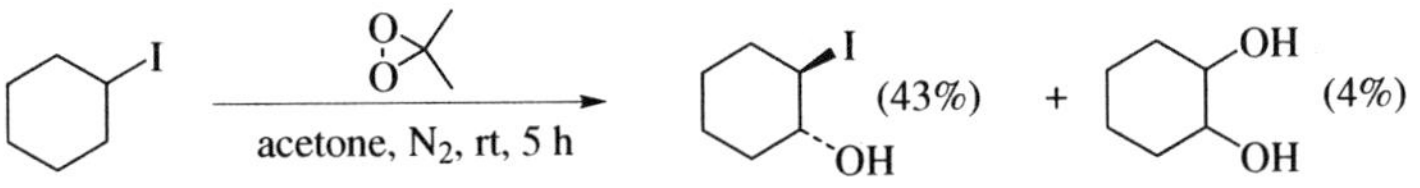

I; acetone, N_2, rt, 5 h; I, OH (43%) + OH, OH (4%) (Eq. 66)

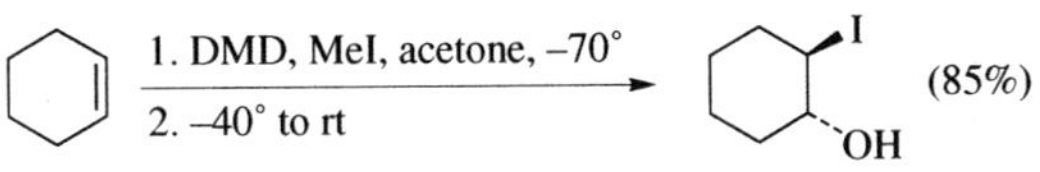

1. DMD, MeI, acetone, –70°; 2. –40° to rt; I, OH (85%) (Eq. 67)

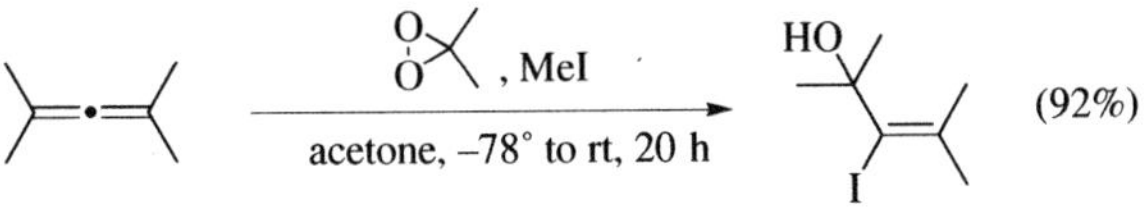

, MeI; acetone, –78° to rt, 20 h; HO, I (92%) (Eq. 68)

Alkanes and Silanes

One of the highlights of dioxirane chemistry is the facile oxidation of C–H and Si–H σ bonds. Some typical examples of these oxidations are collected in the rosette of Scheme 18. Both DMD (isol.) and TFD (isol.) are employed for the oxidation of alkanes; TFD is more effective than DMD. In a few cases in situ generated dioxiranes have also been employed for this purpose.

Alkanes. Usually, alkanes are difficult to functionalize, but dioxiranes, especially TFD (isol.), effect hydroxylation under mild conditions. Their reactivity order follows the sequence primary < secondary < tertiary < benzylic < allylic C–H bonds. Only one example of hydroxylation by TFD (isol.) at a primary position of an unfunctionalized alkane (without secondary and tertiary C–H bonds) appears to have been reported (Eq. 69);[172] in contrast, papers on hydroxylation at a secondary position are relatively abundant. For example, cyclohexane gives cyclohexanone as the only product in high yield (98%) under exceedingly

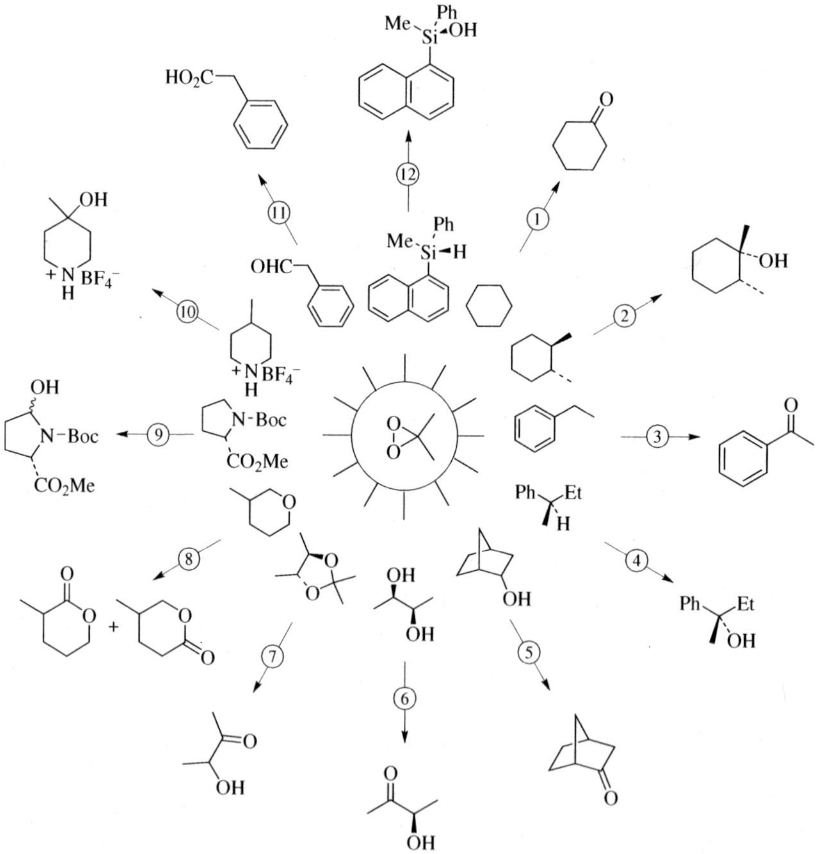

Scheme 18. An overview of dioxirane oxidations of alkanes and silanes.

mild conditions (Eq. 70). The primary oxidation product, namely cyclohexanol, is more reactive toward dioxirane oxidation than cyclohexane. Oxidation of the secondary alcohol may be circumvented by in situ acylation with trifluoroacetic anhydride (Eq. 70). Related cycloalkanes follow this reactivity pattern.[30,172,173]

TFP, $(CF_3CO)_2O$, CH_2Cl_2, 0°, 60 h (> 99%) (Eq. 69)

TFP, $(CF_3CO)_2O$, CH_2Cl_2, 0°, 10 min (> 99%)

TFP/CH_2Cl_2, −22°, 18 min (98%) (Eq. 70)

When *n*-alkanes are used, a mixture of regioisomeric ketones is usually obtained,[172] unless the intermediary secondary alcohols are again protected in situ through acylation (Eq. 71).[173] Oxidation of bicyclic substrates usually affords a mixture of diastereomers (Eq. 72).[30] That the hydroxylation of a tertiary C–H bond is preferred over primary and secondary C–H bonds is exemplified in the oxidation of *cis*-1,2-dimethylcyclohexane by either DMD (isol.)[174] or TFD (isol.) (Eq. 73).[30] The resulting tertiary alcohol of this C–H insertion also demonstrates that oxygen transfer takes place stereoselectively, i.e., with complete retention at the stereogenic center. Absolute stereoretention has been rigorously confirmed by employing optically active 2-phenylbutane as substrate (Eq. 74).[31]

TFP, $(CF_3CO)_2O$, CH_2Cl_2, 0°, 40 min (58%) + (42%) (Eq. 71)

TFP/CH_2Cl_2, −22°, 90 min (69%) + (6%) + (14%) (Eq. 72)

TFP/CH_2Cl_2, −22°, 4 min (95%)

acetone, 22°, dark, 18 h (ca. 100%) (Eq. 73)

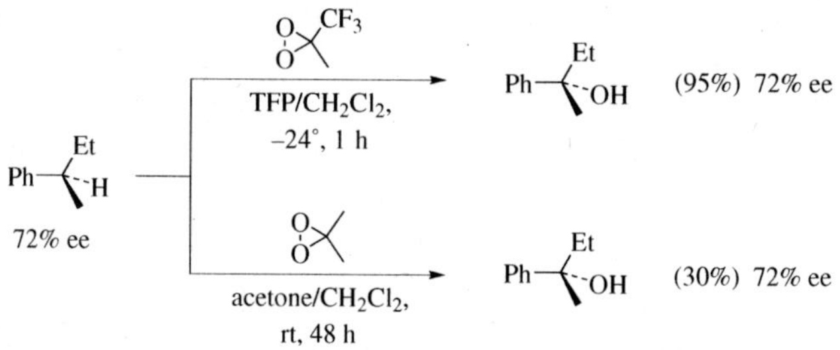

(Eq. 74)

Benzylic C–H bonds are particularly reactive toward dioxirane oxidation, with numerous examples documented in the literature.[175] A preparatively useful approach is shown in Eq. 75, in which a benzhydryl C–H bond is oxyfunctionalized.[176]

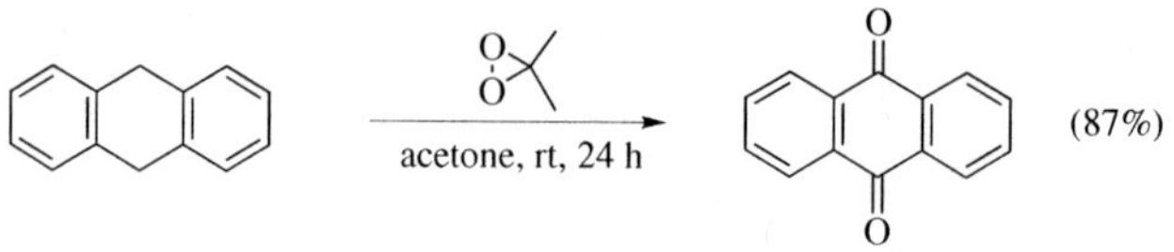

(Eq. 75)

The oxidation of alkanes by dioxiranes is a convenient and useful method in organic synthesis. For example, the polycyclic substrate in Eq. 76[177] is hydroxylated in near quantitative yield by DMD (isol.). Such a transformation would be difficult to realize with conventional oxidants. Similarly, all four bridgehead positions in adamantane may be hydroxylated by TFD (isol.) on repetitive oxidation, affording the tetrahydroxy derivative.[178]

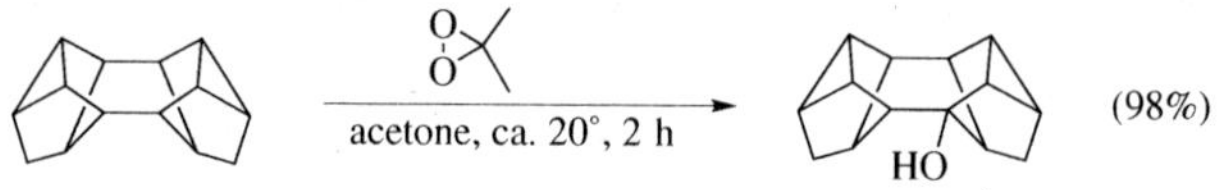

(Eq. 76)

Tertiary C–H bonds in the side chains of several steroids have also been selectively hydroxylated (Eq. 77).[179] In the absence of such C–H bonds, the tertiary C–H bond at the junction of the A and B rings is hydroxylated (Eq. 78).[180] This chemoselectivity derives presumably from steric factors, since the tertiary C–H bond in the side chain is sterically more exposed.

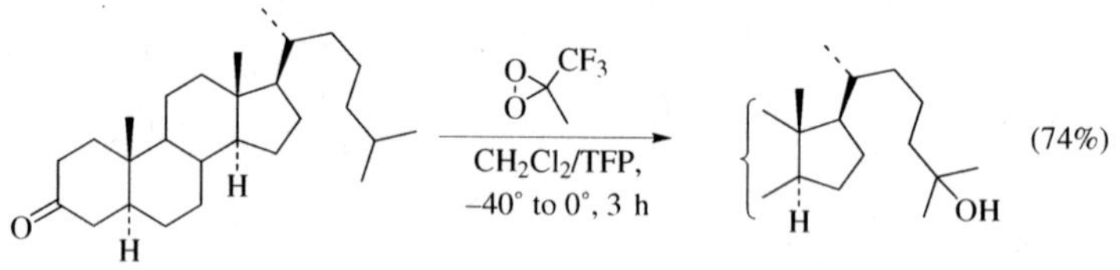

(Eq. 77)

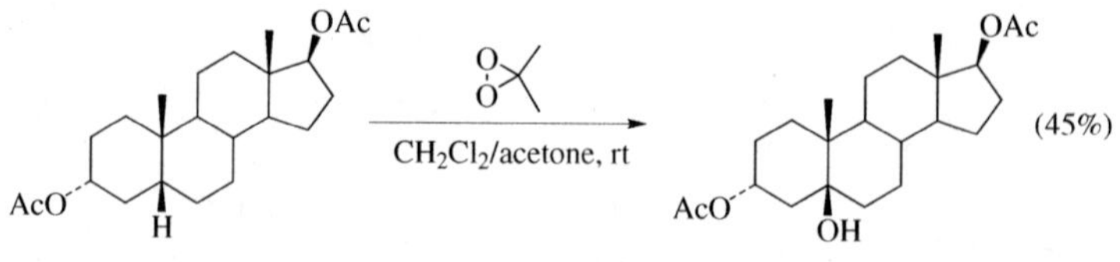

(Eq. 78)

Carbon-hydrogen bonds adjacent to functional groups on the carbon atom are usually more reactive toward dioxirane oxidation, as has already been illustrated for alcohols. Primary alcohols, although much less reactive compared to secondary alcohols, are oxidized to aldehydes and/or carboxylic acids (Eq. 79).[29]

CH_2Cl_2/TFP, –20°, 90 min; CHO (5%) + CO_2H (83%) (Eq. 79)

The same reaction may be responsible for the slow decomposition of DMD (isol.) in methanol.[181] Consequently, primary and secondary alcohols are not recommended as solvents for conducting oxidations with DMD (isol.) and especially TFD (isol.).

The oxidation of secondary alcohols, which is actually a very facile reaction, leads to the corresponding ketone as the product; expectedly, the tertiary hydroxy functionality is not oxidized (Eq. 80).[182] The reactivity difference between primary and secondary alcohols may be exploited for chemoselective oxidation, as shown in Eq. 81.[183] Such oxidative transformations of vicinal diols to the corresponding α-hydroxy carbonyl products are particularly useful in organic synthesis, since the latter comprise valuable building blocks. In view of the mild oxidation conditions, this transformation has been employed for the preparation of optically active α-hydroxy ketones from the corresponding diols, as exemplified in Eq. 82.[184] Since the requisite enantioenriched diols may be readily obtained by a Sharpless dihydroxylation, this oxidation constitutes a convenient and effective entry into non-racemic α-hydroxy ketones.

CH_2Cl_2/acetone, 0°, 2 h; (94%) (Eq. 80)

acetone, 18-22 h; (100%) (Eq. 81)

96% ee; TFP/CH_2Cl_2, 0°, 100 min; (88%) > 92% ee (Eq. 82)

Oxidation of C–H bonds is usually more difficult than epoxidation; however, steric effects can cause allylic C–H oxidation to compete efficiently with epoxidation.[185] In the reaction shown in Eq. 83 the large silyl group directs the DMD oxidation preferably toward C–H insertion (for a detailed discussion see ref. 15).

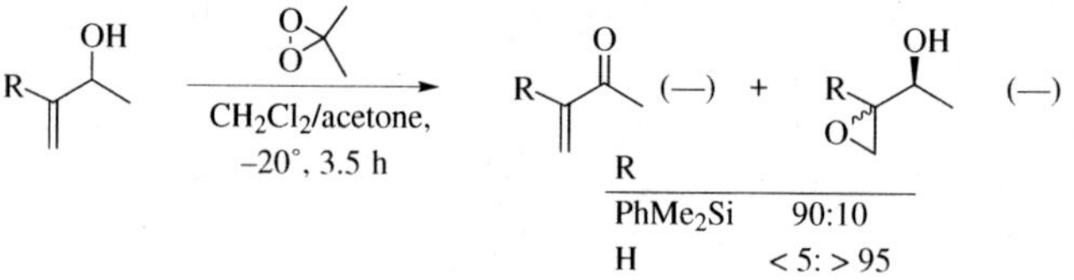

(Eq. 83)

R	
$PhMe_2Si$	90:10
H	< 5: > 95

Conversion of the alcohol into the corresponding ether derivative reduces reactivity; however, DMD (isol.), and even more so TFD (isol.), oxidize such substrates to their respective carbonyl products (Eq. 84).[186] Since even ethers, such as diethyl ether and tetrahydrofuran, may be cleaved by dioxiranes, they are not recommended as solvents for dioxirane oxidations. Indeed, for the successful preparation of TFD (isol.), use of ether-free 1,1,1-trifluoropropan-2-one (TFP) is essential.[187] Nevertheless, when properly controlled, oxidation of ethers provides a useful method for the deprotection of the benzyl group in carbohydrates (Eq. 85).[188,189]

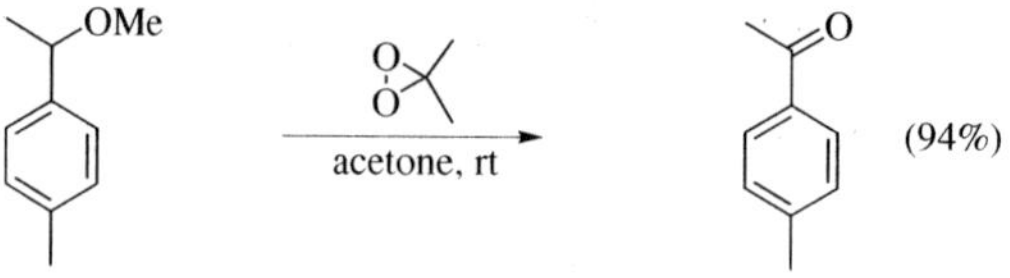

(Eq. 84)

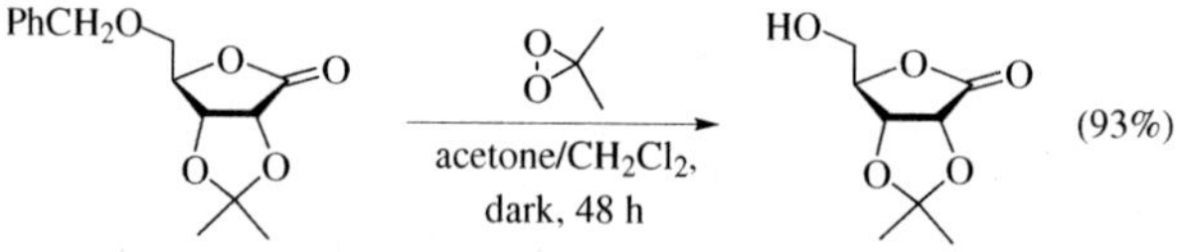

(Eq. 85)

Acetals of vicinal diols are also subject to α-oxidation by dioxirane, especially with the more reactive TFD (isol.). For example, oxidative cleavage of an acetal functionality to the corresponding α-hydroxy ketone (Eq. 86)[190] constitutes a convenient deprotection protocol, coupled with an alcohol oxidation. Like the oxidation of vicinal diols, the second stereogenic center is preserved. Although further oxidation of α-hydroxy ketones to 1,2-diketones is possible, such reactions are sluggish because of electronic reasons, in that the α-carbonyl group deactivates the C–H bond. Thus, 1,2-diketones are not usually formed in appreciable amounts in the dioxirane oxidation of vicinal diols or their acetals.[184,190]

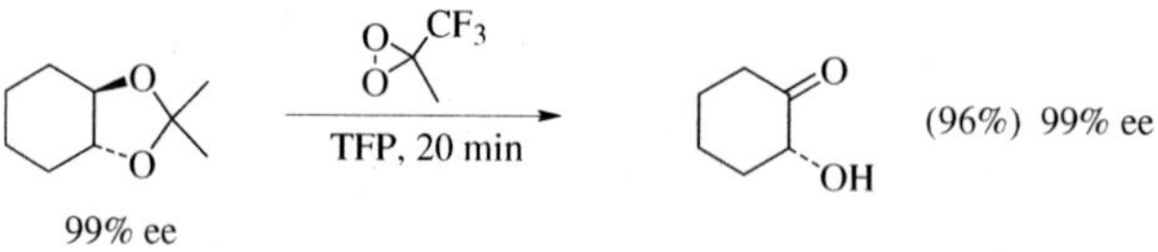

(Eq. 86)

Dioxirane oxidation of the acetals of aldehydes leads to esters (Eq. 87);[189,191] similarly, oxidation of orthoformates furnishes carbonates (Eq. 88).[191]

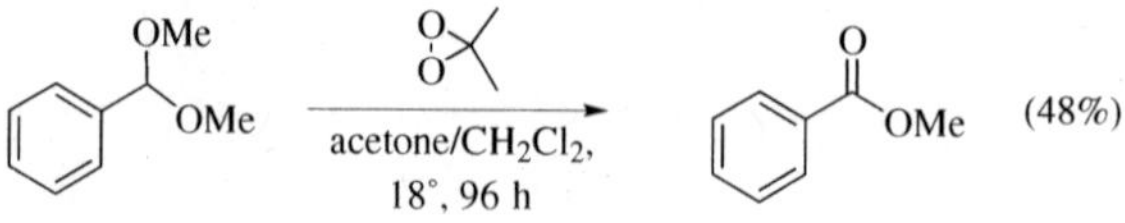

(Eq. 87)

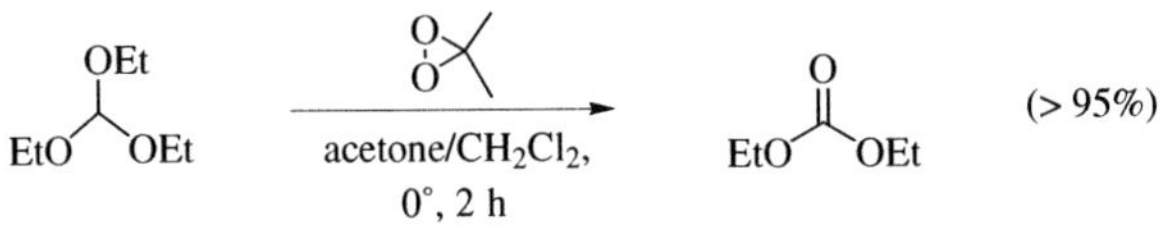
(Eq. 88)

Direct oxidation of the C–H bond in an amine is not feasible given the much higher reactivity of the nitrogen atom; however, protection of the amino group as the ammonium salt (strongly acidic conditions must be used to tie up all of the amine) or as an amide will suppress the nitrogen oxidation effectively (for a detailed discussion of chemoselective dioxirane oxidation see ref. 15). A typical example is shown in Eq. 89.[192] Unlike for alcohols, the α-hydroxylation of an amine is rare. One such example is shown in Eq. 90,[93] for which the hydroxylation is made possible by protection of the amine as a carbamate.

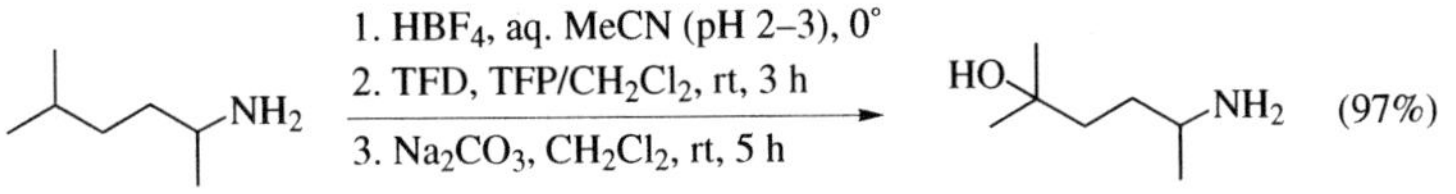
(Eq. 89)

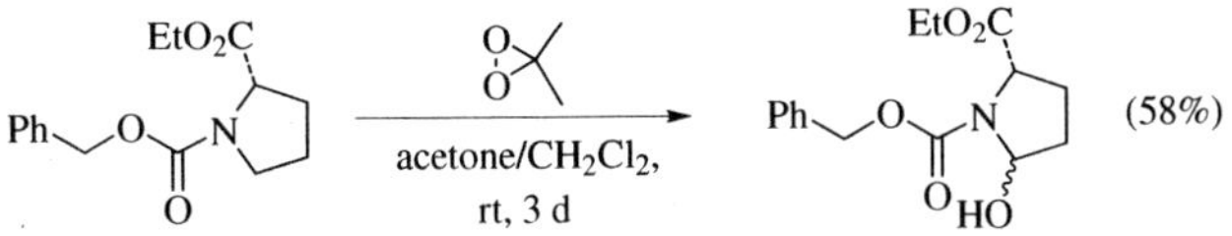
(Eq. 90)

The acidic C–H bond of 1,3-dicarbonyl compounds is also reactive toward dioxirane (both isolated and in situ generated) oxidation (Eq. 91).[193] Although the oxidation appears like a C–H insertion, the possibility that epoxidation of the enolate is involved cannot be ruled out. The fact that this oxidation may be catalyzed by either $Ni(OAc)_2$ or $Ni(acac)_2$ implies involvement of enolate intermediates.[194]

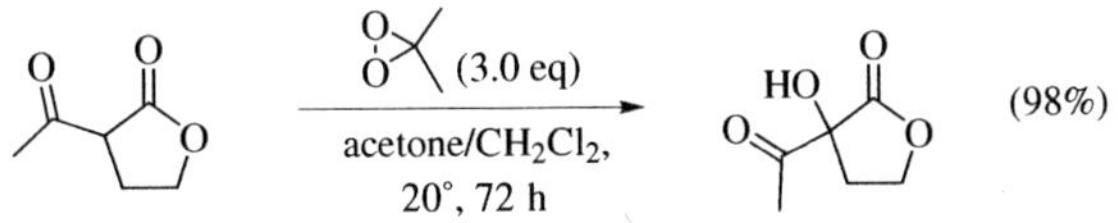
(Eq. 91)

As is evident from the above discussion, C–H oxidation is a highly chemoselective reaction. The chemoselectivity is mainly governed by the reactivity of the chemically different C–H bonds, and sometimes by steric factors when the reactivities are similar. In the special situation illustrated in Eq. 92,[195] an in situ generated intramolecular dioxirane chemoselectively oxidizes the C–H bond at the δ site rather than the usually more reactive tertiary hydrogen (γ site) because of a more favorable concerted six-membered cyclic transition state.[195] Moreover, the equatorial C–H bond is preferentially oxidized such that the trans product dominates.

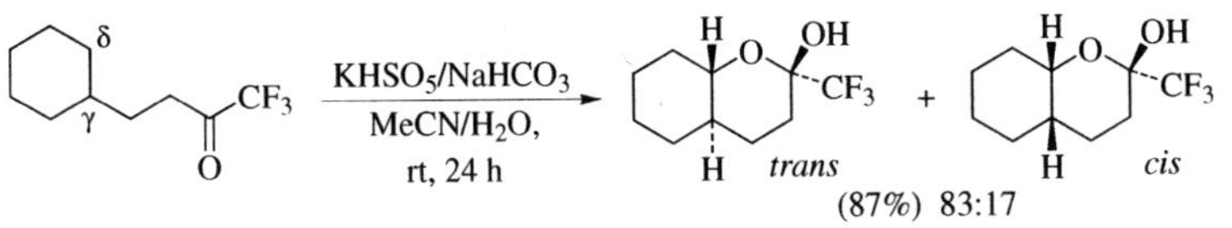
(Eq. 92)

As C–H oxidation by dioxiranes is a stereoselective reaction, an attractive opportunity arises to carry out enantioselective C–H oxidations by employing optically active dioxiranes; however, such asymmetric C–H oxidations by dioxiranes are still largely unexplored. The only known example of enantioselective C–H oxidation appears to be the oxidation of vicinal diols by an in situ generated fructose-derived dioxirane.[196,197] Through either the desymmetrization of meso-diols (Eq. 93) or the kinetic resolution of racemic diols (Eq. 94), enantioenriched α-hydroxy ketones may be obtained in up to 71% ee.[197] The desymmetrization of the acetals of meso-diols leads to higher enantioselectivities compared with that of meso-diols, but the conversion is lower because of their reduced reactivity.[197]

(3.0 eq), $KHSO_5$, K_2CO_3, buffer (pH 10.5)/MeCN, 0°, 3 h

meso → (95%) 58% ee (Eq. 93)

(3.0 eq), $KHSO_5$, K_2CO_3, buffer (pH 10.5)/MeCN, 0°, 3 h

rac → (31%) 71% ee (Eq. 94)

Since presently not much is known about enantioselective C–H oxidation with optically active dioxiranes, more research in this important area is needed. Analogous to the recent development of dioxirane-mediated asymmetric epoxidations, we expect progress in asymmetric C–H functionalization with chiral dioxiranes in the near future. The problem resides in designing more persistent and reactive optically active ketones as the dioxirane precursors.

Silanes. The Si–H bond in silanes is weaker than the C–H bond in alkanes; therefore, the oxidation of silanes is more facile. Nevertheless, only a few examples of silane oxidation by dioxiranes are known. Oxidation of dimethylphenylsilane by TFD (isol.)[39] or DMD (isol.)[198] affords the silanol in high yield, as shown in Eq. 95. As in the case of C–H oxidation, TFD is significantly more reactive than DMD toward the Si–H bond. The mild and neutral conditions lead exclusively to silanol product without any formation of disiloxane.

$$\mathrm{Ph{-}SiMe_2{-}H} \xrightarrow[\text{methyl(trifluoromethyl)dioxirane, TFP/CH}_2\text{Cl}_2,\ -20^\circ,\ <1\text{ min, or dimethyldioxirane, acetone, }0^\circ,\ 0.5\text{ h}]{} \mathrm{Ph{-}SiMe_2{-}OH}\ (>98\%) \quad \text{(Eq. 95)}$$

Like C–H oxidation, Si–H oxidation of silanes by dioxirane is also stereoselective. As displayed in Eq. 96, the original configuration of the silane is preserved during the oxidation.[39] The hydroxylation of silanes has also been applied to organometallic substrates (see the following section).

$$\text{1-Naphthyl-Si(Ph)(Me)H (96.5\% ee)} \xrightarrow[\text{TFP/CH}_2\text{Cl}_2,\ -20^\circ,\ <1\text{ min}]{\text{methyl(trifluoromethyl)dioxirane}} \text{1-Naphthyl-Si(Ph)(Me)OH}\ (>98\%)\ 97\%\ \text{ee} \quad \text{(Eq. 96)}$$

Organometallic Compounds

Oxidation of organometallic compounds by dioxiranes is usually more complex since both the metal center as well as the organic ligands may be oxidized. The examples presented in this section include both types of oxidations. Unless redox chemistry at the metal center complicates matters, the more electron-rich organic ligand in the organometallic complex usually undergoes direct oxidation (epoxidation, heteroatom oxidation, σ-bond insertion) by the dioxirane more readily than the metal center. In the case of highly reactive electron-rich alkenes, direct non-selective epoxidation prevails.

Dioxirane oxidation of the metal center leads to a higher oxidation state of the metal. For example, the DMD oxidation of a manganese(II) porphyrin complex at subambient temperature leads to the manganese(IV) derivative in quantitative yield (Eq. 97).[199] Analogously, manganese(III) and iron complexes are oxidized under similar conditions.[199] The use of DMD (isol.) as a stoichiometric oxygen donor in the Jacobsen–Katsuki epoxidation with the manganese-salen catalyst enables the enantioselective epoxidation of prochiral olefins under homogeneous conditions.[200–203] Since the oxidation of the manganese is much faster than the epoxidation of the olefin, good to excellent enantiomeric excesses are obtained for the epoxides.[200–203]

$$\mathrm{Mn^{II}(TPP)} \xrightarrow[\text{acetone, }-50^\circ\text{ to }-20^\circ]{\text{dimethyldioxirane}} \mathrm{O{=}Mn^{IV}(TPP)}\ (100\%) \quad \text{(Eq. 97)}$$

TPP = Tetraphenylporphyrin

Dioxirane oxidation of metal-carbene complexes usually leads to demetallation.[204–207] A specific example of such an oxidation by DMD (isol.) is shown in Eq. 98.[204] When the ligands of the metal complex contain a functionality that is more prone to oxidation, demetallation may be suppressed (Eq. 99).[208] With even an excess of DMD (2.2 equiv.), a mixture of sulfoxide and the sulfone products is formed exclusively (93% yield) without demetallation.[208]

$$(CO)_5Cr{=}C(Ph)(OEt) \xrightarrow[\text{acetone, 20°, 3 h}]{\text{DMD}} PhCO_2Et\ (97\%) \quad \text{(Eq. 98)}$$

$$(CO)_5Cr{=}C(NC_4H_8)(CH_2)_3SPh \xrightarrow[\text{acetone, –78°, 10 min}]{\text{DMD (1.2 eq)}} (CO)_5Cr{=}C(NC_4H_8)(CH_2)_3SOPh\ (80\%) + (CO)_5Cr{=}C(NC_4H_8)(CH_2)_3SO_2Ph\ (20\%) \quad \text{(Eq. 99)}$$

In an unusual example, the niobium complex shown in Eq. 100 is oxidized by TFD (isol.) to the metallaoxirane.[209] Selective oxidative decarbonylation reactions of several rhenium and molybdenum carbonyl complexes by DMD (isol.) have also been observed; a particular case for rhenium is given in Eq. 101.[210]

$$\text{Me}_3\text{Si-Cp–Nb(H)=C=CPh}_2 \xrightarrow[\text{TFP/Et}_2\text{O, rt}]{\text{TFD}} \text{Me}_3\text{Si-Cp–Nb(H)(O–C=CPh}_2) \ (—) \quad \text{(Eq. 100)}$$

$$\text{Cp*Re(CO)}_3 \xrightarrow[\text{acetone, 0°}]{\text{DMD}} \text{Cp*ReO}_3\ (74\%) \quad \text{(Eq. 101)}$$

The dioxirane oxidation of ligands in organometallic complexes is more abundant. The chemical nature of the ligand determines whether the oxidation takes place at a π bond (epoxidation), at a lone pair (heteroatom oxidation), or at a σ bond (Si–H insertion).

Organometallic substrates with ligands that contain a reactive C=C double bond may undergo epoxidation. For example, the tungsten complex shown in Eq. 102 is epoxidized by DMD (isol.) in quantitative yield at ambient temperature.[211] Similarly, titanium enolates are functionalized by DMD (isol.) to the corresponding α-hydroxy ketones after acidic workup (Eq. 103).[212] When enantiomerically pure enolates bearing titanium TADDOLates as chiral ligands are subjected to this oxidation, enantiomerically enriched α-hydroxy ketones are obtained.[212] If the metal complex is sufficiently robust, even electron-poor double bonds may be epoxidized under more strenuous conditions, as illustrated for the ferrocene derivative shown in Eq. 104.[213]

$$\text{Tp*W(O)}_2\text{CH}_2\text{C(CH}_3\text{)=CH}_2 \xrightarrow[\text{acetone/CH}_2\text{Cl}_2\text{, 20°, 45 min}]{\text{DMD}} \text{Tp*W(O)}_2\text{CH}_2\text{C(CH}_3\text{)(O)CH}_2\ (100\%) \quad \text{(Eq. 102)}$$

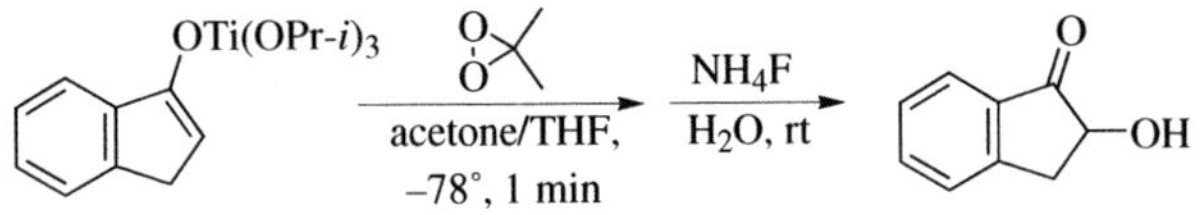

(Eq. 103)

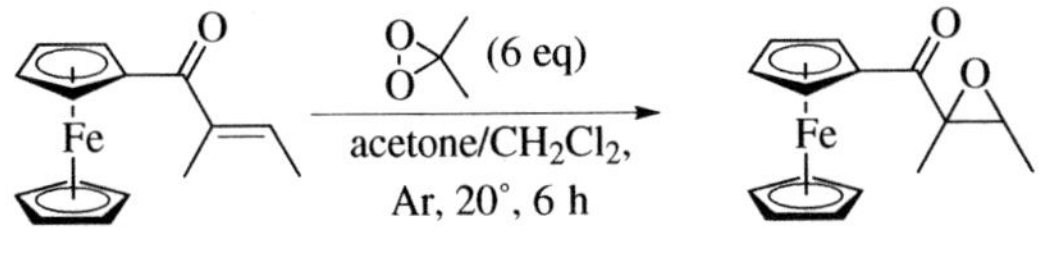

(Eq. 104)

When the C=C double bond is, however, directly coordinated to the metal center, the reactivity drops. For example, oxidation of an iron complex by DMD (isol.) takes place only at the more electron-rich furan ring (Eq. 105).[214] Thus, the iron-tricarbonyl fragment may be utilized as an oxidatively resistant protecting group for the 1,3-diene functionality.

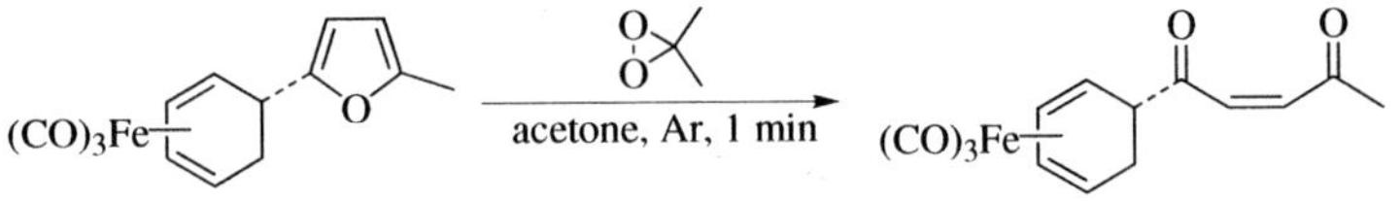

(Eq. 105)

Dioxirane oxidation of cyclopentadiene, ligands widely used in organometallic complexes, has not been reported. Apparently, the complexed cyclopentadiene ligand resists dioxirane oxidation. In contrast, the metal-coordinated triple bond in a manganese-acetylene complex is oxidized by DMD (isol.) to a manganese-carbene complex, as illustrated in Eq. 106.[215]

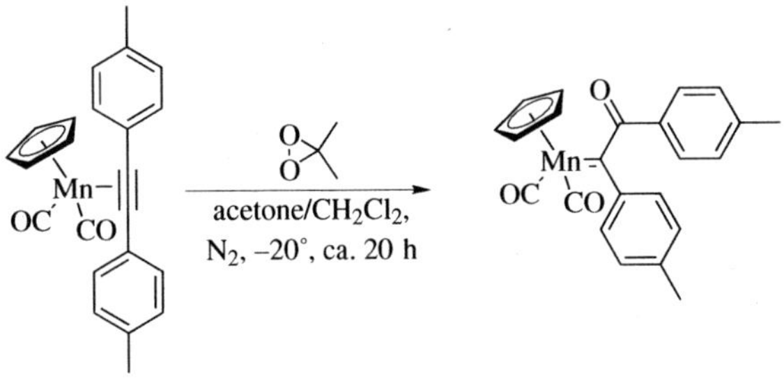

(Eq. 106)

For ligands with a heteroatom functionality (sulfur, phosphorus, or nitrogen), the heteroatom is usually the preferred site of dioxirane oxidation. These oxidations usually follow the general trends presented in the section on Heteroatom Substrates (see above); sulfides are oxidized to sulfoxides (Eq. 107)[127] and/or sulfones (Eq. 108),[208] and phosphines to phosphine oxides (Eq. 109).[216]

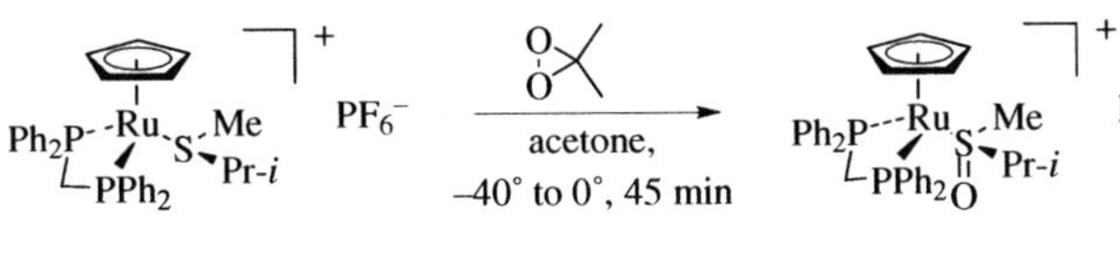

(Eq. 107)

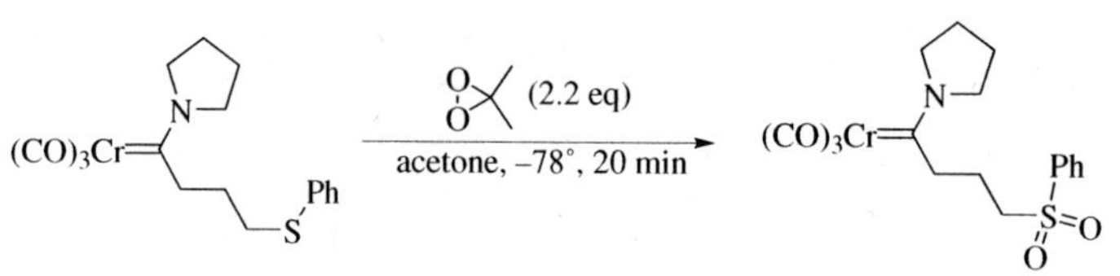

(Eq. 108)

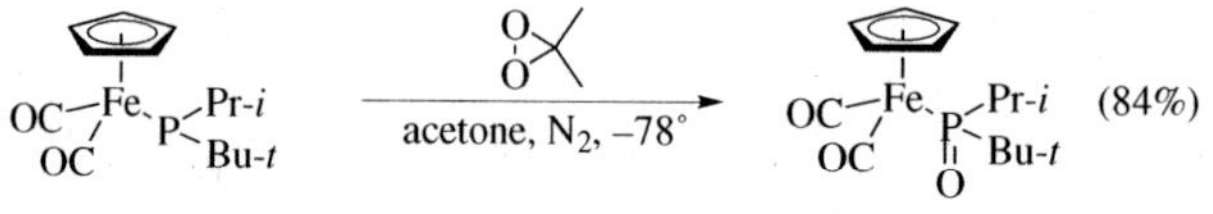

(Eq. 109)

A tertiary amine ligand affords the *N*-oxide with DMD (isol.), which may eliminate hydroxylamine on warming to room temperature, thus generating the vinyl group in the final product (Eq. 110).[217] This result illustrates that the nitrogen functionality is more readily oxidized than an alkenyl double bond.

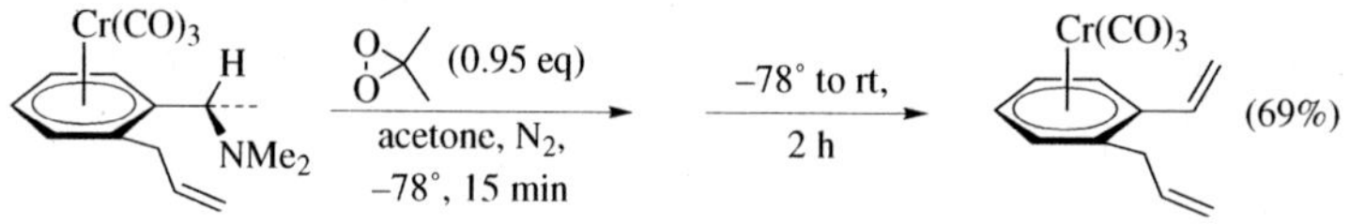

(Eq. 110)

Notably, oxidation of a molybdenum complex having a molybdenum-phosphorus triple bond occurs at the trivalent phosphorus ligand, affording the corresponding complex with a P=O functionality (Eq. 111).[218]

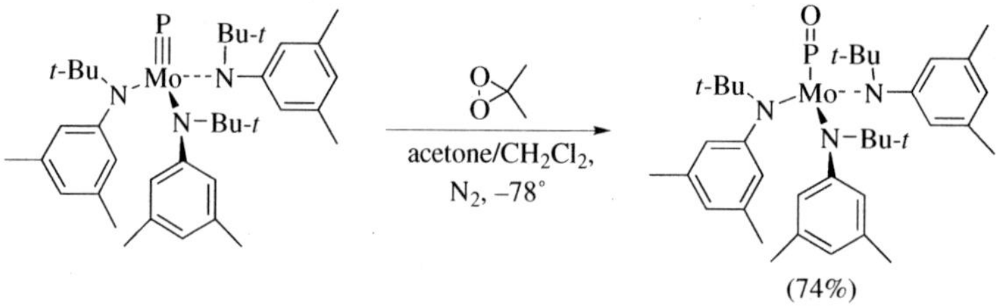

(Eq. 111)

DMD (isol.) mediated oxidation of a ruthenium complex having both sulfide and double-bond functionalities reveals once again that the sulfur atom is more prone to oxidation than the C=C double bond, even though the sulfide functionality is coordinated to the metal center (Eq. 112).[219] The corresponding epoxide may only be obtained once the sulfur atom has been functionalized.[219]

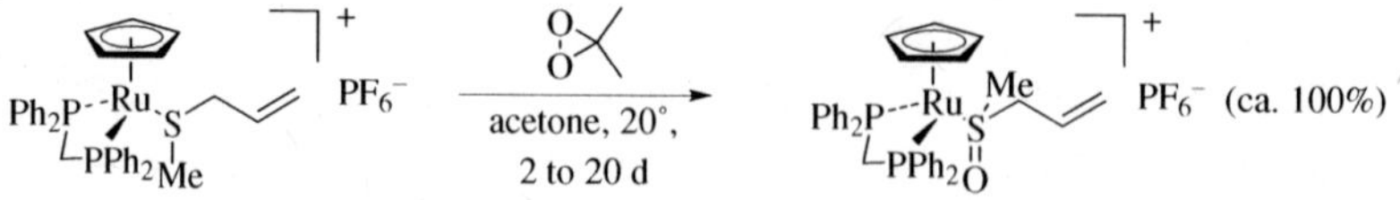

(Eq. 112)

Insertion of oxygen into Si–H with DMD or TFD when Si–H is a component of an organometallic substrate has also been documented. For example, organometallic complexes with silane ligands are successfully hydroxylated by

DMD (isol.).[220–227] A preparatively valuable example of the regioselective double hydroxylation of a ferriodisilane is shown in Eq. 113.[227]

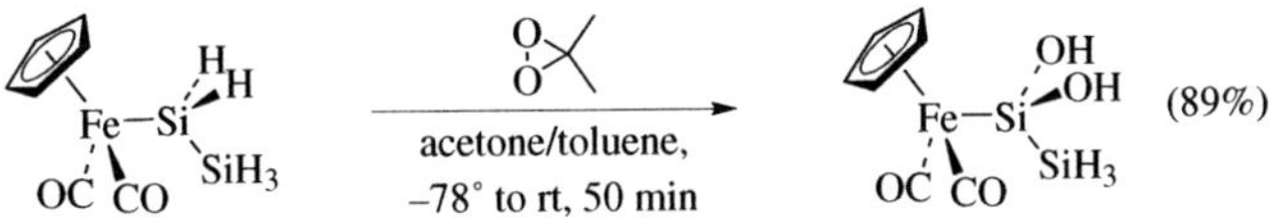

acetone/toluene, −78° to rt, 50 min (89%) (Eq. 113)

Insertion into a CH bond of a ligand belonging to an organometallic compound is much more difficult. The only case known to date is the TFD (isol.) oxidation of the tungsten-boron complex in Eq. 114,[211] in which hydroxylation of the pyrazole ligand occurs.

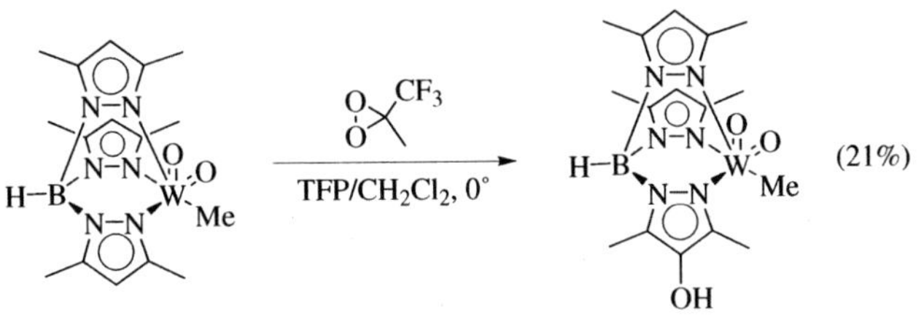

TFP/CH_2Cl_2, 0° (21%) (Eq. 114)

COMPARISON WITH OTHER METHODS

The three reaction types of dioxiranes presented in this chapter, namely π-bond oxidation (epoxidation) of allenes, acetylenes, and arenes, lone-pair oxidation of heteroatom substrates (N, S, P heteroatoms), and σ-bond oxidation (CH/SiH insertions) of alkanes and silanes, are quite different in their nature. Consequently, in comparing the dioxirane performance with other oxidants, it is convenient and even essential, to deal with these three classes separately in this subsection. It should, however, be kept in mind that the general features are quite similar, such that considerable overlap in the oxidative behavior exists for these substrates.

Allenes, Acetylenes, and Arenes

The dioxirane oxidation of cumulenes, acetylenes, and aromatic compounds all entail initial formation of epoxides. Some of these epoxides are rather labile and the final oxidation product may be structurally altered. Such functionalizations are to be classified as epoxidation reactions, which have been thoroughly covered in a previous chapter on dioxirane chemistry.[15] The interested reader should consult that coverage for details; herein we reiterate only the more specifically applicable features in regard to oxidants other than dioxiranes.

As cumulenes and arenes are more sluggishly epoxidized than alkenes, potent oxidizing agents must be employed. For example, perhydrates (hexafluoroacetone/H_2O_2),[228,229] oxaziridines,[230–232] and the Payne oxidation reagents (MeCN/H_2O_2/HO^-)[233–235] are hardly suitable. In addition, oxidations catalyzed by most transition metals (Co, Cr, Mn, Mo, Ti,V, W) are relatively ineffective for these substrates. An exception is rhenium, which in the form of methyltrioxorhenium

(MTO), efficiently oxidizes cumulenes and arenes, with the advantage that the MTO/H_2O_2 oxidant operates catalytically.[40]

It is most unfortunate that the usually highly efficient enantioselective Jacobsen–Katsuki epoxidation with chiral manganese-salen complexes is not applicable to functionalize cumulenes and arenes to the corresponding optically active oxidation products. Similarly, the optically active ketones (such as Shi's fructose-derived ketone[132,236]) employed in the catalytic, enantioselective, in situ mode of epoxidation are also ineffective for these substrates because of their low oxidative reactivity.

Peracids (most frequently *m*CPBA) are usually employed for the oxidation of cumulenes, acetylenes, and arenes, but as already pointed out,[77–86] isolated dioxiranes are more advantageous for the preparation of labile epoxides. The disadvantage of peracids resides in the fact that acid-sensitive substrates and/or products must be avoided. When the acidity is buffered, the substrate and resulting epoxide must resist hydrolysis. Such limitations are not an issue when isolated dioxiranes are used, but the in situ mode of generating dioxiranes is subject to the same disadvantages as for peracids. The benefits of dioxirane chemistry should be conspicuous for the oxidation of allenes, acetylenes, and arenes.

Heteroatom Substrates

Of the substrates considered in this chapter, those with heteroatoms are the easiest to oxidize, such that many oxidizing agents are available. For some heteroatoms, particularly divalent sulfur/selenium and more so trivalent phosphorus compounds, even H_2O_2 without activation will do, although the rate of oxidation is relatively slow. In the context of reactivity, dioxiranes present no definite advantages as heteroatom oxidants over the traditional ones such as peracids[237] and transition-metal catalysts.[238] (For a detailed comparison of the reactivity and selectivity of the methyltrioxorhenium (MTO) catalyst with dioxiranes, see a recent review.[40])

On the contrary, overoxidation by the more reactive dioxiranes may be a more serious problem to control. Whereas sp^3-type (amines, hydroxylamine, hydrazines) and sp^2-type (imines, oximes, hydrazones, heteroarenes) nitrogen-containing substances are readily oxidized to a plethora of products, the direct oxidation of the sp-type nitrogen atom in nitriles to the corresponding nitrile oxides is still a difficult task even for the highly reactive TFD. Similarly, the oxidative functionalization of amides and imides lacks suitable oxidizing agents, since neither dioxiranes nor traditional oxidants serve this purpose.

A unique chemical property of dioxiranes is their propensity to oxidize oxygen-type nucleophiles (e.g., HO_2^-, RO_2^-, $RC(O)O_2^-$, ClO^-) to molecular oxygen; the latter is formed in the singlet-excited state, namely singlet oxygen.[26,27] This unusual transformation appears not to have an equivalent among other oxidants. It is a consequence of the high electrophilic character of the dioxiranes, which makes them amenable for attack by the oxygen-centered nucleophile on the peroxide bond of the dioxirane. Some amine *N*-oxides[165] also engage in this type of reaction and are deoxygenated into singlet-excited molecular oxygen and the free amine, again a unique chemical behavior of dioxiranes.

As for enantioselective oxidations, specifically sulfoxidation, the chiral dioxiranes, such as Shi's fructose-based dioxirane,[132,236] are inferior to the asymmetric oxygen transfer catalyzed by transition metals, namely the $Ti(OR)_4/t\text{-}BuO_2H$ oxidant (Kagan sulfoxidation[239]). The ability, however, to achieve the asymmetric efficiency delivered by oxidative enzymes[240] and microorganisms[241,242] is still a formidable task in oxidation chemistry, particularly for chiral dioxiranes. Nevertheless, sulfoxides of high enantiomeric purity may be obtained through the sequence of desymmetrizing a prochiral sulfide by complexation with a transition metal based chiral auxiliary, followed by DMD oxidation, and final removal of the chiral auxiliary.[126] Such methodology should be able to compete in efficacy with the established protocols such as the Kagan enantioselective sulfoxidation.[239]

The dioxirane-related oxaziridines, which in optically active form deliver sulfoxides with enantioselectivities up to 98% ee, are effective for asymmetric sulfoxidation.[243,244] Oxaziridinium salts also show promise and offer potential, but, as yet, enantioselectivity of only about 35% ee has been achieved.[245]

Alkanes and Silanes

Indisputably, the greatest challenge in oxidation chemistry is still the direct functionalization of unactivated C–H bonds. It is especially desirable to carry out such insertion reactions enantioselectively under catalytic conditions. Nature has perfected oxygen-atom insertions into C–H bonds by developing efficacious enzymes for this purpose, namely the oxidases and oxygenases.[246] Along these lines, biomimetic oxidants based on chemical catalysts have been developed,[247] most notably for the remote hydroxylation of steroids.[248–250]

Although as yet the dioxiranes do not offer a general method for the enantioselective functionalization of hydrocarbons, it should be appreciated that these readily accessible oxidants, especially the simple structures DMD and TFD, work as impressively as they do. In this context, we reiterate that such non-metal-catalyzed C–H insertions by dioxiranes may take place highly stereoselectively as, for example, with complete retention of configuration in the hydroxylation of (*R*)-2-phenylbutane (see Scheme 5) by DMD.[37] Indeed, even a few asymmetric C–H oxidations with optically active dioxiranes, such as Shi's fructose-derived system,[196,197] have been reported to occur in substantial enantiomeric excess, under quasi-catalytic conditions. These simple metal-free functionalizations of alkanes approach the efficiency of enzymatic C–H insertions. However, their catalytic efficiency still needs to be improved. The future challenges in dioxirane chemistry lie in enhancing the catalytic reactivity of these oxidants to achieve high enantioselectivity.

There are only a few alternative chemical methods that are competitive with dioxiranes; most comprise metal-catalyzed and radical-type C–H oxidations. One such method is the so-called "Gif oxidation,"[251] which is of limited synthetic utility because complex product mixtures are usually obtained.[252] A detailed comparison of metal-catalyzed C–H insertion with methyltrioxorhenium (MTO) and with dioxiranes has been made recently.[40] Generally, the performance (reactivity, selectivity) of the dioxiranes is better, but the MTO/H_2O_2 oxidant offers excellent catalytic efficiency. An effective nonmetal-type C–H oxidation of alcohols to ketones is catalyzed by the TEMPO nitroxyl radical, which has the advantage over dioxiranes that C=C double bonds may be present, since this reagent does not effect epoxidation.[253–263] In fact, this method may be used for the kinetic resolution of secondary alcohols to afford ee values of up to 98% by engaging chiral binaphthyl-based nitroxyl radicals.[264]

Silanes are more readily oxidized than alkanes, since the Si–H bond (ca. 77 kcal/mol) is considerably weaker than the C–H bond (ca. 99 kcal/mol). The advantages of dioxiranes for oxygen insertion into Si–H bonds has been amply emphasized;[39,198] a competitive alternative is the catalytic MTO/H_2O_2 oxidant.[40] For oxidation of optically active silanes, the urea/H_2O_2 adduct (UHP) should be employed instead of hydrogen peroxide to obtain enantioselectivities comparable to those of dioxiranes.[199]

EXPERIMENTAL CONDITIONS

Caution! The dioxiranes DMD and TFD are volatile peroxides and must be handled with care. The oxidations should be carried out in a hood with good ventilation. Inhalation and direct exposure to skin must be avoided! Although no explosions have been reported for dioxiranes, all safety precautions should be employed!

EXPERIMENTAL PROCEDURES

DMD (isol.), acetone, 22°, 20 h → CO_2H / OH (42%) + (6%) + (trace)

2-Hydroxy-2-methylpropanoic Acid [Oxidation of an Alkyne with DMD (isol.)].[63] To a magnetically stirred solution of 2-butyne (216 mg, 4.00 mmol) in acetone (5.0 mL) in a 250-mL flask, was added a solution of DMD in acetone (140 mL, 0.060 M, 8.40 mmol) at room temperature (ca. 22°). The progress of the reaction was followed by GLC analysis, which indicated the presence of three products in the ratio of 10 : 15 : 75. After 20 hours, the excess acetone was removed on a rotary evaporator (20°, 15 mmHg), the dark yellow residue (25 mL) was subjected to fractional distillation (80°, 5 mmHg) to afford a colorless material which solidified. The solid was recrystallized from CH_2Cl_2/hexane to give 174 mg (42%) of 2-hydroxy-2-methylpropanoic acid as colorless needles, mp 77–79°; ^1H NMR ($CDCl_3$) δ 1.50 (s, 6H), 6.30 (br s, 2H); ^{13}C NMR ($CDCl_3$): δ 27.0, 72.2, 181.4; EIMS m/z (%): 89 (5), 59 (100), 45 (7), 44 (4), 43 (53).

The yellow distillate contained two other products, which were separated by preparative GLC. One of the products was obtained in trace amount and identified as 1-oxiranylethanone: ^{1}H NMR ($CDCl_3$) δ 2.06 (s, 3H), 2.90 (dd, $J = 5.7$, 2.5 Hz, 1H), 3.01 (dd, $J = 5.7$, 4.7 Hz, 1H), 3.40 (dd, $J = 4.6$, 2.5 Hz, 1H); ^{13}C NMR ($CDCl_3$) δ 23.7, 45.8, 53.7, 205.5. EIMS m/z (%): 87 (M + H, 1), 86 (M^+, 17), 85 (13), 71 (18), 55 (7), 53 (1), 44 (3), 43 (100).

The other product, 2,2,5,5-tetramethyl-1,3-dioxolane-4-one, was isolated by distillation (80°, 5 mmHg) as a colorless liquid (18 mg, 6%); IR (KBr) 2990, 2936, 1797, 1466, 1380, 1301, 1192, 1076, 1015, 931, 868, 838 cm^{-1}; ^{1}H NMR ($CDCl_3$) δ 1.48 (s, 6H), 1.58 (s, 6H); ^{13}C NMR ($CDCl_3$) δ 26.5, 28.6, 77.2, 109.3, 175.7; EIMS m/z (%): 130 (1.3), 129 (M–CH_3, 22), 101 (20), 100 (8), 59 (81), 58 (39), 43 (100).

TFD (isol.)
CF_3COMe, CH_2Cl_2, 0°, 3 min
(83%) + (13%)

***cis*-Bicyclo[5.3.0]decan-2-one [Oxidation of an Alkyne with TFD (isol.)].[64]** A 25-mL flask was charged with cyclodecyne (136 mg, 1.00 mmol) and a trifluoroacetone solution of TFD (isol.) (20.0 mL, 0.010 M, 2.00 mmol) at 0°. After magnetic stirring for 3 minutes, the volatiles were removed on a rotary evaporator (10°, 15 mmHg), and the residue was purified by column chromatography on silica gel, to give the title compound (126 mg, 83%); ^{13}C NMR (50 MHz, $CDCl_3$) δ 21.0, 21.7, 22.1, 23.1, 28.0, 29.7, 43.4, 58.3, 59.2, 206.5. Further elution resulted in isolation of *cis*-bicyclo[4.4.0]decan-2-one (22 mg, 13%): ^{13}C NMR (50 MHz, $CDCl_3$) δ 24.5, 25.4, 26.2, 27.8, 32.5, 32.2, 40.4, 43.2, 54.6, 214.0.

DMD (isol.)
acetone, 20°
(92%)

6-Hydroxy-2,2-dimethyl-3-oxacyclohexanone [Oxidation of an Allene with DMD (isol.)].[57] A 25-mL flask at 20° was charged with 5-methyl-3,4-hexadien-1-ol (22 mg, 0.20 mmol) and a solution of DMD (12.0 mL, 0.100 M, 1.20 mmol) under vigorous stirring. Removal of the solvent on a rotary evaporator (20°, 15 mmHg), followed by chromatographic purification on silica gel, afforded the title compound as a colorless liquid (26 mg, 92%); IR 3425, 1720, 1158, 1076 cm^{-1}; ^{1}H NMR (300 MHz, $CDCl_3$) δ 1.37 (s, 3H), 1.39 (s, 3H), 1.99 (m, 1H), 2.51 (m, 1H), 3.20–3.80 (br s, 1H), 3.87 (ddd, $J = 13.0$, 5.0, 2.0 Hz, 1H), 4.05 (ddd, $J = 13.0$, 12.0, 4.0 Hz, 1H), 4.55 (dd, $J = 12.0$, 7.0 Hz, 1H); ^{13}C NMR (75 MHz, $CDCl_3$) δ 22.6, 23.8, 36.3, 59.0, 70.1, 80.3, 212.0. EIMS m/z (%): 145 (14), 127 (9), 116 (10), 99 (2), 87 (11), 85 (5), 83 (100), 71 (3). HRMS calcd for $C_7H_{13}O_3$ (M + H), 145.0860, found 145.0865.

DMD (isol.), acetone, 20° (96%)

6-Hydroxy-5,5-dimethyl-3-oxacyclohexanone [Oxidation of an Allene with DMD (isol.)].[57] A 25-mL flask was charged at 20° with 2,2-dimethyl-3,4-pentadien-1-ol (20 mg, 0.180 mmol) and a solution of DMD (18.0 mL, 0.100 M, 1.80 mmol) under vigorous magnetic stirring. Removal of the excess solvent on a rotary evaporator (20°, 15 mmHg), followed by chromatographic purification on silica gel afforded 6-hydroxy-5,5-dimethyl-3-oxacyclohexanone as a colorless liquid (25 mg, 96%); IR 3460, 1727, 1248, 1106 cm^{-1}; ^{1}H NMR (300 MHz, $CDCl_3$) δ 0.94 (s, 3H), 1.08 (s, 3H), 3.40–3.50 (br s, 1H), 3.63 (dd, $J = 15.0$, 12.0 Hz, 2H), 4.00 (dd, $J = 14.4$, 1.2 Hz, 1H), 4.02 (br s, 1H), 4.13 (dd, $J = 14.4$, 0.5 Hz, 1H); ^{13}C NMR (75 MHz, $CDCl_3$) δ 17.7, 22.6, 42.8, 72.7, 76.4, 81.0, 206.9. EIMS m/z (%): 145 (23), 144 (16), 127 (4), 101 (5), 85 (37), 71 (100). HRMS calcd for $C_7H_{13}O_3$ (M + H), 145.0860, found 145.0865.

DMD (isol.), acetone, K_2CO_3, 20°, 20 min (95%)

2,5-Hexamethylene-1,4-dioxaspiro[2.2]pentane [Diepoxidation of a Cyclic Allene with DMD (isol.)].[16] To a stirred solution of DMD (0.100 M, 4.44 mmol) in acetone (40.0 mL) dried over K_2CO_3, was added the cyclic allene (112 mg, 0.900 mmol) at 20°. Stirring was continued for 20 minutes at the same temperature. The solvent was removed on a rotary evaporator (20°, 15 mmHg), and the product was separated from the K_2CO_3 by triturating with ether (3 × 10 mL). The combined ether triturates were filtered, dried ($MgSO_4$), and concentrated (20°, 15 mmHg) to give the title compound (135 mg, 95%) as a colorless oil; IR 1626, 1605 cm^{-1}; ^{1}H NMR (300 MHz, $CDCl_3$) δ 1.35–1.42 (m, 6H), 1.49–1.54 (m, 2H), 1.69–1.74 (m, 2H), 2.12–2.18 (m, 2H), 3.75 (dd, $J = 6.4$ Hz, 2H); ^{13}C NMR (75 MHz, $CDCl_3$) δ 20.1, 24.7, 27.4, 60.1, 84.3; EIMS m/z (%): 154 (8), 130 (29), 98 (100), 82 (83), 69 (65). HRMS calcd for $C_9H_{14}O_2$, 154.0994, found 154.1000.

DMD-d_6 (isol.), acetone-d_6, N_2, –78°, 30 min (100%)

2,3-Epoxy-2,3-dihydro-2,3-dimethylbenzo[*b*]furan [Epoxidation of a Benzofuran with DMD-d_6 (isol.)].[87] A 5-mm NMR tube was charged with an acetone-d_6 solution of 2,3-dimethylbenzo[*b*]furan (113 µL, 0.220 M, 25 µmol) at −78° under a N_2 atmosphere. By means of a syringe, a well-dried (over 4 Å molecular sieves) DMD-d_6 (isol.) solution in acetone-d_6 (500 µL, 0.0500 M, 25 µmol) was added rapidly at −78°. After 30 minutes, the NMR tube was submitted to low-temperature (−50°) ^{13}C-NMR spectroscopy, which revealed that the olefinic carbon resonances in 2,3-dimethylbenzo[*b*]furan were replaced by the characteristic epoxide resonances for the product. At temperatures higher than 0°, complete decomposition of the epoxide occurred within 30 minutes.

DMD (isol.), acetone, 22°, 1 h (97%)

1,2-Epoxyacenaphthene [Epoxidation of an Arene with DMD (isol.)].[265] To a magnetically stirred solution of acenaphthylene (611 mg, 4.02 mmol) in acetone (5.0 mL) was added an acetone solution of DMD (66.0 mL, 0.0620 M, 4.09 mmol) at room temperature (ca. 20°). The progress of the reaction was monitored by GLC, which indicated that acenaphthylene was converted into its 1,2-epoxide within one hour. Removal of the solvent on a rotary evaporator (20°, 15 mmHg) afforded a white solid, which was taken up into CH_2Cl_2 (30 mL) and dried over Na_2SO_4. After removal of the drying agent, the solvent was removed on a rotary evaporator first at 20°, 15 mmHg and subsequently at 20°, 5 mmHg, to give the analytically pure oxide (654 mg, 97%), mp 83–84°; ^{1}H NMR ($CDCl_3$) δ 4.81 (s, 2H), 7.39–7.77 (m, 6H).

$KHSO_5$, Me_2CO, CH_2Cl_2, phosphate buffer, *n*-Bu_4NHSO_4, KOH, 0-10°, 5.5 h (84%)

Bisbenzo[3′,4′]cyclobuta[1′,2′:1,2:1″,2″:3,4]biphenyleno[1,8*b*-*b*:2,3-*b*′:4,4*a*-*b*″]trisoxirene [Epoxidation of an Arene with DMD (in situ)].[67] To a solution of tris(benzocyclobutadieno)cyclohexatriene (1.50 g, 5.06 mmol) in an acetone/CH_2Cl_2 mixture (350 mL, 5 : 2 v/v) contained in a 1000-mL three-necked flask were added phosphate buffer (50 mL) and tetra-*n*-butylammonium hydrogen sulfate (200 mg, 0.590 mmol). A solution of potassium monoperoxysulfate (46.0 g, 30.2 mmol) in water (225 mL) was added dropwise under vigorous magnetic stirring at 0–10° over 1.5 hours. The pH was maintained at 7.5–8.5 by the dropwise addition of an aqueous solution of KOH (2–3%). The reaction mixture was stirred for an additional 4 hours and then mixed with an equal volume of ice-cold water. The reaction mixture was extracted with CH_2Cl_2 (1 × 150 mL), the extract was washed with ice-cold water (3 × 100 mL), and the combined organic layers were dried (K_2CO_3). The solvent was removed on a rotary evaporator (20°, 15 mmHg) and the solid residue was purified by preparative TLC on silica gel with CH_2Cl_2/hexane (1 : 1) as eluent. Recrystallization from the same solvent mixture gave the oxide as colorless plates (1.48 g, 84%), mp 180–182°;

IR (KBr) 1613, 1512, 1495, 1463, 1430, 1339, 1261, 1154, 1094, 1003, 918, 861, 802, 751 cm^{-1}; ^{1}H NMR (400 MHz, $CDCl_3$) δ 7.38 (m, 6H), 7.46 (m, 6H); ^{13}C NMR (100 MHz, $CDCl_3$) δ 76.3, 122.6, 131.5, 143.71. EIMS m/z (%): 348 (M^+, 11), 316 (8), 261 (10), 248 (9), 176 (21), 175 (100), 174 (11), 156 (61), 135 (22), 127 (20), 123 (30), 121 (22), 107 (18), 85 (11), 73 (38).

t-BuO(C=O)NH–CH(CO$_2$Me)–CH$_2$–(indol-3-yl) → [DMD (isol.); acetone, CH$_2$Cl$_2$, 10°, 2 d] → t-BuO(C=O)NH–CH(CO$_2$Me)–CH$_2$–(2-oxoindolin-3-yl) (91%)

Methyl Boc-β-(2,3-dihydro-2-oxo-indol-3-yl)alaninate [Oxidation of an Indole with DMD (isol.)].[93] A 10-mL flask equipped with a magnetic stirring bar was charged with a solution of Boc-Trp-OMe (292 mg, 0.910 mmol) in CH_2Cl_2 (5.0 mL). After cooling to 10° by means of an ice bath, a freshly prepared acetone solution of DMD (23.0 mL, 0.100 M, 2.30 mmol) was added. Stirring was continued for 2 days at 10°. The solvent was removed (10°, 15 mmHg) and the residue was purified by flash column chromatography on silica gel, with a mixture of EtOAc and hexane as eluent, to afford two diastereoisomers (A : B = 1 : 1, 91% yield). Diastereomer A: $[\alpha]^{25}{}_D - 221.4°$ (c 1.0, $CHCl_3$); ^{1}H NMR (200 MHz, $CDCl_3$) δ 1.38 (s, 9H), 2.39–2.57 (m, 2H), 3.76 (s, 3H), 4.19–4.32 (m, 1H), 5.44 (s, 1H), 6.59–6.62 (m, 1H), 6.63–6.84 (m, 1H), 7.09–7.20 (m, 1H), 7.21–7.31 (m, 2H); ^{13}C NMR (50 MHz, $CDCl_3$): δ 28.2, 41.4, 52.3, 59.8, 81.2, 84.3, 110.5, 119.4, 123.2, 130.3, 148.3, 154.1, 173.9, 175.2; EIMS m/z (%): 334 (M^+, 18). Anal. Calcd for $C_{17}H_{22}N_2O_5$: C, 61.06, H, 6.63, N, 8.38. Found: C, 61.0, H, 6.59, N, 8.15.

Diastereomer B: $[\alpha]^{25}{}_D + 83.6°$ (c 1.0, $CHCl_3$); ^{1}H NMR (200 MHz, $CDCl_3$) δ 1.42 (s, 9H), 2.57–2.64 (m, 2H), 3.64 (s, 3H), 4.58–4.78 (m, 1H), 5.30–5.45 (m, 1H), 6.88–7.05 (m, 2H), 7.12–7.32 (m, 2H); ^{13}C NMR (50 MHz, $CDCl_3$) δ 28.2, 42.1, 52.2, 61.8, 81.6, 98.6, 115.1, 121.9, 123.2, 130.2, 149.1, 154.5, 172.4, 173.9; EIMS m/z (%): 334 (M^+, 16). Anal. Calcd for $C_{17}H_{22}N_2O_5$: C, 61.06, H, 6.63, N, 8.38. Found: C, 61.10, H, 6.63, N, 8.19.

n-Bu–NH_2 → [DMD (isol.); acetone, 22°, dark, 30 min] → n-Bu–NO_2 (84%)

1-Nitrobutane [Oxidation of a Primary Aliphatic Amine with DMD (isol.)].[266] A 150-mL flask was charged with a solution of n-butylamine (52 mg, 0.700 mmol) in acetone (5.0 mL), and an acetone solution of DMD (95.0 mL, 0.050 M, 4.80 mmol). The mixture was stirred at room temperature (ca. 20°) for 30 minutes in the dark. The solvent was removed on a rotary evaporator (20°, 20 mmHg), to afford the title compound (62 mg, 84%).

NH2 / O2N / NO2 — DMD (isol.), acetone, 22°, 12 h → NO2 / O2N / NO2 (94%)

1,3,5-Trinitrobenzene [Oxidation of a Primary Aromatic Amine with DMD (isol.)].[101] To a stirred solution of 3,5-dinitroaniline (30 mg, 0.165 mmol) in acetone (5.0 mL) was added an acetone solution of DMD (30.0 mL, 0.0600 M, 1.80 mmol) at room temperature (ca. 20°). After the reaction mixture was stirred for 12 hours, excess solvent was removed on a rotary evaporator (20°, 15 mmHg), and the residue was purified by preparative TLC on silica gel with CH_2Cl_2/hexane (1 : 1) as eluent. The product streak was scraped from the plate and extracted from the silica gel with CH_2Cl_2 (15 mL) that contained 5% methanol. Evaporation of the volatiles under reduced pressure (20°, 15 mmHg) gave 1,3,5-trinitrobenzene (33 mg, 94%), mp 121–122°.

NH2 / OMe — $KHSO_5$, (phosphate buffer), Me_2CO, n-Bu_4NHSO_4, pH 7.5-8.5, 0°, KOH, 45 min → NO2 / OMe (100%)

***o*-Nitroanisole [Oxidation of a Primary Aromatic Amine with DMD (in situ)].**[267] A 500-mL, three-necked round-bottomed flask, fitted with two addition funnels and a pH electrode, was charged with *o*-anisidine (1.10 mL, 10.0 mmol), CH_2Cl_2 (100 mL), acetone (100 mL), an aqueous solution of sodium phosphate (50 mL, 0.080 M), and tetra-*n*-butylammonium hydrogen sulfate (170 mg, 0.500 mmol). In one of the addition funnels was placed an aqueous solution of $KHSO_5$ (150 mL, 20.0 g, 32.0 mmol), and in the other an aqueous solution of KOH (100 mL, 150 mL). After the mixture was cooled to 0°, the aqueous solution of $KHSO_5$ was added dropwise over 30 minutes, while maintaining the pH between 7.5–8.5 by dropwise addition of an aqueous solution of KOH (2.00 N). After addition, the mixture was stirred at the same temperature for 15 minutes and then treated with 1 mL of methyl sulfide to destroy residual peroxide. The suspended material was removed by filtration and the organic layer was washed with water (50 mL), dried ($MgSO_4$), and concentrated (20°, 15 mmHg). The residue was purified by column chromatography on silica gel (50 g) with CH_2Cl_2 as eluent to afford *o*-nitroanisole (1.50 g, 100%).

OH / N–H — DMD (isol.), acetone, 0°, 30 min → OH / N–O• (100%)

1-Oxyl-2,2,6,6-tetramethyl-4-hydroxypiperidine [Oxidation of a Hindered Secondary Amine with DMD (isol.)].[97] To a magnetically stirred solution of 2,2,6,6-tetramethylpiperidinol (312 mg, 2.00 mmol) in acetone (20 mL) in a

100-mL flask was slowly added a pale yellow stock solution of DMD (60.0 mL, 0.0670 M, 4.00 mmol) at 0° (ice bath). The reaction mixture turned to deep yellow within 10 minutes. After stirring was continued for another 30 minutes, the solvent was removed on a rotary evaporator (20°, 15 mmHg) to afford the nitroxide (354 mg, 100%) as a bright yellow powder, mp 71–72°.

$KHSO_5$, (phosphate buffer) / Me_2CO, n-Bu_4NHSO_4, pH 7.5-8.5 / 22°, KOH, 2 h (93%)

Pyridine *N*-Oxide, Method A [Oxidation of Pyridine with DMD (in situ)].[121] To a 500-mL, three-necked flask, equipped with a mechanical stirrer, was added pyridine (1.00 g, 12.6 mmol), acetone (5 mL, 68.0 mmol), and phosphate buffer (50 mL). An aqueous solution of potassium monoperoxysulfate (100 mL, 18.3 g, 29.8 mmol) was added dropwise by means of an addition funnel. Simultaneously, an aqueous solution of KOH (1.00 N) was added in portions to maintain the pH at 7.5–8.0. After completion of the addition, the reaction mixture was stirred for 2 hours, then extracted with CH_2Cl_2 (4 × 30 mL). The combined extracts were dried ($MgSO_4$) and concentrated (20°, 15 mmHg). The residue was crystallized from a CH_2Cl_2/hexane mixture to afford the pyridine oxide (1.10 g, 93%) as a white crystalline solid, mp 64–65° (lit.[268] mp 65–66°); ^{1}H NMR (60 MHz, $CDCl_3$) δ 7.30–7.50 (m), 8.3–8.5 (m).

DMD (isol.) / acetone, hexane, 22° (88%)

Pyridine *N*-Oxide, Method B [Oxidation of Pyridine with DMD (isol.)].[121] To a stirred mixture of an acetone solution of pyridine (0.50 mL, 0.200 M, 7.90 mg, 0.10 mmol) and a hexane solution of decane (0.50 mL, 0.100 M) was added an acetone solution of DMD (1.00 mL, 0.116 M, 0.116 mmol) at room temperature (ca. 20°). The solvent was removed (20°, 15 mmHg) leaving pyridine *N*-oxide (8.4 mg, 88%).

DMD (isol.) / acetone, –20°, 36 h (100%)

Thiophene 1,1-Dioxide [Oxidation of Thiophene with DMD (isol.)].[141] A 50-mL flask was charged with an acetone solution of thiophene (10.0 mL, 84 mg, 1.00 mmol) at −20° and an acetone solution of DMD (isol.) (30 mL, 0.100 M, 3.00 mmol) was added rapidly under magnetic stirring at −20°. Stirring was continued at the same temperature for 36 hours. The solvent and unreacted DMD were removed (below −40°, 5 mmHg) to afford pure thiophene 1,1-dioxide as colorless plates (116 mg, ca. 100%); UV ($CHCl_3$) 245 (870) and 288 (1070) nm; IR (neat) 1152, 1306, 1327, 1530, 3100, 3175 cm^{-1}; ^1H NMR (300 MHz, $CDCl_3$, −40°) δ 6.53–6.61 (m, 2H), 6.75–6.83 (m, 2H); ^{13}C NMR (100 MHz, $CDCl_3$, −40°) δ 129.3, 113.1; HRMS calcd for $C_4H_4O_2S$, 115.9932, found 115.9931.

Me(Et)S=N–OAc → [DMD (isol.); acetone, rt, 2 h] → Me(Et)S(O⁻)=N–OAc (86%)

S-Ethyl-S-methyl-N-(acetyl)sulfoximine [Oxidation of a Sulfilimine with DMD (isol.)].[148] A 25-ml flask, supplied with a magnetic stirrer, was charged with a solution of the substituted sulfilimine (20 mg, 0.150 mmol) in acetone (5 mL). An acetone solution of DMD (4.0 mL, 0.080 M, 0.32 mmol) was added dropwise at 0°. The reaction mixture was stirred for 2 hours at room temperature; the reaction progress was monitored by TLC (silica gel). After complete consumption of the sulfilimine, the solvent was removed (20°, 15 mmHg) and the residue was purified by column chromatography on silica gel with a mixture of Et_2O/MeOH (95 : 5) as eluent, affording the sulfoximine (20 mg, 86%) as an oil; ^1H NMR (300 MHz, $CDCl_3$) δ 1.40 (t, $J = 7.0$ Hz, 3H), 2.40 (s, 3H), 3.30 (s, 3H), 3.50 (q, $J = 7.0$ Hz, 2H), 7.20–7.80 (m, 4H). Anal. Calcd for $C_5H_{11}NO_2S$: C, 40.25, H, 7.43, N, 9.39. Found: C, 40.01, H, 7.33, N, 9.46.

MeS–Ph → [DMD (isol.); acetone, 22°] → MeS(=O)–Ph (65%)

Methyl Phenyl Sulfoxide [Oxidation of a Thioether with DMD (isol.)].[121] A 10-mL flask was charged with a solution of phenyl methyl sulfide (13.6 mg, 0.11 mmol) in acetone (0.50 mL) and an acetone solution of DMD (0.58 mL, 0.189 M, 0.110 mmol) at ca. 20°. The reaction mixture was stirred at room temperature (ca. 22°) until consumption of the sulfide was complete as determined by GC analysis. The solvent was removed under reduced pressure (20°, 15 mmHg) and the crude product was purified by preparative TLC on silica gel by elution with a mixture of hexane and EtOAc to afford the solid phenyl methyl sulfoxide (11 mg, 65%).

O_2N–C_6H_4–O–P(=S)(OEt)$_2$ → [DMD (isol.); acetone, CH_2Cl_2, 22°, 5 min] → O_2N–C_6H_4–O–P(=O)(OEt)$_2$ (100%)

Diethyl 4-Nitrophenylphosphate [Oxidation of a Thiophosphate with DMD (isol.)].[150] An acetone solution of DMD (200 mL, 0.100 M, 20.0 mmol), dried over 4 Å molecular sieves, was added rapidly to a magnetically stirred dry CH_2Cl_2 solution of *O,O*-diethyl *O*-(4-nitrophenyl)thiophosphate (10.0 mL, 29 mg, 0.100 mmol) at room temperature (ca. 20°). After standing for 5 minutes, the crude reaction mixture was diluted with pentane (10 mL) and dried ($MgSO_4$). The drying agent was removed by filtration and washed with pentane (5 mL), and the filtrate was concentrated on a rotary evaporator (20°, 15 mmHg), affording the title compound (27 mg) in quantitative yield; ^{1}H NMR (300 MHz, $CDCl_3$) δ 1.38 (t, $J = 7.0$ Hz, 3H), 1.39 (t, $J = 7.0$ Hz, 3H), 4.25 (q, $J = 7.0$ Hz, 2H), 4.32 (q, $J = 7.0$ Hz, 2H), 7.38 (dd, $J = 9.0$ Hz, 1.0 Hz, 2H), 8.24 (dd, $J = 9.0$, 1.0 Hz, 2H); ^{13}C NMR (300 MHz, $CDCl_3$) δ 15.9, 16.1, 65.2, 65.3, 120.4, 125.0, 148.2, 155.7; ^{31}P NMR δ −6.6.

Ph Ph Ph Se Ph —DMD (isol.) / acetone, CH_2Cl_2, −50°→ Ph Ph Ph Se(=O) Ph (100%)

Tetraphenylselenophene 1-Oxide [Oxidation of a Selenophene with DMD (isol.)].[155] A cold (−50°) acetone solution of DMD (11.50 mL, 0.086 M, 1.00 mmol) was added to a cold (−50°), vigorously stirred dry CH_2Cl_2 solution of tetraphenylselenophene (2.0 mL, 435 mg, 1.00 mmol). After complete addition, the solvent was removed (−40°, 0.001 mmHg), to afford the title compound (451 mg) in quantitative yield; IR (KBr) 3056, 1596, 1573, 1487, 1444, 817, 788, 761, 743, 710, 693 cm^{-1}; ^{1}H NMR (400 MHz, $CDCl_3$) δ 6.91 (d, $J = 7.2$ Hz, 4H), 7.10 (t, $J = 7.2$ Hz, 4H), 7.16 (t, $J = 7.2$ Hz, 2H), 7.19–7.27 (m, 6H), 7.29–7.36 (m, 4H); ^{13}C NMR (100 MHz, $CDCl_3$) δ 128.0, 128.1, 128.6, 128.8, 129.4, 129.5, 131.9, 134.7, 147.2, 150.0; ^{77}Se NMR (76 MHz, $CDCl_3$) δ 1014. Anal. Calcd for $C_{28}H_{20}OSe$: C, 74.50, H, 4.47. Found: C, 73.98, H, 4.44.

t-Bu Se Se Bu-*t* —DMD (isol.) / acetone, CH_2Cl_2, 0-22°, 30 min→ *t*-Bu O=Se Se=O Bu-*t* (ca. 100%)

1,6-Di-*tert*-butyl-2,2,5,5-tetramethyl-7,8-diselenabicyclo[4.1.1]octane 7-*endo*, 8-*endo*-Dioxide [Oxidation of a Selenoether with DMD (isol.)].[134] To a stirred solution of the starting diselenetane (65 mg, 0.160 mmol) in CH_2Cl_2 (10 mL) was added an acetone solution of DMD (5.0 mL, 0.082 M, 0.410 mmol) in three portions at 0°. The mixture was warmed to room temperature (ca. 20°) and magnetically stirred for 30 minutes. The solvent was removed (20°, 20 mmHg) to give spectroscopically pure title compound (70 mg, ca. 100%) as a colorless powder, which decomposed above 80°; IR (KBr) 824 cm^{-1}; ^{1}H NMR (400 MHz, $CDCl_3$) δ 1.23–1.30 (m, 2H), 1.47 (s, 6H), 1.51 (s, 18H), 1.73 (s, 6H), 4.30–4.37 (m, 2H); ^{13}C NMR (100 MHz, $CDCl_3$) δ 28.4, 31.6, 35.4, 38.8, 43.9, 49.0, 97.3. Anal. Calcd for $C_{18}H_{34}Se_2O_2$: C, 49.09, H, 7.78. Found: C, 48.63, H, 7.74.

Ph_3P —DMD (isol.) / acetone, 22°→ $Ph_3P{=}O$ (ca. 100%)

Triphenylphosphine Oxide [Oxidation of a Phosphine with DMD (isol.)].[121] To a stirred solution of triphenylphosphine (26 mg, 0.100 mmol) in acetone (0.50 mL) was added a freshly prepared acetone solution of DMD (0.50 mL, 0.185 M, 6.9 mg, 0.0900 mmol) at room temperature (ca. 20°). Capillary GC was used to monitor the reaction progress by injecting 1.0-μL aliquots of the reaction mixture at intervals of 15–30 minutes (the peak areas of the triphenylphosphine and its oxide product were compared). Removal of the solvent (20°, 15 mmHg) afforded triphenylphosphine oxide (27.8 mg, ca. 100%).

$$PhN^{+}(Me)_2-O^{-} \xrightarrow[\text{CDCl}_3,\ 20^\circ]{\text{DMD (isol.)}} PhNMe_2 + {}^1O_2\ (0.33\%)$$

Singlet-Oxygen Generation by Oxidation of *N*,*N*-Dimethylaniline *N*-Oxide with DMD (isol.).[165] To a stirred solution of *N*,*N*-dimethylaniline *N*-oxide in $CDCl_3$ (1.0 mL, 0.50 mM) was added a $CDCl_3$ solution of DMD (0.080 M, 3 equiv.) at 20°. The reaction mixture was magnetically stirred at this temperature for 10 minutes. The consumption of the dioxirane was monitored by means of the peroxide test (KI/HOAc), while the amount of singlet oxygen (0.33%) was determined by its characteristic IR chemiluminescence at 1268 nm using a photodiode detector.

DMD (isol.), acetone, CH_2Cl_2, 0° (ca. 100%)

[(2*S*,4*S*,5*S*)-5-Acetylamino-4-benzoyloxy-2-methoxycarbonyltetrahydropyran-2-yl] Propen-2-yl *N*-Acetyl-2′,3′-di-*O*-acetyl-5′-cytidylate [Oxidation of a Phosphite to a Phosphate with DMD (isol.)].[158] To a cold (0°) stirred solution of the starting phosphite (9.7 mg, 0.0120 mmol) in CH_2Cl_2 (1 mL) was added an acetone solution of DMD (162 μL, 0.083 M, 0.0310 mmol) at 0°. After 10 minutes, the reaction mixture was concentrated on a rotary evaporator (0°, 15 mmHg) to give the title compound (9.9 mg, 100%) as a colorless foam (the product was contaminated with the α-linked diastereomer); ^1H NMR (500 MHz, $CDCl_3$) δ 1.89 (s, 4.3H), 1.91 (s, 3.2H), 2.02 (s, 6.6H), 2.08 (s, 12.0H), 2.15 (s, 15.2H), 2.22 (s, 4.4H), 2.25 (s, 3.7H), 2.74 (dd, J = 13.5, 4.8 Hz, 1H), 2.92 (dd, J = 13.7, 4.5 Hz, 1.4H), 3.32 (dd, J = 14.9, 4.4 Hz, 0.2H), 3.76–3.80

(m, 1.8H), 3.80 (m, 1.8H), 3.82 (s, 3.4H), 3.85–3.90 (m, 1.8H), 4.18–4.22 (m, 2.6H), 4.29–4.49 (m, 10.2H), 4.54–4.63 (m, 4.0H), 4.69 (t, $J = 7.4$ Hz, 2.0H), 5.26–5.40 (m, 6.6H), 5.44 (d, $J = 3.6$ Hz, 2.5H), 5.47–5.54 (m, 1.5H), 5.64 (dd, $J = 4.8, 4.0$ Hz, 1.0H), 5.89–6.02 (m, 2.6H), 6.24 (d, $J = 2.8$ Hz, 1.0H), 6.37 (d, $J = 6.2$ Hz, 0.9H), 6.67 (d, $J = 7.7$ Hz, 0.8H), 6.95 (d, $J = 8.7$ Hz, 0.8H), 7.36–7.48 (m, 7.9H), 7.53–7.58 (m, 2.5H), 7.83 (d, $J = 7.3$ Hz, 0.5H), 7.92–8.02 (m, 5.4H), 8.56 (d, $J = 7.7$ Hz, 1.1H), 8.90 (s, 1.2H), 9.12 (s, 0.9H); ^{31}P NMR (203 Hz, $CDCl_3$) δ −4.54, −4.60.

I, SO_2CF_3 —DMD (isol.), acetone, Et_2O, rt, 1 h→ SO_2CF_3 (58%)

1-[(Trifluoromethyl)sulfonyl]methylcyclohexene [Oxidation of an Iodoalkane with DMD (isol.)].[169] A 10-mL flask, equipped with a magnetic stirring bar, was charged at 25° with 356 mg (1.00 mmol) of *trans*-1-iodo-2-[(trifluoromethyl)sulfonyl]methylcyclohexane in ether (5 mL) and an acetone solution of DMD (20.0 mL, 0.100 M, 2.00 mmol). After the mixture was stirred at 25° for 1 hour, the solvent was removed (10°, 15 mmHg) and the product was dried to give the title alkene (132 mg, 58%) as a colorless oil.

I —DMD (isol.), acetone, N_2, 22°, 5 h→ I, OH (43%) + OH, OH (4%)

***trans*-2-Iodocyclohexanol [Oxidation of Iodocyclohexane with DMD (isol.)].**[168] A 25-mL, round-bottomed flask was charged with an acetone solution of DMD (11.0 mL, 0.090 M, 1.0 mmol) at ca. 20° under a N_2 atmosphere. While stirring magnetically, the cyclohexyl iodide (210 mg, 1.00 mmol) was added. After five hours, GLC analysis indicated formation of *trans*-2-iodocyclohexanol (102 mg, 43%) and 1,2-cyclohexanediol (5 mg, 4%).

O, O, =N−NMe_2, 93% ee —DMD (isol.), acetone, 0°, 2 h→ O, O, CN (92%) 93% ee

(*S*)-4-Cyano-2,2-dimethyl-1,3-dioxalane [Conversion of a Hydrazone into a Nitrile with DMD (isol.)].[113] A cold (0°) solution of DMD (0.100 M, 1.00 mmol) in acetone (10.0 mL) was added to a cold (0°) acetone solution of (+)-2,3-*O*-isopropylidene-D-glyceraldehyde *N*,*N*-dimethylhydrazone (5 mL, 86 mg, 0.50 mmol, enantiomeric purity 93%) with vigorous magnetic stirring. The reaction progress was monitored by GLC analysis, which indicated that the starting material was converted into the nitrile product within 2 hours. Removal of the acetone on a rotary evaporator (20°, 15 mmHg) afforded the title compound (58 mg, 92%); $[\alpha]_D$ + 1.36 (c 1.33, $CHCl_3$), 93% optically pure.

PhCHMeCN + [1-chloro-3-nitrobenzene] —t-BuOK, THF/DMF, Ar→ [intermediate] —DMD (isol.), H_2O, DMF/THF/acetone, −70°, 5 min→ [phenol] (87%) + [nitro compound] (3%)

2-(2-Chloro-4-hydroxyphenyl)-2-phenylpropionitrile [Tandem Nuleophilic Addition/Conversion of a Nitrobenzene into a Phenol with DMD (isol.)].[110] An oven dried, 100-mL, three-necked, round-bottom flask, equipped with a magnetic stirring bar, was charged with t-BuOK (123 mg, 1.10 mmol) and THF (20 mL) at −70° under an argon gas atmosphere. A solution of 2-phenylpropionitrile (131 mg, 1.00 mmol) and 1-chloro-3-nitrobenzene (157 mg, 1.00 mmol) in DMF (1.0 mL) was added at −70° within 2 minutes via syringe. The resulting mixture was magnetically stirred for 5 minutes, and a precooled (−70°) acetone solution of DMD (14.5 mL, 1.20 mmol, 0.0830 M) was added in one portion. After 5 minutes, H_2O (18.0 μL, 1.00 mmol) was added. The mixture was stirred for an additional 5 minutes, hydrolyzed with saturated aqueous NH_4Cl (1.0 mL), raised to 20°, and dried over $MgSO_4$. The drying agent was removed by filtration, washed with THF (3 × 20 mL), and the solvent was evaporated (30°, 12 mmHg). The residue was purified by chromatography on silica gel (4 : 1 hexane/EtOAc, followed by 2 : 1 hexane/EtOAc as eluents) to give the nitro compound (8.6 mg, 3%) as a minor product, and the title phenol (223 mg, 87%) as colorless flakes, mp 180–182°; IR (KBr) 3375, 2236, 1607, 1575, 1495, 1430, 1312, 1291, 1215 cm^{-1}; ^1H NMR (200 MHz, $CDCl_3$) δ 2.09 (s, 3H), 5.69 (br s, 1H), 6.81–6.89 (dd, $J = 8.6$, 2.6 Hz, 1H), 6.91–6.94 (d, $J = 2.6$ Hz, 1H), 7.20–7.39 (m, 5H), 7.42–7.50 (d, $J = 8.6$ Hz, 1H); ^{13}C NMR (50 MHz, $CDCl_3$) δ 29.6, 44.8, 113.9, 118.9, 122.0, 125.8, 127.5, 128.6, 128.8, 129.0, 135.3, 141.3, 156.5; EIMS m/z (%): 257 (M^+), 242 (100), 222, 215, 207, 206, 195, 177, 165, 152, 89, 77. Anal. Calcd for $C_{15}H_{12}ClNO$: C, 69.91, H, 4.69, N, 5.43, Cl, 13.76. Found: C, 69.74, H, 4.43, N, 5.29, Cl, 13.83.

PhC(O)C(=PPh_3)C(O)OMe —DMD (isol.), acetone, CH_2Cl_2, rt, 1 h→ PhC(O)C(OH)₂C(O)OMe (ca.100%)

Methyl 3-Phenyl-2,2-dihydroxy-3-oxopropionate [Oxidation of a Phosphorane to a Ketone Hydrate with DMD (isol.)].[159] A 25-mL, round-bottomed flask was charged with a solution of methyl-3-oxo-3-phenyl-2-(triphenylphosphoranylidene)propionate (219 mg, 0.500 mmol) in CH_2Cl_2 (2.0 mL). Under vigorous magnetic stirring, an acetone solution of DMD (15.0 mL, 1.5 mmol, 0.100 M) was added and the stirring continued at room temperature for one hour until all the starting material had been consumed as monitored by TLC. The reaction mixture was concentrated (20°, 15 mmHg) and the residue was purified by flash column chromatography on silica gel (hexane/EtOAc, 1 : 1 as eluent) to give the product as a yellow oil (105 mg, ca. 100%); IR (neat) 3600–3300, 3060, 2940, 1760, 1750, 1690, 1600, 1450, 1440, 1230, 1130, 1100, 1010 cm^{-1}; ^{1}H NMR ($CDCl_3$) δ 3.66 (s, 3H), 5.81 (br s, 2H), 7.38–7.44 (m, 2H), 7.54–7.58 (m, 1H), 8.05–8.08 (m, 2H); ^{13}C NMR ($CDCl_3$) δ 53.4, 91.9, 128.6, 129.0, 129.8, 131.1, 170.1, 191.4; HRMS (M + H) calcd for $C_{10}H_{10}O_5$, 211.0603, found 211.0606.

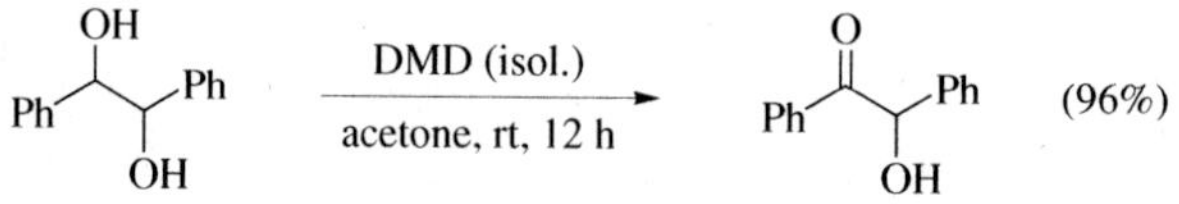

DMD (isol.), acetone, rt, 12 h (96%)

Benzoin. Method A [Oxidation of a Benzyl Alcohol to an Aryl Ketone with DMD (isol.)].[269] A 25-mL flask was charged with hydrobenzoin (214 mg, 1.00 mmol) in acetone (1.0 mL) at room temperature, and then a solution (at ca. 20°) of DMD (1.50 mmol, 0.080 M) in acetone (19.0 mL) was added rapidly under vigorous magnetic stirring. The solvent was removed by distillation (20°, 15 mmHg) on a Vigreux column, and the residue was purified by flash column chromatography on silica gel with 1 : 1 hexane/EtOAc as eluent, to afford the benzoin (204 mg, 96%).

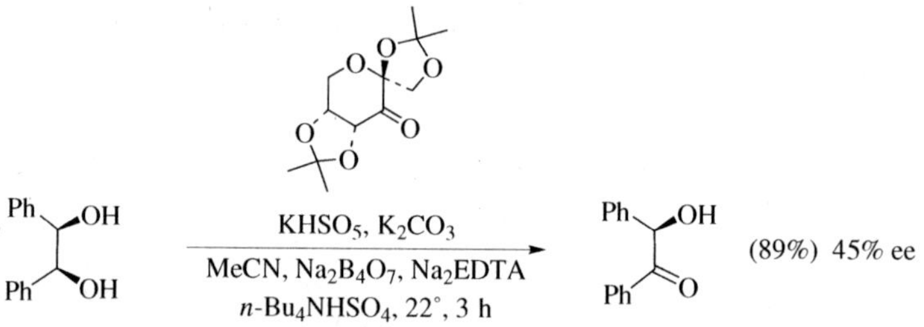

$KHSO_5$, K_2CO_3, MeCN, $Na_2B_4O_7$, Na_2EDTA, n-Bu_4NHSO_4, 22°, 3 h (89%) 45% ee

(*R*)-Benzoin. Method B [Catalytic Asymmetric Oxidation of Hydrobenzoin].[197] To a MeCN solution of *meso*-hydrobenzoin (1.5 mL, 21.4 mg, 0.100 mmol) was added the ketone catalyst 1,2:4,5-bis-*O*-(1-methylethylidene)-β-D-*erythro*-2,3-hexodiulo-2,6-pyranose (77.5 mg, 0.300 mmol), Bu_4NHSO_4 (1.5 mg, 4.0 μmol), and $Na_2B_4O_7$ (1.0 mL, 0.050 M) in aqueous Na_2EDTA (4×10^{-4} M) while stirring magnetically at 0°. Solutions of potassium monoperoxy sulfate (92.0 mg, 0.150 mmol) and K_2CO_3 (87.0 mg, 0.630 mmol), each in an aqueous solution (0.65 mL) of Na_2EDTA (4×10^{-4} M), were added simultaneously using syringes over a period of 2 hours. The mixture was stirred for another hour and then diluted with H_2O (20 mL), extracted with ether (3 × 20 mL), washed with H_2O (2 × 10 mL), and dried over $MgSO_4$. After removal of the solvent on a rotary evaporator (20°, 20 mmHg), the residue was purified by column chromatography (silica gel) to give the recovered ketone (40–60%) and benzoin (18.9 mg, 89%), with an ee value of 45% for the R enantiomer.

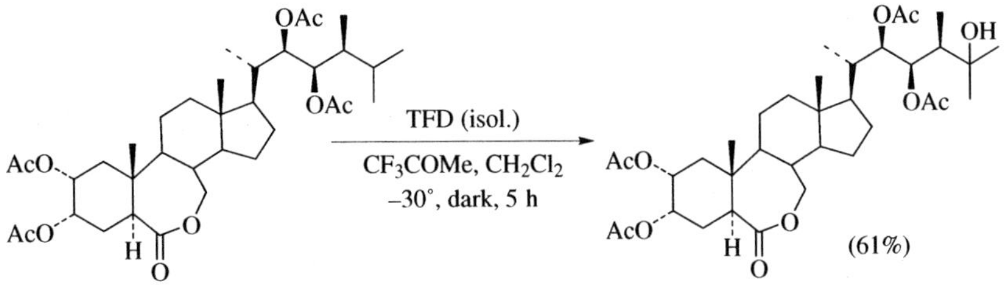

TFD (isol.)
CF_3COMe, CH_2Cl_2
–30°, dark, 5 h
(61%)

2,3,22,23-Tetra-*O*-acetyl-25-hydroxybrassinolide [Hydroxylation of a Tertiary Carbon Center with TFD (isol.)].[270] To a stirred solution of 2,3,22,23-tetra-*O*-acetylbrassinolide (20 mg, 0.0300 mmol) in dry CH_2Cl_2 (0.40 mL) was added dropwise a trifluoroacetone solution of TFD (0.20 mL, 0.50 M, 0.10 mmol) at $-30°$. The reaction mixture was stirred magnetically in the dark for 5 hours at $-30°$. The solvent was removed on a rotary evaporator (20°, 15 mmHg), and the residue was purified by flash chromatography on silica gel (2 : 1 hexane/EtOAc as eluent) to give the title compound (12.5 mg, 61%) as colorless needles, mp 226–229°; ^1H NMR (600 MHz, $CDCl_3$) δ 0.73 (s, 3H), 0.98 (s, 3H), 1.03 (d, $J = 6.8$ Hz, 3H), 1.04 (d, $J = 7.3$ Hz, 3H), 1.15 (s, 3H), 1.16 (m, 1H), 1.18 (m, 2H), 1.22 (s, 3H), 1.25 (m, 1H), 1.27 (m, 1H), 1.28 (m, 1H), 1.41 (m, 1H), 1.62 (m, 2H), 1.66 (dq, $J = 7.3$, 1.0 Hz, 1H), 1.68 (m, 1H) 1.73 (m, 1H), 1.75 (s, 1H), 1.76 (m, 1H), 1.92 (m, 1H), 1.93 (m, 1H), 1.98 (m, 1H), 2.00 (s, 6H), 2.02 (s, 3H), 2.09 (m, 1H), 2.11 (s, 3H), 2.29 (ddd, $J = 15.1$, 12.2, 2.4 Hz, 1H), 2.99 (dd, $J = 12.2$, 4.4 Hz, 1H), 4.04 (dd, $J = 12.2$, 9.3 Hz, 1H), 4.12 (dd, $J = 12.2$, 1.0 Hz, 1H), 4.87 (ddd, $J = 12.2$, 4.4, 2.4 Hz, 1H), 5.12 (dd, $J = 9.3$, 1.0 Hz, 1H), 5.37 (m, 1H), 5.49 (dd, $J = 9.3$, 1.0 Hz, 1H); ^{13}C NMR (150 MHz, $CDCl_3$) δ 9.1, 11.7, 12.7, 15.5, 20.8, 21.1, 21.2, 22.2, 24.7, 26.6, 28.1, 28.6, 29.3, 37.1, 38.4, 38.9, 39.2, 39.4, 42.1, 42.5, 43.4, 51.3, 52.4, 58.4, 68.0, 68.9, 70.4, 72.3, 72.4, 75.5, 170.0, 170.2, 170.5, 171.0, 175.0; HRMS (FAB) (M + H) calcd for $C_{36}H_{57}O_{11}$, 665.3901, found, 665.3900.

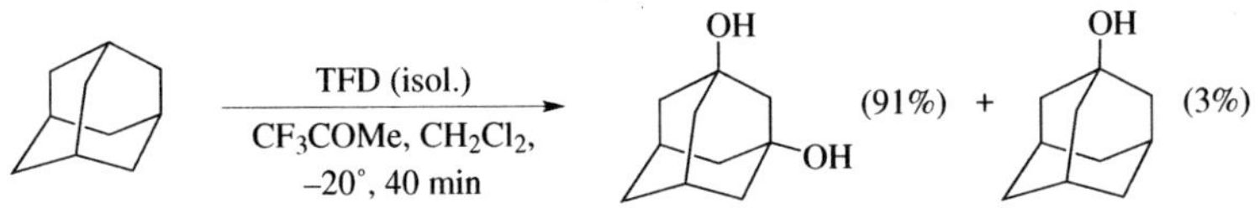

TFD (isol.)
CF_3COMe, CH_2Cl_2,
–20°, 40 min
(91%) + (3%)

1,3-Dihydroxyadamantane [Dihydroxylation of Adamantane with TFD (isol.)].[178] A solution of TFD (4.60 mL, 0.50 M, 2.30 mmol) in trifluoroacetone/CH_2Cl_2 (2 : 1 v/v) at $-20°$ was added to a solution of adamantane (136 mg, 0.100 mmol) in CH_2Cl_2 (5 mL) also at $-20°$, while stirring vigorously magnetically. The progress of the reaction was followed by GLC analysis, which indicated that 97% of the adamantane was converted to its hydroxylated products in 40 minutes. Removal of the solvent on a rotary evaporator ($-20°$, 15 mmHg) afforded a mixture of the 1,3-dihydroxyadamantane (156 mg, 91%) and the monohydroxy adamantane (4.6 mg, 3%).

KHSO$_5$, NaHCO$_3$

MeCN, H$_2$O, Na$_2$EDTA, 22°, 3.5 h

(91%)

Cycloheptanone [Oxidation of a Secondary Alcohol to a Ketone under In Situ Catalytic Conditions].[52] To a vigorously stirred solution of cycloheptanol (38.4 mg, 0.300 mmol) in MeCN (1.5 mL), was added an aqueous Na$_2$EDTA solution (4×10^{-4} M) of 7*H*-dibenzo[*g*,*i*]-1,5-dioxacycloundecin-5,8,11(9*H*)-trione (catalyst, 1.0 mL, 17.6 mg, 0.0600 mmol) at room temperature (ca. 20°). A mixture of KHSO$_5$ (282 mg, 0.600 mmol) and NaHCO$_3$ (156 mg) was added in portions, and consumption of the alcohol was complete after 4 hours as confirmed by GLC analysis. The reaction mixture was poured into water (20 mL), extracted with CH$_2$Cl$_2$ (3 × 30 mL), and dried (Na$_2$SO$_4$). After removal of the solvent on a rotary evaporator (20°, 15 mmHg), the residue was purified by flash column chromatography (Et$_3$N-buffered silica gel) to give cycloheptanone (34.4 mg, 91%).

CO$_2$Me KHSO$_5$, NaHCO$_3$ MeCN, H$_2$O, Na$_2$EDTA, 22°, 24 h CO$_2$Me OH (70%)

Methyl (*S,*S**)-6-Ethyl-2-hydroxytetrahydropyran-2-carboxylate [Hydroxylation of a Secondary Carbon Center under In Situ Catalytic Conditions].**[195] To a magnetically stirred MeCN solution (30 mL) of methyl-2-oxo-octanoate (86 mg, 0.50 mmol) was added an aqueous Na$_2$EDTA solution (20 mL, 4×10^{-4} M), followed by a mixture of KHSO$_5$ (1.54 g, 2.5 mmol) and NaHCO$_3$ (0.65 g) at ca. 20° over a period of 1 hour. After stirring for 24 hours, the reaction mixture was poured into brine (10 mL) and extracted with EtOAc (3 × 30 mL). The combined organic layers were dried (MgSO$_4$), and the solvent was removed on a rotary evaporator (20°, 15 mmHg). The residue was purified by flash column chromatography on silica gel (1 : 4 EtOAc/hexane as eluent) to give the title compound (66 mg, 70%) as a colorless syrup; ^{1}H NMR (300 MHz, CDCl$_3$) δ 0.89 (t, $J = 7.5$ Hz, 3H), 1.20–1.74 (m, 6H), 1.79–1.97 (m, 2H), 3.60 (s, 1H), 3.82 (s, 3H), 3.85 (m, 1H); ^{13}C NMR (68 MHz, CDCl$_3$) δ 9.8, 18.4, 28.8, 30.0, 30.4, 53.1, 72.5, 94.8, 171.7; EIMS m/z (%): 171 (M$^+$ − OH, 14), 130 (12), 129 (100), 111 (41).

Ph Si–H Me DMD (isol.) acetone, CH$_2$Cl$_2$, 0°, 18 min Ph Si–OH Me (98%) 97% ee

(*R*)-Methylphenyl(1-naphthyl)silanol [Hydroxylation of a Silane with TFD (isol.)].[39] A cold (−20°) 1,1,1-trifluoropropanone solution of TFD (80.0 mL, 0.500 M, 4.00 mmol) was rapidly added to a cold (−20°) solution of 96.5% optically pure (*R*)-methylphenyl(1-naphthyl)silane (0.992 g, 4.00 mmol) in dry CH_2Cl_2 (30 mL). Capillary GC analysis indicated complete consumption of the silane immediately on addition of the oxidant. Removal of the solvent on a rotary evaporator (10–20°, 80–100 mmHg) afforded very pure (97% ee) silanol (1.03 g, 98%).

CO, OC, Re, CO → DMD (isol.), acetone, 0° → ReO_3 (74%)

(η^5-Pentamethylcyclopentadienyl)trioxorhenium [Oxidation of a Rhenium Complex with DMD (isol.)].[210] A 100-ml flask, equipped with a magnetic stirring bar, was charged with a chilled (0°) anhydrous acetone solution of $Cp^*Re(CO)_3$ (20 mL, 80 mg, 0.20 mmol) and an acetone solution of DMD (25.0 mL, 0.05 M, 1.25 mmol) was added dropwise at 0°. The reaction terminated within a few minutes under immediate gas evolution, accompanied by a slight darkening of the pale yellow solution. The solvent was removed on a rotary evaporator (20°, 15 mmHg) until a volume of ca. 3 mL remained. Hexane (15 mL) was added to the residue, and the resulting solution cooled to 0° for 3 hours to afford the title compound (54 mg, 74%) as yellow needles; IR (KBr) 913 and 878 cm^{-1}.

$(CO)_5Cr$, OEt, Ph → DMD (isol.), acetone, O_2, −20°, 4 h → Ph—≡—CO_2Et (90%)

Ethyl Phenylpropiolate [Oxidation of a Fischer Carbene Complex with DMD (isol.)].[205] To a vigorously stirred acetone solution of the Fischer carbene complex (10 mL, 99 mg, 0.28 mmol), previously filtered over Celite and protected from light at −20°, was added an acetone solution of DMD (13.6 mL, 0.041 M, 0.56 mmol) dropwise over 4 hours. The reaction progress was monitored by TLC (silica gel), which indicated complete consumption of the complex within minutes. The solvent was evaporated (room temperature, 20 mmHg), the residue taken up in CH_2Cl_2 (10 mL), and the chromium oxides were removed by filtration through Celite. The solvent was removed on a rotary evaporator (room temperature, 20 mmHg) to afford pure ethyl phenylpropiolate (44 mg, 90%).

H, Si—H, SiH_3, Fe, OC, CO → DMD (isol.), acetone, toluene, −78° to 22°, 50 min → OH, Si—OH, SiH_3, Fe, OC, CO (89%)

[Dicarbonyl(η^5-pentamethylcyclopentadienyl)ferrio]-1,1-dihydroxydisilane [Hydroxylation of an Iron-Complexed Silane with DMD (isol.)].[227] A cold (−78°) acetone solution of DMD (11.0 mL, 1.3 M, 0.84 mmol) was added to a solution of $Me_5Cp(CO)_2FeSi_2H_5$ (130 mg, 0.420 mmol) in toluene (5 mL) at −78° while stirring magnetically. After complete addition (ca. 10 minutes), the color of the reaction mixture changed from yellow to orange. Subsequently, the temperature of the reaction mixture was raised to ca. 20° and after 80 minutes a material precipitated. The solvent was removed (20°, 15 mmHg), the residue was washed with pentane (10 mL) and dried over $MgSO_4$ to give the title compound (98 mg, 89%) as a yellow powder, mp 65–66°; IR (toluene) 1931, 1986, 2107, 3479 cm^{-1}; ^{1}H NMR (400 MHz, benzene-d_6) δ 3.57 (s, ^1J (SiH) = 182 Hz, 3H), 2.28 (br s, 2H), 1.58 (s, 15H); ^{13}C NMR (100 MHZ, benzene-d_6) δ 9.7, 95.7, 215.9; ^{29}Si NMR (benzene-d_6) δ −95.26 (s), 96.70 (s). Anal. Calcd for $C_{12}H_{20}FeO_4Si_2$: C, 42.35, H, 5.92. Found C, 42.26, H, 6.01.

TABULAR SURVEY

The oxidation of allenes, alkynes, arenes, heteroatom substrates, alkanes and silanes, and organometallic compounds is presented in the appended tables. The tabular survey covers the literature reported through March, 2005.

The tables are arranged in the order of the discussion in the section on Scope and Limitations. Thus, the data on the oxidation of allenes and alkynes, arenes, heteroatom substrates, alkanes and silanes, and organometallic compounds by isolated dioxiranes (DMD and TFD) are presented in Tables 1A, 2A, 3A–3E, 4A, 5A, 5E, and 6. Oxidations with in situ generated dioxiranes of allenes, alkynes, arenes, heteroatom substrates, and alkanes are shown in Tables 1B, 2B, 3F–H, 4B, and 5C. Regioselective oxidations of alkanes by isolated dioxiranes are compiled in Table 5B. Asymmetric oxidations of alkanes by in situ generated dioxiranes are shown in Table 5D. Miscellaneous oxidations are presented in Table 7.

The entries within each table are arranged in order of increasing carbon number of the substrates. The carbon count is based on the total number of carbon atoms. Yields of products are given in parentheses, and an em-dash (—) indicates that no yield was reported in the original reference. Data on conversion (% conv.) are included in the product column, preferentially in subtables, and labeled as such. Ratios of different products or diastereomers are given without parentheses. For those reactions that were carried out both with and without a co-solvent, the cosolvent is enclosed in parentheses to indicate that its use is optional.

The following abbreviations are used in the tables:

Ac	acetyl
Ad	adamantyl
Bn	benzyl
Bz	benzoyl
Boc	*tert*-butyloxycarbonyl

Cbz	benzyloxycarbonyl
Cp	cyclopentadienyl
Cy	cyclohexyl
de	diastereomeric excess
DEK	diethyl ketone
DMD	dimethyldioxirane
DMD (in situ)	in situ generated dioxirane
DMD (isol.)	isolated dimethyldioxirane in acetone
DMD-d_6 (isol.)	isolated hexadeuterated dimethyldioxirane in acetone-d_6
DMIPS	dimethylisopropylsilyl
DMM	dimethoxymethane
dr	diastereomeric ratio
EDTA	ethylenediaminetetraacetic acid
Na_2EDTA	disodium salt of ethylenediaminetetraacetic acid
F112	1,1,1,2-tetrachlorodifluoroethane
ee	enantiomeric excess
LDA	lithium diisopropylamide
Ms	methanesulfonyl
MOM	methoxymethyl
Naph	naphthyl
NPhth	*N*-phthalimido
Oxone®	potassium monoperoxysulfate ($2KHSO_5{\cdot}KHSO_4{\cdot}K_2SO_4$)
PMB	*p*-methoxybenzyl
PMP	*p*-methoxyphenyl
PG	protecting group
PPTS	pyridinium *p*-toluenesulfonate
TAS	tris(dimethylamino)sulfonium difluorotrimethyl siliconate
TBDPS	*tert*-butyldiphenylsilyl
TBS	*tert*-butyldimethylsilyl
TES	triethylsilyl
Tf	trifluoromethanesulfonyl (trifyl)
TFD	methyl(trifluoromethyl)dioxirane
TFD (in situ)	in situ generated methyl(trifluoromethyl)dioxirane
TFD (isol.)	isolated methyl(trifluoromethyl)dioxirane
TFP	1,1,1-trifluoro-2-propanone
TIPS	triisopropylsilyl
TMP	tetramesitylporphyrin
TMS	trimethylsilyl
Tp	hydridotris(1-pyrazoylborate)
Tp*	3,5-dimethylhydridotris(1-pyrazoylborate)
TPP	tetraphenylporphyrin
TPS	triphenylsilyl
Ts	*p*-toluenesulfonyl

TABLE 1A. OXIDATION OF ALLENES AND ALKYNES BY ISOLATED DIOXIRANES

	Substrate	Conditions	Product(s) and Yield(s) (%)	Refs.
C_4 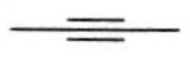		DMD, acetone, rt, 20 h	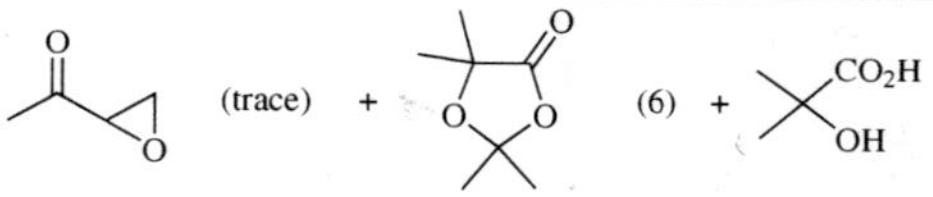 (trace) + (6) + CO_2H, OH (42)	63
C_5 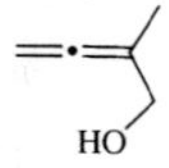	HO	DMD, acetone, rt, 10 min	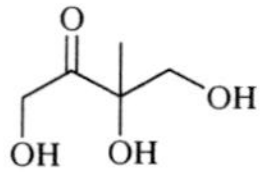 O, OH, OH, OH (89)	56, 57, 60
C_{5-7}	R^1, R^1, R^2, $( )_n$, HO	DMD, acetone, rt, 10 min	**I** or **II**	56
			R^1 | R^2 | n | Product: Me | H | 1 | **I** (55); H | Me | 1 | **II** (89); Me | H | 2 | **I** (92)	
C_6	TMS	DMD, acetone, rt, 140 h	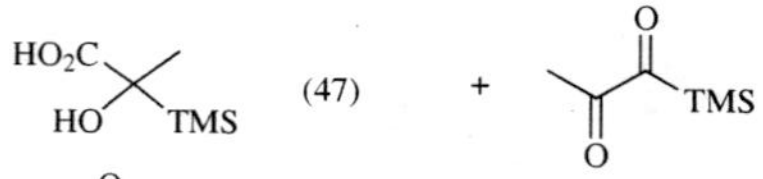 HO_2C, HO, TMS (47) + TMS (—)	63
	HO	DMD, acetone, rt, 5 min	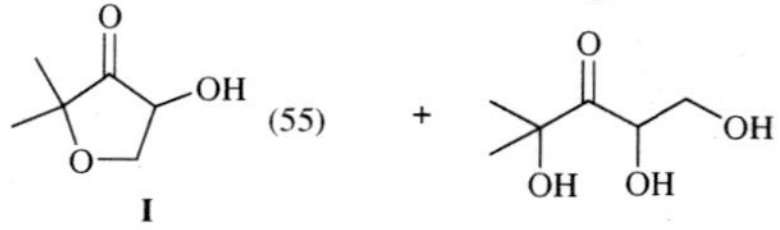 **I** (55) + OH, OH, OH (10)	56, 57, 60
		DMD, acetone, CH_2Cl_2, TsOH, rt, 10 min	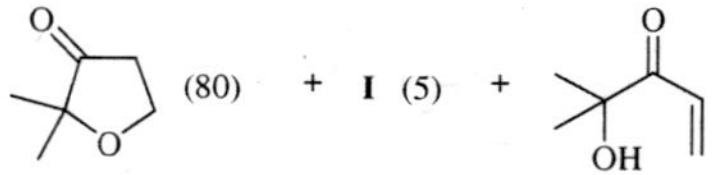 (80) + **I** (5) + OH (10)	56, 57, 60
		1. Lewis acid/ligand (pre-mixed, 9.0 eq), furan 2. DMD (3-5 eq), CH_2Cl_2, 8-10 h	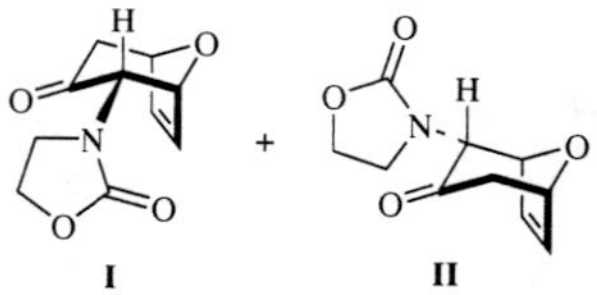 **I** + **II**	271

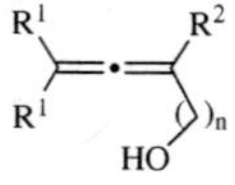

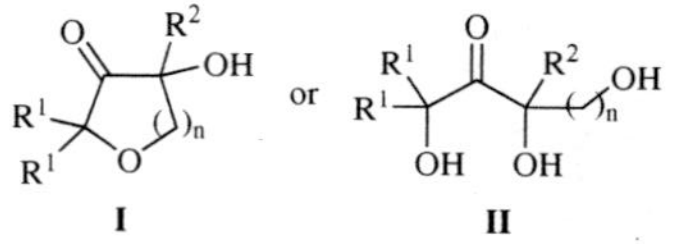

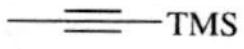

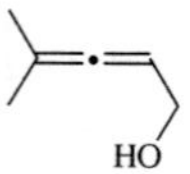

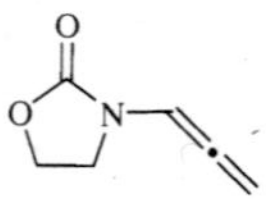

Ligand	eq	Lewis Acid (eq)	Temp	**I**	% ee	**II**	% ee
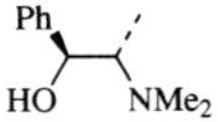 Ph, HO, NMe_2	1.3	$Zn(OTf)_2$ (1.1)	–55°	(0)	—	(48)	8
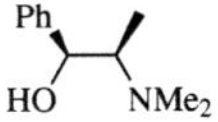 Ph, HO, NMe_2	1.3	$Zn(OTf)_2$ (1.1)	–55°	(0)	—	(58)	2
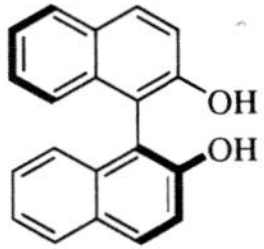 OH, OH	1.3	$Zn(OTf)_2$ (1.1)	–55°	(0)	—	(70)	21
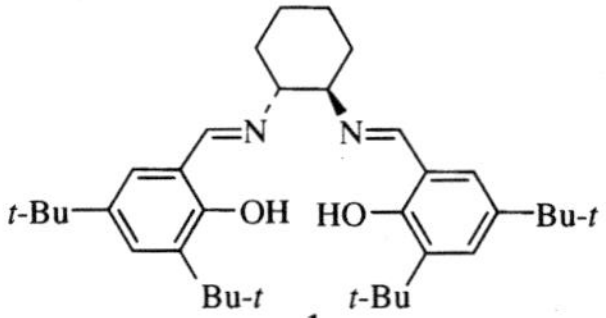 *t*-Bu, OH, HO, Bu-*t*, Bu-*t*, *t*-Bu, N, N **1**	1.0	$Co(OAc)_2{\bullet}4H_2O$ (1.1)	–55°	(0)	—	(—)	—
1	1.0	$Co(OAc)_2{\bullet}4H_2O$ (0.25)	–55°	(0)	—	(46)	1
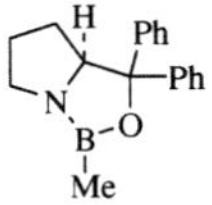 H, Ph, Ph, N, B, O, Me	0.25	—	–78°	(0)	—	(40)	3
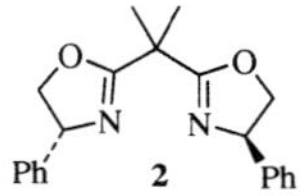 O, N, Ph, O, N, Ph **2**	1.3	$Sn(OTf)_2$ (1.1)	–55°	(0)	—	(48)	5
2	0.32	MgI_2 (0.25)	–78°	(0)	—	(46)	3
2	1.1	$Cu(OTf)_2$ (0.85)	–55°	(62)	78	(0)	—
2	0.32	$Cu(OTf)_2$ (0.25)	–78°	(46)	74	(0)	—

TABLE 1A. OXIDATION OF ALLENES AND ALKYNES BY ISOLATED DIOXIRANES (*Continued*)

Substrate	Conditions	Product(s) and Yield(s) (%)	Refs.
C_6 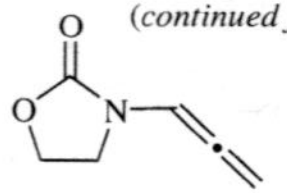 (structure)	*(continued from previous page)*		

Ligand	eq	Lewis Acid (eq)	Temp	I	% ee	II	% ee
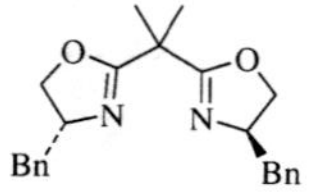 (structure: Bn, Bn)	1.2	$Cu(OTf)_2$ (1.1)	–55°	(53)	22	(0)	—
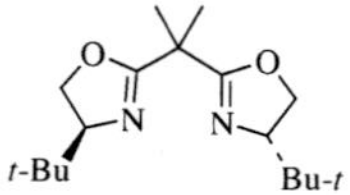 (structure: t-Bu, Bu-t)	0.32	$Cu(OTf)_2$ (0.25)	–78°	(0)	—	(46)	10
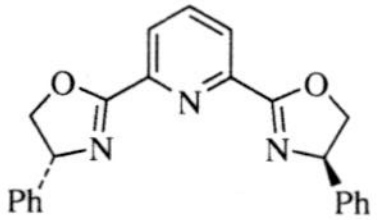 (structure: Ph, Ph)	0.32	$Cu(OTf)_2$ (0.25)	–78°	(54)	61	(0)	—
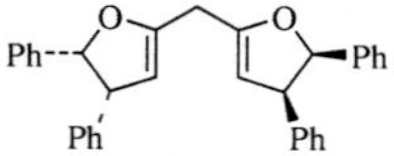 (structure: Ph, Ph, Ph, Ph)	0.32	$Cu(OTf)_2$ (0.25)	–78°	(46)	82	(0)	—
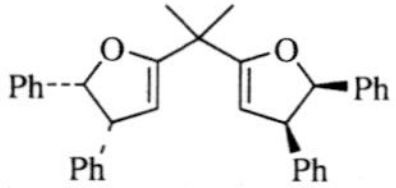 (structure: Ph, Ph, Ph, Ph)	0.32	$Cu(OTf)_2$ (0.25)	–78°	(46)	90	(0)	—
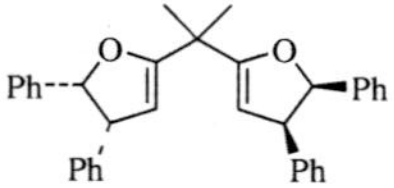 (structure: Ph, Ph, Ph, Ph)	0.12	$Cu(OTf)_2$ (0.10)	–78°	(76)	59	(0)	—
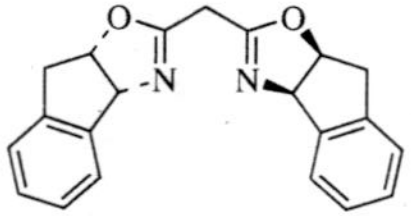 (structure)	0.32	$Cu(OTf)_2$ (0.25)	–78°	(84)	2	(0)	—

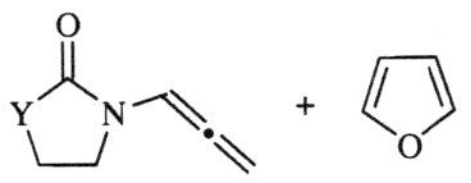

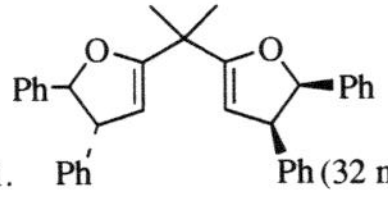

1. Ph, Ph (32 mol%), $CuOTf_2$ (25 mol%)
2. DMD (2-5 eq), acetone/CH_2Cl_2, 8-10 h (syringe pump method)
3. 4 Å MS

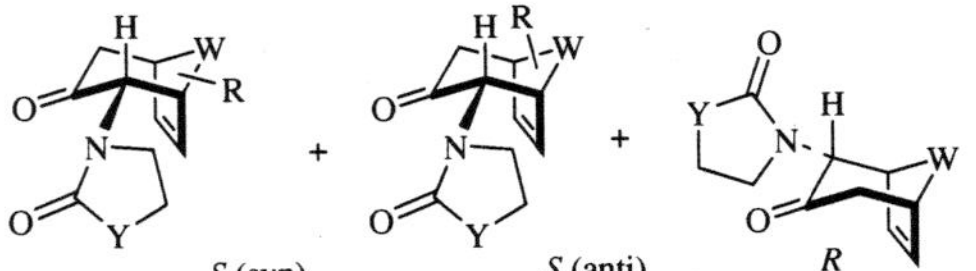

I S (syn) + II S (anti) + III R — 271

I + II + III
W = O; R = H

Y	Additive	I + II	I:II	III	% ee
O	—	(90)	1:1	(0)	92
O	$AgSbF_6$	(91)	1:1	(0)	99
O	—	(0)	—	(87)	82
CH_2	—	(60)	1:1	(0)	58
$(CH_2)_2$	—	(47)	1:1	(0)	14

Y = O + cyclopentadiene — " — **I + II** (90), **I**:**II** = 1:1, 43% ee; W = CH_2; Additive = $AgSbF_6$ — 271

Y = O + 2-R-furan — " — **I + II**, W = O — 271

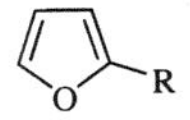

R	Additive	I + II	I:II	% ee
Me	—	(37)	(4:1)	—
Me	$AgSbF_6$	(88)	(20:1)	71
CO_2Me	—	(33)	(100:0)	—
CO_2Me	$AgSbF_6$	(61)	(100:0)	67

Y = O + 2,5-dimethylfuran — " — **I + II** (81), 36% ee; W = O — 271

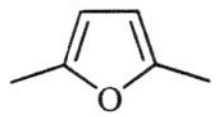

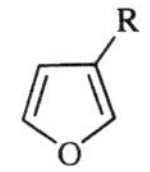

Y = O + 3-R-furan — " — **I + II**, W = O — 271

R	Additive	I + II	I:II	% ee
Me	$AgSbF_6$	(91)	(0:100)	99
Br	$AgSbF_6$	(58)	(1:14)	84
CH_2OTPS	$AgSbF_6$	(66)	(1:2.3)	92

TABLE 1A. OXIDATION OF ALLENES AND ALKYNES BY ISOLATED DIOXIRANES (*Continued*)

Substrate	Conditions	Product(s) and Yield(s) (%)	Refs.
C_6			
CO_2H	DMD (6-10 eq), acetone, rt, 0.5-2 h	(18)	272
	DMD, TsOH, acetone	n = 1, 2 (—)	56
R^2, R^1, R^3, R^3, CO_2H	DMD, acetone	R^1, R^2, OH, R^3, R^3 **I** + R^1, R^2, OH, R^3, R^3 **II**	60
		R^1 R^2 R^3 **I + II** **I:II**	
		Me H H (76) 40:60	
		Me Me Me (96) 50:50	
		n-Pr H Me (84) 40:60	
C_7			
Bu-*t*	DMD, acetone, rt, 20 h	*t*-Bu (6) + OAc (18) + Ac (40) + Ac (4) + *t*-Bu (3) + *t*-Bu, CO_2H, OH (9)	63
HO	DMD, acetone, rt	OH **I** (92)	57
	DMD, acetone, CH_2Cl_2, TsOH, rt, 10 min	(79) + **I** (10)	57

Substrate	Conditions	Product(s) and Yield(s) (%)	Refs.
OHC	DMD, acetone, H_2O	HO, O, O, OH (42)	59
	DMD, acetone, TsOH	O, O, OH (61)	59
	DMD, acetone, MeOH	O, O, OMe (35) + O, OHC (38)	59
CO_2H	DMD, acetone	O, OH, O, O (73)	60
N OH	DMD, acetone	O, O, N, OH (34)	58
HO	DMD, acetone	OH, O, O, **I** (—)	56
	DMD, acetone, rt	**I** (96)	57
OH	DMD, acetone, rt	HO, O, O (48)	57
OMe, HO	DMD, acetone, rt, 15 min	O, OMe, O **I** + HO, O, OMe, O **II** **I** + **II** (85), **I**:**II** = 25:75	56, 57, 60

TABLE 1A. OXIDATION OF ALLENES AND ALKYNES BY ISOLATED DIOXIRANES (*Continued*)

Substrate	Conditions	Product(s) and Yield(s) (%)	Refs.
C_7			
NH_2	DMD, acetone, K_2CO_3	O, O, N, OH (—) + O, OH, OH, CN (53)	58
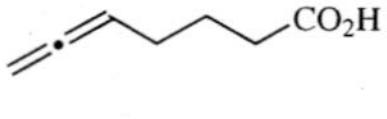 CO_2H	DMD (6-10 eq), K_2CO_3, acetone, rt, 0.5-2 h	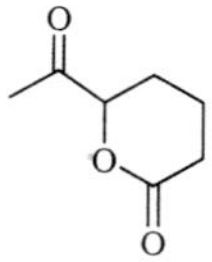 O, O, O (68)	272
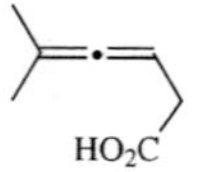 HO_2C	DMD (6-10 eq), acetone, rt, 0.5-2 h	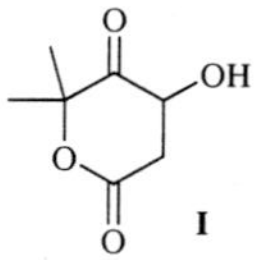 O, OH, O, O **I** (87)	272
	DMD (6-10 eq), $NaHCO_3$, acetone, rt, 0.5-2 h	**I** (52) + O, O, O (10)	272
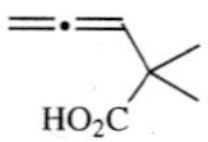 HO_2C	DMD (6-10 eq), acetone, rt, 0.5-2 h	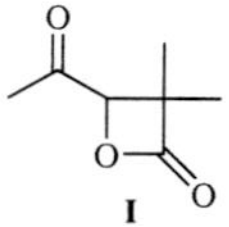 O, O, O **I** (52)	272
	DMD (6-10 eq), acetone $NaHCO_3$, rt, 0.5-2 h	**I** (84)	272
	DMD (6-10 eq), acetone TsOH , rt, 0.5-2 h	**I** (15) + O, O, O, O (43) + O, O, OH, O, O (32)	272

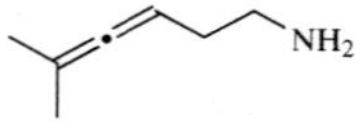

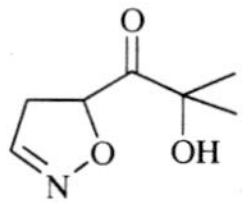

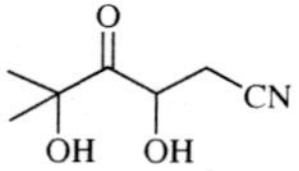

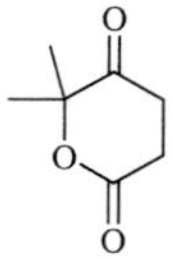

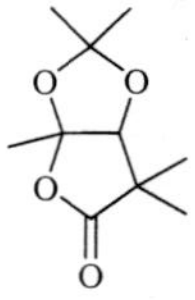

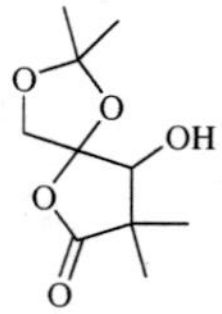

C_{7-8}

n = 1,2

DMD, acetone

(21)

58

C_{7-9}

DMD, acetone

I or II

n		
1	**I**	(87)
2	**II**	(82)
3	**II**	(91)

60

DMD, acetone

R^1	R^2	n	
Me	H	1	(88)
H	Me	1	(48)
H	H	2	(55)
Me	H	2	(75)
H	Me	2	(65)

56

DMD, acetone, rt

R^1	R^2	
H	H	(55)
H	Me	(65)
Me	H	(75)

57

C_{7-11}

DMD, acetone

I + II

55

R^1	R^2	R^3	R^4	Temp	Time	**I + II**	**I:II**
Me	Me	Me	Me	–50°	30 min	(44)	50:50
Me	Me	*n*-Bu	H	rt	10 min	(95)	90:10
Me	Me	*t*-Bu	H	rt	10 min	(84)	100:0
n-Bu	*n*-Bu	H	H	–40°	1.5 h	(80)	50:50
n-Oct	H	H	H	–40°	2.5 h	(50)	83:17

TABLE 1A. OXIDATION OF ALLENES AND ALKYNES BY ISOLATED DIOXIRANES (*Continued*)

Substrate	Conditions	Product(s) and Yield(s) (%)	Refs.
C_8			
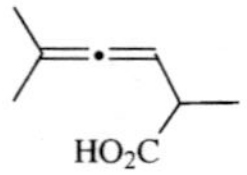 HO_2C	DMD (6-10 eq), acetone, rt, 0.5 to 2 h	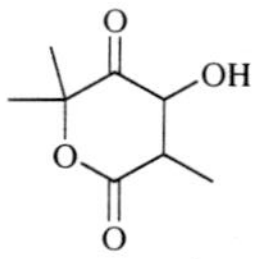 I (92)	272
	DMD (6-10 eq), $NaHCO_3$, acetone, rt, 0.5-2 h	I (54) 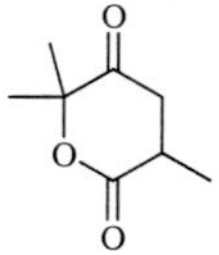 + (28)	272
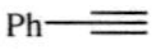 Ph	DMD, acetone, CH_2Cl_2, 0°, 6 h	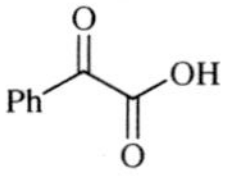 Ph, OH (12) + PhCHO (38)	64
	TFD, TFP, CH_2Cl_2, 0°, 7 min	PhCHO (49)	64
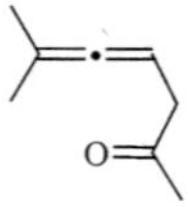	DMD, acetone, CH_2Cl_2, MeOH, 3 Å MS	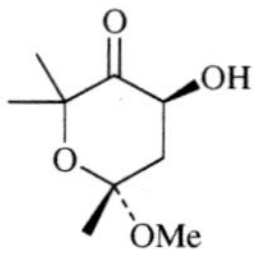 OH, OMe (47) 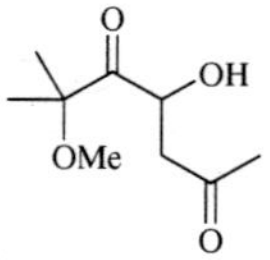 + OH, OMe (27)	59
	DMD, acetone, H_2O	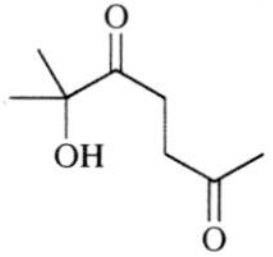 OH (58) 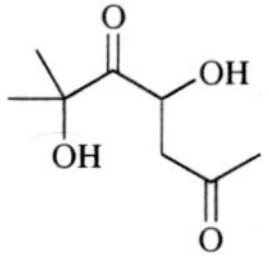 + OH, OH (29)	59
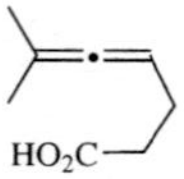 HO_2C	DMD, acetone, H_2O	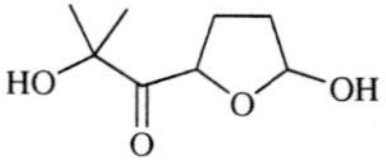 HO, OH (68)	59

Substrate	Conditions	Product(s) and Yield(s) (%)	Refs.
	DMD, acetone, MeOH	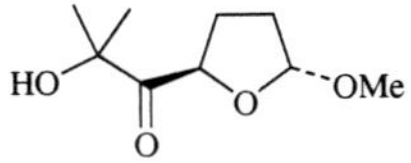 HO, OMe, O (51)	59
	DMD, acetone (dry)	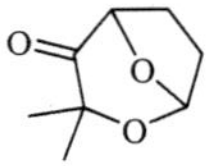 O (17)	59
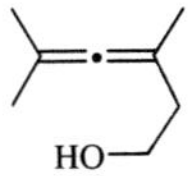 HO	DMD, acetone, rt	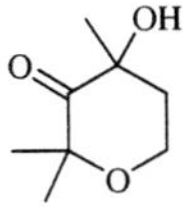 OH, O (65)	57
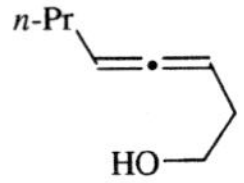 *n*-Pr, HO	DMD, acetone, rt	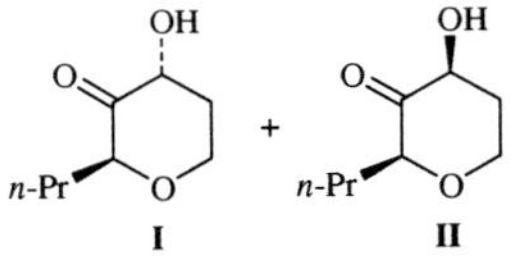 OH, O, *n*-Pr, I + OH, O, *n*-Pr, II; **I** + **II** (67), **I**:**II** = 1.2:1	57
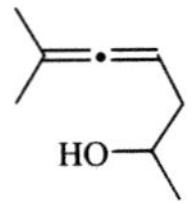 HO	DMD, acetone, rt	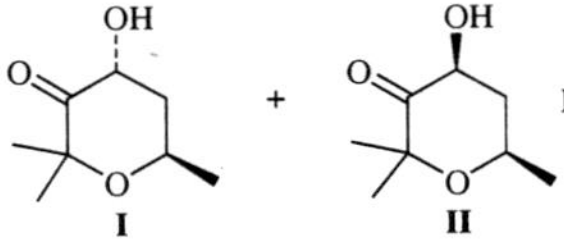 OH, O, I + OH, O, II; **I** + **II** (80), **I**:**II** = 1.3:1	57
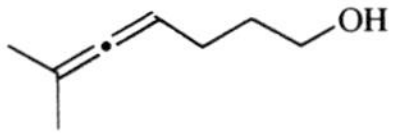 OH	DMD, acetone, rt	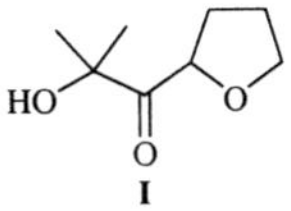 HO, O, O, I (88)	57
	DMD, acetone, CH_2Cl_2, TsOH, rt, 1.7 h	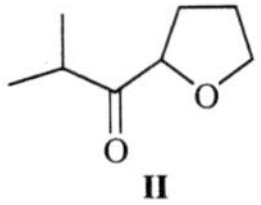 O, O, II + **I**; **I** + **II** (50), **I**:**II** = 5:1	57

TABLE 1A. OXIDATION OF ALLENES AND ALKYNES BY ISOLATED DIOXIRANES (*Continued*)

Substrate	Conditions	Product(s) and Yield(s) (%)	Refs.
C_{8-9}			
CO_2H (n)	DMD (6-10 eq), acetone, rt, 0.5-2 h	OH, O, O (n) **I** n = 1 (82) **II** n = 2 (71)	272
	DMD (6-10 eq), Cs_2CO_3, acetone, rt, 0.5-2 h	**I** + **II** **I** n = 1 (100) **II** n = 2 (83)	272
C_9			
	DMD, acetone, K_2CO_3, rt, 20 min	O, O (95)	16
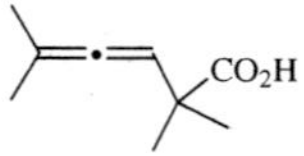 CO_2H	DMD (6-10 eq), acetone, rt, 0.5-2 h	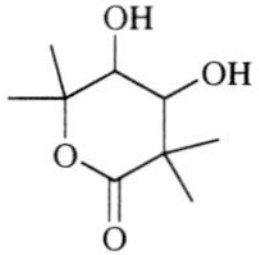 OH, OH, O, O **I** (96)	272
	DMD (6-10 eq), acetone $NaHCO_3$, rt, 0.5-2 h	**I** (44) 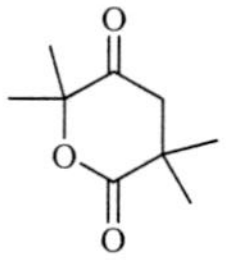 + O, O, O **II** (11) 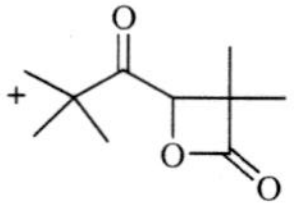 + O, O, O (11)	272
	DMD (6-10 eq), acetone, TsOH, rt, 0.5-2 h	**I** + **II** (88), **I**:**II** = 1:4	272
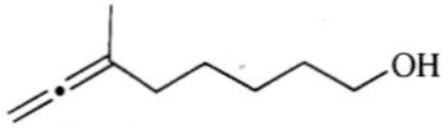 OH	DMD, acetone, rt	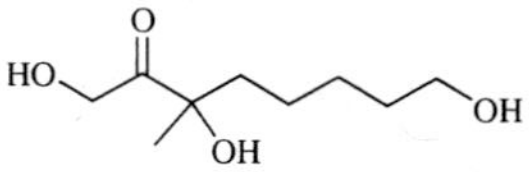 HO, O, OH, OH (66)	57
	DMD, acetone, KOAc, rt	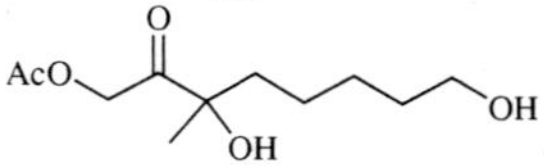 AcO, O, OH, OH (63)	57

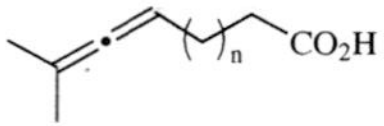

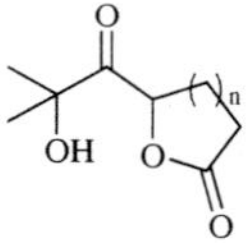

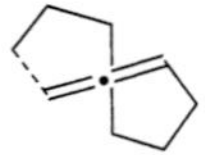

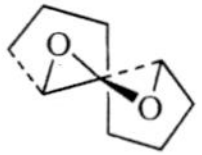

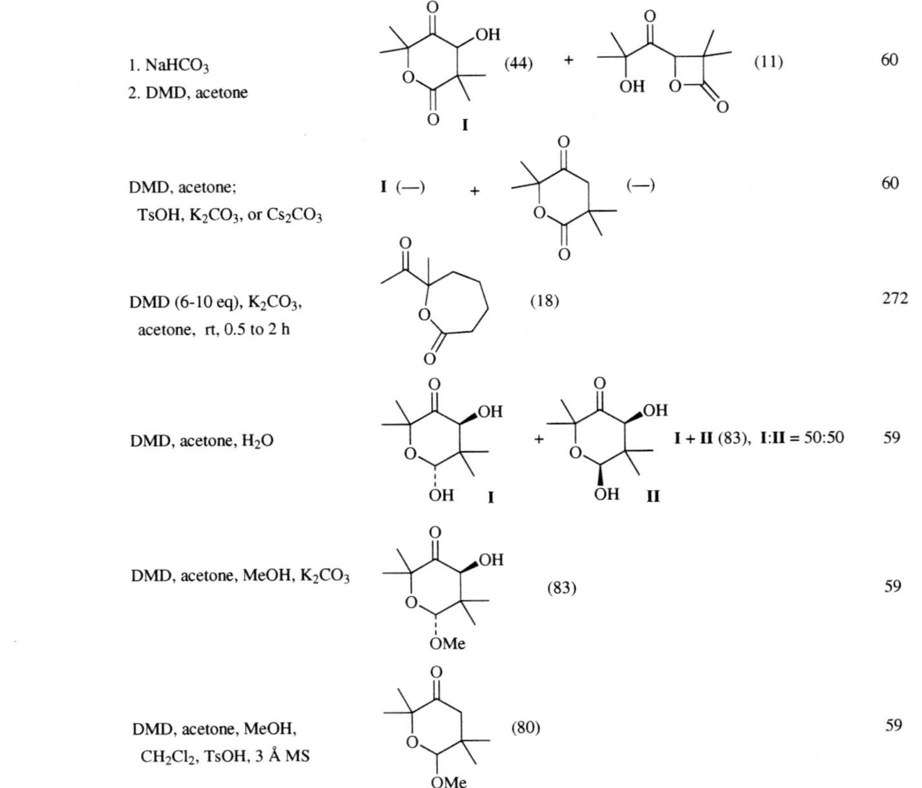

Reactant	Conditions	Product(s) and Yield(s) (%)	Refs.
	1. $NaHCO_3$ 2. DMD, acetone	**I** (44) + (11)	60
	DMD, acetone; TsOH, K_2CO_3, or Cs_2CO_3	**I** (—) + (—)	60
	DMD (6-10 eq), K_2CO_3, acetone, rt, 0.5 to 2 h	(18)	272
	DMD, acetone, H_2O	**I** + **II** (83), **I**:**II** = 50:50	59
	DMD, acetone, MeOH, K_2CO_3	(83)	59
	DMD, acetone, MeOH, CH_2Cl_2, TsOH, 3 Å MS	(80)	59

TABLE 1A. OXIDATION OF ALLENES AND ALKYNES BY ISOLATED DIOXIRANES (*Continued*)

Substrate	Conditions	Product(s) and Yield(s) (%)	Refs.
C_9			
CHO	DMD, acetone (dry), CH_2Cl_2, 3 Å MS	O, O (31) + O, O, O (59)	59
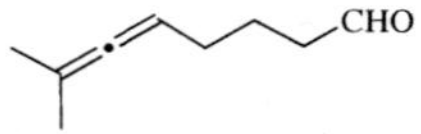 CHO	DMD, acetone, H_2O	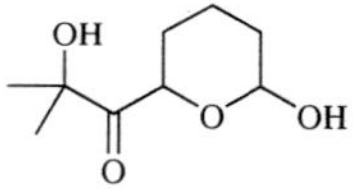 OH, O, O, OH (38)	59
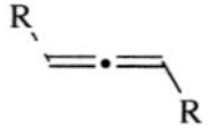 R, R	DMD, acetone, K_2CO_3, rt	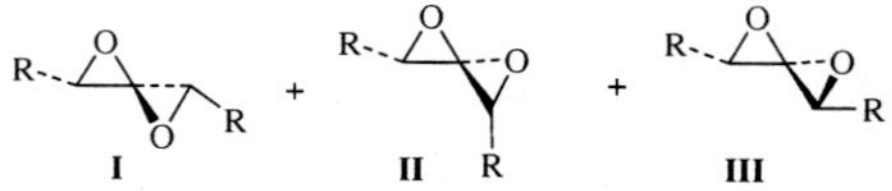 R, O, O, R **I** + R, O, O, R **II** + R, O, O, R **III**	16
		R / Time / **I:II:III**	
		n-Pr / 20 min / 1:1:0.15 / (99)	
		i-Pr / 20 min / 2:1:0 / (75)	
		t-Bu / 25 min / 1:0:0 / (98)	
C_9			
HO	DMD, acetone, rt	OH, O, O (72)	57
C_{9-21}			
O, Y, N, R[1], R[2], R[3] + Z	DMD (2-3 eq), acetone, THF, –40 to 50°	R[1], R[2], R[3], Y, O, N, H, O, Z **I** + R[1], R[2], R[3], Y, O, N, O, Z **II**	273

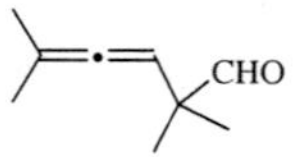

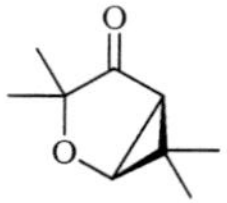

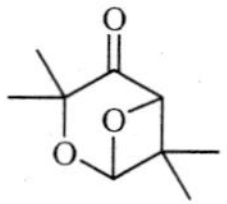

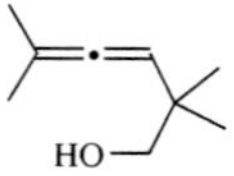

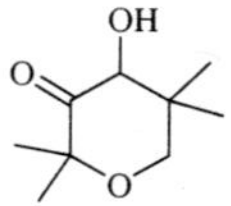

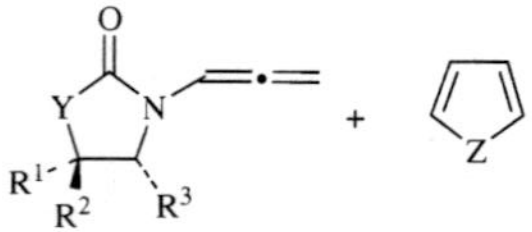

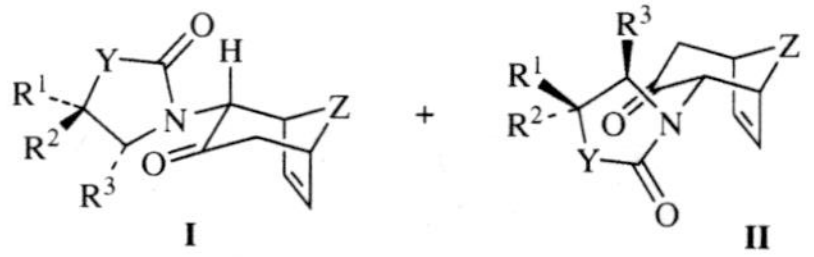

R^1	R^2	R^3	Y	Z	**I + II**	**I:II**
H	H	*i*-Pr	O	O	(70)	55:45
H	H	Ph	O	CH_2	(40)	> 95:5
H	H	$PhCH_2$	O	O	(67)	77:23
Me	H	Ph	NMe	O	(60)	> 95:5
Me	H	Ph	NMe	CH_2	(83)	> 96:4
H	H	Ph_2CH	O	O	(74)	> 95:5
H	H	Ph_2CH	O	CH_2	(62)	93:7
Ph	Ph	*i*-Pr	O	O	(72)	94:6

C_{10}

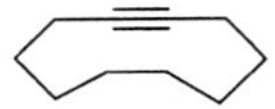

TFD, TFP, CH_2Cl_2, 0°, 3 min — (83) + (12) — 64

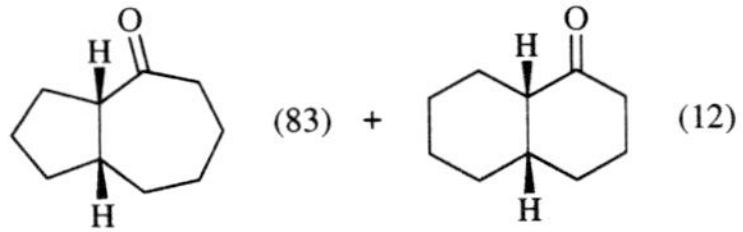

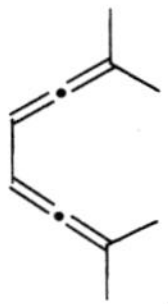

DMD (1 eq), acetone, rt 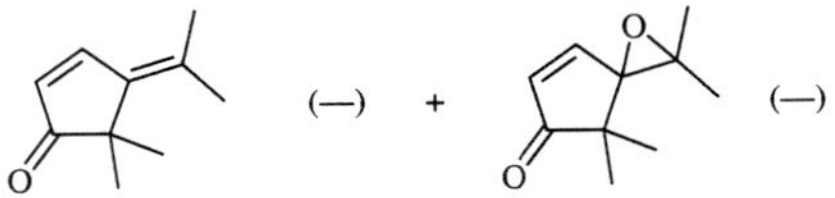 — (—) + (—) — 61

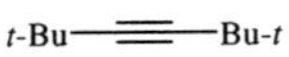

t-Bu—≡—Bu-*t*

DMD, acetone, rt, 140 h — Bu-*t* (7.5) + Bu-*t* (6.2) + Bu-*t* (22.4) — 63

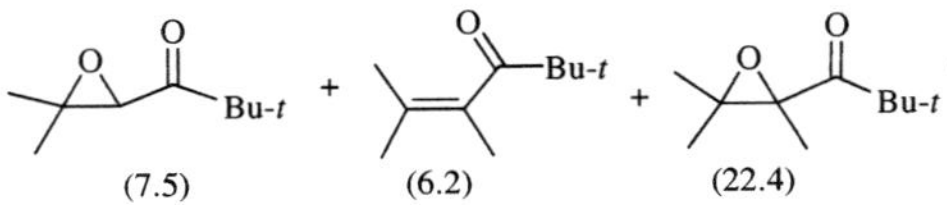

n-Pr, CO_2H

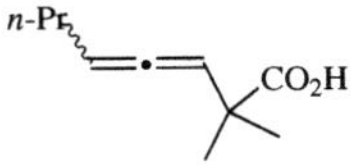

DMD (6-10 eq), acetone, rt, 0.5-2 h — *n*-Pr, OH (83) **I** — 272

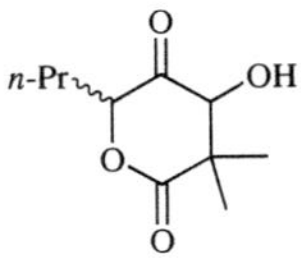

TABLE 1A. OXIDATION OF ALLENES AND ALKYNES BY ISOLATED DIOXIRANES (*Continued*)

Substrate	Conditions	Product(s) and Yield(s) (%)	Refs.
C_{10}			
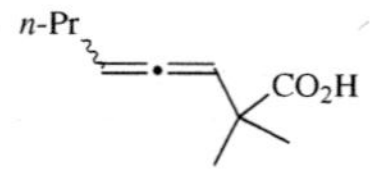 n-Pr, CO_2H	DMD (6-10 eq), Na_2CO_3, acetone, rt, 0.5-2 h	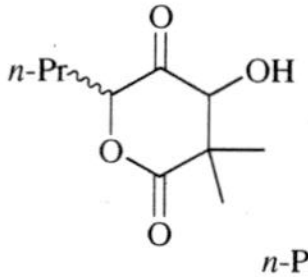 n-Pr, OH (37) 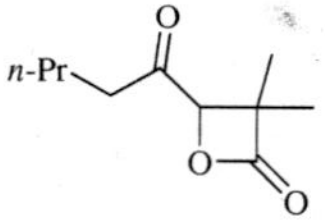 + n-Pr (9) 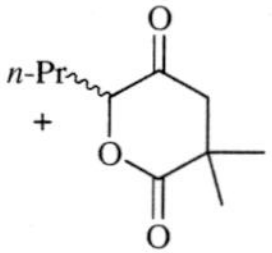 + n-Pr (7)	272
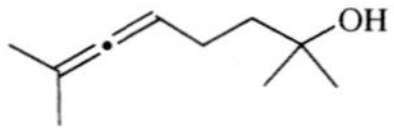 OH	DMD, acetone, rt	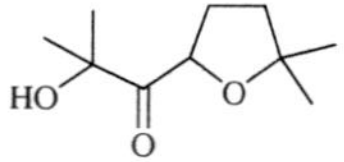 HO (70)	57
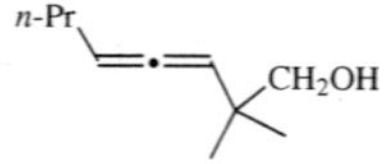 n-Pr, CH_2OH	DMD, acetone, rt	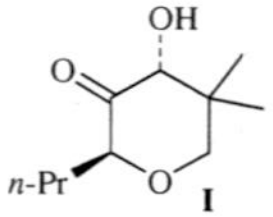 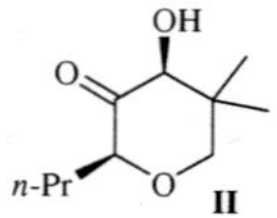 **I** + **II** (75), **I**:**II** = 2.5:1	57
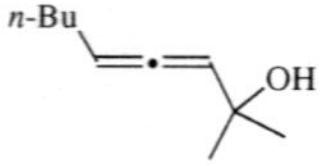 n-Bu, OH	DMD, acetone, rt	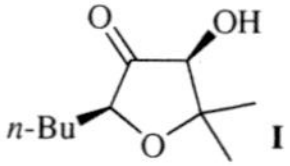 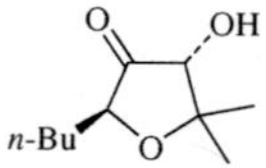 **I** + **II** (56)	56, 57, 60
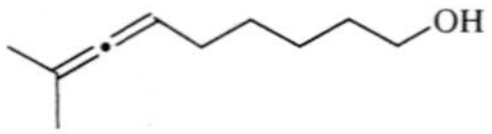 OH	DMD, acetone, rt	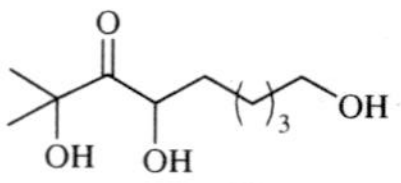 **I** (36) 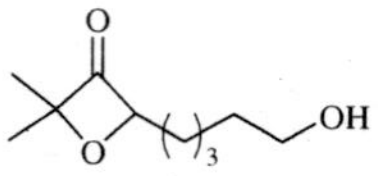 + **II** (23) 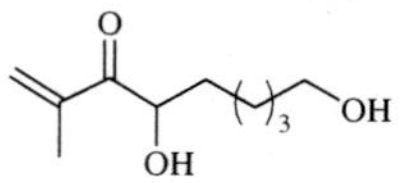 + **III** (13)	56, 57
	DMD (dry), acetone, rt	**I** (5) + **II** (38) + **III** (33)	57

Substrate	Conditions	Product(s) and Yield(s) (%)	Refs.
	DMD, acetone, K_2CO_3, rt, 10 min	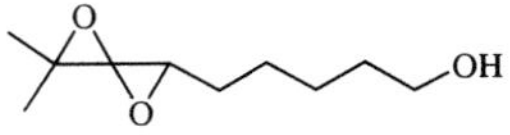 OH (96)	56, 57
	DMD (excess), acetone, rt, 24 h	**II** (—)	61
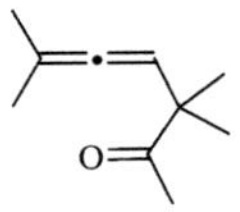	DMD, acetone, CH_2Cl_2, MeOH, 3 Å MS	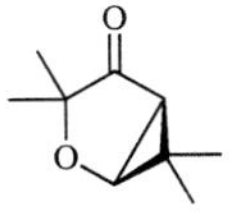 (25) + OH, OMe (60)	59
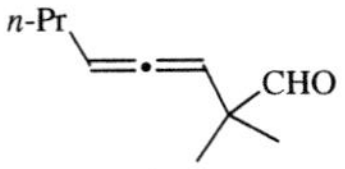 n-Pr, CHO	DMD, acetone, MeOH, CH_2Cl_2	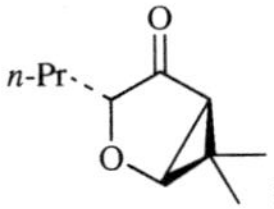 n-Pr **I** (32) 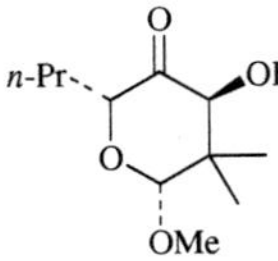 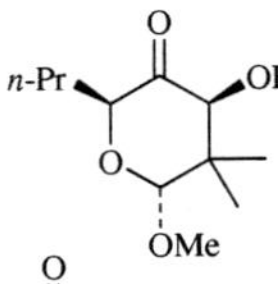 + n-Pr, OH, OMe **II** (—) + n-Pr, OH, OMe **III** (—) **II**:**III** = 80:20	59
	DMD, acetone (dry), CH_2Cl_2, 3 Å MS	**I** 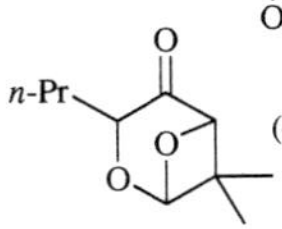 + n-Pr (—)	59
	DMD, acetone, CH_2Cl_2, TsOH, H_2O	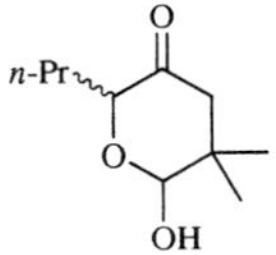 n-Pr, OH (72)	59

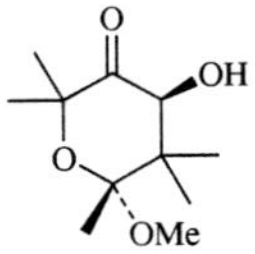

TABLE 1A. OXIDATION OF ALLENES AND ALKYNES BY ISOLATED DIOXIRANES (*Continued*)

Substrate	Conditions	Product(s) and Yield(s) (%)	Refs.
C_{10} 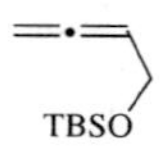 TBSO	DMD, acetone, rt	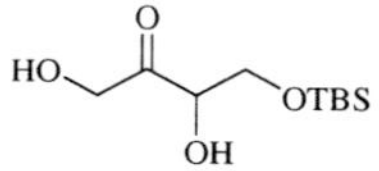 HO, O, OTBS, OH (37)	57
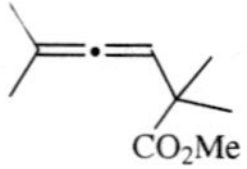 CO_2Me	DMD (6-10 eq), acetone, rt, 0.5-2 h	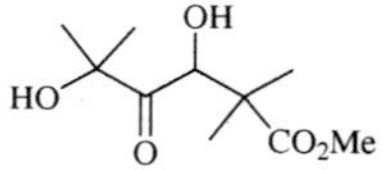 HO, O, OH, CO_2Me (80)	272
C_{11} 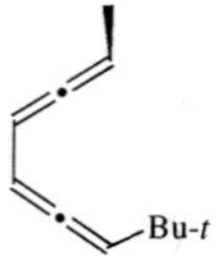 Bu-*t*	DMD, acetone, rt, 24 h	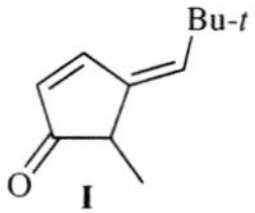 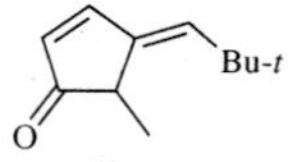 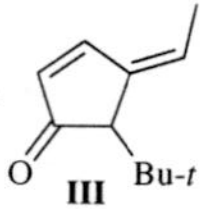 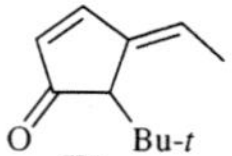 Bu-*t*, O, **I** + Bu-*t*, O, **II** + O, Bu-*t*, **III** + O, Bu-*t*, **IV** **I + II + III + IV** (—), **I:II:III:IV** = 26:30:28:16	61
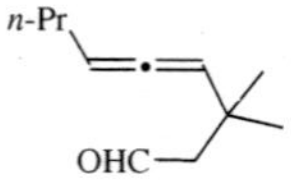 *n*-Pr, OHC	DMD, acetone, H_2O	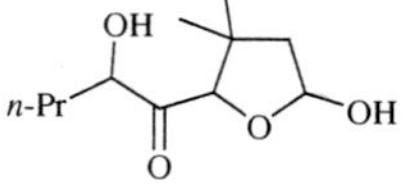 OH, *n*-Pr, O, O, OH (90)	59
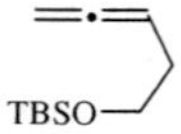 TBSO	DMD, acetone, $MgSO_4$, rt	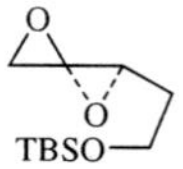  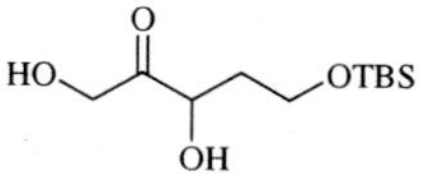 O, O, TBSO, **I** + O, O, TBSO, **II** + HO, O, OTBS, OH, **III** **I + II + III** (100), **I:II:III** = 3.4:1:1.4	57
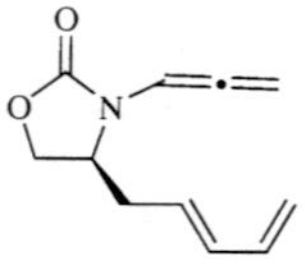 O, O, N	DMD (2-5 eq), acetone, CH_2Cl_2	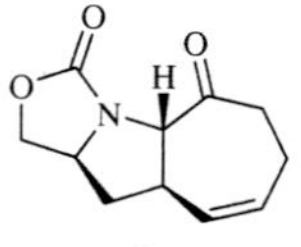 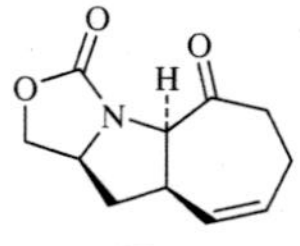 O, O, N, H, O, **I** + O, O, N, H, O, **II**	274

Addition Method	Temp	I + II	I:II
syringe-pumped	rt	(76)	60:40
cannulated	–45°	(75)	62:38

C_{12} TBSO

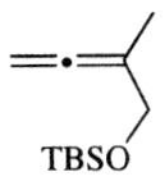

DMD, acetone, rt

TBSO (100) 57

t-Bu C_5H_{11}-n

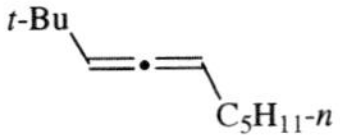

DMD, acetone, K_2CO_3, rt, 45 min

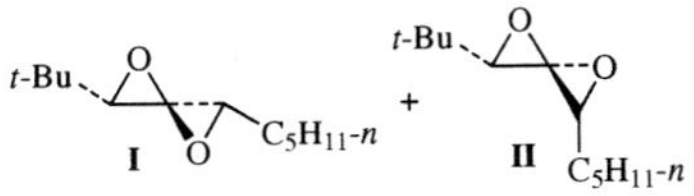

t-Bu C_5H_{11}-n **I** + t-Bu **II** C_5H_{11}-n **I + II** (100), **I:II** = 2.4:1 16

TBSO

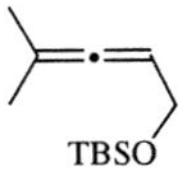

DMD, acetone, K_2CO_3, rt, 12 min

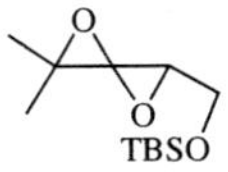

TBSO (100) 57

Ph +

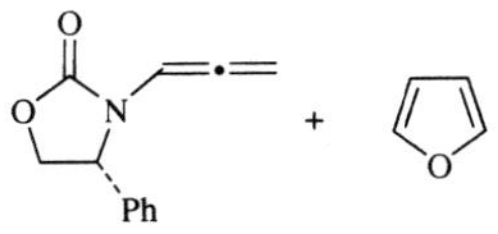

DMD (2-3 eq), additive (2 eq), acetone

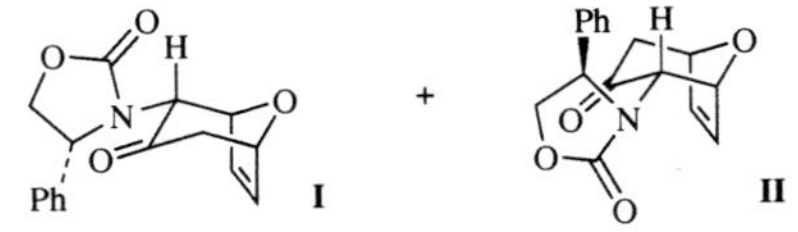

Ph **I** + Ph **II** 273

Temp	Solvent	Additive	I + II	I:II
–40°	CH_2Cl_2	—	(75)	75:25
–40°	Et_2O	—	(77)	75:25
–40°	MeCN	—	(< 10)	—
–40°	THF	—	(80)	75:25
rt	THF	—	(80)	75:25
–78°	THF	—	(70)	82:18
–40°	THF	$LiClO_4$	(81)	75:25
–40°	THF	$MgBr_2$	(< 10)	—
rt	THF	$ZnCl_2$	(40)	90:10
–40°	THF	$ZnCl_2$	(77)	94:6
–78°	THF	$ZnCl_2$	(80)	> 96:4

TABLE 1A. OXIDATION OF ALLENES AND ALKYNES BY ISOLATED DIOXIRANES (*Continued*)

Substrate	Conditions	Product(s) and Yield(s) (%)	Refs.
C_{12-16}	DMD (2-5 eq), acetone, CH_2Cl_2	I + II	274
C_{13} TsHN	DMD, acetone, TsOH	N-Ts (42) + NHTs, OH (—)	58
TBSO	DMD, acetone, $NaHCO_3$, rt	TBSO (93)	57
OTBS	DMD, acetone, $NaHCO_3$, −78° to rt, 5 h	HO, OH, OTBS (45)	57

Addition Method	Additive	Y	n	Temp	I + II	I:II
syringe-pumped	—	O	0	−45°	(82)	> 96:4
syringe-pumped	—	O	0	rt	(77)	> 96:4
cannulated	—	O	0	−45°	(75)	> 96:4
cannulated	—	O	1	−45°	(75)	> 96:4
cannulated	$ZnCl_2$	O	1	−78°	(30)	95:5
syringe-pumped	—	O	2	−45°	(30)	87:13
syringe-pumped	—	O	2	rt	(70)	83:17
syringe-pumped	—	O	3	rt	(57)	70:30
syringe-pumped	—	CH_2	1	rt	(90)	87:13
cannulated	—	CH_2	1	−45°	(85)	93:7
syringe-pumped	—	CH_2	3	rt	(40)	52:48

Substrate	Conditions	Product(s) and Yield(s) (%)	Refs.
C_{13-14} 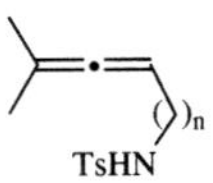 TsHN( )$_n$ allene	DMD (excess), acetone	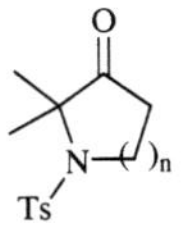 N-Ts pyrrolidinone/piperidinone ( )$_n$; n: 1 (56), 2 (52) 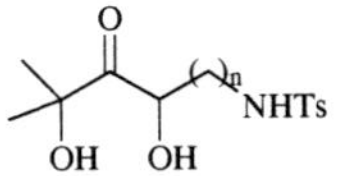 + OH OH ( )$_n$NHTs diol ketone (—)	58
C_{14} Ph—≡—Ph	DMD, acetone, CH_2Cl_2, 0°, 8 h	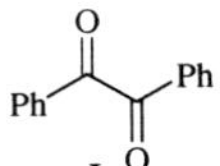 PhCOCOPh **I** (15) + $Ph_2C{=}O$ **II** (29)	64
	TFD, TFP, CH_2Cl_2, 0°, 6 min	**I** (25) + **II** (49)	64
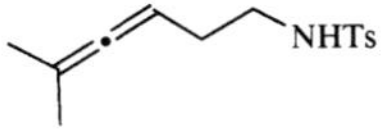 NHTs allene	DMD, acetone, CH_2Cl_2, TsOH	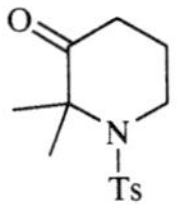 N-Ts piperidinone (2) 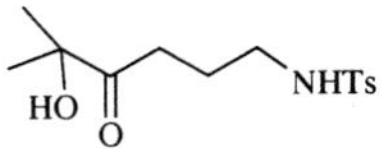 + HO ketone NHTs (33)	58
	DMD, acetone, K_2CO_3	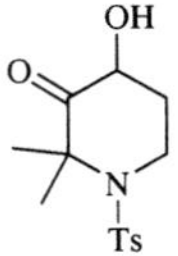 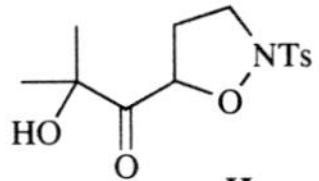 OH N-Ts piperidinone **I** (—) + HO ketone NTs isoxazolidine **II** (47) **I**:**II** = 1:1.6	58
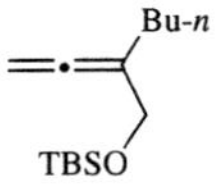 Bu-*n*, TBSO allene	DMD, acetone, $NaHCO_3$, –50 to 10°	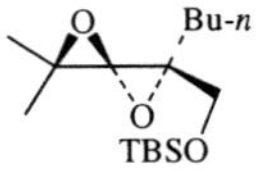 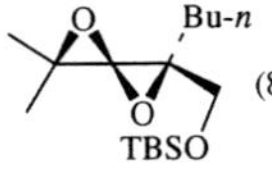 Bu-*n* TBSO spirodiepoxide **I** + Bu-*n* TBSO spirodiepoxide **II** (85) **I**:**II** = 1.1:1	57
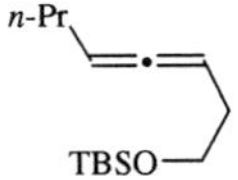 *n*-Pr, TBSO allene	DMD, acetone, $NaHCO_3$, –40 to 20°, 6 h	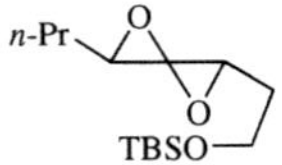 *n*-Pr TBSO spirodiepoxide (—)	57
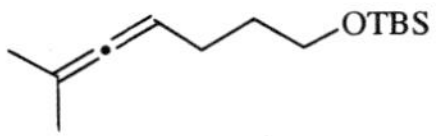 OTBS allene	DMD, acetone, K_2CO_3, rt	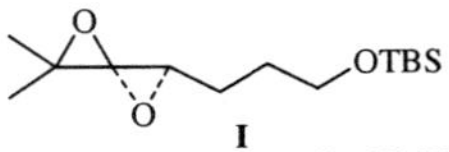 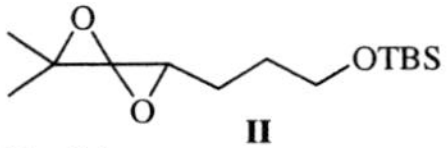 OTBS spirodiepoxide **I** + OTBS spirodiepoxide **II**; **I** + **II** (91), **I**:**II** = 9:1	57

TABLE 1A. OXIDATION OF ALLENES AND ALKYNES BY ISOLATED DIOXIRANES (*Continued*)

Substrate	Conditions	Product(s) and Yield(s) (%)	Refs.
C_{14} OTBS	DMD, acetone, rt	OTBS (—) dr 67:33	57
C_{14-18} RO, O, N, n	DMD (2-3 eq), acetone, CH_2Cl_2, –45°	I + II (RO, O, H, N, n)	274
		R / n / I / II: *t*-Bu / 0 / (80) / (14); Bn / 0 / (75) / (—); *t*-Bu / 1 / (< 5) / (—); Bn / 1 / (< 5) / (—)	
C_{15} NHTs	DMD, acetone, $NaHCO_3$	HO, O, N–Ts **I** (64)	58
	DMD, acetone, TsOH	O, N–Ts **II** (81) + **I** (—)	58
	DMD, acetone, K_2CO_3	**II** (67) + HO, O, NTs, O (13)	58

Substrate	Conditions	Product(s) and Yield(s) (%)	Refs.
C_{16} OTBS 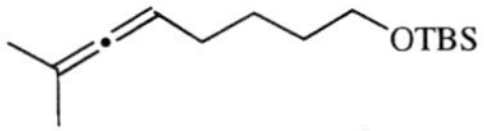	DMD, acetone, rt	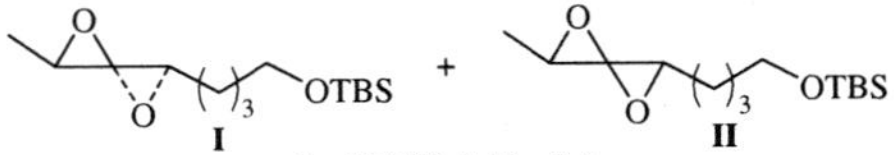 I (OTBS) + II (OTBS); I + II (98), I:II = 9:1	57
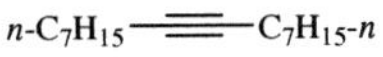 n-C_7H_{15}—≡—C_7H_{15}-n	DMD, acetone, CH_2Cl_2, 0°, 8 h	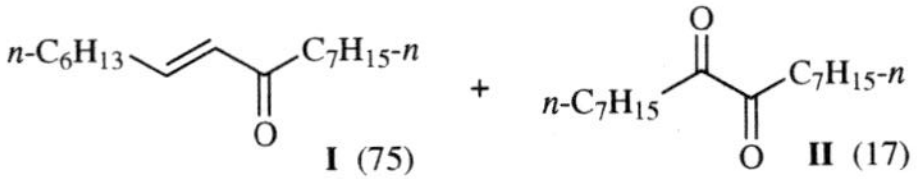 n-C_6H_{13}, C_7H_{15}-n, I (75) + n-C_7H_{15}, C_7H_{15}-n, II (17)	64
	TFD, TFP, CH_2Cl_2, 0°, 7 min	I (74) + II (18)	64
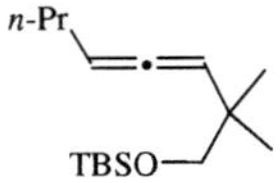 n-Pr, TBSO	DMD, acetone, $NaHCO_3$, rt	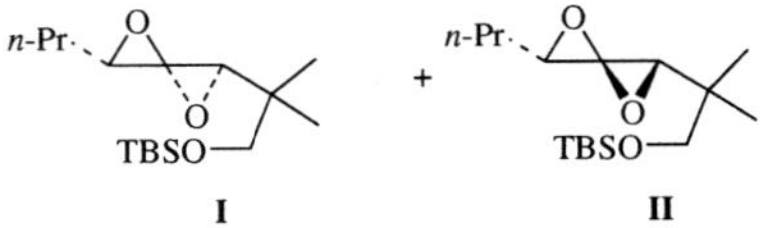 n-Pr, TBSO, I + n-Pr, TBSO, II; I + II (100), I:II = 2.2:1	57
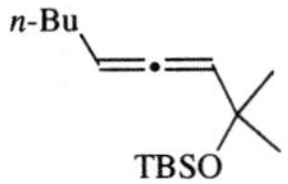 n-Bu, TBSO	DMD, acetone, rt, 1 h	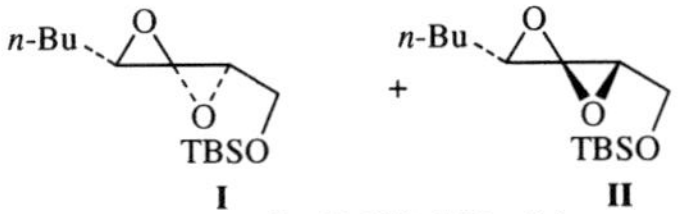 n-Bu, TBSO, I + n-Bu, TBSO, II; I + II (62), I:II = 2:1	57
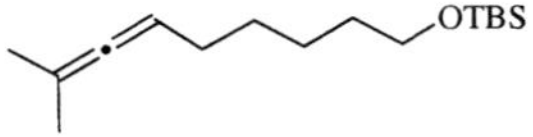 OTBS	DMD, acetone, rt	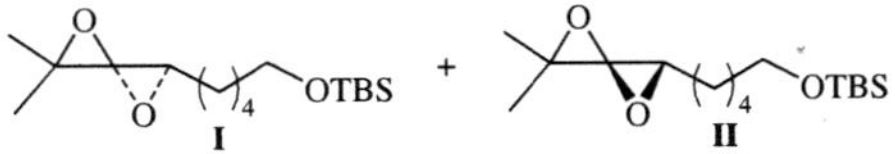 I (OTBS) + II (OTBS); I + II (100), I:II = 9:1	57
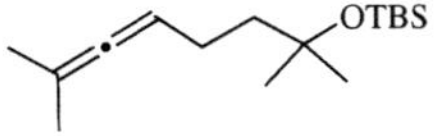 OTBS	DMD, acetone, rt	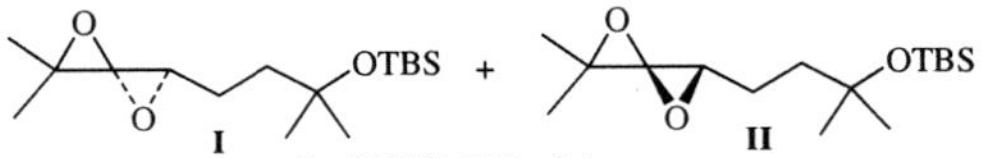 I (OTBS) + II (OTBS); I + II (98), I:II = 9:1	57

TABLE 1A. OXIDATION OF ALLENES AND ALKYNES BY ISOLATED DIOXIRANES (*Continued*)

Substrate	Conditions	Product(s) and Yield(s) (%)	Refs.
C_{16}			
O, N, OH, Ph	DMD (2-5 eq), acetone, –40°, 2 h	O, Ph, N, O, HO, O **I** (88)	275
	DMD (2-5 eq), acetone, PPTS (0.5 eq), –40°, 2 h	**I** + O, Ph, N, O, O **II** **I + II** (70), **I:II** = 75:25	275
O, O, N, Pr-*i*, O	DMD (2-5 eq) cannulated, acetone/CH_2Cl_2, –78°, 5-15 min	O, O, N, O, *i*-Pr, O (65) dr 95:5	276
C_{17}			
OH, O, O, N, Ph	1. DMD, acetone 2. TBDPSCl, imadazole, CH_2Cl_2, 2 h	O, Ph, N, O, TBDPSO, O (46) Temp / ratio at C-2: rt 1:1; –45° 2:1; –78° 2:1	275
O, BnO, N	DMD (2-5 eq) syringe-pumped, acetone, CH_2Cl_2, –45°	BnO, O, H, O, N **I** + BnO, O, H, O, N **II** **I + II** (65), **I:II** = 53:47	274

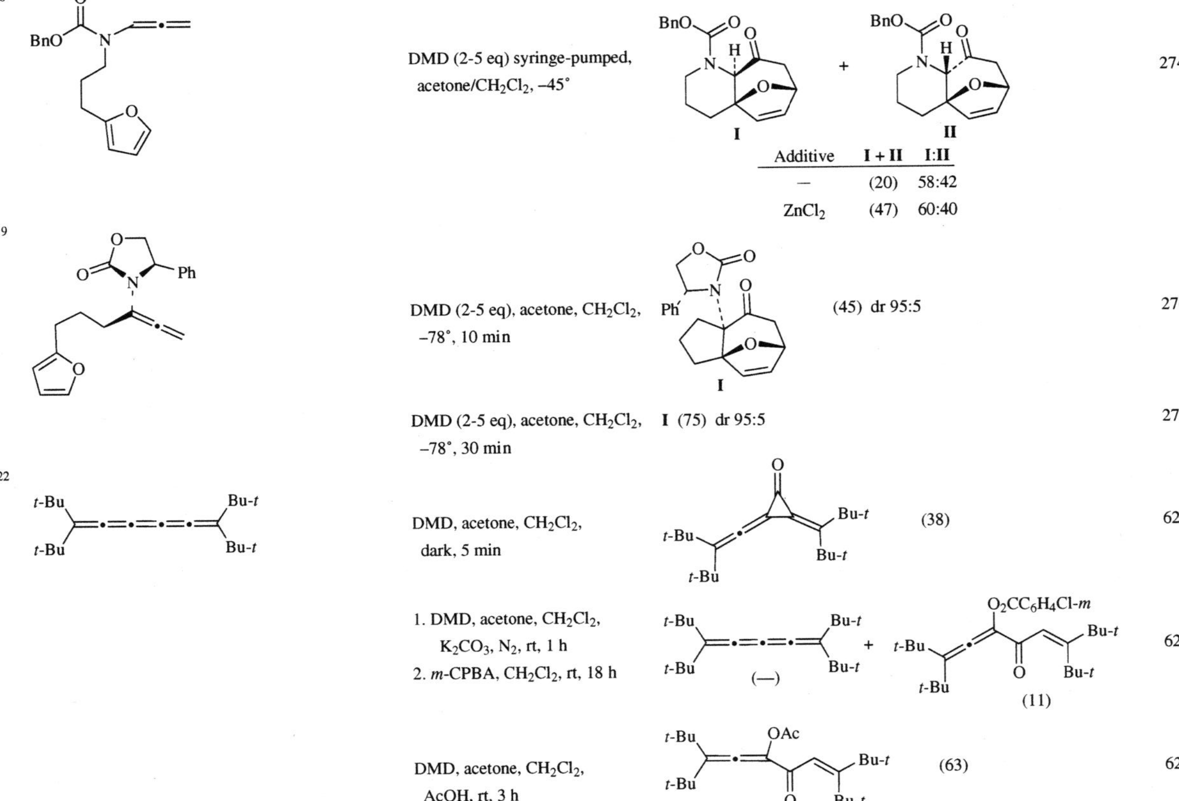

Reactant	Conditions	Product(s) and Yield(s) (%)	Refs.
C_{18}	DMD (2-5 eq) syringe-pumped, acetone/CH_2Cl_2, –45°	I + II	274
		Additive: —, I + II (20), I:II 58:42	
		Additive: $ZnCl_2$, I + II (47), I:II 60:40	
C_{19}	DMD (2-5 eq), acetone, CH_2Cl_2, –78°, 10 min	I (45) dr 95:5	276
	DMD (2-5 eq), acetone, CH_2Cl_2, –78°, 30 min	**I** (75) dr 95:5	276
C_{22}	DMD, acetone, CH_2Cl_2, dark, 5 min	(38)	62
	1. DMD, acetone, CH_2Cl_2, K_2CO_3, N_2, rt, 1 h 2. *m*-CPBA, CH_2Cl_2, rt, 18 h	(—) + $O_2CC_6H_4Cl$-*m* (11)	62
	DMD, acetone, CH_2Cl_2, AcOH, rt, 3 h	OAc (63)	62

TABLE 1A. OXIDATION OF ALLENES AND ALKYNES BY ISOLATED DIOXIRANES (*Continued*)

Substrate	Conditions	Product(s) and Yield(s) (%)	Refs.
C_{22-26} 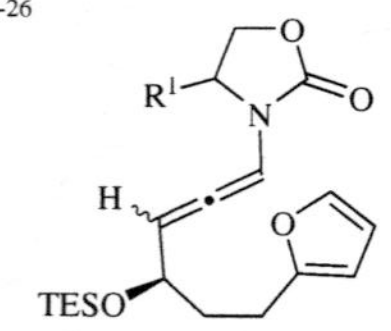	DMD (2.5 eq), acetone, CH_2Cl_2, –78°, 5-15 min	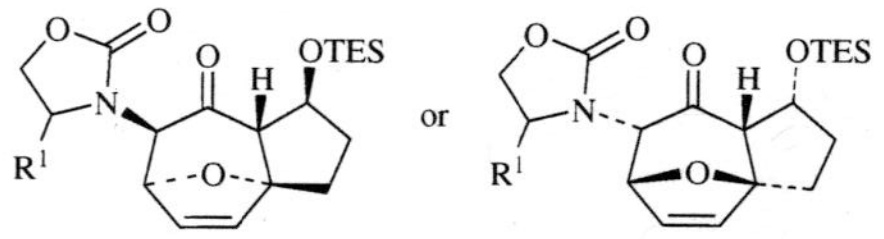 I or II	276
C_{25} 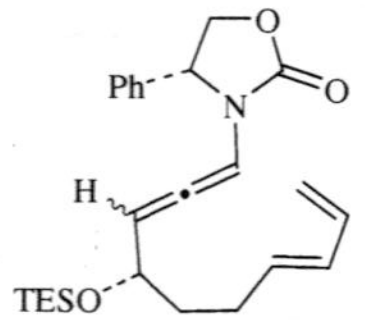	DMD (2.5 eq), acetone, CH_2Cl_2, –78°, 5-15 min	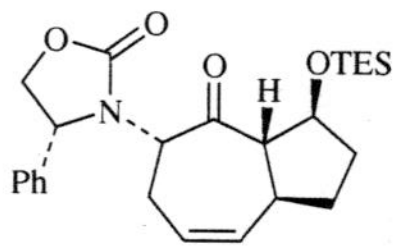	276

C_{22-26} substrate:

R^1	Allene isomer	I	II	dr
(*R*)-Bn	P/M 3:1	(65)	(—)	71:29
(*S*)-Ph	P/M 1:1	(—)	(60)	90:10
(*S*)-Pr-*i*	P	(—)	(60)	95:5
(*S*)-Pr-*i*	M	(—)	(60)	95:5

C_{25} substrate:

Allene isomer	Yield (%)	dr
M	(30)	< 5:95
P	(34)	< 5:95

C_{25-26}

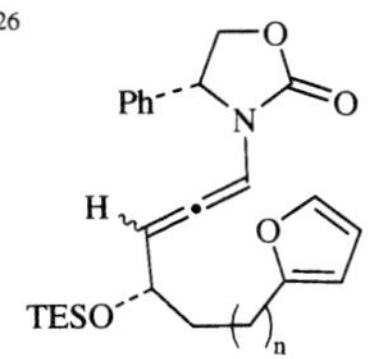

n	Allene isomer
1	P
1	M
2	P/M 1:1
3	P/M 1:1

DMD (2.5 eq), acetone, CH_2Cl_2, –78°, 5-15 min

I or II

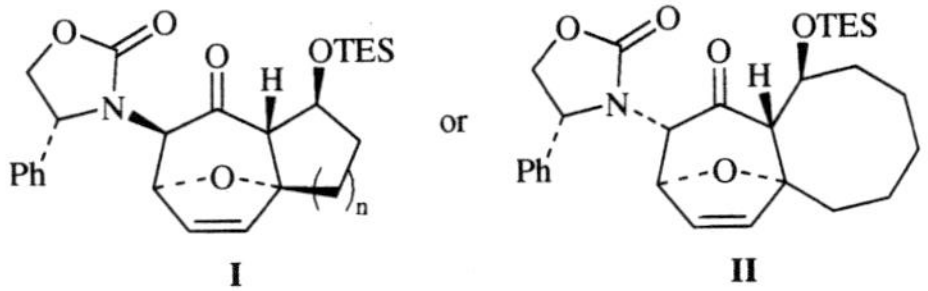

I	II	dr
(60)	(—)	90:10
(75)	(—)	90:10
(65)	(—)	93:7
(—)	55	< 5:95

276

C_{28-32}

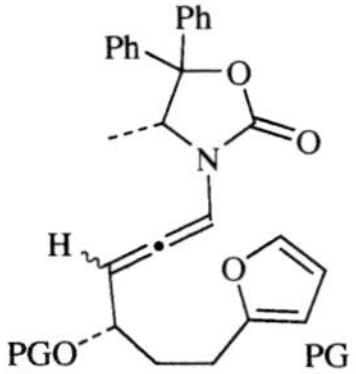

PG	Allene isomer
TES	P/M 2.5:1
Ac	P/M 2.5:1

DMD (2.5 eq), acetone, CH_2Cl_2, –78°, 5-15 min

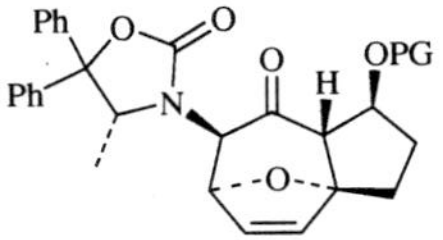

(78)
(83)

276

TABLE 1B. OXIDATION OF ALLENES AND ALKYNES BY IN SITU GENERATED DIOXIRANES

Substrate	Conditions	Product(s) and Yield(s) (%)	Refs.
C_6			
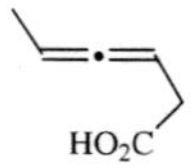 HO_2C	Oxone®, $NaHCO_3$, acetone, CH_2Cl_2, H_2O, rt, 2 h	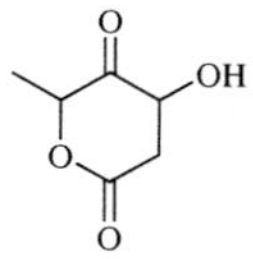 O, OH, O, O (42)	272
C_7			
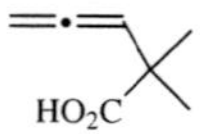 HO_2C	Oxone®, $NaHCO_3$, acetone, CH_2Cl_2, H_2O, rt, 2 h	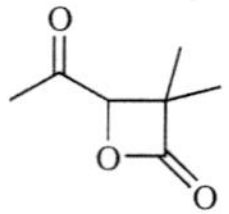 O, O, O (72)	272
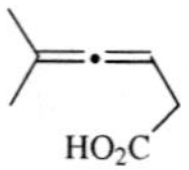 HO_2C	Oxone®, $NaHCO_3$, acetone, CH_2Cl_2, H_2O, rt, 2 h	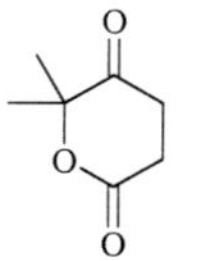 O, O, O (67)	272
C_8			
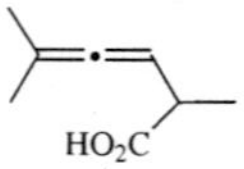 HO_2C	Oxone®, $NaHCO_3$, acetone, CH_2Cl_2, H_2O, rt, 2 h	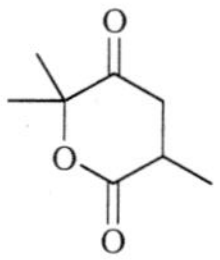 O, O, O (74)	272
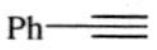 Ph—≡	Oxone®, acetone, phosphate buffer (pH 7.5), CH_2Cl_2, Bu_4NHSO_4, 10°, 40 h	PhCHO **I** (6) + $PhCO_2H$ **II** (15) + $PhCH_2CO_2H$ **III** (52)	64
	Oxone®, acetone, phosphate buffer (pH 7.5), CH_2Cl_2, Bu_4NHSO_4, 10°, 8 h	**I** (25) + **II** (20) + **III** (19)	64

Substrate	Conditions	Product(s) and Yield(s) (%)	Refs.
C_9 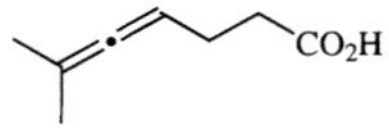	Oxone®, $NaHCO_3$, acetone, CH_2Cl_2, H_2O, rt, 2 h	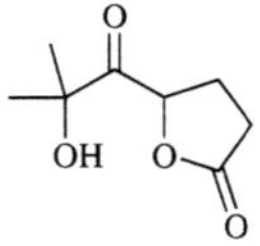 (—)	272
C_{10} 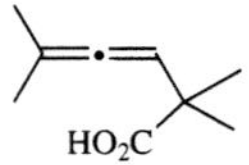	Oxone®, $NaHCO_3$, acetone, CH_2Cl_2, H_2O, rt, 2 h	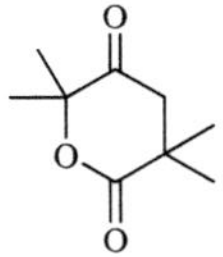 (76)	272
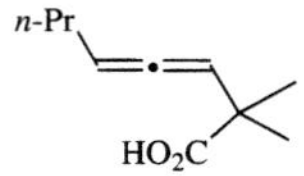	Oxone®, $NaHCO_3$, acetone, CH_2Cl_2, H_2O, rt, 2 h	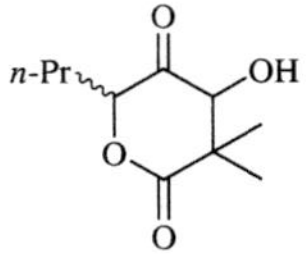 (53) + 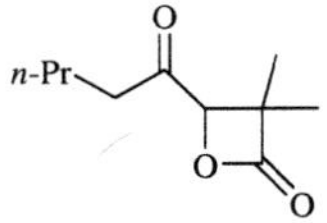 (25)	272
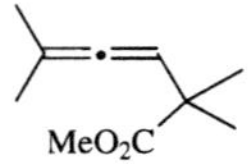	Oxone®, $NaHCO_3$, acetone, CH_2Cl_2, H_2O, rt, 2 h	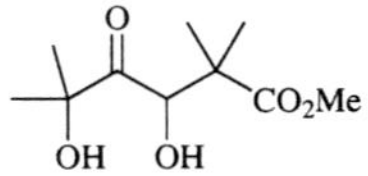 (—)	272
C_{14} Ph—≡—Ph	Oxone®, acetone, phosphate buffer (pH 7.5), CH_2Cl_2, Bu_4NHSO_4, 5°, 60 h	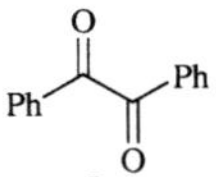 **I** (15) + $Ph_2C{=}O$ **II** (45) + Ph_2CHCO_2H **III** (14)	64
	Oxone®, acetone, phosphate buffer (pH 7.5), CH_2Cl_2, Bu_4NHSO_4, 5°, 12 h	**I** (12) + **II** (44) + **III** (6)	64

TABLE 1B. OXIDATION OF ALLENES AND ALKYNES BY IN SITU GENERATED DIOXIRANES (*Continued*)

Substrate	Conditions	Product(s) and Yield(s) (%)	Refs.
C_{16}			
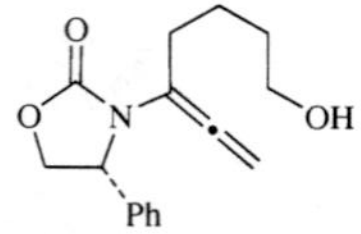 O, N, OH, Ph	Oxone®, acetone, CH_2Cl_2, H_2O (1:1:2), rt, 2-3 h	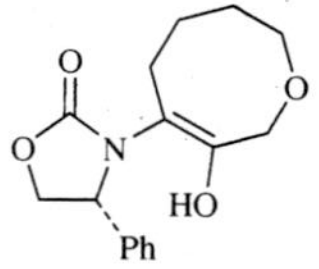 O, N, O, HO, Ph (40)	275
 n-C_7H_{15}—≡—C_7H_{15}-n	Oxone®, acetone, phosphate buffer (pH 7.5), CH_2Cl_2, Bu_4NHSO_4, 5°, 16 h	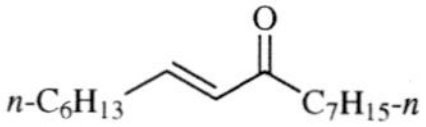 n-C_6H_{13}, O, C_7H_{15}-n (56) 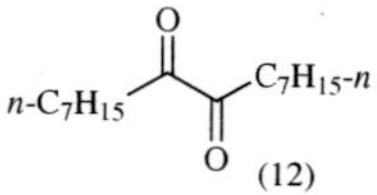 + n-C_7H_{15}, O, O, C_7H_{15}-n (12)	64

TABLE 2A. OXIDATION OF ARENES AND HETEROARENES BY ISOLATED DIOXIRANES

Substrate	Conditions	Product(s) and Yield(s) (%)	Refs.
C_4	1. DMD, acetone, 0°, 30 min 2. $Ph_3PC(Me)CHO$, CH_2Cl_2, acetone, 0° to rt, 3 h	OHC, CHO (84)	75
C_{4-10} R^1, R^2	DMD, acetone, rt	O, O, R^1 R^2 (> 95) R^1 / R^2: H / H; Me / H; Me / Me; H / $AcOCH_2$; H / $AcOCH_2$; $AcOCH_2$ / $CH(OMe)_2$; H / $C(Me)_2OTMS$	73
C_5	1. DMD, acetone, 0°, 30 min 2. $Ph_3PCHCHO$, CH_2Cl_2, acetone, 0° to rt, 3 h	O, CHO (73)	75
C_{5-6} R^1, OH, R^2	DMD, acetone, rt	HO, O, R^1, R^2 (> 95) R^1 / R^2: H / H; H / Me; Me / H; CH_2OH / H	73
C_6	TFD, TFP, F113, 0°, 6 h	OHC, CHO **I** + OHC, CHO **II** **I** + **II** (4), **I**:**II** = 1:1.5	65
OH, OH	TFD, acetone, TFP, –20°, 1 h	O, OH, OH, O (70)	277

TABLE 2A. OXIDATION OF ARENES AND HETEROARENES BY ISOLATED DIOXIRANES (*Continued*)

	Substrate	Conditions	Product(s) and Yield(s) (%)	Refs.
C_6	OH, OH, Br + NaCl	DMD, acetone, 10% H_2SO_4, rt, 1 min	Cl, OH, OH, Br (75)	278
C_{6-8}	OR, OR + HX	DMD, acetone, rt, 1 min	OR, OR, X **I** R X: H Br (95); H Cl (90); Me Br (95); Me Cl (95)	278
	OR, OR + NaX	DMD, acetone, 10% H_2SO_4, rt, 1 min	**I** R X: H Br (95); H Cl (90); Me Br (95); Me Cl (95); Me I (95)	278
	OR, OR, X + HX	DMD, acetone, rt, 1 min	X, OR, OR, X **I** R X: H Br (78); H Cl (85); Me Br (85); Me Cl (90)	278
	OR, OR, X + NaX	DMD, acetone, 10% H_2SO_4, rt, 1 min	**I** R X: H Br (78); H Cl (85); Me Br (85); Me Cl (90); Me I (90)	278

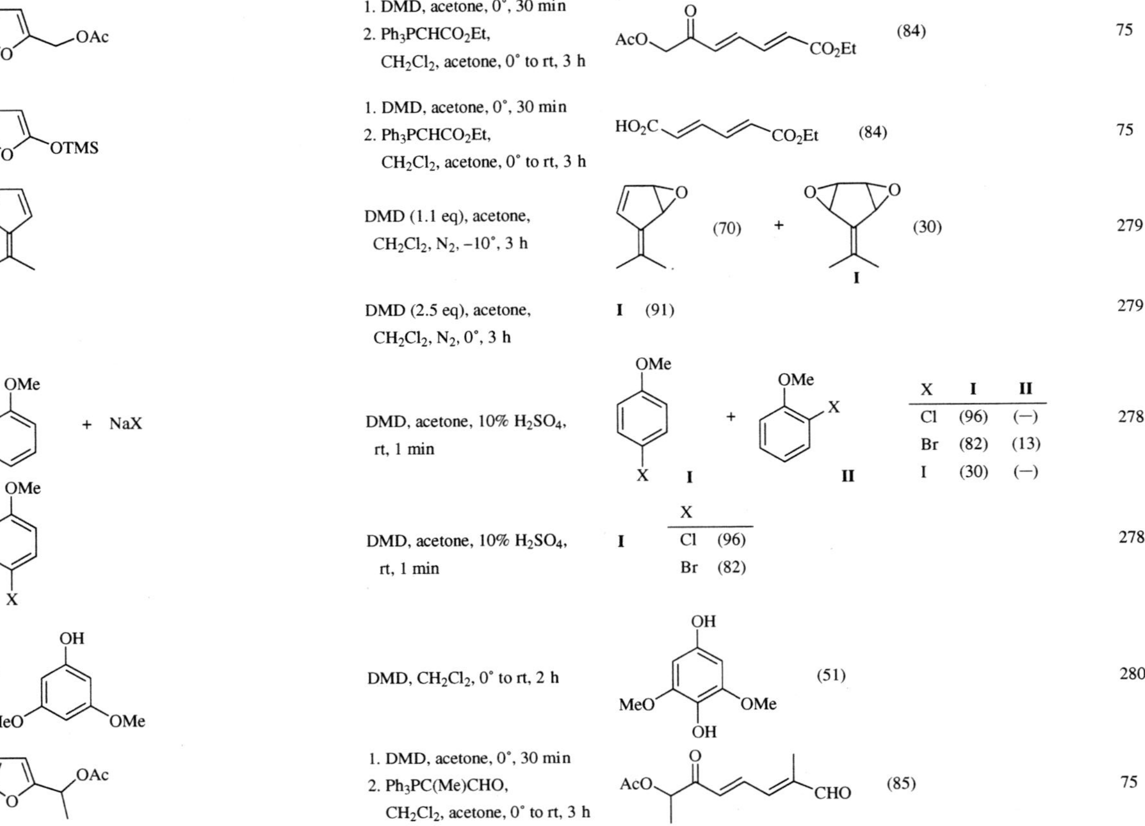

Reactant	Conditions	Product(s) and Yield(s) (%)	Refs.
C_7	1. DMD, acetone, 0°, 30 min 2. Ph_3PCHCO_2Et, CH_2Cl_2, acetone, 0° to rt, 3 h	(84)	75
	1. DMD, acetone, 0°, 30 min 2. Ph_3PCHCO_2Et, CH_2Cl_2, acetone, 0° to rt, 3 h	(84)	75
	DMD (1.1 eq), acetone, CH_2Cl_2, N_2, –10°, 3 h	(70) + **I** (30)	279
	DMD (2.5 eq), acetone, CH_2Cl_2, N_2, 0°, 3 h	**I** (91)	279
+ NaX	DMD, acetone, 10% H_2SO_4, rt, 1 min	**I** + **II** X: Cl, I (96), II (—) X: Br, I (82), II (13) X: I, I (30), II (—)	278
	DMD, acetone, 10% H_2SO_4, rt, 1 min	**I** X: Cl (96) X: Br (82)	278
C_8	DMD, CH_2Cl_2, 0° to rt, 2 h	(51)	280
	1. DMD, acetone, 0°, 30 min 2. $Ph_3PC(Me)CHO$, CH_2Cl_2, acetone, 0° to rt, 3 h	(85)	75

TABLE 2A. OXIDATION OF ARENES AND HETEROARENES BY ISOLATED DIOXIRANES (*Continued*)

Substrate	Conditions	Product(s) and Yield(s) (%)	Refs.
C_8 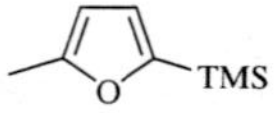 TMS	DMD, acetone, rt	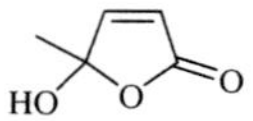 HO, O, O (>95)	73
C_{8-14} 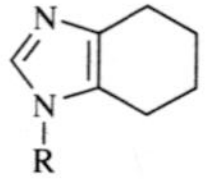 N, N, R	DMD (1.5 eq), acetone, CH_2Cl_2, rt, 1.5 h	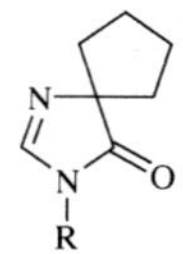 N, N, R, O R: Bn (73); Me (70); MOM (60); SO_2NMe_2 (—)	281
C_9 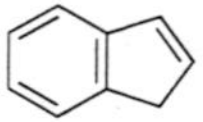	DMD, acetone, 1 h	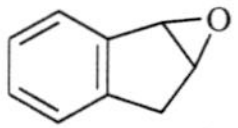 O (96)	265
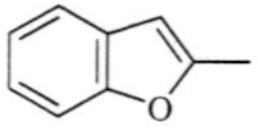 O	DMD, acetone, CH_2Cl_2, N_2, 20°, 19 h	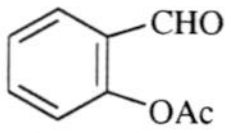 CHO, OAc (35)	89
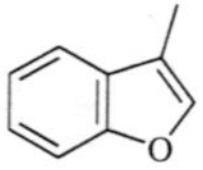 O	DMD-d_6, acetone-d_6, N_2, –55°, 4 h	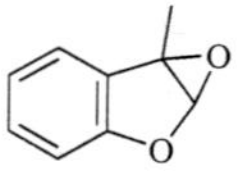 O, O (—)	89
C_{9-10} 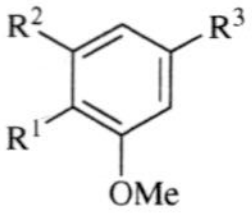 R^2, R^3, R^1, OMe	DMD, acid, acetone, 0°, N_2	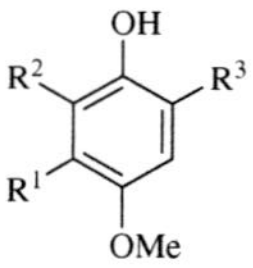 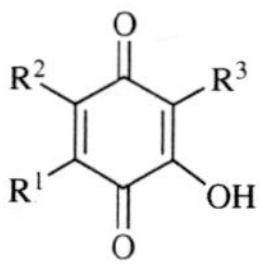 **I** 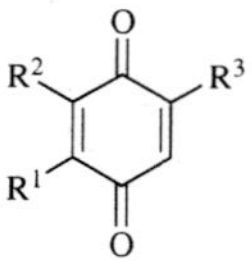 (R^2, OH, R^3, R^1, OMe) + **II** (R^2, O, R^3, R^1, O) + **III** (R^2, O, R^3, R^1, O, OH)	282

R^1	R^2	R^3	Acid	Time	I	II	III
MeO	H	Me	—	360 min	(—)	(3)	(11)
MeO	H	Me	H_2SO_4	60 min	(—)	(8)	(2)
H	MeO	Me	—	300 min	(—)	(—)	(12)
H	MeO	Me	H_2SO_4	45 min	(1)	(35)	(5)
H	MeO	Me	$H_3PMo_{12}O_{40}$	300 min	(1)	(38)	(2)
Me	MeO	MeO	—	30 min	(26)	(—)	(—)
Me	MeO	MeO	H_2SO_4	60 min	(—)	(50)	(—)
Me	MeO	MeO	$H_3PMo_{12}O_{40}$	30 min	(—)	(23)	(—)
Me	MeO	MeO	H_3PO_4	30 min	(—)	(51)	(—)
Me	MeO	MeO	CF_3CO_2H	30 min	(—)	(46)	(—)
Me	MeO	MeO	AcOH	30 min	(6)	(16)	(—)
MeO	MeO	Me	—	390 min	complex mixture		
MeO	MeO	Me	H_2SO_4	240 min	(11)	(14)	(—)

C_{9-17}

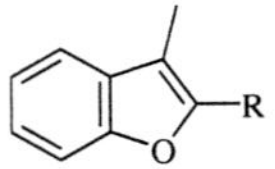

DMD, acetone, CH_2Cl_2, N_2

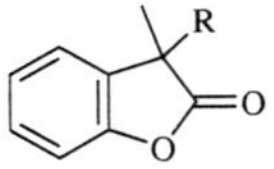

R	Temp	Time	
H	0°	3 h	(98)
CH_2OH	0°	3 h	(89)
CO_2Et	20°	48 h	(72)
Ph(MeO)CH	0°	9 h	(77)

89

C_{10}

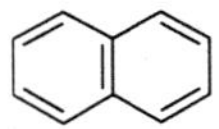

TFD, CH_2Cl_2, TFP, –20°, 30 min 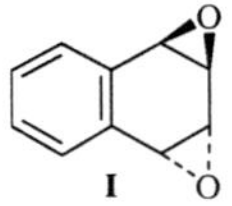 — **I** (82) — 65

TFD, CH_2Cl_2, TFP, –22°, 40 min — **I** (95) — 65

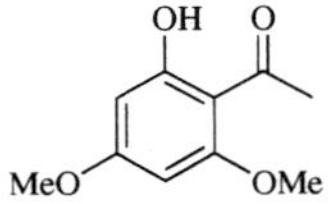

DMD, CH_2Cl_2, 0° to rt, 10 min 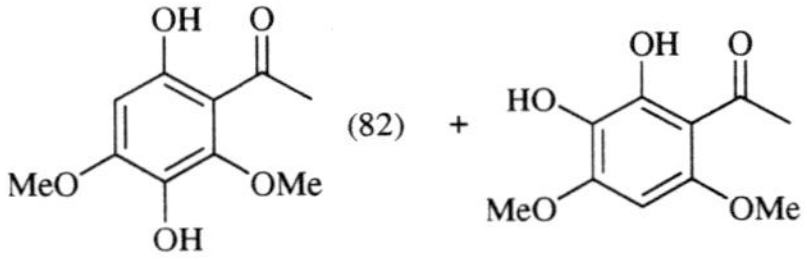 — (82) + (17) — 280

TABLE 2A. OXIDATION OF ARENES AND HETEROARENES BY ISOLATED DIOXIRANES (*Continued*)

Substrate	Conditions	Product(s) and Yield(s) (%)	Refs.
C_{10} 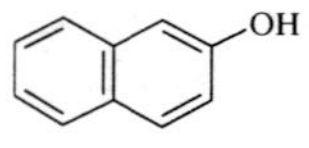 OH	DMD, acetone	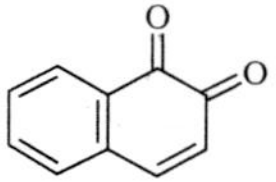 O O (38) **I**	17
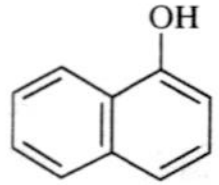 OH	DMD, acetone	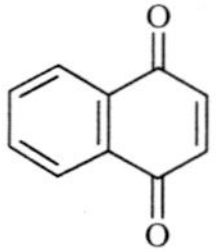 O O (17) + **I** (14)	17
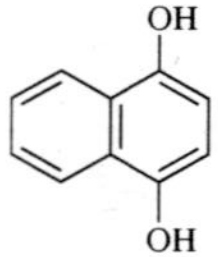 OH OH	DMD, acetone, Ar, 20°, 24 h	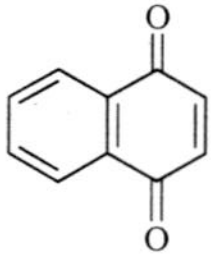 O O (27) 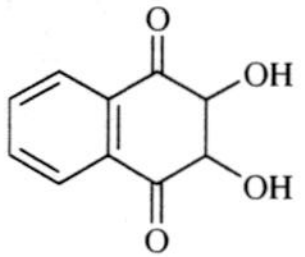 + O OH OH O (51)	66
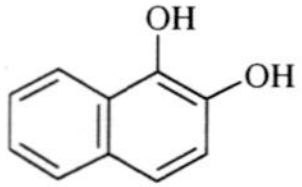 OH OH	DMD, acetone	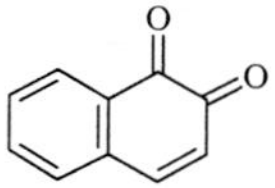 O O (100)	17
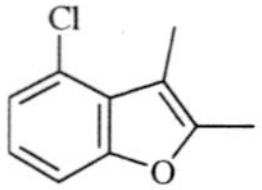 Cl O	DMD, acetone, N_2, –40°, 11 h	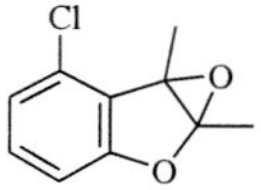 Cl O O (41)	79
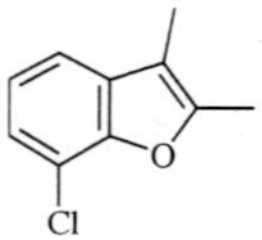 O Cl	DMD, acetone, N_2, –40°, 9 h	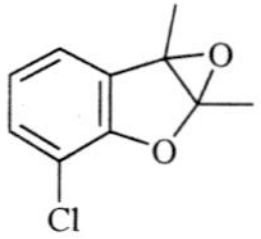 O O Cl (72)	79

Reactant	Conditions	Product(s) and Yield(s) (%)	Refs.
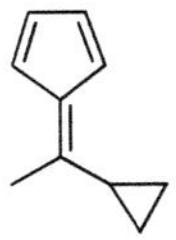	DMD (1.0 eq), acetone, CH_2Cl_2, N_2, –10°, 3 h	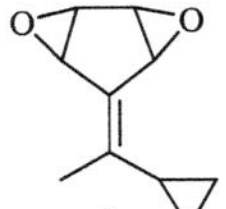 **I** (13) + 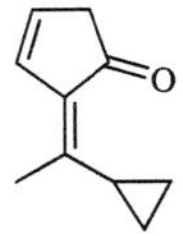 (72)	279
	DMD (2.5 eq), acetone, CH_2Cl_2, N_2, 0°, 3 h	**I** (85)	279
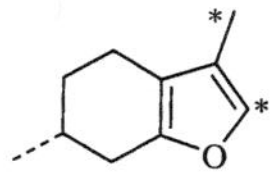 * = ^{13}C	DMD-d_6, acetone-d_6, –40°	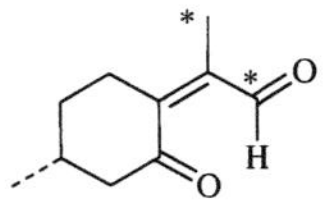 H (—)	283
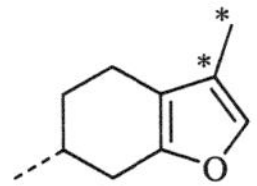 * = ^{13}C	DMD-d_6, acetone-d_6, –40 to –20°	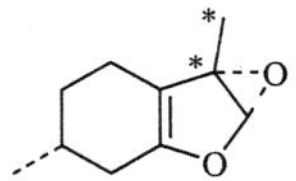 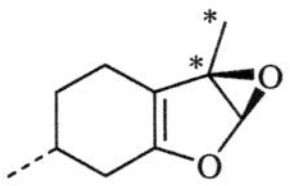 (—) + (—)	283
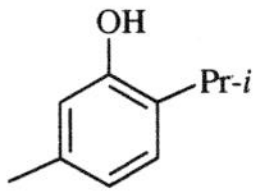 OH, Pr-*i*	DMD, acetone	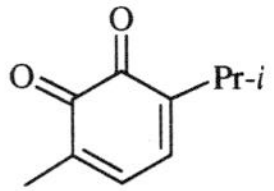 Pr-*i* (5) 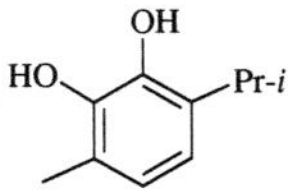 + HO, OH, Pr-*i* (15) 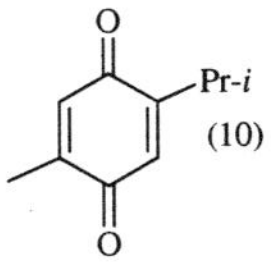 + Pr-*i* (10) 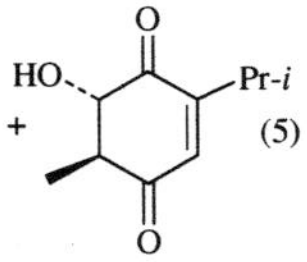 + HO, Pr-*i* (5) 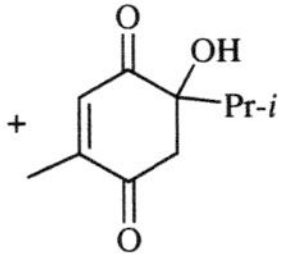 + OH, Pr-*i* (5)	17
H, * = ^{13}C	DMD-d_6, acetone-d_6, –40 to –20°	(—) + (—)	283

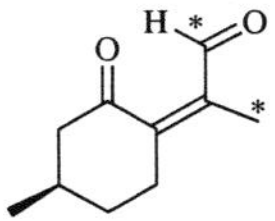

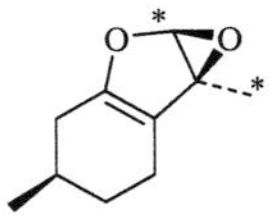

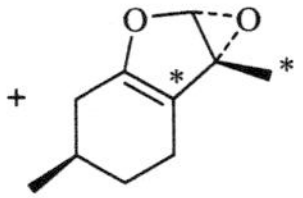

TABLE 2A. OXIDATION OF ARENES AND HETEROARENES BY ISOLATED DIOXIRANES (*Continued*)

Substrate	Conditions	Product(s) and Yield(s) (%)	Refs.
C_{10} 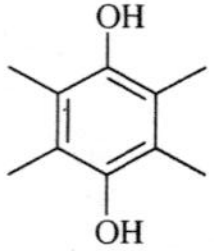 OH, OH	DMD, acetone, Ar, 20°, 1.5 h	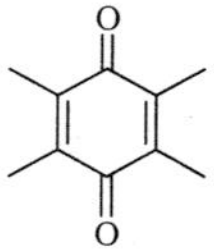 (72) 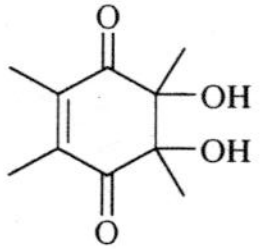 + OH, OH (11)	66
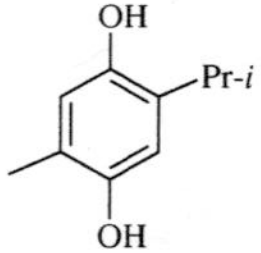 OH, Pr-*i*, OH	DMD, acetone	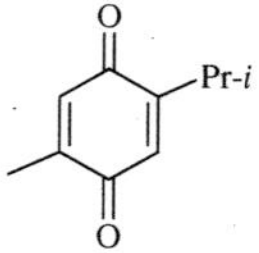 Pr-*i* (77)	17
C_{10-11} 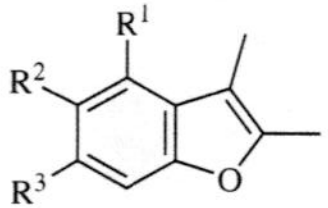 R^1, R^2, R^3	DMD, acetone, CH_2Cl_2, N_2, –78 to –20°, 0.5 to 3 h	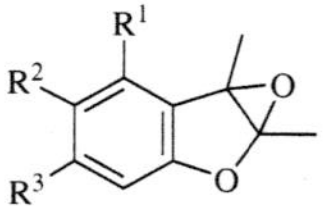 (100) R^1 R^2 R^3: H H H; Me H H; H Me H; H H Me	77, 79
C_{10-12} 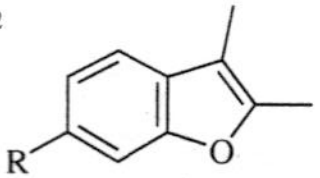 R	DMD-d_6, acetone-d_6, N_2, –78°, 0.5 h	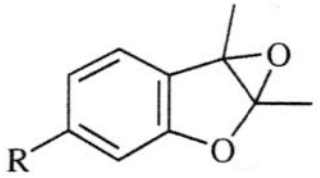 R = H, OAc (100)	87
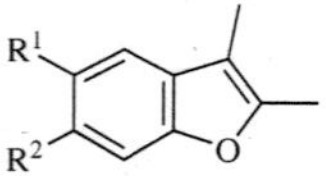 R^1, R^2	DMD, acetone, N_2, –40°, 11-12 h	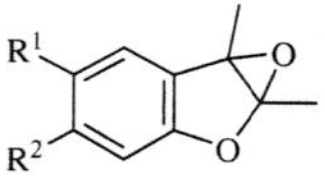 (100) R^1 R^2: Cl H; H Cl; H Ac	79
C_{10-18} 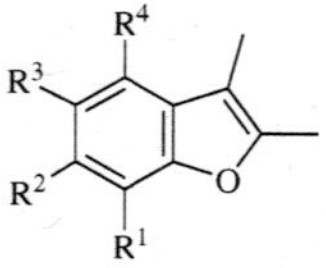 R^1, R^2, R^3, R^4	DMD, acetone, CH_2Cl_2, N_2	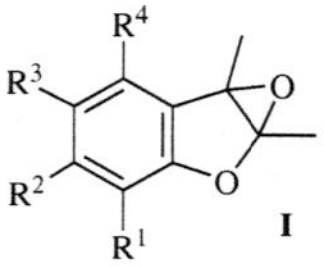 **I** 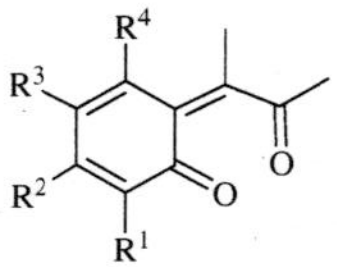 + **II**	82

R^1	R^2	R^3	R^4	Temp	Time	% Convn	Products
Cl	H	H	H	–20°	10 h	> 95	**I**
H	Cl	H	H	–35°	6 h	79	**I**
H	H	Cl	H	–20°	9 h	71	**I**
H	H	H	Cl	–20°	8 h	92	**I**
H	H	H	H	–40°	7 h	> 95	**I**
Me	H	H	H	–45°	3 h	95	**I**
H	Me	H	H	–50°	2 h	> 95	**I+II** (31:69)
H	H	Me	H	–40°	3 h	> 95	**I**
H	H	H	Me	–35°	4 h	> 95	**I**
H	MeO	H	H	–60°	0.5 h	> 95	**II**
H	H	*t*-Bu	H	–30°	2 h	> 95	**I**
H	*t*-Bu	H	*t*-Bu	–45°	3 h	> 95	**I**

C_{11}

Substrate	Conditions	Products	Refs.
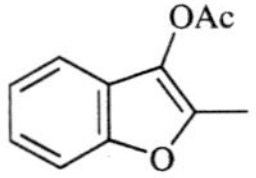 OAc	DMD, acetone, CH_2Cl_2, 0°, 9 h 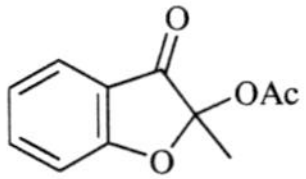	OAc (99)	83
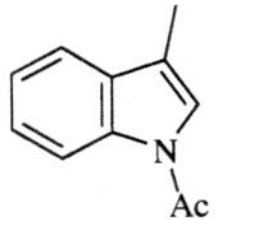 N, Ac	DMD, acetone, CH_2Cl_2, –78°	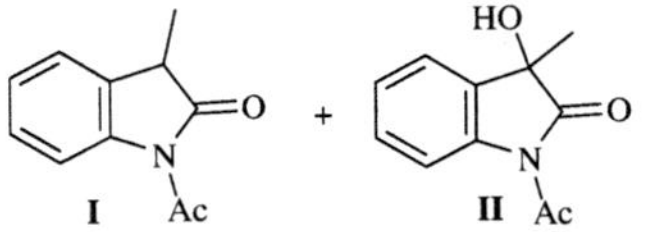 O, N, Ac, **I** + HO, O, N, Ac, **II**; **I + II** (> 98), **I**:**II** = 11:1	92
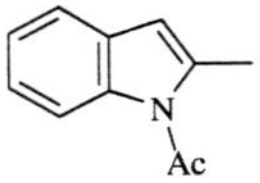 N, Ac	DMD, acetone, CH_2Cl_2, –78°	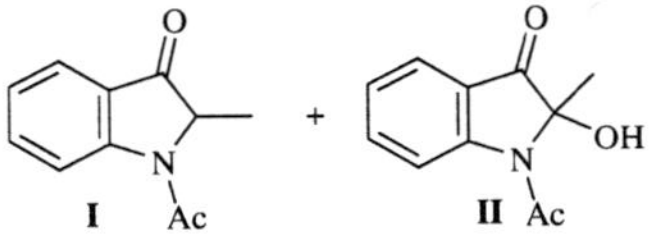 O, N, Ac, **I** + O, OH, N, Ac, **II**; **I + II** (> 98), **I**:**II** = 10:1	92
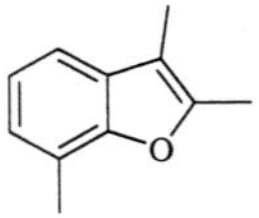 O	DMD, acetone, CH_2Cl_2, N_2, –78 to –20°, 3 h 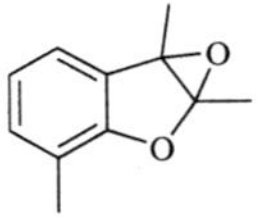	O, O (80)	79

TABLE 2A. OXIDATION OF ARENES AND HETEROARENES BY ISOLATED DIOXIRANES (*Continued*)

Substrate	Conditions	Product(s) and Yield(s) (%)	Refs.
C_{11}			
	1. DMD, acetone, 0°, 30 min 2. $Ph_3PCHCHO$, CH_2Cl_2, acetone, 0° to rt, 3 h	CHO (78)	75
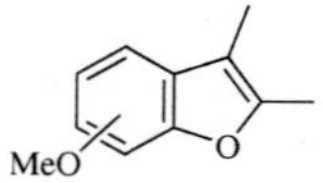 MeO	DMD, acetone, –70 to –20°, 3 to 5 h	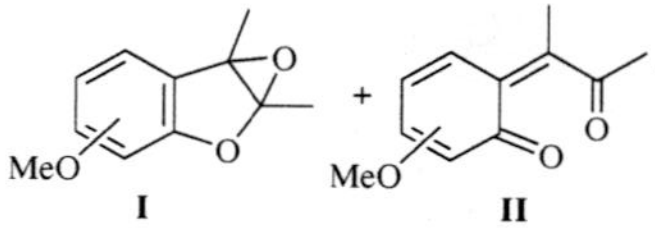 MeO I + MeO II 4-MeO I (100) II (—) 5-MeO I (34) II (66) 6-MeO I (—) II (100) 7-MeO I (57) II (43)	81
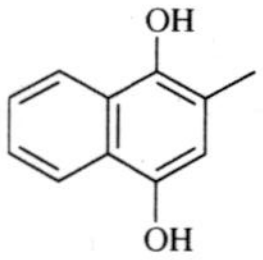 OH, OH	DMD, acetone, Ar, 20°, 24 h	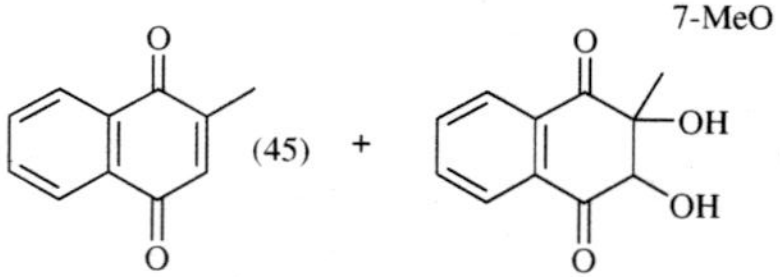 (45) + OH, OH (28)	66
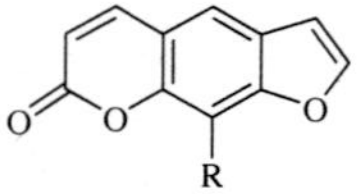 R	TFD, TFP, CH_2Cl_2, N_2, –78 to –40°	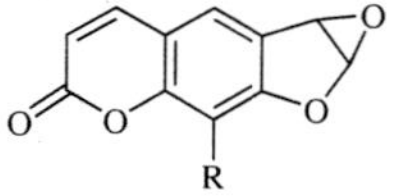 R R = H, Time 30 min (100) R = MeO, Time 2 h (100)	84
C_{12}			
	DMD, acetone, 20°, 24 h	(29)	284
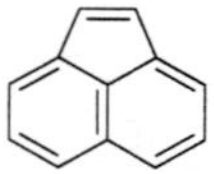	DMD, acetone, rt, 1 h	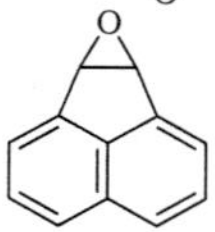 (97)	265
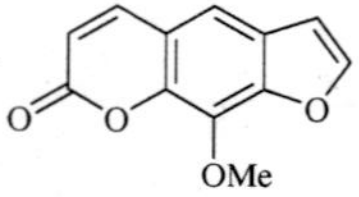 OMe	DMD, acetone, MeOH, rt, 19 h	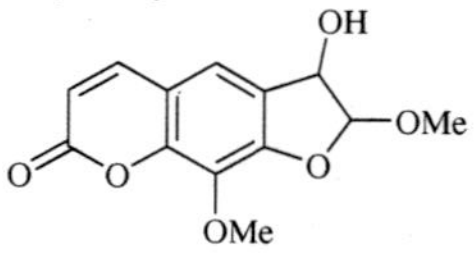 OH, OMe, OMe (83)	85

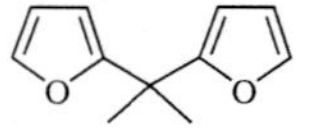

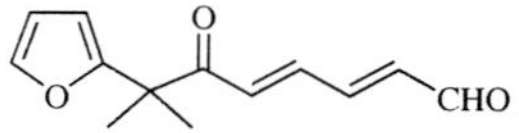

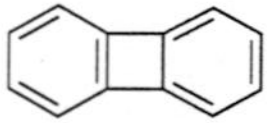

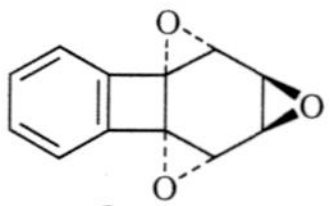

Substrate	Conditions	Product(s) and Yield(s) (%)	Refs.
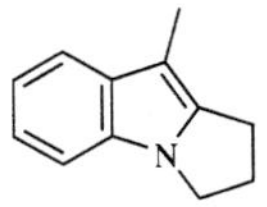	DMD (1 eq), acetone, CH_2Cl_2, Ar, –40 to 22°, 3 h	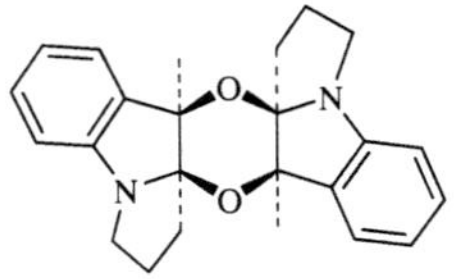 (—)	285
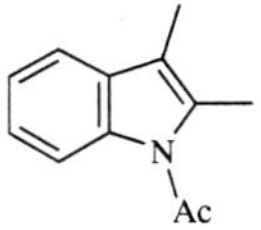	DMD, acetone, 0°, 0.5 h	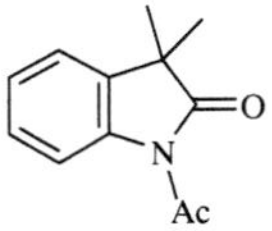 (89)	91
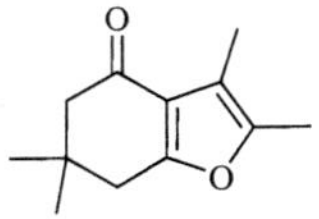	DMD-d_6, acetone-d_6, –40°, 1.5 h	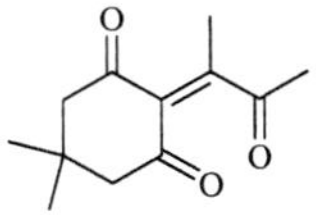 (> 95)	74
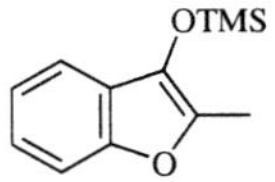	DMD, acetone, CH_2Cl_2, N_2, –70 to –30°, 0.5-1 h	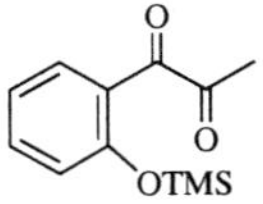 OTMS (72) 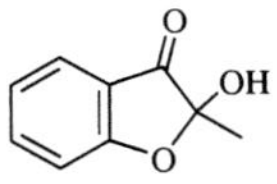 + OH (28)	78
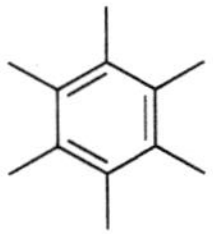	DMD (4 eq), acetone, rt, 24 h	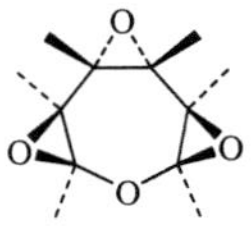 **I** 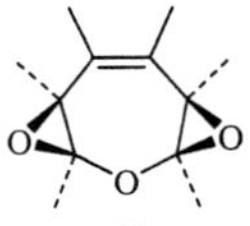 (—) + **II** 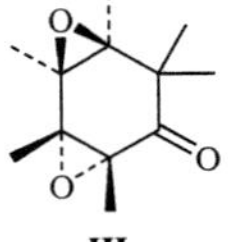 (—) + **III** 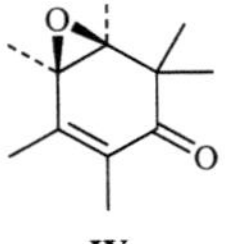 (—) + **IV** (—)	286, 287

TABLE 2A. OXIDATION OF ARENES AND HETEROARENES BY ISOLATED DIOXIRANES (*Continued*)

Substrate	Conditions	Product(s) and Yield(s) (%)	Refs.
C_{12} 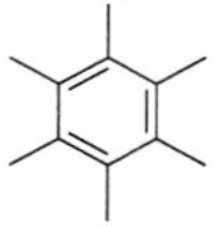	DMD (5 eq), acetone, rt, 48 h	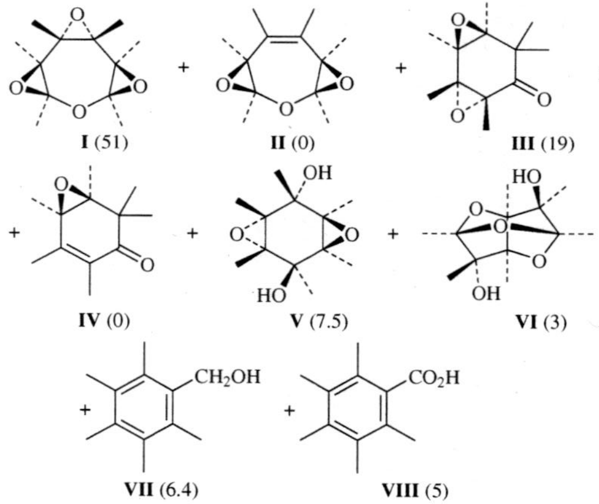 **I** (51) + **II** (0) + **III** (19) + **IV** (0) + **V** (7.5) + **VI** (3) + **VII** (6.4) (CH_2OH) + **VIII** (5) (CO_2H)	286, 287
	DMD (2 eq), acetone, rt, 5 h	**I** (5) + **II** (29) + **IV** (5) + **III** (traces)	286, 287
	DMD (6 eq), acetone, –25°, 72 h	**I** (—) + **II** (—) + **IV** (—) + **V** (—) + **III** (traces) + **VII** (traces)	286, 287
	DMD (6 eq), acetone, $NaHCO_3$, rt, dark, 64 h	**I** (—) + **II** (—) + **III** (—), **I**:**II**:**III** = 58:27:15	286, 287
	DMD (> 6 eq), acetone, $NaHCO_3$, rt, dark, > 76 h	**I** (—) + **III** (—), **I**:**III** = 90:10	286, 287
	Dry DMD (6 eq), acetone, rt, 96 h	**I** (—) + **II** (—) + **VII** (—), **I**:**II**:**VII** = 58:15:27	286, 287
	DMD (6 eq), acetone, TsOH, rt, 96 h	**I** (—) + **III** (—) + **V** (traces) + **VII** (traces), **I**:**III** = 15:85	286, 287

Carbons	Substrate	Conditions	Product(s) and Yield(s) (%)	Refs.
	CO_2Me, NH, N, H, N, *t*-BuO, O	DMD, acetone, CH_2Cl_2, 10°, 2 d	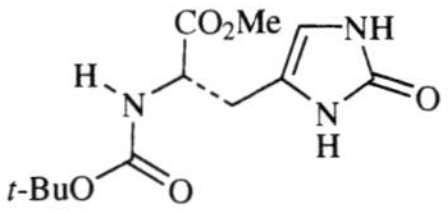 CO_2Me, NH, H, N, N, H, O, *t*-BuO, O (55)	93
	R^1, CH_2R^2, CH_2R^3, N, Ac	DMD-d_6, acetone-d_6, –40°, 10-30 min	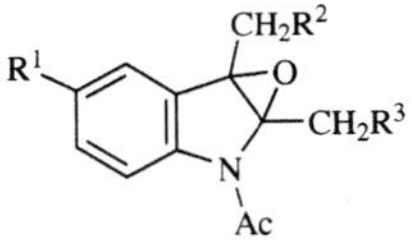 R^1, CH_2R^2, O, CH_2R^3, N, Ac (<95)	288
	R^1, N, $COCH_2R^2$	DMD, acetone, CH_2Cl_2, –78°	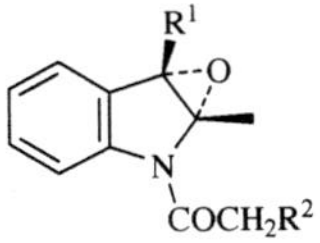 R^1, O, N, $COCH_2R^2$	92
C_{12-17}	R^1, CH_2R^2, CH_2R^3, N, Ac	DMD-d_6, acetone-d_6, –70 to –40°, 10-30 min	R^1, CH_2R^2, O, CH_2R^3, N, Ac (>95)	90
C_{13}	O, N, O, O	DMD, acetone, rt	H, O, O, N, O, O (> 95)	73

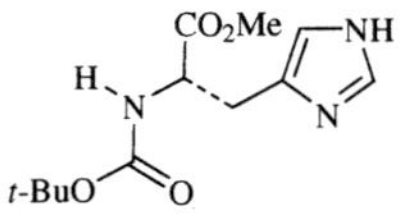

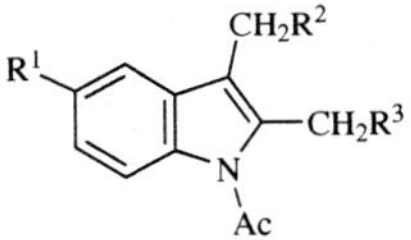

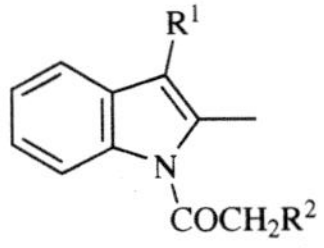

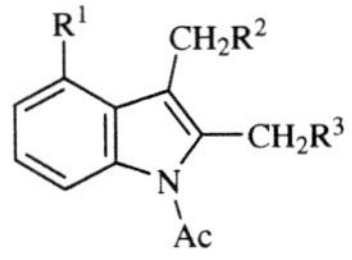

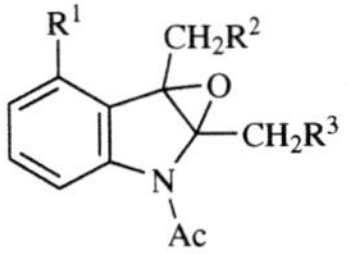

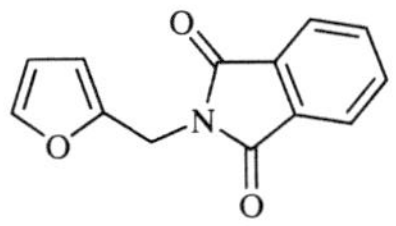

Ref. 288:

R^1	R^2	R^3
H	H	H
H	—$(CH_2)_2$—	
CO_2Et	—$(CH_2)_2$—	

Ref. 92:

R^1	R^2	Time	
Me	Cl	2 h	(> 95)
Me	H	1 h	(> 97)
Et	H	1.5 h	(> 98)
i-Pr	H	1 h	(> 98)
t-Bu	H	1 h	(—)

Ref. 90:

R^1	R^2	R^3
H	H	H
H	—CH_2—	
MeO	H	H
H	—$(CH_2)_2$—	
CO_2Et	—$(CH_2)_2$—	

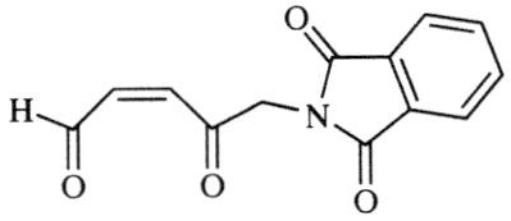

TABLE 2A. OXIDATION OF ARENES AND HETEROARENES BY ISOLATED DIOXIRANES (*Continued*)

Substrate	Conditions	Product(s) and Yield(s) (%)	Refs.
C_{13} 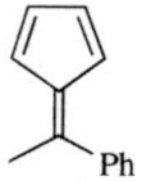 Ph	DMD (1.0 eq), acetone, CH_2Cl_2, N_2, –10°, 3 h	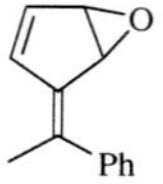 Ph (67) 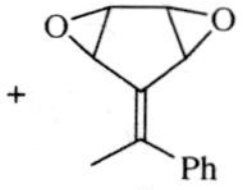 + Ph **I** (14)	279
	DMD (3.1 eq), acetone, CH_2Cl_2, N_2, 0°, 4 h	**I** (89)	279
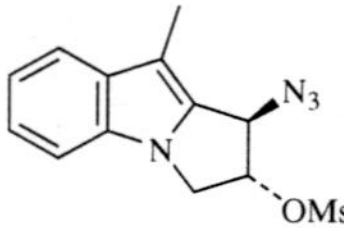 N_3, OMs	DMD, acetone, H_2O	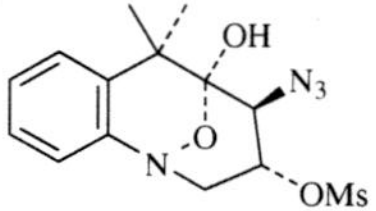 OH, N_3, OMs (59)	289
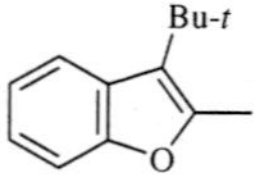 Bu-*t*	DMD, acetone, CH_2Cl_2, N_2, 0°, 3 h	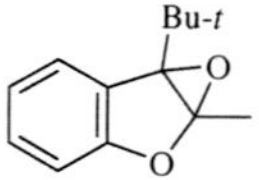 Bu-*t* (> 95)	89
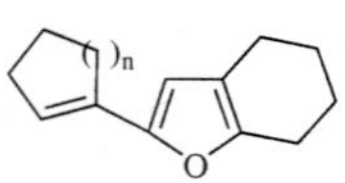 n = 1, 2, 3	DMD, acetone, CH_2Cl_2, –20°	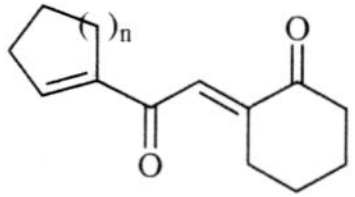 (100)	290
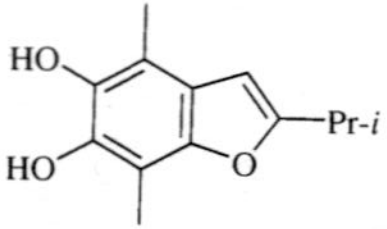 HO, HO, Pr-*i*	DMD, acetone, CH_2Cl_2, –78°, 20 min	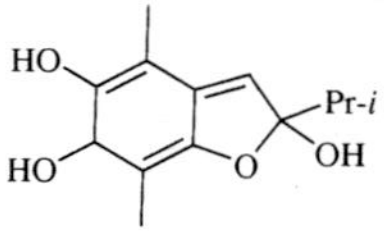 HO, HO, Pr-*i*, OH (73)	291
C_{14} 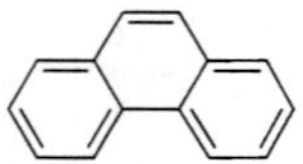	DMD, acetone, MeCN, 26°	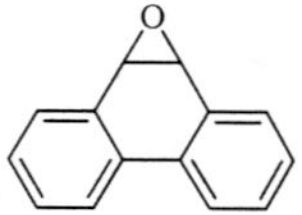 **I** (83)	121

Substrate	Conditions	Product(s) and Yield(s) (%)	Refs.
	Ethylmethyldioxirane, 2-butanone, MeCN, rt	**I** (—)	121
	TFD, TFP, –20°, 5 min	**I** (74)	173
	DMD, acetone, 22°, 20 h	**I** (—)	173
	TFD, CH_2Cl_2, TFP, –20°, 8 min	**I** (96)	65
	TFD, CH_2Cl_2, TFP, 0°, 30 min	(80)	65
OH	DMD, acetone	O O (77)	17
O O O	DMD, acetone, CH_2Cl_2, N_2, –78 to –20°, 16 h	O O O O (100)	84
O O O	DMD, acetone, CH_2Cl_2, N_2, –78° to –20°, 22 h	O O O O (100)	84
O_2N N Ac	DMD, acetone, CH_2Cl_2, 20°	O_2N O N Ac (99)	91

TABLE 2A. OXIDATION OF ARENES AND HETEROARENES BY ISOLATED DIOXIRANES (*Continued*)

Substrate	Conditions	Product(s) and Yield(s) (%)	Refs.
C_{14} 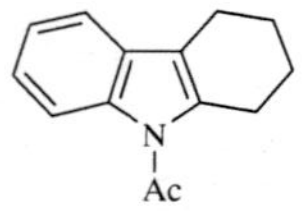 N Ac	DMD, acetone, 0°, 0.5 h	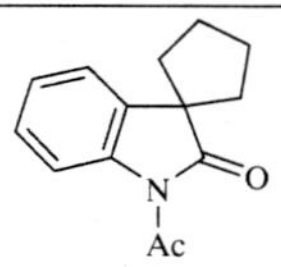 N Ac O (75) **I**	91
	DMD, acetone, CH_2Cl_2, –78°	**I** (100)	92
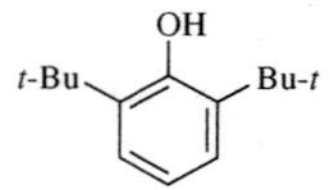 OH *t*-Bu Bu-*t*	TFD, TFP, CH_2Cl_2, N_2, 0°, 1 min	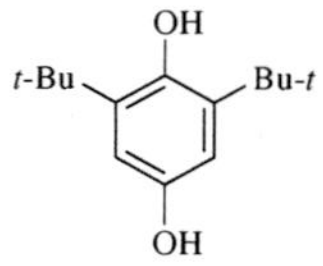 OH *t*-Bu Bu-*t* OH **I** (4) 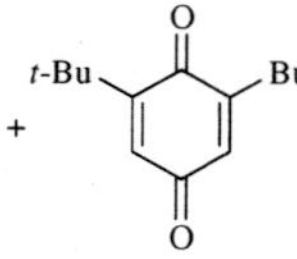 + O *t*-Bu Bu-*t* O **II** (23) 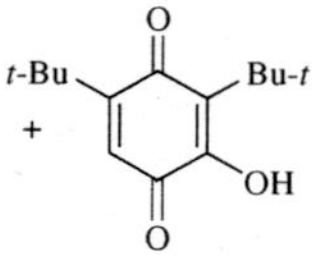 + O *t*-Bu Bu-*t* OH O **III** (67)	277
	DMD, acetone, CH_2Cl_2, N_2, 0°, 48 h	**II** (19) + **III** (13) 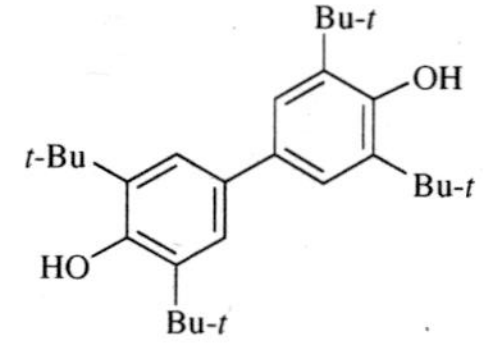 + Bu-*t* OH *t*-Bu Bu-*t* HO Bu-*t* (2)	277
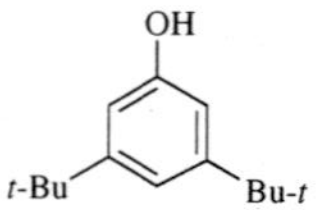 OH *t*-Bu Bu-*t*	DMD, acetone	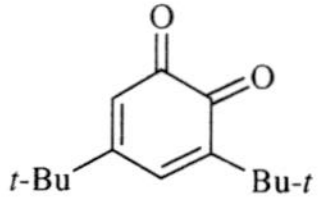 O O *t*-Bu Bu-*t* (46) 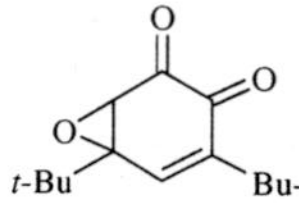 + O O O *t*-Bu Bu-*t* (20)	17
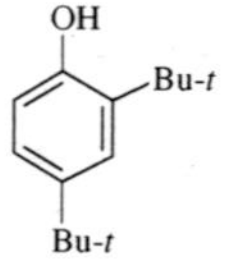 OH Bu-*t* Bu-*t*	DMD, acetone	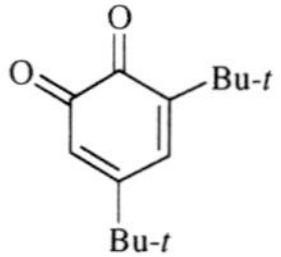 O O Bu-*t* Bu-*t* (55)	17

Substrate	Conditions	Product(s) and Yield(s) (%)	Refs.
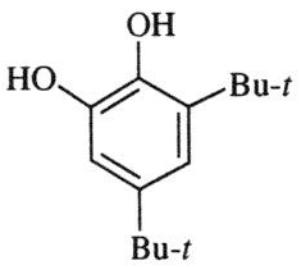 HO, OH, Bu-t, Bu-t	DMD, acetone	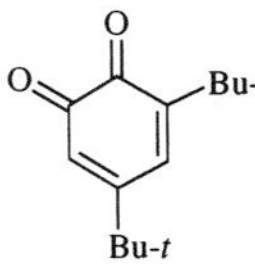 O, O, Bu-t, Bu-t (100)	17
C_{15} 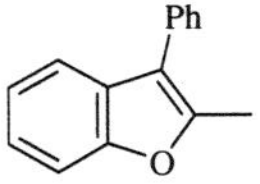 Ph	DMD, acetone, CH_2Cl_2, N_2, –20°, 2 d	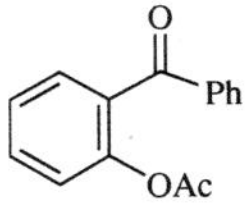 Ph, OAc **I** (75) 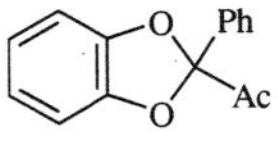 + O, O, Ph, Ac **II** (14)	292
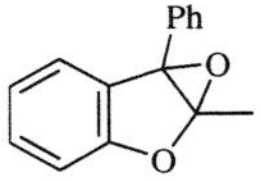 Ph, O, O	DMD, acetone, CH_2Cl_2, N_2, –20°, 1 d	**I** (74) + **II** (13)	292
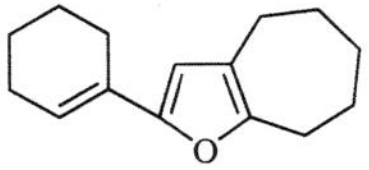 O	DMD, acetone, CH_2Cl_2, –20°	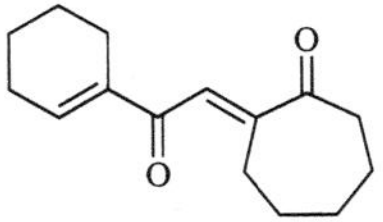 O, O (—)	290
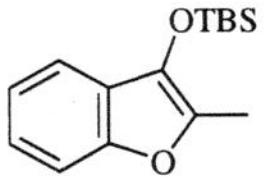 OTBS, O	DMD, acetone, CH_2Cl_2, –78°, 1 h	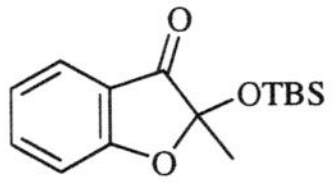 O, OTBS, O (74) 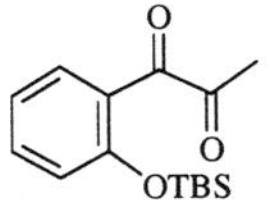 + O, O, OTBS (13)	83
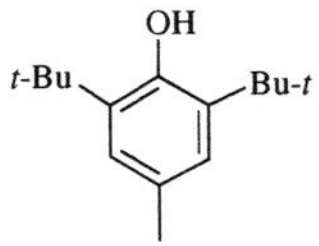 OH, t-Bu, Bu-t	DMD, acetone	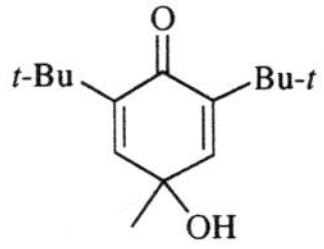 O, t-Bu, Bu-t, OH (13)	17

TABLE 2A. OXIDATION OF ARENES AND HETEROARENES BY ISOLATED DIOXIRANES (*Continued*)

Substrate	Conditions	Product(s) and Yield(s) (%)	Refs.
C_{15-16} Ph; R; O	DMD, acetone, –70 to –20°, 4 h	Ph; R; O; O; I + Ph; O; O; II R / I / II H / (100) / (—) MeO / (—) / (100)	80
	DMD, acetone, N_2, –78 to –10°	I R / Time H / 7 h / (100) MeO / 1 h / (100)	293
N; O; OR	DMD, acetone, CH_2Cl_2, N_2, –78 to –20°, 3 h	O; N; O; OR R *t*-Bu (95) $(CH_2)_2TMS$ (93)	294
C_{15-17} R^1; R^2; O; NH; Ph; O; O	DMD, acetone, CH_2Cl_2, –78°, 10-20 min; 20°, 60-140 min	R^1; R^2; O; O; Ph; O; OH; N; O R^1 / R^2 H / H / (98) Me / H / (97) Me / Me / (98)	295
R^2; R^3; R^1; O	DMD, acetone, N_2	R^2; R^3; R^1; O; O; I + R^2; R^3; R^1; OH; O; II	86

R^1	R^2	R^3	Temp	Time	I	II
H	H	H	–78° to –10°	7 h	(> 95)	(—)
MeO	H	NO_2	–78° to –20°	3 h	(—)	(> 95)
MeO	H	H	–78° to –20°	1 h	(—)	(> 95)
Ac	H	H	–78° to 0°	14 h	(—)	(—)
MeO	MeO	H	–78° to –20°	1 h	(—)	(> 95)

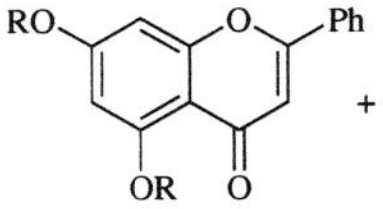

RO, OR, O, Ph + HCl | DMD, acetone, rt | RO, Cl, Cl, OR, O, Ph | 278

R	
H	(72)
Me	(87)

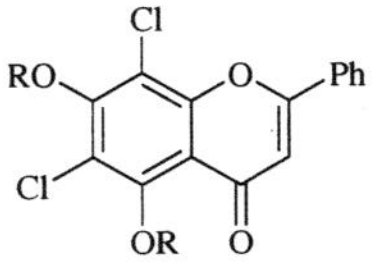

C_{16}

Reactant	Conditions	Product(s)	Refs.
	DMD, acetone, 0-5°, 12 h	I (—)	69
	TFD (1.1 eq), CH_2Cl_2, TFP, –20°, 5 min	I (1.2)	65
	TFD (2.2 eq), CH_2Cl_2, TFP, –20°, 5 min	I (2.4)	65
O	DMD, acetone, 0-5°, 12 h	O, O (7)	69
Ac, O	DMD, acetone, CH_2Cl_2, N_2, –78 to –20°, 5 h	Ac, O, O (100)	84

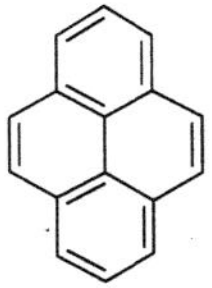

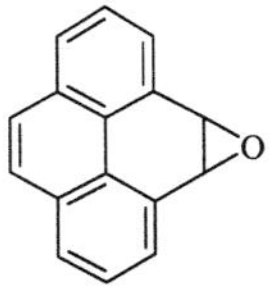

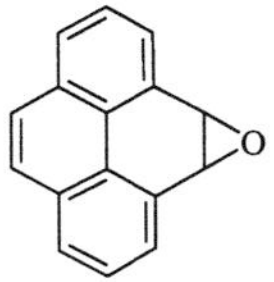

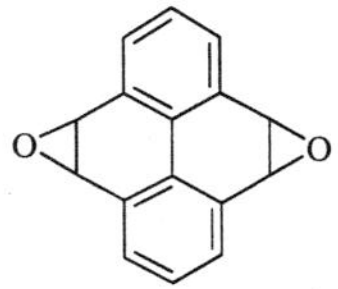

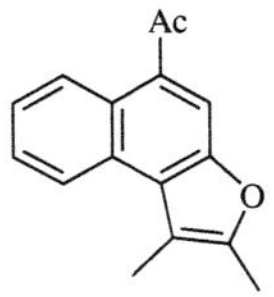

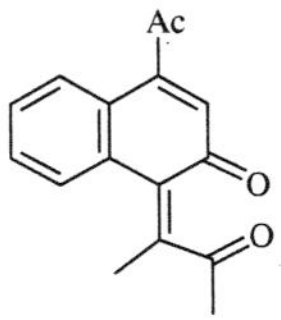

TABLE 2A. OXIDATION OF ARENES AND HETEROARENES BY ISOLATED DIOXIRANES (*Continued*)

	Substrate	Conditions	Product(s) and Yield(s) (%)	Refs.
C_{16} 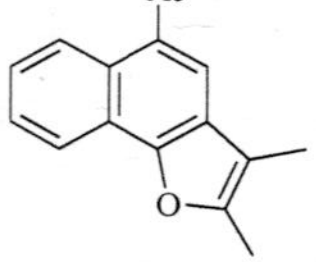	Ac	DMD, acetone, CH_2Cl_2, N_2, –78 to –20°, 7 h	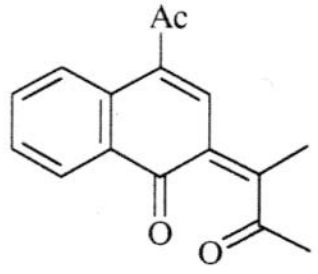 Ac; O; O (100)	84
	OMe O; Ph	DMD, acetone, rt	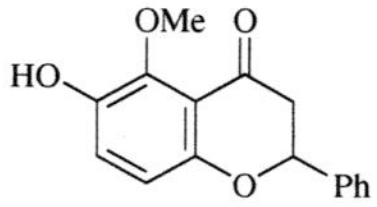 OMe O; HO; Ph (35-40)	68
		DMD, acetone, 2 N HCl, –30°	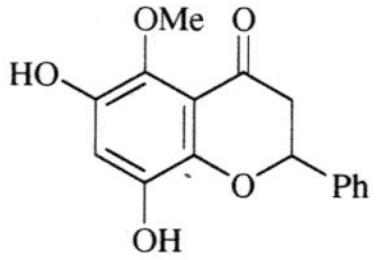 OMe O; HO; OH; Ph (100)	68
	O; OMe	DMD, acetone, 2 N HCl, –30°	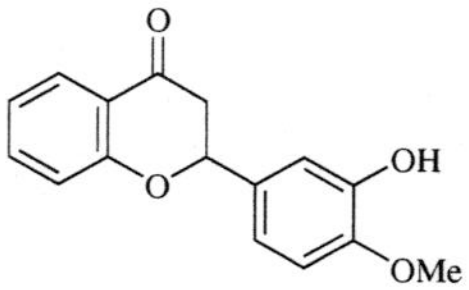 O; OH; OMe (100)	68
	MeO_2C; N; N; Bn	DMD (1.5 eq), acetone, CH_2Cl_2, rt, 2 h	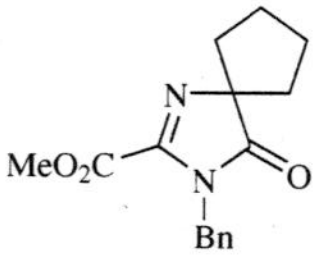 MeO_2C; N; N; Bn; O (40)	281

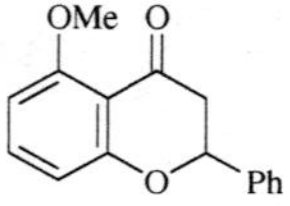

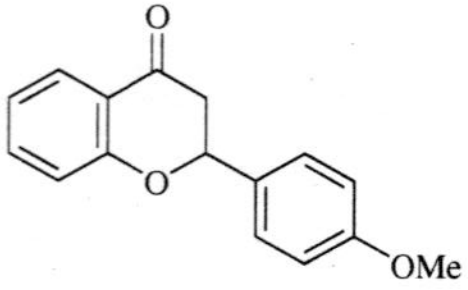

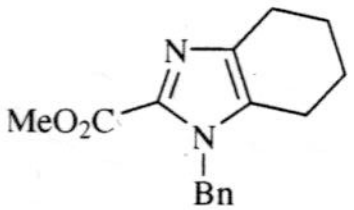

Substrate	Conditions	Product(s) and Yield(s) (%)	Refs.
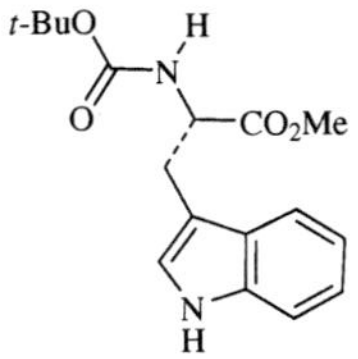	DMD, acetone, CH_2Cl_2, 10°, 2 d	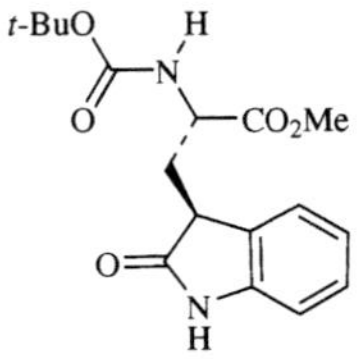 (45) 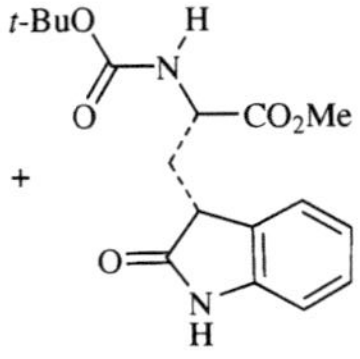 + (45)	93
C_{17}	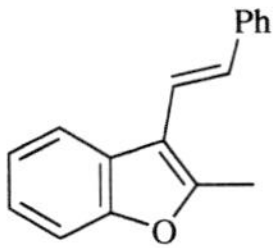 DMD, acetone, CH_2Cl_2, N_2, –70 to –20°, 3 h	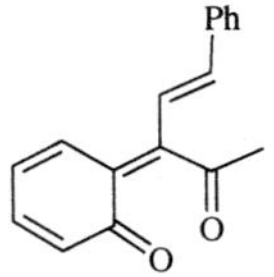 (100)	83
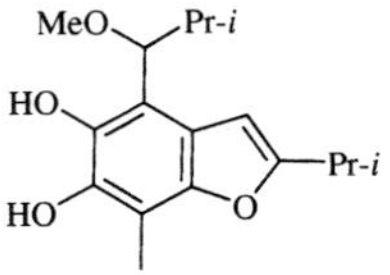	DMD, acetone, CH_2Cl_2, –78°, 20 min	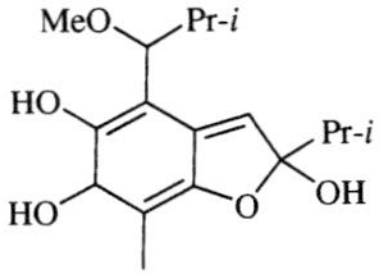 (65)	291
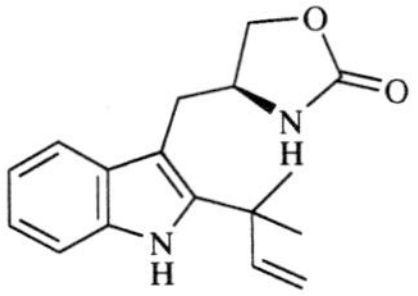	DMD, acetone	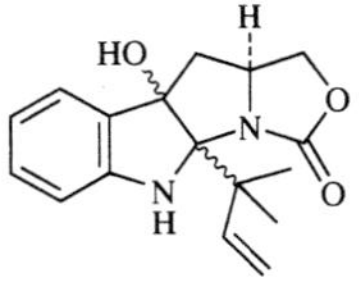 (—)	296
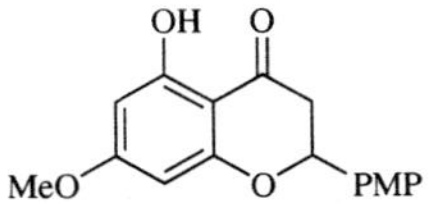	DMD, acetone, 0-22°, 20 min	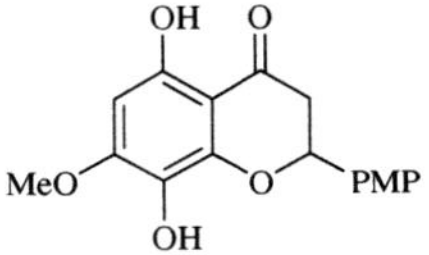 (88)	280

TABLE 2A. OXIDATION OF ARENES AND HETEROARENES BY ISOLATED DIOXIRANES (*Continued*)

Substrate	Conditions	Product(s) and Yield(s) (%)	Refs.
$C_{17\text{-}19}$ 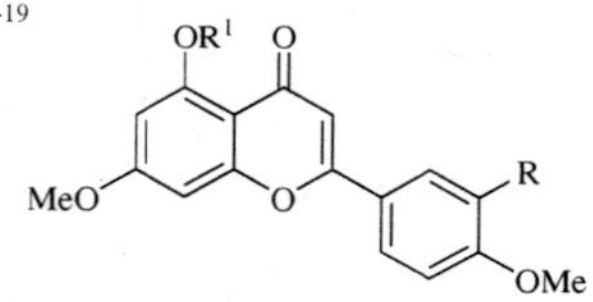 OR[1], O, MeO, O, R, OMe	1. DMD, CH_2Cl_2, 2 h 2. Me_2SO_4, acetone, K_2CO_3, 60°, 1 h	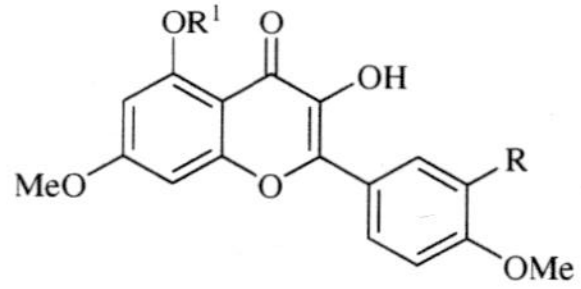 OR[1], O, OH, MeO, O, R, OMe R / R[1]: H / H (85); Me / H (55); Me / Me (38)	280
C_{18} 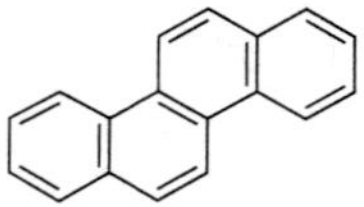	DMD, acetone, rt, dark, 72 h	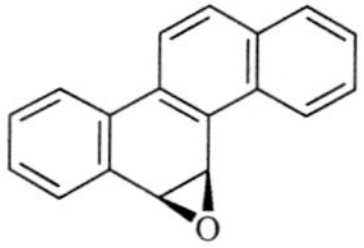 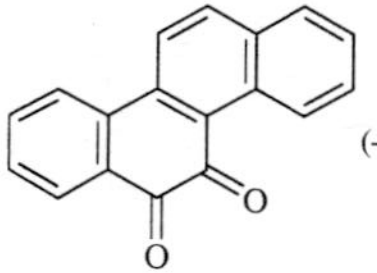 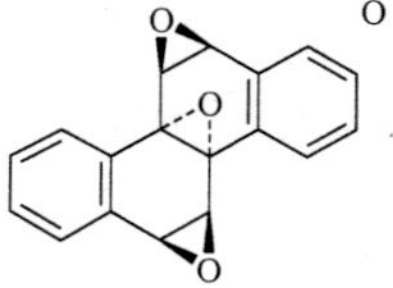 O (—) + O, O (—) + O, O, O (8)	70
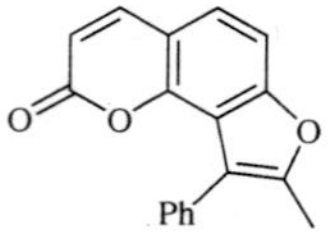 O, O, O, Ph	DMD, acetone, CH_2Cl_2, N_2, –78 to –20°, 14 h	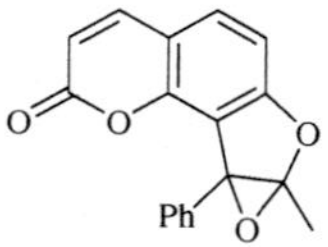 O, O, O, Ph, O (100)	84
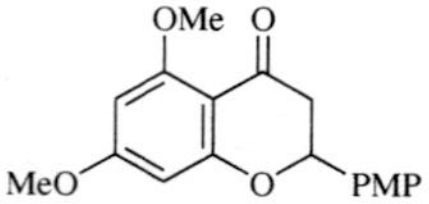 OMe, O, MeO, O, PMP	DMD, acetone, rt	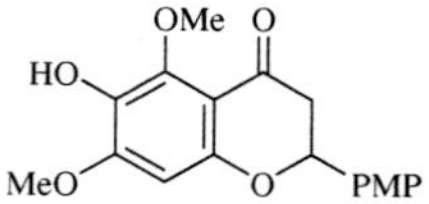 OMe, O, HO, MeO, O, PMP (35-40)	68
	DMD, acetone, 2 N HCl, –30°	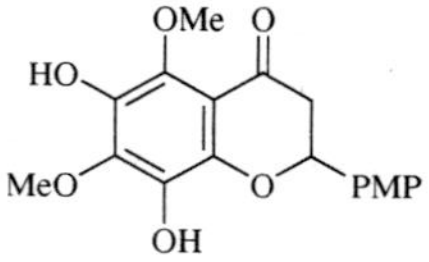 OMe, O, HO, MeO, O, PMP, OH (100)	68

Substrate	Conditions	Product(s) and Yield(s) (%)	Refs.
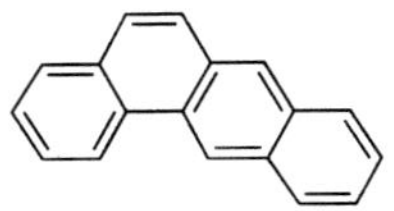	DMD, acetone, 0-5°, 12 h	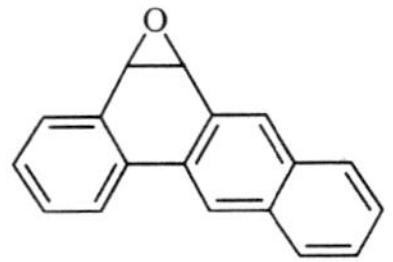 (—)	69
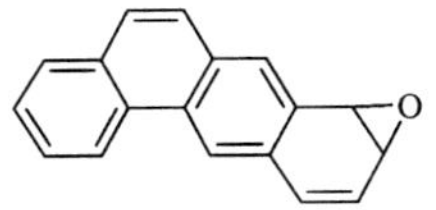	DMD, acetone, 0-5°, 12 h	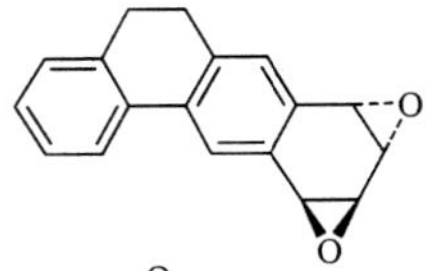 (97)	69
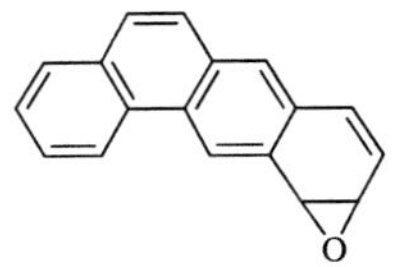	DMD, acetone, 0-5°, 12 h	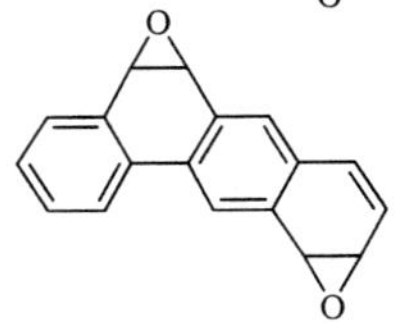 (18)	69
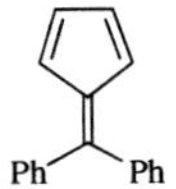 Ph Ph	DMD (1.0 eq), acetone, CH_2Cl_2, N_2, –10°, 2 h	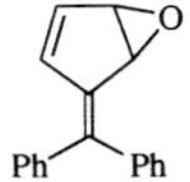 Ph Ph (61) 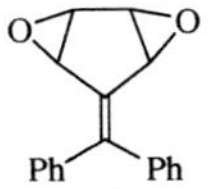 + Ph Ph **I** (14)	279
	DMD (2.2 eq), acetone, CH_2Cl_2, N_2, 20°, 5 min	**I** (100)	279
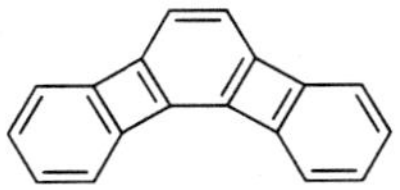	DMD, acetone, rt, 30 min	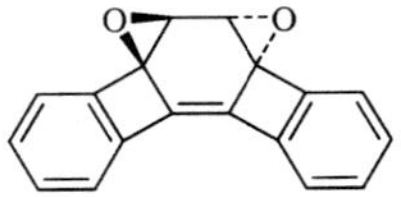 (100)	297
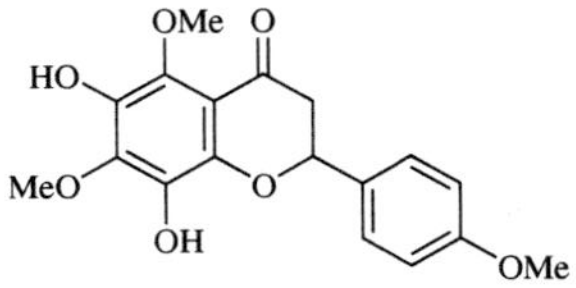 HO, OMe, MeO, OH, OMe	DMD, acetone, 2 N HCl, –30°	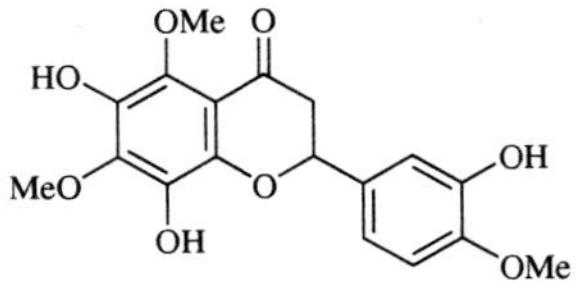 HO, OMe, MeO, OH, OH, OMe (100)	68

TABLE 2A. OXIDATION OF ARENES AND HETEROARENES BY ISOLATED DIOXIRANES (*Continued*)

Substrate	Conditions	Product(s) and Yield(s) (%)	Refs.
C_{18-19} OH, O, MeO, OMe, R, OMe	DMD, CH_2Cl_2, 0° to rt, 1 h	OH, O, MeO, OMe, OH, R, OMe; R: H (76), OMe (65)	280
OH, O, MeO, MeO, O, R, OMe	DMD, CH_2Cl_2, 1 h	OH, O, MeO, MeO, O, OH, R, OMe; R: H (90), OMe (90)	280
C_{19} N, Me	DMD, acetone, CH_2Cl_2, –78°	**I** HO, N, Me, O + **II** O, OH, N, Me; **I** + **II** (> 98), **I**:**II** = 10:1	92
OMe, O, MeO, O, OMe, OMe	DMD, CH_2Cl_2, N_2, 30 min	OMe, O, OH, MeO, O, OMe, OMe (86)	280
OH, O, OMe, MeO, O, OMe, OMe	DMD, CH_2Cl_2, 50 min	OH, O, OMe, MeO, O, OH, OMe, OMe (89)	280

Substrate	Conditions	Product(s) and Yield(s) (%)	Refs.
	DMD, HCl, acetone, –30°, 1 min	(96)	278
	DMD (3.0 eq), HCl, –30°, 1 min	(96)	280
	DMD (40 eq), HCl, –30°, 1 min	(90)	280
C_{21}			
	DMD, acetone, 20°, 6-8 h	(100)	298
	DMD, acetone	(—)	296

TABLE 2A. OXIDATION OF ARENES AND HETEROARENES BY ISOLATED DIOXIRANES (*Continued*)

Substrate	Conditions	Product(s) and Yield(s) (%)	Refs.
C_{21} 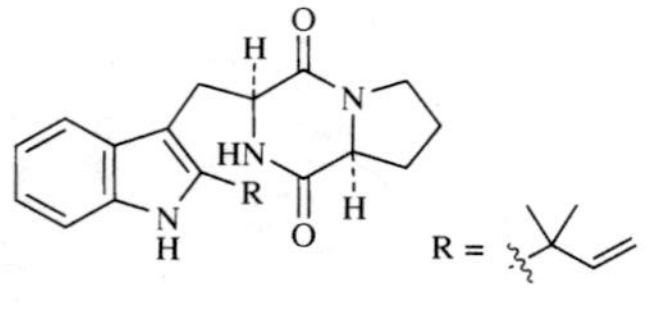 R = (1,1-dimethylallyl)	DMD (4 eq), acetone, CH_2Cl_2, –78 to 0°, 25 h	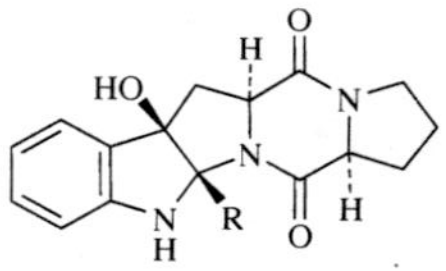 (64) 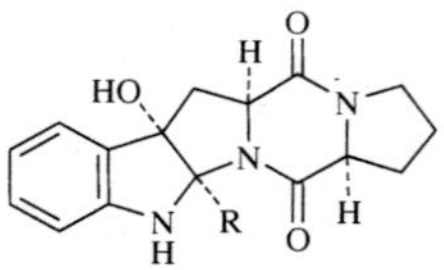 + (12)	296
C_{22} 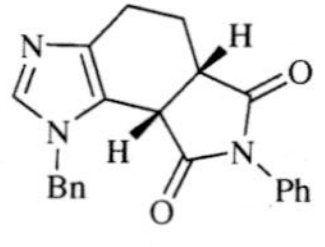	DMD (1.5 eq), acetone, CH_2Cl_2, rt, 2 h	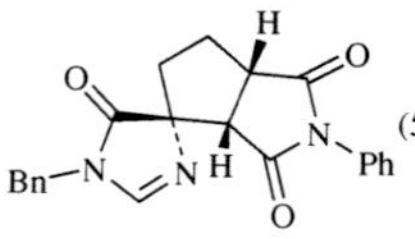 (56) + 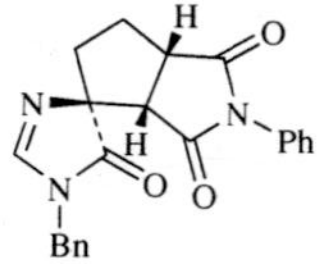 (27)	281
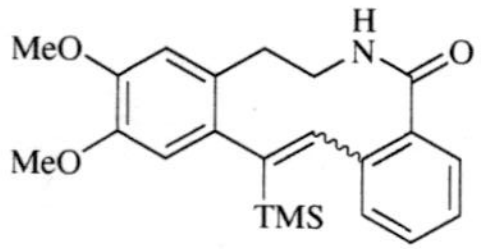	TFD, TFP, CH_2Cl_2, rt, 9 h	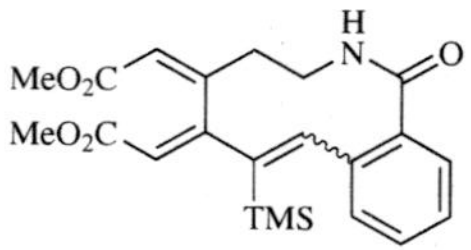 (90)	299, 300
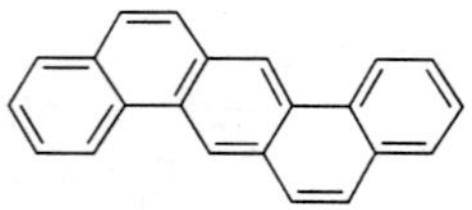	DMD, acetone, 0-5°, 12 h	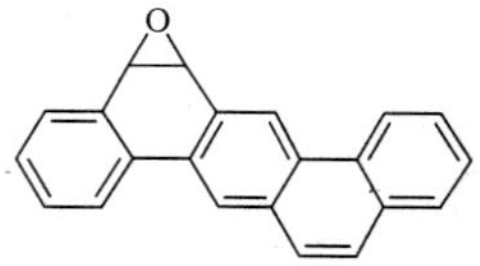 (—)	69
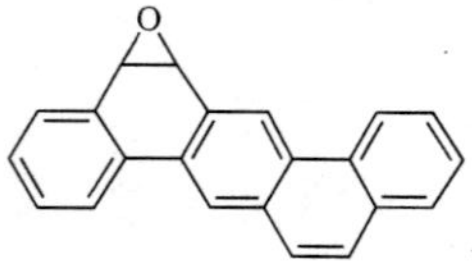	DMD, acetone, 0-5°, 12 h	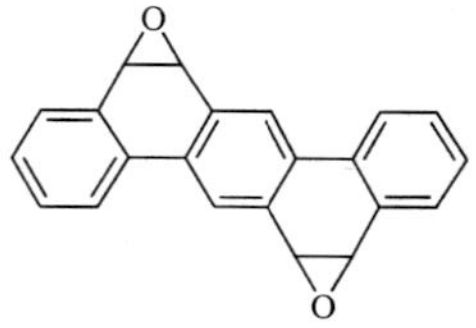 (16)	69

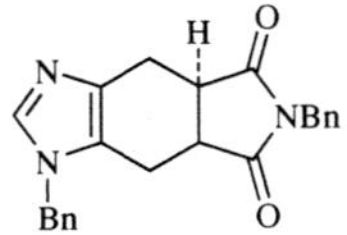

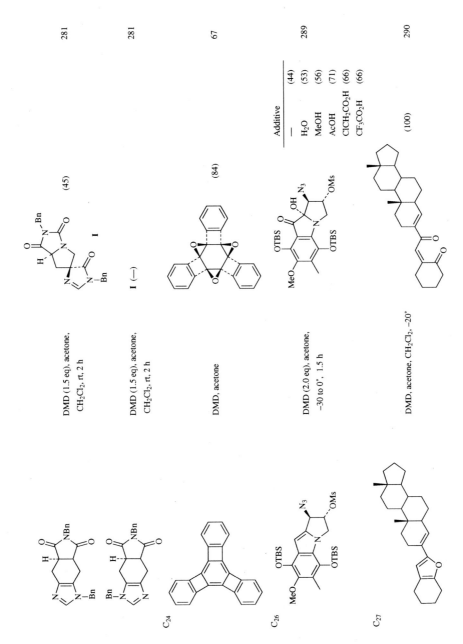

Formula	Reactant	Conditions	Product(s) and Yield(s) (%)	Refs.
		DMD (1.5 eq), acetone, CH_2Cl_2, rt, 2 h	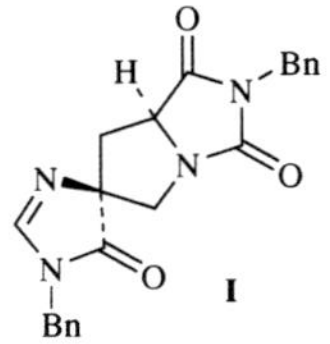 **I** (45)	281
		DMD (1.5 eq), acetone, CH_2Cl_2, rt, 2 h	**I** (—)	281
C_{24} 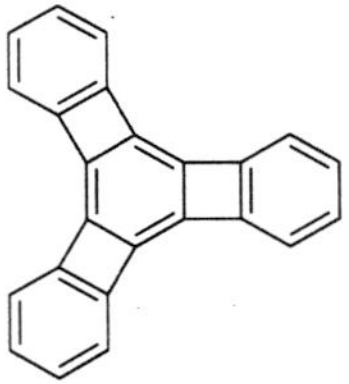		DMD, acetone	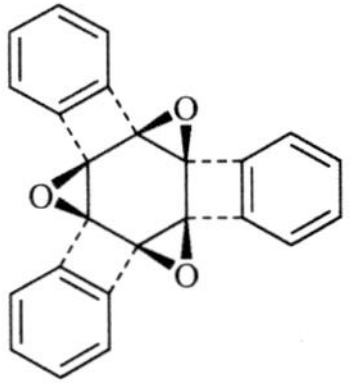 (84)	67
C_{26} 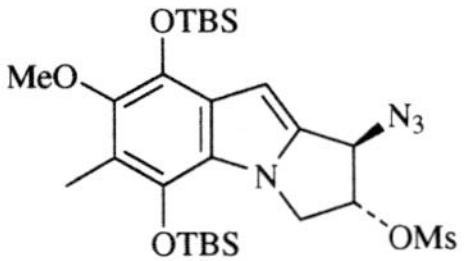		DMD (2.0 eq), acetone, –30 to 0°, 1.5 h	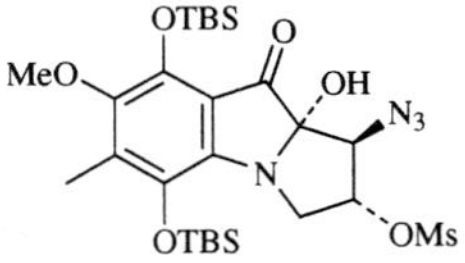 Additive: — (44); H_2O (53); MeOH (56); AcOH (71); $ClCH_2CO_2H$ (66); CF_3CO_2H (66)	289
C_{27} 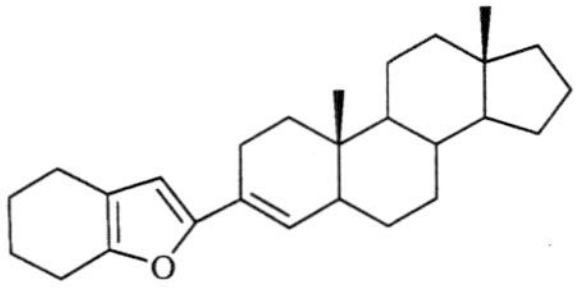		DMD, acetone, CH_2Cl_2, –20°	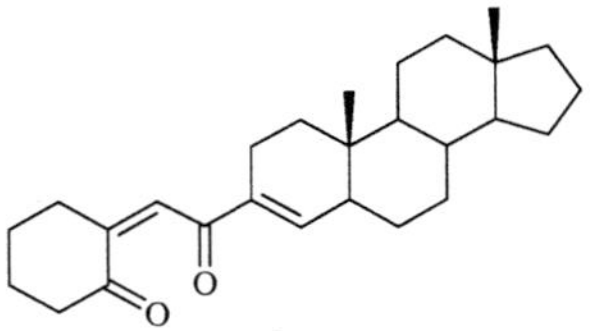 (100)	290

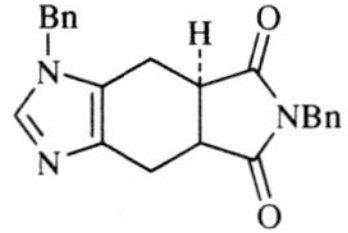

TABLE 2A. OXIDATION OF ARENES AND HETEROARENES BY ISOLATED DIOXIRANES (*Continued*)

Substrate	Conditions	Product(s) and Yield(s) (%)	Refs.
C_{28} 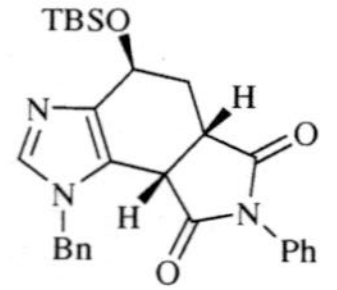 TBSO, N, N, Bn, H, H, O, O, N, Ph	DMD (1.5 eq), acetone, CH_2Cl_2, –78°, 2 h	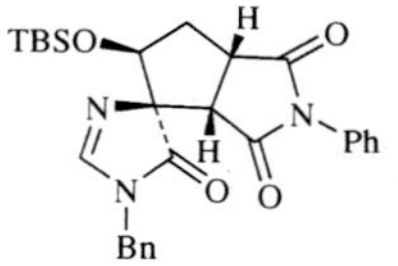 TBSO, H, O, N, N, Ph, H, O, O, N, Bn (32)	281
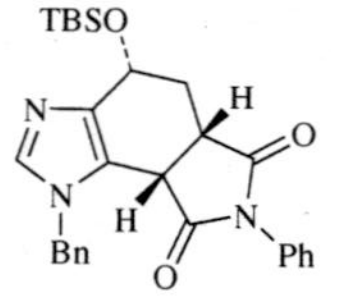 TBSO, N, N, Bn, H, H, O, O, N, Ph	DMD (1.5 eq), acetone, CH_2Cl_2, –78°, 2 h	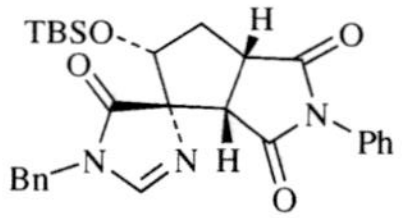 TBSO, O, H, O, N, Ph, Bn–N, N, H, O (44)	281
C_{29} 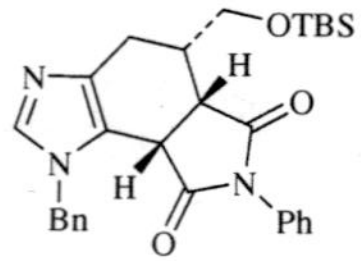 OTBS, N, N, Bn, H, H, O, O, N, Ph	DMD (1.5 eq), acetone, CH_2Cl_2, rt, 1 h	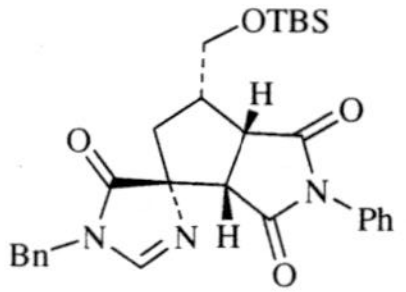 OTBS, H, O, O, N, Ph, Bn–N, N, H, O (60)	281
C_{30} 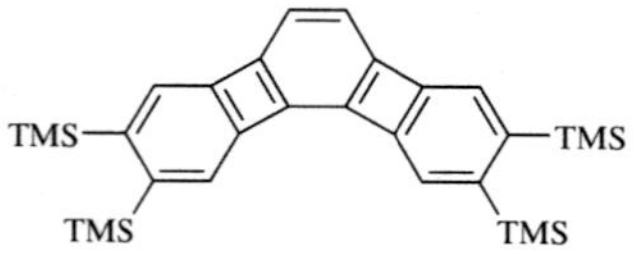 TMS, TMS, TMS, TMS	DMD, acetone, rt, 1 h	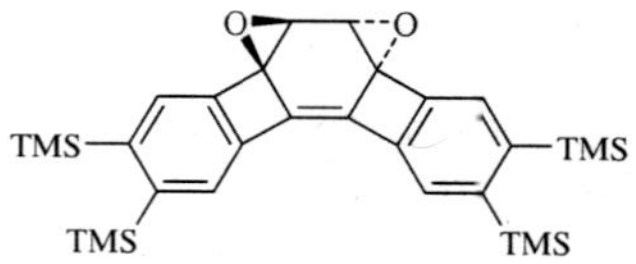 O, O, TMS, TMS, TMS, TMS (100)	297
	DMD, acetone, rt, 6 h	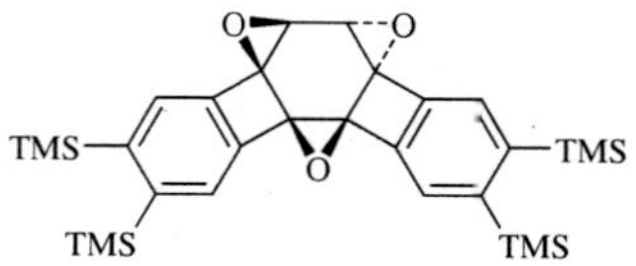 O, O, O, TMS, TMS, TMS, TMS (26)	297

Substrate	Conditions	Product(s) and Yield(s) (%)	Refs.
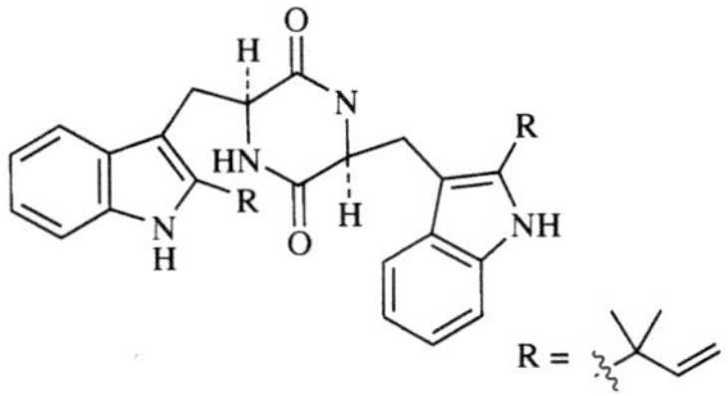 C_{32} R = [structure]	DMD (4 eq), acetone, CH_2Cl_2, –78 to 0°, 25 h	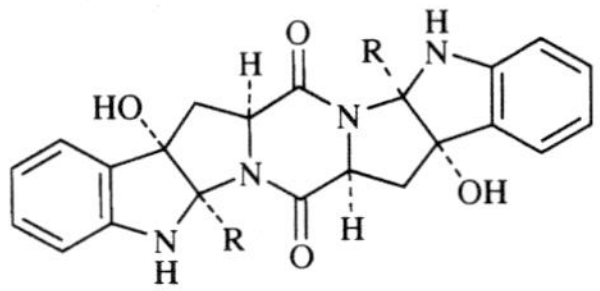 (18) 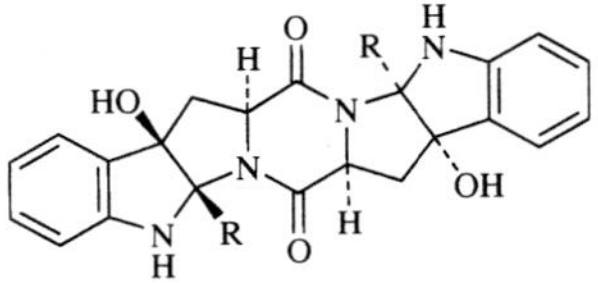 + (40) 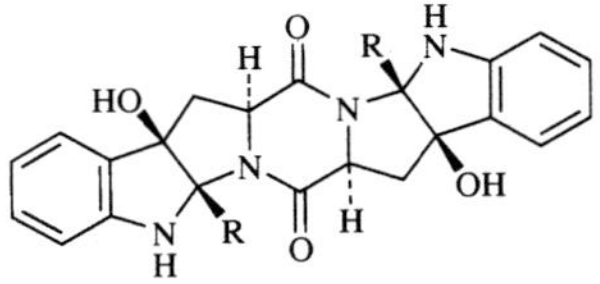 + (21)	296
C_{35}	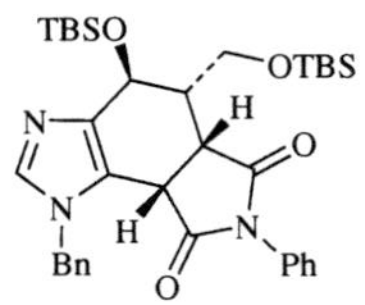 DMD (1.5 eq), acetone, CH_2Cl_2, rt, 2 h	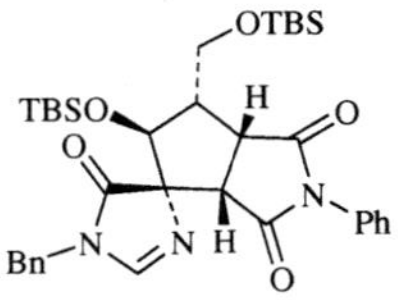 (56) 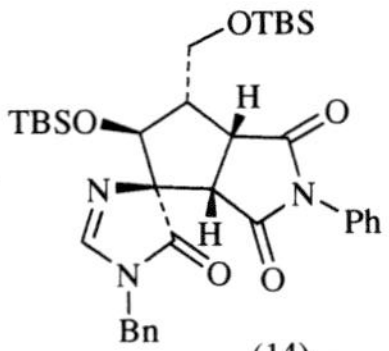 + (14)	281

TABLE 2A. OXIDATION OF ARENES AND HETEROARENES BY ISOLATED DIOXIRANES (*Continued*)

Substrate	Conditions	Product(s) and Yield(s) (%)	Refs.
C_{43-50} R²O, OR², OR², OR², OR¹	DMD, acetone, –15°, 0.6 h	R²O, HO, OR², OR², OR², OR¹ R¹ H, R² Bn (10-15) R¹ Bn, R² Bn (13)	301
C_{43-53} BnO, R², OBn, OBn, OBn, OR¹	DMD, acetone	BnO, HO, R², OBn, OBn, OBn, OR¹	301

R¹	R²	Temp	Time	Yield (%)
H	Br	–40°	7.5 h	(33-36)
H	CF_3CO	–30°	36 h	(37)
Bn	CHO	–40°	7.5 h	(38)
Bn	CO_2H	–15°	0.6 h	(30-40)
Bn	CH_2OAc	–40°	7.5 h	(34)
Bn	CF_3CO	–30°	20 h	(49)

Reactant	Conditions	Product(s) and Yield(s) (%)	Refs.
C_{60} 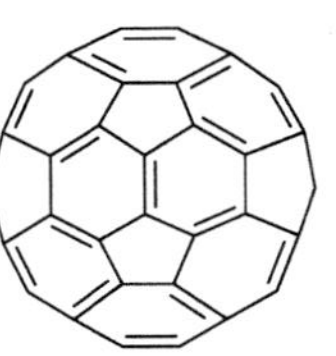	TFD (4 eq), TFP, 1,2-dichlorobenzene, 0°, 6 min	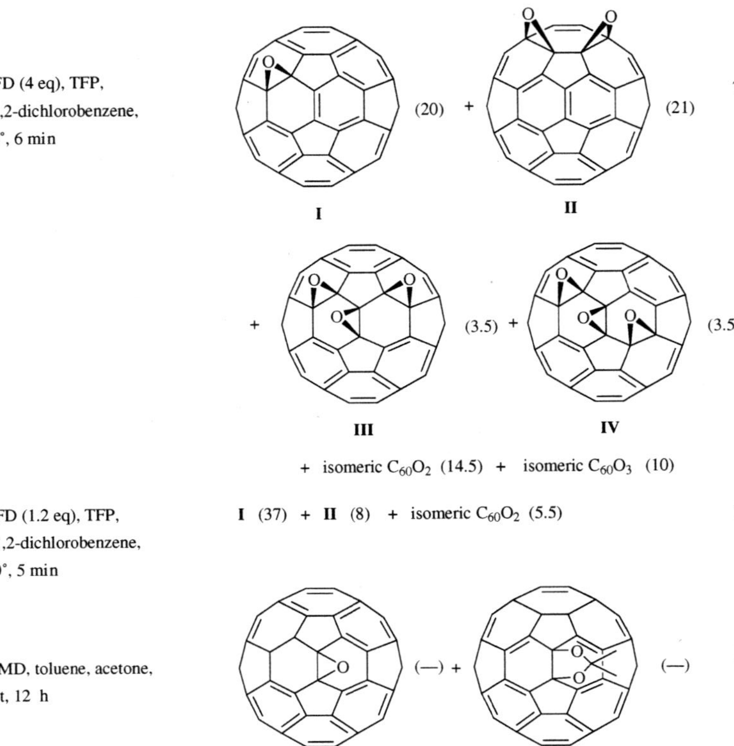 I (20) + II (21) + III (3.5) + IV (3.5) + isomeric $C_{60}O_2$ (14.5) + isomeric $C_{60}O_3$ (10)	72
	TFD (1.2 eq), TFP, 1,2-dichlorobenzene, 0°, 5 min	**I** (37) + **II** (8) + isomeric $C_{60}O_2$ (5.5)	72
	DMD, toluene, acetone, rt, 12 h	(—) + (—)	71

TABLE 2B. OXIDATION OF ARENES AND HETEROARENES BY IN SITU GENERATED DIOXIRANES

Substrate	Conditions	Product(s) and Yield(s) (%)	Refs.
C_{13-14} R-substituted N-Boc indole	1. $(i\text{-Pr})_3B$, LDA, THF 2. Oxone®, NaOH, acetone, THF, H_2O, $NaHSO_3$	R-substituted N-Boc oxindole R: 5-Me (89); 4-Cl (70); 7-Me (63); 5-Br (62); H (78)	302
C_{15-16} 2-(4-R-phenyl)chroman-4-one R: H; OMe; H; OMe	Oxone®, acetone/H_2O (v/v 1:1), NaX, rt X (eq), Time: Br (1), 3 h; Br (1), 0.3 h; Cl (1), 1 h; Cl (4), 0.3 h	X-substituted products Product: 6-Br (97); 3',6-Br_2 (74); 6-Cl (85); 3',6-Cl_2 (55)	303
C_{16} 7-MeO-2-Ph-chroman-4-one	Oxone®, acetone/H_2O (v/v 1:1), NaX, rt X (eq), Time: Br (10), 0.3 h; Br (5), 0.5 h; Cl (4), 6 h; Cl (5), 0.5 h	6-X-7-MeO-2-Ph-chroman-4-one **I** + 6,8-X_2-7-MeO-2-Ph-chroman-4-one **II** **I**, **II**: (97), (—); (—), (93); (38), (30); (—), (93)	303

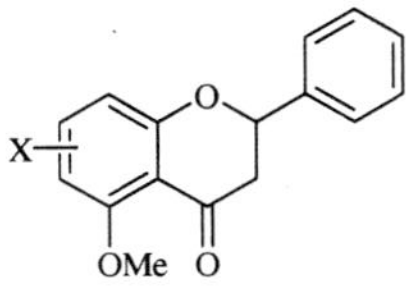

X, OMe, O

Oxone®, acetone/H_2O, NaX, rt

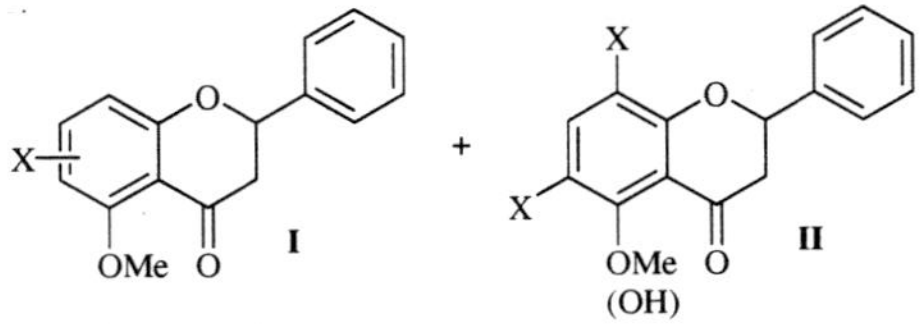

X, OMe, O, **I** + X, X, OMe (OH), O, **II**

303

Acetone/H_2O	X (eq)	Time	Product
5:1	Br (1)	0.5 h	**I**, 6-Br (19) + 8-Br (79)
1:1	Br (1)	0.25 h	**II**, 5-OMe (58) + **II**, 5-OH, Br_2 (36)
5:1	Cl (1)	2 h	**I**, 8-Cl (59) + **II**, Cl_2 (16)
1:1	Cl (3)	0.25 h	**II**, Cl_2 (98)

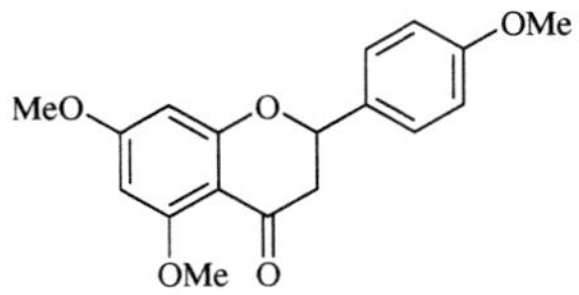

MeO, OMe, OMe, O

Oxone®, acetone/H_2O (v/v 1:1), NaX (1 eq), rt

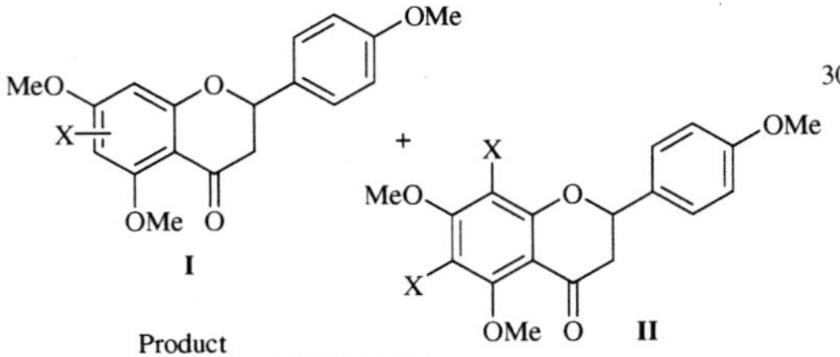

MeO, X, OMe, O, OMe, **I** + MeO, X, X, OMe, O, OMe, **II**

303

Acetone/H_2O	X	Time	Product
5:1	Br	45 min	**I**, 6-Br (98)
1:1	Br	45 min	**II**, Br_2 (98)
1:1	Cl	60 min	**I**, 6-Cl (30) + 8-Cl (65)

TABLE 3A. NITROGEN OXIDATION BY ISOLATED DIOXIRANES

Substrate	Conditions	Product(s) and Yield(s) (%)	Refs.
C_0			
NO_2^-	DMD, acetone, rt, 0.5 h	NO_3^- (—)	163
C_4			
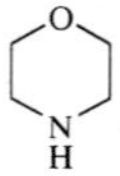 (morpholine, N-H)	DMD, acetone, 0°, 20-30 min	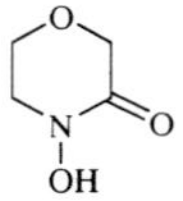 (4-hydroxymorpholin-3-one, N-OH, C=O) (67)	99
n-$BuNH_2$	DMD (x eq), acetone, rt	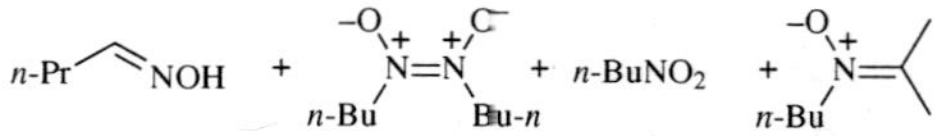 n-Pr–CH=NOH (I) + n-Bu–N(–O⁻)(+)=N(+)(–O⁻)–Bu-n (II) + n-$BuNO_2$ (III) + n-Bu–N(+)(–O⁻)=C(CH₃)₂ (IV)	114
	x = 5, Additives —:	I (20)[a], II (34)[a], III <5[b], IV (25)	
	x = 6, Additives $NaHCO_3$:	I (24), II (36), III (trace), IV (15)	
	x = 6, Additives K_2CO_3:	I (16), II (58), III (trace), IV (—)	
	DMD, acetone, dark, rt, 30 min	**III** (84)	266
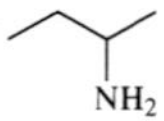 (sec-butylamine, NH_2)	DMD, acetone, dark, rt, 30 min	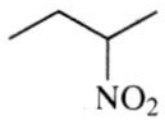 (2-nitrobutane, NO_2) (87)	266
t-BuNH	DMD, acetone, dark, rt, 30 min	t-$BuNO_2$ (90)	266
$C_{4\text{-}5}$ (cyclic amine, N-H, ( )$_n$)	DMD, acetone, 0°, 20 to 30 min	(N-hydroxy lactam, N-OH, C=O, ( )$_n$) n = 1 (27); n = 2 (72)	99
$C_{4\text{-}6}$ (pyrazine, R^1, R^2, R^3, R^4)	DMD, acetone, rt, 18-48 h	(pyrazine N-oxide, N^+–O^-)	304
		R^1 = H, R^2 = Cl, R^3 = H, R^4 = H (91)	
		R^1 = Me, R^2 = Cl, R^3 = Me, R^4 = H (81)	
		R^1 = H, R^2 = Cl, R^3 = Cl, R^4 = H (40)	

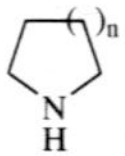

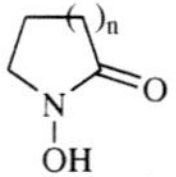

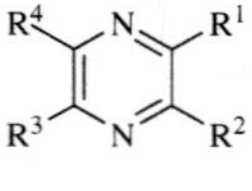

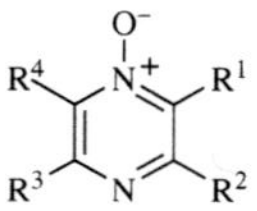

Substrate	Conditions	Product(s) and Yield(s) (%)	Refs.
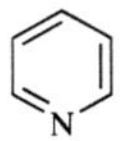 C_5 pyridine	DMD, acetone, hexane, rt	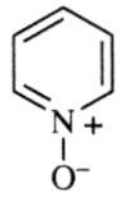 pyridine N-oxide **I** (75)	121
	DMD, acetone, 0°, < 1 h	**I** (100)	95
	Cyclohexanone dioxirane, cyclohexanone, –20°, 5 min	**I** (100)	98
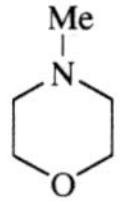 N-methylmorpholine	DMD, acetone, 0°, < 1 h	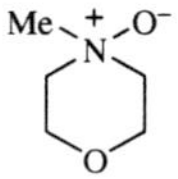 N-methylmorpholine N-oxide (100)	95
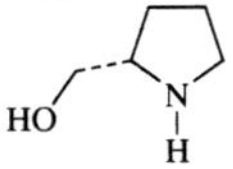 2-(hydroxymethyl)pyrrolidine	DMD, acetone, –78°, 30 min	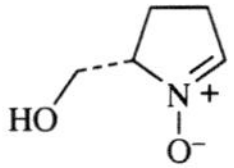 2-(hydroxymethyl)-1-pyrroline N-oxide (32)	305
C_{5-9} R-pyridine	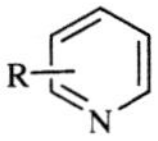 DMD, acetone	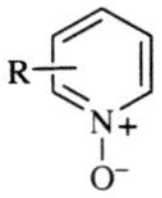 R-pyridine N-oxide (—)	21, 306

R	k_{rel}
3-Br	0.104±0.0050
H	3.12±0.12
4-CN	0.0352±0.0035
2-Me	3.12±0.12
3-Me	9.34±0.47
4-Me	10.6±0.53
3,4-Me_2	23.0±1.4
3,5-Me_2	19.7±1.2
2,6-Me_2	0.0415±0.0042
2,4,6-Me_3	0.0923±0.0092
1,2-CH=CHCH=CH–	1.08
2,3-CH=CHCH=CH–	5.59±0.28
4-CF_3	0.017±0.002
4-OMe	5.36±0.27
4-Ph	1.01±0.06

TABLE 3A. NITROGEN OXIDATION BY ISOLATED DIOXIRANES (*Continued*)

Substrate	Conditions	Product(s) and Yield(s) (%)	Refs.
C_{5-12} 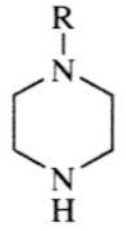 R–N(piperazine)N–H	DMD, acetone, 0°, 20-30 min	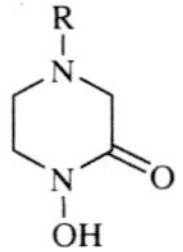 R–N(piperazinone, C=O)N–OH; R: CHO (89), $PhCH_2O_2C$ (67), 4-AcC_6H_4 (85)	99
C_6 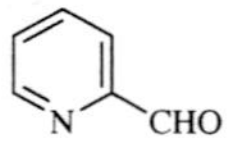 2-pyridinecarboxaldehyde (N, CHO)	DMD, acetone, 20°, 1 h	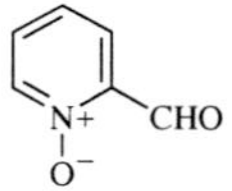 2-CHO pyridine N^+–O^- (60)	307
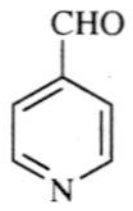 4-pyridinecarboxaldehyde (CHO, N)	DMD, acetone, 20°, 2 h	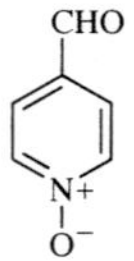 4-CHO pyridine N^+–O^- (98)	307
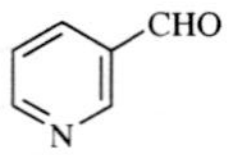 3-pyridinecarboxaldehyde (CHO, N)	DMD, acetone, 20°, 1 h	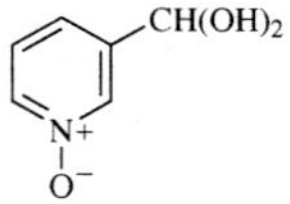 3-$CH(OH)_2$ pyridine N^+–O^- (98)	307
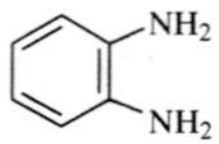 1,2-$C_6H_4(NH_2)_2$	DMD, acetone, 20°, 1 h	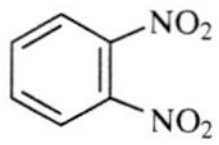 1,2-$C_6H_4(NO_2)_2$ (100)	308
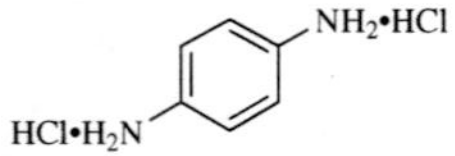 $HCl•H_2N$–C_6H_4–$NH_2•HCl$ (para)	DMD, acetone, H_2O, rt, 23 h	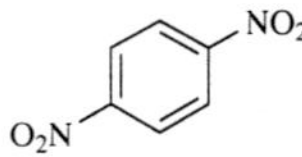 O_2N–C_6H_4–NO_2 (para) (82)	101

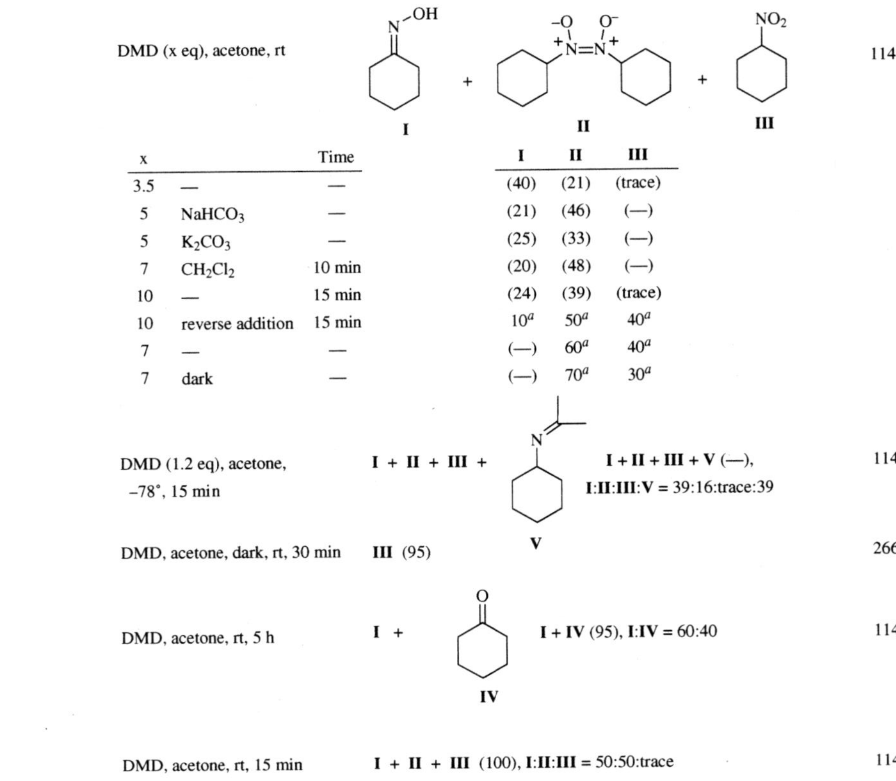

Substrate	Conditions	Product(s) and Yield(s) (%)	Refs.
Cyclohexylamine	DMD (x eq), acetone, rt	I + II + III	114
Cyclohexanone oxime	DMD, acetone, rt, 5 h	I + IV; I + IV (95), I:IV = 60:40	114
N-Cyclohexylhydroxylamine	DMD, acetone, rt, 15 min	I + II + III (100), I:II:III = 50:50:trace	114

x		Time	I	II	III
3.5	—	—	(40)	(21)	(trace)
5	$NaHCO_3$	—	(21)	(46)	(—)
5	K_2CO_3	—	(25)	(33)	(—)
7	CH_2Cl_2	10 min	(20)	(48)	(—)
10	—	15 min	(24)	(39)	(trace)
10	reverse addition	15 min	10[a]	50[a]	40[a]
7	—	—	(—)	60[a]	40[a]
7	dark	—	(—)	70[a]	30[a]

DMD (1.2 eq), acetone, −78°, 15 min — **I** + **II** + **III** + **V**; **I** + **II** + **III** + **V** (—), **I**:**II**:**III**:**V** = 39:16:trace:39 — 114

DMD, acetone, dark, rt, 30 min — **III** (95) — 266

TABLE 3A. NITROGEN OXIDATION BY ISOLATED DIOXIRANES (*Continued*)

Substrate	Conditions	Product(s) and Yield(s) (%)	Refs.
C_6			
[structure] CO_2Me, NH_2	DMD, acetone, –45° to rt	[structure] CO_2Me, H–N–OH (82)	100
[structure] NH_2, NH_2	DMD, acetone, 0°, 1 h	[structure] (86)	308
[structure] NH_2•HCl	DMD, acetone, rt, 1 h	[structure] **I** + [structure] NO_2 **II** **I** + **II** (—), **I**:**II** = 90:10	114
	DMD, acetone, H_2O, rt, 30 min	**I** + **II** (—), **I**:**II** = 26:53	114
H_2N–[structure]–NH_2	DMD, acetone, H_2O, rt, 24 h	O_2N–[structure]–NO_2 (20)	101
[structure] NH_2, NH_2	DMD, acetone, 0°, 1 h	[structure] NO_2, NO_2 (13) + [structure] (49)	308
HCl•H_2N–[structure]–NH_2•HCl	DMD, acetone, H_2O, rt, 23 h	O_2N–[structure]–NO_2 (60)	101
C_{6-7}			
[structure] NH_2, R^1, R^2, R^3, R^4, R^5	DMD, acetone, dark, 22°	[structure] NO_2, R^1, R^2, R^3, R^4, R^5	101, 266

R^1	R^2	R^3	R^4	R^5	Time	
H	H	Cl	H	H	30 min	(97)
Cl	H	Cl	H	Cl	10 h	(98)
F	H	H	H	F	10 h	(96)
H	H	H	H	H	30 min	(97)
NH_2	H	H	H	H	6 h	(85)
NO_2	H	H	H	H	6 h	(65)
H	NO_2	H	H	H	30 min	(97)
H	H	NO_2	H	H	30 min	(98)
H	NO_2	H	NO_2	H	overnight	(94)
H	H	Me	H	H	30 min	(98)
H	H	CF_3	H	H	2 h	(93)
H	H	CN	H	H	30 min	(90)
H	H	CO_2H	H	H	30 min	(95)
H	H	MeO	H	H	30 min	(94)
H	H	Ac	H	H	30 min	(95)

C_7

Substrate	Conditions	Product (Yield)	Refs.
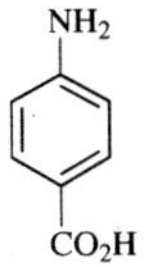 4-$NH_2C_6H_4CO_2H$	Cyclohexanone dioxirane, acetone, cyclohexanone, CH_2Cl_2, −10°, 20 min 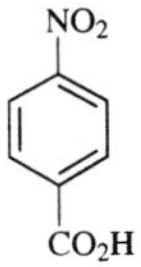	4-$NO_2C_6H_4CO_2H$ (91)	98
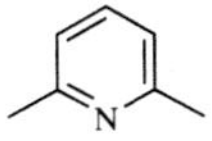 2,6-dimethylpyridine	DMD, acetone, 0°, < 1 h	2,6-dimethylpyridine N-oxide (100)	95
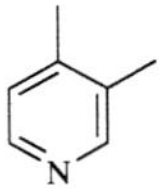 3,4-dimethylpyridine	DMD, acetone, 0°, < 1 h	3,4-dimethylpyridine N-oxide (100)	95

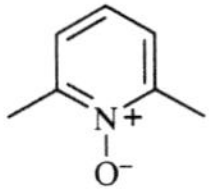

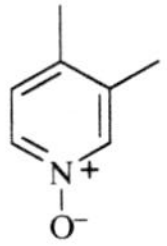

TABLE 3A. NITROGEN OXIDATION BY ISOLATED DIOXIRANES (*Continued*)

Substrate	Conditions	Product(s) and Yield(s) (%)	Refs.
C_7 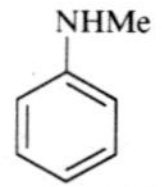 NHMe	DMD, acetone, 0°, 5 min	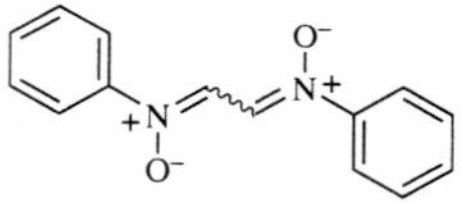 (57)	309
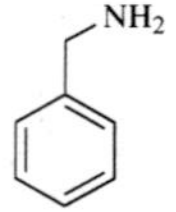 NH_2	DMD (6 eq), acetone, rt	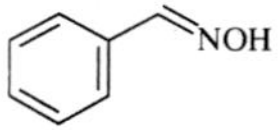 NOH; Additives / E:Z: — (60) —; $NaHCO_3$ (69) 9:1; K_2CO_3 (65) 8:1	114
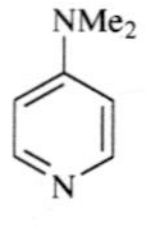 NMe_2 N	DMD, acetone	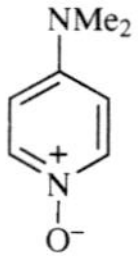 NMe_2 + N O⁻ (—)	21
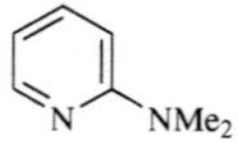 N NMe_2	DMD, acetone	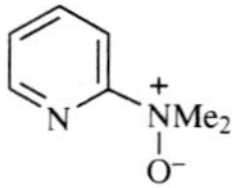 N + NMe_2 O⁻ (—)	21
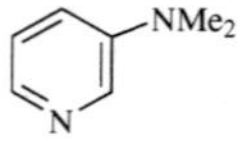 NMe_2 N	DMD, acetone	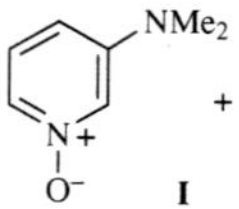 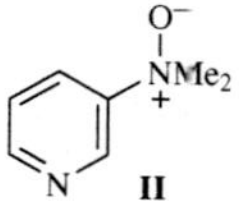 NMe_2 N+ O⁻ **I** + O⁻ NMe_2 + N **II** **I** + **II** (—), **I**:**II** = 17:83	21
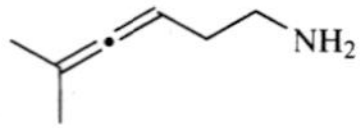 NH_2	DMD (x eq), acetone, rt	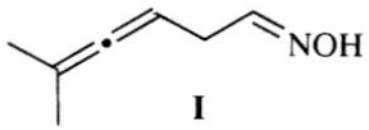 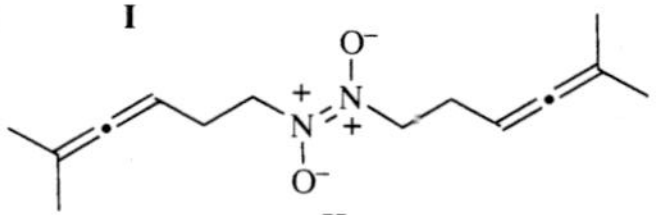 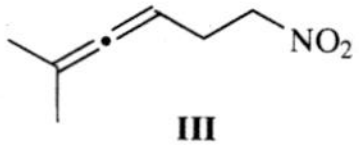 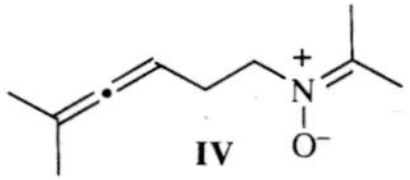 NOH **I** + O⁻ + N = N + O⁻ **II** + NO_2 **III** + + N O⁻ **IV**	114

x	Additives	I	II	III	IV
5	—	(20)	(5)	$< 5^b$	(28)
6	$NaHCO_3$	(36)	(9)	$< 5^b$	(—)
7	K_2CO_3	(40)	(8)	$< 5^b$	(—)

Substrate	Conditions	Product(s) and Yield(s) (%)	Refs.
NH_2 (norbornylamine)	DMD, acetone, rt, 34 min	NO_2 (81)	101
NH_2•HCl (norbornylamine hydrochloride)	1. DMD, acetone, H_2O, 21°, 3 min 2. rt, 7 h	NO_2 (58)	101
n-BuN=	DMD, acetone, rt, 15 min	*n*-PrCH=NOH (**I**) + [*n*-Bu$\overset{+}{N}$(O⁻)]$_2$ (**II**) + *n*-Bu$\overset{+}{N}$(O⁻)=CMe_2 (**III**) **I + II + III** (—), **I:II:III** = 44:30:26	114

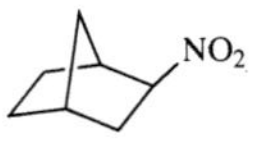

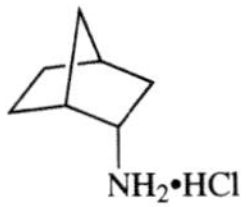

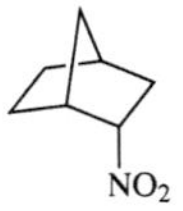

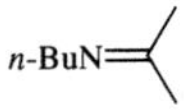

$C_{7\text{-}15}$

Substrate	Conditions	Product	Refs.
N, N, N, R (benzotriazole)	DMD (x eq), acetone, CH_2Cl_2, rt	O⁻, N+, N, N, R	310

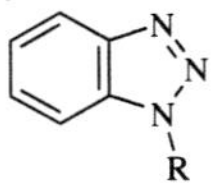

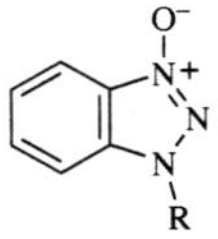

R	x	Time	
Me	2.0	24 h	(90)
CH_2CN	3.0	60 h	(40)
Et	2.0	24 h	(74)
CH_2CCH	3.0	60 h	(35)
n-Pr	2.0	24 h	(76)
CH_2CO_2Et	3.0	36 h	(68)
n-C_6H_{13}	2.0	34 h	(81)
Bz	2.0	48 h	(0)
Bn	3.0	60 h	(92)
CH_2OPh	3.0	36 h	(91)
CH_2CH_2Ph	2.2	48 h	(92)
CH(Me)Ph	2.0	48 h	(73)
n-C_9H_{19}	3.0	48 h	(90)

TABLE 3A. NITROGEN OXIDATION BY ISOLATED DIOXIRANES (*Continued*)

Substrate	Conditions	Product(s) and Yield(s) (%)	Refs.
C_8			
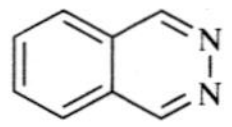	DMD, acetone	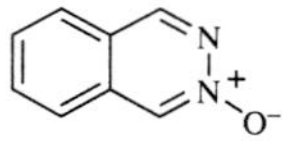 N^+–O^- (—)	21
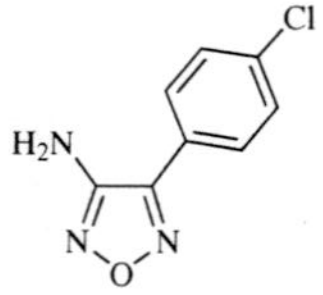 H_2N, Cl	DMD, acetone	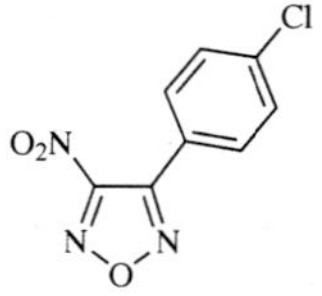 O_2N, Cl (—)	149
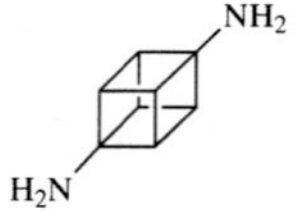 NH_2, H_2N	DMD, acetone, CH_2Cl_2, rt, 48 h	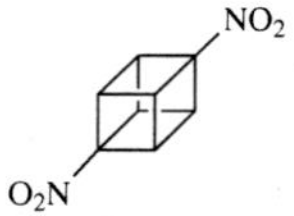 NO_2, O_2N (80)	101
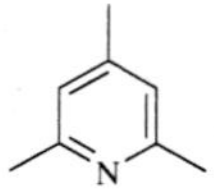 N	DMD, acetone, 0°, < 1 h	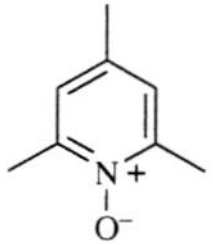 N^+–O^- (100)	95
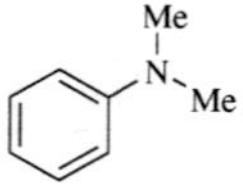 Me–N–Me	DMD, acetone, 0°, < 1 h	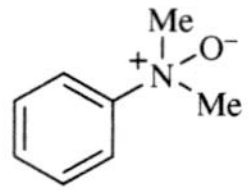 Me, ^+N, O^-, Me (100)	95
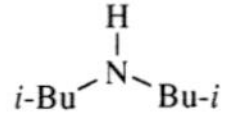 *i*-Bu–N(H)–Bu-*i*	DMD, acetone, 0°, 10 min	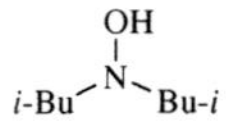 *i*-Bu–N(OH)–Bu-*i* (97)	96
C_{8-9}			
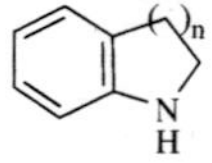 ( )n, NH	DMD, acetone, 0°, 20-30 min	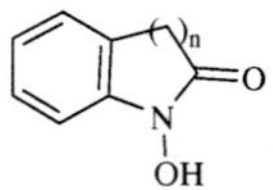 ( )n, =O, N–OH n = 1 (54) n = 2 (83)	99

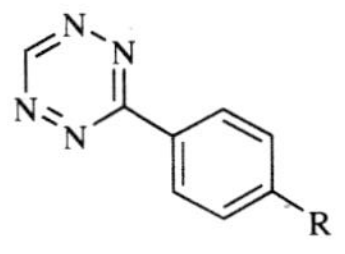

TFD, CH_2Cl_2, TFP, 0°, 20 min

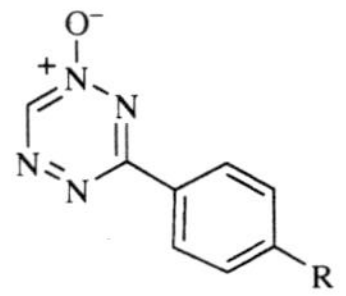

R	
Cl	(74)
H	(68)
Me	(90)
MeO	(73)

311

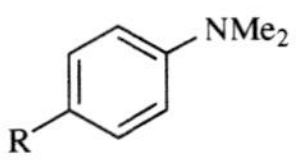

DMD, acetone, 0-5°, dark

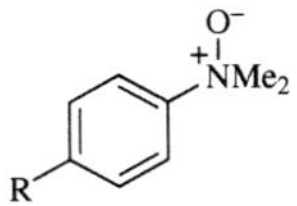

(—) R = Cl, NO_2, H, MeO

25

C_{8-11}

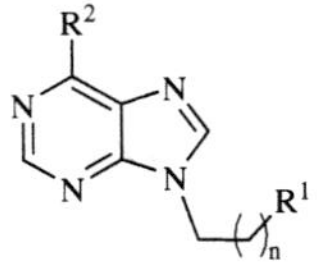

DMD (1.2 eq), acetone, CH_2Cl_2, rt, 8 h

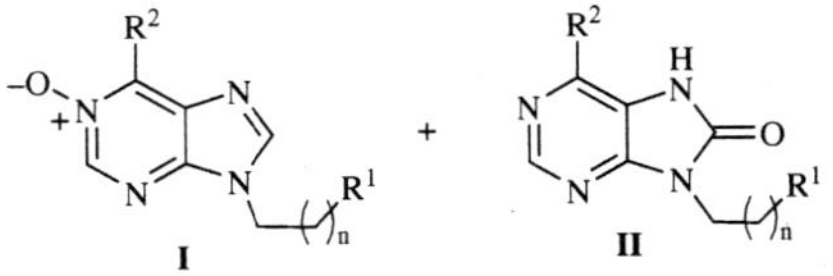

I + II

312

Ar =

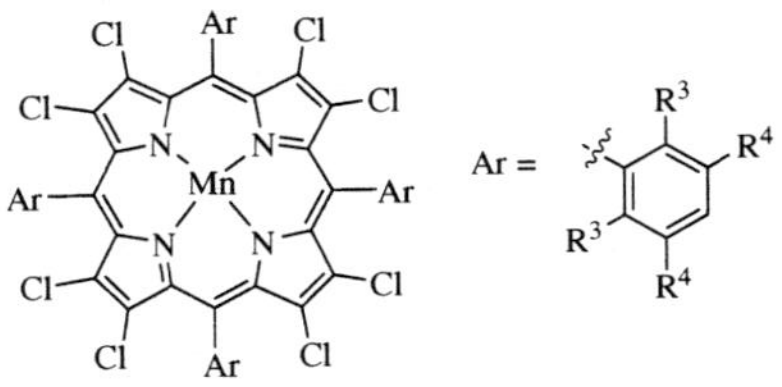

Catalyst **A**: R^3 = OMe, R^4 = Cl

Catalyst **B**: R^3 = Cl, R^4 = H

R^1	R^2	n	Catalyst	I	II
OH	NH_2	3	**A**	(—)	(98)
OH	NH_2	3	**B**	(93)	(—)
OH	H	3	**A**	(98)	(—)
OH	NH_2	2	**A**	(92)	(—)
OH	NH_2	2	**B**	(94)	(—)
OH	NH_2	4	**B**	(95)	(—)
OH	NH_2	4	**A**	(15)	(30)
Me	NH_2	4	**A**	(95)	(—)
Me	NH_2	4	**B**	(97)	(—)
OAc	NH_2	3	**A**	(91)	(—)
OAc	NH_2	3	**B**	(93)	(—)

TABLE 3A. NITROGEN OXIDATION BY ISOLATED DIOXIRANES (*Continued*)

Substrate	Conditions	Product(s) and Yield(s) (%)	Refs.
C_{8-12} Boc-N(H)-CH(R)-$(CH_2)_n$-CO_2Me	TFD (x eq), TFP, CH_2Cl_2, –20 to 0°	Boc-N(OH)-CH(R)-$(CH_2)_n$-CO_2Me R, n, x, Time: H, 0, 2.0, 5 h (82); *i*-Pr, 0, 3.1, 6 h (75); H, 4, 2.0, 8 h (74)	313
C_9			
1-allylbenzotriazole	DMD (1.5 eq), acetone, CH_2Cl_2, 0°, 24 h; rt, 12 h	1-allylbenzotriazole 3-oxide (28) + 1-(oxiranylmethyl)benzotriazole (46)	310
isoquinoline	DMD, acetone, 0°, < 1 h	isoquinoline N-oxide (100)	95
quinoline bis-epoxide	DMD, acetone, 10°, 3 h	quinoline bis-epoxide N-oxide (99)	314
quinoline bis-epoxide (stereoisomer)	DMD, acetone, 10°, 3 h	quinoline bis-epoxide N-oxide (93)	314
1,2,3,4-tetrahydroisoquinoline	DMD, acetone, 0°, 20-30 min	3,4-dihydroisoquinoline N-oxide (48)	99
Ph-CH_2-N(Me)-Me	DMD, acetone, 0°, < 1 h	Ph-CH_2-N^+(Me)(Me)-O^- (100)	95

Reactant	Conditions	Product(s) and Yield(s) (%)	Refs.
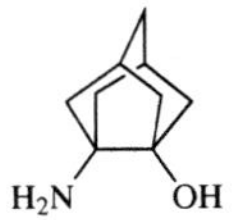 H_2N, OH	1. DMD (3 eq), acetone, CH_2Cl_2, –25° 2. rt, 12-15 h	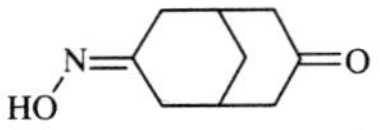 HO, N=, =O (92)	315
	1. DMD (7 eq), acetone, CH_2Cl_2, –25° 2. rt, 12 to 15 h	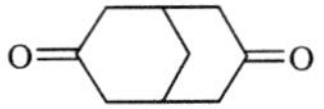 O=, =O (100)	315
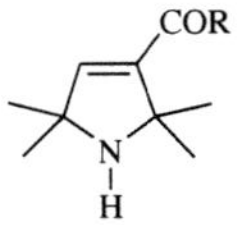 COR, N, H	DMD (2 eq), acetone, 0°, 30 min	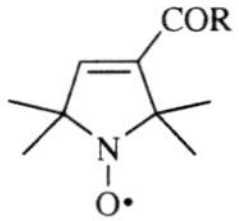 COR, N, O• R: OH (94); NH_2 (98)	97
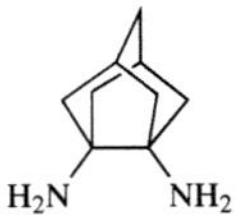 H_2N, NH_2	1. DMD (6 eq), acetone, CH_2Cl_2, –25° 2. rt, 12 to 15 h	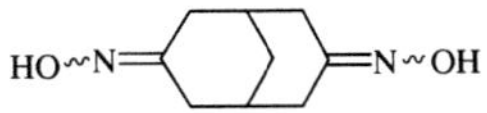 HO~N=, =N~OH (100)	315
	1. DMD (10 eq), acetone, CH_2Cl_2, –25° 2. rt, 12 to 15 h	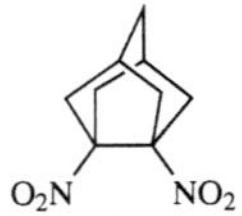 O_2N, NO_2 **I** (96)	315
	1. DMD (20 eq), acetone, CH_2Cl_2, –25° 2. rt, 12 to 15 h	**I** 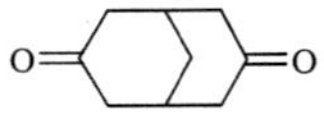 + O=, =O **II** **I** + **II** (100), **I**:**II** = 80:20	315
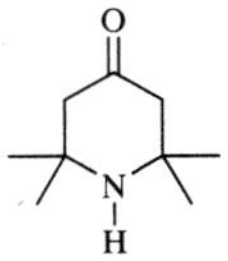 O, N, H	DMD (2 eq), acetone, 0°, 30 min	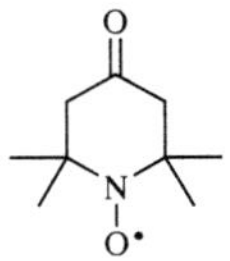 O, N, O• (98)	97

TABLE 3A. NITROGEN OXIDATION BY ISOLATED DIOXIRANES (*Continued*)

Substrate	Conditions	Product(s) and Yield(s) (%)	Refs.
C_9			
CONH$_2$ N H	DMD, acetone, 0°, 2 h	CONH$_2$ N OH (—)	96
	DMD (2 eq), acetone, 0°, 30 min	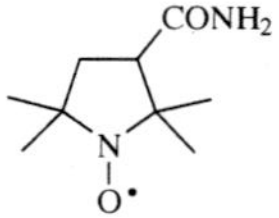 CONH$_2$ N O• (100)	97
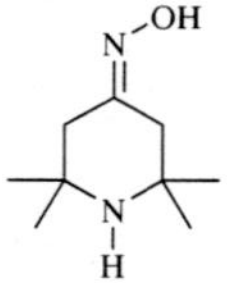 N OH N H	DMD (2 eq), acetone, 0°, 30 min	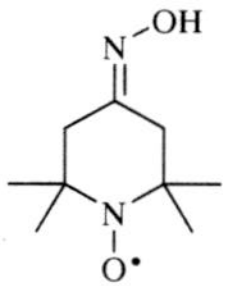 N OH N O• (99)	97
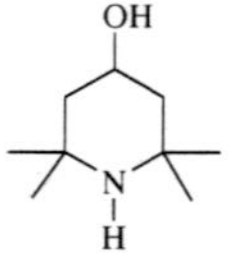 OH N H	DMD, acetone, 0°, 2 h	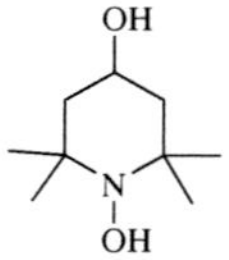 OH N OH (99)	96
	DMD (2 eq), acetone, 0°, 30 min	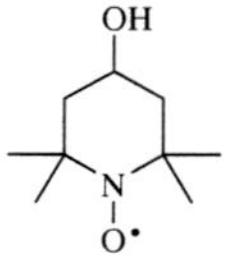 OH N O• (100)	97
C_{10}			
NCO OCN	DMD, acetone, H_2O, dark, N_2, rt, 1.5 h	NO$_2$ O$_2$N (85)	316

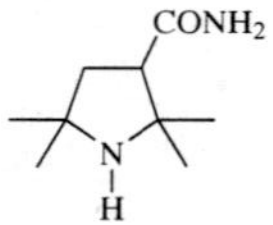

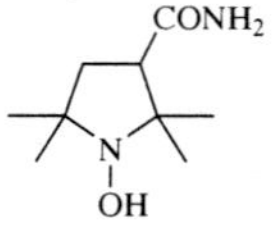

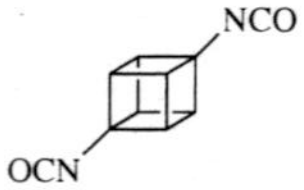

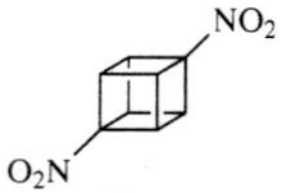

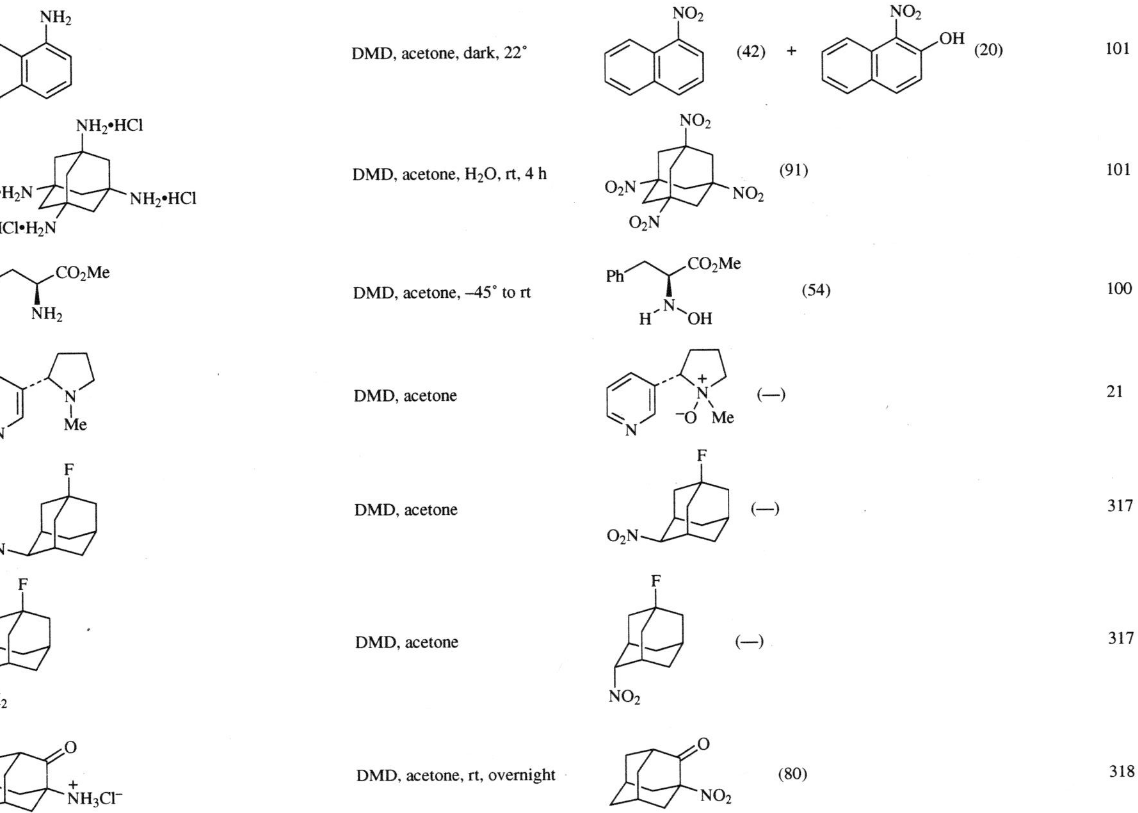

Conditions	Product(s) and Yield(s) (%)	Refs.
DMD, acetone, dark, 22°	(42) + (20)	101
DMD, acetone, H_2O, rt, 4 h	(91)	101
DMD, acetone, –45° to rt	(54)	100
DMD, acetone	(—)	21
DMD, acetone	(—)	317
DMD, acetone	(—)	317
DMD, acetone, rt, overnight	(80)	318

TABLE 3A. NITROGEN OXIDATION BY ISOLATED DIOXIRANES (*Continued*)

Substrate	Conditions	Product(s) and Yield(s) (%)	Refs.
C_{10}  NH_2	DMD, acetone, dark, rt, 30 min	 NO_2 (95)	266
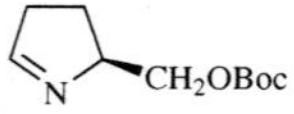 CH_2OBoc, N	TFD (1.2 eq), TFP, CH_2Cl_2, −78°, 1 h	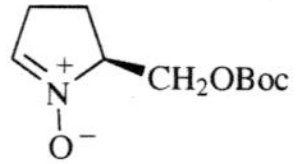 CH_2OBoc, N^+, O^- (100)	319
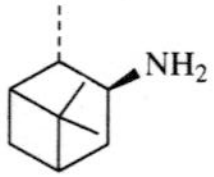 NH_2	DMD, acetone, CH_2Cl_2, rt, 1 h	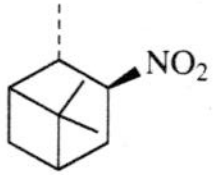 NO_2 (100)	22
	DMD (insufficient amount), acetone, CH_2Cl_2, rt, 35 min	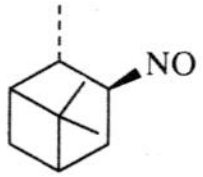 NO (10) 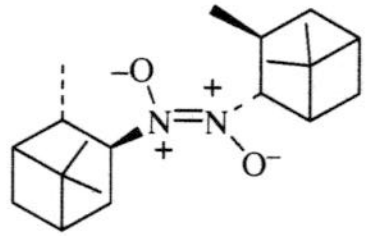 + $-O$, $N^+=N^+$, O^- (90)	22
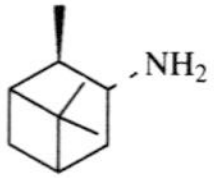 NH_2	DMD, acetone, CH_2Cl_2, rt, 1 h	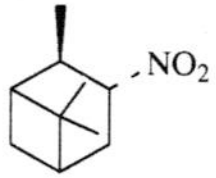 NO_2 (96)	22
	DMD (insufficient amount), acetone, CH_2Cl_2, rt, 35 min	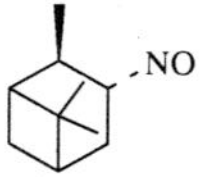 NO (22) 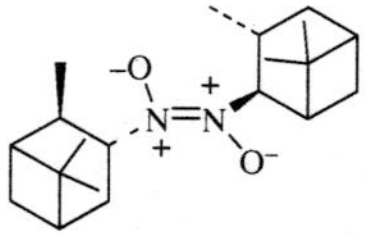 + $-O$, $N^+=N^+$, O^- (78)	22
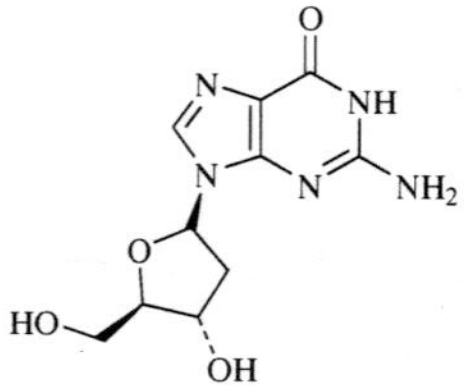 O, N, NH, N, NH_2, O, HO, OH	DMD (2.5 eq), acetone, 90°, 5 h	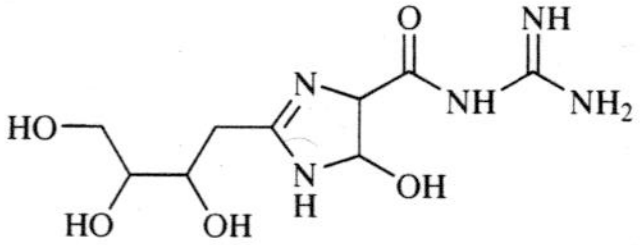 HO, HO, OH, N, N H, OH, O, NH, NH, NH_2 (35)	320

Substrate	Conditions	Product(s) and Yield(s) (%)	Refs.
O, N, NH, N, NH$_2$, N, O, OH, HO, OH	DMD (2.5 eq), acetone, 90°, 5 h	N, O, NH, NH, NH$_2$, N, H, OH (25)	320
H, N, O	DMD, acetone, 0°, 2 h	HO, N, O (96)	96
	DMD (2 eq), acetone, 0°, 30 min	•O, N, O **I** (100)	97
	Cyclohexanone dioxirane, acetone, cyclohexanone, 20°, 10 min	**I** (88)	98
	Acetone, cyclohexanone, –20°, 10 min	**I** (88)	98
n-$C_{10}H_{21}NH_2$	DMD (6 eq), acetone, rt	n-C_9H_{19}CH=NOH **I** + n-$C_{10}H_{21}$N(O$^-$)=N$^+$(O$^-$)$C_{10}H_{21}$-n **II** + n-$C_{10}H_{21}NO_2$ **III** + n-$C_{10}H_{21}$N$^+$(O$^-$)=C(CH$_3$)$_2$ **IV**	114

Additives	I	II	III	IV
—	(51)	(18)	(8)	(3)
$NaHCO_3$	(43)	(13)	(trace)	(17)
K_2CO_3	(73)	(18)	(trace)	(—)

TABLE 3A. NITROGEN OXIDATION BY ISOLATED DIOXIRANES (*Continued*)

Substrate	Conditions	Product(s) and Yield(s) (%)	Refs.
C_{10-23}			
N-H oxazolidine, $(CH_2)_x$, $(CH_2)_yCO_2R$	DMD, acetone, 0°, 30 min	I (N–O•) + II (N–OH)	321
x = 0, y = 2, R = Me		I (—), II (95)	
x = 1, y = 3, R = Me		I (—), II (95)	
x = 6, y = 3, R = Me		I (95), II (—)	
x = 6, y = 3, R = $[(CH_2)_2O]_3Me$		I (95), II (—)	
C_{11}			
Cl, N, N-H, $CONH_2$, NO_2	DMD, acetone, 0°, 10 min	Cl, N, N-OH, $CONH_2$, NO_2 (—)	322
N, Ph	DMD, acetone, 0°, < 1 h	$^-$O, N+, Ph (100)	95
R, N, Bu-*t*	DMD, acetone, CH_2Cl_2, 0°, 2 h	I ($^+$N, Bu-*t*, O$^-$) + II (R, CHO)	102
R = Br		I (17), II (56)	
R = Cl		I (15), II (47)	
R = NO_2		I (23), II (31)	
R = H		I (12), II (36)	

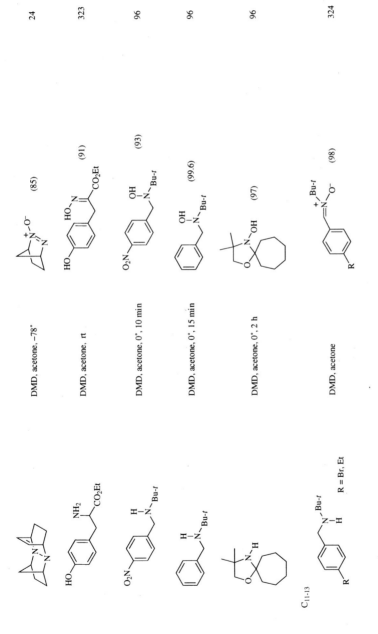

Substrate	Conditions	Product(s) and Yield(s) (%)	Refs.
N, N	DMD, acetone, –78°	N^+–O^-, N (85)	24
HO, NH_2, CO_2Et	DMD, acetone, rt	HO, HO–N, CO_2Et (91)	323
O_2N, H, N, Bu-*t*	DMD, acetone, 0°, 10 min	O_2N, OH, N, Bu-*t* (93)	96
H, N, Bu-*t*	DMD, acetone, 0°, 15 min	OH, N, Bu-*t* (99.6)	96
O, N–H	DMD, acetone, 0°, 2 h	O, N–OH (97)	96
C_{11-13}			
R, N, Bu-*t*, H; R = Br, Et	DMD, acetone	R, N^+, Bu-*t*, O^- (98)	324

TABLE 3A. NITROGEN OXIDATION BY ISOLATED DIOXIRANES (*Continued*)

Substrate	Conditions	Product(s) and Yield(s) (%)	Refs.
C_{11-17}			
R[1]-C6H4-CH2-NH-R[2]	DMD, acetone, 0°, 10 min	R[1]-C6H4-CH=N(+)(O−)-R[2] R[1] R[2] Cl *t*-Bu (99) F *t*-Bu (99) NO_2 *t*-Bu (95) H *t*-Bu (96) CF_3 *t*-Bu (98) Me *t*-Bu (99) MeO *t*-Bu (98) *t*-Bu *t*-Bu (99) Cl Ph (96) H Ph (98) H Bn (96) H 1-Ad (99)	309
C_{12}			
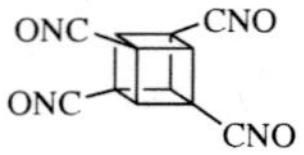 ONC, ONC, CNO, CNO (tetraisocyanatocubane)	DMD, acetone, H_2O, overnight	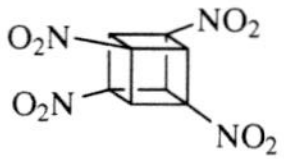 O_2N, O_2N, NO_2, NO_2 (tetranitrocubane) (>30)	325
N, N (phenanthroline)	DMD, acetone	N-oxide, N-oxide (—)	21
N, N (phenanthroline)	DMD, acetone	N-oxide, N (—)	21
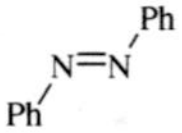 Ph-N=N-Ph	DMD, acetone, dark, rt, 30 min	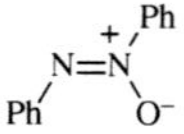 Ph-N=N(+)(O−)-Ph (96)	266

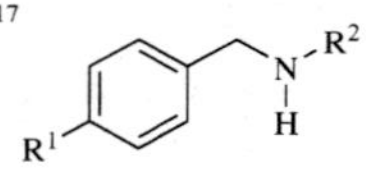

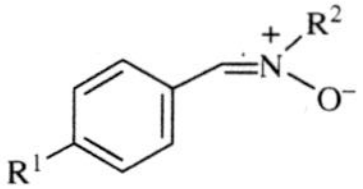

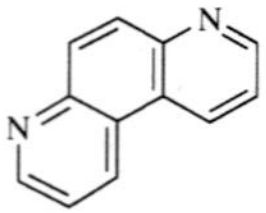

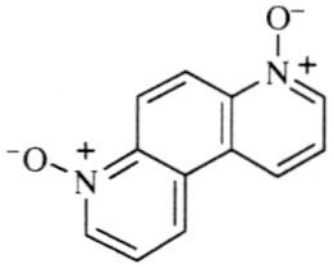

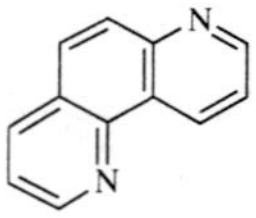

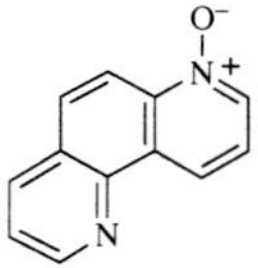

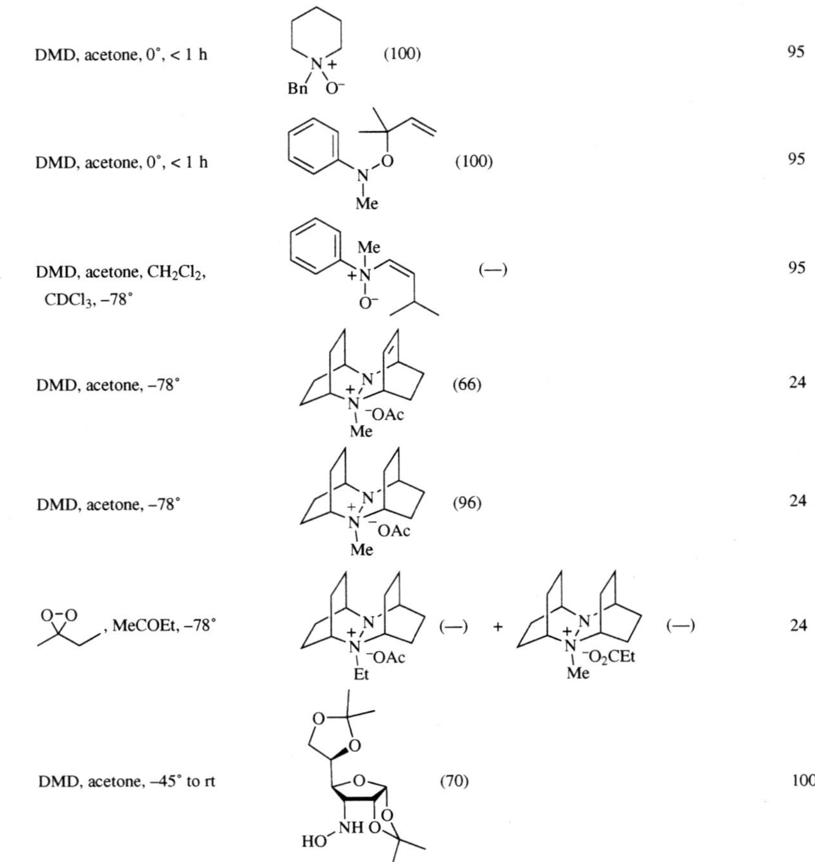

Conditions	Product(s) and Yield(s) (%)	Refs.
DMD, acetone, 0°, < 1 h	(100)	95
DMD, acetone, 0°, < 1 h	(100)	95
DMD, acetone, CH_2Cl_2, $CDCl_3$, −78°	(—)	95
DMD, acetone, −78°	(66)	24
DMD, acetone, −78°	(96)	24
(dioxirane), MeCOEt, −78°	(—) + (—)	24
DMD, acetone, −45° to rt	(70)	100

TABLE 3A. NITROGEN OXIDATION BY ISOLATED DIOXIRANES (*Continued*)

Substrate	Conditions	Product(s) and Yield(s) (%)	Refs.
C_{12} **I**	DMD, acetone, rt, 6 h	**I** (—) + **II** (—) **I**:**II** = 90:10	114
	DMD, acetone, 0°, 10 min	(82)	96
	TFD (5 eq), TFP, CH_2Cl_2, –20 to 0°, 6 h	(57) + (21)	313
n-Bu_3N	DMD, acetone, 0°, < 1 h	*n*-$Bu_3\overset{+}{N}$–O^- (100)	95
C_{13}	DMD, acetone, CH_2Cl_2, 0°, 2 h	(90) + (2)	102
	DMD, acetone, 0°, < 1 h	(100)	95
	DMD, acetone, CH_2Cl_2, 0°, 2 h	(39)	102

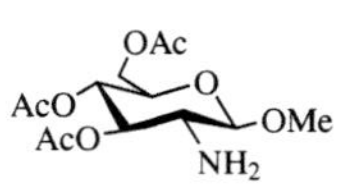

Substrate	Conditions	Product(s) and Yield(s) (%)	Refs.
(tri-O-acetyl methyl glycoside, 2-NH$_2$)	DMD, acetone, –45° to rt	(2-N(H)OH product) (76)	100

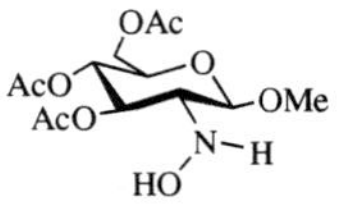

C_{13-16}

Substrate	Conditions	Product(s) and Yield(s) (%)	Refs.
(pentamethylbenzylidene imine, =N–R)	DMD, acetone, CH_2Cl_2, 0°, 2 h	I (nitrone) + II (pentamethylbenzaldehyde, CHO)	102

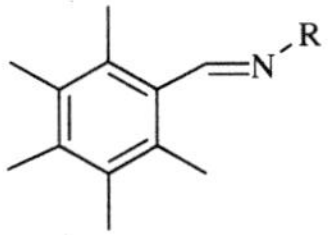

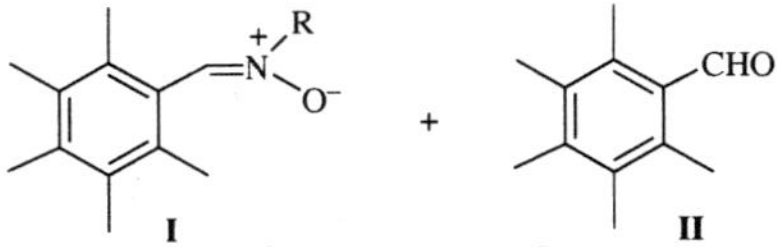

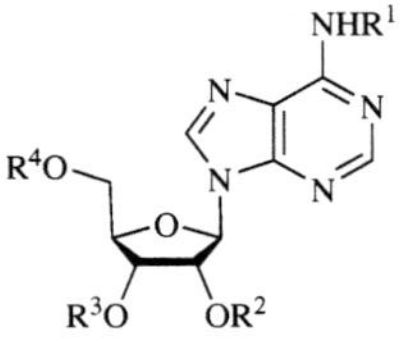

R	I	II
Me	(71)	(28)
t-Bu	(65)	(30)

Substrate	Conditions	Product(s) and Yield(s) (%)	Refs.
(adenosine derivative: NHR[1], R^4O, R^3O, OR^2)	DMD (1.2 eq), acetone, CH_2Cl_2, rt, 8 h	I (N-oxide, N^+–O^-) + II (8-oxo, N–H)	312

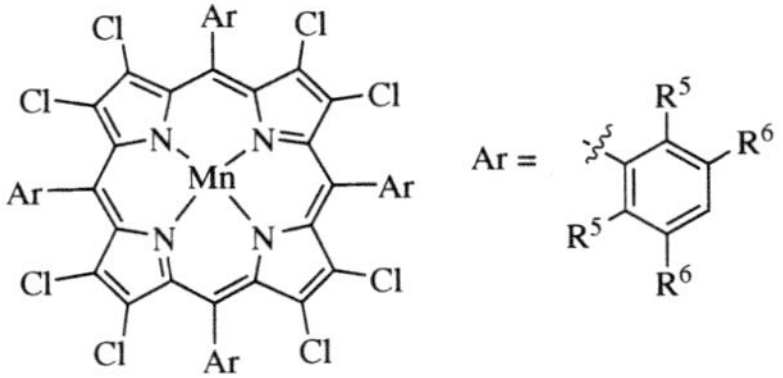

Catalyst: Mn porphyrin with Cl substituents, Ar =

Catalyst **A**: R^5 = OMe, R^6 = Cl

Catalyst **B**: R^5 = Cl, R^6 = H

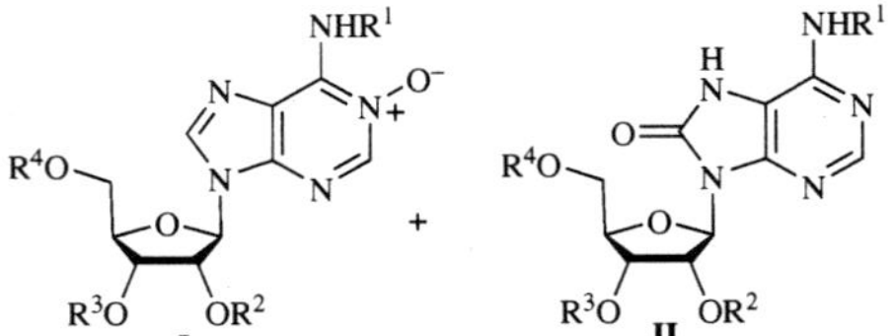

R^1	R^2	R^3	R^3	Catalyst	I	II
Ac	Ac	Ac	Ac	**A**	(93)	(—)
Ac	Ac	Ac	Ac	**B**	(93)	(—)
H	—C(Me)$_2$—		H	**A**	(25)	(70)
H	—C(Me)$_2$—		H	**A**	(—)	(76)
H	—C(Me)$_2$—		H	**B**	(88)	(—)
H	—C(Me)$_2$—		H	**B**	(81)	(—)

TABLE 3A. NITROGEN OXIDATION BY ISOLATED DIOXIRANES (*Continued*)

	Substrate	Conditions	Product(s) and Yield(s) (%)	Refs.
C_{14}	Bn–NH–Bn	DMD, acetone, 0°, 15 min	Bn–N(OH)–Bn **I** (98)	96
		Cyclohexanone dioxirane, cyclohexanone, 20°, 5 min	**I** (68) + Bn–N⁺(O⁻)=CH–Ph **II** (24)	98
		Cyclohexanone dioxirane (2 eq), cyclohexanone, 20°, 5 min	**II** (65)	98
	N-Bn, 4-ethylidenepiperidine	DMD, acetone, 0°, < 1 h	N-oxide (Bn, O^-) (100)	95
	N-Bn, 3-ethyl-1,2,5,6-tetrahydropyridine	DMD, acetone, 0°, < 1 h	N-oxide (Bn, O^-) (100)	95
	N-Bn, 3-ethylpiperidine	DMD, acetone, 0°, < 1 h	N-oxide (Bn, O^-) (100)	95
	CbzHN, H; HO_2C ... NH_2	1. $SOCl_2$, MeOH 2. DMD, acetone, –78°	CbzHN, H; HO_2C ... $N^+(O^-)=CMe_2$ (48)	326
		1. $SOCl_2$, MeOH, –5° to rt, overnight 2. $NaHCO_3$, rt 3. DMD, acetone, –78°, 5-10 min	CbzHN, H; MeO_2C ... $N^+(O^-)=CMe_2$ (48)	327

Substrate	Conditions	Product(s) and Yield(s) (%)	Refs.
C_{14-18} RO_2C–CH(NHCbz)–$(CH_2)_n$–CH_2NH_2 n R 2 Me 3 *t*-Bu 3 *t*-Bu	DMD, acetone	RO_2C–CH(NHCbz)–$(CH_2)_n$–$CH_2N^+(O^-)$=CMe_2 (—)	328
C_{14-19} fluorenone imine, N–R	DMD, acetone, CH_2Cl_2, 0°, 2 h	fluorenone nitrone, ^-O–N^+–R **I** + fluorenone **II** R I II Me (91) (5) Et (92) (9) *i*-Pr (31) (12) *t*-Bu (4) (14) Ph (43) (—)	102
C_{15} oxime ketone (HO–N=, =O)	1. DMD (7 eq), acetone, CH_2Cl_2, –25° 2. rt, 12-15 h	O_2N, OH **I** (90)	316
	1. DMD (3 eq), acetone, CH_2Cl_2, –25° 2. rt, 12-15 h	**I** (86)	316
	1. DMD (10 eq), acetone, CH_2Cl_2, –25° 2. rt, 12-15 h	**I** (100)	316
dioxime (HO~N=, =N~OH)	1. DMD (6 eq), acetone, CH_2Cl_2, –25° 2. rt, 12-15 h	**I** (95)	316
	1. DMD (20 eq), acetone, CH_2Cl_2, –25° 2. rt, 12-15 h	**I** (100)	316

TABLE 3A. NITROGEN OXIDATION BY ISOLATED DIOXIRANES (*Continued*)

Substrate	Conditions	Product(s) and Yield(s) (%)	Refs.
C_{15} 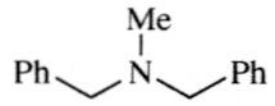 Ph, Me, N, Ph	DMD, acetone, 0°, < 1 h	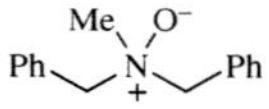 Me, O^-, Ph, N^+, Ph (—)	95
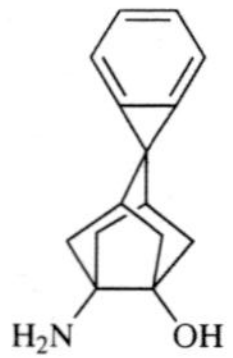 H_2N, OH	1. DMD (3 eq), acetone, CH_2Cl_2, –25° 2. rt, 12-15 h	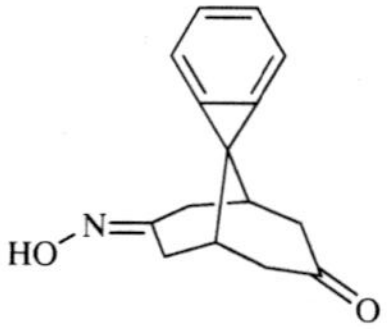 HO, N=, O (100)	316
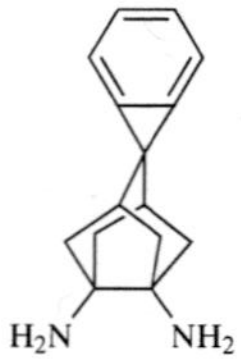 H_2N, NH_2	1. DMD (6 eq), acetone, CH_2Cl_2, –25° 2. rt, 12-15 h	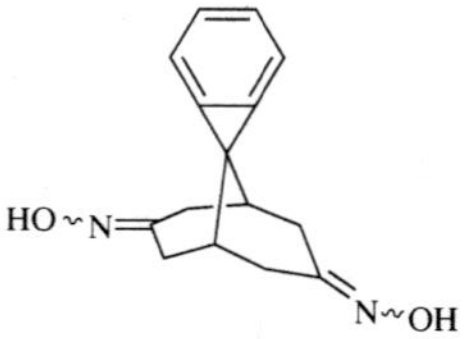 HO~N=, N~OH (100)	316
	1. DMD (12 eq), acetone, CH_2Cl_2, –25° 2. rt, 12-15 h	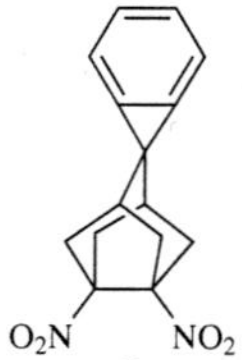 O_2N, NO_2 **I** (96)	316
C_{16} 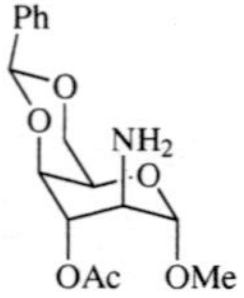 Ph, O, O, NH_2, O, OAc, OMe	DMD, acetone, –45° to rt	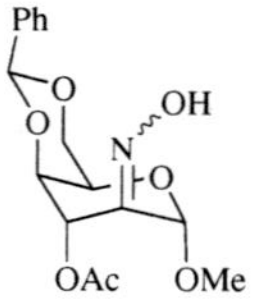 Ph, O, O, N, OH, O, OAc, OMe (—)	100

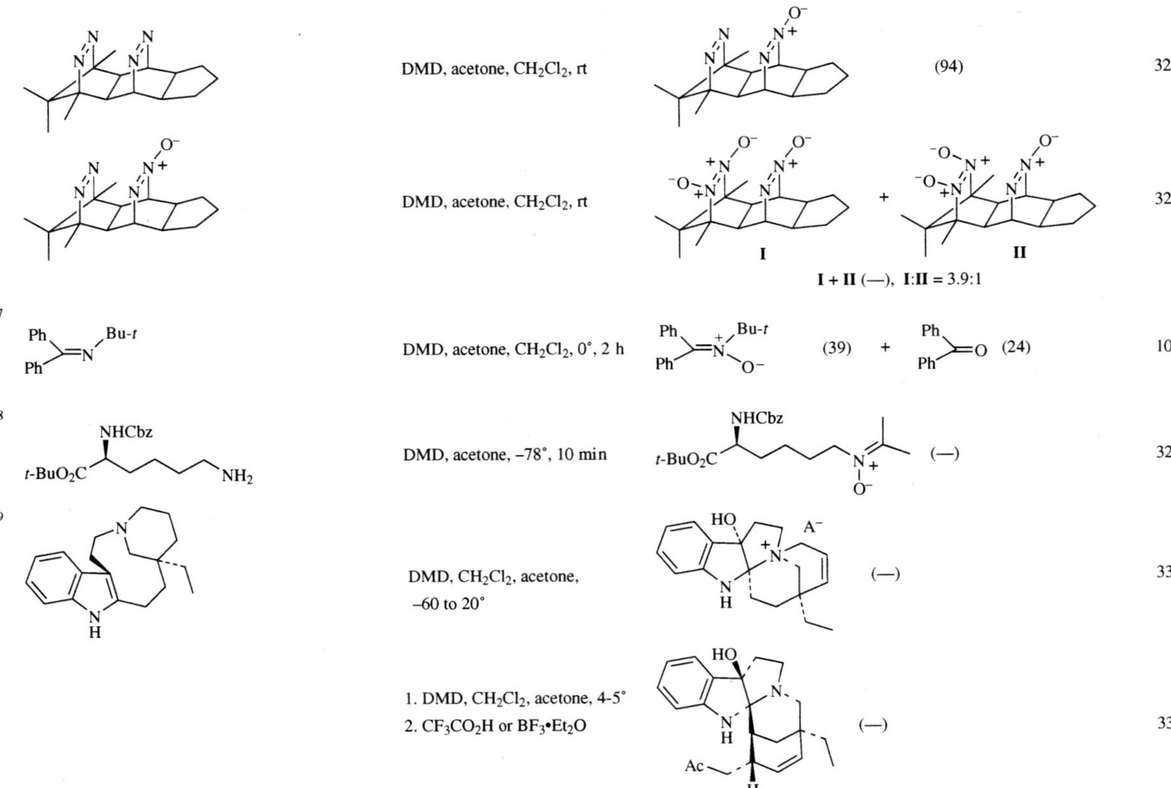

Substrate	Conditions	Product(s) and Yield(s) (%)	Refs.
	DMD, acetone, CH_2Cl_2, rt	(94)	329
	DMD, acetone, CH_2Cl_2, rt	**I** + **II**; **I** + **II** (—), **I**:**II** = 3.9:1	329
C_{17}	DMD, acetone, CH_2Cl_2, 0°, 2 h	(39) + (24)	102
C_{18}	DMD, acetone, –78°, 10 min	(—)	326
C_{19}	DMD, CH_2Cl_2, acetone, –60 to 20°	(—)	330
	1. DMD, CH_2Cl_2, acetone, 4-5° 2. CF_3CO_2H or BF_3•Et_2O	(—)	330

TABLE 3A. NITROGEN OXIDATION BY ISOLATED DIOXIRANES (*Continued*)

Substrate	Conditions	Product(s) and Yield(s) (%)	Refs.
C_{20}			
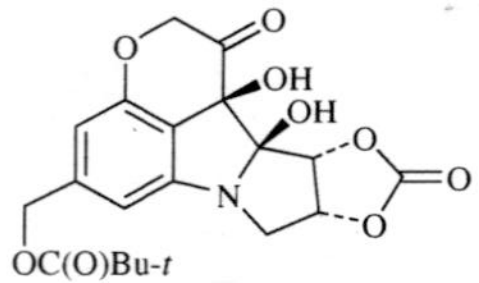 OC(O)Bu-*t* (structure)	DMD, acetone	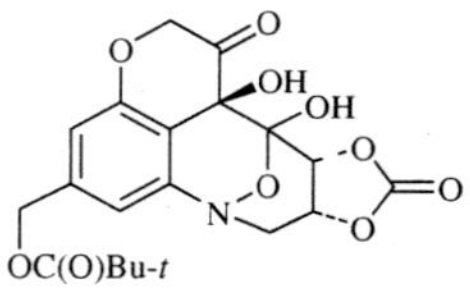 OC(O)Bu-*t* (structure) (—)	331
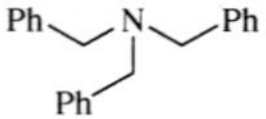 Ph, N, Ph, Ph (structure)	DMD, acetone, –78°	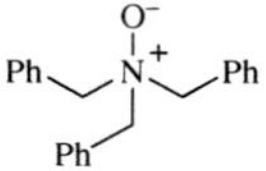 O^-, Ph, N^+, Ph, Ph (structure) (—)	95
	DMD, acetone, 0°, < 1 h	PhCHO (31) 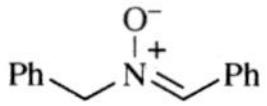 + Ph, N^+, O^-, Ph (structure) (38)	95
 H_2N, BzO, OMe, OBz (structure)	DMD, acetone, –45° to rt	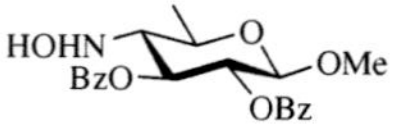 HOHN, BzO, OMe, OBz (structure) (75)	100
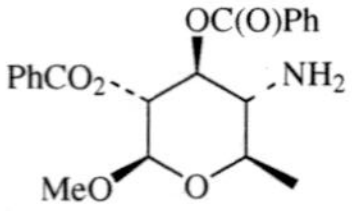 $PhCO_2$, OC(O)Ph, NH_2, MeO (structure)	DMD, acetone, –45° to rt	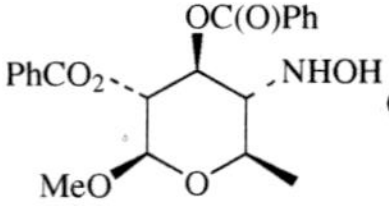 $PhCO_2$, OC(O)Ph, NHOH, MeO (structure) (14) 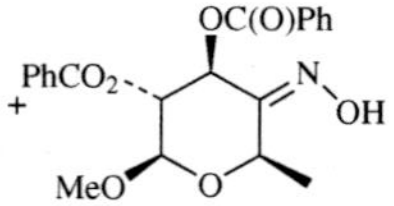 + $PhCO_2$, OC(O)Ph, N, OH, MeO (structure) (—)	332

Substrate	Conditions	Product(s) and Yield(s) (%)	Refs.
NH_2•HCl	1. DMD, acetone, MeOH, rt, overnight 2. Ac_2O, pyridine, rt, overnight	NO_2, OH, H (3) + AcO, OH, H, NO_2 (9) + AcO, OH, H, NO_2 (18)	333
CO_2Me	DMD, CH_2Cl_2, acetone, 4-5°	^-O, N^+, HO, CO_2Me (—)	330
	1. DMD, CH_2Cl_2, acetone, 4-5° 2. CF_3CO_2H or BF_3•Et_2O	HO, MeO_2C (—) + MeO_2C, HO (—)	330
C_{23} CO_2Me	DMD, acetone, 0°, 20 min	CO_2Me, N^+, O^- (64)	334

TABLE 3A. NITROGEN OXIDATION BY ISOLATED DIOXIRANES (*Continued*)

	Substrate	Conditions	Product(s) and Yield(s) (%)	Refs.
C_{23} 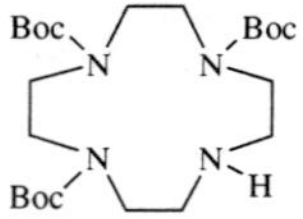	Boc, Boc, Boc, N, N, N, N–H	DMD, acetone, 0° to rt, 4 h	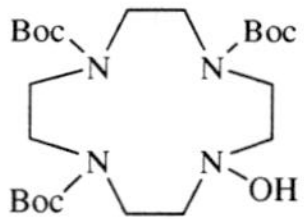 Boc, Boc, Boc, N, N, N, N–OH (81)	335
C_{24} 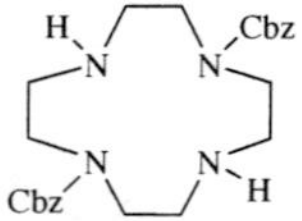	H, Cbz, Cbz, N, N, N, N–H	DMD, acetone, 0° to rt, 70 min	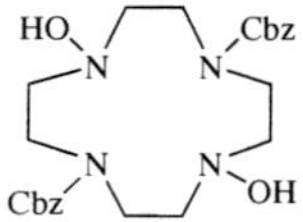 HO, Cbz, Cbz, N, N, N, N–OH (46)	335
C_{26} 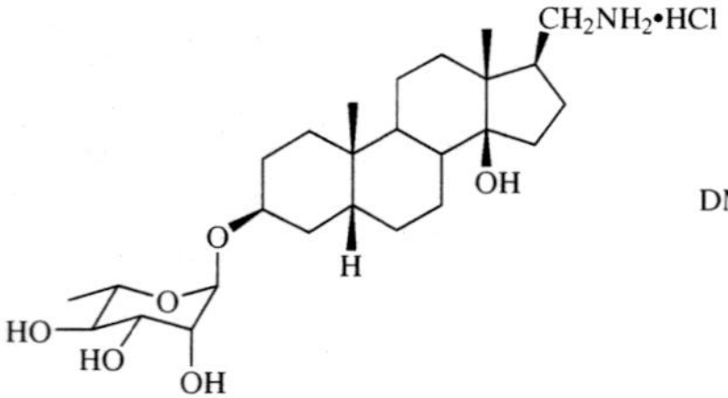	CH_2NH_2•HCl, OH, H, O, HO, HO, OH	DMD, acetone, MeOH, 30 min	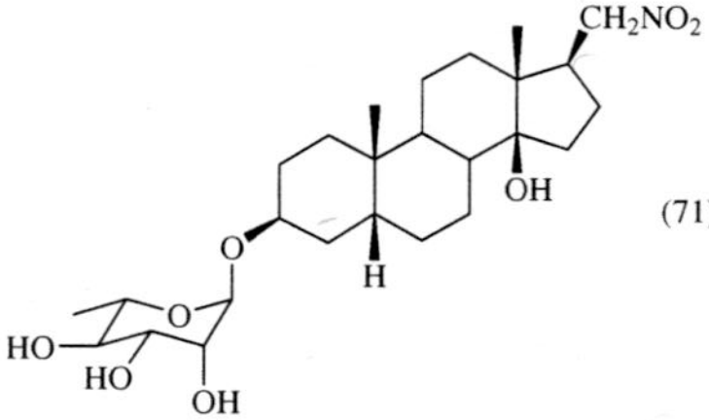 CH_2NO_2, OH, H, O, HO, HO, OH (71)	336
	Ph_3P=N, Cl, N, O, N	DMD, acetone	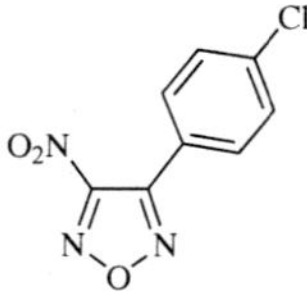 O_2N, Cl, N, O, N (—) + Ph_3PO (—)	149
	OH, MOMO, O, NCO_2Me, MeO_2C, N, PMB	TFD, CH_2Cl_2, –22°	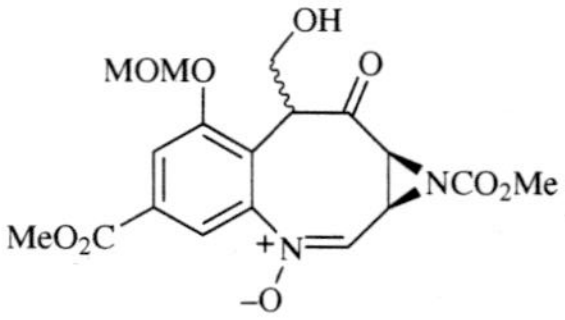 OH, MOMO, O, NCO_2Me, MeO_2C, +N=, –O (—)	337

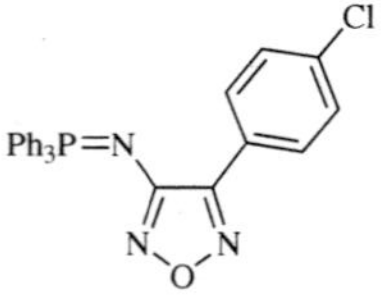

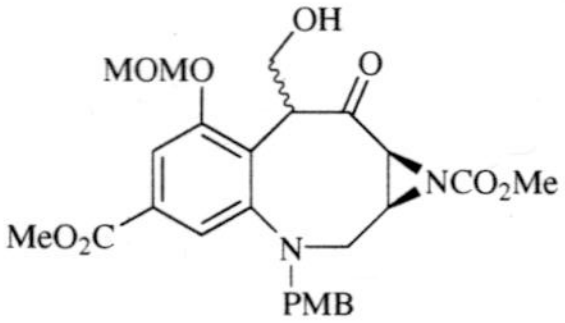

Formula	Substrate	Conditions	Product(s) and Yield(s) (%)	Refs.
C_{27}	OMe, O, OH, HO, Ac, OH, O, NH$_2$, HO	DMD, acetone, –45° to rt	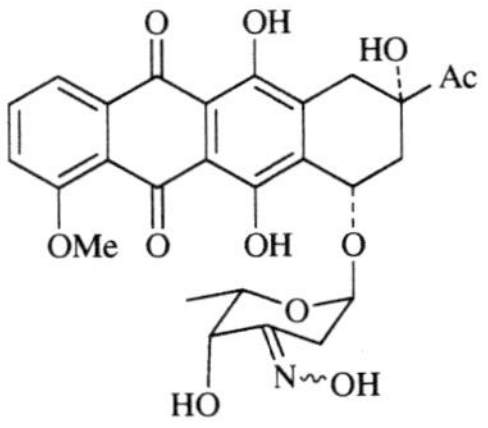 OMe, O, OH, HO, Ac, OH, O, N〰OH, HO (—)	100
	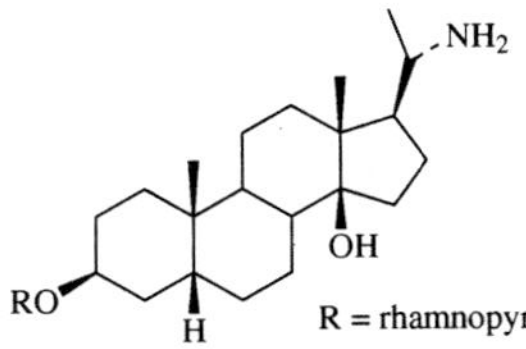 NH_2, OH, RO, H; R = rhamnopyranosyl (2-L)	DMD, acetone, MeOH, rt, overnight	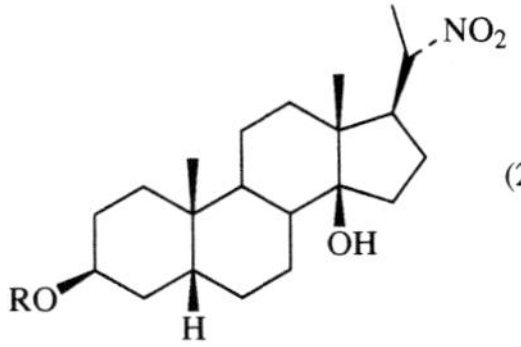 NO_2, OH, RO, H (21)	333
	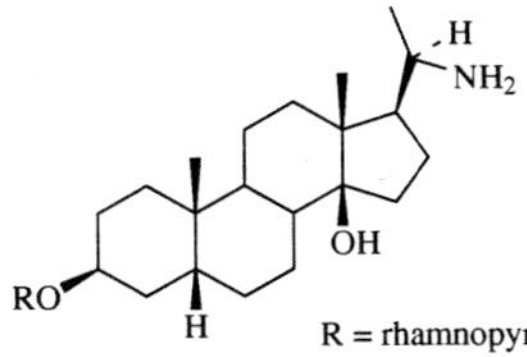 H, NH_2, OH, RO, H; R = rhamnopyranosyl (2-L)	DMD, acetone/MeOH, rt, overnight	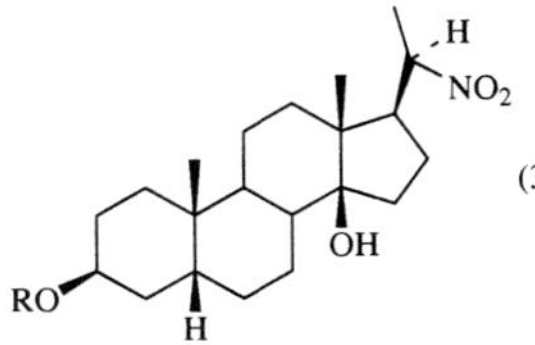 H, NO_2, OH, RO, H (36)	333
	CH_2OAc, $AcOCH_2$, $CH_2N(Pr\text{-}i)_2$, $AcOCH_2$, CH_2OAc	DMD, acetone, overnight	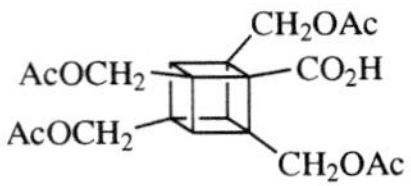 CH_2OAc, $AcOCH_2$, CO_2H, $AcOCH_2$, CH_2OAc (—)	325
C_{29} 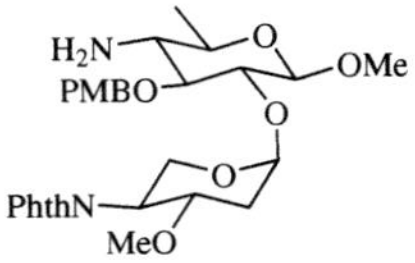	H_2N, PMBO, O, OMe, O, PhthN, MeO	DMD, acetone, –45° to rt	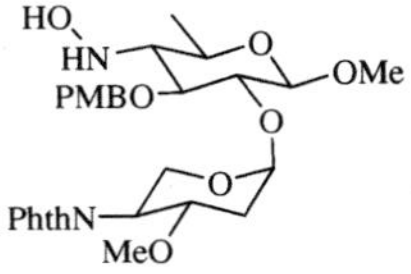 HO, HN, PMBO, O, OMe, O, PhthN, MeO (80)	100

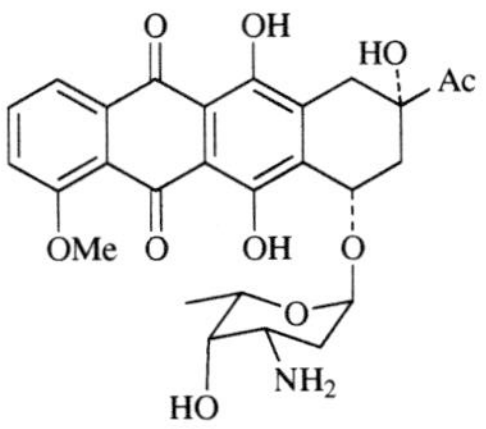

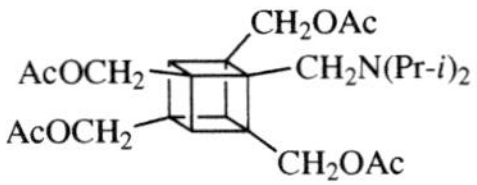

TABLE 3A. NITROGEN OXIDATION BY ISOLATED DIOXIRANES (*Continued*)

Substrate	Conditions	Product(s) and Yield(s) (%)	Refs.
C_{31} 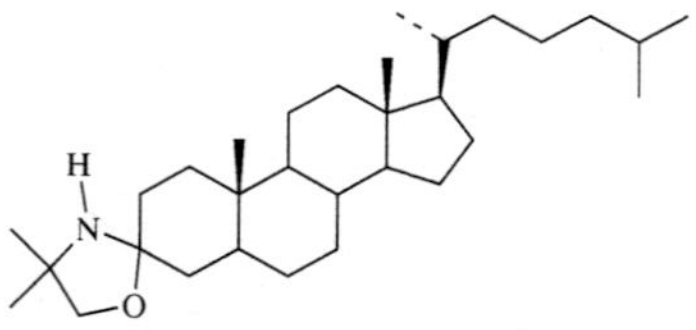	DMD, acetone, 0°, 2 h	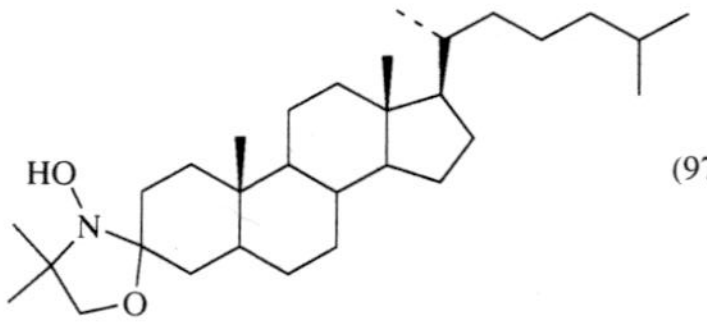 (97)	96
C_{32} 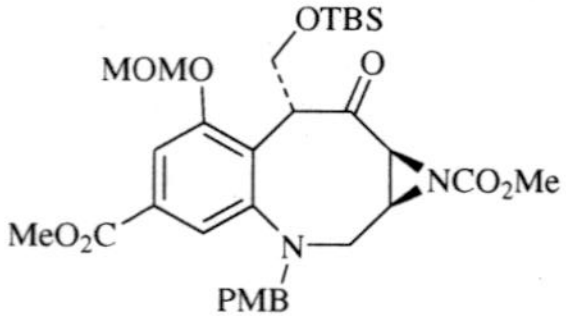	DMD, CH_2Cl_2, sat K_2CO_3, 0° to rt, 7 h	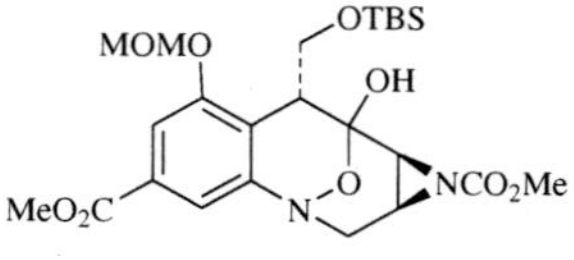 I (30-50)	337
	DMD, CH_2Cl_2, 0°-rt	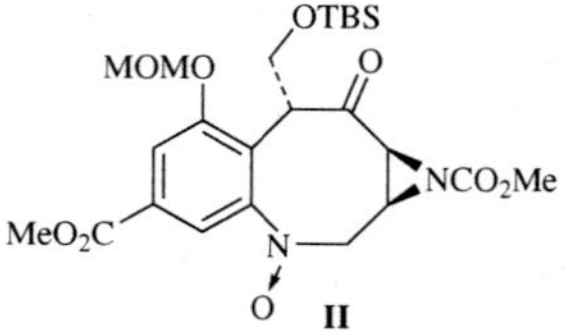 II (90)	337
	DMD, CH_2Cl_2, sat $NaHCO_3$, 0° to rt	**I**:**II** = 1:1	337
C_{34} 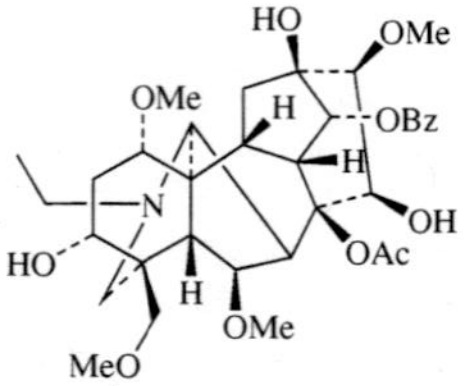	DMD, acetone, –20°	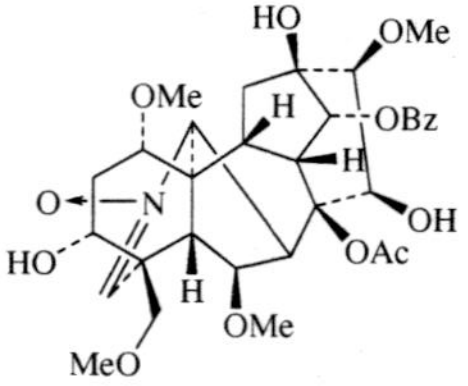 (65)	338

Substrate	Conditions	Product(s) and Yield(s) (%)	Refs.
C_{38} NHOBn, OMe, OBn, OAc, OAc	DMD, acetone, –78° to rt, 24 h	NO_2, OMe, OH, OAc, OAc (83)	339
C_{39} NH_2, OH, RO, H; R = tris-β-D-digitoxosyl	DMD, acetone, CH_2Cl_2, rt, 15 min	NO_2, OH, RO, H (42)	333
C_{45} OMOM, Boc, Boc, Boc, BocHN	DMD, acetone, HCl, Ar, 1 h	OH, OH, NH, H_2N (73)	340
C_{47} OMOM, MOMO, Boc, Boc, Boc, NHBoc	1. DMD, acetone, N_2, 0°, 15 min 2. $NaCNBH_3$, AcOH, rt, 2 h	OMOM, MOMO, OH, Boc, Boc, Boc, NHBoc (33)	340

TABLE 3A. NITROGEN OXIDATION BY ISOLATED DIOXIRANES (*Continued*)

Substrate	Conditions	Product(s) and Yield(s) (%)	Refs.
C_{50} 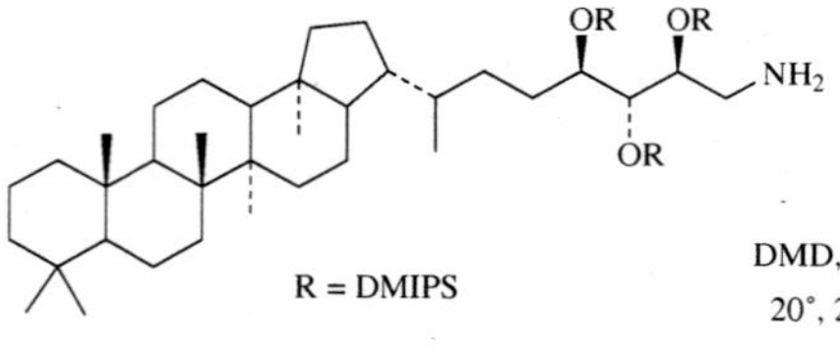 R = DMIPS	DMD, acetone, CH_2Cl_2, 20°, 2.5 h	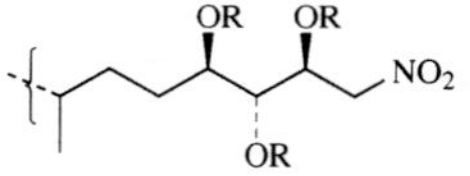 (—) + 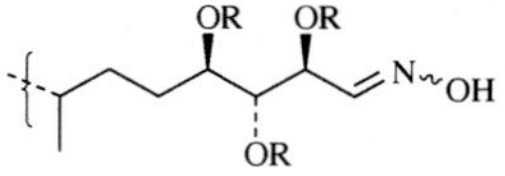 (—) + 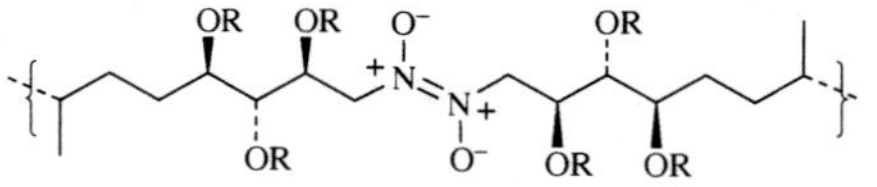 (—)	341
C_{53} 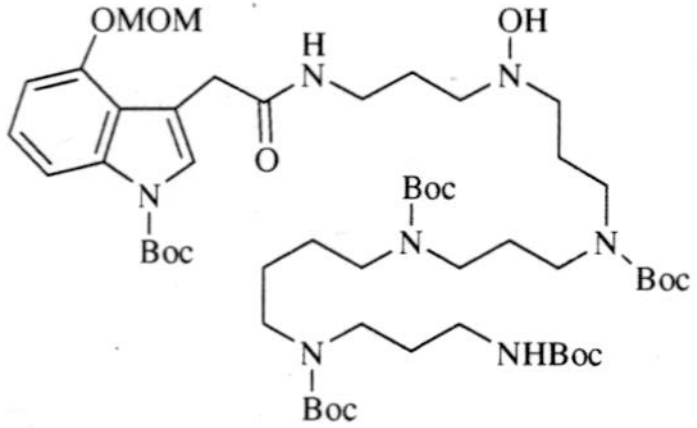	DMD, acetone, HCl, Ar, 1 h	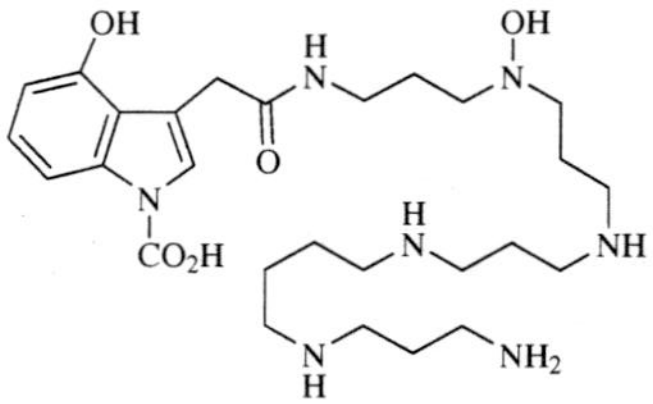 (—)	340

[a] The two products were obtained as a mixture.

[b] This value is the ratio of this compound in the crude products.

TABLE 3B. SULFUR AND SELENIUM OXIDATION BY ISOLATED DIOXIRANES

	Substrate	Conditions	Product(s) and Yield(s) (%)	Refs.
C_0	$^{34}S_8$	DMD, acetone, CH_2Cl_2, 0°, 1 h	$^{34}S_2O$ (—)	342
	S_8	1. DMD, CH_2Cl_2, benzene, acetone, 0°, 1 h 2. 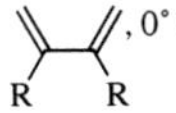 2,3-R-1,3-butadiene, 0° to rt	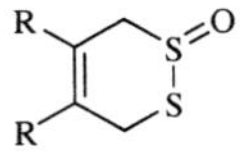 4,5-R-3,6-dihydro-1,2-dithiin 1-oxide; R = Me (33), Ph (27)	343
		DMD, CH_2Cl_2, benzene, acetone, 0°, 1 h	S_2O (—)	343
C_{2-5}	RSH	1. DMD, acetone, CH_2Cl_2, N_2, –40°, 1 h 2. Air	RSO_2H; R = Et (73), *n*-Pr (74), *i*-Pr (90), *n*-Bu (84), *t*-Bu (95), *n*-C_5H_{11} (96)	115
C_3	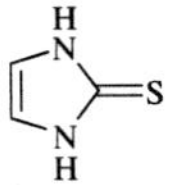 imidazole-2-thione	DMD, acetone, MeOH, rt	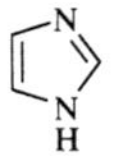 imidazole (81)	142
	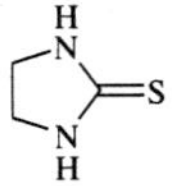 imidazolidine-2-thione	DMD, acetone, MeOH, rt	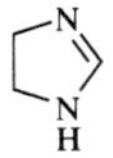 2-imidazoline (53) 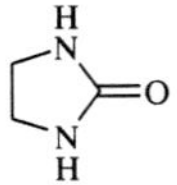 + imidazolidin-2-one (15)	142
C_4	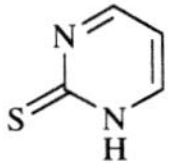 pyrimidine-2(1H)-thione	DMD, acetone, CH_2Cl_2, EtOH, rt	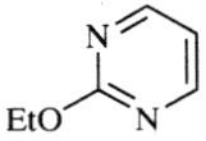 2-EtO-pyrimidine (53)	143
		DMD, acetone, CH_2Cl_2, rt	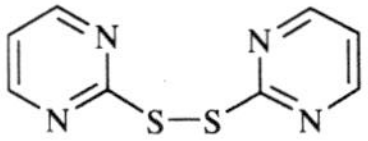 di(pyrimidin-2-yl) disulfide (77)	143

TABLE 3B. SULFUR AND SELENIUM OXIDATION BY ISOLATED DIOXIRANES (*Continued*)

Substrate	Conditions	Product(s) and Yield(s) (%)	Refs.
C_4 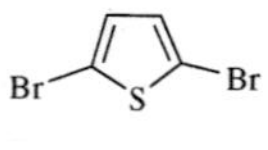 2,5-dibromothiophene	DMD, acetone, several days	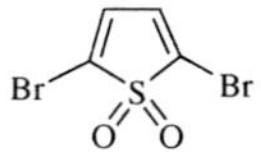 2,5-dibromothiophene 1,1-dioxide (27)	344
 thiophene	DMD, acetone, temp 1; solvent removal, temp 2	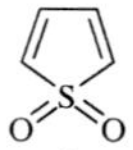 I 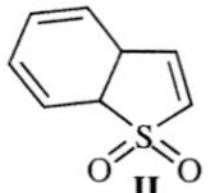 (—) + II (—)	141, 345, 346,
	Temp 1 / Time / Temp 2	I:II	
	−18° / 2 h / < −15°	88:12	
	−25° / 36 h / < −25°	91:9	
	−25° / 6 h / < −40°	100:0	
$C_{4\text{-}13}$ 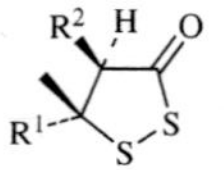 1,2-dithiolan-3-one (R^1, R^2)	DMD, acetone, CH_2Cl_2, −78°	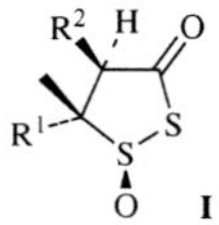 I 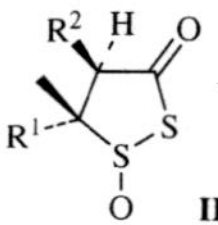 (—) + II (—)	137

R^1	R^2	I:II
H	H	2:1
Me	NHAc	8:1
Me	$NHSO_2C_6H_4Br$-4	18:1
Me	$NHCOC_6H_3(NO_2)_2$-3,5	5:1
Me	NHBz	2:1
Me	NHTs	15:1
Me	NHCOBn	8:1

Substrate	Conditions	Product(s) and Yield(s) (%)	Refs.
C_{4-30}	DMD, acetone, CH_2Cl_2, rt		138
C_{5-9}	DMD, acetone, CH_2Cl_2, rt	I + II	143
C_{5-19}	DMD, acetone, rt		148
C_7	DMD (1-2 eq), acetone, rt	(—)	347
	DMD (4 eq), acetone, rt, 2-4 h	(100)	347

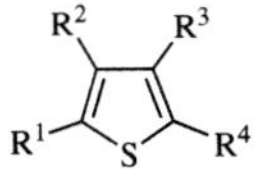

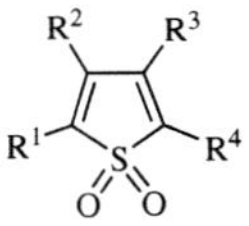

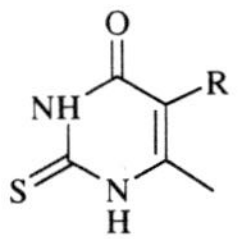

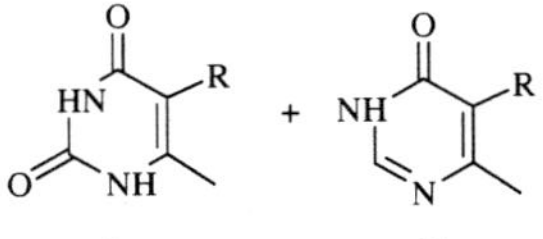

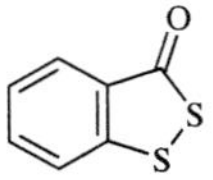

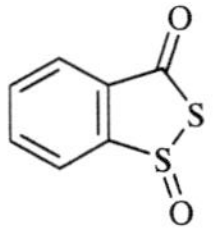

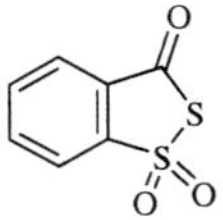

R^1	R^2	R^3	R^4	
Br	H	H	Br	(27)
Et	H	H	H	(—)
Me	H	H	Me	(93)
Et	H	H	Ac	(53)
Bn	H	H	Bn	(93)
Ph	Ph	Ph	Ph	(99)
PhCO	Ph	Ph	Bz	(76)

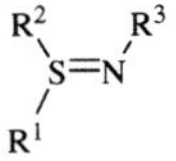

R	I	II
H	(55)	(32)
i-Bu	(40)	(37)

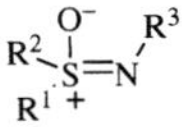

R^1	R^2	R^3	Time	
Me	Et	Ac	2 h	(86)
Me	Et	Ts	4 h	(86)
Me	4-ClC_6H_4	Ts	6 h	(49)
Me	Ph	Ts	6 h	(72)
Me	4-MeC_6H_4	Ts	6 h	(77)
Me	4-$MeOC_6H_4$	Ts	6 h	(86)
Ph	Ph	Ts	6 h	(64)

TABLE 3B. SULFUR AND SELENIUM OXIDATION BY ISOLATED DIOXIRANES (*Continued*)

Substrate	Conditions	Product(s) and Yield(s) (%)	Refs.
C_7			
$PhSCD_3$	TFD, vigorous conditions	$PhS(O)CD_3$ (—) + $PhSO_2CD_3$ (—)	348
$4\text{-}RC_6H_4SOMe$	DMD, acetone	$4\text{-}RC_6H_4SO_2Me$ (—)	349
$PhSMe$ **I**	DMD, acetone, rt	$PhS(O)Me$ (65)	121
	DMD, $CDCl_3$, minutes	$PhS(O)Me$ (—)	156, 157
	TFD, solvent, 0°	$PhS(O)Me$ **II** + $PhSO_2Me$ **III** (—)	348

R	k_R/k_H
Cl	0.7185
NO_2	0.2762
H	1.0000
Me	1.337
MeO	1.857

I:TFD	Solvent	**I**:**II**:**III**	**III**:**II**
1:1	CH_2Cl_2	42:12:46	3.8
10:1	CH_2Cl_2	93:3:4	1.3
2:1	CH_2Cl_2	68:9:23	2.6
2:1	CH_2Cl_2/acetone (1:19)	71:8:21	2.6
2:1	CH_2Cl_2/CH_3CN (1:19)	68:9:23	2.6
2:1	CH_2Cl_2/CF_3CH_2OH (1:7)	66:18:16	0.9

Substrate	Conditions	Product(s) and Yield(s) (%)	Refs.
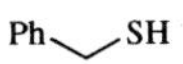 Ph SH	1. DMD, acetone, CH_2Cl_2, N_2, –40°, 1 h 2. Air	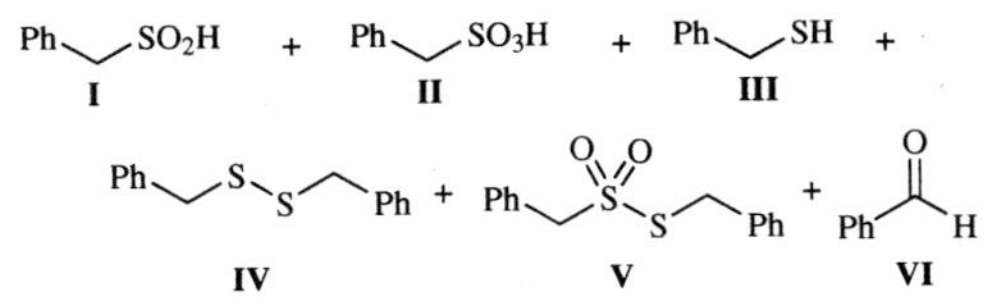 Ph SO_2H (**I**) + Ph SO_3H (**II**) + Ph SH (**III**) + Ph S S Ph (**IV**) + Ph S(O,O) S Ph (**V**) + Ph(C=O)H (**VI**) **I** + **II** + **III** + **IV** + **V** + **VI** (—), **I**:**II**:**III**:**IV**:**V**:**VI** = 71:5:8:7:7:2	115
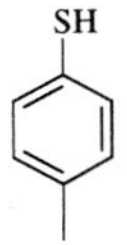 SH	1. DMD, acetone, CH_2Cl_2, N_2, –40°, 1 h 2. Air	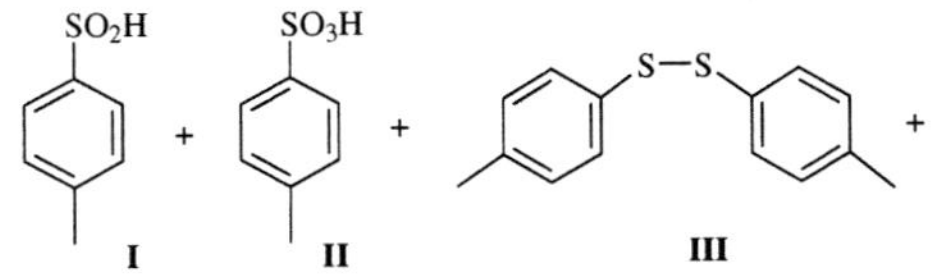 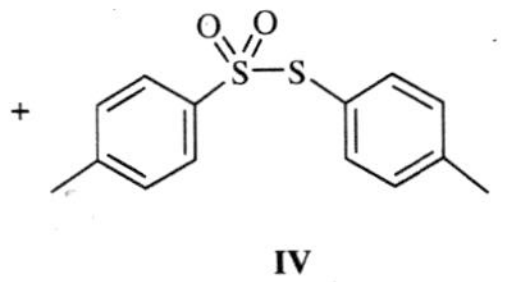 SO_2H (**I**) + SO_3H (**II**) + S—S (**III**) + S(O,O)—S (**IV**) **I** + **II** + **III** + **IV** (—), **I**:**II**:**III**:**IV** = 18:29:33:20	115
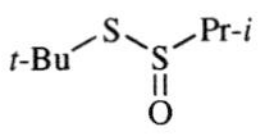 *t*-Bu S S(O) Pr-*i*	DMD, acetone, N_2, –78° to rt	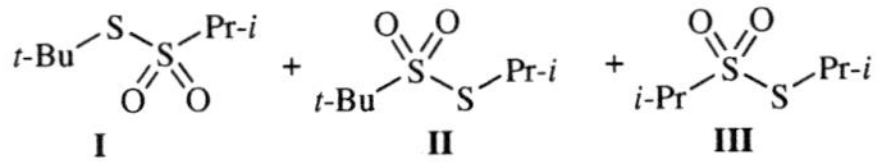 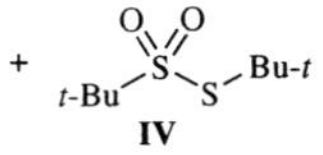 *t*-Bu S S(O,O) Pr-*i* (**I**) + *t*-Bu S(O,O) S Pr-*i* (**II**) + *i*-Pr S(O,O) S Pr-*i* (**III**) + *t*-Bu S(O,O) S Bu-*t* (**IV**) **I** + **II** + **III** + **IV** (—), **I**:**II**:**III**:**IV** = 55:30:10:5	350

TABLE 3B. SULFUR AND SELENIUM OXIDATION BY ISOLATED DIOXIRANES (*Continued*)

Substrate	Conditions	Product(s) and Yield(s) (%)	Refs.
C_7			
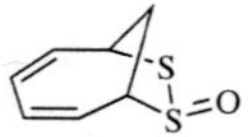 S=O	DMD (1.5 eq), CH_2Cl_2, –50 to 30°, 5 h	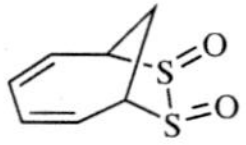 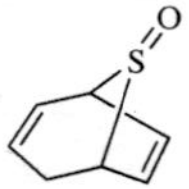 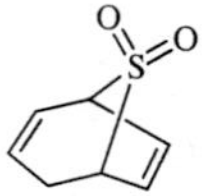 S=O S=O (48) 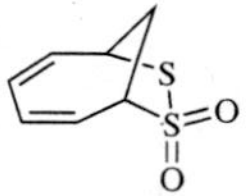 + S S=O O (trace) + S=O (trace) + O S=O I (trace)	351
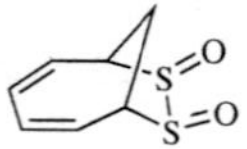 S=O S=O	DMD (2.2 eq), CH_2Cl_2, 0°, 3 h	**I** (88)	351
 S S	DMD, CH_2Cl_2, 0°	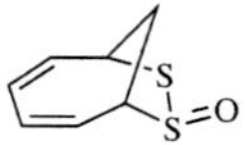 S S=O (98)	351
$C_{7\text{-}8}$			
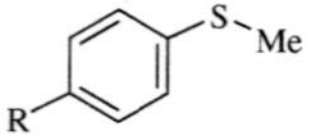 R S Me	DMD, acetone	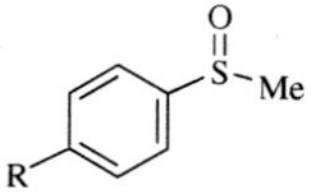 R S(=O) Me (—)	349

R	k_R/k_H
Br	0.6356
Cl	0.6598
NO_2	0.2093
H	1.0000
Me	1.237
MeO	1.548

C_7

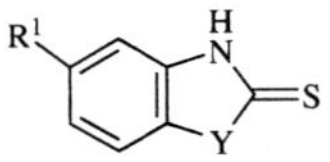

DMD, CH_2Cl_2, acetone, rt

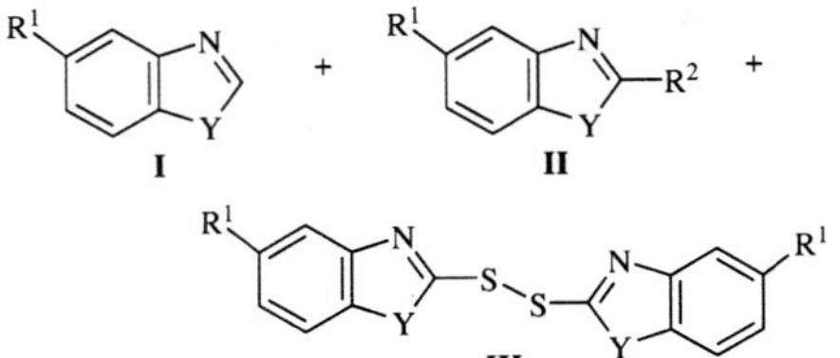

I + II + III

142

R^1	R^2	Y	I	II	III
H	$AcCH_2$	O	(18)	(37)	(42)
H	—	S	(—)	(—)	(95)
Me	OH	NH	(55)	(18)	(—)

DMD, acetone, CH_2Cl_2, MeOH, rt

I + II + III + 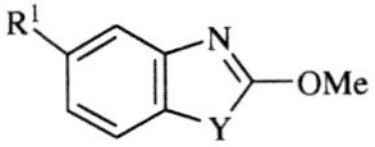 IV

142

R^1	Y	I	II	III	IV
H	O	(11)	(13)	(0)	(72)
H	S	(—)	(—)	(88)	(—)
Me	NH	(20)	(—)	(—)	(74)

DMD, acetone, R^2NH_2, CH_2Cl_2, rt

I + 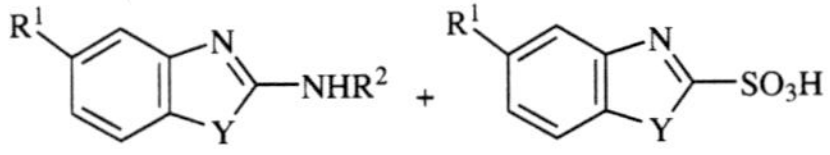 V + VI

142

R^1	R^2	Y	I	V	VI
H	H	O	(20)	(68)	(—)
H	H	S	(—)	(—)	(9)
Me	H	NH	(—)	(75)	(—)
Me	Et	NH	(21)	(62)	(—)
Me	*n*-Pr	NH	(29)	(54)	(—)
Me	*n*-Bu	NH	(35)	(57)	(—)

TABLE 3B. SULFUR AND SELENIUM OXIDATION BY ISOLATED DIOXIRANES (*Continued*)

Substrate	Conditions	Product(s) and Yield(s) (%)	Refs.
C_{7-13} R^1–S–R^2	TFD (x eq), TFP, 0°	R^1–S(=O)–R^2 **I** + R^1–S(=O)₂–R^2 **II**	352

—ArN(Me)Ar— = (N-methyl diarylamine bridge: two benzene rings linked through N–Me)

Solvent	R^1	R^2	x	% Conv.	**I**	**II**	**II:I**
CH_2Cl_2	Ph	Me	0.5	31	(27)	(73)	2.7
CCl_4	Ph	Me	0.5	25	(20)	(80)	4.0
CH_2Cl_2/acetone (1:19)	Ph	Me	0.5	29	(28)	(72)	2.6
CH_2Cl_2/MeCN (1:19)	Ph	Me	0.5	32	(28)	(72)	2.6
CH_2Cl_2/CF_3CH_2OH (1:19)	Ph	Me	0.5	34	(53)	(47)	0.9
CH_2Cl_2/CF_3CO_2H (2 eq)	Ph	Me	0.5	40	(50)	(50)	1.0
CH_2Cl_2/CF_3CO_2H (10 eq)	Ph	Me	0.5	41	(66)	(34)	0.53
CH_2Cl_2/CF_3CO_2H (20 eq)	Ph	Me	0.5	53	(73)	(27)	0.36
CH_2Cl_2/CF_3CO_2H (1:19)	Ph	Me	0.5	46	(100)	(—)	0.0
CH_2Cl_2	Ph	Me	1.0	58	(21)	(79)	3.8
CH_2Cl_2/DMSO (1 eq)	Ph	Me	1.0	42	(28)	(72)	2.6
CH_2Cl_2/DMSO (3 eq)	Ph	Me	1.0	27	(32)	(68)	2.1
CH_2Cl_2/DMSO (5 eq)	Ph	Me	1.0	23	(40)	(60)	1.5
CH_2Cl_2	4-NCC_6H_4	Me	0.5	31	(45)	(54)	1.2
CH_2Cl_2/CF_3CH_2OH (1:19)	4-NCC_6H_4	Me	0.5	44	(77)	(23)	0.3
CH_2Cl_2	Ph	Ph	0.5	45	(36)	(64)	1.8
CCl_4	Ph	Ph	0.5	22	(28)	(72)	2.6
CH_2Cl_2/CF_3CH_2OH (1:19)	Ph	Ph	0.5	30	(61)	(39)	0.6
CH_2Cl_2	—ArN(Me)Ar—		0.5	21	(31)	(69)	2.2
CCl_4	—ArN(Me)Ar—		0.5	18	(13)	(87)	6.7
CH_2Cl_2/CF_3CH_2OH (1:19)	—ArN(Me)Ar—		0.5	25	(87)	(13)	0.15
CH_2Cl_2	—ArN(Me)Ar—		1.0	41	(27)	(73)	2.7
CH_2Cl_2/CH_3CO_2H (15:1)	—ArN(Me)Ar—		1.0	52	(42)	(58)	1.4
CH_2Cl_2/CH_3CO_2H (1.7:1)	—ArN(Me)Ar—		1.0	75	(69)	(31)	0.45
CH_2Cl_2/*t*-BuOH (1.7:1)	—ArN(Me)Ar—		1.0	64	(35)	(65)	1.9

C_{7-14}

Dioxirane (x eq), CH_2Cl_2, TFP, acetone, 0°

Dioxirane	R^1	R^2	x	% Conv.	**I**	**II**	**II:I**
TFD	Ph	CF_3	0.5	48	(> 99)	(—)	(—)
TFD	4-ClC_6H_4	Me	0.5	30	(38)	(62)	1.6
TFD	4-$O_2NC_6H_4$	Me	0.5	35	(47)	(51)	1.1
DMD	Ph	Me	1.0	95	(95)	(5)	< 0.1
DMD	Ph	Me	0.5	55	(100)	(—)	0.0
TFD	Ph	Me	1.0	58	(21)	(79)	3.8
TFD	Ph	Me	0.5	31	(27)	(73)	2.7
TFD	Ph	Me	0.33	20	(35)	(65)	1.9
TFD	Ph	Me	0.33[a]	20	(40)	(60)	1.5
TFD	Ph	Me	0.2	13	(38)	(61)	1.6
TFD	Ph	Me	0.1	7	(43)	(57)	1.3
TFD	4-NCC_6H_4	Me	0.5	34	(46)	(55)	1.2
TFD	4-MeC_6H_4	Me	0.5	27	(24)	(76)	3.2
TFD	4-$MeOC_6H_4$	Me	0.5	30	(29)	(71)	2.4
DMD	*n*-Bu	*n*-Bu	1.0	90	(94)	(6)	< 0.1
TFD	*n*-Bu	*n*-Bu	1.0	63	(29)	(71)	2.5
TFD	*n*-Bu	*n*-Bu	0.33	22	(41)	(59)	1.4
TFD	*n*-Bu	*n*-Bu	0.1	7	(43)	(57)	1.3
TFD	Ph	Ph	0.5	45	(36)	(64)	1.8
DMD	—ArN(Me)Ar—		0.5	49	(100)	(—)	0.0
TFD	—ArN(Me)Ar—		1.0	41	(27)	(73)	2.7
TFD	—ArN(Me)Ar—		0.5	21	(31)	(69)	2.2
TFD	—ArN(Me)Ar—		0.33	19	(34)	(66)	1.9
DMD	Bn	Bn	1.0	93	(96)	(4)	< 0.1
TFD	Bn	Bn	1.0	63	(27)	(73)	2.7
TFD	Bn	Bn	0.5	30	(36)	(64)	1.8
TFD	Bn	Bn	0.33	22	(41)	(59)	1.4
TFD	Bn	Bn	0.1	7	(43)	(57)	1.3

TABLE 3B. SULFUR AND SELENIUM OXIDATION BY ISOLATED DIOXIRANES *(Continued)*

Substrate	Conditions	Product(s) and Yield(s) (%)	Refs.
C_{7-14} R^1SR^2	TFD (x eq), TFP, CH_2Cl_2	$R^1S(O)R^2$ (I) + $R^1SO_2R^2$ (II)	352
C_7 PhSMe	TFD (1.0 eq), TFP, solvent, additive, 0°		352
PhSCD$_3$ (I) + PhS(O)Me (II)	Dioxirane (**III**), solvent, 0°	PhS(O)CD$_3$ (IV) + PhSO$_2$CD$_3$ (V) + PhSO$_2$Me (VI) (—)	352

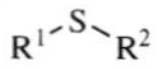

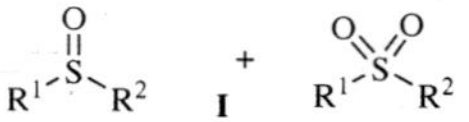

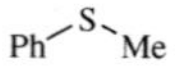

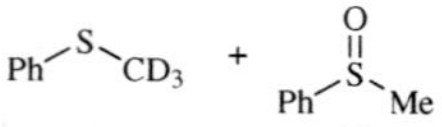

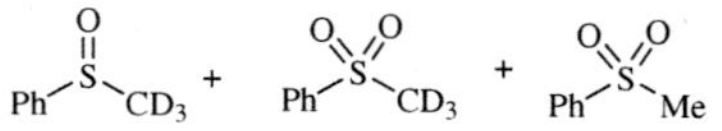

R^1	R^2	x	Temp	% Conv.	I	II	II:I
Ph	Me	0.5	0°	31	(27)	(73)	2.7
Ph	Me	0.5	–40°	31	(32)	(68)	2.1
Ph	Me	0.5	–80°	35	(49)	(51)	1.1
Ph	Me	0.33	0°	20	(35)	(65)	1.9
Ph	Me	0.33	–40°	22	(41)	(59)	1.4
Ph	Me	0.33	–80°	24	(54)	(46)	0.8
Bn	Bn	0.5	0°	30	(36)	(64)	1.8
Bn	Bn	0.5	–40°	34	(41)	(59)	1.4
Bn	Bn	0.5	–80°	38	(58)	(42)	0.7

Additive[b]	Eq	Solvent	% Convn	I	^{18}O	II	^{18}O	II:I
$CF_3C(^{18}OH)_2CH_3$	100	CH_2Cl_2/MeCN (1:4)	56	(34)	(23)	(66)	(12)	1.7
$H_2{}^{18}O$	100	CH_2Cl_2/CF_3CH_2OH (1:4)	41	(33)	(11)	(67)	(6)	2.0
$H_2{}^{18}O$, CF_3CO_2H	100, 2.5	CH_2Cl_2/CF_3CH_2OH (1:4)	67	(47)	(21)	(53)	(5)	1.1
$CF_3C^{18}O_2H$	200	CH_2Cl_2	54	(94)	(1.1)	(6)	(0.5)	< 0.1

Dioxirane	Solvent	I:II:III (molar)	I:IV:V	II:VI	I:II	VI:IV
DMD	CH_2Cl_2/acetone (1:1)	1:1:1	13:80:7	93:7	13.4	0.1
TFD	CH_2Cl_2	1:1:1	58:11:31	66:34	2.1	2.8
TFD	CH_2Cl_2	1:3:1	73:9:18	80:20	2.3	2.0
TFD	CH_2Cl_2/CF_3CH_2OH (1:9)	1:1:1	53:21:26	70:30	2.4	1.2

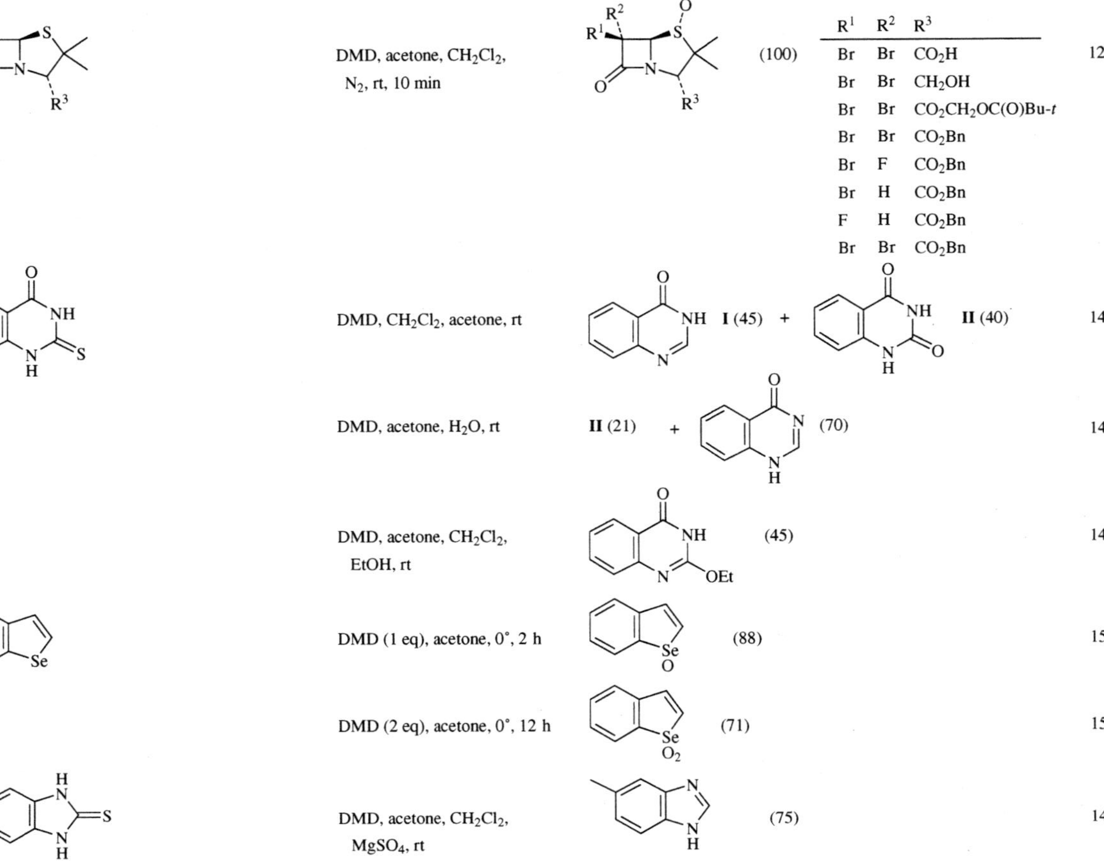

Substrate	Conditions	Product(s) and Yield(s) (%)	Refs.
C_{7-15}	DMD, acetone, CH_2Cl_2, N_2, rt, 10 min	(100)	123
C_8	DMD, CH_2Cl_2, acetone, rt	**I** (45) + **II** (40)	143
	DMD, acetone, H_2O, rt	**II** (21) + (70)	143
	DMD, acetone, CH_2Cl_2, EtOH, rt	(45)	143
	DMD (1 eq), acetone, 0°, 2 h	(88)	153
	DMD (2 eq), acetone, 0°, 12 h	(71)	153
	DMD, acetone, CH_2Cl_2, $MgSO_4$, rt	(75)	142

R^1	R^2	R^3
Br	Br	CO_2H
Br	Br	CH_2OH
Br	Br	$CO_2CH_2OC(O)Bu$-*t*
Br	Br	CO_2Bn
Br	F	CO_2Bn
Br	H	CO_2Bn
F	H	CO_2Bn
Br	Br	CO_2Bn

TABLE 3B. SULFUR AND SELENIUM OXIDATION BY ISOLATED DIOXIRANES (*Continued*)

Substrate	Conditions	Product(s) and Yield(s) (%)	Refs.
C_8			
t-Bu–S(=O)–S–Bu-t	DMD, acetone, N_2, –78°, 1 h	t-Bu–S(O_2)–S–Bu-t (83) + t-BuS_xO_y (—)	350
t-Bu–S(O_2)–S–Bu-t	DMD, acetone, N_2, –78°	t-Bu–S(O_2)–S(O_2)–Bu-t (63)	350
C_{8-10}			
cis-diol cyclohexadiene–CH_2–S–R, R = Me, Et, n-Pr	DMD, acetone, 0°	sulfoxide diastereomer (45) + sulfoxide diastereomer (45)	353
C_{8-18}			
R^2–P(=S)(R^1)–R^3	DMD, acetone, CH_2Cl_2, rt, 5 min	R^2–P(=O)(R^1)–R^3 (100)	150
2,5-R-selenophene	DMD (2 eq), acetone, 1-38 h	2,5-R-selenophene 1,1-dioxide (—)	154
C_9			
cis-diol cyclohexadiene–S-(1,3-dithian-2-yl)...	DMD, acetone, 0°	S-oxide diastereomer (45) + S-oxide diastereomer (45)	353

R^1	R^2	R^3
MeO	MeO	1,3,4-$Cl_3C_6H_2O$
MeO	MeO	3-Me-4-$O_2NC_6H_3O$
EtO	EtO	1,4-Cl_2-4-BrC_6H_2O
EtO	EtO	4-$O_2NC_6H_4O$
n-Bu	n-Bu	n-Bu
Ph	Ph	Ph

Temp	R
rt	CO_2Me
0° to rt	Bz

C_{9-15}

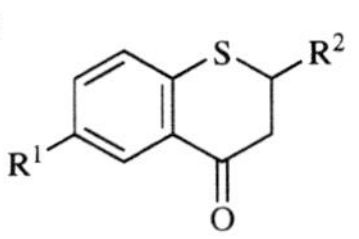

R^1	R^2	x	% Conv.	I	cis:trans I	II
H	H	1.42	86	(93)	—	(4.8)
H	H	3.25	100	(—)	—	(96)
Cl	Me	2.25	98	(86)	—	(1.3)
Cl	Me	3.17	100	(0)	60:40	(86)
H	Me	2.01	91	(81)	—	(14)
H	Me	3.00	100	(—)	59:41	(85)
Me	Me	1.80	100	(88)	—	(8.9)
Me	Me	3.50	100	(—)	61:39	(84)
H	Ph	2.12	100	(77)	—	(16)
H	Ph	4.10	100	(—)	70:30	(92)

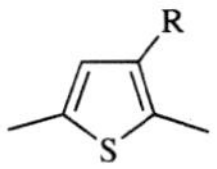

DMD (x eq), acetone, CH_2Cl_2, 0-5° — I + II — 354

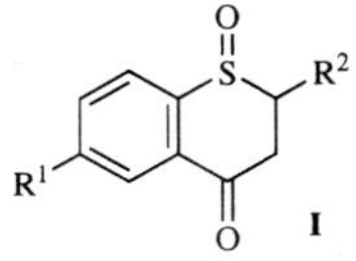

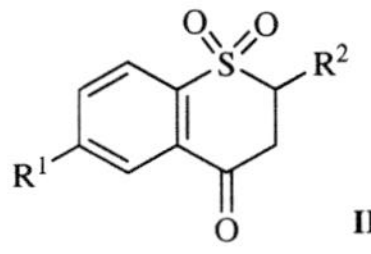

DMD, acetone, CH_2Cl_2, 5°, 3 h — 139

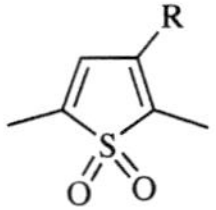

R	
$SnMe_3$	(65)
$OSi(i\text{-}Pr)_3$	(25)

C_{10}

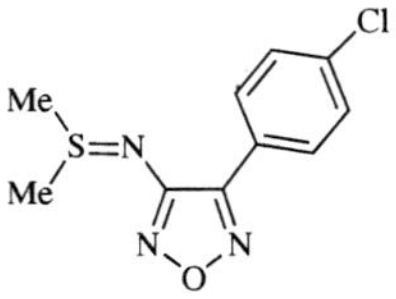

DMD, acetone, rt, 16 h 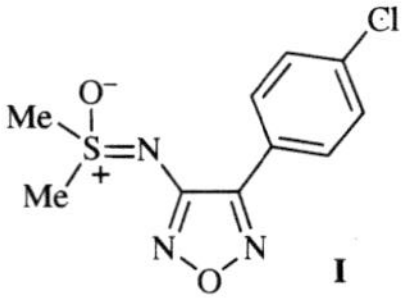 — I (63) + O_2N II (37) — 149

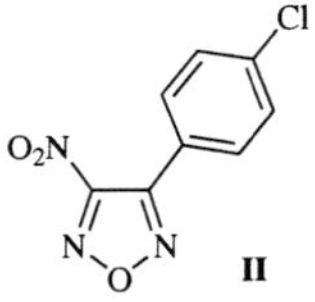

Wet DMD, acetone, rt — **I** (36) + **II** (64) — 149

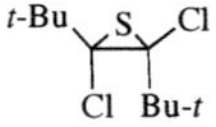

DMD, acetone, CH_2Cl_2 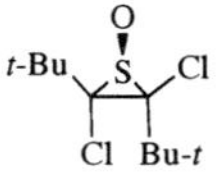 — (100) — 355

TABLE 3B. SULFUR AND SELENIUM OXIDATION BY ISOLATED DIOXIRANES (*Continued*)

Substrate	Conditions	Product(s) and Yield(s) (%)	Refs.
$C_{10\text{-}15}$			
[thiochromen-4-one, 3-R]	DMD, acetone, cyclohexanone, –20°, 10 min	[sulfone] R: Me (89); Ph (99)	356
[1,5-benzothiazepin-4-one, R^1, R^2]	DMD (x eq), acetone, CH_2Cl_2, 0 to 5°	I + II + III	357

R^1	R^2	x	I + II	III	I:II
Me	H	1.41	(71)	(3)	23:77
Me	Me	1.80	(66)	(11)	50:50
2-furyl	H	2.40	(88)	(6)	0:100
2-furyl	H	2.85	(58)	(35)	0:100
Ph	H	2.08	(63)	(8)	0:100
Ph	H	3.73	(0)	(98)	—

Substrate	Conditions	Product(s) and Yield(s) (%)	Refs.
C_{11}			
[2-thienyl 4-chlorophenyl ketone]	DMD, acetone, H_2O, rt	[hydroxy dihydrothiophene dioxide] (50)	358
[Ph, t-Bu pentathiepane]	DMD, CH_2Cl_2, acetone, –20° to rt	(25) + (17) + (9) + (12) + (9) + (8) + (1)	359

C_{12}

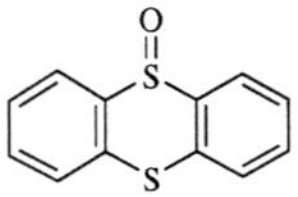

Dioxirane, acetone, CH_2Cl_2, additive, 1 h

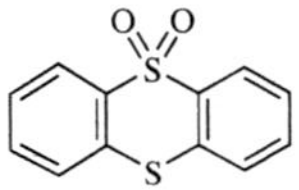

I 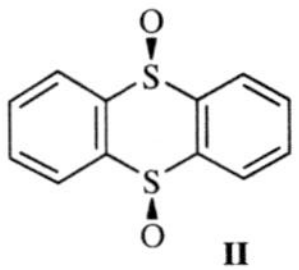 + **II** 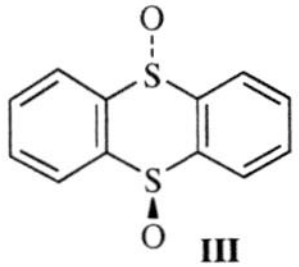 + **III** 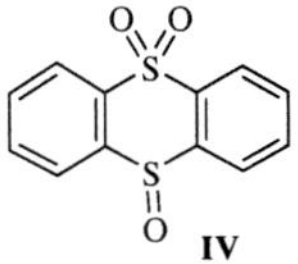 + **IV**

360, 361

Dioxirane	Additive	Temp	% Conv.	**I**	**II**	**III**	**IV**
TFD	—	–78°	21	(0.8)	(3.1)	(96)	(0.2)
TFD	—	0°	20	(2.8)	(3.9)	(85)	(8.1)
DMD	—	–50°	52	(6.8)	(3.1)	(90)	(0.1)
DMD	—	0°	56	(12)	(3.9)	(84)	(0.3)
DMD	MeOH	0°	35	(8.8)	(160	(75)	(—)
DMD	AcOH	0°	35	(6.5)	(130	(80)	(—)
DMD	CF_3CO_2H	0°	53	(1.6)	(5.7)	(93)	(—)
Methyl(isopropyl)dioxirane	—	–40°	44	(4.3)	(1.2)	(94)	(—)
Methyl(isopropyl)dioxirane	—	0°	44	(13)	(2.50	(84)	(—)
Methyl(isopropyl)dioxirane	CF_3CO_2H	0°	42	(—)	(2.5)	(98)	(—)
Cyclohexanone dioxirane	—	–70°	46	(5.8)	(11)	(82)	(1.9)
Cyclohexanone dioxirane	—	0°	46	(12)	(12)	(76)	(0.9)
Cyclohexanone dioxirane	CF_3CO_2H	0°	45	(1.0)	(13)	(86)	(—)

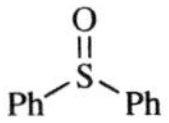

Ph–S(O)–Ph

Phenyl(trifluoromethyl)-dioxirane, MeCN, 20°, 30 min — Ph–SO_2–Ph **I** (—) — 362

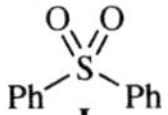

Diphenyldioxirane, 20°, 1.5 h — **I** (19) — 363

Bis(4-methoxy)phenyldioxirane, 20°, 1.5 h — **I** (42) — 363

TABLE 3B. SULFUR AND SELENIUM OXIDATION BY ISOLATED DIOXIRANES (*Continued*)

Substrate	Conditions	Product(s) and Yield(s) (%)	Refs.
C_{12}			
Ph–S–S–Ph	DMD, acetone, CH_2Cl_2, N_2, 30°	PhS(O)(O)–S–Ph	5
3,5-di-*t*-Bu-selenophene (*t*-Bu, Se, Bu-*t*)	DMD, CH_2Cl_2, acetone, –50°	*t*-Bu, Se=O, Bu-*t* (100)	155
Ph–S–Ph	Diphenyldioxirane, 20°, 1.5 h	PhS(=O)Ph **I** (2.2)	363
	Bis(4-methoxyphenyl)dioxirane, 20°, 1.5 h	**I** (37)	363
t-Bu, *t*-Bu, S=O, S, S	DMD, acetone, –18°	**I** + **II**; **I** + **II** (100), **I**:**II** = 7:1	364
O, S–S, O, N, NH, O, HO	DMD, acetone, rt, 2h	O, S–S=O, O, N, NH, O, HO (96)	365
$C_{12\text{-}32}$			
R^1, R^2, R^3, R^4, Se	DMD (2 eq), acetone, 0° to rt, 1-38 h	R^1, R^2, R^3, R^4, SeO_2	153

R^1	R^2	R^3	R^4	
t-Bu	H	*t*-Bu	H	(97)
Me	Ph	Ph	Me	(97)
4-ClC_6H_4	4-ClC_6H_4	4-ClC_6H_4	4-ClC_6H_4	(69)
Ph	Ph	Ph	Ph	(97)
4-MeC_6H_4	4-MeC_6H_4	4-MeC_6H_4	4-MeC_6H_4	(89)
4-$MeOC_6H_4$	4-$MeOC_6H_4$	4-$MeOC_6H_4$	4-$MeOC_6H_4$	(99)

C_{13}

Substrate	Conditions	Product (Yield)	Refs.
PhSCH$_2$SPh	DMD, acetone, cyclohexanone, –20°, 10 min	PhSO$_2$CH$_2$SO$_2$Ph (—)	5
2-thioxo-5-(C_8H_{17}-*n*)-6-methyl-pyrimidin-4-one	DMD, acetone, CH_2Cl_2, EtOH, rt	2-EtO-5-(C_8H_{17}-*n*)-6-methyl-pyrimidin-4-one (91)	143
2-Cl-3-PhS-spiro ketal cyclopentenone	DMD (2 eq), acetone, 20°, 30 min	2-Cl-3-PhSO$_2$-spiro ketal cyclopentenone (95)	366

C_{13-14}

R^1	R^2
F	*t*-$BuCO_2CH_2$
Br	Bn
Cl	Bn
F	Bn

Substrate	Conditions	Product (Yield)	Refs.
penam (R^1, R^2)	DMD, acetone, CH_2Cl_2, N_2, rt, 10 min	penam sulfoxide (100)	123

C_{13-15}

R^1	R^2
H	H
Me	H
MeO	H
Me	Me
MeO	Me

Substrate	Conditions	Product (Yield)	Refs.
4-$R^1C_6H_4$SC(O)C_6H_4-4-R^2	DMD, acetone, CH_2Cl_2, N_2, –30°, 2-4 h	4-$R^1C_6H_4$SO$_2$C(O)C_6H_4-4-R^2 (—)	367

TABLE 3B. SULFUR AND SELENIUM OXIDATION BY ISOLATED DIOXIRANES (*Continued*)

Substrate	Conditions	Product(s) and Yield(s) (%)	Refs.
C_{13-15}			
$4\text{-}R^1C_6H_4C(O)SC_6H_4\text{-}4\text{-}R^2$	DMD, acetone, CH_2Cl_2, N_2, 30°	$4\text{-}R^1C_6H_4C(O)SO_2C_6H_4\text{-}4\text{-}R^2$ (100) R^1 R^2: H H; H Me; H MeO; Me Me; Me MeO	367
C_{14}			
PhS–C≡C–SPh	DMD, acetone, 20°, 1 h	$PhSO_2$–C≡C–SO_2Ph (—)	133
$(NC)_2C{=}C(Ph)CH{=}C(SMe)_2$	DMD, acetone, cyclohexanone, –20°, 10 min	$(NC)_2C{=}C(Ph)CH{=}C(SO_2Me)_2$ (—)	368
PMP–S–S–PMP	DMD, acetone	PMP–S(O)–S–PMP (—)	135
PMP–S–S–S–PMP	DMD, acetone-d_6, –20°	PMP–S(O)–S–S–PMP **I** (—) + PMP–SO_2–S–S–PMP **II** (—) + PMP–SO_2–S–S(O)–PMP **III** (—)	135
PMP–S(O)–S–S–PMP	DMD, acetone-d_6, –20°	**I** (—) + **II** (—) + **III** (—) + PMP–S(O)–S–S(O)–PMP **IV** + other products (—)	135
PMP–S–S(O)–S–PMP	DMD, acetone, CH_2Cl_2, rt	PMP–S–S–PMP **I** (—) + PMP–S–S–S–PMP **II**	135

Substrate	Conditions	Product(s) and Yield(s) (%)	Refs.
	TFD, acetone, TFP or CH_2Cl_2, TFP, rt	**I** (—) + **II** (—)	135
	TFD, acetone, TFP, –80°	**I** (—)	135
PMP–SO_2–S–S–PMP	DMD, acetone-d_6, –20°	**III** (—) + other products	135
C_6H_4Me-4 sulfoxide (OH)	DMD, acetone, rt, 5 h	$SO_2C_6H_4Me$-4 sulfone (OH) (—)	369
C_{15}			
2-Phenylthiochromen-4-one	DMD (1.2 eq), solvent, 0°, 24 h	**I** + **II**	354
Me–S(=NTs)–C_6H_4Me-4, 80% ee	DMD, acetone, 0° to rt	Me–$S^+(O^-)$(=NTs)–C_6H_4Me-4 (97), 80% ee	148

Solvent	% Conv.	**I:II**
CCl_4	29	7:93
CCl_4/acetone (9:1)	25	12:88
acetone	15	35:65
CCl_4/MeOH (1:1)	29	48:52
$CHCl_3$	21	60:40
CCl_4/AcOH (1:1)	25	66:34

TABLE 3B. SULFUR AND SELENIUM OXIDATION BY ISOLATED DIOXIRANES (*Continued*)

Substrate	Conditions	Product(s) and Yield(s) (%)	Refs.
C_{15}	DMD, acetone, CH_2Cl_2, ROH, rt	I + II	117
	DMD (4 eq), CH_2Cl_2, acetone, –78°, 1 h; rt, 4 h	(28) + (24)	370
	DMD, CH_2Cl_2, acetone, –20°	+ (—)	370
	DMD (4 eq), CH_2Cl_2, acetone, –20°	(—) + other products	370
C_{15-21}	DMD, acetone, rt, 3 h	I + II + III + IV	136

ROH	I	II
—	(95)	(—)
MeOH	(22)	(70)
EtOH	(18)	(65)
n-BuOH	(20)	(75)

R	I + II	I:II	III	IV
t-Bu	(31)	50:50	(4)	(52)
1-Ad	(38)	50:50	(12)	(25)

Substrate	Conditions	Product(s) and Yield(s) (%)	Refs.
 C_{16} 4-MeC$_6$H$_4$S–C≡C–SC$_6$H$_4$Me-4	DMD, acetone, 20°, 3 h	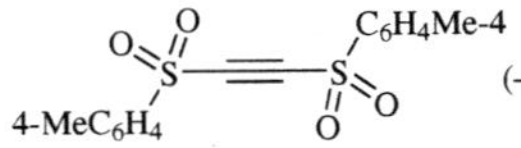 4-MeC$_6$H$_4$SO$_2$–C≡C–SO$_2$C$_6$H$_4$Me-4 (—)	133
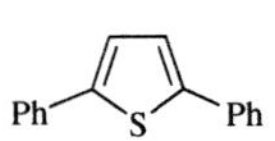 2,5-Diphenylthiophene	DMD (1 eq), acetone, CH_2Cl_2, 20°	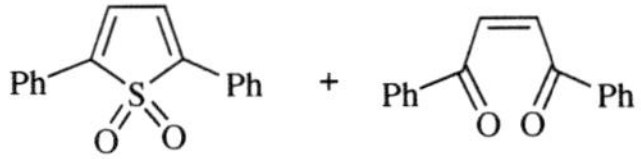 I + II; I + II (40), I:II = 75:25	140
	DMD (2 eq), acetone, CH_2Cl_2, 0°, 4 h	I (85)	140
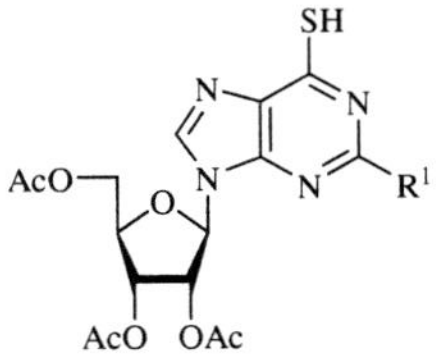 AcO, AcO, OAc, SH, R[1]	DMD, acetone, CH_2Cl_2, R^2NH_2, rt	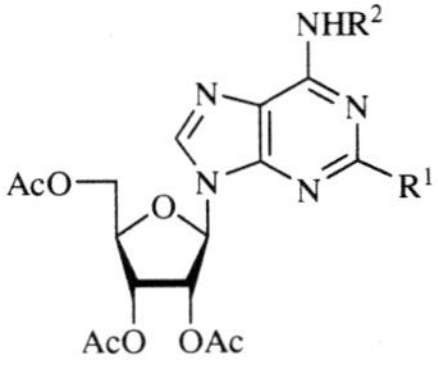 AcO, AcO, OAc, NHR[2], R[1]	116, 117, 118, 119, 120, 144
		R^1 / R^2: H / H (75); H / Me (55); NHAc / H (65); NHAc / Me (63); NHAc / Et (67); NHAc / 4-MeC$_6$H$_4$ (69); SH / H (53); SH / Me (46)	
BocNH, CH$_2$OAc, CO$_2$Me	DMD, acetone, dark, 0-5°, 40-60 min	I + II (—), I:II = 47:53	124
t-Bu, Bu-t, S=O	DMD, acetone, CH_2Cl_2, 0° to rt, 1 h	(98)	134

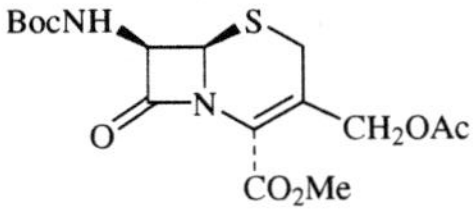

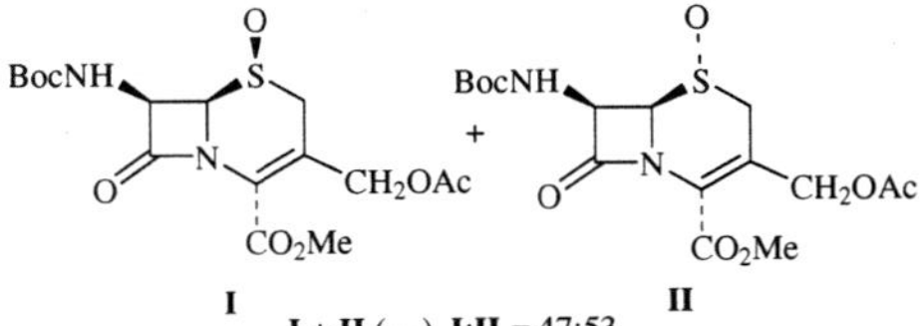

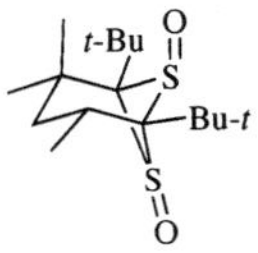

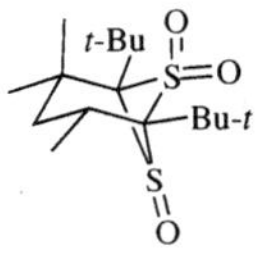

TABLE 3B. SULFUR AND SELENIUM OXIDATION BY ISOLATED DIOXIRANES (*Continued*)

Substrate	Conditions	Product(s) and Yield(s) (%)	Refs.
C_{16-17}			
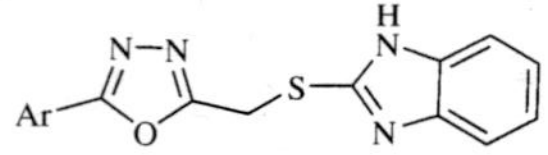 Ar	DMD (2-4 eq), acetone, rt, 30 min	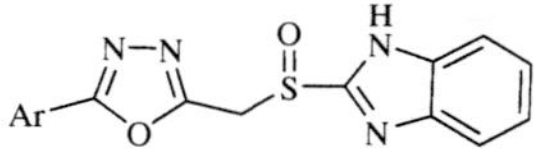 Ar: 4-ClC_6H_4 (74); 4-$MeOC_6H_4$ (75)	372
	DMD (5-8 eq), acetone, rt, 30 min	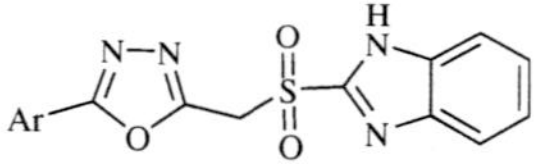 Ar: 4-ClC_6H_5 (90); 4-$MeOC_6H_5$ (91)	372
C_{16-18}	DMD, acetone, CH_2Cl_2, rt	n: 3 (63); 4 (93)	138
C_{16-12} R[1]: BocNH, R[2]: $MeCO_2CH_2$, R[3]: Me; R[1]: BnCONH, R[2]: Me, R[3]: $AcOCH_2$; R[1]: $PhOCH_2CONH$, R[2]: Cl, R[3]: Bn	DMD, acetone, 0-5°; dark, 40-60 min	(100)	124
C_{17} t-Bu, Bu-t	1. DMD, acetone, CH_2Cl_2, 0° 2. rt, 1 h	(98)	134

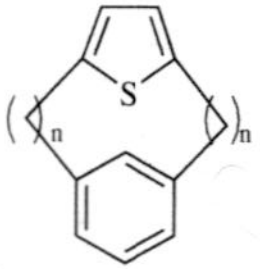

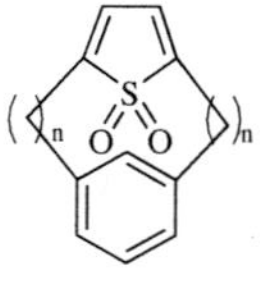

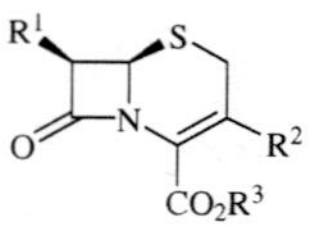

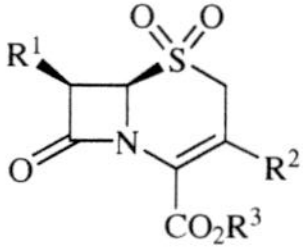

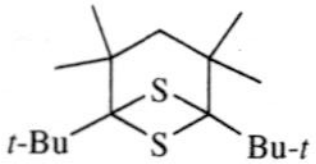

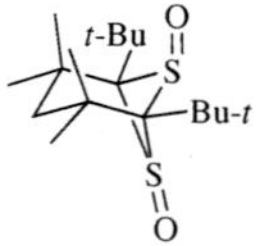

Substrate	Conditions	Product(s) and Yield(s) (%)	Refs.
C_{18-19} CbzNH, S, R, O, N, CO_2Me	DMD, acetone, 0 to 5°; dark, 40-60 min	CbzNH, O, O, S, R, O, N, CO_2Me	124
		R: CH=CH_2 (95); CH_2CO_2Me (100)	
C_{18} S, HN, N, AcHN, N, N, O, OAc, AcO, OAc	DMD, acetone, CH_2Cl_2, ROH, rt	**I** (O, HN, N, AcHN, N, N) + **II** (OR, N, N, AcHN, N, N)	117
		R / I / II: — (95) (–); Me (22) (70); Et (18) (65); *n*-Bu (20) (75)	
t-Bu, Y, Y, Bu-*t*	DMD, acetone, CH_2Cl_2	*t*-Bu, O, Y, Y, O, Bu-*t*	134
		Y / Temp / Time: S, 0°, 1 h (89); Se, 0° to rt, 30 min (100)	
C_{19} S=O, *t*-Bu, Bu-*t*, Bu-*t*	DMD, acetone, –78° to rt, overnight	*t*-Bu, O, O, S, O, Bu-*t*, Bu-*t* (9) + O, O, S, *t*-Bu, Bu-*t* (29)	147
S, *t*-Bu, Bu-*t*, Bu-*t*	DMD, acetone	**I** (O, S, *t*-Bu, Bu-*t*, Bu-*t*) + **II** (S=O, *t*-Bu, Bu-*t*, Bu-*t*) **I + II** (100)	147
		Temp / **I**:**II**: 20°, 20:1; –78°, 100:0	

TABLE 3B. SULFUR AND SELENIUM OXIDATION BY ISOLATED DIOXIRANES (*Continued*)

Substrate	Conditions	Product(s) and Yield(s) (%)	Refs.
C_{20} SPh, TMS	1. DMD (1 eq), CH_2Cl_2, −78° to rt, 6 h 2. KOAc, AcOH, −78° to rt 3. KOH, EtOH, H_2O, rt	(52) + O^-, S^+, Ph (40)	373
C_{20-29} R^1, S, N, O, R^2, CO_2R^3	DMD, acetone, 0 to 5°; dark, 4 days	R^1, O, O, S, N, O, O, R^2, CO_2R^3	124

R^1	R^2	R^3	Yield (%)	dr
BnCONH	Me	$AcOCH_2$	(85)	83:17
$PhOCH_2CONH$	Cl	Ph_2CH	(95)	—
BnCONH	Me	$AcOCH_2$	(85)	83:17
$PhOCH_2CONH$	Cl	Ph_2CH	(95)	—

Substrate	Conditions	Product(s) and Yield(s) (%)	Refs.
C_{21-31} R^1, O, R^2, SPh, Ar, O	1. DMD, acetone, rt 2. Toluene, heating	R^1, O, R^2, Ar, O **I** + R^1, O, R^2, SOPh, Ar, O **II**	374

R^1	R^2	Ar	I	II
H	H	Ph	(38)	(35)
H	H	$4\text{-}MeOC_6H_4$	(65)	(—)
MeO	H	Ph	(52)	(—)
H	H	$2,4\text{-}(MeO)_2C_6H_3$	(85)	(—)
MeO	H	$4\text{-}MeOC_6H_4$	(67)	(15)
H	H	$2,4,6\text{-}(MeO)_3C_6H_2$	(82)	(—)
H	$4\text{-}MeOC_6H_4$	$2,4,6\text{-}(MeO)_3C_6H_2$	(61)	(—)

Substrate	Conditions	Product(s) and Yield(s) (%)	Refs.
C_{22}	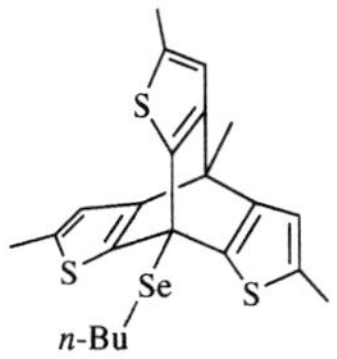 DMD, acetone, –78°	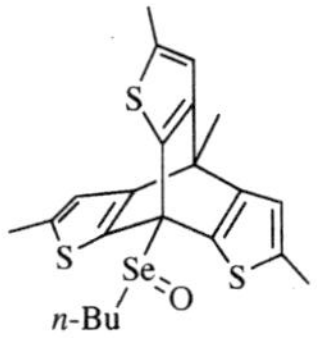 (—)	152
	1. DMD, acetone, –78° 2. rt	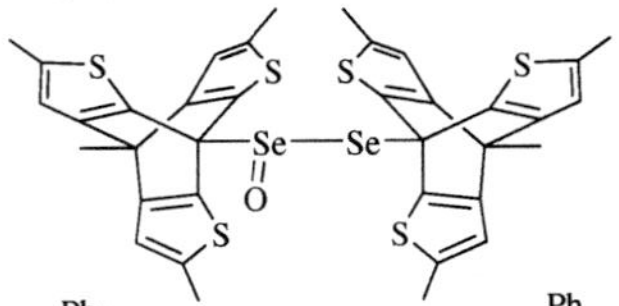 (70)	152
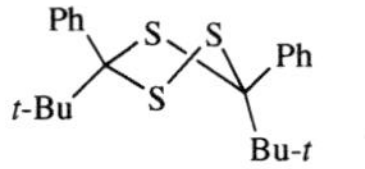	DMD (1.2 eq), acetone, CH_2Cl_2, –20°	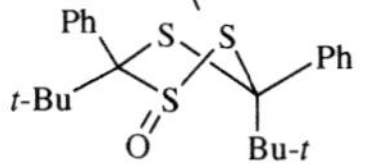 (54) 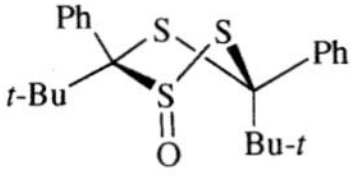 + (39)	375
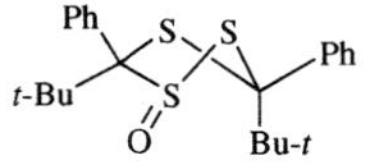	DMD (2 eq), acetone, CH_2Cl_2, –20°, 1.5 h	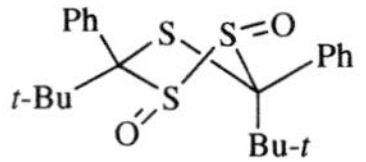 (100)	375
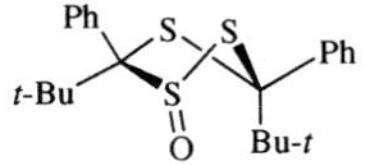	DMD (4 eq), acetone, CH_2Cl_2, –20°, 2.5 h	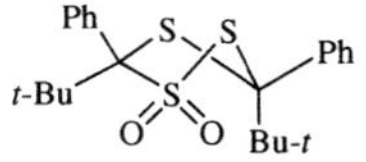 (100)	375
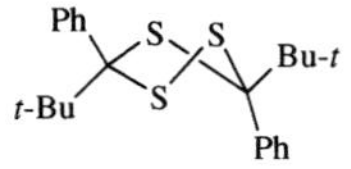	DMD (1.2 eq), acetone, CH_2Cl_2, –20°, 1 h	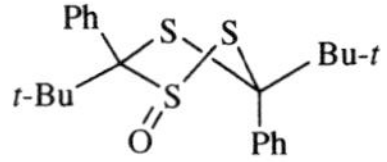 (88) 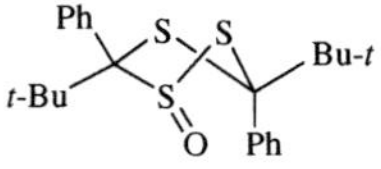 + (4)	376
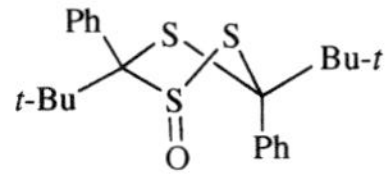	DMD (4 eq), acetone, CH_2Cl_2, –20°, 7 h	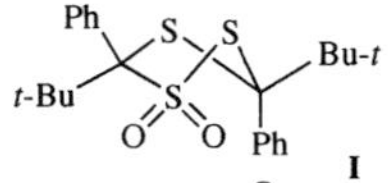 **I** (71) 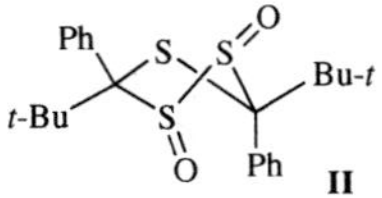 + **II** (20)	376
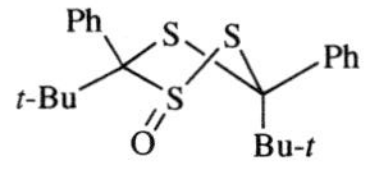	DMD (2 eq), acetone, CH_2Cl_2, –20°, 2 h	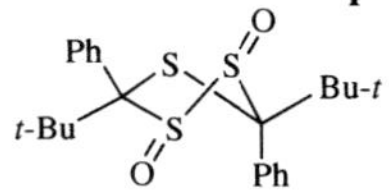 (70) + **I** (21) + **II** (4)	376

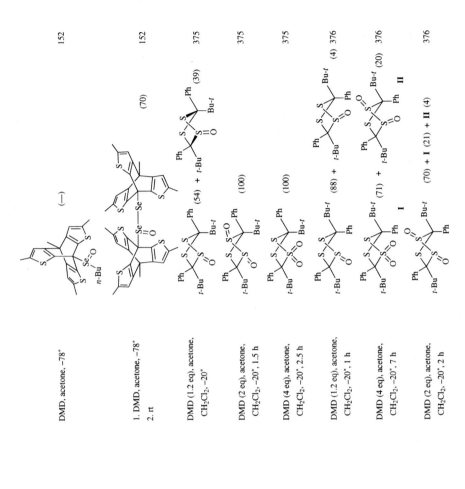

TABLE 3B. SULFUR AND SELENIUM OXIDATION BY ISOLATED DIOXIRANES (*Continued*)

Substrate	Conditions	Product(s) and Yield(s) (%)	Refs.
C_{22}			
Ph, Bu-*t*, t-Bu, Ph thiirane	DMD (1 eq), acetone, CH_2Cl_2, 0°, 10 min	Ph, Bu-*t*, *t*-Bu, Ph thiirane S-oxide (100)	376
PhS, SPh, Ph	DMD, acetone, –78° to rt	PhO_2S, SO_2Ph, Ph (87)	377
C_{22-34}			
Ph, R, Ph, R (S, S, S, S)	DMD (4 eq), acetone, CH_2Cl_2, –20°, 1 h	Ph, R, Ph, R (S=O, O=S); R: *t*-Bu (42); 1-Ad (60)	267
C_{23}			
$PhOCH_2CONH$, S=O, $CO_2CH_2C_6H_4NO_2$-4	DMD, acetone, 0-5°, dark, 40-60 min	$PhOCH_2CONH$, SO_2, $CO_2CH_2C_6H_4NO_2$-4 (95)	124
C_{23-28}			
S, Ar^2, N–Ac, Ar^1	DMD, acetone, CH_2Cl_2, rt, 16 h	SO_2, Ar^2, N–Ac, Ar^1 (see below)	378

Ar^1	Ar^2	(%)
Ph	Ph	(84)
Ph	2-$AcOC_6H_4$	(81)
Ph	4-$AcOC_6H_4$	(88)
Ph	4-$AcOC_6H_4$	(83)
4-(*i*-Pr)C_6H_4	2-$AcOC_6H_4$	(94)
3-ClC_6H_4	2-$AcOC_6H_4$	(91)
4-ClC_6H_4	2-$AcOC_6H_4$	(94)
2,4-$Cl_2C_6H_3$	2-$AcOC_6H_4$	(78)
3,4-$Cl_2C_6H_3$	2-$AcOC_6H_4$	(81)
4-ClC_6H_4	4-$AcOC_6H_4$	(84)
2,4-$Cl_2C_6H_3$	4-MeC_6H_4	(78)

	Substrate	Conditions	Product(s) and Yield(s) (%)	Refs.
C_{24}	CONPh$_2$, SOPh	DMD, acetone, cyclohexanone, –20°, 10 min	CONPh$_2$, SO$_2$Ph (—)	379
	1-Ad, Se, Ad-1	DMD, CH_2Cl_2, acetone, –50°	1-Ad, Se, O, Ad-1 (100)	155
	t-Bu, Bu-*t*, O=S, S=O, Bu-*t*, Bu-*t*	DMD (1.0 eq), acetone, 0°	(19) + (38)	380
		DMD (2.8 eq), acetone, 0°	(100)	380
C_{26}	*n*-C_9H_{19}, S, P, OBn, OBn, F, F	DMD, acetone, rt, 3 h	*n*-C_9H_{19}, O, P, OBn, OBn, F, F (97)	151
C_{28}	Ph, Ph, Ph, Ph, Se	DMD, CH_2Cl_2, acetone, –50°	Ph, Ph, Ph, Ph, Se, O (100)	155

TABLE 3B. SULFUR AND SELENIUM OXIDATION BY ISOLATED DIOXIRANES (*Continued*)

	Substrate	Conditions	Product(s) and Yield(s) (%)	Refs.
C_{29}	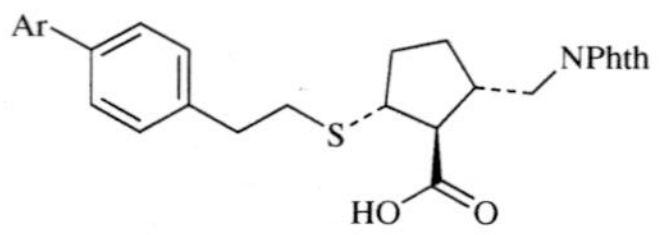 Ar; NPhth; S; HO; O Ar = Ph, 4-FC_6H_4, 4-ClC_6H_4	DMD, acetone, 0° to rt	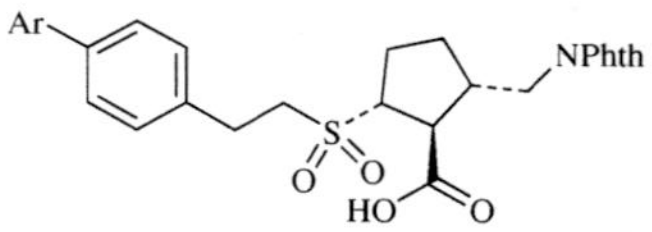 Ar; NPhth; S; O; O; HO; O (—)	371
C_{36}	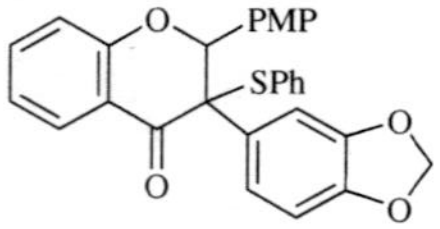 O; PMP; SPh; O; O; O	1. DMD, acetone, rt 2. Toluene, heating	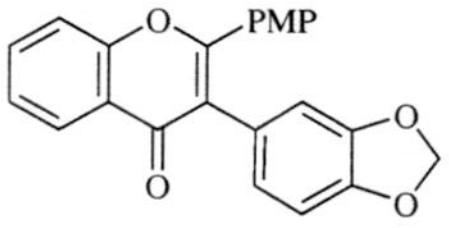 O; PMP; O; O; O (41)	374
	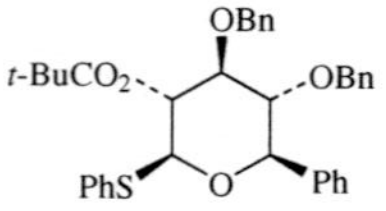 OBn; t-$BuCO_2$; OBn; PhS; O; Ph	DMD, acetone, CH_2Cl_2, 0°, 30 min	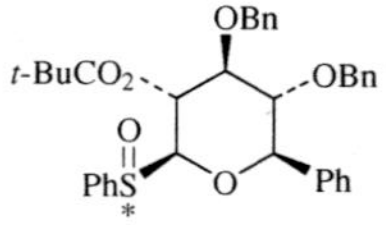 OBn; t-$BuCO_2$; OBn; O; PhS*; O; Ph (77) dr 2.5:1	125
C_{38}	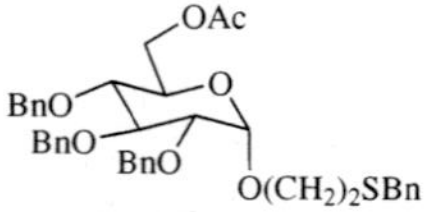 OAc; BnO; O; BnO; BnO; $O(CH_2)_2SBn$	DMD, acetone, –80° to rt, 2 h	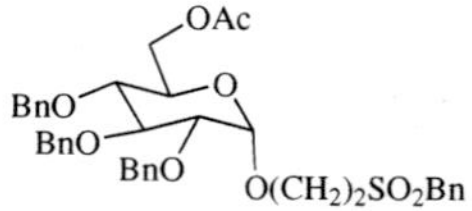 OAc; BnO; O; BnO; BnO; $O(CH_2)_2SO_2Bn$ (100)	381
C_{65}	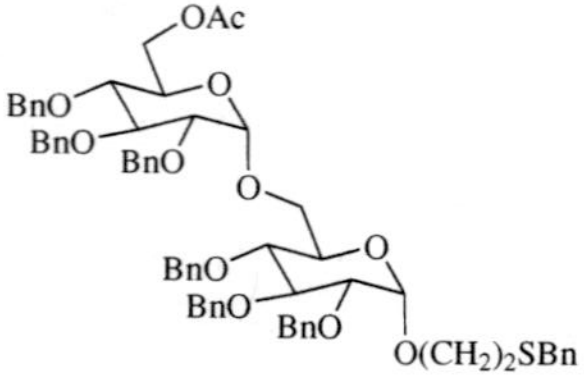 OAc; BnO; O; BnO; BnO; O; BnO; O; BnO; BnO; $O(CH_2)_2SBn$	DMD, acetone, –80° to rt, 2 h	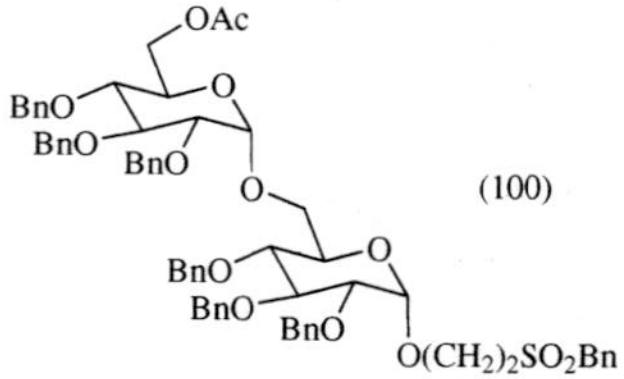 OAc; BnO; O; BnO; BnO; O; BnO; O; BnO; BnO; $O(CH_2)_2SO_2Bn$ (100)	381

C_{92}	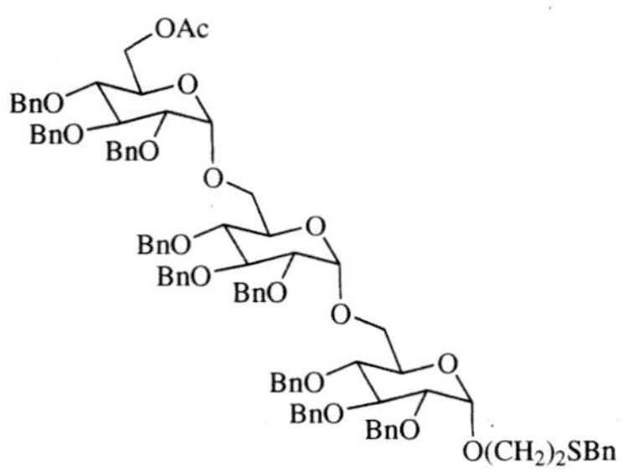 OAc, BnO, BnO, BnO, O, BnO, BnO, BnO, O, BnO, BnO, BnO, $O(CH_2)_2SBn$	DMD, acetone, –80° to rt, 2 h	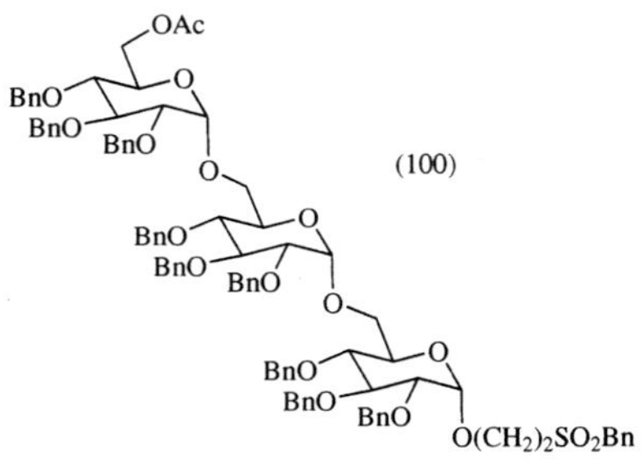 OAc, BnO, BnO, BnO, O, BnO, BnO, BnO, O, BnO, BnO, BnO, $O(CH_2)_2SO_2Bn$ (100)	381

[a] Methyl(trifluoromethyl)dioxirane (TFD) was added dropwise.

[b] The percent isotopic oxygen labeling for these compounds was as follows : $CF_3C(^{18}OH)_2CH_3$: 49%; $H_2{}^{18}O$: 98%; $CF_3C^{18}O_2H$: 49%.

TABLE 3C. PHOSPHORUS OXIDATION BY ISOLATED DIOXIRANES

Substrate	Conditions	Product(s) and Yield(s) (%)	Refs.
C_{18}			
Ph_3P	DMD, acetone	$Ph_3P{=}O$ **I** (100)	121
	DMD, acetone, –70 to –50°	**I** (—)	156
	DMD, $CDCl_3$	**I** (—)	157
C_{34}			
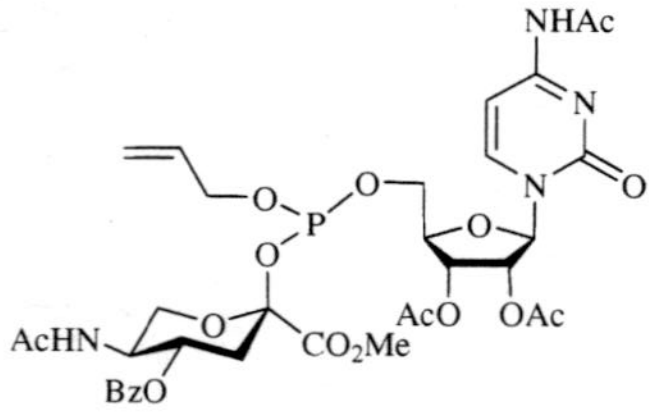	DMD, acetone, CH_2Cl_2, 0°, 10 min	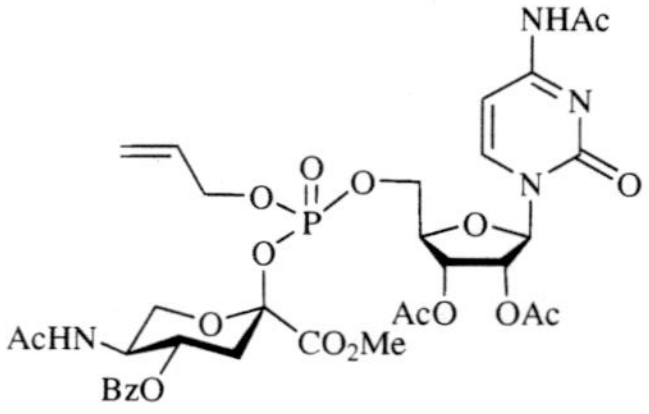 (100)	158
$C_{38\text{-}44}$			
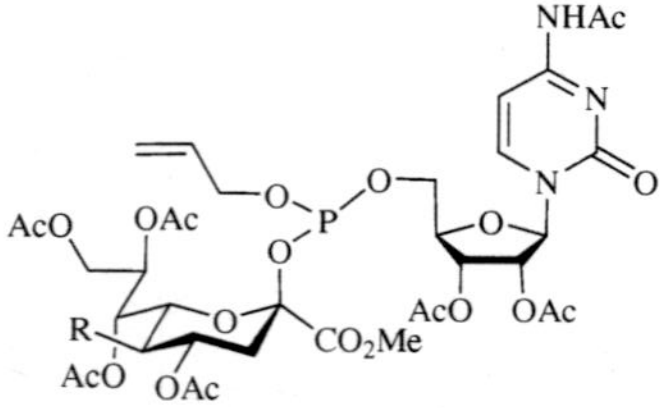	DMD, acetone, CH_2Cl_2, rt, 10 min	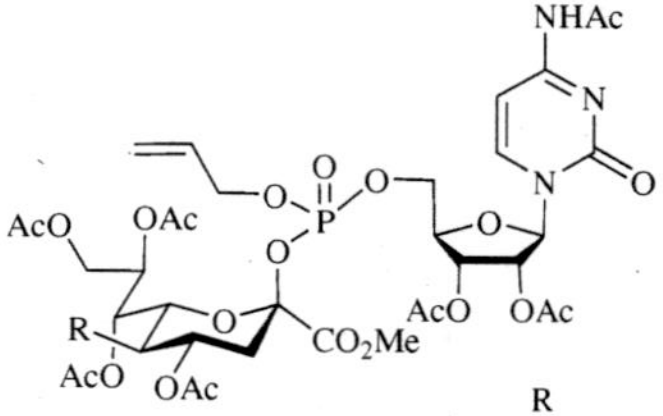	158

R	
NHAc	(98)
OAc	(100)
NHCbz	(99)

	Substrate	Conditions	Product(s) and Yield(s) (%)	Refs.
C_{39}		1. Methyl glycolate, tetrazole, MeCN, 0° 2. DMD, acetone, CH_2Cl_2, rt	(78)	382
		2-Ethyl-2-methyldioxirane, CH_2Cl_2, rt, 2 h	(>23)	383
C_{53}		DMD, acetone, –40°	(80)	384

TABLE 3D. OXYGEN OXIDATION BY ISOLATED DIOXIRANES

	Substrate	Conditions	Product(s) and Yield(s) (%)	Refs.
C_0	HSO_5^-	DMD, acetone/H_2O, $NaHCO_3$, 20°	O_2 (—) + HSO_4^- (—)	160
C_1	$F_2C{=}O$	$F_2C(O_2)$, CsF, –50°, 20 h	$CF_3O(CF_2O)_nOC(O)F$ (—) n = 1, 2, 3	385
	$F_2{}^{*}C{=}O$, * = ^{13}C	$F_2C(O_2)$, CsF, –50°, 16 h	$^{*}CF_3OOC(O)F$ (—) + $^{*}CF_3O(CF_2O)_nOC(O)F$ (—) n = 1, 2, 3	385
C_3	methyl(trifluoromethyl)dioxirane	TFD, CH_2Cl_2, n-Bu_4NI, 0°, 5 min	O_2 (96)	164
	dimethyldioxirane	DMD, acetone, dark, rt, 5-10 d	3,3,6,6-tetramethyl-1,2,4-trioxane (—)	121
		BF_3, acetone, ether, 0°	$MeCO_2Me$ (—)	121
$C_{5\text{-}8}$	pyridine N-oxide (substituents R^1–R^5)	DMD, $CDCl_3$, 20°	$^1O_2$ **I**	165

R^1	R^2	R^3	R^4	R^5	
H	H	H	H	H	(—)
Me	H	H	H	Me	(—)
H	Me	Me	H	H	(—)
Me	H	Me	H	Me	(—)
H	H	Me_2N	H	H	(30)

Substrate	Conditions	Product(s) and Yield(s) (%)	Refs.
C_{5-12}			
	TFD, TFA, 20°	I (R1 R2 R3 R4 R5: H H H H H (4.8); Me H H H Me (1.3); H Me Me H H (0.2); Me H Me H Me (0.6); H H Me_2N H H (5))	165
$R\text{-}N^+(O^-)$ ring with Y	DMD, $CDCl_3$, 20°	I (R, Y: Me, O (0.08-0.60); $PhCH_2$, CH_2 (0.04-0.09))	165
C_6			
$TMS^{18}O^{18}OSO_3TMS$	Cyclohexanone, CH_2Cl_2, He, –80 to –10°, 10 h	$^{32}O_2$ (—) + $^{34}O_2$ (—) + $^{36}O_2$ (—) **I**:**II**:**III** = 92.7:6.4:0.9 (I, II, III)	386
C_7			
4-(NMe_2)pyridine	DMD, acetone, 0°	$^1O_2$ (—) + 4-(NMe_2)pyridine **I** (—) + 4-(NMe_2)pyridine N-oxide **II** (—) **I**:**II** = 16:84	26
4-(NMe_2)pyridine N-oxide	DMD, acetone	$^1O_2$ (—) + 4-(NMe_2)pyridine (—)	26
C_8			
$PhMe_2N^+\text{—}O^-$	DMD, $CDCl_3$, 20°	$^1O_2$ (0.33)	165
C_9			
Isoquinoline N-oxide	TFD, TFA, 20°	$^1O_2$ (0.02)	165

TABLE 3D. OXYGEN OXIDATION BY ISOLATED DIOXIRANES (*Continued*)

Substrate	Conditions	Product(s) and Yield(s) (%)	Refs.
C_9 N–O•	DMD, acetone, air or N_2, 20°, 2-4 h	N–OMe **I** (> 98) + N–OCH_2COMe (< 1)	166
	TFD, TFA, 0°, 2-4 h	$^1O_2$ (> 96)	166
C_{12} N–O•	TFD, CH_2Cl_2, 0°, 2 h	N–OMe **I** (8)	23
	TFD, hν, CH_2Cl_2, –10°	**I** (39) + N–OCF_3 (20)	23
$Bu_3\overset{+}{N}–O^-$	TFD, TFA, 20°	**I** (0.12-0.88)	165
C_{15} Ph, N–O•	Dioxirane, acetone, 0°	Ph, N–OMe **I** + Ph, N–H **II** + Ph, N^+–O^- **III** + Ph, N^+–O^- **IV**	166

Dioxirane	Time	I	II	III	IV
DMD	3 h	(16)	(8.5)	(18)	(12)
TFD	30 min	(12)	(14)	(19)	(21)

Reactant	Conditions	Product(s) and Yield(s) (%)	Refs.
C_{16}			
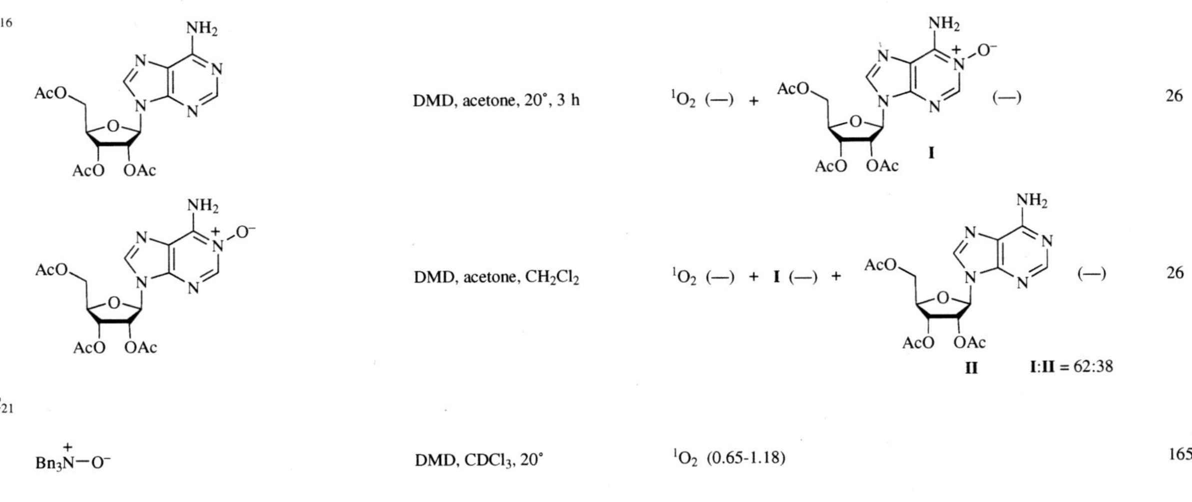	DMD, acetone, 20°, 3 h	$^1O_2$ (—) + I (—)	26
	DMD, acetone, CH_2Cl_2	$^1O_2$ (—) + **I** (—) + II (—) **I**:**II** = 62:38	26
C_{21}			
$Bn_3\overset{+}{N}-O^-$	DMD, $CDCl_3$, 20°	$^1O_2$ (0.65-1.18)	165

TABLE 3E. HALOGEN OXIDATION BY ISOLATED DIOXIRANES

Substrate		Conditions	Product(s) and Yield(s) (%)	Refs.
C_0				
LiI		TFD, TMSCl, CH_2Cl_2, 0°, 20 min	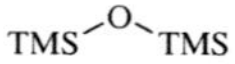 TMS–O–TMS (—)	164
C_1				
CH_3I	Alkene (0.5 equiv)	1. DMD (1 eq), acetone, –70°, 1 h 2. Alkene, –70° to rt, 20 h		
	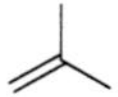		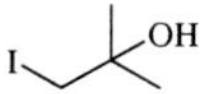 I, OH (78)	170
			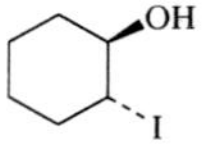 OH, I (85)	170
	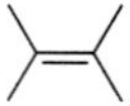		 I, OH (0)	170
			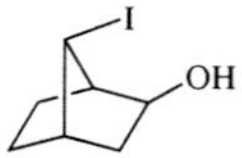 I, OH (72)	170
	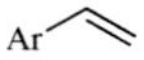 Ar		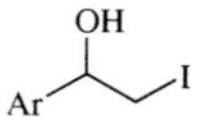 OH, Ar, I Ar: 4-ClC_6H_4 (82); Ph (85); 4-$CF_3C_6H_4$ (44); 4-MeC_6H_4 (75); 4-$MeOC_6H_4$ (60)	170
	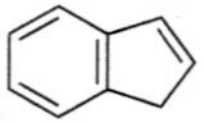		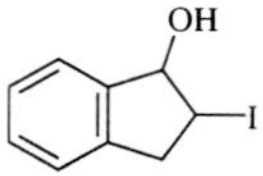 OH, I (82)	170
	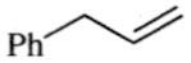 Ph		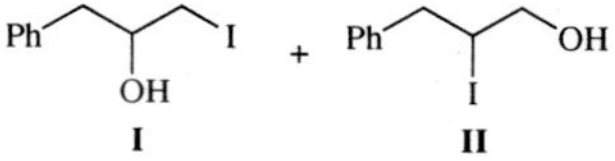 Ph, OH, I **I** + Ph, I, OH **II** **I + II** (85), **I:II** = 65:35	170

Substrate	Conditions	Product(s) and Yield(s) (%)	Refs.
Ph (α-methylstyrene)		Ph, OH, I (72)	170
Et, C_5H_{11}-*n*		HO, C_5H_{11}-*n*, Et, I **I** + I, C_5H_{11}-*n*, Et, OH **II**; **I** + **II** (85), **I**:**II** = 1:1	170
Ph, Ph		I, OH, Ph, Ph (0)	170
	1. DMD (1 eq), acetone, –70°, 1 h 2. (tetramethylallene), –70° to rt, overnight	I, OH (94)	171
C_6 PhI	DMD, acetone, quinoline, 22°, 4 h	PhI=O (—)	35
	DMD, acetone, AcOH, N_2, rt, 5 h	$PhI(OAc)_2$ (56)	168
Cyclohexyl iodide	DMD, acetone, N_2, rt, 5 h	I, OH (43) + OH, OH (4)	168
	DMD, acetone, AcOH, N_2, rt, 5 h	I, OAc (35)	168
C_{6-7} 4-R-C_6H_4I	DMD, acetone, N_2, rt, 5 h	4-R-C_6H_4IO **I** + 4-R-$C_6H_4IO_2$ **II**	168

R	I	II
H	(26)	(13)
MeO	(—)	(46)

TABLE 3E. HALOGEN OXIDATION BY ISOLATED DIOXIRANES (*Continued*)

Substrate	Conditions	Product(s) and Yield(s) (%)	Refs.
C_{7-8} 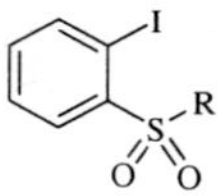 I, SO_2R (R)	DMD, acetone, CH_2Cl_2, 0°, 8 h	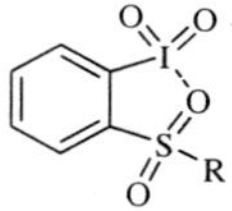 R: OMe (89); OEt (84)	387
C_8 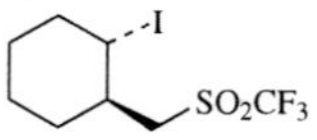 I, SO_2CF_3	DMD, acetone, ether, rt, 1 h	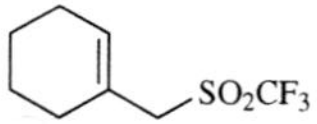 SO_2CF_3 (58)	169
C_{9-13} 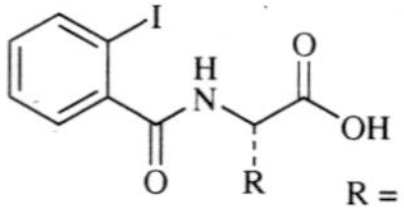 I, H, N, O, OH, R; R = H, Me, *i*-Pr, *i*-Bu	DMD, acetone, rt, 8 h	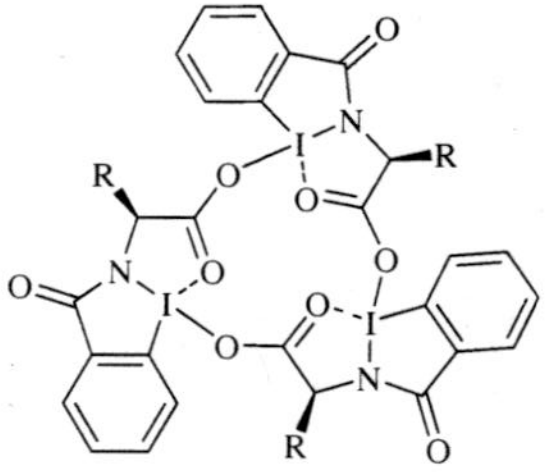 R, O, N, I (76-90)	388
C_{10} 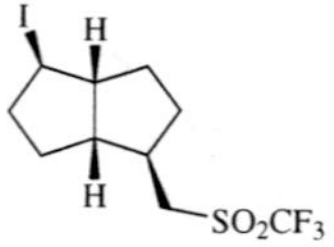 I, H, H, SO_2CF_3	DMD, acetone, ether, 0°	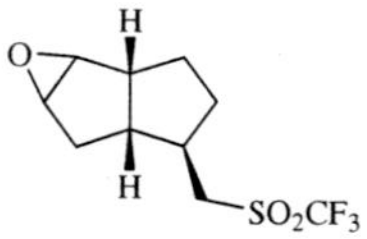 O, H, H, SO_2CF_3 (77)	169
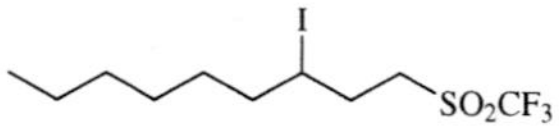 I, SO_2CF_3	DMD, acetone, ether, rt, 1 h	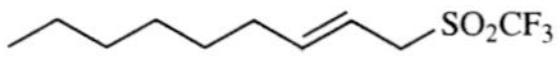 SO_2CF_3 (80)	169

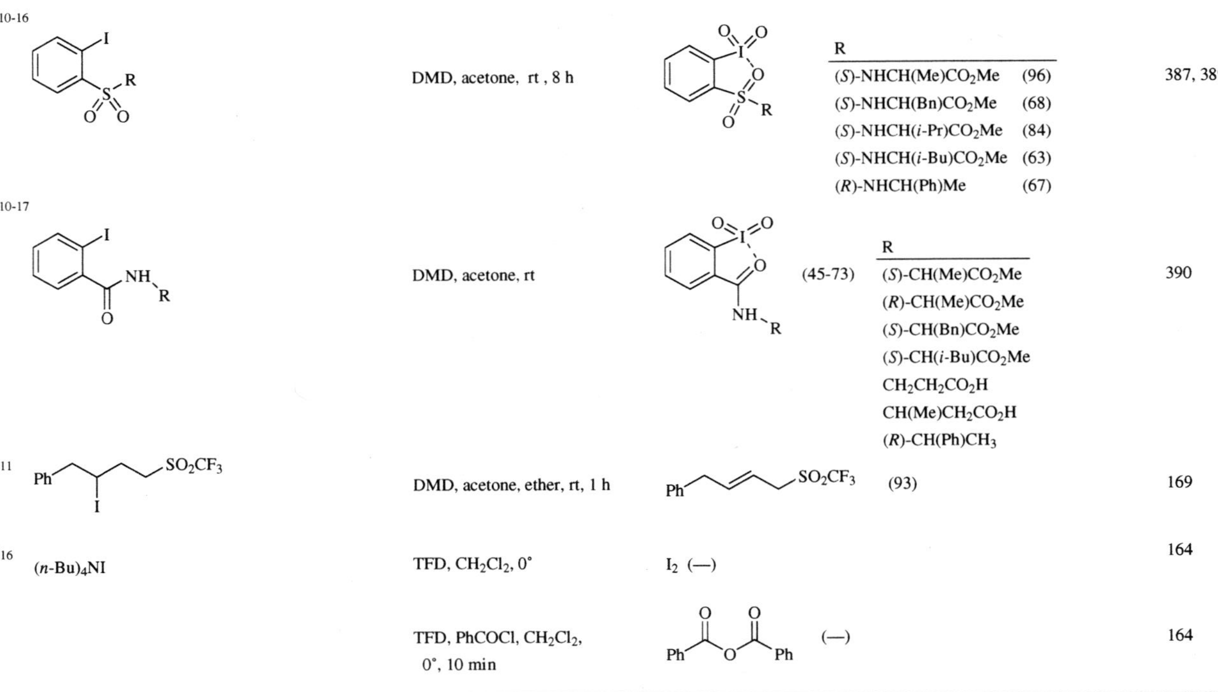

Substrate	Conditions	Product(s) and Yield(s) (%)	Refs.
C_{10-16}	DMD, acetone, rt, 8 h	R: (*S*)-NHCH(Me)CO_2Me (96); (*S*)-NHCH(Bn)CO_2Me (68); (*S*)-NHCH(*i*-Pr)CO_2Me (84); (*S*)-NHCH(*i*-Bu)CO_2Me (63); (*R*)-NHCH(Ph)Me (67)	387, 389
C_{10-17}	DMD, acetone, rt	(45-73) R: (*S*)-CH(Me)CO_2Me; (*R*)-CH(Me)CO_2Me; (*S*)-CH(Bn)CO_2Me; (*S*)-CH(*i*-Bu)CO_2Me; $CH_2CH_2CO_2H$; CH(Me)CH_2CO_2H; (*R*)-CH(Ph)CH_3	390
C_{11}	DMD, acetone, ether, rt, 1 h	(93)	169
C_{16} $(n\text{-Bu})_4NI$	TFD, CH_2Cl_2, 0°	I_2 (—)	164
	TFD, PhCOCl, CH_2Cl_2, 0°, 10 min	(—)	164

TABLE 3F. NITROGEN OXIDATION BY IN SITU GENERATED DIOXIRANES

Substrate	Conditions	Product(s) and Yield(s) (%)	Refs.
C_{4-11} $R^1R^2CHNH_2$	Oxone® (x eq), acetone, H_2O, $NaHCO_3$, rt	$R^1R^2C{=}NOH$ (I) + azoxy dimer (II) + $R^1R^2CHNO_2$ (III) + oxaziridine (IV)	114

R^1	R^2	x	Solvent	I	II	III	IV
n-Pr	H	4.5	CH_2Cl_2	(24)	(30)	6[a]	27[a]
n-Pr	H	4.5	—	(40)	(40)	12[a]	(—)
—$(CH_2)_5$—		4.5	CH_2Cl_2	(10)	(43)	(10)	(10)
—$(CH_2)_5$—		8	CH_2Cl_2	12[a]	60[a]	14[a]	14[a]
—$(CH_2)_5$—		16	CH_2Cl_2	12[a]	50[a]	18[a]	20[a]
—$(CH_2)_5$—		8	—	2[a]	26[a]	72[a]	(—)
—$(CH_2)_5$—		16	—	(—)	14[a]	86[a]	(—)
—$(CH_2)_5$—		20	—	(—)	(11)	(65)	(—)
Ph	H	6	CH_2Cl_2	(55)	(—)	(—)	38[a]
Ph	H	6	—	(52)	(—)	(—)	(trace)
n-C_9H_{19}	H	4.5	CH_2Cl_2	(15)	(35)	5[a]	(37)
n-C_9H_{19}	H	4.5	—	(—)	(64)	(9)	13[a]

Substrate	Conditions	Product(s) and Yield(s) (%)	Refs.
C_5 pyridine	Oxone®, acetone, H_2O (pH 7.5-8.0), 2 h	pyridine N-oxide I (93)	121
	Oxone®, acetone, 50°, 16 h	I (63) + cyclic triperoxide (4)	121
	Oxone®, acetone	I (70-80)	121

Substrate	Conditions	Product(s) and Yield(s) (%)	Refs.
C_6			
NH_2 (benzene ring)	Oxone®, cyclohexanone, buffer (pH)	**I** pH: 7.0 (72); 7.5 (85); 8.0 (94); 8.5 (96); 9.0 (95); 9.5 (86)	391
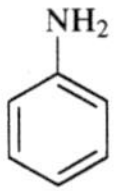	Oxone®, acetone, CH_2Cl_2, phosphate buffer (pH 7.5-8.5), aq. KOH, $(n\text{-Bu})_4NHSO_4$, 0°, 45 min	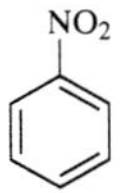 NO_2 (benzene ring) **I** (17.5) 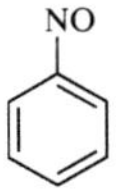 + NO (benzene ring) (17.5)	267
	Oxone®, THF, acetone, CH_2Cl_2, phosphate buffer (pH 7.5-8.5), aq. KOH, $(n\text{-Bu})_4NHSO_4$, 0°, 15 min	**I** (78)	267
	Oxone®, 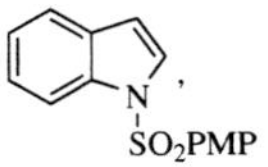 (indole, N–SO_2PMP), acetone, CH_2Cl_2, phosphate buffer (pH 7.5-8.5), aq. KOH, $(n\text{-Bu})_4NHSO_4$, 0°, 45 min	**I** (98)	267
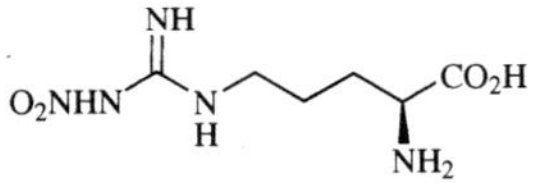 O_2NHN–C(=NH)–NH–$(CH_2)_3$–CH(NH_2)–CO_2H	Oxone®, acetone, phosphate buffer (pH 8.0), KOH, 5°, 7 h	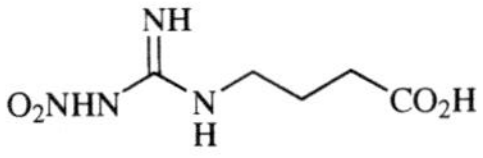 O_2NHN–C(=NH)–NH–$(CH_2)_3$–CO_2H (34)	112
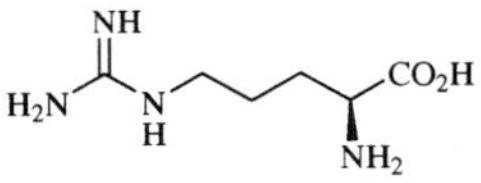 H_2N–C(=NH)–NH–$(CH_2)_3$–CH(NH_2)–CO_2H	Oxone®, acetone, phosphate buffer (pH 8.0), KOH, 5°, 7 h	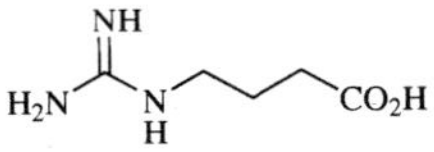 H_2N–C(=NH)–NH–$(CH_2)_3$–CO_2H (57)	112

TABLE 3F. NITROGEN OXIDATION BY IN SITU GENERATED DIOXIRANES (*Continued*)

Substrate	Conditions	Product(s) and Yield(s) (%)	Refs.
C_7			
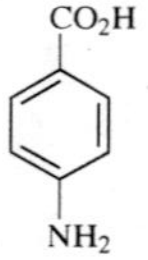 4-$H_2NC_6H_4CO_2H$	Oxone®, acetone, H_2O, NaOH, $NaHCO_3$, EDTA, 2-18°, 4 h	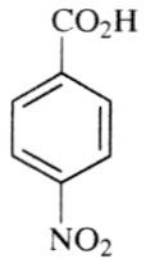 4-$O_2NC_6H_4CO_2H$ (73)	392
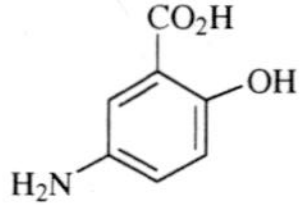 4-amino-2-hydroxybenzoic acid (CO_2H, OH, H_2N)	Oxone®, acetone, H_2O, NaOH, $NaHCO_3$, EDTA, 2-18°, 4 h	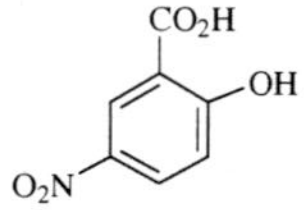 2-hydroxy-4-nitrobenzoic acid (CO_2H, OH, O_2N) (74)	392
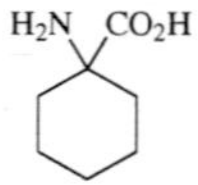 1-aminocyclohexanecarboxylic acid (H_2N, CO_2H)	Oxone®, acetone, phosphate buffer (pH 8.0), KOH, 5°, 7 h	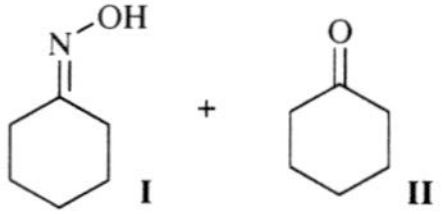 cyclohexanone oxime (N–OH) **I** + cyclohexanone **II**; **I** + **II** (88), **I**:**II** = 3:1	112
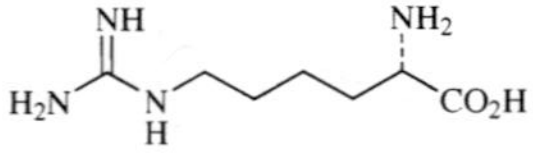 $H_2NC(=NH)NH(CH_2)_3CH(NH_2)CO_2H$	Oxone®, acetone, phosphate buffer (pH 8.0), KOH, 5°, 7 h	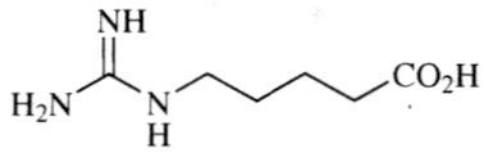 $H_2NC(=NH)NH(CH_2)_3CO_2H$ (56)	112
C_{7-9}			
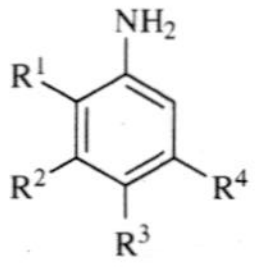 2-R^1-3-R^2-4-R^3-5-R^4-C_6HNH_2 (aniline, NH_2)	Oxone®, acetone, CH_2Cl_2, phosphate buffer (pH 7.5-8.5), aq. KOH, $(n\text{-Bu})_4NHSO_4$, 0°, 45 min	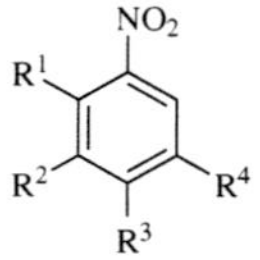 2-R^1-3-R^2-4-R^3-5-R^4-C_6HNO_2 (nitrobenzene, NO_2); see table below	267

R^1	R^2	R^3	R^4	
MeO	H	H	H	(100)
H	MeO	H	H	(73)
H	—OCH_2O—		H	(57)
MeO	H	MeO	H	(85)
MeO	H	H	MeO	(51)
H	MeO	MeO	MeO	(52)
H	MeO	CO_2Me	H	(78)

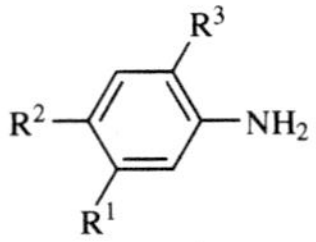

Oxone® (1-2 eq), CH_2Cl_2, H_2O 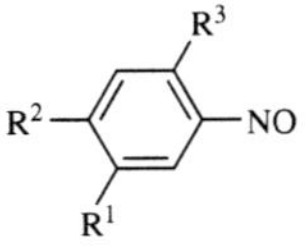 — 393

R^1	R^2	R^3	Time	Yield
H	CO_2Me	H	0.5 h	(100)
H	CO_2Me	H	1 h	(100)
CO_2Me	H	H	3 h	(100)
CO_2H	H	H	3.5 h	(96)
CN	H	H	3 h	(96)
H	Br	H	3.5 h	(70)
H	Et	H	12 h	(65)
H	SO_3H	H	4 h	(—)
H	SO_3NBu_4	H	0.5 h	(97)
H	CH_2OH	H	1 h	(—)
H	CH_2CO_2H	H	4 h	(—)
H	CH_2NHBoc	H	9.5 h	(60)
CO_2Me	H	Me	3.5 h	(96)

C_8

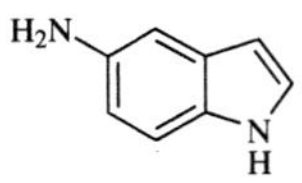

Reactant	Conditions	Product(s) and Yield(s) (%)	Refs.
H_2N (5-aminoindole)	Oxone®, acetone, CH_2Cl_2, phosphate buffer (pH 7.5-8.5), aq. KOH, $(n\text{-Bu})_4NHSO_4$, 0°, 45 min	O_2N (5-nitroindole) (40) + ON (5-nitrosoindole) (18)	267
N–N, N=N, Ph (phenyltetrazine)	Oxone®, TFP, CH_2Cl_2, H_2O, $NaHCO_3$, 0°, 80 min	O^-, N^+, Ph (47)	311
NH_2, Ph, CO_2H	Oxone®, acetone, phosphate buffer (pH 8.0), KOH, 5°, 7 h	Ph=NHOH **I** + $PhCO_2H$ **II** **I + II** (74), **I**:**II** = 3:1	112

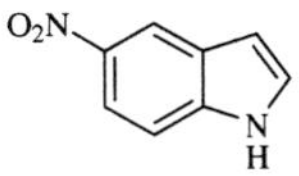

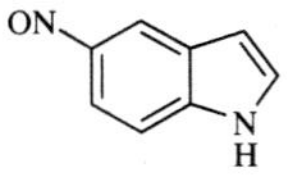

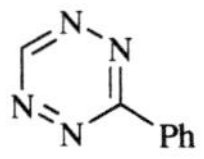

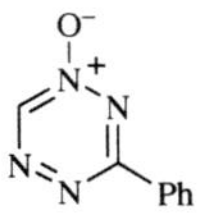

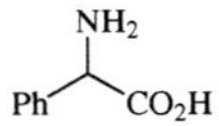

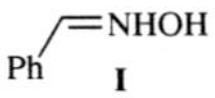

TABLE 3F. NITROGEN OXIDATION BY IN SITU GENERATED DIOXIRANES (*Continued*)

	Substrate	Conditions	Product(s) and Yield(s) (%)	Refs.
C_8	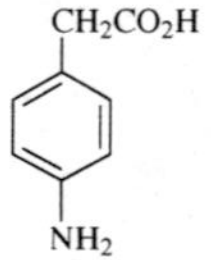 CH_2CO_2H, NH_2	Oxone®, acetone, H_2O, NaOH, $NaHCO_3$, EDTA, 2-18°, 4 h	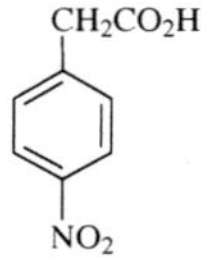 CH_2CO_2H, NO_2 (81)	392
	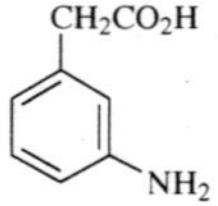 CH_2CO_2H, NH_2	Oxone®, acetone, H_2O, NaOH, $NaHCO_3$, EDTA, 2-18°, 4 h	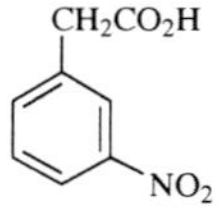 CH_2CO_2H, NO_2 (81)	392
	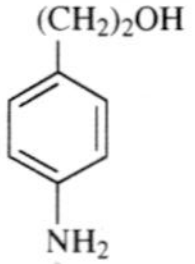 $(CH_2)_2OH$, NH_2	Oxone®, acetone, H_2O, NaOH, $NaHCO_3$, EDTA, 2-18°, 4 h	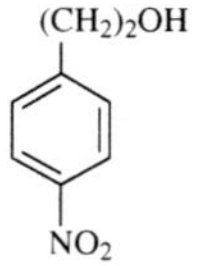 $(CH_2)_2OH$, NO_2 (73)	392
	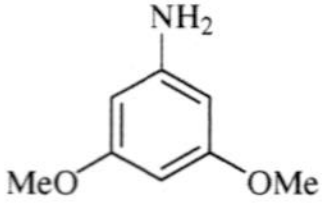 NH_2, MeO, OMe	Oxone®, acetone, CH_2Cl_2, phosphate buffer (pH 7.5-8.5), aq. KOH, $(n\text{-}Bu)_4NHSO_4$, 0°, 45 min	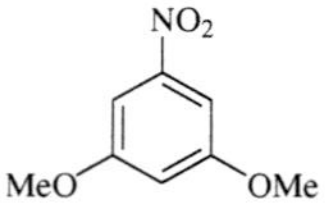 NO_2, MeO, OMe (20) 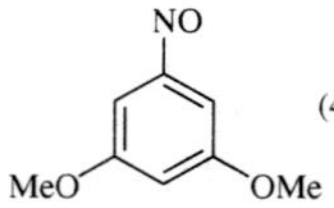 + NO, MeO, OMe (40)	267
	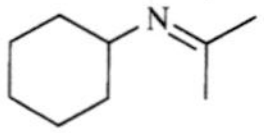 N	Oxone®, acetone, H_2O, $NaHCO_3$, rt	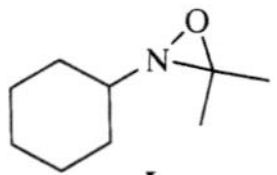 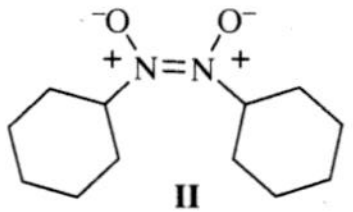 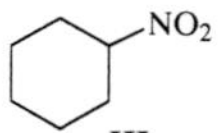 O, N **I** + ^{-}O, O^{-}, $^{+}N{=}N^{+}$ **II** + NO_2 **III**; **I + II + III** (—), **I:II:III** = 73:5:22	114

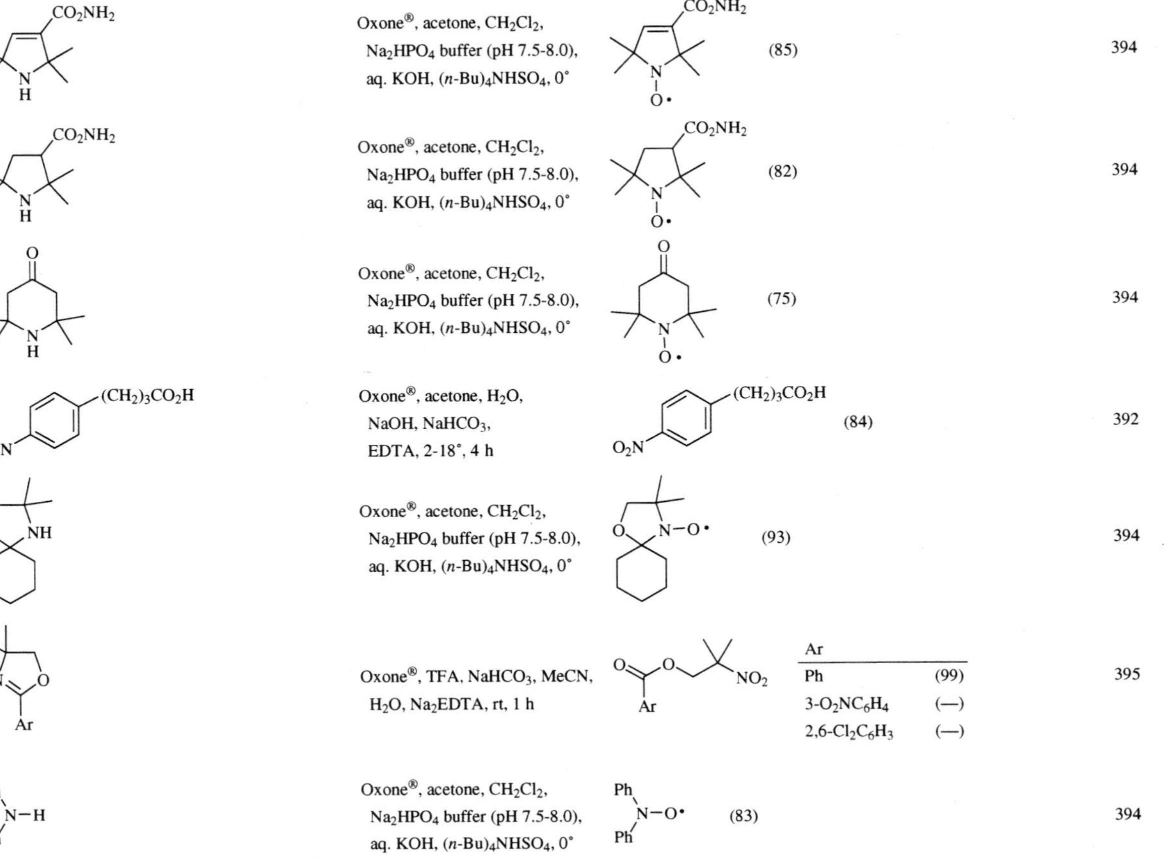

Substrate	Conditions	Product(s) and Yield(s) (%)	Refs.
	Oxone®, acetone, CH_2Cl_2, Na_2HPO_4 buffer (pH 7.5-8.0), aq. KOH, $(n\text{-Bu})_4NHSO_4$, 0°	(85)	394
	Oxone®, acetone, CH_2Cl_2, Na_2HPO_4 buffer (pH 7.5-8.0), aq. KOH, $(n\text{-Bu})_4NHSO_4$, 0°	(82)	394
	Oxone®, acetone, CH_2Cl_2, Na_2HPO_4 buffer (pH 7.5-8.0), aq. KOH, $(n\text{-Bu})_4NHSO_4$, 0°	(75)	394
C_{10}	Oxone®, acetone, H_2O, NaOH, $NaHCO_3$, EDTA, 2-18°, 4 h	(84)	392
	Oxone®, acetone, CH_2Cl_2, Na_2HPO_4 buffer (pH 7.5-8.0), aq. KOH, $(n\text{-Bu})_4NHSO_4$, 0°	(93)	394
C_{11}	Oxone®, TFA, $NaHCO_3$, MeCN, H_2O, Na_2EDTA, rt, 1 h	Ar: Ph (99); $3\text{-}O_2NC_6H_4$ (—); $2,6\text{-}Cl_2C_6H_3$ (—)	395
C_{12}	Oxone®, acetone, CH_2Cl_2, Na_2HPO_4 buffer (pH 7.5-8.0), aq. KOH, $(n\text{-Bu})_4NHSO_4$, 0°	(83)	394

TABLE 3F. NITROGEN OXIDATION BY IN SITU GENERATED DIOXIRANES (*Continued*)

Substrate	Conditions	Product(s) and Yield(s) (%)	Refs.
C_{14}			
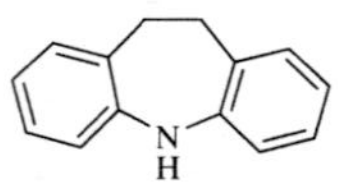 N–H (structure)	Oxone®, acetone, CH_2Cl_2, Na_2HPO_4 buffer (pH 7.5-8.0), aq. KOH, $(n\text{-Bu})_4NHSO_4$, 0°	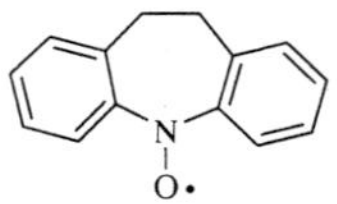 N–O• (structure) (89)	394
 N–H, N–H (structure)	Oxone®, acetone, CH_2Cl_2, Na_2HPO_4 buffer (pH 7.5-8.0), aq. KOH, $(n\text{-Bu})_4NHSO_4$, 0°	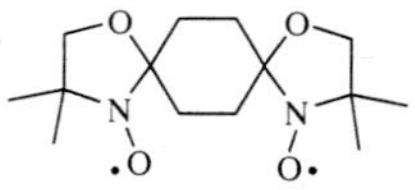 •O–N, N–O• (structure) (90)	394
C_{15}			
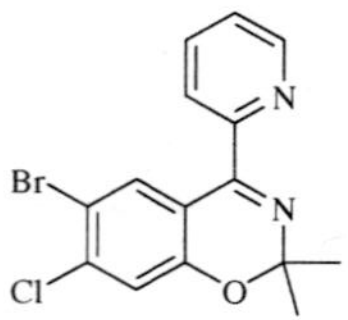 Br, Cl, N, O (structure)	Oxone® (2.3 eq), acetone, H_2O, CH_2Cl_2, buffer salt (x eq)	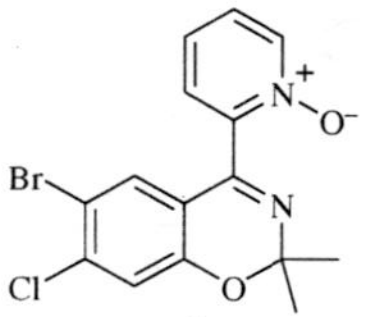 I 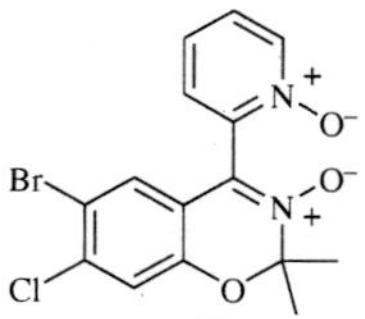 + II (structures)	396

Buffer salt	x	I	II
$NaHCO_3$	6.4	(76)	(4)
Na_2HPO_4	6.4	(79)	(2)
Na_2HPO_4	8.0	(86)	(3)
Na_2HPO_4	9.2	(84)	(3)
K_2HPO_4	6.4	(79)	(2)

Substrate	Conditions	Product(s) and Yield(s) (%)	Refs.
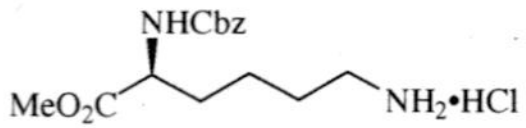 MeO_2C, NHCbz, NH_2•HCl (structure)	Oxone®, acetone, H_2O, $NaHCO_3$, Ar, rt, 30 min	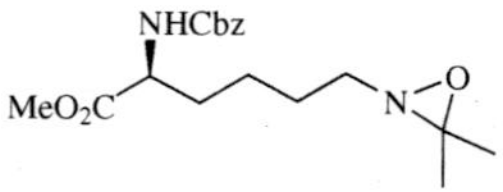 MeO_2C, NHCbz, N–O (structure) (67)	327

[a] This value is the ratio of the products in the crude mixture.

TABLE 3G. SULFUR OXIDATION BY IN SITU GENERATED DIOXIRANES

Substrate	Conditions	Product(s) and Yield(s) (%)	Refs.
C_4 (3-bromothiolane, Br wedge)	Oxone®, acetone, H_2O, 0-5°, 75 min	3-bromothiolane S-oxide (67)	397
C_{7-13} R^1-S-R^2	Oxone®, acetone, bovine serum albumin, $NaHCO_3$, Na_2EDTA, buffer (pH 7.2-7.8), 4°	$R^1S(O)R^2$ **I** + $R^1S(O)R^2$ **II** + $R^1SO_2R^2$ **III**	130, 131

R^1	R^2	Time	I + II	III	I:II
Ph	Me	180 min	(98)	(—)	53.5:46.5
Bz	Me	180 min	(85)	(—)	62:38
Ph	Et	60 min	(51)	(—)	49.5:50.5
4-MeC_6H_4	Me	60 min	(77)	(—)	66:34
4-MeC_6H_4	Et	105 min	(68)	(—)	82:18
Ph	*i*-Pr	120 min	(56)	(5)	10.5:89.5
Ph	*t*-Bu	120 min	(70)	(—)	13.5:86.5
4-MeC_6H_4	*i*-Pr	120 min	(50)	(11)	64.5:35.5
4-MeC_6H_4	*t*-Bu	85 min	(40)	(14)	45.5:54.5
Ph	*c*-C_6H_{11}	25 min	(45)	(5)	24:76
Ph	Bz	180 min	(30)	(14)	84:16

Substrate	Conditions	Product(s) and Yield(s) (%)	Refs.
C_{8-13} R^1-S-R^2	Oxone®, $PhCOCF_3$, bovine serum albumin, $NaHCO_3$, Na_2EDTA, buffer (pH 7.2-7.8), 4°	**I + II + III**	130, 131

R^1	R^2	Time	I + II	III	I:II
4-MeC_6H_4	Me	135 min	(80)	(—)	45.5:54.5
4-MeC_6H_4	Et	180 min	(78)	(5)	52:48
Ph	*i*-Pr	120 min	(96)	(—)	18:82
Ph	*t*-Bu	90 min	(37)	(10)	14.5:85.5
Ph	Bz	150 min	(22)	(3)	86.5:13.5

TABLE 3G. SULFUR OXIDATION BY IN SITU GENERATED DIOXIRANES (*Continued*)

Substrate	Conditions	Product(s) and Yield(s) (%)	Refs.
C_{8-9} 4-$R^1C_6H_4SR^2$	Oxone® (x eq), acetone, buffer (pH 7.5-8.0), $NaHCO_3$, EDTA	4-$R^1C_6H_4S(O)R^2$ (I) + 4-$R^1C_6H_4SO_2R^2$ (II)	398

R^1	R^2	x	Time	Temp	I	II
H	CH_2CO_2H	0.65	5 min	0-2°	(98.6)	(—)
H	CH_2CO_2H	1.35	1-2 h	rt	(—)	(96.7)
CO_2H	Me	0.65	5 min	0-2°	(76.8)	(—)
CO_2H	Me	1.35	1-2 h	rt	(—)	(92.6)
H	$(CH_2)_2OH$	0.65	5 min	0-2°	(81.5)	(—)
H	$(CH_2)_2OH$	1.35	1-2 h	rt	(—)	(92.8)
CH_2OH	Me	0.65	5 min	0-2°	(46.6)	(—)
CH_2OH	Me	1.35	1-2 h	rt	(—)	(83.4)
H	Z-CH=$CHCO_2H$	0.65	5 min	0-2°	(95.0)	(—)
H	Z-CH=$CHCO_2H$	1.35	1-2 h	rt	(—)	(35)

Substrate	Conditions	Product(s) and Yield(s) (%)	Refs.
R^1SR^2	Oxone®, cyclohexanone, bovine serum albumin, $NaHCO_3$, Na_2EDTA, buffer (pH 7.2-7.8), 4°	$R^1S(O)R^2$ (I) + $R^1S(O)R^2$ (II)	130, 131

R^1	R^2	Time	I + II	I:II
4-MeC_6H_4	Me	240 min	(60)	53:47
Ph	*i*-Pr	120 min	(46)	42:58

Substrate	Conditions	Product(s) and Yield(s) (%)	Refs.
C_{8-14} R^1SR^2	Oxone®, TFP, bovine serum albumin, $NaHCO_3$, Na_2EDTA, buffer (pH 7.2-7.8), 4°	I + II + $R^1SO_2R^2$ (III)	130, 131

R^1	R^2	Time	I + II	III	I:II
4-MeC_6H_4	Me	10 min	(78)	(5)	52:48
4-MeC_6H_4	Et	60 min	(66)	(12)	80.5:19.5
Ph	*i*-Pr	5 min	(67)	(—)	5.5:94.5
Ph	*t*-Bu	5 min	(55)	(—)	16.5:83.5
Ph	Bz	5 min	(20)	(21)	86:14
4-MeC_6H_4	Bz	5 min	(95)	(—)	69:31

C_9

t-Bu, *t*-Bu C=S | Oxone®, acetone, benzene, H_2O, 18-crown-6, $KHCO_3$, N_2, rt, 4 h | *t*-Bu, *t*-Bu C=S→O (33) | 146

C_{9-15}

4-$R^1C_6H_4SCH_2COR^2$ | Oxone®, bovine serum albumin, $NaHCO_3$, Na_2EDTA, buffer (pH 7.2-7.8), 4° | 4-$R^1C_6H_4S(O)CH_2COR^2$ **I** + 4-$R^1C_6H_4SO_2CH_2COR^2$ **II** | 130, 131

R^1	R^2	Time	I	% ee	II
H	Me	40 min	(100)	6	(—)
Me	Me	30 min	(100)	72	(—)
H	*i*-Pr	10 min	(84)	35	(—)
Me	*i*-Pr	10 min	(94)	82	(—)
Me	*t*-Bu	30 min	(83)	79	(—)
Me	Ph	20 min	(59)	9	(10)

C_{10}

Adamantanethione | Oxone®, acetone, benzene, H_2O, 18-crown-6, $KHCO_3$, N_2, rt, 4 h | Adamantanone (20) + adamantanethione S-oxide (=SO) (29) | 146

C_{11}

3-TsO-tetrahydrothiophene | Oxone®, acetone, H_2O, 0°, 40 min | 3-TsO-tetrahydrothiophene S-oxide (94) + diastereomeric S-oxide (6) | 122

TABLE 3G. SULFUR OXIDATION BY IN SITU GENERATED DIOXIRANES (*Continued*)

Substrate	Conditions	Product(s) and Yield(s) (%)	Refs.
C_{12} 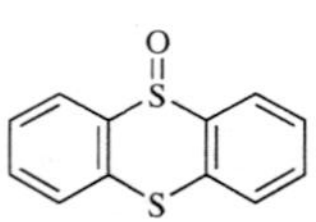	Oxone®, acetone, phosphate buffer (pH 7.5), 18-crown-6, EDTA, 0° to rt, 24 h	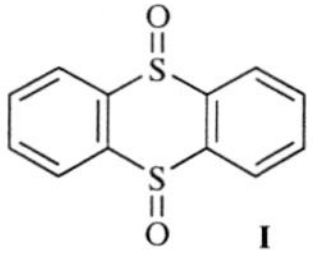 I (4) + 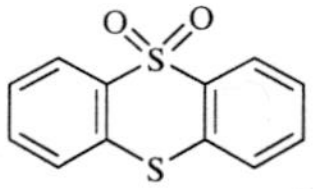 II (16) + 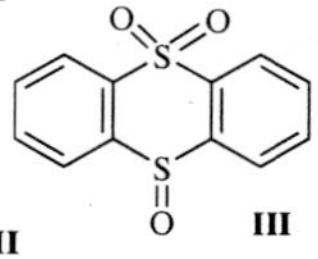 III (6)	399
	Oxone®, carbonyl compound, 18-crown-6, CH_2Cl_2, H_2O, 0°	I + II + III	400
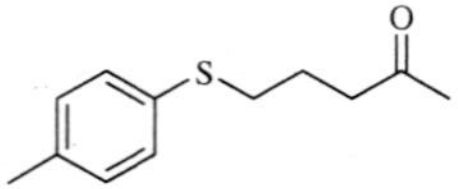	Oxone®, bovine serum albumin, $NaHCO_3$, Na_2EDTA, buffer (pH 7.2-7.8), 4°, 140 min	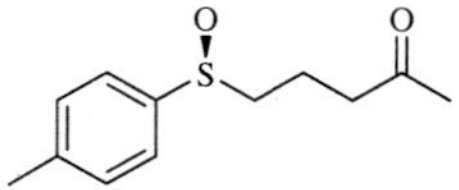 (63) 84% ee	130, 131
C_{13} 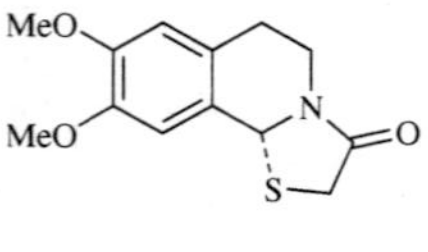	Oxone®, $CHCl_3$, MeOH H_2O, 0°, 30 min	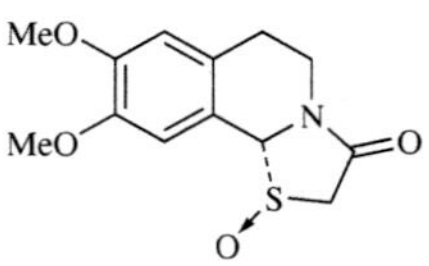 (68)	401

Ketone or aldehyde	I + II + III	I:II:III
acetone	(10.1)	59:15:26
t-BuCOMe	(5.33)	61:31:8
cyclohexanone	(1.88)	58:30:12
t-BuCHO	(5.04)	49:3:48

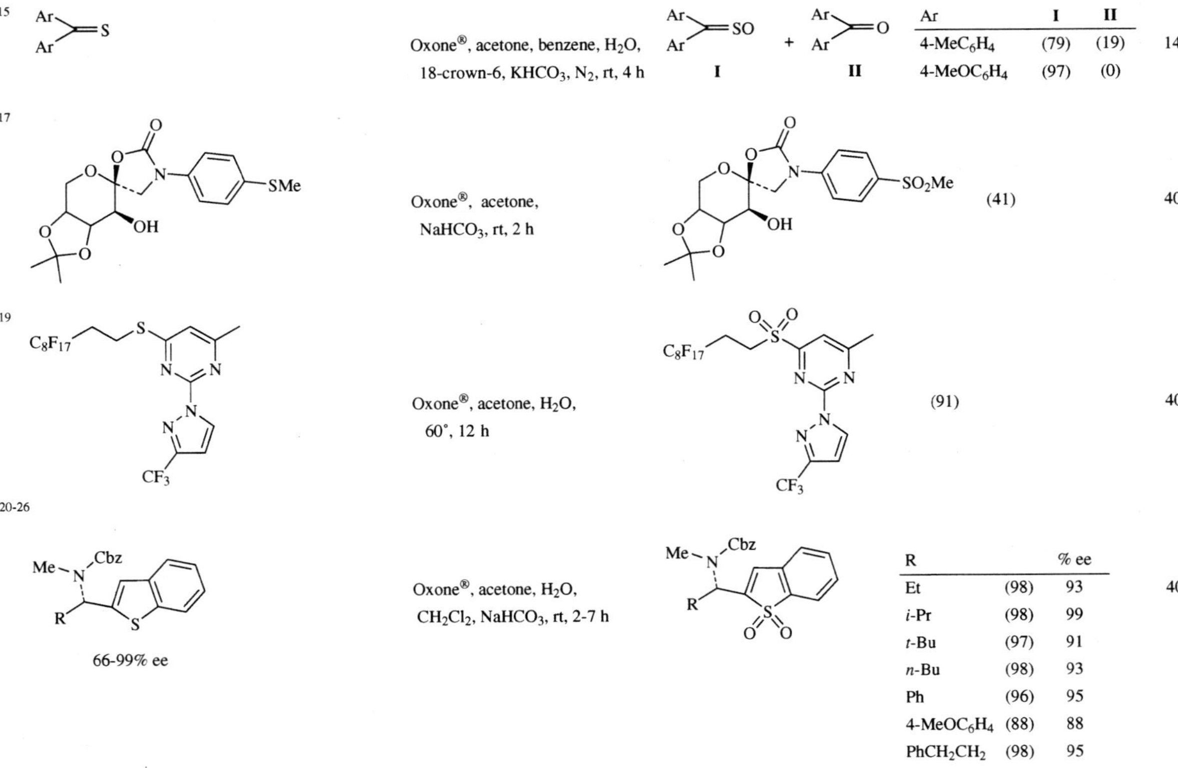

	Substrate	Conditions	Product(s) and Yield(s) (%)	Refs.
C_{15}		Oxone®, acetone, benzene, H_2O, 18-crown-6, $KHCO_3$, N_2, rt, 4 h	I + II	146
C_{17}		Oxone®, acetone, $NaHCO_3$, rt, 2 h	(41)	402
C_{19}		Oxone®, acetone, H_2O, 60°, 12 h	(91)	403
C_{20-26}	66-99% ee	Oxone®, acetone, H_2O, CH_2Cl_2, $NaHCO_3$, rt, 2-7 h		404

Ar	I	II
4-MeC_6H_4	(79)	(19)
4-$MeOC_6H_4$	(97)	(0)

R		% ee
Et	(98)	93
i-Pr	(98)	99
t-Bu	(97)	91
n-Bu	(98)	93
Ph	(96)	95
4-$MeOC_6H_4$	(88)	88
$PhCH_2CH_2$	(98)	95

TABLE 3G. SULFUR OXIDATION BY IN SITU GENERATED DIOXIRANES (*Continued*)

	Substrate	Conditions	Product(s) and Yield(s) (%)	Refs.
C_{27} 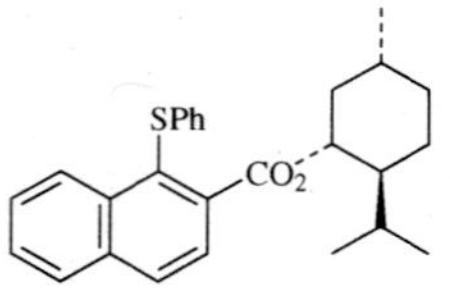	SPh, CO_2	Oxone®, acetone, MeCN, H_2O, $NaHCO_3$, 5°, 20 h	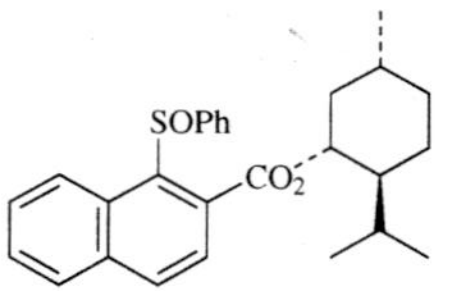 SOPh, CO_2 (51) dr 58:42	405
C_{29} 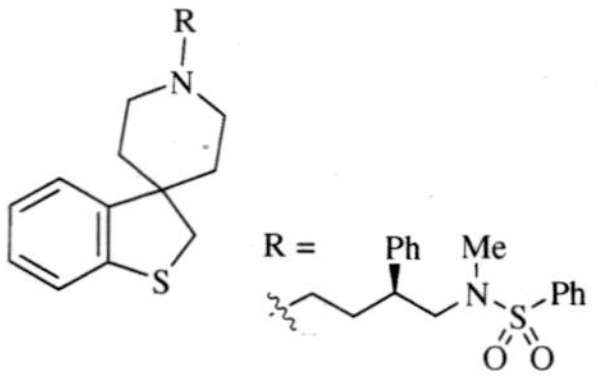	R–N, S; R = Ph, Me, N–S(O)(O)–Ph	Oxone® (1.2 eq), MeOH, –20°, 2-5 min	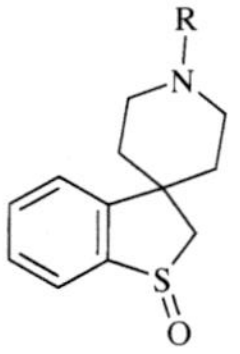 R–N, S=O (—)	406
		Oxone® (3 eq), MeOH, rt	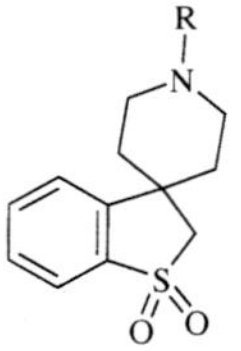 R–N, S(=O)(=O) (—)	406
C_{31} 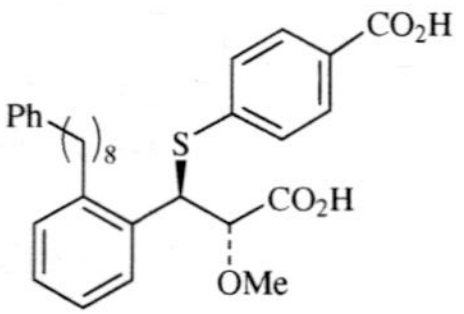	Ph–( )$_8$, S, CO_2H, CO_2H, OMe	Oxone®, acetone, buffer (pH 7.5-8.0), $NaHCO_3$, EDTA, rt	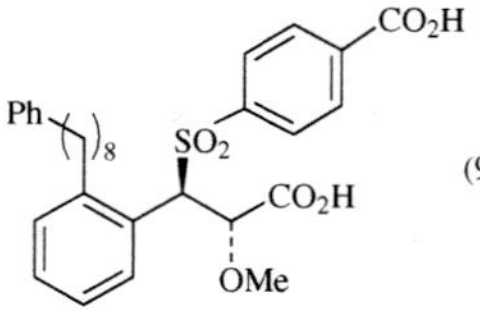 Ph–( )$_8$, SO_2, CO_2H, CO_2H, OMe (96)	398

TABLE 3H. OXIDATION OF OTHER HETEROATOMS BY IN SITU GENERATED DIOXIRANES

Substrate	Conditions	Product(s) and Yield(s) (%)	Refs.
C_0			
Cl^-	Oxone®, ketone, buffer (pH 9.0)	OCl^- (—) Ketone / k_{rel}: — / < 0.1 acetone / 1.0 cyclohexanone / 6.1 *N*,*N*-dimethyl-4-oxopiperidinium nitrate / 1300	161
ONO_2^-	$AcCO_2Me$, phosphate buffer (pH 7.4)	NO_2^- (—) **I** + NO_3^- (—) **II**	163
	AcCOMe, phosphate buffer (pH 7.4)	**I** (—) + **II** (—)	163
	Oxone®, ketone, buffer (pH = 9.0)	SO_4^{2-} (—) + O_2 (—) Ketone / k_{rel}: — / < 0.1 acetone / 1.0 cyclohexanone / 9.4 *N*,*N*-dimethyl-4-oxopiperidinium nitrate / 1400	161
C_{18}			
Ph_3P	Oxone®, acetone, phosphate buffer, KOH, < 0°, 10 min	$Ph_3P = 0$ (—)	121

TABLE 4A. C=Y OXIDATION BY ISOLATED DIOXIRANES

Substrate	Conditions	Product(s) and Yield(s) (%)	Refs.
C_{3-9}			
$RCH_2CH(NO_2)CH_3$	1. *t*-BuOK, THF, 20°, 5 min 2. DMD, H_2O, acetone, 20°, 5 min	$RCH_2C(O)CH_3$ R: OH (—); SPh (25); SO_2Ph (—)	407
C_5			
$EtO_2CC(O)CH{=}N_2$	DMD, acetone, rt, min	$EtO_2CC(O)CH(OH)_2$ (100)	103
$(EtO)_2P(O)CH{=}N_2$	DMD, acetone, rt	$(EtO)_2P(O)CH(OH)_2$ (100)	408
C_{5-9}			
R—N=•=S	1. DMD, acetone, N_2, rt, 15 min 2. Isopropylamine, 0°, 1.5 h	RNHC(O)NH-*i*-Pr R: *n*-Bu (71); Ph (89); Bn (84); $BnCH_2$ (67)	409
$R^1R^2CHNO_2$	1. *t*-BuOK, THF, 20°, 5 min 2. DMD, H_2O, acetone, 20°, 5 min	$R^1R^2C{=}O$ R^1 / R^2: H / $(CH_2)_2CO_2Me$ (73) Et / $(CH_2)_2CN$ (86) Et / $(CH_2)_2COMe$ (99) Et / $(CH_2)_2CO_2Me$ (90) $(CH_2)_2CO_2Me$ / $(CH_2)_2CO_2Me$ (83)	107
C_{5-23}			
$R^1C(O)C(=N_2)C(O)R^2$	DMD, acetone, rt	$R^1C(O)C(O)C(O)R^2$ **I** or $R^1C(O)C(OH)_2C(O)R^2$ **II**	105

R_1	R^2	Time	Product	
Me	Me	16 h	**II**	(95)
Me	EtO	10 h	**I**	(100)
—$(CH_2)_3$—		26 h	**I**	(93)
EtO	EtO	29 h	**II**	(100)
—$CH_2C(Me)_2CH_2$—		20 h	**II**	(89)
—(1,2-C_6H_4)—		20 h	**II**	(100)
Me	Ph	24 h	**II**	(98)
OEt	Ph	35 h	**I**	(94)
Ph	Ph	30 h	**II**	(100)
(–)-menthO	(–)-menthO	10 h	**II**	(100)

(–)-menthO =

C_6

DMD, acetone, rt, 2 h — **I** + **II** (—), **I**:**II** = 35:65 — 106

DMD, acetone, rt, min — Y: S (88); O (98) — 103

1. BuLi, THF, Ar, –70°, 5 min
2. DMD, acetone, H_2O, THF, Ar, –70°, 6 min

(15) + (15) + (11) + (5) — 407

DMD, H_2O, –70°, 15 min — (28) + (29) — 410

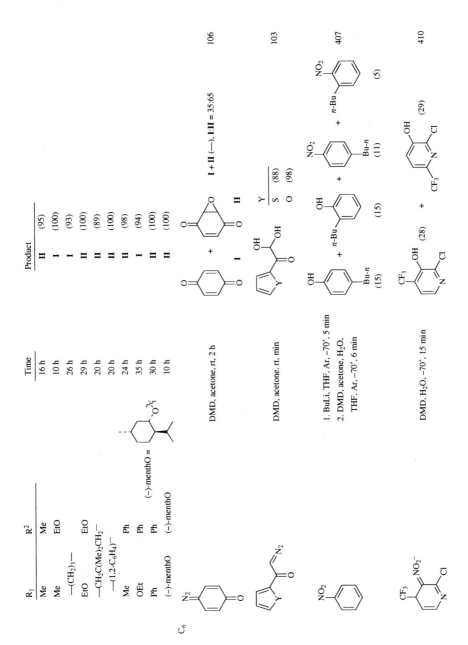

TABLE 4A. C=Y OXIDATION BY ISOLATED DIOXIRANES (*Continued*)

Substrate	Conditions	Product(s) and Yield(s) (%)	Refs.
C_7			
N_2	DMD, acetone, rt, min	OH, OH (100)	103
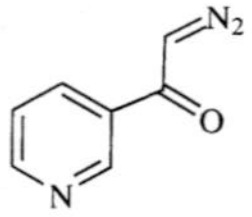 N_2	DMD, acetone, rt, min	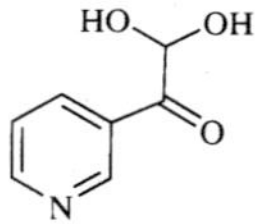 HO, OH (100)	103
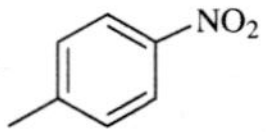 NO_2	1. *n*-BuLi, THF, Ar, –70°, 5 min 2. DMD, acetone, H_2O, –70°, 5 min	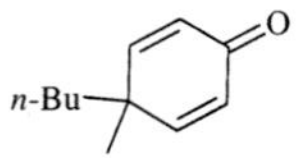 *n*-Bu (16) + Bu-*n*, OH (7)	111
 NOH	DMD, acetone, rt, 1 h	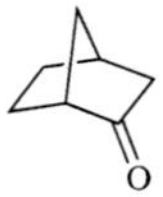 (85)	411
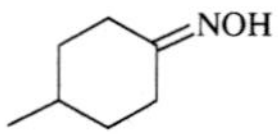 NOH	DMD, acetone, 0°, 30 min	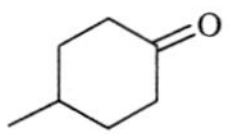 (80)	411
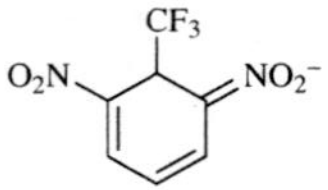 O_2N, CF_3, NO_2^-	DMD, H_2O, –70°, 15 min	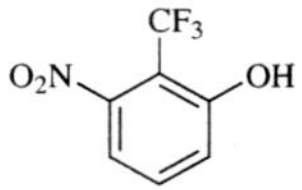 O_2N, CF_3, OH (47) + O_2N, OH, CF_3 (18) + O_2N, OH, CF_3 (13)	410
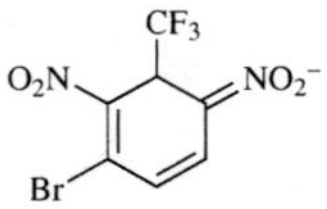 O_2N, CF_3, NO_2^-, Br	DMD, H_2O, –70°, 15 min	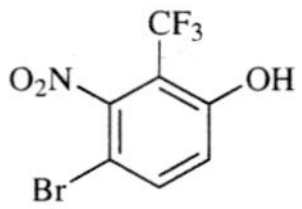 O_2N, CF_3, OH, Br (29) + O_2N, CF_3, OH, Br (6) + O_2N, OH, CF_3, Br (11)	410

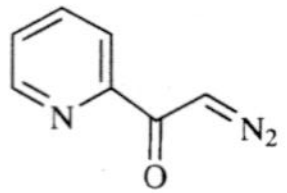

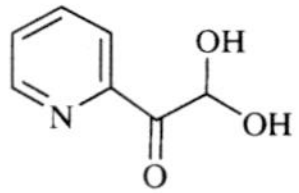

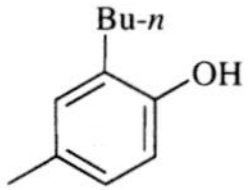

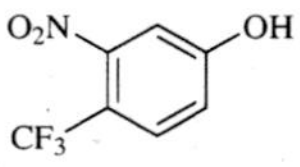

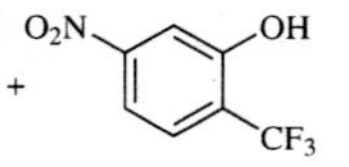

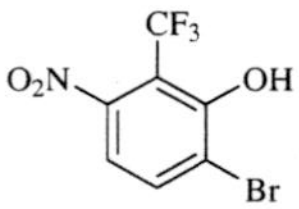

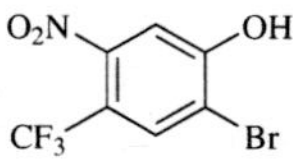

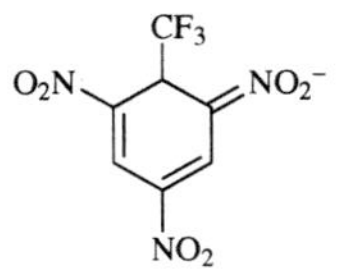

C_{7-23}

DMD, H_2O, –70°, 15 min 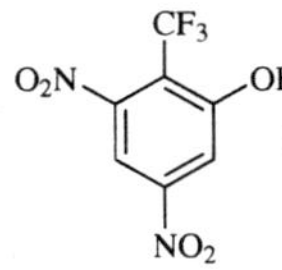 — (22) + 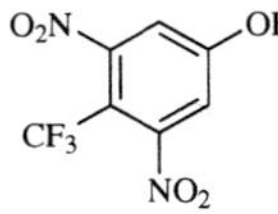 (31) — 410

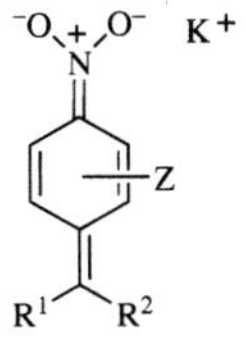

DMD, acetone, THF, Ar, 20°, 5 min 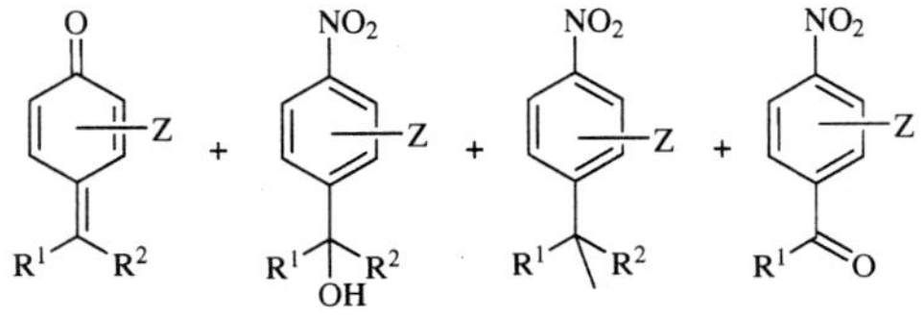 — I + II + III + IV — 413

R^1	R^2	Z	I	II	III	IV
Me	Me	H	(—)	(85)	(—)	(—)
H	SO_2Ph	H	(—)	(—)	(—)	(60)
Me	SO_2Ph	H	(—)	(—)	(—)	(33)
Ph	CO_2Me	H	(—)	(63)	(—)	(—)
Ph	Ph	2-Cl	(58)	(30)	(—)	(—)
Ph	Ph	3-Cl	(61)	(10)	(—)	(—)
Ph	Ph	2-I	(33)	(38)	(—)	(—)
Ph	Ph	H	(51)	(28)	(18)	(—)
—(9-fluorenyl)—		H	(—)	(99)	(—)	(—)
Ph	Ph	3-CN	(64)	(—)	(—)	(—)
Ph	Ph	2-MeO	(58)	(36)	(—)	(—)
Ph	Ph	3-MeO	(44)	(13)	(30)	(—)
Ph	4-ClC_6H_4	3-MeO	(40)	(26)	(33)	(—)
Ph	1-Naph	H	(75)	(20)	(—)	(—)
Ph	Ph	$(CH)_4$	(91)	(—)	(—)	(—)

TABLE 4A. C=Y OXIDATION BY ISOLATED DIOXIRANES (*Continued*)

Substrate	Conditions	Product(s) and Yield(s) (%)	Refs.
$C_{7\text{-}17}$			
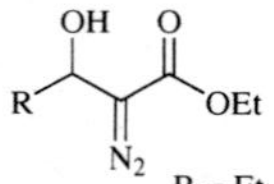 OH O R OEt N_2 R = Et, *i*-Bu, *c*-C_5H_9, *n*-C_6H_{13}, Ph, 3-BrC_6H_4, 4-FC_6H_4, 2-$MeOC_6H_4$, 4-$MeOC_6H_4$, 3-$CF_3C_6H_4$, 4-PhC_6H_4	DMD, acetone, –35°, 15-30 min	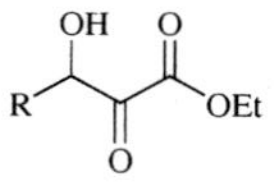 OH O R OEt O (—)	412
C_8			
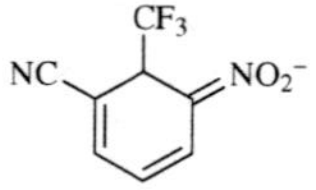 CF_3 NC NO_2^-	DMD, H_2O, –70°, 15 min	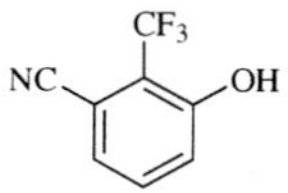 CF_3 NC OH (22) 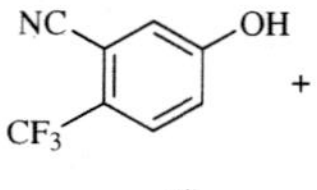 + NC OH CF_3 (4) 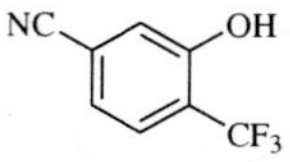 + NC OH CF_3 (23)	410
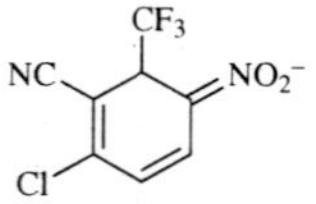 CF_3 NC NO_2^- Cl	DMD, H_2O, –70°, 15 min	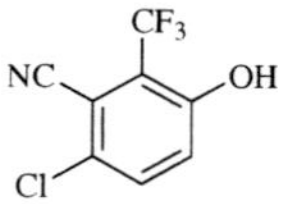 CF_3 NC OH Cl (47) 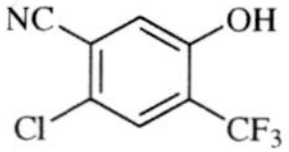 + NC OH Cl CF_3 (16)	410
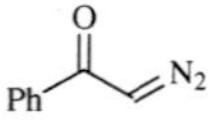 O Ph N_2	DMD, acetone, rt, min	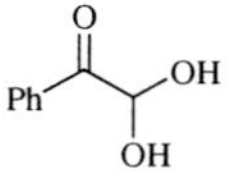 O Ph OH OH (100)	103
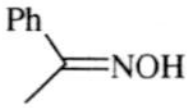 Ph NOH	DMD, acetone, rt, 24 h	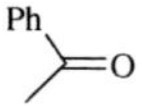 Ph O (95)	411
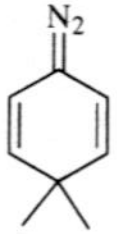 N_2	DMD, acetone, CH_2Cl_2, 0°, 1 min	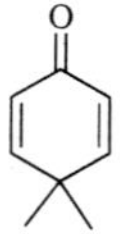 O (95) 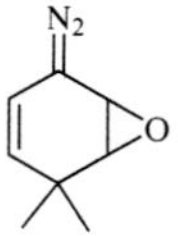 + N_2 O (5)	106
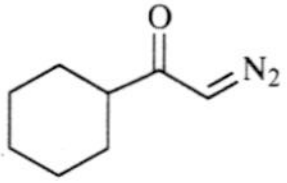 O N_2	DMD, acetone, rt, min	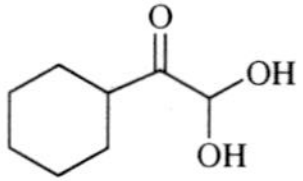 O OH OH (96)	103

Substrate	Conditions	Product(s) and Yield(s) (%)	Refs.
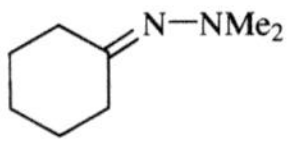	DMD, acetone, CH_2Cl_2, 20°, 3 min	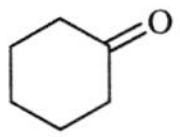 (89) 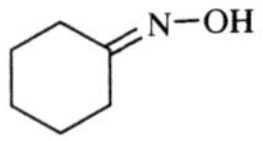 + $N-OH$ (10)	414
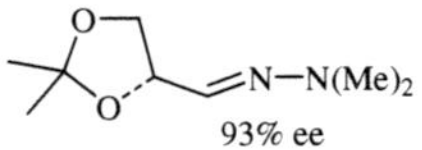 $N-N(Me)_2$ 93% ee	DMD, acetone, 0°, 2-3 min	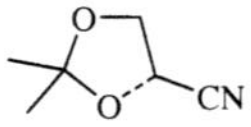 CN (92) 93% ee	113
C_9			
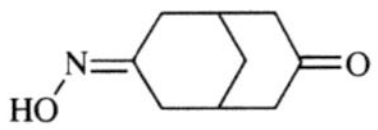 HO N	1. DMD (3 eq), acetone, CH_2Cl_2, –25° 2. rt, 12-15 h	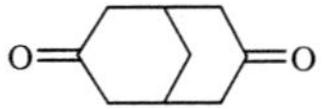 I (100) 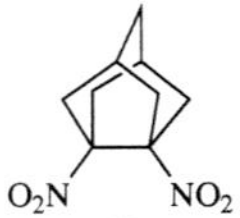 O_2N NO_2 II (0)	315
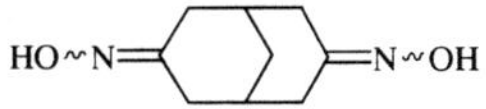 HO~N N~OH	1. DMD (6 eq), acetone, CH_2Cl_2, –25° 2. rt, 12-15 h	**I** + **II** (100), **I**:**II** = 10:90	315
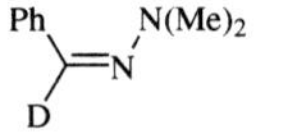 Ph D $N(Me)_2$	DMD, acetone, 0°, 2-3 min	Ph—≡N (97)	113
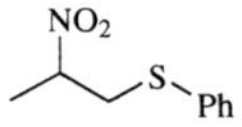 NO_2 S Ph	1. *t*-BuOK, THF, rt, 5 min 2. DMD, acetone, H_2O, –70°, 7 min	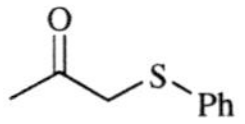 O S Ph (31)	407
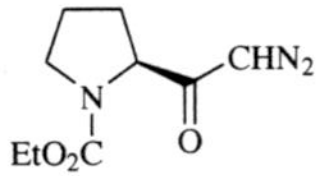 CHN_2 EtO_2C	DMD, acetone	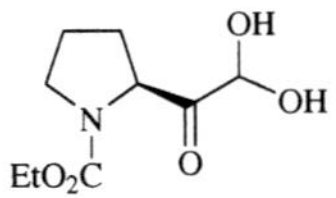 OH OH EtO_2C (—)	104
C_{9-10}			
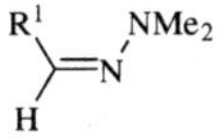 R^1 H NMe_2	DMD, acetone, 0°, 2-3 min	R^1—≡N; R^1: 4-ClC_6H_4 (98), 3-$O_2NC_6H_4$ (97), Ph (97), 4-MeC_6H_4 (98)	113
C_{9-14}			
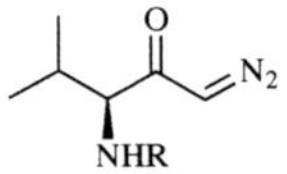 N_2 NHR	DMD, acetone	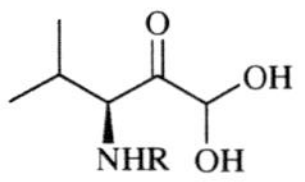 NHR OH OH (—); R: CO_2Et, phthaloyl, CO_2Bn	104

TABLE 4A. C=Y OXIDATION BY ISOLATED DIOXIRANES (*Continued*)

Substrate	Conditions	Product(s) and Yield(s) (%)	Refs.
C_{10} 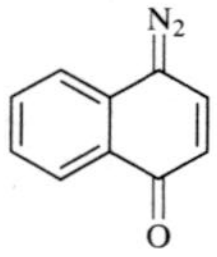 N_2 O	DMD, acetone, rt, 2 h	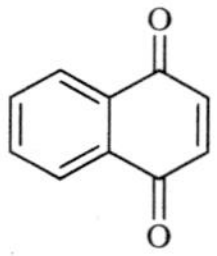 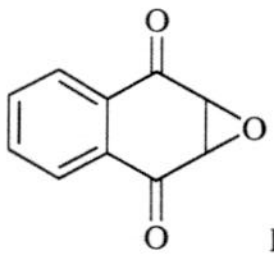 O O I + O O O II I + II (—), I:II = 70:30	106
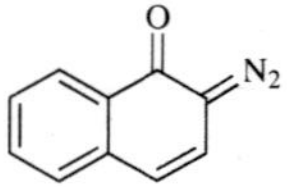 O N_2	DMD, acetone, rt, 2 h	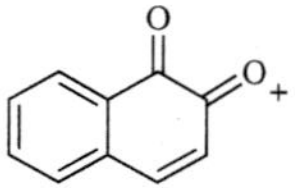 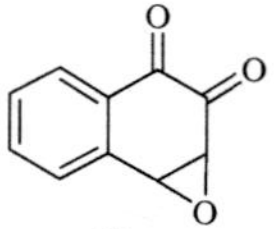 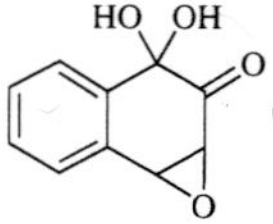 O O I + O O O II + HO OH O O III (—) I + II + III (—), I:(II+III) = 30:70	106
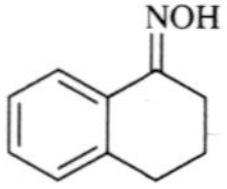 NOH	DMD, acetone, rt, 24 h	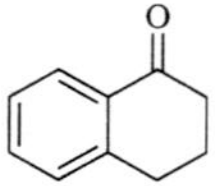 O (97)	411
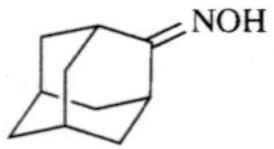 NOH	DMD, acetone, 0°, 5 min	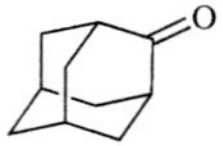 O (100)	411
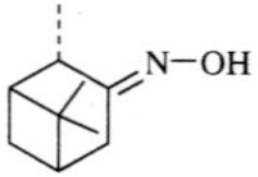 N—OH	DMD, acetone, rt, 0.5 h	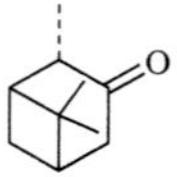 O (31)	22
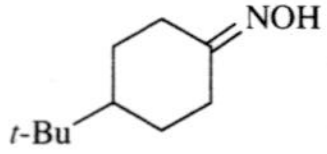 NOH *t*-Bu	DMD, acetone, 0°, 30 min	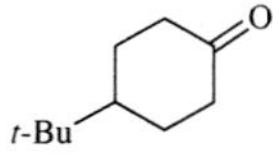 O *t*-Bu (98)	411
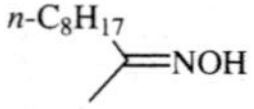 *n*-C_8H_{17} =NOH	DMD, acetone, 0°, 10 min	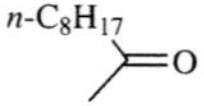 *n*-C_8H_{17} =O (94)	411

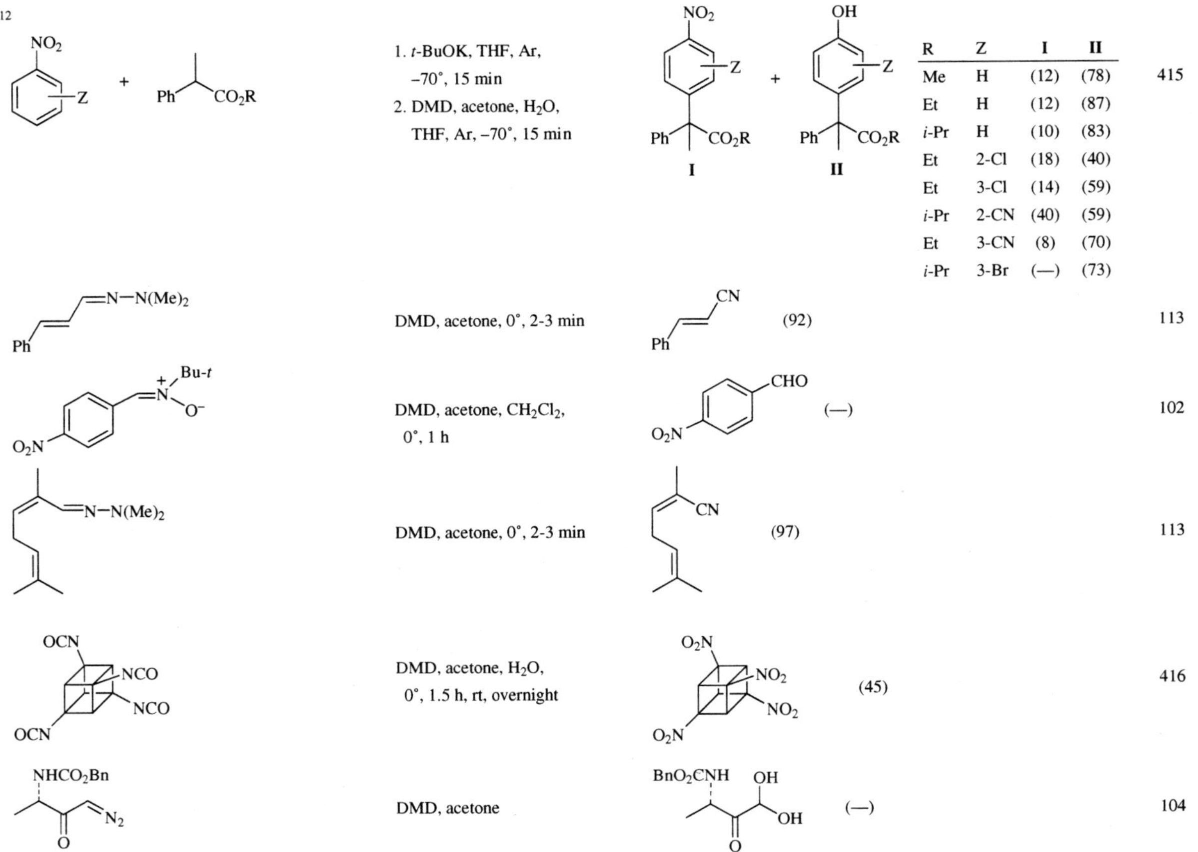

R	Z	I	II
Me	H	(12)	(78)
Et	H	(12)	(87)
i-Pr	H	(10)	(83)
Et	2-Cl	(18)	(40)
Et	3-Cl	(14)	(59)
i-Pr	2-CN	(40)	(59)
Et	3-CN	(8)	(70)
i-Pr	3-Br	(—)	(73)

TABLE 4A. C=Y OXIDATION BY ISOLATED DIOXIRANES (*Continued*)

Substrate	Conditions	Product(s) and Yield(s) (%)	Refs.
C_{12} N_2, O	DMD, acetone, rt	OH, OH, O (100)	103
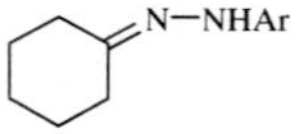 N—NHAr	DMD, acetone (or TFD, TFP), CH_2Cl_2	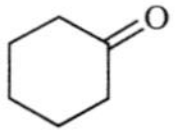 O Ar / Dioxirane / Temp / Time 2,4-$(O_2N)_2C_6H_3$ DMD 2° 6 h (92) 2,4-$(O_2N)_2C_6H_3$ TFD 0° 90 min (95) 4-$O_2NC_6H_4$ DMD 20° 3 h (91) 4-$O_2NC_6H_4$ TFD 0° 30 min (92) Ph DMD 20° 30 min (94)	414
O, N-CO_2Bu-*t*, N_2, O	DMD, acetone	O, N-CO_2Bu-*t*, OH, OH, O (—)	104
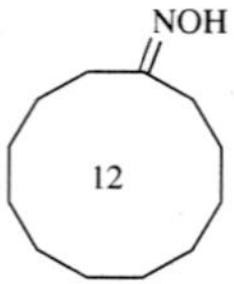 NOH, 12	DMD, acetone, rt, 1 h	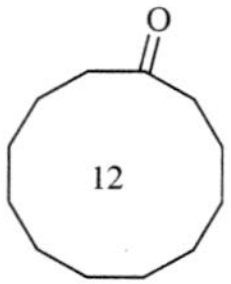 O, 12 (100)	411
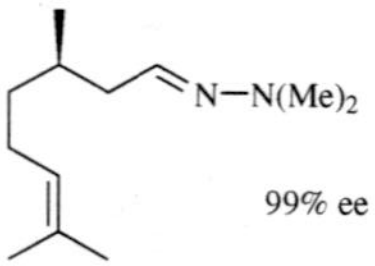 N—N(Me)$_2$ 99% ee	DMD, acetone, 0°, 2-3 min	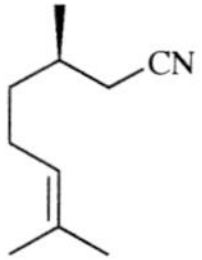 CN (92) 98% ee	113
C_{12-15} O, N_2, NHR	DMD, acetone	O, OH, NHR, OH (—) R: CO_2Bu-*t*; CO_2Bn	104

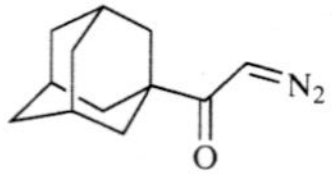

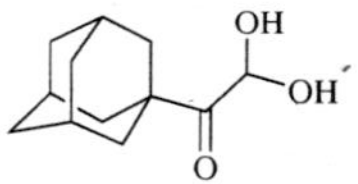

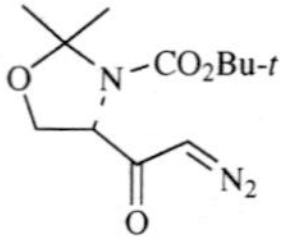

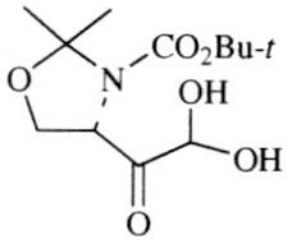

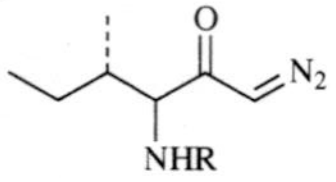

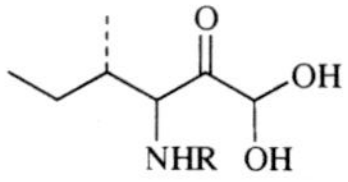

Substrate	Conditions	Product(s) and Yield(s) (%)	Refs.
C_{13}	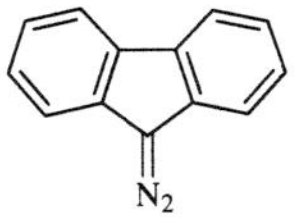 DMD, acetone, rt	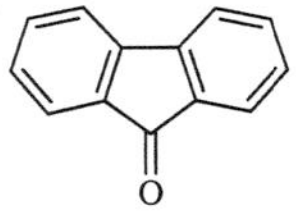 (100)	106
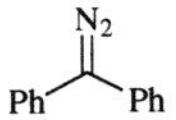	DMD, acetone, rt	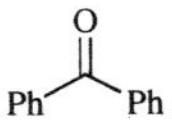 (100)	106
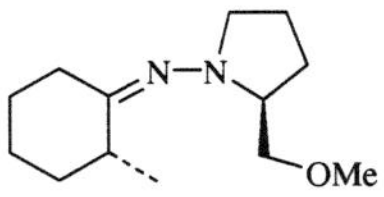	DMD, acetone, CH_2Cl_2, 0°, 5 min	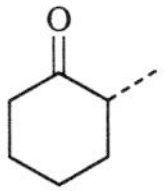 (85) 92% ee 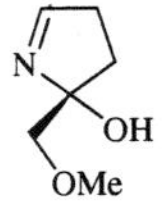 + (30)	414
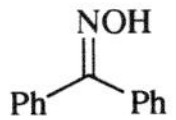	1. *t*-BuOK, THF, 20°, 5 min 2. DMD, acetone, 20°, 5 min	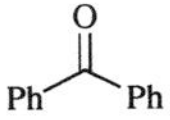 **I** (76)	111
	DMD, acetone, rt, 48 h	**I** (97)	411
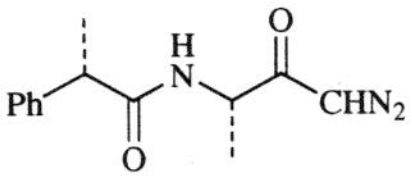	DMD, acetone	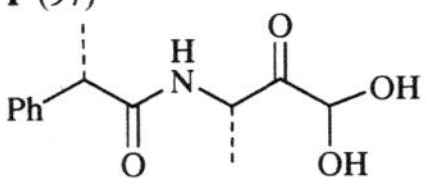 (—)	104
C_{13-15}	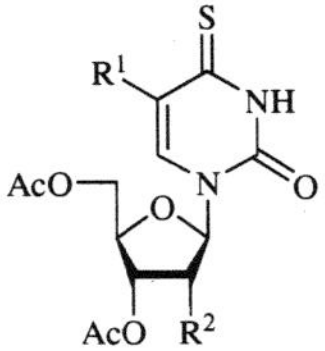 DMD, acetone, CH_2Cl_2, R^3NH_2, rt	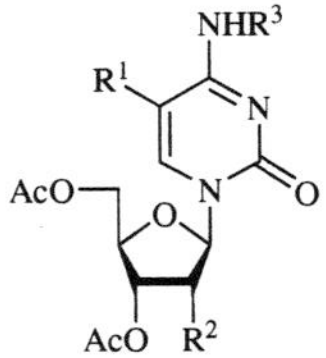 see table below	144

R^1	R^2	R^3	
H	H	H	(81)
H	H	Me	(58)
H	H	Et	(78)
H	H	4-MeC_6H_4	(79)
Me	H	H	(77)
Me	H	Me	(73)
Me	H	Et	(61)
Me	H	4-MeC_6H_4	(68)
H	OAc	H	(90)
H	OAc	Me	(72)
H	OAc	Et	(72)
H	OAc	4-MeC_6H_4	(76)
H	OAc	2,6-$(Me)_2C_6H_3$	(81)

TABLE 4A. C=Y OXIDATION BY ISOLATED DIOXIRANES (*Continued*)

Substrate	Conditions	Product(s) and Yield(s) (%)	Refs.
C_{14} 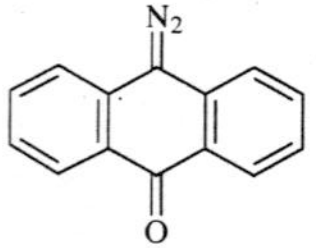 N_2 O	DMD, acetone, rt	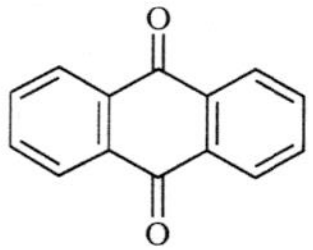 O O (100)	106
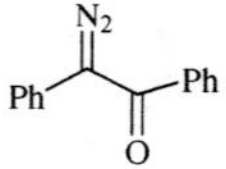 N_2 Ph Ph O	DMD, acetone, rt	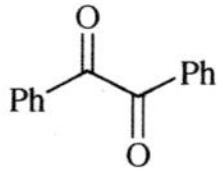 O Ph Ph O (100)	106
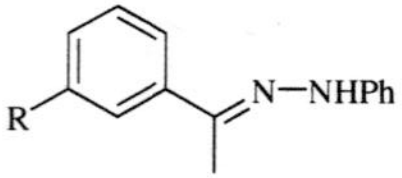 R N—NHPh	DMD, acetone, CH_2Cl_2, 20°	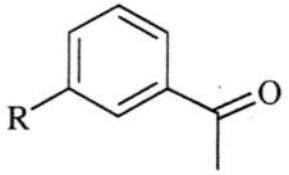 R O R / Time NO_2 / 2 h / (82) H / 30 min / (92)	414
C_{15} 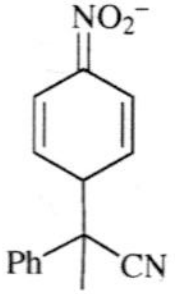 NO_2^- Ph CN	DMD, H_2O	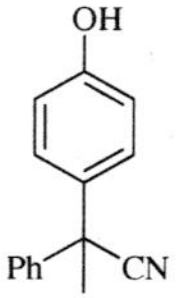 OH Ph CN (81)	417
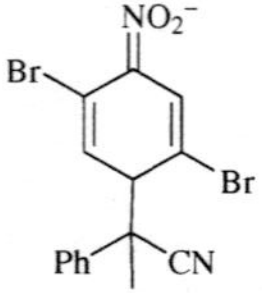 NO_2^- Br Br Ph CN	DMD, H_2O, DMF, THF, acetone, Ar, –70°, 5 min	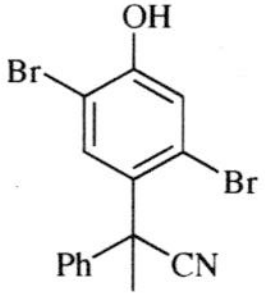 OH Br Br Ph CN (72)	110
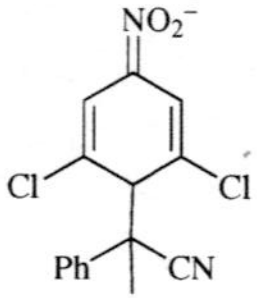 NO_2^- Cl Cl Ph CN	DMD, H_2O, DMF, THF, acetone, Ar, –70°, 5 min	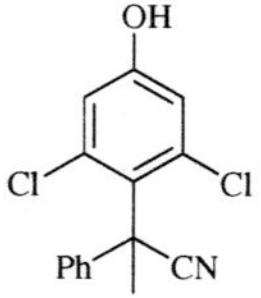 OH Cl Cl Ph CN (69-77)	110

Reactant	Conditions	Product(s) and Yield(s) (%)	Refs.
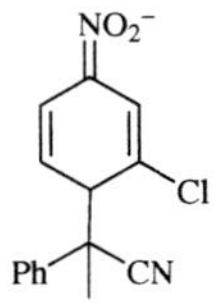 NO_2^- / Cl / Ph, CN	DMD, H_2O, DMF, THF, acetone, Ar, –70°, 5 min	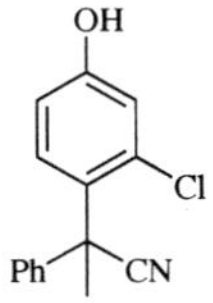 OH / Cl / Ph, CN (87) 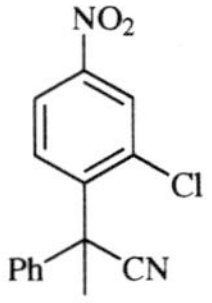 + NO_2 / Cl / Ph, CN (3)	110
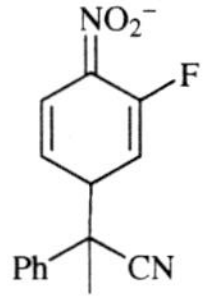 NO_2^- / F / Ph, CN	DMD, H_2O, DMF, THF, acetone, Ar, –70°, 5 min	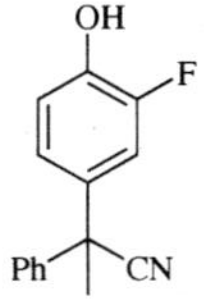 OH / F / Ph, CN (81) 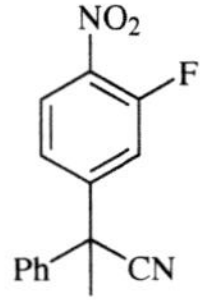 + NO_2 / F / Ph, CN (7)	110
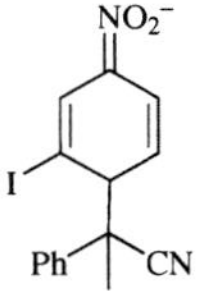 NO_2^- / I / Ph, CN	DMD, acetone, THF, Ar, –70°, 5 min	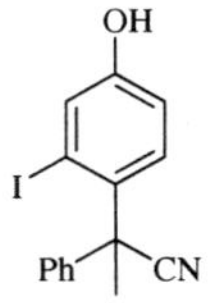 OH / I / Ph, CN **I** (25)	407
	DMD, acetone, H_2O, THF, Ar, –70°, 5 min	**I** (—)	407
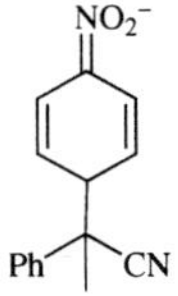 NO_2^- / Ph, CN	DMD, H_2O (x eq), DMF, THF, acetone, Ar, –70°, 5 min	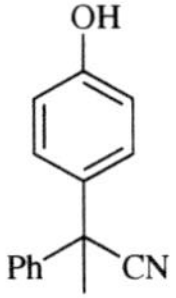 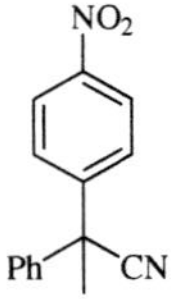 OH / Ph, CN **I** + NO_2 / Ph, CN **II**	110

x	I	II
0	(47)	(6)
0.5	(71)	(5)
1.0	(77-83)	(4-6)

TABLE 4A. C=Y OXIDATION BY ISOLATED DIOXIRANES (*Continued*)

Substrate	Conditions	Product(s) and Yield(s) (%)	Refs.
C_{15} 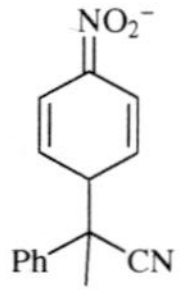 NO_2^- Ph CN	Dioxirane, THF, Ar, –70°, 5 min	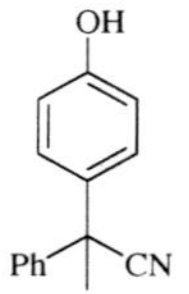 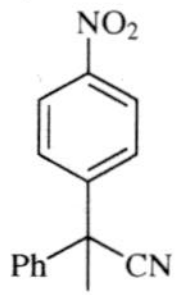 OH Ph CN **I** + NO_2 Ph CN **II**	111
	Dioxirane / Solvent: DMD acetone —; DMD acetone H_2O; DMD acetone MeOH; TFD TFP —; TFD TFP H_2O	**I** / **II**: (trace) (trace); (77-83) (4-6); (33-63) (4-5); (38) (10); (32) (9)	
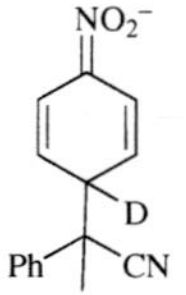 + NO_2^- D Ph CN	DMD, acetone, THF, Ar, –70°, 5 min	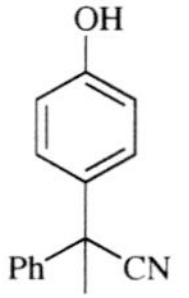 OH Ph CN (—) $K_H/K_D = 1.01 \pm 0.01$	111
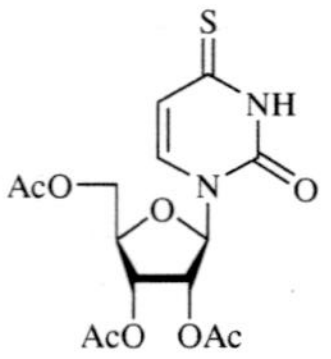 S NH O N AcO O AcO OAc	DMD, acetone, CH_2Cl_2, additive, rt	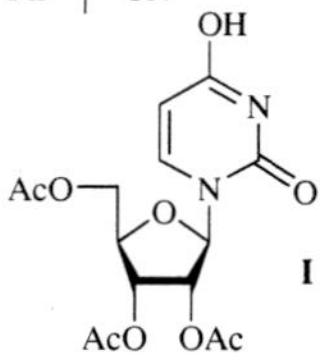 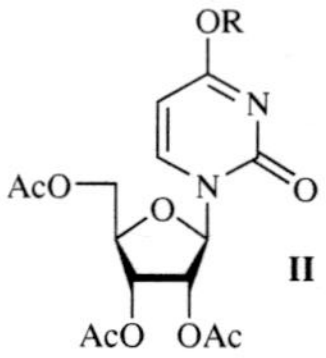 OH N O N AcO O AcO OAc **I** + OR N O N AcO O AcO OAc **II**	144

Additive	R	I	II
—	H	**I** + **II** (95)	**I**:**II** = 1:1
MeOH	Me	(22)	(70)
EtOH	Et	(18)	(65)
n-PrOH	*n*-Pr	(15)	(70)
n-BuOH	*n*-Bu	(20)	(75)

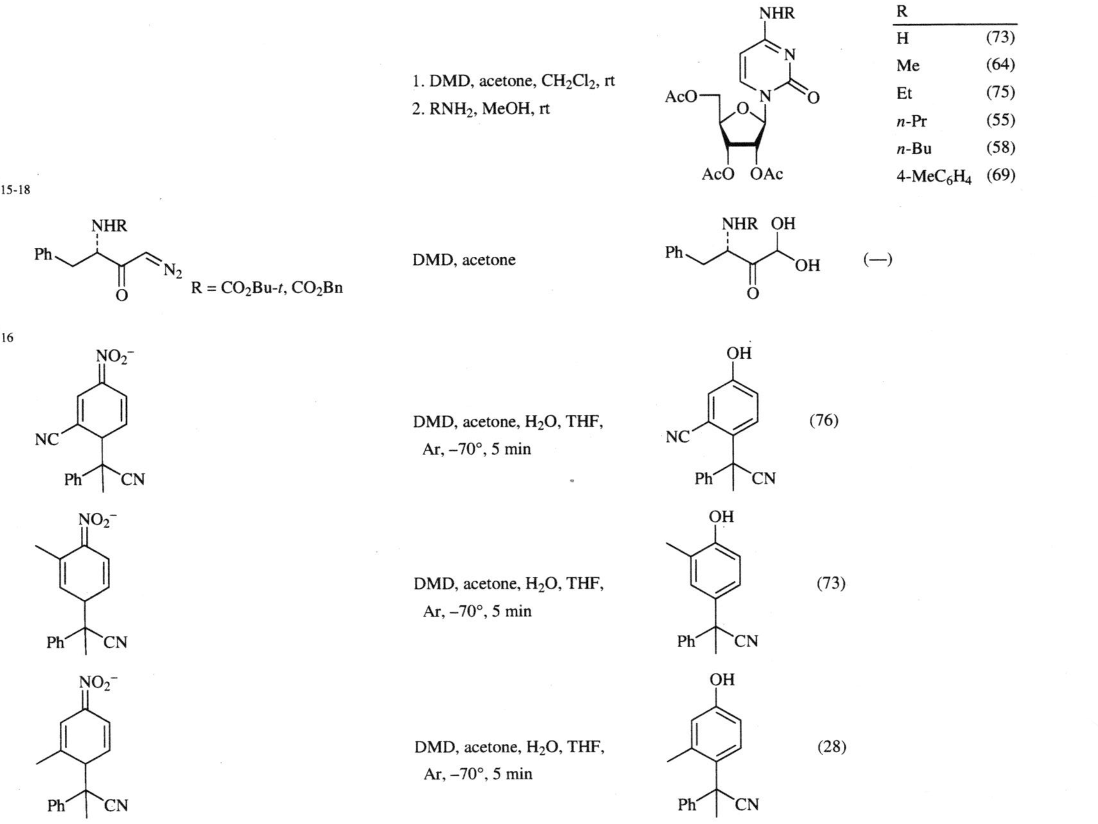

Substrate	Conditions	Product(s) and Yield(s) (%)	Refs.
C_{15-18}			
$R = CO_2Bu$-*t*, CO_2Bn	DMD, acetone	(—)	104
	1. DMD, acetone, CH_2Cl_2, rt 2. RNH_2, MeOH, rt	R: H (73); Me (64); Et (75); *n*-Pr (55); *n*-Bu (58); 4-MeC_6H_4 (69)	144
C_{16}			
	DMD, acetone, H_2O, THF, Ar, –70°, 5 min	(76)	407
	DMD, acetone, H_2O, THF, Ar, –70°, 5 min	(73)	407
	DMD, acetone, H_2O, THF, Ar, –70°, 5 min	(28)	407

TABLE 4A. C=Y OXIDATION BY ISOLATED DIOXIRANES (*Continued*)

Substrate	Conditions	Product(s) and Yield(s) (%)	Refs.
C_{17}			
BnO_2CHN, t-BuO_2C, O, N_2	DMD, acetone	BnO_2CHN, t-BuO_2C, O, OH, OH (—)	104
NHTs, O, P—OMe, OMe, Ph, N_2	DMD, acetone, rt, 2 h	NHTs, O, P—OMe, OMe, Ph, O (97)	418
C_{18}			
NHAc, H, N, S, AcO, O, AcO, OAc	DMD, acetone, CH_2Cl_2, $MeNH_2$, rt	NHAc, MeNH, N, AcO, O, AcO, OAc (83)	116, 117, 118, 119, 120, 144
C_{19}			
NO_2^-, Ph, CN	DMD, acetone, H_2O, THF, Ar, –70°, 5 min	OH, Ph, CN (65)	407
$NO_2^-M^+$, Ph, Ph	DMD, acetone, solvent, Ar, 20°, 5 min	**I** (O, Ph, Ph) + **II** (NO_2, Ph, Ph, OH) + **III** (NO_2, Ph, Ph)	27

M^+	Solvent	% Conv.	**I:II:III**
Li^+	THF	83	17:63:16
Na^+	THF	88	33:47:16
K^+	THF	95	48:26:17
$ClMg^+$	THF	96	0:80:3
$t\text{-}Bu_4N^+$	THF	97	0:95:0
K^+	DMF	93	29:60:8
K^+	THF/DMF (3:1)	91	37:45:14
K^+	toluene	90	51:33:15

$C_{21\text{-}22}$

Me–N⁺(OTAS)= ... Z, CO_2Me

DMD, acetone, CH_2Cl_2, TsOH, rt, 10 min

OH, Z, CO_2Me **I** + NO_2, Z, CO_2Me **II**

Z	**I**	**II**
H	(82)	(12)
2-Cl	(74)	(8)
3-Cl	(64)	(15)
3-CN	(69)	(—)

417

C_{22}

PhO_2S, NO_2, $OCOCH_2CCl_3$, $OCOCH_2CCl_3$, $OCOCH_2CCl_3$

1. *N,N,N',N'*-Tetramethylguanidine, CH_2Cl_2, 0°
2. DMD, acetone

O, OH, $OCOCH_2CCl_3$, $OCOCH_2CCl_3$, $OCOCH_2CCl_3$ (78) 419

C_{23}

N–CO_2Bn, O, H N, O, N_2, Ph

DMD, acetone

N–CO_2Bn, O, H N, O, OH, OH, Ph (—) 104

TABLE 4A. C=Y OXIDATION BY ISOLATED DIOXIRANES (*Continued*)

Substrate	Conditions	Product(s) and Yield(s) (%)	Refs.
C_{24}			
O_2N, H, OMOM, H, OCH_2PMP	1. *t*-BuOK, THF, H_2O 2. DMD, acetone	O, H, OMOM, H, OCH_2PMP (72)	109
C_{26}			
OH, O, CN, PPh_3	DMD, MeOH, acetone, rt	O, O, OH, OMe, O (32) + O, O, OH (37)	420
O, CN, PPh_3	DMD, MeOH, acetone, rt	O, OMe, O (85)	420
O, CN, PPh_3	DMD, MeOH, acetone, rt	O, OMe, O (56)	420
$C_{26\text{-}28}$			
R^1, O, O, OR^2, PPh_3	DMD, acetone, CH_2Cl_2, rt, 1 h	R^1, O, O, OR^2, HO, OH; R^1 / R^2: Me / *t*-Bu (100); $Cl(CH_2)_2$ / *t*-Bu (> 82); Ph / Me (100)	159
C_{27}			
NNHPh, AcO	DMD, acetone, CH_2Cl_2, 20°, 15 min	O, AcO (96)	414

	CN, PPh_3	DMD, MeOH, acetone, rt	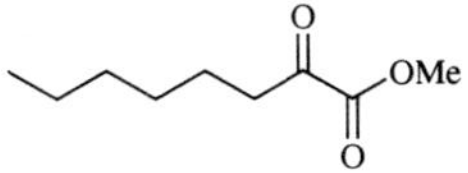 OMe (82)	420
		DMD, $BnNH_2$, CH_2Cl_2, acetone, –78°	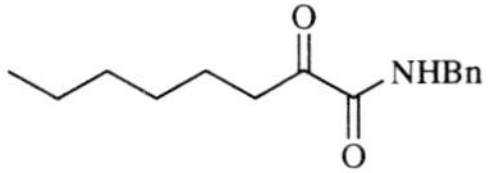 NHBn (81)	420
		DMD, H_2N–CH(Bu-*i*)–CO_2Me, CH_2Cl_2, acetone, –78°	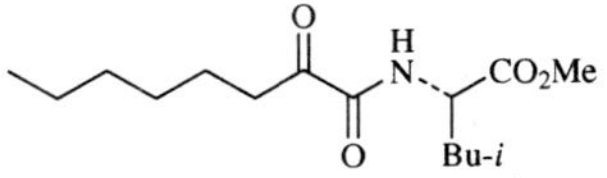 H, N, CO_2Me, Bu-*i* (72)	420
C_{28} 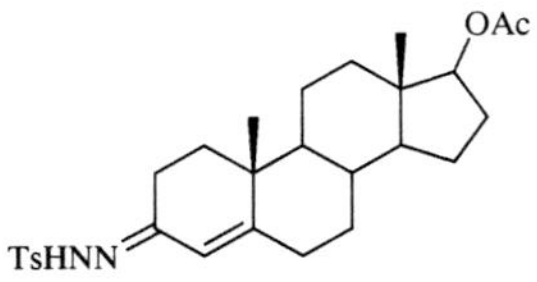	TsHNN, OAc	DMD, acetone, CH_2Cl_2, 20°, 1 h	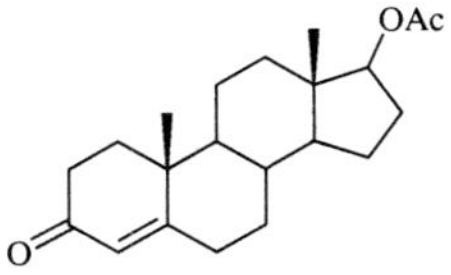 OAc (94)	414
C_{28-32} 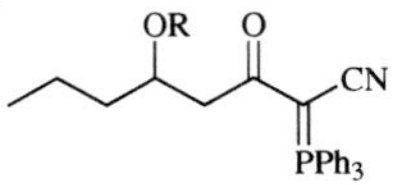	OR, O, CN, PPh_3	DMD, MeOH, acetone, rt	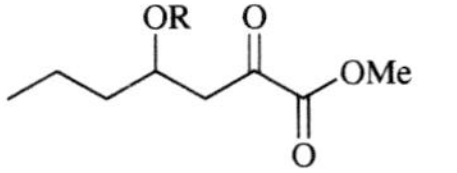 OR, O, OMe; R: Ac (82), *t*-Bu (89), TBS (88)	420
C_{30} 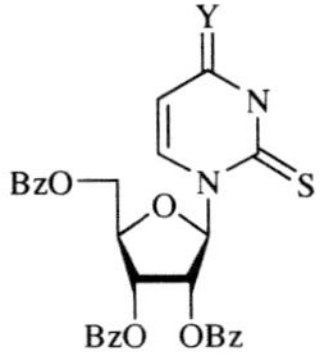	BzO, Y, N, S, BzO, OBz	DMD, acetone, CH_2Cl_2, additive, rt	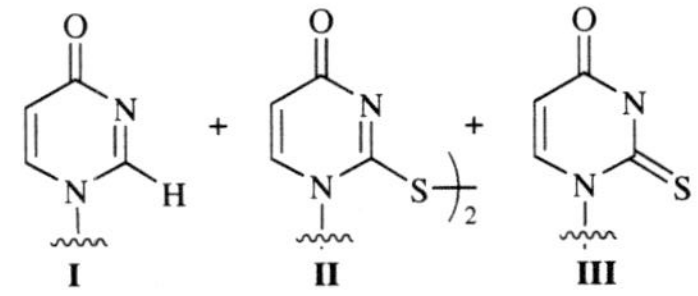 **I** + **II** + **III**	144

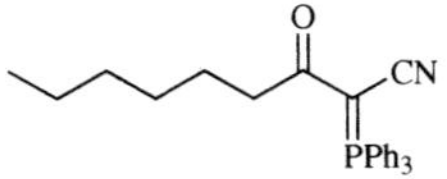

Y	Additive	I	II	III
O	—	(43)	(20)	(—)
O	ROH	(37)	(26)	(—)
O	RNH_2	(24)	(15)	(—)
S	—	(38)	(22)	(8)

TABLE 4A. C=Y OXIDATION BY ISOLATED DIOXIRANES (*Continued*)

Substrate	Conditions	Product(s) and Yield(s) (%)	Refs.
C_{30}			
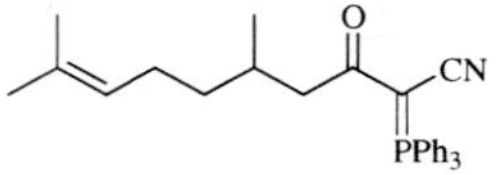 O, CN, PPh_3	DMD (2.4 eq), MeOH, acetone, 0°	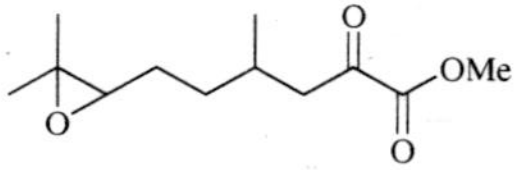 O, OMe, O, O (79)	420
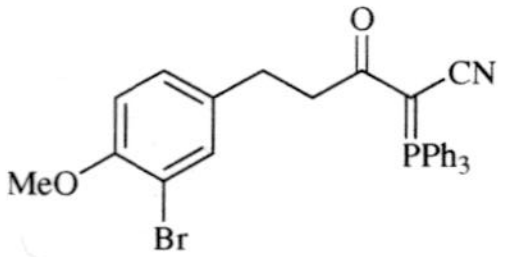 MeO, Br, O, CN, PPh_3	DMD, MeOH, acetone, rt	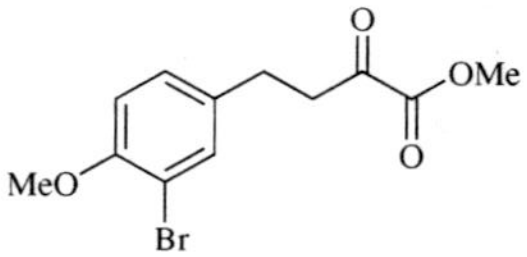 MeO, Br, O, OMe, O (86)	420
C_{36}			
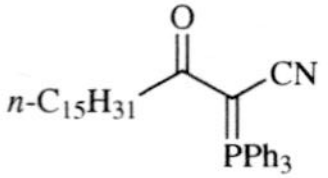 n-$C_{15}H_{31}$, O, CN, PPh_3	DMD, acetone, MeOH	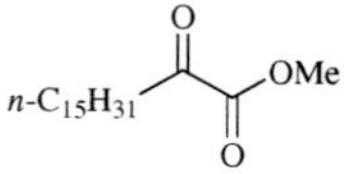 n-$C_{15}H_{31}$, O, OMe, O (86)	159
C_{38}			
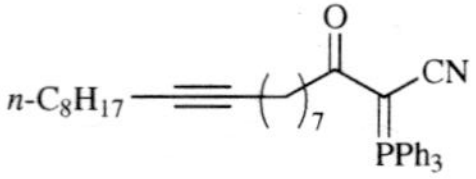 n-C_8H_{17}, ( )$_7$, O, CN, PPh_3	DMD, MeOH, acetone, 0°	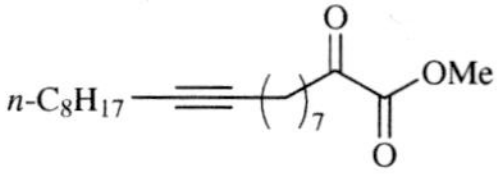 n-C_8H_{17}, ( )$_7$, O, OMe, O (87)	420

TABLE 4B. C=Y OXIDATION BY IN SITU GENERATED DIOXIRANES

Substrate	Conditions	Product(s) and Yield(s) (%)	Refs.
C_{4-13}			
$R^1C(=NOH)R^2$	Oxone®, wet alumina, microwave	$R^1C(=O)R^2$	421

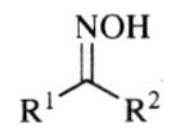

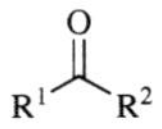

R^1	R^2	Yield (%)
Me	n-Pr	(84)
—$(CH_2)_5$—		(87)
Ph	H	(95)
2-$O_2NC_6H_4$	H	(99)
4-$O_2NC_6H_4$	H	(98)
Ph	Me	(89)
4-MeC_6H_4	H	(94)
4-$MeOC_6H_4$	H	(90)
4-$MeOC_6H_4$	Me	(82)
Ph	Ph	(86)

Substrate	Conditions	Product(s) and Yield(s) (%)	Refs.
C_{5-11}			
2-(P)-S-pyrimidin-4(3H)-one (R¹, R²)	Oxone® (3 eq), dioxane, H_2O, 12 h	uracil (R¹, R²)	422

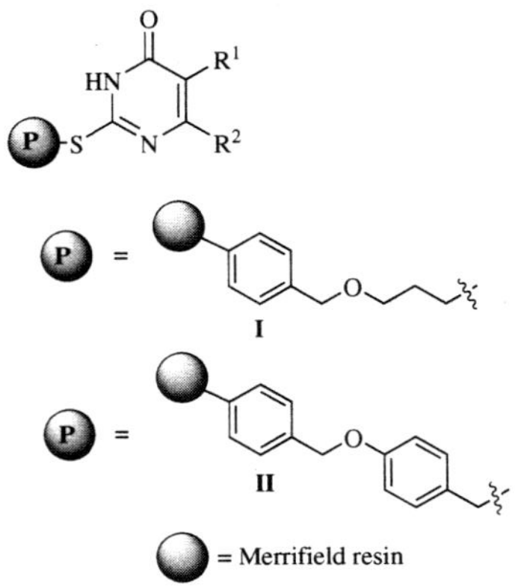

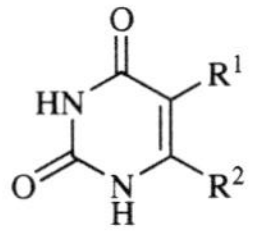

R^1	R^2	I	II
H	Me	(98)	(95)
Me	Me	(83)	(78)
H	CH_2CO_2Et	(78)	(65)
H	Bn	(85)	(70)

Substrate	Conditions	Product(s) and Yield(s) (%)	Refs.
(same)	Oxone® (3 eq), MeOH, 12 h	2-MeO-pyrimidin-4(3H)-one (R¹, R²)	422

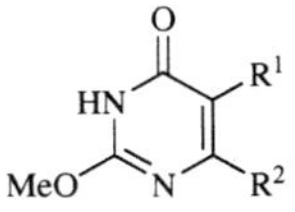

R^1	R^2	I	II
H	Me	(95)	(96)
Me	Me	(72)	(70)
H	CH_2CO_2Et	(67)	(56)
H	Bn	(77)	(73)

(P) = Merrifield resin–C_6H_4–CH_2–O–$CH_2CH_2CH_2$– **I**

(P) = Merrifield resin–C_6H_4–CH_2–O–C_6H_4–CH_2– **II**

● = Merrifield resin

TABLE 4B. C=Y OXIDATION BY IN SITU GENERATED DIOXIRANES (*Continued*)

Substrate	Conditions	Product(s) and Yield(s) (%)	Refs.
C_{14} 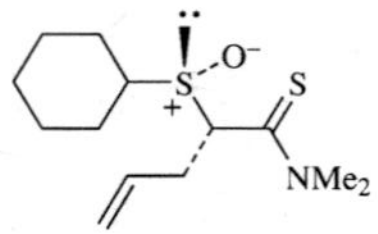 S, O⁻, +, NMe₂	Oxone®, acetone, $NaHCO_3$ H_2O, 0°, 1 h	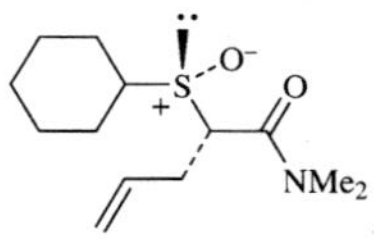 O⁻, +, O, NMe₂ (83)	423
C_{30} 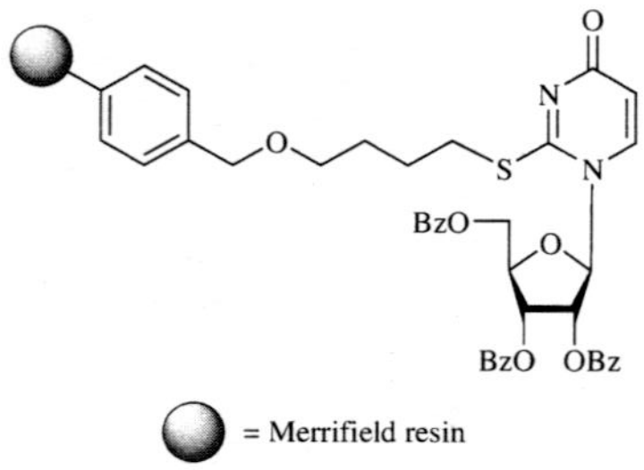 O, N, S, N, BzO, O, BzO OBz; = Merrifield resin	Oxone®, dioxane, H_2O, 12 h	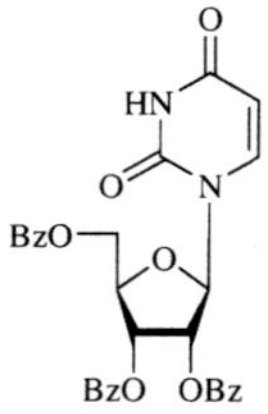 O, HN, O, N, BzO, O, BzO OBz (40)	422

TABLE 5A. C–H OXIDATION BY ISOLATED DIOXIRANES

Substrate	Conditions	Product(s) and Yield(s) (%)	Refs.
C_2 CH_3CHO	DMD, acetone, rt, 2 h	CH_3CO_2H (—)	121
	DMD, 2-Me-quinoline, CF_3CO_2H, 0°, 8 h	CH_3CO_2H (77) + 2-methylquinoline N-oxide (7) + 2,4-dimethylquinoline (4.1) + 4-acetyl-2-methylquinoline (4.2)	35
C_3 EtCHO	DMD, acetone, rt, 2 h	$EtCO_2H$ (—)	121
i-PrOH	TFD, TFP, CH_2Cl_2, –20°, 8 min	acetone (> 88)	29
C_{3-11} 2-R^1-cyclic acetal (ring size n, R^2)	DMD, acetone, 20°	$R^1CO_2(CH_2)_nCH(R^2)OH$ (—) + $R^1CO_2CH(R^2)(CH_2)_nOH$ (—)	424

R^1	R^2	n	k (x 10^3 l mol^{-1} s^{-1})
H	H	1	4.83 ± 0.35
n-Pr	H	1	33.0
i-Pr	H	1	24.8 ± 1.2
H	Me	2	0.61 ± 0.06
i-Pr	H	2	0.74 ± 0.01
i-Pr	Me	2	1.36 ± 0.09
Ph	H	1	45.0 ± 2.2
Ph	H	2	3.2 ± 0.05
Ph	Me	2	9.25 ± 0.42

TABLE 5A. C–H OXIDATION BY ISOLATED DIOXIRANES (*Continued*)

Substrate	Conditions	Product(s) and Yield(s) (%)	Refs.
C_4			
	DMD, 2-Me-quinoline, CF_3CO_2H, acetone, 0°, 8 h	**I** (23) + **II** (3.2) + 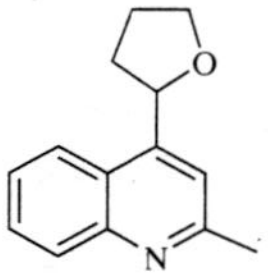 (2.8)	35
	DMD, acetone, CH_2Cl_2, 0°, 70 h	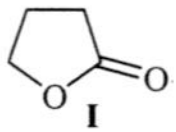 **I** (33) + 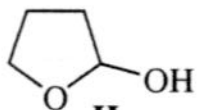 **II** (13)	191
	TFD, TFP, CH_2Cl_2, 0°, 10 min	**I** (65) + **II** (30)	191
 OH	TFD, TFP, CH_2Cl_2, –20°, 10 min	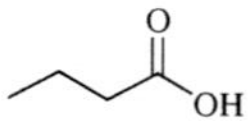 O, OH (81)	29
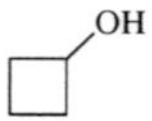 OH	TFD, TFP, CH_2Cl_2, –20°, 20 min	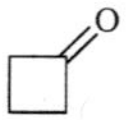 O (96) **I**	29
	TFD, TFP, CH_2Cl_2, –20°, 4 min	**I** (59)	29
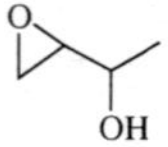 O, OH	TFD, TFP, CH_2Cl_2, –20°, 12 min	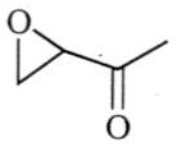 O, O (92)	29
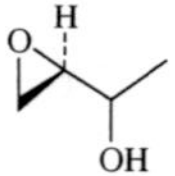 H, O, OH	TFD, TFP, CH_2Cl_2, –20°, 12 min	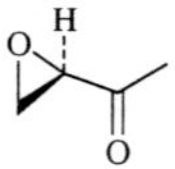 H, O, O (86)	29

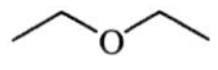 O	DMD, 2-Me-quinoline, CF_3CO_2H, acetone, 0°, 8 h	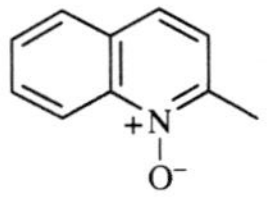 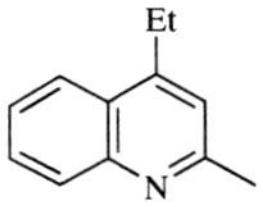 ^{+}N–O^{-} (12) 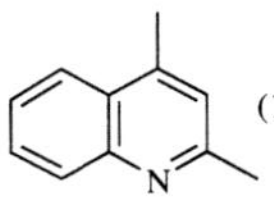 + N (1.2) + Et, N (3.2)		35
	TFD, TFP, –20°, 30 min	EtOH + CH_3CHO + CH_3CO_2H + AcOEt + AcOMe + CF_3CO_2H + CF_3CO_2Me + CF_3CO_2Et + $AcOCF_3$	(—)	187
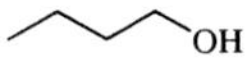 OH	TFD, TFP, CH_2Cl_2, –20°, 10 min	n-PrCO_2H (89) 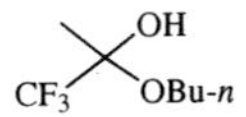 + OH, CF_3, OBu-*n* (8)		29
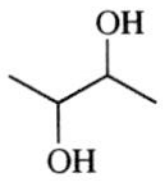 OH, OH	DMD, acetone, rt, overnight	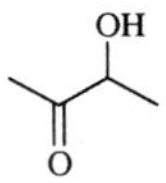 OH, O (85) 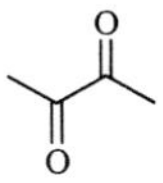 + O, O (5)		269
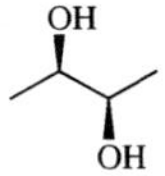 OH, OH	DMD, CH_2Cl_2, acetone, 0°, 4 h	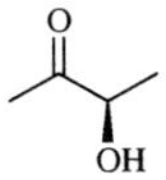 O, OH (92) > 98% ee		184

TABLE 5A. C–H OXIDATION BY ISOLATED DIOXIRANES (*Continued*)

Substrate	Conditions	Product(s) and Yield(s) (%)	Refs.
C_5 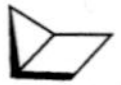 x M	DMD, acetone; or TFD, TFP	 OH (I) + OH, OH (II)	425

x	Dioxirane (M)	Temp	Time	% Conv.	I	II
0.181	DMD (0.018)	20°	60 min	40	(89)	(11)
0.181	DMD (0.018)	20°	150 min	75	(70)	(30)
0.181	DMD (0.002)	20°	15 min	60	(> 99)	(—)
0.181	DMD (0.002)	20°	27 min	82	(> 99)	(—)
0.028	DMD (0.024)	20°	150 min	70	(> 99)	(—)
0.028	DMD (0.002)	20°	120 min	25	(> 99)	(—)
0.033	DMD (0.002)	20°	90 min[a]	35	(> 99)	(—)
0.209	TFD (0.002)	0°	5 min	95	(79)	(21)
0.002	TFD (0.001)	0°	30 min	35	(92)	(8)

Substrate	Conditions	Product(s) and Yield(s) (%)	Refs.
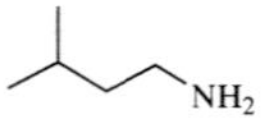 NH_2	1. aq. HBF_4, MeCN, (pH 2-3), 0° 2. TFD, CH_2Cl_2, rt, 3 h 3. Na_2CO_3, CH_2Cl_2, rt, 5 h	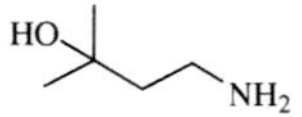 HO, NH_2 (90)	192
	1. aq. HBF_4, MeCN, (pH 2-3), 0° 2. TFD, CH_2Cl_2, rt, 10 h 3. Na_2CO_3, CH_2Cl_2, rt, 5 h	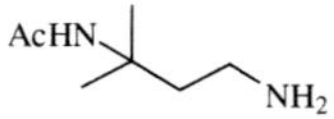 AcHN, NH_2 (93)	192
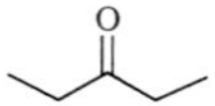 O	TFD, TFP, Et_2O, –20°	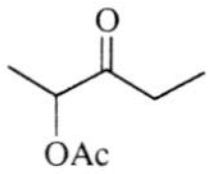 O, OAc (—) + other products	187
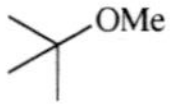 OMe	TFD, CCl_4, 0°, 10 min	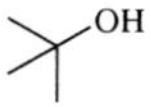 OH (90)	191
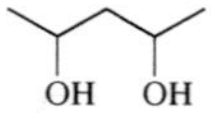 OH OH	DMD, acetone, rt, overnight	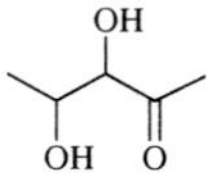 OH, OH O (95)	269

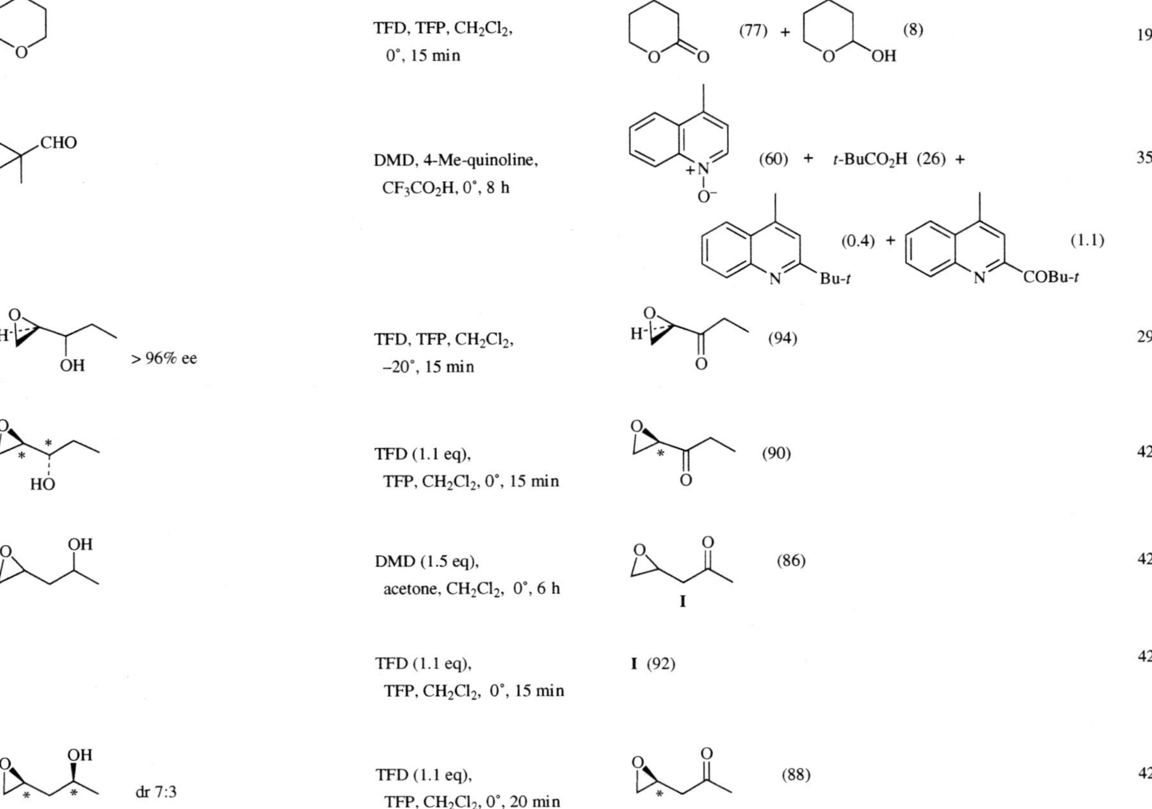

Substrate	Conditions	Product(s) and Yield(s) (%)	Refs.
(structure)	TFD, TFP, CH_2Cl_2, 0°, 15 min	(structure) (77) + (structure) (8)	191
(structure)	DMD, 4-Me-quinoline, CF_3CO_2H, 0°, 8 h	(structure) (60) + t-$BuCO_2H$ (26) + (structure) (0.4) + (structure) (1.1)	35
(structure) > 96% ee	TFD, TFP, CH_2Cl_2, –20°, 15 min	(structure) (94)	29
(structure)	TFD (1.1 eq), TFP, CH_2Cl_2, 0°, 15 min	(structure) (90)	426
(structure)	DMD (1.5 eq), acetone, CH_2Cl_2, 0°, 6 h	(structure) **I** (86)	426
	TFD (1.1 eq), TFP, CH_2Cl_2, 0°, 15 min	**I** (92)	426
(structure) dr 7:3	TFD (1.1 eq), TFP, CH_2Cl_2, 0°, 20 min	(structure) (88)	426

TABLE 5A. C–H OXIDATION BY ISOLATED DIOXIRANES (*Continued*)

Substrate	Conditions	Product(s) and Yield(s) (%)	Refs.
C_5			
OH, syn/anti 6:4	DMD (1.5 eq), acetone, CH_2Cl_2, 0°, 3 h	O **I** (96)	426
	TFD (1.1 eq), TFP, CH_2Cl_2, 0°, 20 min	**I** (96)	426
C_{5-8}			
n = 1, 2, 3, 4	DMD, acetone, rt, 1-2 d	OH n (—)	427
R^1 O R^3 R^2 O n	DMD, acetone, 20°	R^1 O R^2 (—) + O R^3 OH n (—)	424

R^1	R^2	R^3	n	k (x 10^3 l mol^{-1} s^{-1})
Me	Me	H	1	0.38 ± 0.05
Me	Me	CH_2OH	1	0.54 ± 0.04
Me	Me	Me	2	0.42 ± 0.01
—$(CH_2)_5$—		H	1	9.25 ± 0.42

Substrate	Conditions	Product(s) and Yield(s) (%)	Refs.
C_{5-13}			
O O R^1 R^3 R^2	DMD, catalyst, acetone, H_2O, 20°	O O R^1 R^3 R^2 OH	194

R_1	R^2	R^3	Catalyst	Time	% Conv.
OMe	H	OMe	—	24 h	15
OMe	H	OMe	$Ni(OAc)_2$	16 h	> 95
—$O(CH_2)_2$—		Me	—	3.5 h	90
—$O(CH_2)_2$—		Me	$Ni(acac)_2$	3.5 h	> 95
Me	H	OEt	—	24 h	35
Me	H	OEt	$Ni(OAc)_2$	24 h	> 95
—$O(CH_2)_2$—		OEt	—	5 h	46
—$O(CH_2)_2$—		OEt	$Ni(OAc)_2$	5 h	84
—$(CH_2)_3$—		OEt	—	3.5 h	75
—$(CH_2)_3$—		OEt	$Ni(acac)_2$	3.5 h	> 95
OEt	Me	OEt	—	120 h	—
OEt	Me	OEt	$Ni(OAc)_2$	12 h	47
—$(CH_2)_4$—		Et	—	3 h	88
—$(CH_2)_4$—		Et	$Ni(OAc)_2$	3 h	90
Me	Bn	OEt	—	4.25 h	11
Me	Bn	OEt	$Ni(acac)_2$	4.25 h	78

C_6

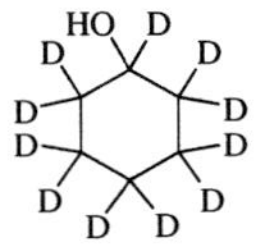

Substrate	Conditions	Product (%)	Refs.
HO, D (cyclohexanol-d_{11})	TFD, TFP, CH_2Cl_2, –12.5°	cyclohexanone-d_{10} (—) $k_H/k_D = 1.6{\pm}0.15$	29
2-acetylbutyrolactone	DMD (3.0 eq), acetone, 20°, 72 h	2-acetyl-2-hydroxybutyrolactone (98)	193
3-R^2-4-R^1-cyclohexanol; R^1 = OH, R^2 = N_3; R^1 = N_3, R^2 = OH	DMD, acetone, rt, 24 h	3-R^2-4-R^1-cyclohexanone (> 95)	428

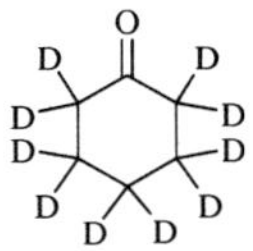

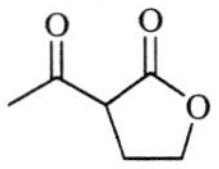

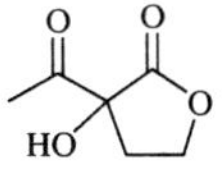

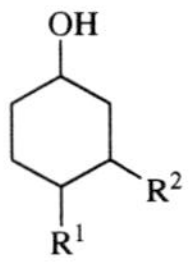

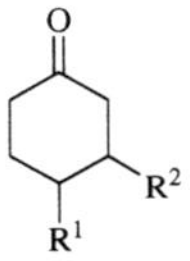

TABLE 5A. C–H OXIDATION BY ISOLATED DIOXIRANES (*Continued*)

Substrate	Conditions	Product(s) and Yield(s) (%)	Refs.
C_6 	TFP, $(CF_3CO)_2O$, CH_2Cl_2, 0°, 10 min	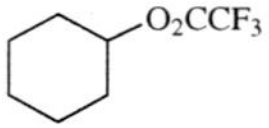 O_2CCF_3 (> 99)	172
	DMD, CCl_3Br, acetone, (O_2)	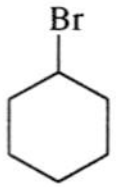 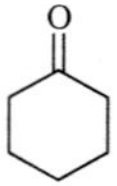 Br (—) + O (—) **I**	429
	TFD, TFP, CH_2Cl_2, –22°, 18 min	**I** (98)	30, 173
	TFD, TFP, –20°, 30 min	**I** (95)	173
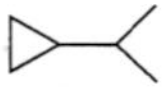	DMD, acetone, rt	 OH (30-50)	429
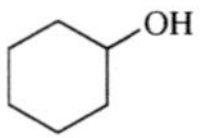 OH	TFD, TFP, CH_2Cl_2, –20°, 20 min	**I** (97)	29
	DMD, acetone, O_2, 22°, 6 h	**I** (—)	35
	DMD, acetone, Ar, 22°, 6 h	**I** (45) + AcOMe (27.2) + MeOH (13.5) + $AcOCH_2Ac$ (12.1) + CH_4 (—)	34, 35
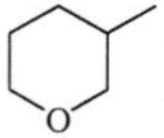 O	DMD, acetone, CH_2Cl_2, 18°, 96 h	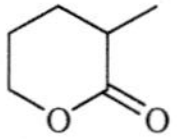 O O (9) 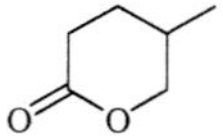 **I** + O O (17) **II**	191
	TFD, TFP, CH_2Cl_2, –10°, 15 min	**I** (32) + **II** (49)	191

Substrate	Conditions	Product(s) and Yield(s) (%)	Refs.
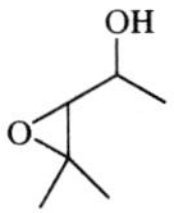	DMD (3.0 eq), acetone, 20°, 72 h	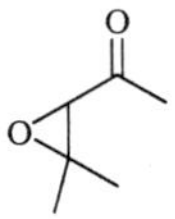 (—)	430
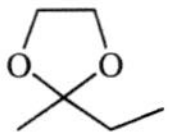	TFD, TFP, CH_2Cl_2, 0°, 30 min	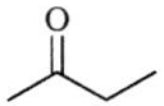 (62)	191
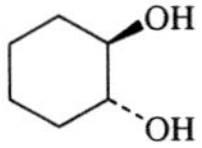 95% ee	DMD, CH_2Cl_2, acetone, 0°, 8 h	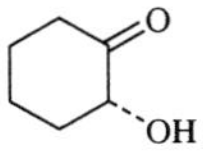 (> 96) 94% ee	184
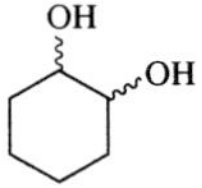	DMD, acetone, rt, overnight	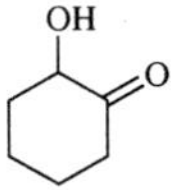 (80-90)	269
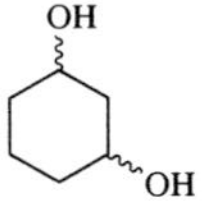	DMD, acetone, rt, overnight	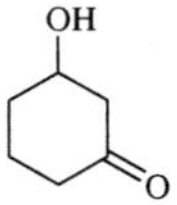 (92)	269
	DMD, acetone, rt, overnight	 (53) + 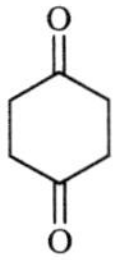 (37)	269
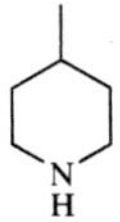	1. aq. HBF_4, MeCN, (pH 2-3), 0° 2. TFD, CH_2Cl_2, rt, 15 h 3. Na_2CO_3, CH_2Cl_2, rt, 5 h	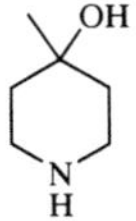 (54)	192
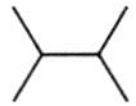	TFD, TFP, CH_2Cl_2, –22°, 3 min	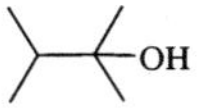 (98)	30

TABLE 5A. C–H OXIDATION BY ISOLATED DIOXIRANES (*Continued*)

Substrate	Conditions	Product(s) and Yield(s) (%)	Refs.
C_6			
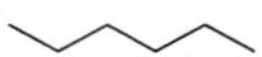	TFP, $(CF_3CO)_2O$, CH_2Cl_2, 0°, 40 min	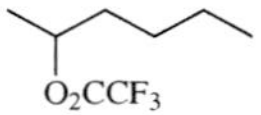 O_2CCF_3 (58) 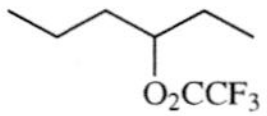 + O_2CCF_3 (42)	172
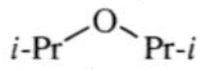 *i*-Pr–O–Pr-*i*	DMD, acetone, Ar, 22°, 6 h	CH_4 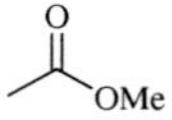 (—) + OMe (38.4) + MeOH (12.1)	34, 35
	DMD, acetone, DEK, Ar, 22°, 6 h	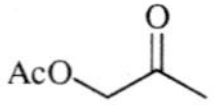 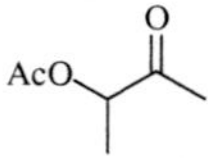 AcO (2.1) + AcO (9.6)	35
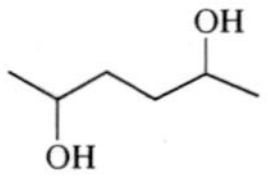 OH OH	DMD, acetone, rt, overnight	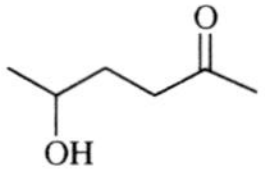 OH (50) 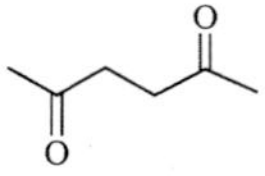 + (25)	269
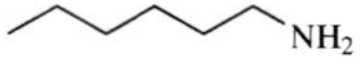 NH_2	1. aq. HBF_4, MeCN (pH 2.0-3.0), 0° 2. TFD, CH_2Cl_2, rt, 8 h 3. Na_2CO_3, CH_2Cl_2, rt, 5 h	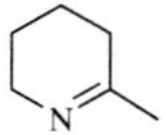 N (88)	192
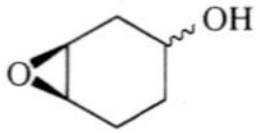 O OH	DMD (1.6 eq), acetone, CH_2Cl_2, 0°, 6 h	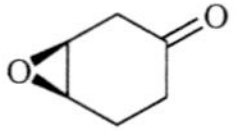 O O (83) **II**	426
	TFD (1.1 eq), TFP, CH_2Cl_2, 0°, 20 min	**II** (88)	426
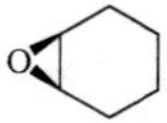 O	TFD (3.0 eq), acetone, 0°, 3 h	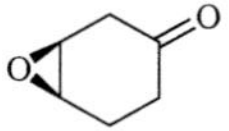 O O (67)	426

C_{6-8}

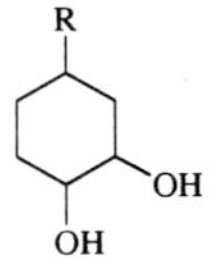

DMD, acetone, rt — 428

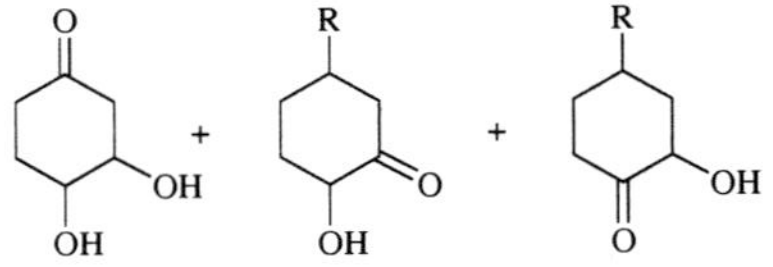

R	Time	I	II	III
OH	24 h	(50)	(0)	(0)
OMe	—	(0)	(0)	(0)
OAc	24 h	(0)	(50)	(50)

C_{6-16}

Dioxirane (x eq), 0°  — 431

R	Dioxirane	x	Solvent	Time	% Conv.	I	II
Me	DMD	1	CH_2Cl_2, acetone	3 h	60	(—)	(98)
Me	DMD	3	CH_2Cl_2, acetone	6 h	95	(3)	(95)
Me	DMD	3	acetone	6 h	95	(65)	(20)
Ph	DMD	3	acetone	6 h	95	(70)	(30)
Me	TFD	3	CH_2Cl_2, TFP	0.5 h	95	(74)	(—)
Me	TFD	3	acetone	0.3 h	95	(90)	(—)
Ph	TFD	3	acetone	0.3 h	95	(96)	(—)

C_7

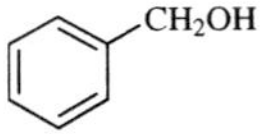

CH_2OH — TFD, TFP, CH_2Cl_2, –20° — 29

Time	I	II
30 min	(43)	(2)
60 min	(48)	(11)
90 min	(5)	(83)

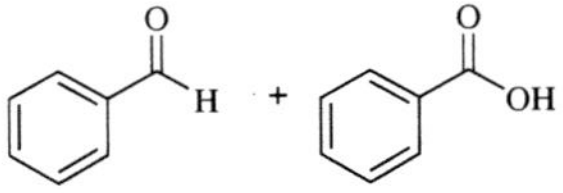

CH_2OH + CD_2OH — DMD, acetone, 0° — (—) + (—) — 432

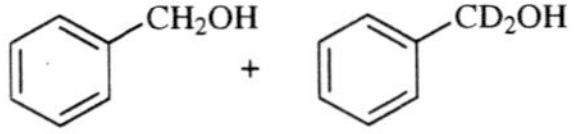

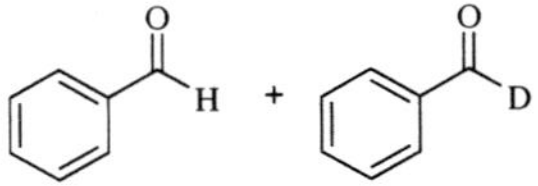

% Conv.	k_H/k_D
8	4.5 ± 0.2
23	4.6 ± 0.2

TABLE 5A. C–H OXIDATION BY ISOLATED DIOXIRANES (*Continued*)

Substrate	Conditions	Product(s) and Yield(s) (%)	Refs.
C_7			
CHDOH	DMD, acetone, 0°	H (—) + D (—); % Conv. 7, k_H/k_D 4.4 ± 0.2; % Conv. 18, k_H/k_D 4.5 ± 0.2	432
	TFD (1.1 eq), TFP, CH_2Cl_2, 0°, 20 min	O (40) + O (4)	433
NH$_2$	1. aq. HBF_4, MeCN, (pH 2.0-3.0), 0° 2. TFD, CH_2Cl_2, rt, 3 h 3. Na_2CO_3, CH_2Cl_2, rt, 5 h	HO, NH$_2$ (97)	192
	1. aq. HBF_4, MeCN, (pH 2.0-3.0), 0° 2. TFD, CH_2Cl_2, rt, 10 h 3. Na_2CO_3, CH_2Cl_2, rt, 5 h	AcHN, NH$_2$ (96)	192
NH$_2$	1. aq. HBF_4, MeCN, (pH 2.0-3.0), 0° 2. TFD, CH_2Cl_2, rt, 8 h 3. Na_2CO_3, CH_2Cl_2, rt, 5 h	N (94)	192
OH, NO$_2$	DMD, acetone, rt, overnight	O, NO$_2$ (75)	434
	TFD, $(CF_3CO)_2O$, CH_2Cl_2, 0°, 40 min	O_2CCF_3 (5) + O_2CCF_3 (95)	172

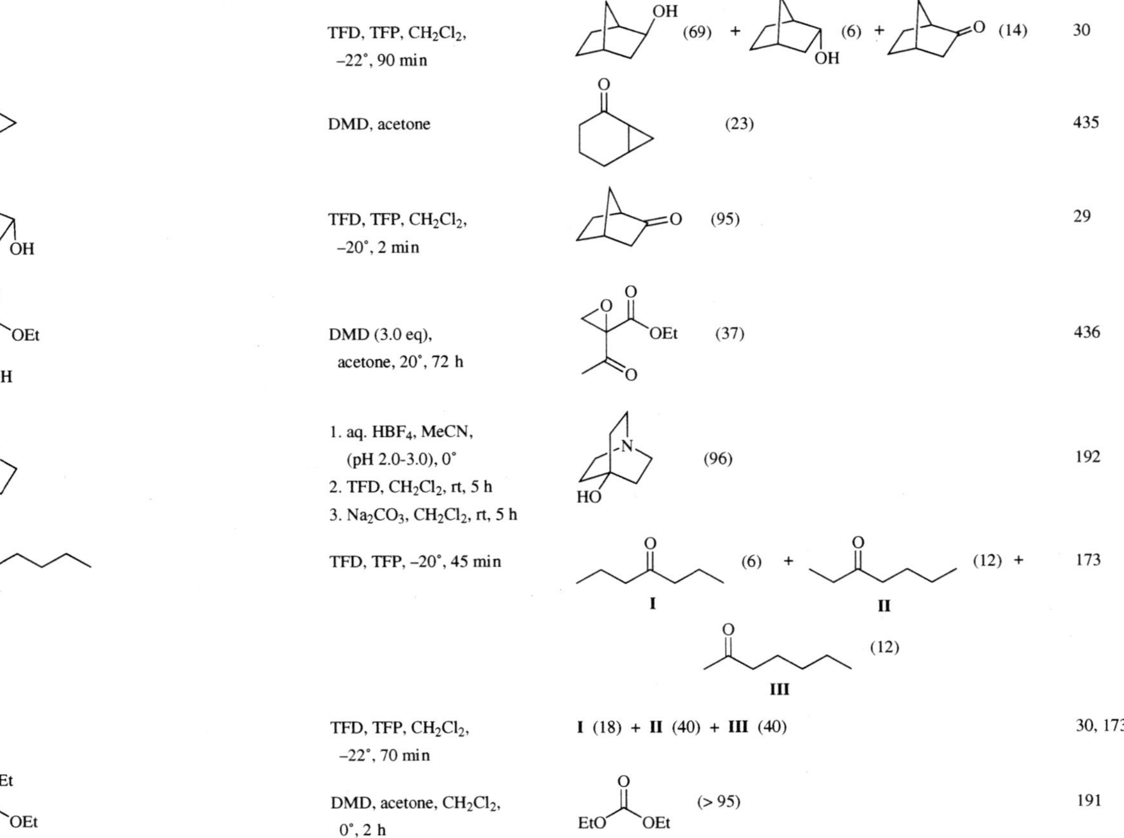

Substrate	Conditions	Product(s) and Yield(s) (%)	Refs.
	TFD, TFP, CH_2Cl_2, –22°, 90 min	OH (69) + OH (6) + O (14)	30
	DMD, acetone	O (23)	435
OH	TFD, TFP, CH_2Cl_2, –20°, 2 min	O (95)	29
O, OEt, OH	DMD (3.0 eq), acetone, 20°, 72 h	O, O, OEt, O (37)	436
N	1. aq. HBF_4, MeCN, (pH 2.0-3.0), 0° 2. TFD, CH_2Cl_2, rt, 5 h 3. Na_2CO_3, CH_2Cl_2, rt, 5 h	N, HO (96)	192
	TFD, TFP, –20°, 45 min	O, **I** (6) + O, **II** (12) + O, **III** (12)	173
	TFD, TFP, CH_2Cl_2, –22°, 70 min	**I** (18) + **II** (40) + **III** (40)	30, 173
OEt, EtO, OEt	DMD, acetone, CH_2Cl_2, 0°, 2 h	O, EtO, OEt (> 95)	191

TABLE 5A. C–H OXIDATION BY ISOLATED DIOXIRANES (*Continued*)

Substrate	Conditions	Product(s) and Yield(s) (%)	Refs.
C_7			
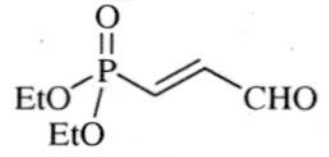 EtO, EtO, O, P, CHO	DMD, acetone, rt, 4 d	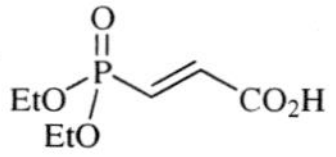 EtO, EtO, O, P, CO_2H (100)	437
	TFD, TFP, CH_2Cl_2, –22°, 8 min	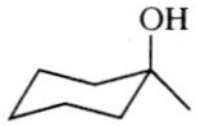 OH (90) 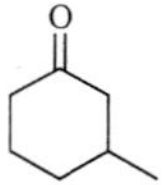 + O (8)	30
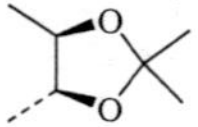 O, O 98% ee	TFD, TFP, CH_2Cl_2, 0°, 40 min	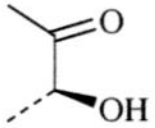 O, OH (> 96) 98% ee	190
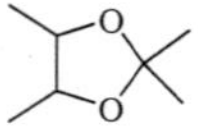 O, O	TFD, TFP, CH_2Cl_2, 0°, 20 min	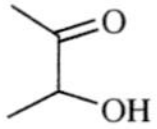 O, OH (> 94)	190
C_{7-8}			
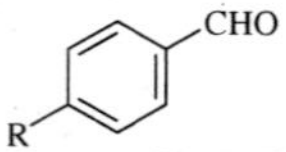 R, CHO R = Br, NO_2, H, CN, Me, MeO	DMD, acetone, rt, dark, 18 h	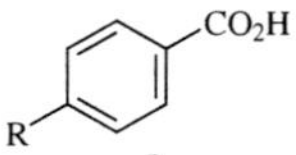 R, CO_2H **I** (98)	438
R = Br, NO_2, H, MeO	DMD, acetone, rt, dark, N_2 or Ar	**I** (52 ± 8)	438
C_{7-10}			
 R	1. TFD, CH_2Cl_2, –40° 2. $(CF_3CO)_2O$	  R, $OCOCF_3$ + R, $OCOCF_3$	439

R	k_{eq}/k_{ax} C-3	k_{eq}/k_{ax} C-4
TMS	2.92	2.61
t-Bu	1.35	1.10
Me	1.34	0.90
CF_3	0.44	0.52

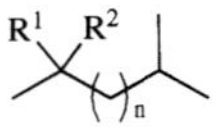

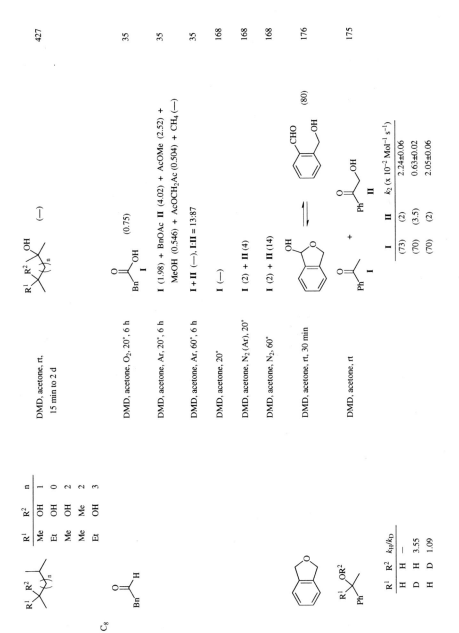

R^1	R^2	n
Me	OH	1
Et	OH	0
Me	OH	2
Me	Me	2
Et	OH	3

DMD, acetone, rt, 15 min to 2 d — (—) — 427

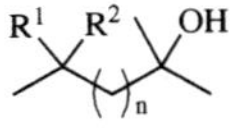

C_8

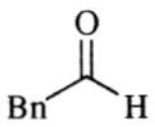

Conditions	Product(s) and Yield(s) (%)	Refs.
DMD, acetone, O_2, 20°, 6 h	**I** (0.75)	35
DMD, acetone, Ar, 20°, 6 h	**I** (1.98) + BnOAc **II** (4.02) + AcOMe (2.52) + MeOH (0.546) + $AcOCH_2Ac$ (0.504) + CH_4 (—)	35
DMD, acetone, Ar, 60°, 6 h	**I** + **II** (—), **I**:**II** = 13:87	35
DMD, acetone, 20°	**I** (—)	168
DMD, acetone, N_2 (Ar), 20°	**I** (2) + **II** (4)	168
DMD, acetone, N_2, 60°	**I** (2) + **II** (14)	168

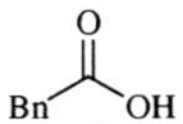

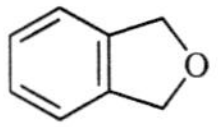

DMD, acetone, rt, 30 min 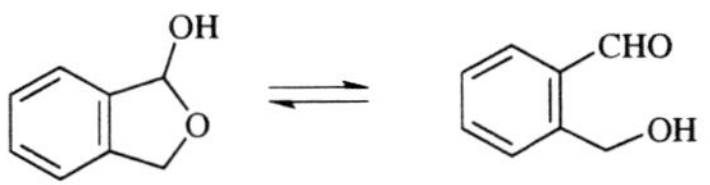 — (80) — 176

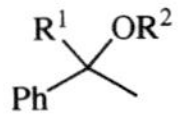

DMD, acetone, rt — **I** + **II** — 175

R^1	R^2	k_H/k_D
H	H	—
D	H	3.55
H	D	1.09

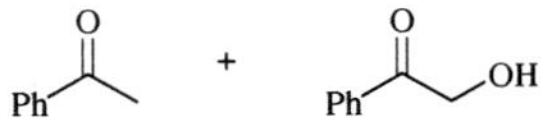

I	II	k_2 (x 10^{-2} Mol^{-1} s^{-1})
(73)	(2)	2.24±0.06
(70)	(3.5)	0.63±0.02
(70)	(2)	2.05±0.06

TABLE 5A. C–H OXIDATION BY ISOLATED DIOXIRANES (*Continued*)

Substrate	Conditions	Product(s) and Yield(s) (%)	Refs.
C_8			
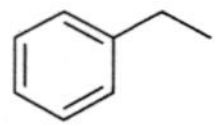	TFD, TFP, $(CF_3CO)_2O$, CH_2Cl_2, 0°, 20 min	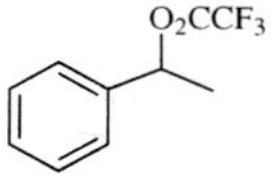 O_2CCF_3 (9)	172
	TFD, TFP, CH_2Cl_2, 0°, 80 min	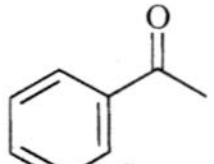 O **I** (9)	30
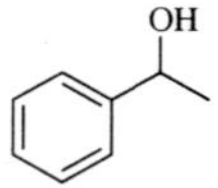 OH	TFD, TFP, CH_2Cl_2, –20°, 20 min	**I** (97)	29
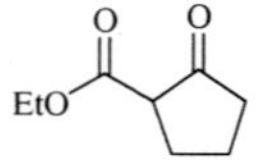 EtO O O	DMD (3.0 eq), acetone, 20°, 72 h	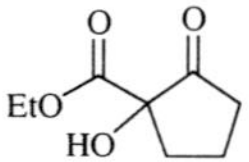 EtO O O HO (100)	193
	TFP, $(CF_3CO)_2O$, CH_2Cl_2, 0°, 2 min	 O_2CCF_3 (50) 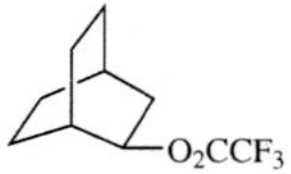 + O_2CCF_3 (50)	172
	TFD, TFP, CH_2Cl_2, –22°, 2 min	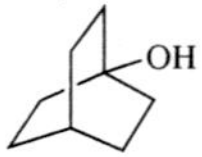 OH (56) 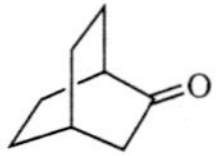 + O (41)	30
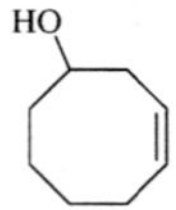 HO	DMD (2.1 eq), acetone, CH_2Cl_2, 20°, 24 h	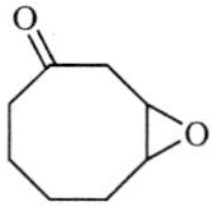 O O (> 95)	440
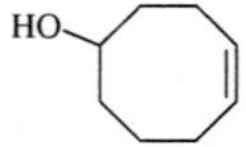 HO	DMD (2.2 eq), acetone, CH_2Cl_2, 20°, 24 h	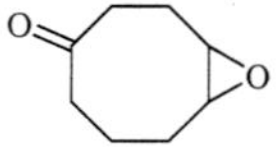 O O (85)	440

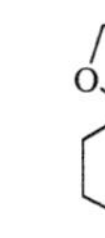	DMD, acetone, CH_2Cl_2, 0°, 24 h	I (33)	191
	TFD, TFP, CH_2Cl_2, 0°, 2 h	I (98)	191
	TFD, TFP, CH_2Cl_2, 0°, 15 min	I (96)	182
	DMD, acetone, CH_2Cl_2, 0°, 4 h	I (65)	182
	DMD, acetone, 22°, dark, 18 h	I (45)	174
	TFD, TFP, CH_2Cl_2, −22°, 10 min	I (90) + (5) + (4)	30
	DMD, acetone, 22°, dark, 18 h	I (100)	174
	TFD, TFP, CH_2Cl_2, −22°, 4 min	I (95)	30
	DMD, acetone/solvent (v/v = 1:1), rt	(—)	441

Solvent	k_2 (x 10^{-3} M^{-1} s^{-1})
acetone	1.37±0.03
2-butanone	1.29±0.01
EtOAc	1.02±0.02
CH_2Cl_2	3.08±0.06
$CHCl_3$	4.44±0.13
$CDCl_3$	4.05±0.04

TABLE 5A. C–H OXIDATION BY ISOLATED DIOXIRANES (*Continued*)

Substrate	Conditions	Product(s) and Yield(s) (%)	Refs.
C_8			
	TFP, $(CF_3CO)_2O$, CH_2Cl_2, 0°, 60 h	$CH_2O_2CCF_3$ (> 99)	172
OH	TFD, TFP, CH_2Cl_2, –20°, 20 min	O (99)	29
NH_2	1. aq. HBF_4, MeCN, (pH 2.0-3.0), 0° 2. TFD, CH_2Cl_2, rt, 8 h 3. Na_2CO_3, CH_2Cl_2, rt, 5 h	N (88)	192
OH + OH, D, CD_3	DMD, acetone, 0°	O (—) + O, CD_3 (—)	432
		% Conv. / k_H / k_D: 6 / 1.00 ± 0.02; 11 / 0.99 ± 0.02; 18 / 1.01 ± 0.02; 32 / 0.98 ± 0.02	
	TFD (1.2 eq), TFP, CH_2Cl_2, 0°, 35 min	O (38)	433
HO, O	DMD (1.5 eq), acetone, CH_2Cl_2, 0°, 6 h	O, O **I** (77)	427
	TFD (1.1 eq), TFP, CH_2Cl_2, 0°, 10 min	**I** (86)	427

Substrate	Conditions	Product(s) and Yield(s) (%)	Refs.
C_8 HO	DMD (1.5 eq), acetone, CH_2Cl_2, 0°, 6 h	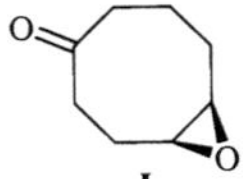 I (93)	427
	TFD (1.1 eq), TFP, CH_2Cl_2, 0°, 10 min	**I** (94)	427
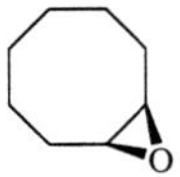	TFD (4.0 eq), acetone, 0°, 2 h	(57) + (19)	427
OH	DMD (1.6 eq), acetone, CH_2Cl_2, 0°, 3 h	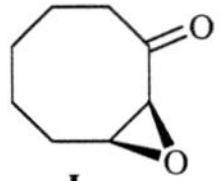 I (91)	427
	TFD (1.1 eq), TFP, CH_2Cl_2, 0°, 10 min	**I** (94)	427
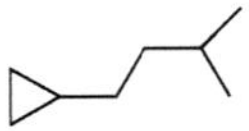	TFD (1.1 eq), TFP, CH_2Cl_2, 0°, 30 min	OH (39) + O (4) + OH (4)	427
C_{8-9} OH, R	DMD, acetone, rt, 3 h	R, O, **I** + R, O, OH, **II**	175

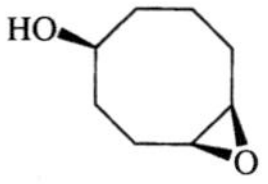

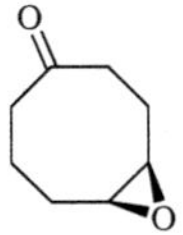

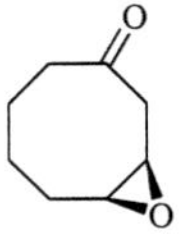

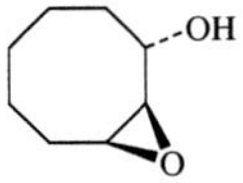

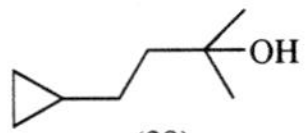

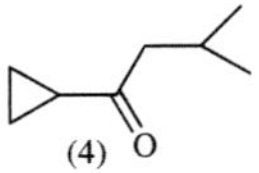

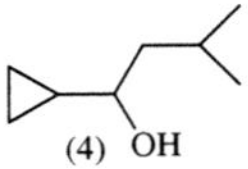

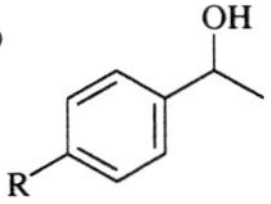

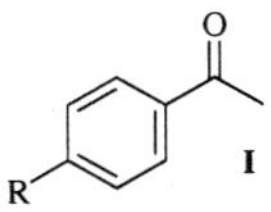

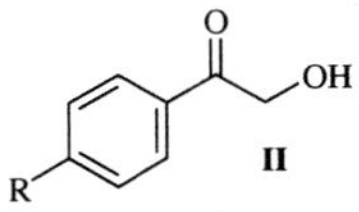

R	I	II	k_2 (x 10^{-2} Mol^{-1} s^{-1})
Br	(96)	(3)	1.51±0.04
Cl	(97)	(3)	1.59±0.04
F	(97)	(2)	1.85±0.04
H	(98)	(2)	2.24±0.06
Me	(97)	(2)	2.89±0.06
MeO	(97)	(3)	3.56±0.08
CN	(97)	(2)	0.76±0.04

TABLE 5A. C–H OXIDATION BY ISOLATED DIOXIRANES (*Continued*)

Substrate	Conditions	Product(s) and Yield(s) (%)	Refs.
C_{8-13}			
$R^1C(OR^3)HR^2$	DMD, acetone, rt	$R^1C(=O)R^2$	175
		R^1 / R^2 / R^3	
		Ph / Me / H (98)	
		Ph / Et / H (92)	
		Ph / *n*-Pr / H (92)	
		Ph / *i*-Pr / H (90)	
		Ph / *t*-Bu / H (90)	
		Ph / Ph / H (96)	
		Ph / *c*-C_3H_5 / H (92)	
		Bn / Me / H (85)	
		Ph / Me / Me (90)	
		Ph / Et / Me (81)	
		Ph / *n*-Pr / Me (86)	
		Ph / *i*-Pr / Me (48)	
		Ph / *t*-Bu / Me (24)	
		Ph / Ph / Me (84)	
		Ph / *c*-C_3H_5 / Me (88)	
		Bn / Me / Me (80)	
		Ph / Me / TMS (95)	
		Ph / Me / Ac (96)	
C_{8-16}			
$PhCH_2R$	DMD, acetone, dark, rt, 3 d	$PhC(=O)R$	442
		R	
		Me (6.9)	
		Et (22)	
		t-Bu (2.7)	
		Ph (3.2)	
		$PhCH_2$ (22)	
		$Ph(CH_2)_2$ (28)	
		$Ph(CH_2)_2$ (99)	
		$Ph(CH_2)_3$ (16)	

C$_9$

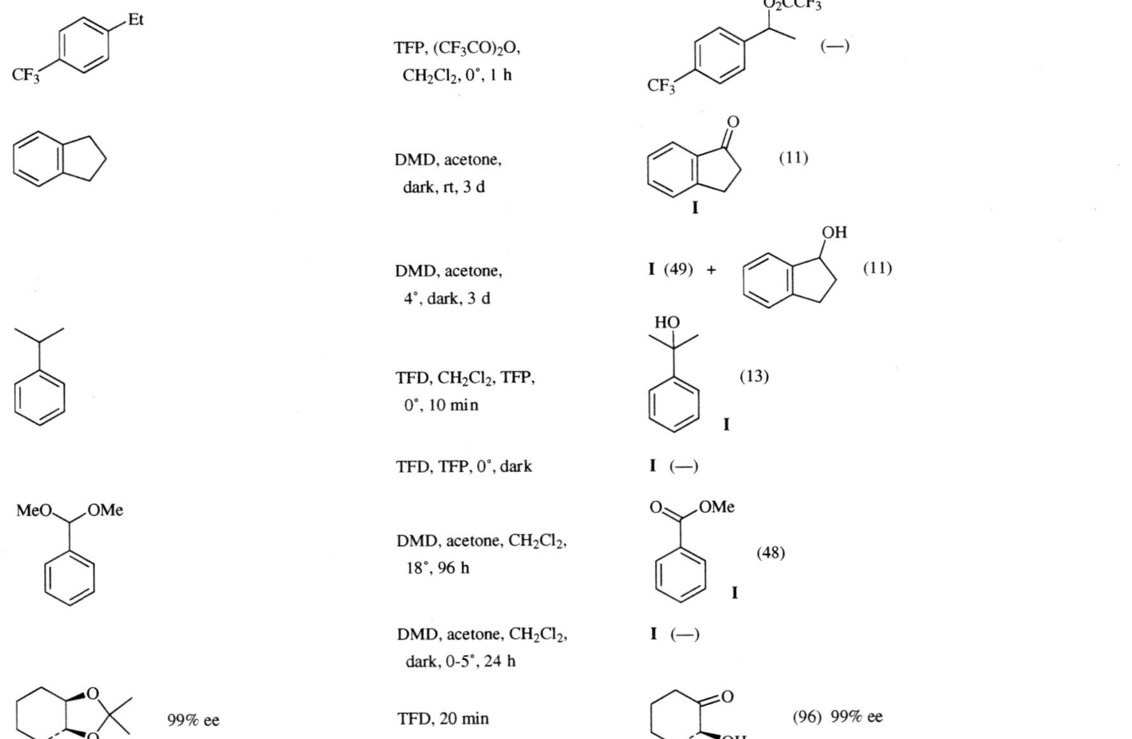

Substrate	Conditions	Product(s) and Yield(s) (%)	Refs.
$CF_3C_6H_4Et$ (4-Et)	TFP, $(CF_3CO)_2O$, CH_2Cl_2, 0°, 1 h	$4-CF_3C_6H_4CH(CH_3)O_2CCF_3$ (—)	172
Indane	DMD, acetone, dark, rt, 3 d	1-Indanone **I** (11)	442
	DMD, acetone, 4°, dark, 3 d	**I** (49) + 1-Indanol (11)	442
Cumene	TFD, CH_2Cl_2, TFP, 0°, 10 min	$PhC(CH_3)_2OH$ **I** (13)	30
	TFD, TFP, 0°, dark	**I** (—)	443
$PhCH(OMe)_2$	DMD, acetone, CH_2Cl_2, 18°, 96 h	$PhCO_2Me$ **I** (48)	191
	DMD, acetone, CH_2Cl_2, dark, 0-5°, 24 h	**I** (—)	189
Cyclohexane-1,2-diol acetonide, 99% ee	TFD, 20 min	2-Hydroxycyclohexanone (96) 99% ee	190

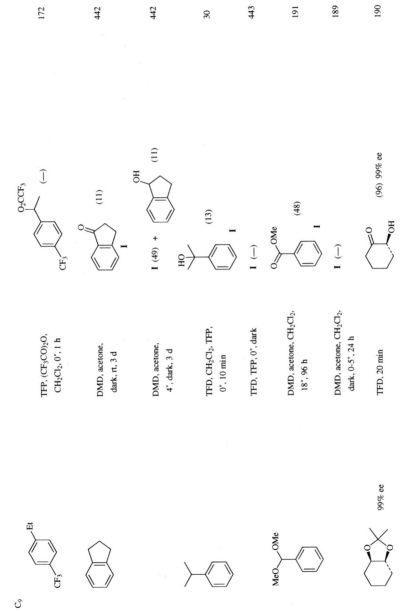

TABLE 5A. C–H OXIDATION BY ISOLATED DIOXIRANES (*Continued*)

Substrate	Conditions	Product(s) and Yield(s) (%)	Refs.
C_9 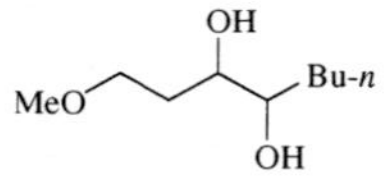	DMD (6 eq), acetone, rt, 72 h	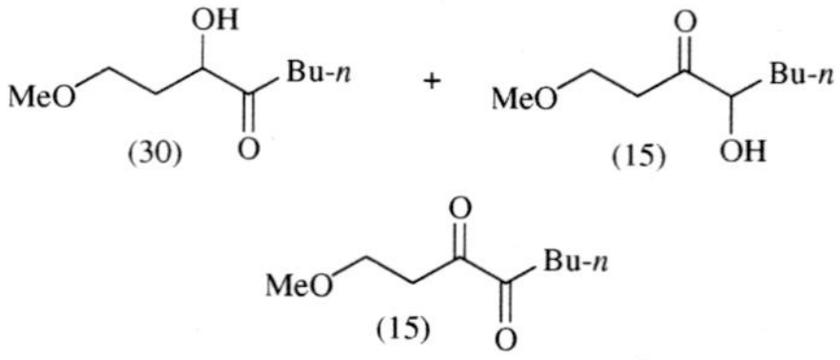 (30) + (15) + (15)	428
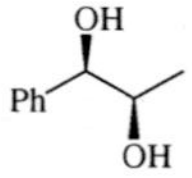 89% ee	DMD, CH_2Cl_2, acetone, 0°, 22 h	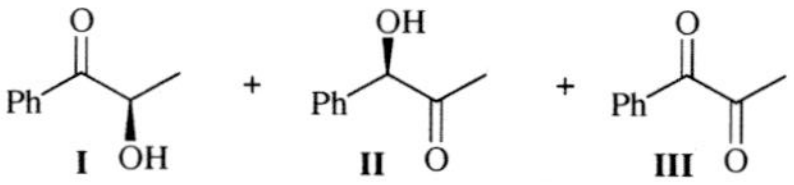 I (42) 86% ee + II (34) 82% ee + III (12)	184
	TFD, CH_2Cl_2, TFP, 0°, 40 min	**I** (41) 85% ee + **II** (34) 82% ee + **III** (12)	184
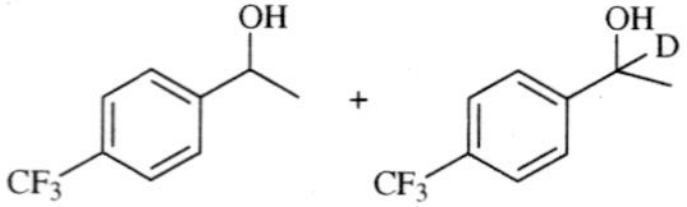 +	DMD, acetone, 0°	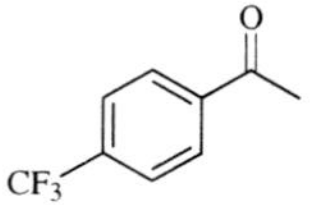 (—)	432
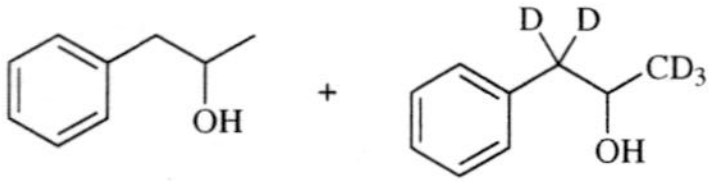 +	DMD, acetone, 0°	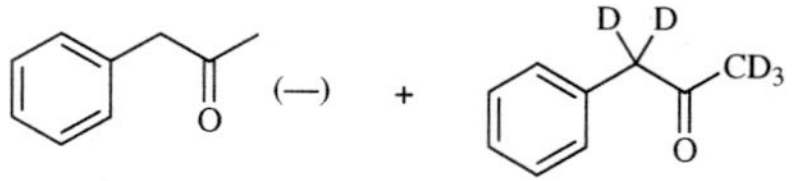 (—) + (—)	432

% Conv.	k_H/k_D
20	4.8 ± 0.2
31	4.8 ± 0.2

% Conv.	k_H/k_D
4	0.98 ± 0.2
8	0.98 ± 0.2
17	1.01 ± 0.2
35	0.98 ± 0.2

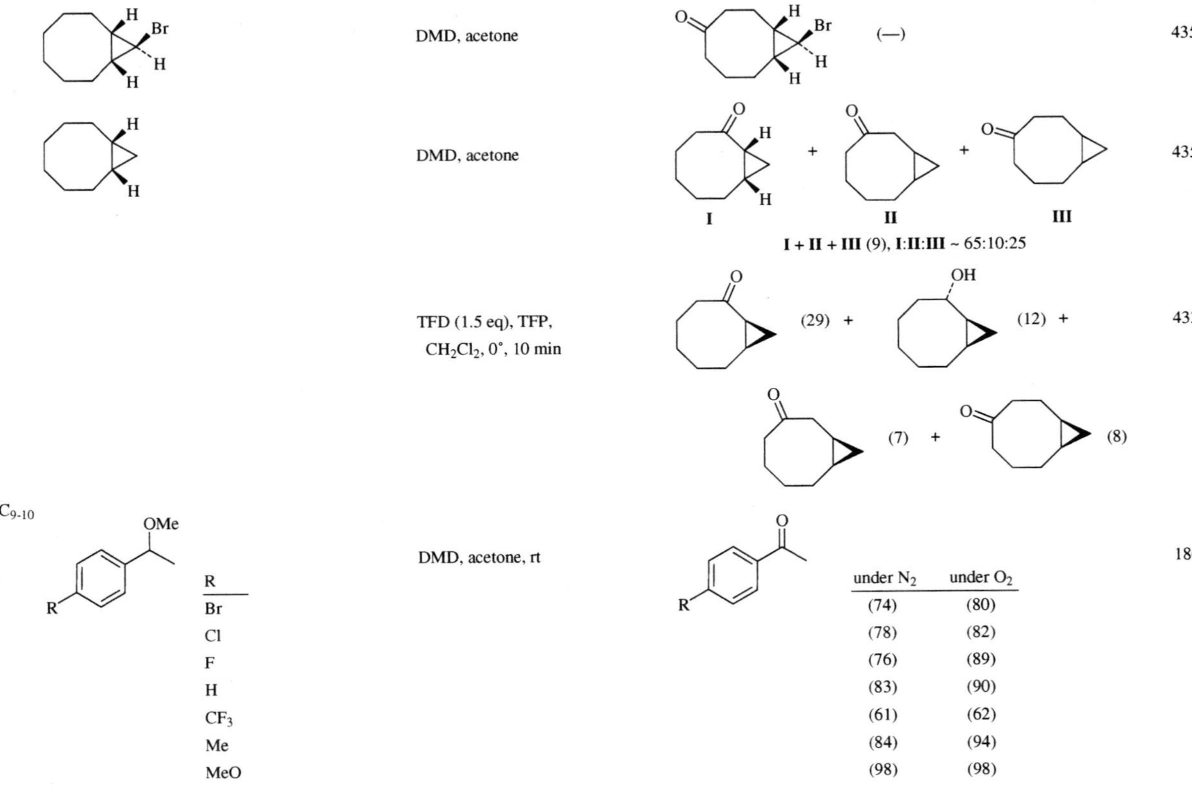

Substrate	Conditions	Product(s) and Yield(s) (%)	Refs.
(8-bromobicyclo[6.1.0]nonane)	DMD, acetone	(ketone) (—)	435
(bicyclo[6.1.0]nonane)	DMD, acetone	**I** + **II** + **III** (9), **I**:**II**:**III** ~ 65:10:25	435
	TFD (1.5 eq), TFP, CH_2Cl_2, 0°, 10 min	(29) + (12) + (7) + (8)	433
C_{9-10} (1-(4-R-phenyl)ethyl methyl ether)	DMD, acetone, rt	(4-R-acetophenone)	186

R	under N_2	under O_2
Br	(74)	(80)
Cl	(78)	(82)
F	(76)	(89)
H	(83)	(90)
CF_3	(61)	(62)
Me	(84)	(94)
MeO	(98)	(98)

TABLE 5A. C–H OXIDATION BY ISOLATED DIOXIRANES (*Continued*)

Substrate	Conditions	Product(s) and Yield(s) (%)	Refs.
C_{9-15}			
4-R-cumene	DMD, acetone, rt	2-(4-R-phenyl)propan-2-ol (—) R / k_{rel} I / 0.14±0.01 H / 1.00±0.00 OH / 10.17±0.50 Me / 1.91±0.01 MeO / 3.58±0.10 Ac / 0.047±0.002 PhO / 7.05±0.40	444
C_{9-19}			
PhCH(R[1])R[2]	DMD, acetone, dark, rt, 3 d	PhC(OH)(R[1])R[2] R[1] / R[2] Me / Me (13.4) Ph / Me (1.0) Ph / *t*-Bu (0.5) Ph / Ph (17.5)	442
C_{10}			
1-(benzofuran-2-yl)ethanol	DMD, acetone, CH_2Cl_2, N_2, 0°, 12 h	2-acetylbenzofuran (38)	89
tetralin	DMD, acetone, dark, rt, 3 d	1-tetralone (13)	442
3-methylisochroman	DMD (2 eq), acetone, rt, 30 min	1-hydroxy-3-methylisochroman (**I**) + 3-methylisochroman-1-one (**II**) **I** + **II** (100), **I**:**II** = 70:30	176
	DMD (3 eq), acetone, rt, 60 min	**II** (100)	176

Substrate	Conditions	Product(s) and Yield(s) (%)	Refs.
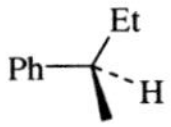 Ph, Et, H 72% ee	TFD, TFP, (CH_2Cl_2), –24°, 1 h	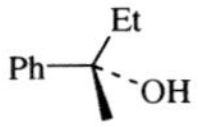 Ph, Et, OH (95) 72% ee	31
	DMD, acetone, CH_2Cl_2, rt, 48 h	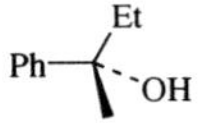 Ph, Et, OH (30) 72% ee	31
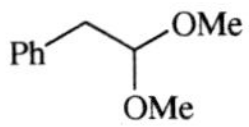 Ph, OMe, OMe	DMD, acetone, CH_2Cl_2, dark, 0-5°, 43 h	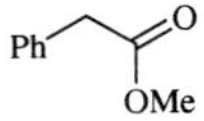 Ph, O, OMe (—)	189
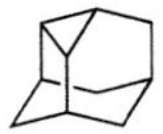	DMD, acetone, 20°, 2 h	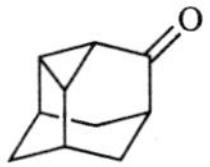 O (29) 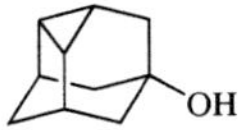 + OH (21)	177
	Cyclohexanone dioxirane, cyclohexanone, CH_2Cl_2, 10°, 6 h	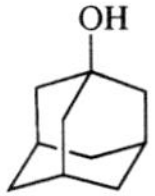 OH **I** (42) 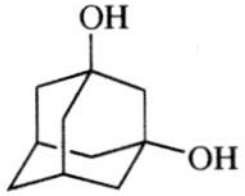 + OH, OH (57)	98
	DMD, acetone, 22°, dark, 18 h	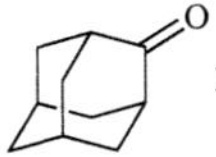 **I** (87) 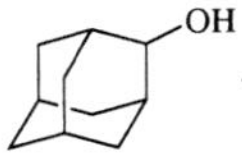 + OH **II** + O **III** **II + III** (2.6)	174
	DMD, acetone, $(CBrCl_3)$	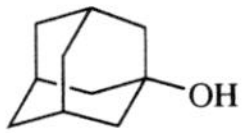 OH (—) + other products	178
	TFD, TFP, CH_2Cl_2, –22°, 1 min	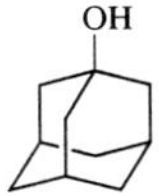 OH (92) 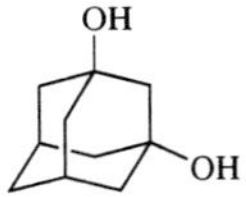 + OH, OH (5)	30
	TFD (2.3 eq), TFP, CH_2Cl_2, –20°, 40 min	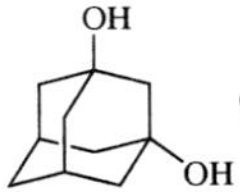 OH, OH (91) 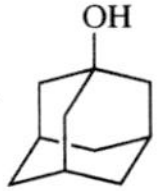 + OH (3)	178

TABLE 5A. C–H OXIDATION BY ISOLATED DIOXIRANES (*Continued*)

Substrate	Conditions	Product(s) and Yield(s) (%)	Refs.
C_{10}			
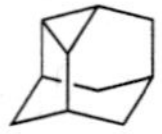	TFD (6 eq), TFP, CH_2Cl_2, –20°, 2 h	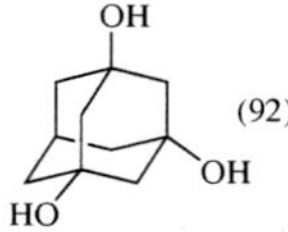 (92) 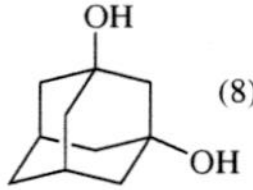 + (8)	178
	TFD (2 eq), TFP, CH_2Cl_2, 0°, 1.5 h	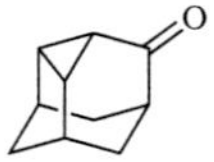 **I** (32) + 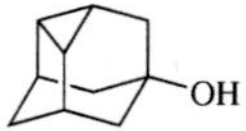 **II** (48)	433
	TFD (4 eq), TFP, CH_2Cl_2, 0°, 1.5 h	**I** (66) + **II** (< 3) + 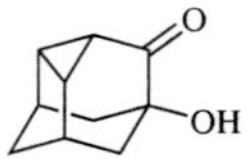 (25)	433
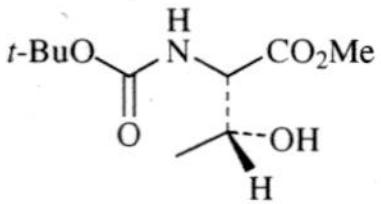 *t*-BuO, N, H, CO_2Me, O, OH, H	DMD, acetone, CH_2Cl_2, rt, 3 d	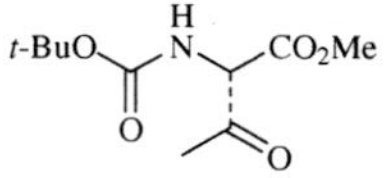 *t*-BuO, N, H, CO_2Me, O, O (61)	93
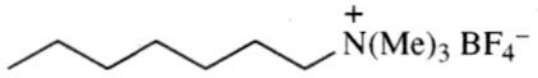 $\overset{+}{N}(Me)_3\ BF_4^-$	TFD, CH_2Cl_2, MeCN (pH 2.0-3.0), 0°, 15 h	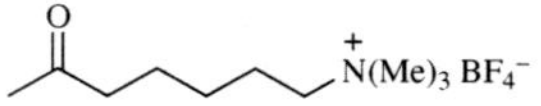 **I** 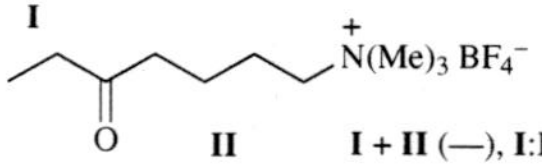 $\overset{+}{N}(Me)_3\ BF_4^-$ + **II** $\overset{+}{N}(Me)_3\ BF_4^-$; **I** + **II** (—), **I**:**II** = 60:40	192
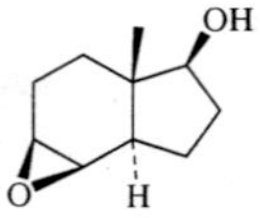 OH, O, H	DMD (1.5 eq), acetone, CH_2Cl_2, 0°, 2.5 h	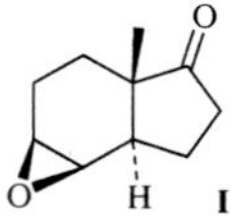 **I** (96)	427
	TFD (1.1 eq), TFP, CH_2Cl_2, 0°, 15 min	**I** (96)	427

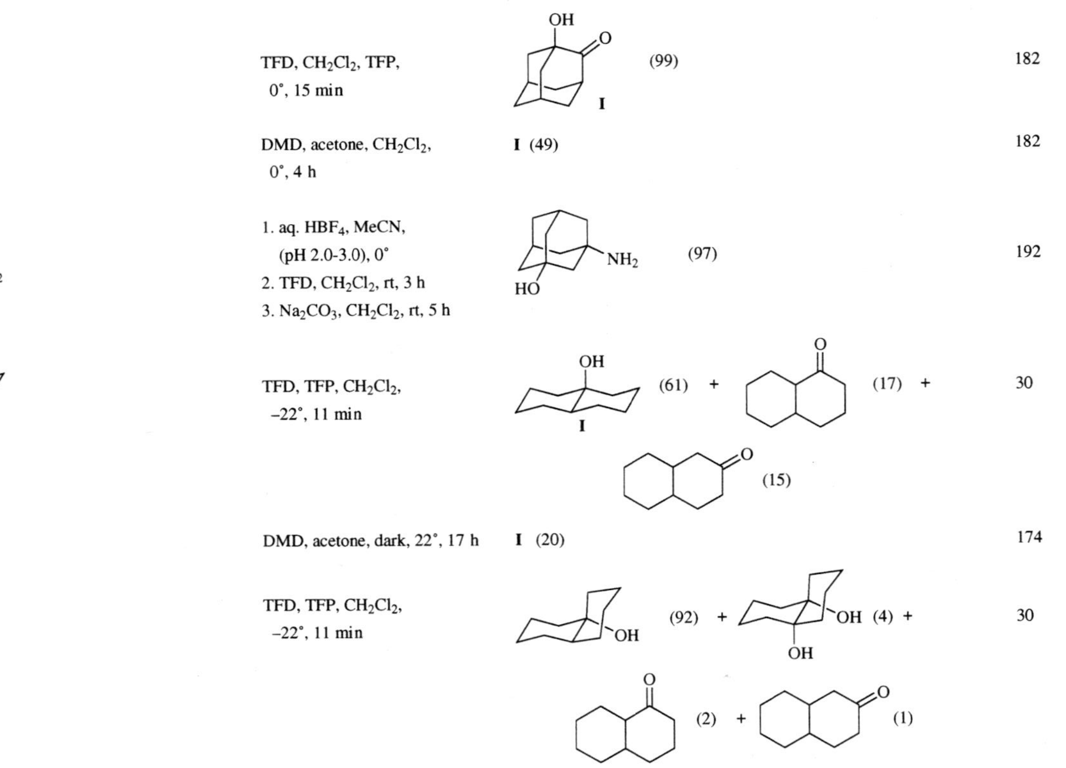

Conditions	Product(s) and Yield(s) (%)	Refs.
TFD, CH_2Cl_2, TFP, 0°, 15 min	**I** (99)	182
DMD, acetone, CH_2Cl_2, 0°, 4 h	**I** (49)	182
1. aq. HBF_4, MeCN, (pH 2.0-3.0), 0° 2. TFD, CH_2Cl_2, rt, 3 h 3. Na_2CO_3, CH_2Cl_2, rt, 5 h	(97)	192
TFD, TFP, CH_2Cl_2, –22°, 11 min	**I** (61) + (17) + (15)	30
DMD, acetone, dark, 22°, 17 h	**I** (20)	174
TFD, TFP, CH_2Cl_2, –22°, 11 min	(92) + (4) + (2) + (1)	30

TABLE 5A. C–H OXIDATION BY ISOLATED DIOXIRANES (*Continued*)

Substrate	Conditions	Product(s) and Yield(s) (%)	Refs.
C_{10}			
	DMD, acetone, 22°, dark, 17 h	OH (84)	174
OH	DMD, acetone, rt, 8 h	O (99)	22
OH, OH	DMD, acetone/CH_2Cl_2, 0°, 2 h	OH, O (94)	182
O, O	TFD, 15 min	O, OH (98)	190
R	DMD, acetone, rt	R, OH (—)	444

R	k_2 (10^{-3} M^{-1} s^{-1})
Br	0.290±0.011
Cl	0.213±0.007
H	2.978±0.099
OH	1.430±0.063
CO_2H	0.957±0.049
Ac	0.406±0.006
CO_2Et	0.557±0.04

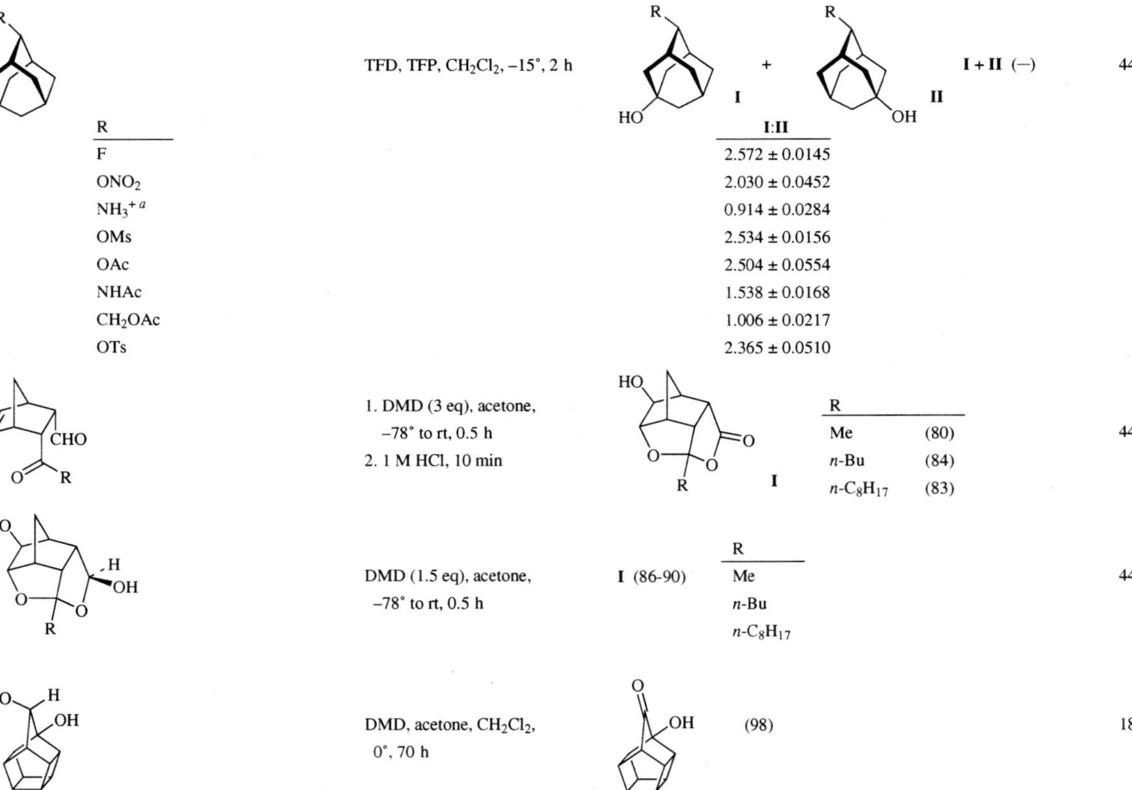

(R-substituted adamantane)	TFD, TFP, CH_2Cl_2, –15°, 2 h	I + II (—)	445

R	I:II
F	2.572 ± 0.0145
ONO_2	2.030 ± 0.0452
NH_3^+ [a]	0.914 ± 0.0284
OMs	2.534 ± 0.0156
OAc	2.504 ± 0.0554
NHAc	1.538 ± 0.0168
CH_2OAc	1.006 ± 0.0217
OTs	2.365 ± 0.0510

1. DMD (3 eq), acetone, –78° to rt, 0.5 h
2. 1 M HCl, 10 min

R	I
Me	(80)
n-Bu	(84)
n-C_8H_{17}	(83)

446

DMD (1.5 eq), acetone, –78° to rt, 0.5 h — **I** (86-90)

R
Me
n-Bu
n-C_8H_{17}

446

C_{11}

DMD, acetone, CH_2Cl_2, 0°, 70 h — (98) — 182

TABLE 5A. C–H OXIDATION BY ISOLATED DIOXIRANES (*Continued*)

Substrate	Conditions	Product(s) and Yield(s) (%)	Refs.
C_{11} 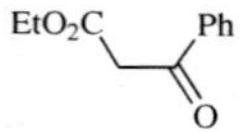 EtO$_2$C, Ph, O	DMD, acetone, CH_2Cl_2, $MgSO_4$	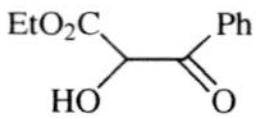 EtO$_2$C, Ph, HO, O (97)	206
Ph	DMD, acetone, atmos., rt, 2 h	HO, Ph **I** + O, Ph **II** + OH, Ph **III** + O, Ph **IV** + OH, O, Ph **V** + OAc, Ph **VI**	38
	Atmos.	I / II / III + IV + V / VI	
	—	(0.3) / (2.1) / (—) / (—)	
	—	(0.3) / (1.7) / (0.05) / (0.14)	
	O_2	(0.3) / (2.2) / (0.08) / (0.01)	
	Ar	(0.3) / (1.7) / (0.05) / (0.09)	
OH, Ph	DMD	O, Ph (83)	38
OH, Ph	DMD	O, Ph (5) + OH, O, Ph (25)	38

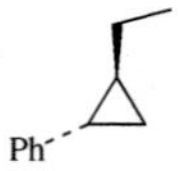

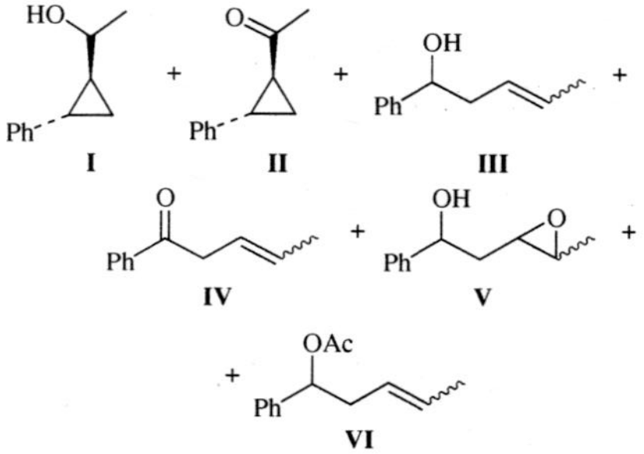

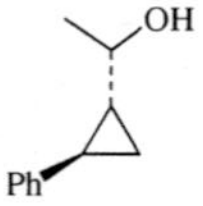

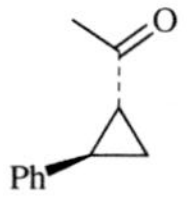

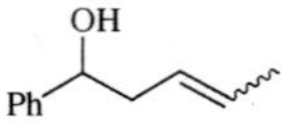

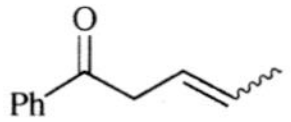

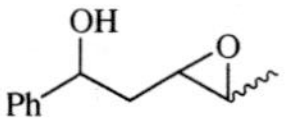

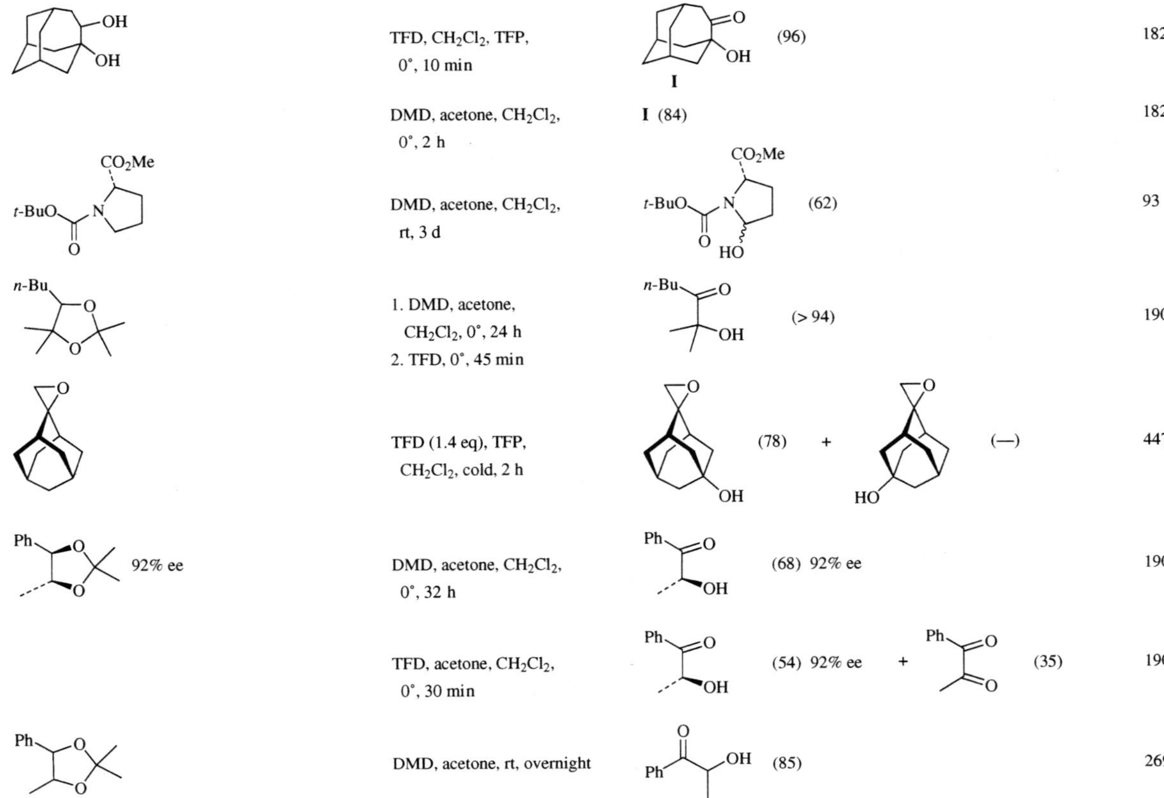

Substrate	Conditions	Product(s) and Yield(s) (%)	Refs.
	TFD, CH_2Cl_2, TFP, 0°, 10 min	**I** (96)	182
	DMD, acetone, CH_2Cl_2, 0°, 2 h	**I** (84)	182
	DMD, acetone, CH_2Cl_2, rt, 3 d	(62)	93
	1. DMD, acetone, CH_2Cl_2, 0°, 24 h 2. TFD, 0°, 45 min	(> 94)	190
	TFD (1.4 eq), TFP, CH_2Cl_2, cold, 2 h	(78) + (—)	447
C_{12} 92% ee	DMD, acetone, CH_2Cl_2, 0°, 32 h	(68) 92% ee	190
	TFD, acetone, CH_2Cl_2, 0°, 30 min	(54) 92% ee + (35)	190
	DMD, acetone, rt, overnight	(85)	269

TABLE 5A. C–H OXIDATION BY ISOLATED DIOXIRANES (*Continued*)

Substrate	Conditions	Product(s) and Yield(s) (%)	Refs.
C_{12}			
[structure] CH_2OH	DMD, acetone, rt, 3 d	[structure] CO_2H (—)	287
BocHN, OMe, O [structure]	DMD, CH_2Cl_2, acetone, rt, 3 d	BocHN, O, O [structure] (42)	93, 448
BocHN, OMe, O [structure]	DMD (6 eq), acetone, CH_2Cl_2, 20°, 120 h	BocHN, O, O [structure] (15)	310
Ph, O, O [structure]	DMD, acetone, 20°	Ph, O, O, OH [structure] (—) + Ph, O, O, OH [structure] (—) $k = 4.52 \pm 0.27 \times 10^3$ l mol^{-1} s^{-1}	424
n-C_6H_{13}–O–C_6H_{13}-n	TFD, TFP, –20°, 30 min	n-$C_6H_{13}OH$ (—) + n-$C_5H_{11}CHO$ (—) + n-$C_5H_{11}CO_2H$ (—) + n-$C_5H_{11}CO_2C_6H_{13}$-n (—) + HO, n-$C_6H_{13}O$, CF_3 [structure] (—) + n-$C_5H_{11}CO_2Me$ (—) + $CF_3CO_2C_6H_{13}$-n (—)	187
[structure]	TFD (1.2 eq), TFP, CH_2Cl_2, 0°, 20 min	HO [structure] (> 90)	433

Substrate	Conditions	Product(s) and Yield(s) (%)	Refs.
C_{12-14}			
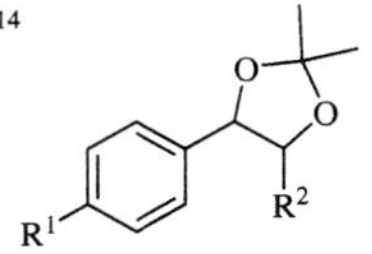 4-$R^1C_6H_4$-substituted 2,2-dimethyl-1,3-dioxolane bearing R^2	DMD (x eq), acetone, CH_2Cl_2, rt	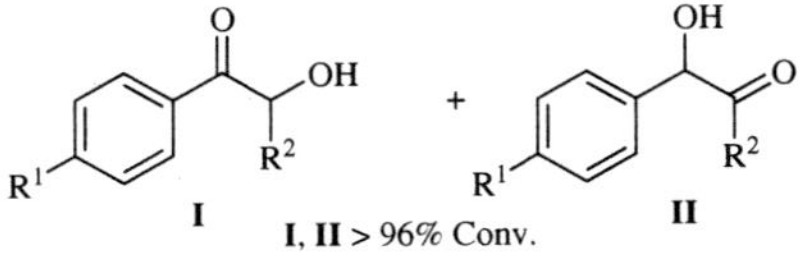 I + II; I, II > 96% Conv.	449

R^1	R^2	x	Time	I	II
H	Me	3	12 h	(85)	(8)
NO_2	Me	5	24 h	(70)	(15)
MeO	Me	3	12 h	(> 96)	(—)
MeO	CO_2Me	5	48 h	(72)	(—)

Substrate	Conditions	Product(s) and Yield(s) (%)	Refs.
C_{13}			
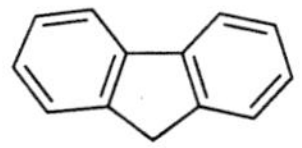 Fluorene	DMD, acetone, dark, rt, 3 d	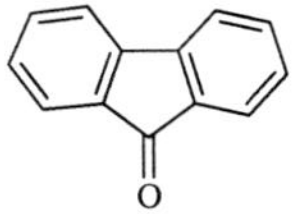 Fluorenone (8.2)	442
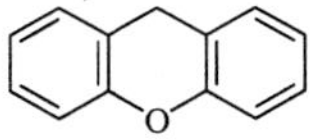 Xanthene	DMD, acetone, rt, 8 h	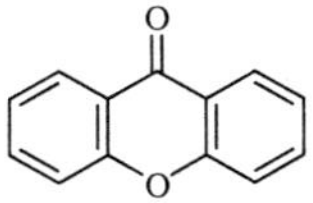 Xanthone (95)	176
 $PhCH_2Ph$	DMD, acetone, rt, 24 h	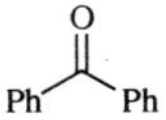 $PhC(O)Ph$ (92)	176
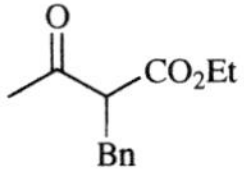 $MeC(O)CH(Bn)CO_2Et$	DMD (4.0 eq), acetone, 20°, 72 h	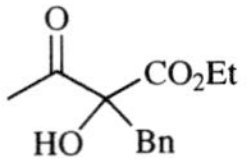 $MeC(O)C(OH)(Bn)CO_2Et$ (100)	193
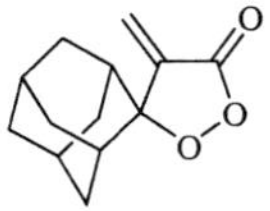 Adamantane-spiro α-methylene-1,2-dioxolan-3-one	DMD, CH_2Cl_2, –20°, 4 h	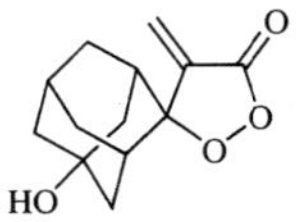 HO-substituted adamantane-spiro α-methylene-1,2-dioxolan-3-one (70)	436

TABLE 5A. C–H OXIDATION BY ISOLATED DIOXIRANES (*Continued*)

Substrate	Conditions	Product(s) and Yield(s) (%)	Refs.
C_{13}			
O, O; 97% ee	TFD, 0°, 35 min	O, OH (> 70) 96% ee	190
 O	TFD (1.4 eq), TFP, CH_2Cl_2, cold, 2 h	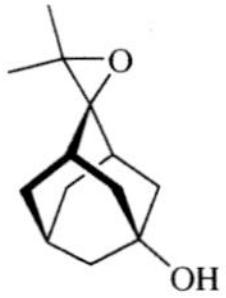 O, OH (42)  + O, HO (28)	447
C_{14}			
	DMD, acetone, rt, 24 h	O, O (87)	176
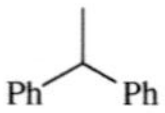 Ph, Ph	TFD, TFP, dark, 0°	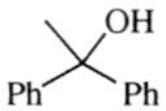 OH, Ph, Ph (—)	443
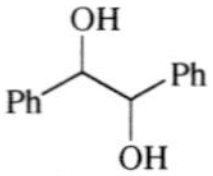 OH, Ph, Ph, OH	DMD, acetone, rt, overnight	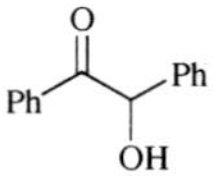 O, Ph, Ph, OH (> 96)	269
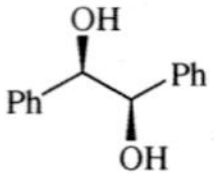 OH, Ph, Ph, OH; 96% ee	DMD, CH_2Cl_2, acetone, 0°, 48 h	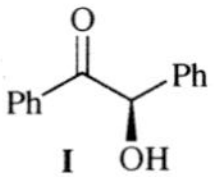 O, Ph, Ph, OH **I** (46) > 92% ee	184
	TFD, CH_2Cl_2, TFP, 0°, 100 min	**I** (88) > 92% ee	184

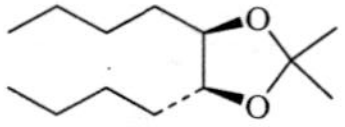

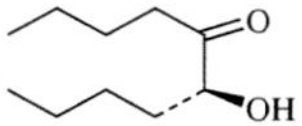

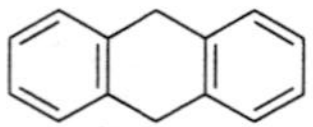

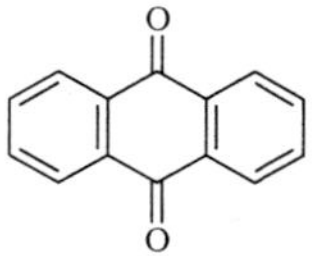

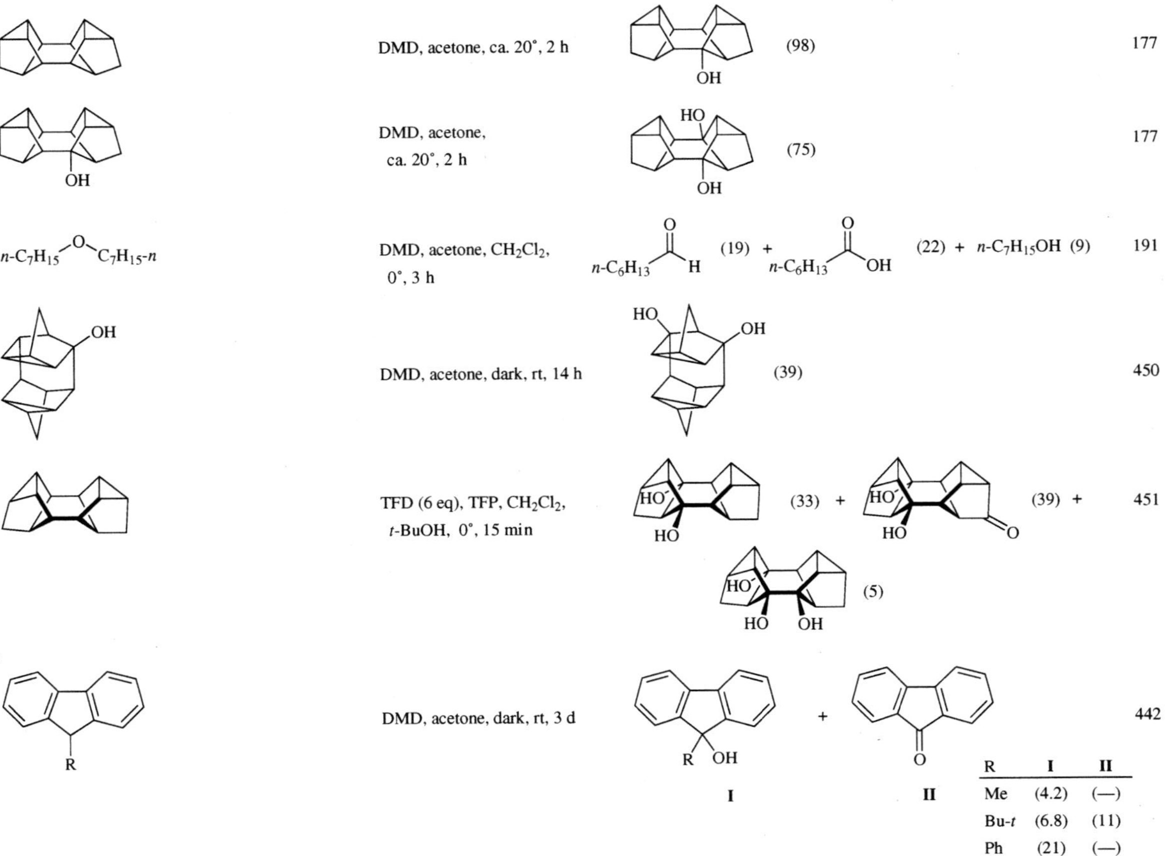

R	I	II
Me	(4.2)	(—)
Bu-t	(6.8)	(11)
Ph	(21)	(—)

TABLE 5A. C–H OXIDATION BY ISOLATED DIOXIRANES (*Continued*)

Substrate	Conditions	Product(s) and Yield(s) (%)	Refs.
C_{15}			
O, Ph, H	DMD, acetone, CH_2Cl_2, rt	O, Ph, OH (27) + O, Ph (14) + O, OH, Ph (10)	452
O, O, Ph, Ph	DMD (4.0 eq), acetone, N_2, 20°, 96 h	O, O, O, Ph, Ph (15) + O, O, Ph, Ph, HO, OH (85)	193
BnO, O, O, O, O	DMD, acetone, CH_2Cl_2, dark, 48 h	HO, O, O, O, O (93)	188
EtO_2C, Ph, O, N, O	DMD, acetone, CH_2Cl_2, rt, 3 d	EtO_2C, Ph, O, N, O, HO (58)	93
BocHN, O, N, H, CO_2Me	DMD, acetone, CH_2Cl_2, rt, 3 d	BocHN, O, N, H, O, O (35)	93, 448

Substrate	Conditions	Product(s) and Yield(s) (%)	Refs.
3-Phenylisochroman-1-one (Ph)	DMD (xs), acetone, CH_2Cl_2, rt	2-(2-oxo-2-phenylethyl)benzoic acid (CO_2H, Ph) (100)	449
3-Aryl isochroman (R)	DMD (x eq), acetone, CH_2Cl_2, rt	I (OH) + II + III (CO_2H)	449

R	x	Time	% Conv.	**I**	**II**	**III**
3'-F	1	8 h	60	(46)	(14)	(—)
3'-F	3	8 h	> 95	(—)	(45)	(34)
4'-F	1	12 h	60	(52)	(8)	(—)
4'-F	3	8 h	> 95	(—)	(58)	(25)
4'-NO_2	1	12 h	63	(38)	(25)	(—)
4'-NO_2	3	18 h	> 95	(—)	(68)	(17)
H	1	6 h	75	(59)[c]	(16)	(—)
H	3	12 h	> 95	(—)	(56)	(34)
3'-OMe	3	5 h	90	(—)	(56)	(34)
3'-OMe	5	12 h	> 95	(—)	(—)	(70)
4'-OMe	1	4 h	65	(—)	(32)[d]	(—)
4'-OMe	3	10 h	> 95	(19)[d]	(30)	(65)

TABLE 5A. C–H OXIDATION BY ISOLATED DIOXIRANES (*Continued*)

Substrate	Conditions	Product(s) and Yield(s) (%)	Refs.
C_{16} PMP	DMD, acetone, rt	PMP (40)	68
Ph, Ph, OH	DMD, acetone, rt, 11 h	Ph, Ph (100)	453
BnO	DMD, CH_2Cl_2, acetone, dark, 20°, 45 h	HO (35)	189
BnO	DMD, acetone, CH_2Cl_2, dark, rt, 48 h	HO (89)	188
HO, BnO, OH	DMD, acetone, CH_2Cl_2, dark, rt, 48 h	OH, HO, OH, BnO, OH (90)	188
BocHN, CO_2Et	DMD, acetone, CH_2Cl_2, rt, 3 d	BocHN, CO_2Et, HO (38)	93, 448
Ph, Ph 97% ee	TFD, 50 min	Ph, Ph, OH (92) 97% ee	190

C_{17-21}

DMD, acetone, CH_2Cl_2, rt

R^1	R^2	
H	OAc	(63)
OAc	H	(57)

452

C_{18}

DMD, acetone, CH_2Cl_2, rt, 3 d

(33) + (1)

93, 448

DMD, CH_2Cl_2, acetone, 20°, 40 h

(69)

179

C_{19}

DMD, acetone, CH_2Cl_2, 0°

(47) + (20)

454

DMD, acetone, 60 h

I + II + III

455

Temp	I	II	III
20°	(76)	(20)	(—)
30°	(46)	(7)	(11)

TABLE 5A. C–H OXIDATION BY ISOLATED DIOXIRANES (*Continued*)

Substrate	Conditions	Product(s) and Yield(s) (%)	Refs.
C_{19-23} OR, Ar =	DMD, acetone, CH_2Cl_2, dark, rt, 48 h	OH; R: $4-BrC_6H_4CH_2$ (85), Bn (87), $4-CNC_6H_4CH_2$ (87), 2-naphthyl-CH_2 (90)	188
C_{20} Ar =	1. DMD, acetone, –20°, 2 d 2. Ac_2O, pyridine	OAc, **I** (80)	456
Ar =	1. DMD, acetone, –20°, 2 d 2. Ac_2O, pyridine	**I** (75)	456
AcO	DMD, acetone, 20°, 22 h	AcO, OH (74)	457
AcO	DMD, acetone, 20°, 22 h	AcO, OH (80)	457

	Substrate	Conditions	Product(s) and Yield(s) (%)	Refs.
C_{21}	BnO, OBn, O	DMD, CH_2Cl_2, acetone, dark, 22°, 48 h	HO, OBn, O (34)	189
	Ac, H, H, O, H	DMD, CH_2Cl_2, acetone, rt, 60 h	Ac, H, H, O, OH (40) + Ac, H, OH, O, H (11)	458
C_{21-23}	R^1, R^2, H, H, H, AcO, H	DMD (4 eq), acetone, CH_2Cl_2, rt, 24 h	R^1, R^2, H, H, OH, AcO, H	459
			R^1 / R^2: C=O (—); H / H (43); OAc / H (—); Et / H (36); *i*-Pr / H (36)	
C_{22}	O, Ar, H, H, Ar, O, OAc; Ar = 3,4-methylenedioxyphenyl (O, O)	DMD, acetone, –20°, 2 d	HO, O, Ar, H, H, Ar, O, OAc (66)	456
	H, H, H, AcO, H	DMD (4 eq), acetone, CH_2Cl_2, rt, 24 h	OH, H, H, H, AcO, H (45)	459

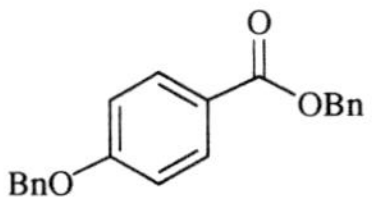

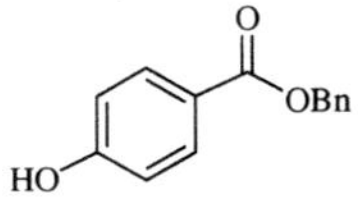

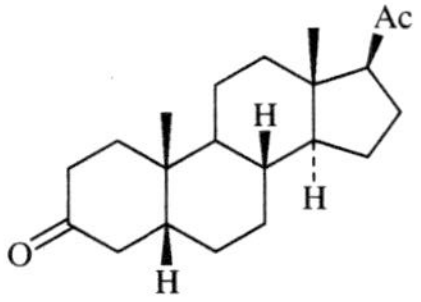

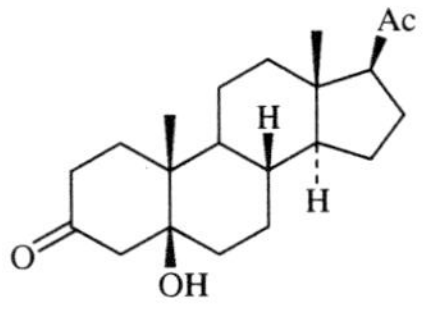

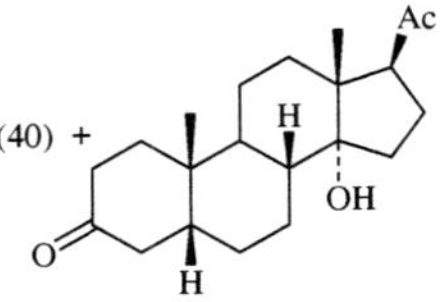

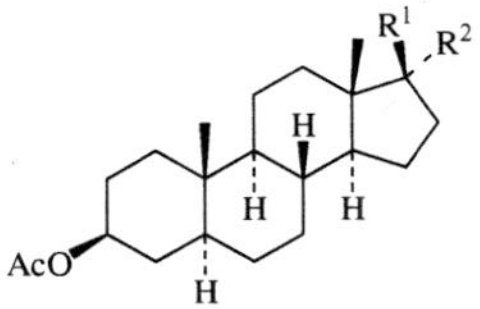

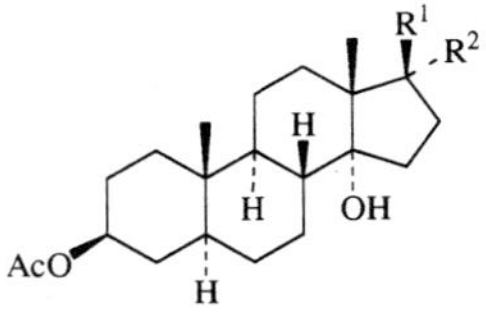

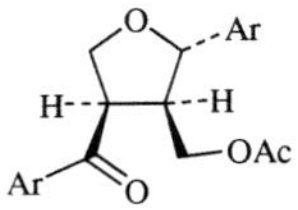

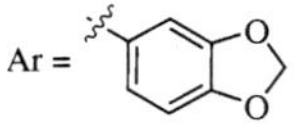

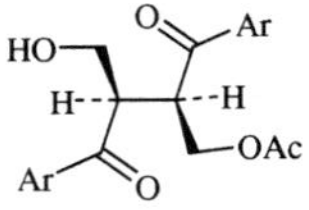

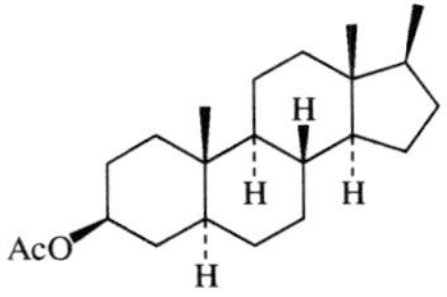

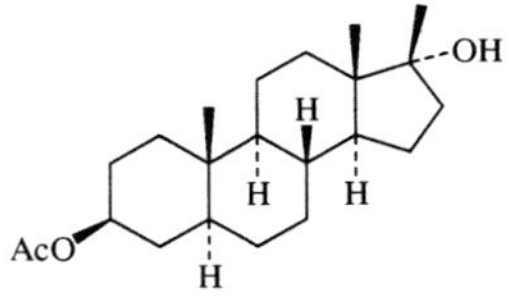

TABLE 5A. C–H OXIDATION BY ISOLATED DIOXIRANES (*Continued*)

Substrate	Conditions	Product(s) and Yield(s) (%)	Refs.
C_{22} 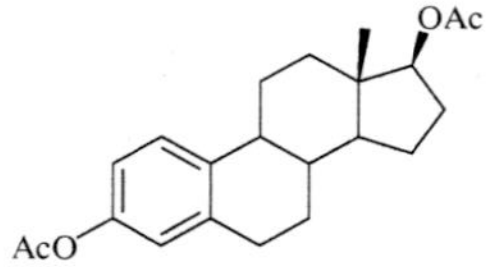 AcO, OAc	DMD, acetone	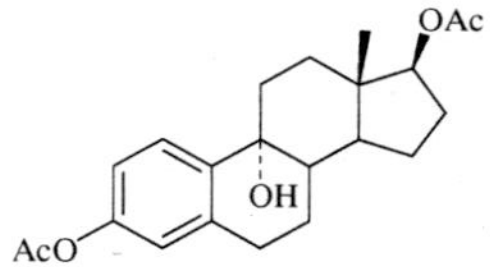 AcO, OAc, OH (75)	460
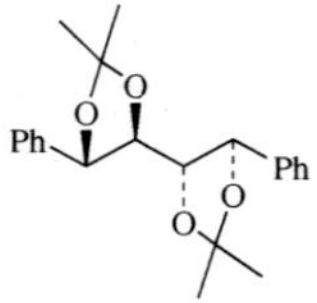 Ph, O, O, Ph, O, O	TFD, TFP, CH_2Cl_2, 0°, 2.5 h	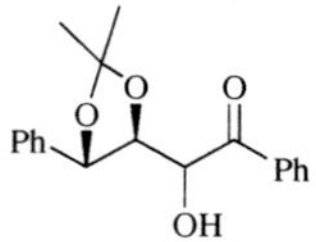 Ph, O, O, O, Ph, OH (35)	190
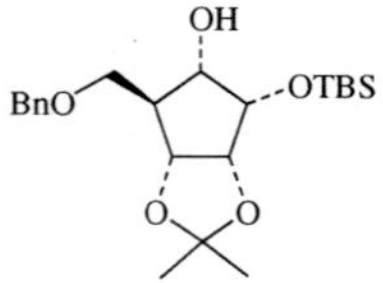 OH, BnO, OTBS, O, O	DMD, acetone, CH_2Cl_2, dark, rt, 48 h	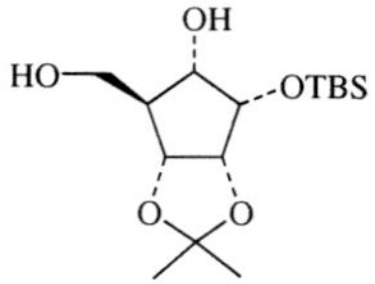 OH, HO, OTBS, O, O (88)	188
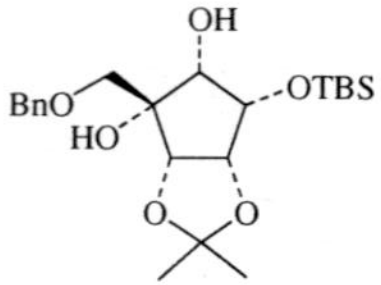 OH, BnO, HO, OTBS, O, O	DMD, acetone, CH_2Cl_2, dark, rt, 48 h	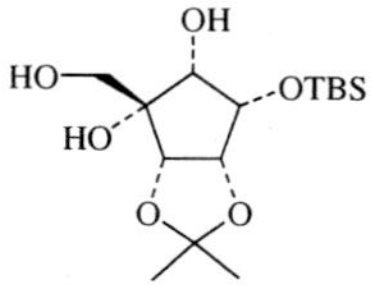 OH, HO, HO, OTBS, O, O (85)	188
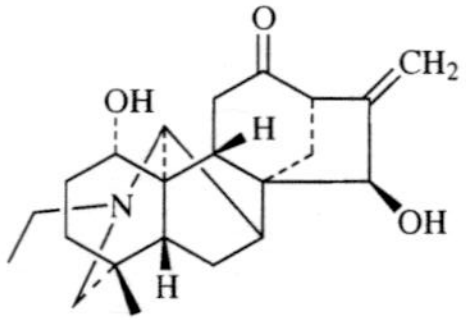 O, CH₂, OH, H, N, OH, H	DMD, acetone, *t*-BuOH, –20°	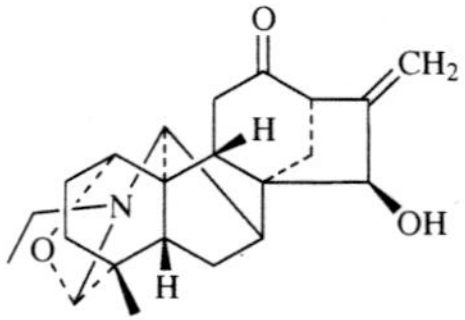 O, CH₂, H, N, O, OH, H (19)	338

C_{22-34}

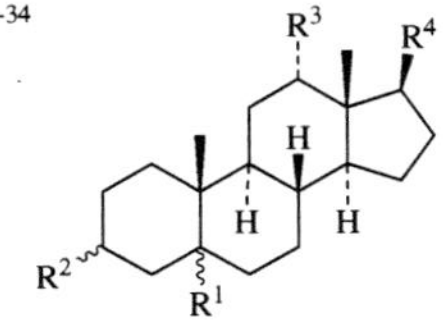

DMD, acetone, rt, 18-54 h 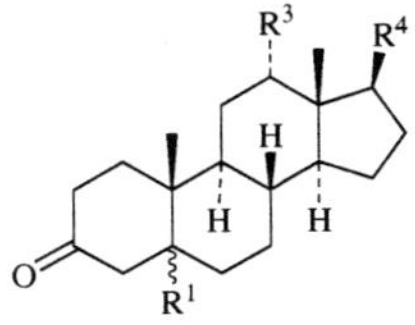 — 461

R^1	R^2	R^3	R^4	Yield
α-H	β-MeO	H	Ac	(95)
α-H	β-BnO	H	Ac	(82)
β-H	α-MeO	OAc	$CH(Me)(CH_2)_2CO_2Me$	(95)
β-H	α-MeO	H	$CH(Me)(CH_2)_3CH(Me)_2$	(57)
α-H	α-MeO	H	$CH(Me)(CH_2)_3CH(Me)_2$	(75)
α-H	β-MeO	H	$CH(Me)(CH_2)_3CH(Me)_2$	(78)
α-H	α-BnO	H	$CH(Me)(CH_2)_3CH(Me)_2$	(88)
α-H	β-BnO	H	$CH(Me)(CH_2)_3CH(Me)_2$	(91)

C_{23}

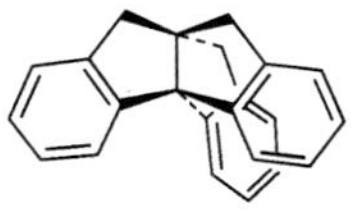

DMD, acetone, 4°, 120 h 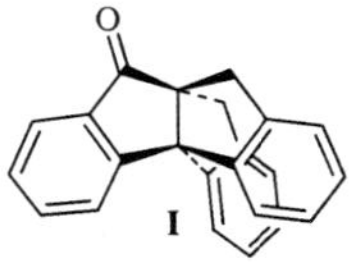 — I (37) — 462

TFD, TFP, CH_2Cl_2, 0°, 40 min — I (76) + 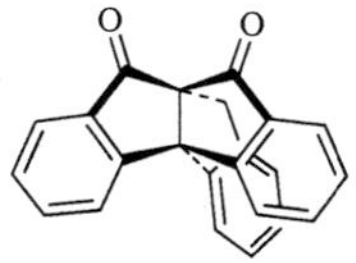 (19) — 462

TABLE 5A. C–H OXIDATION BY ISOLATED DIOXIRANES (*Continued*)

Substrate	Conditions	Product(s) and Yield(s) (%)	Refs.
C_{23}			
	DMD, acetone, 4°, 96 h	O (28) + HO (16) I II	462
	TFD, TFP, CH_2Cl_2, –20°, 2.5 h	**I** (16) + **II** (15) 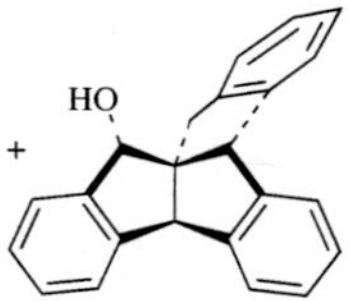 + HO (8)	462
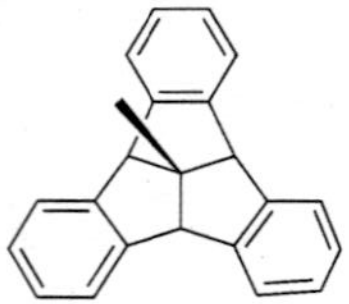	DMD, acetone, 20°, 120 h	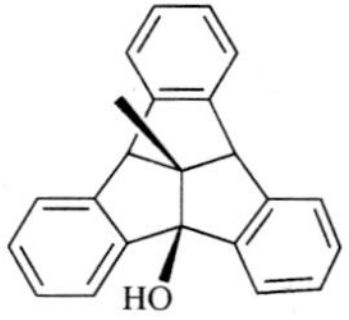 HO **I** (10)	462
	TFD, TFP, CH_2Cl_2, –20°, 40 min	**I** (35)	462
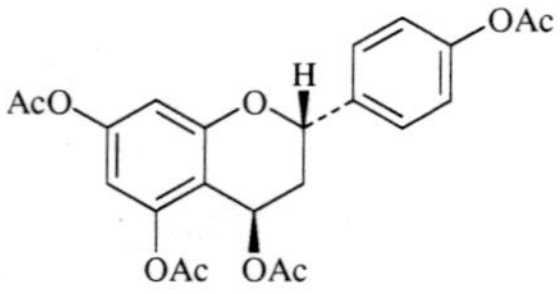 AcO, H, OAc, OAc OAc	DMD, acetone, CH_2Cl_2, rt	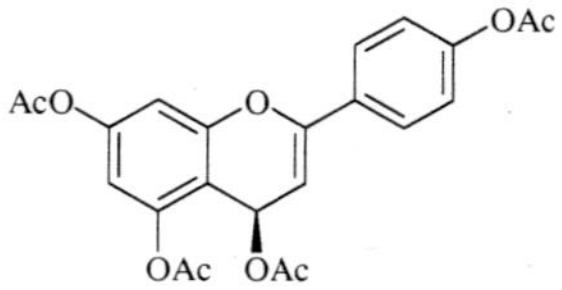 AcO, OAc, OAc OAc (—)	452

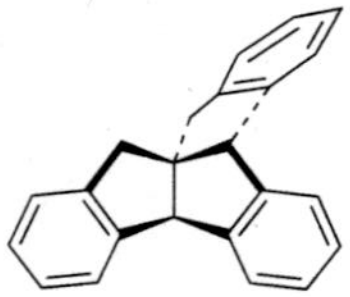

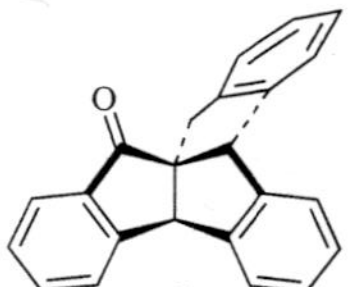

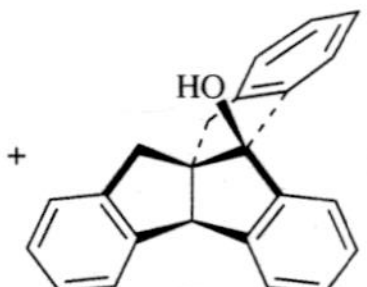

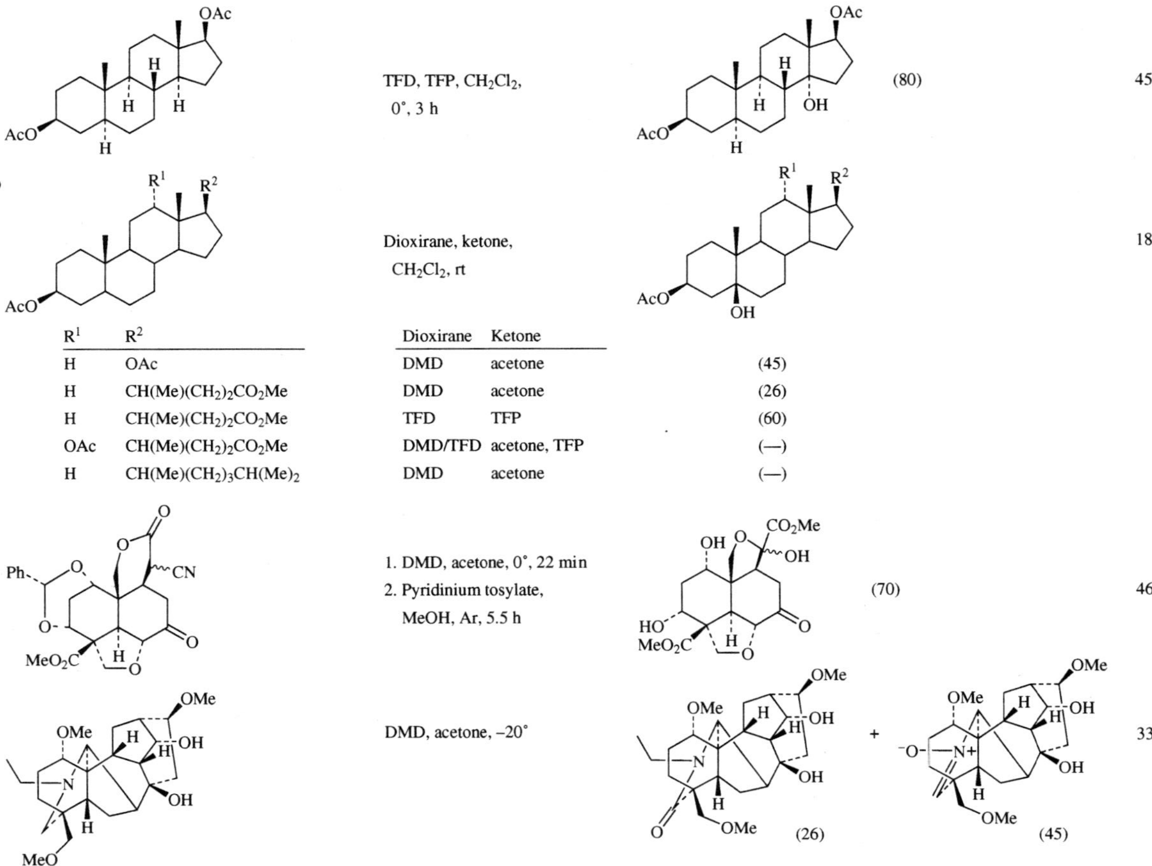

R^1	R^2	Dioxirane	Ketone	
H	OAc	DMD	acetone	(45)
H	$CH(Me)(CH_2)_2CO_2Me$	DMD	acetone	(26)
H	$CH(Me)(CH_2)_2CO_2Me$	TFD	TFP	(60)
OAc	$CH(Me)(CH_2)_2CO_2Me$	DMD/TFD	acetone, TFP	(—)
H	$CH(Me)(CH_2)_3CH(Me)_2$	DMD	acetone	(—)

TABLE 5A. C–H OXIDATION BY ISOLATED DIOXIRANES (*Continued*)

Substrate	Conditions	Product(s) and Yield(s) (%)	Refs.
C_{24}			
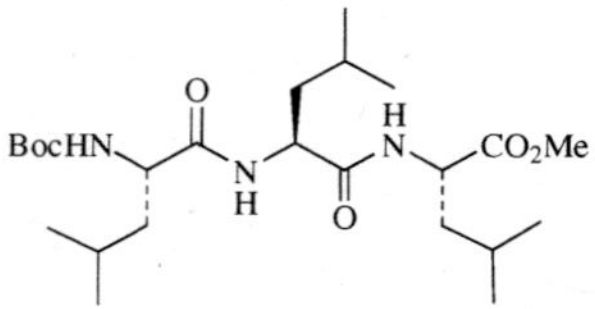 BocHN, N-H, H-N, CO_2Me, O, O	DMD, acetone, rt, 3 d	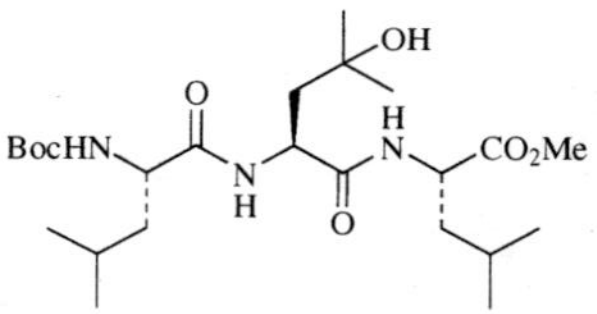 BocHN, N-H, H-N, CO_2Me, OH, O, O (40) 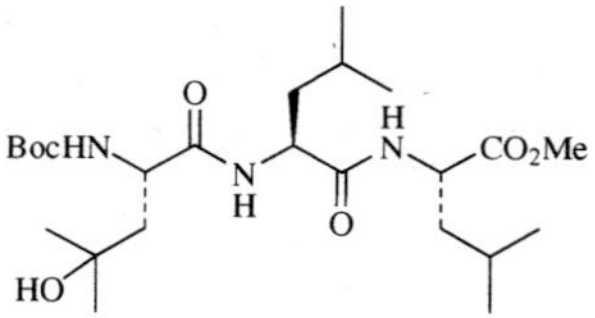 + BocHN, N-H, H-N, CO_2Me, HO, O, O (13)	93
C_{25}			
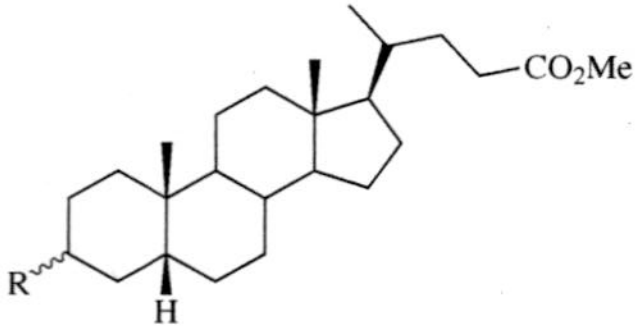 R, H, CO_2Me	DMD (2 eq), $CHCl_3$, CH_2Cl_2, acetone, rt, 0.5 h	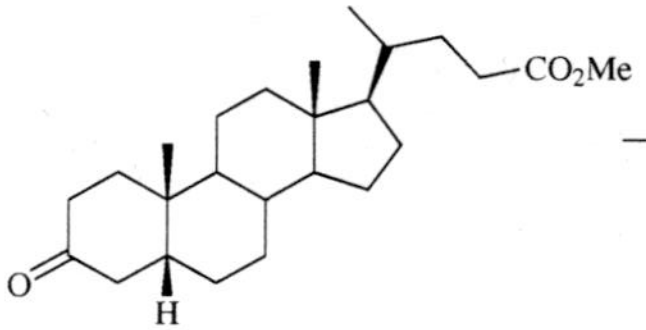 O, H, CO_2Me R: α-OH (90); β-OH (83)	464
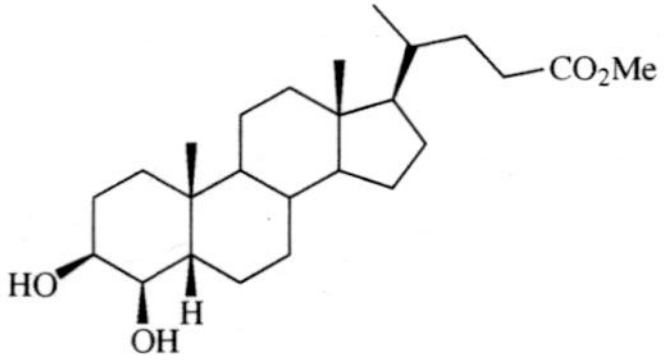 HO, OH, H, CO_2Me	DMD (2 eq), $CHCl_3$, CH_2Cl_2, acetone, rt, 1.5 h	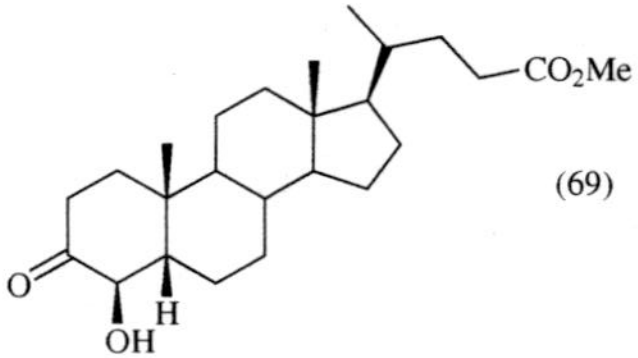 O, OH, H, CO_2Me (69)	464

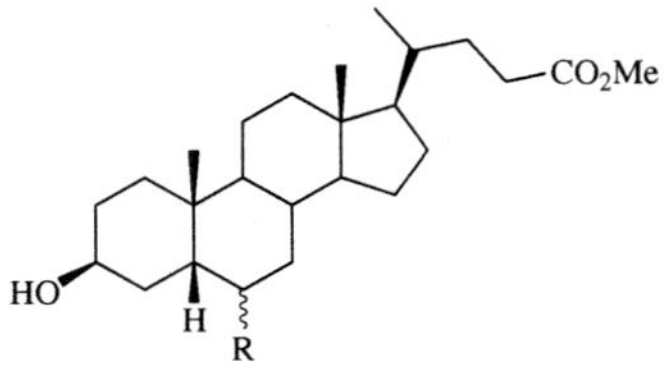

DMD (2 eq), $CHCl_3$,
CH_2Cl_2, acetone, rt, 1.5 h

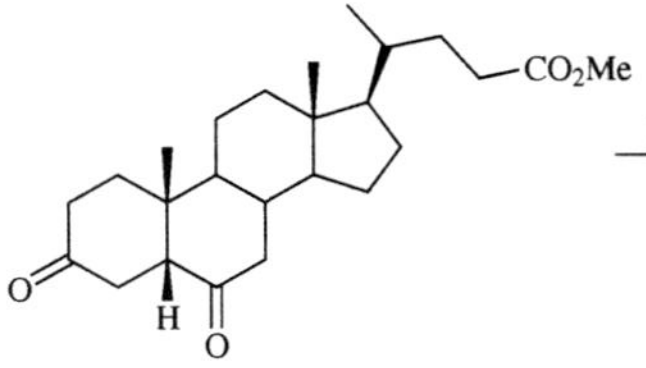

CO_2Me

R		
α-OH	(87)	464
β-OH	(84)	

DMD (2 eq), $CHCl_3$,
CH_2Cl_2, acetone, rt, 6 h

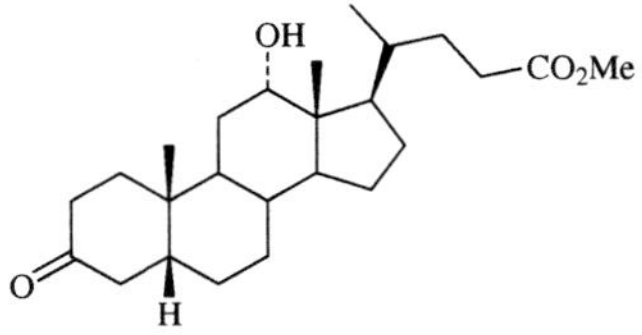

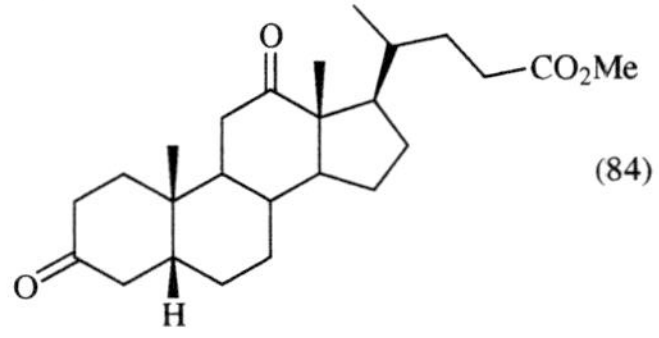

CO_2Me

(84) 464

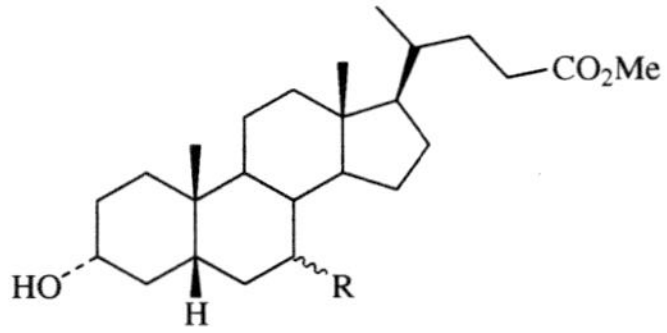

DMD (2 eq), $CHCl_3$,
CH_2Cl_2, acetone, rt, 1 h

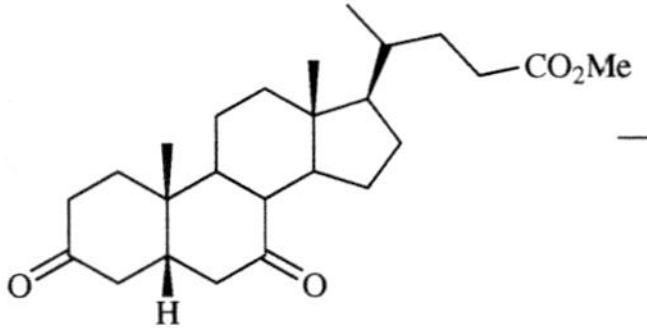

CO_2Me

R		
α-OH	(80)	464
β-OH	(84)	

DMD (2 eq), $CHCl_3$,
CH_2Cl_2, acetone, rt, 0.5 h

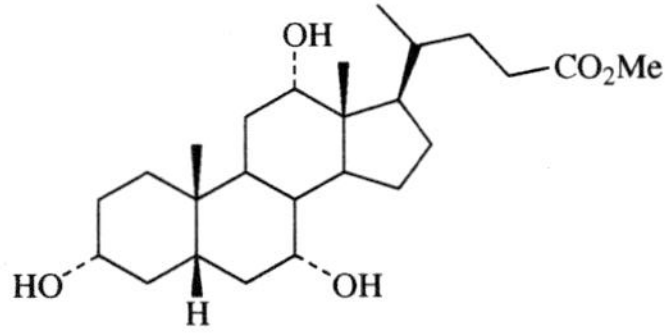

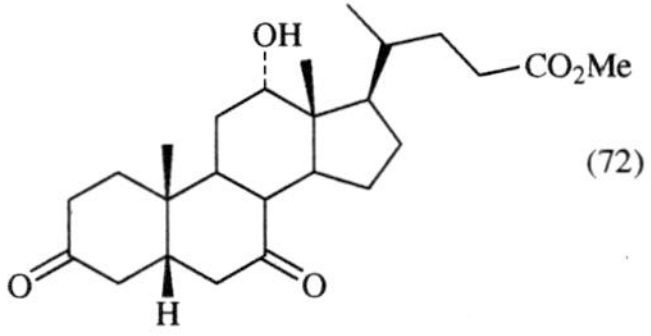

CO_2Me

(72) 464

TABLE 5A. C–H OXIDATION BY ISOLATED DIOXIRANES (*Continued*)

Substrate	Conditions	Product(s) and Yield(s) (%)	Refs.
C_{25}			
HO, H, OH, OH, CO_2Me	DMD (2 eq), $CHCl_3$, CH_2Cl_2, acetone, rt, 1.5 h	O, H, OH, O, CO_2Me (83)	464
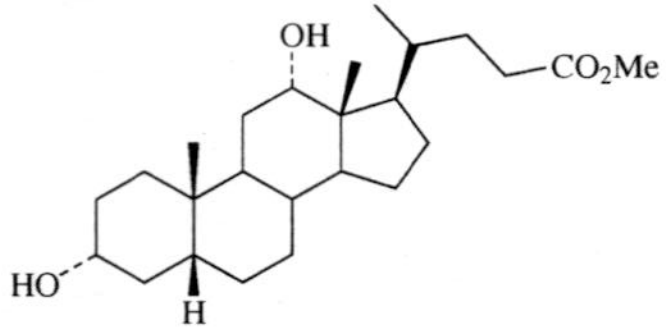 OH, HO, H, CO_2Me	DMD (2 eq), $CHCl_3$, CH_2Cl_2, acetone, rt, 0.5 h	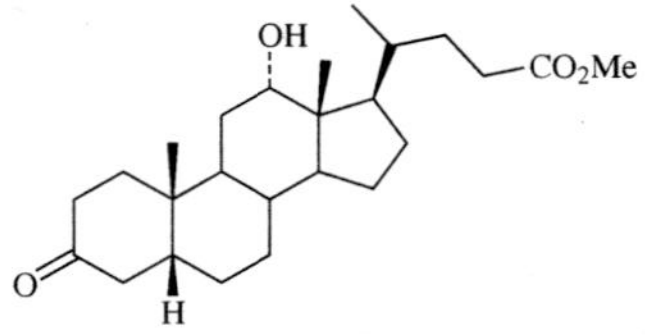 OH, O, H, CO_2Me (87)	464
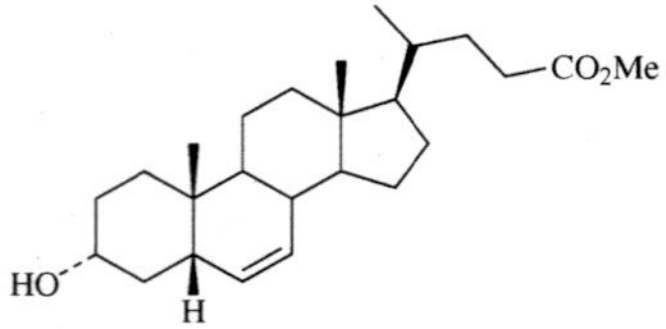 HO, H, CO_2Me	DMD (2 eq), $CHCl_3$, CH_2Cl_2, acetone, rt, 1 h	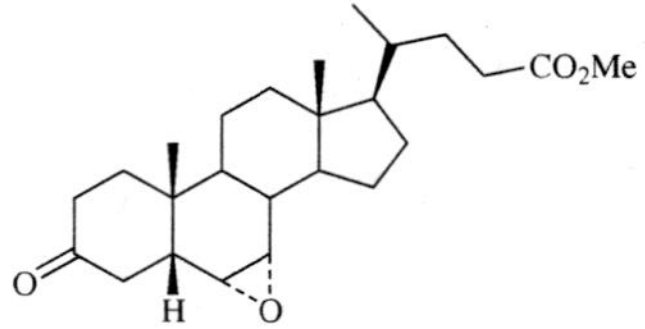 O, H, O, CO_2Me (55)	464
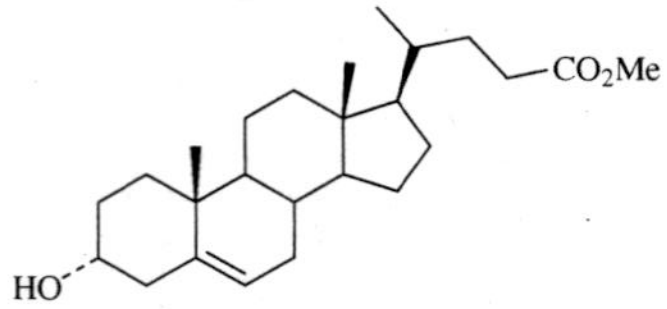 HO, CO_2Me	DMD (2 eq), $CHCl_3$, CH_2Cl_2, acetone, rt, 1 h	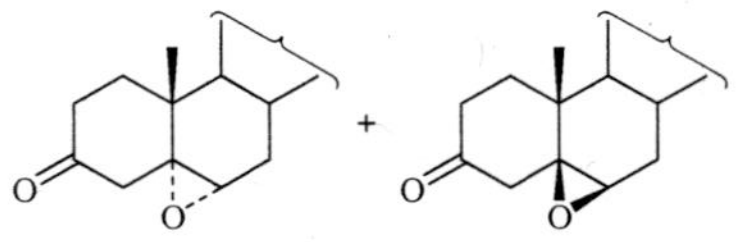 O, O (**I**) + O, O (**II**) **I + II** (83), **I**:**II** = 3:1	464

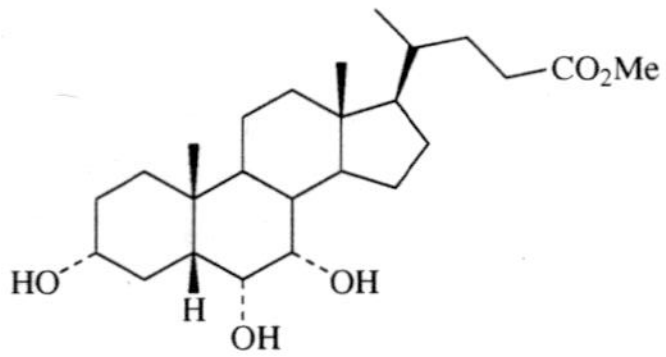

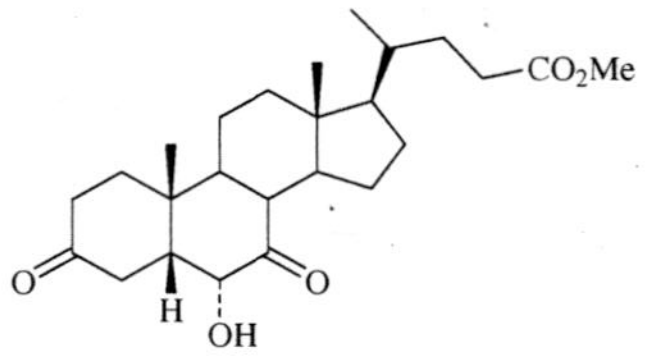

Substrate	Conditions	Product(s) and yield(s) (%)	Refs.
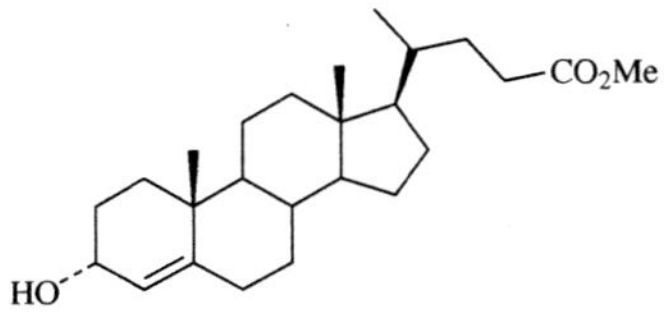	DMD (2 eq), $CHCl_3$, CH_2Cl_2, acetone, rt, 1 h	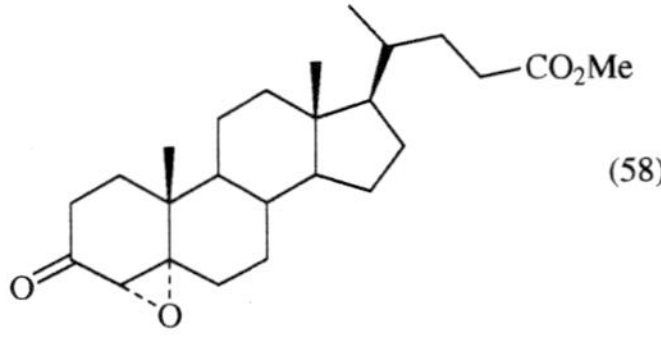 (58)	464
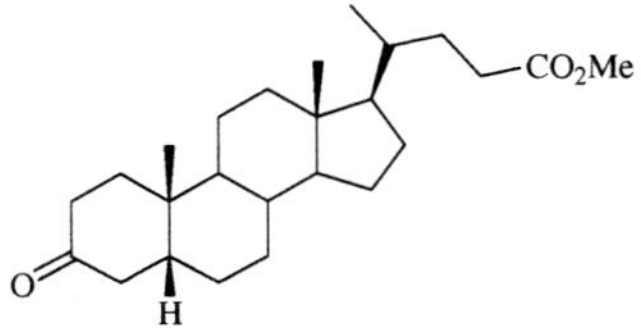	DMD (2 eq), $CHCl_3$, CH_2Cl_2, rt, 36 h	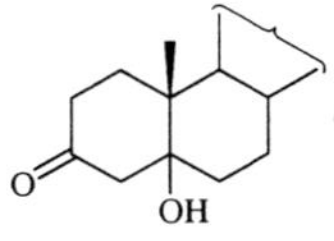 (17) + 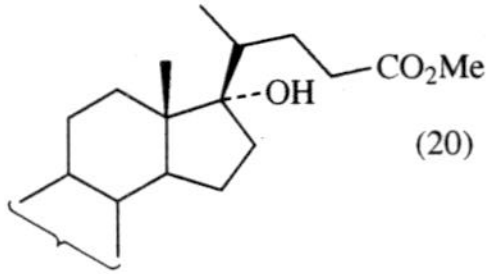 (20) + 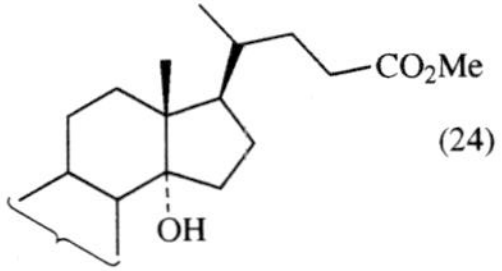 (24)	465
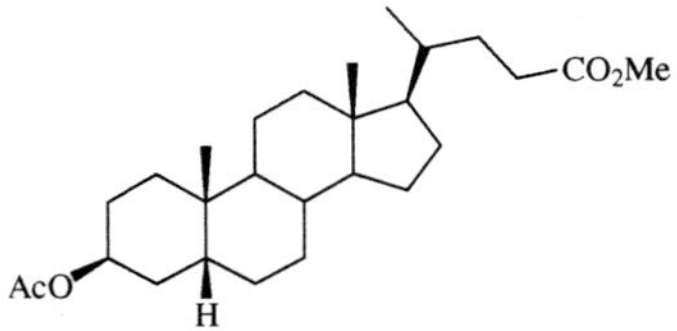 C_{27}	DMD (2 eq), $CHCl_3$, CH_2Cl_2, rt, 36 h	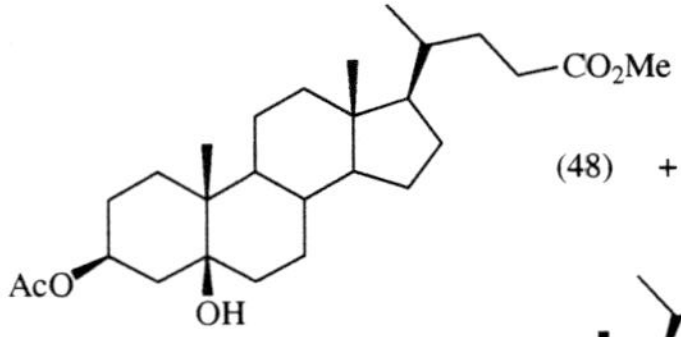 (48) 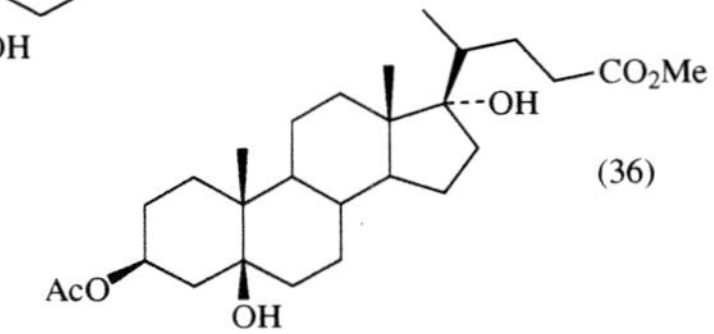 + (36)	466

TABLE 5A. C–H OXIDATION BY ISOLATED DIOXIRANES (*Continued*)

Substrate	Conditions	Product(s) and Yield(s) (%)	Refs.
C_{27}			
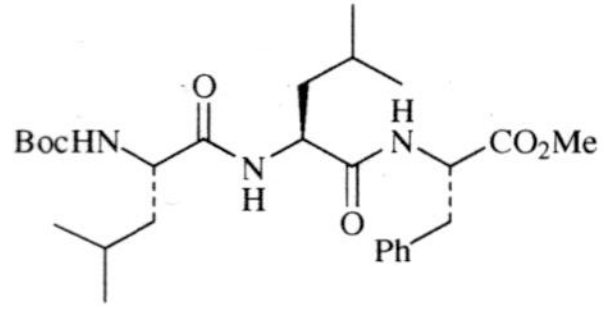 BocHN, O, N H, H N, O, Ph, CO_2Me	DMD, acetone, rt, 3 d	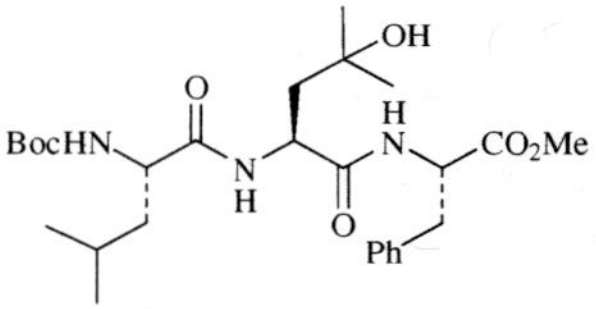 BocHN, O, N H, OH, H N, O, Ph, CO_2Me (43)	93
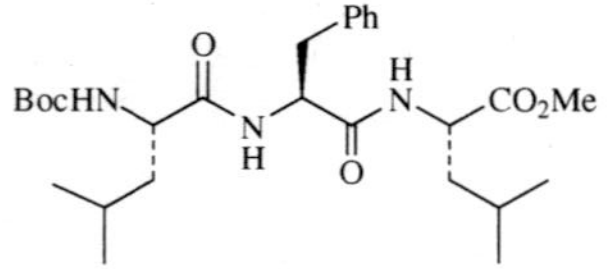 BocHN, O, N H, Ph, H N, O, CO_2Me	DMD, acetone, rt, 3 d	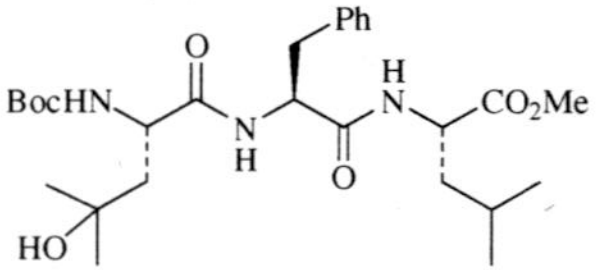 BocHN, HO, O, N H, Ph, H N, O, CO_2Me (41)	93
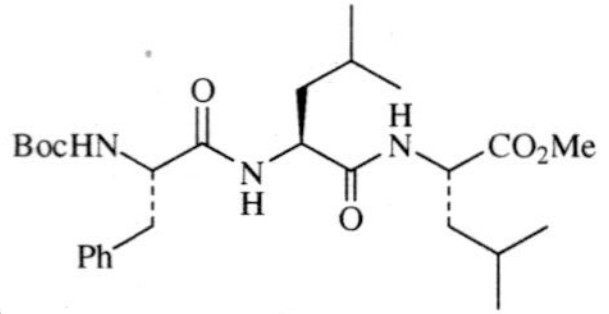 BocHN, Ph, O, N H, H N, O, CO_2Me	DMD, acetone, rt, 3 d	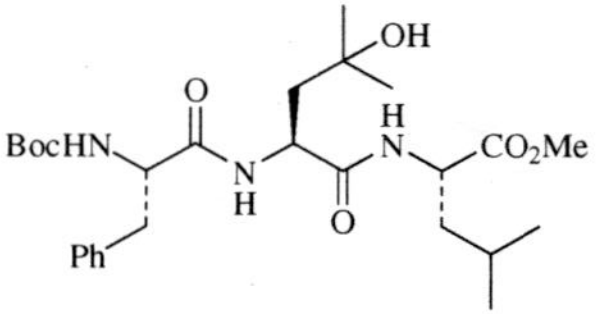 BocHN, Ph, O, N H, OH, H N, O, CO_2Me (38)	93
$C_{27\text{-}29}$			
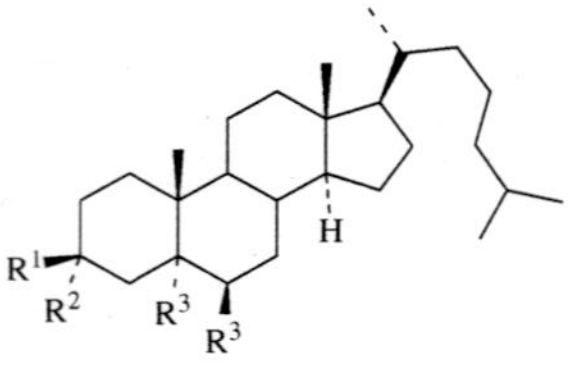 R^1, R^2, R^3, R^3, H	DMD, CH_2Cl_2, acetone, 20°, 24 h	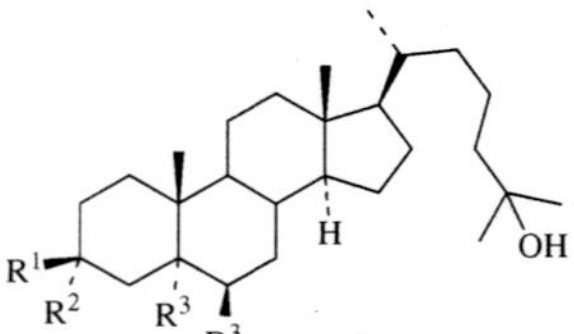 R^1, R^2, R^3, R^3, H, OH **I** R^1 R^2 R^3 C=O H (29) H OAc Br (44) H OAc H (24)	179
	TFD, CH_2Cl_2, TFP, –40 to 0°, 3 h	**I** R^1 R^2 R^3 C=O H (74) H OAc Br (62) H OAc Br (66)	179

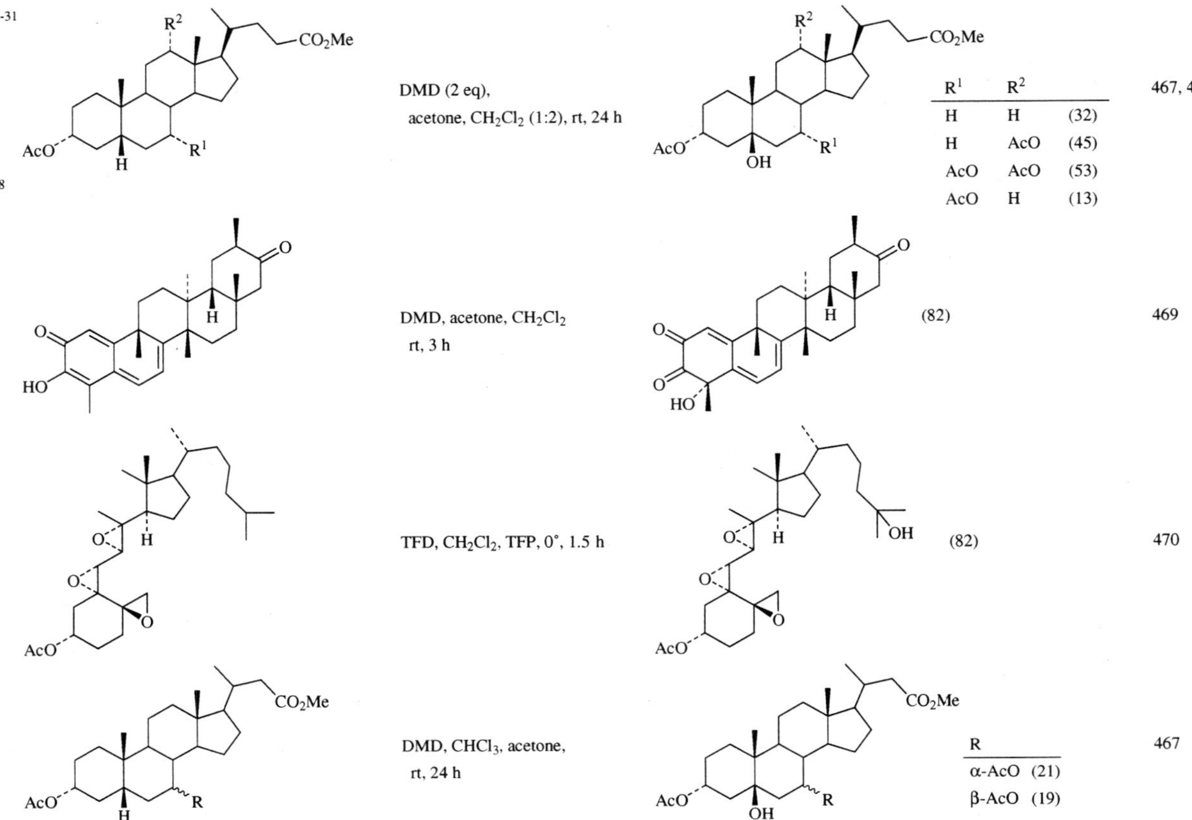

C27-31
R2
CO2Me
AcO
H
R1
DMD (2 eq),
acetone, CH2Cl2 (1:2), rt, 24 h
R2
CO2Me
AcO
OH
R1
R1 R2
H H (32)
H AcO (45)
AcO AcO (53)
AcO H (13)
467, 468
C28
O
HO
H
DMD, acetone, CH2Cl2
rt, 3 h
O
O
O
HO
H
(82)
469
O
O
H
AcO
TFD, CH2Cl2, TFP, 0°, 1.5 h
O
O
H
OH
AcO
(82)
470
CO2Me
AcO
H
R
DMD, CHCl3, acetone,
rt, 24 h
CO2Me
AcO
OH
R
R
α-AcO (21)
β-AcO (19)
467

TABLE 5A. C–H OXIDATION BY ISOLATED DIOXIRANES (*Continued*)

Substrate	Conditions	Product(s) and Yield(s) (%)	Refs.
C_{29}			
OAc, CO_2Me, H, H, H, AcO, H	DMD, CH_2Cl_2, acetone, rt, 48 h	OAc, CO_2Me, H, H, H, AcO, OH (35)	458
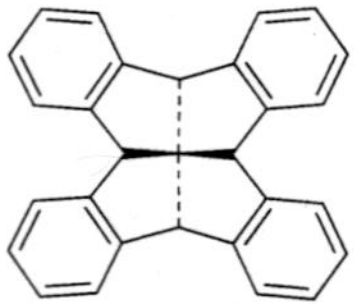	TFD, CH_2Cl_2, TFP, *t*-BuOH, –10°, 25 min	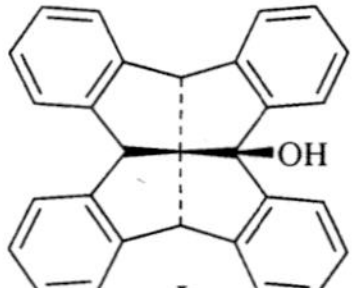 OH **I** (56)	462
	TFD, CH_2Cl_2, TFP, 0°, 20 min	**I** (44)	443
	TFD, CH_2Cl_2, TFP, *t*-BuOH, –10°, 25 min	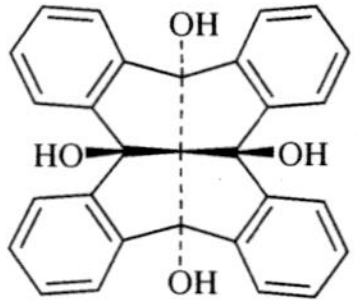 OH, HO, OH, OH (54)	462
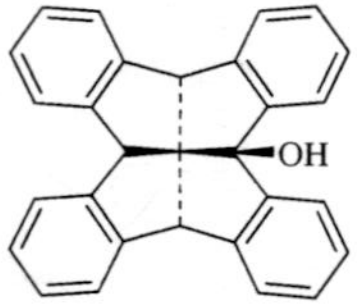 OH	TFD, CH_2Cl_2, TFP, 0°, 15 min	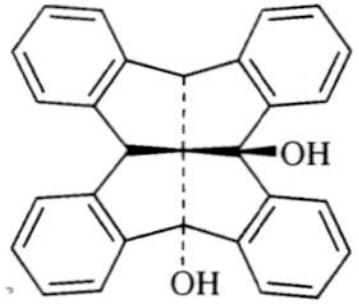 OH, OH (53)	443

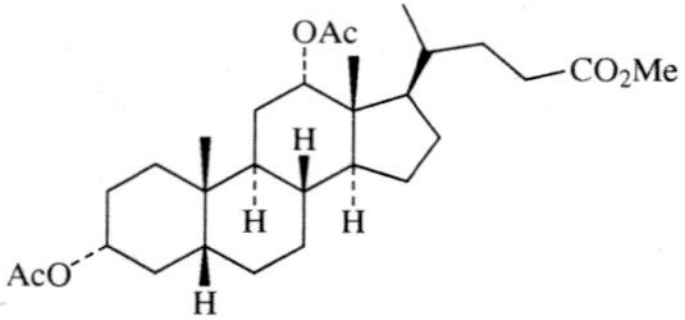

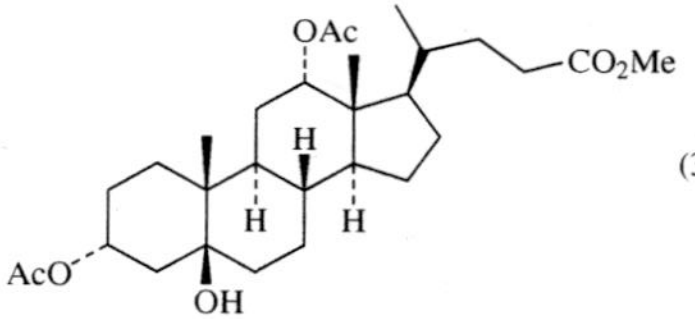

Substrate	Conditions	Product	(Yield, %)	Ref.
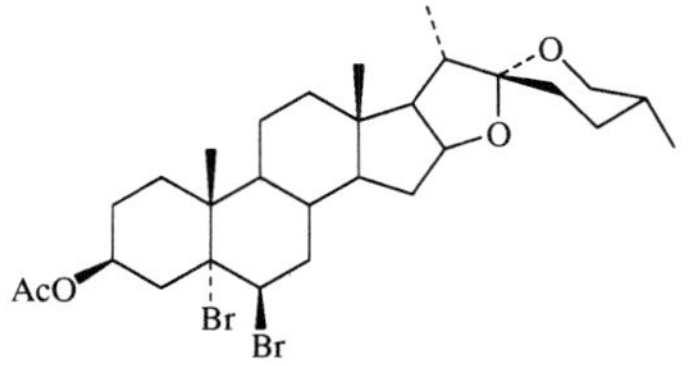 AcO, Br, Br, O, O	DMD, acetone, rt, 2 h	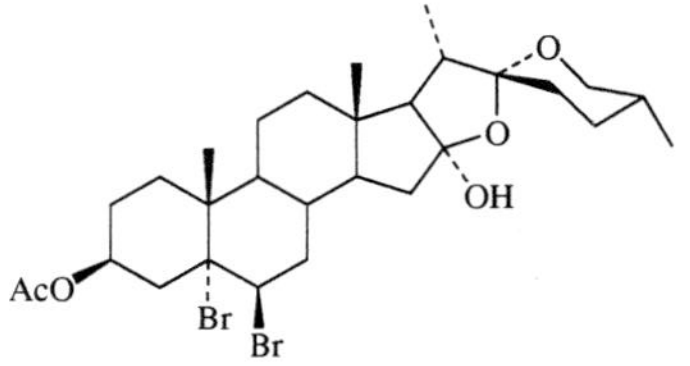 AcO, Br, Br, O, O, OH	(> 90)	471
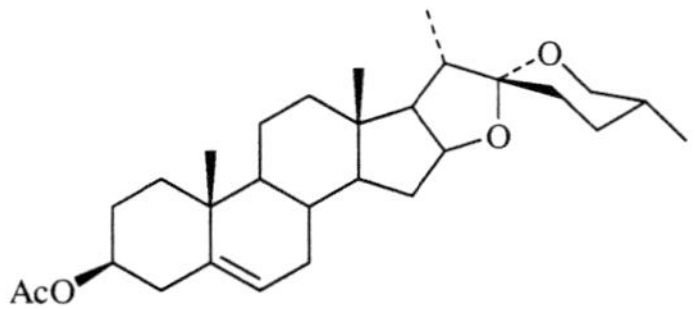 AcO, O, O	DMD, acetone, rt, 3 h	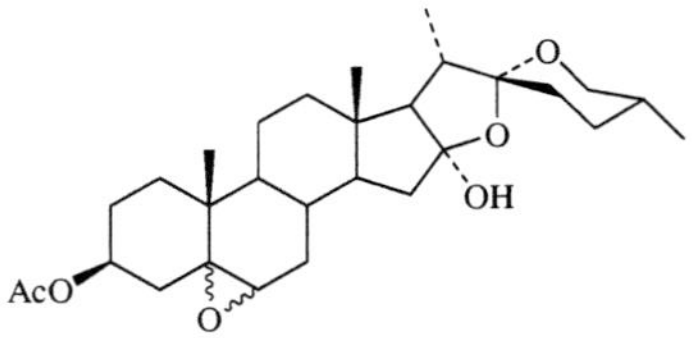 AcO, O, O, O, OH	(92)	470
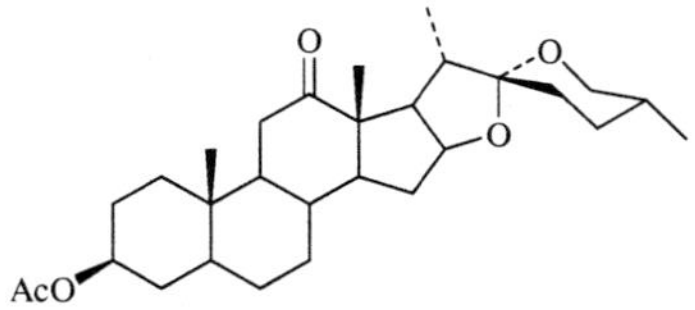 AcO, O, O, O	DMD, acetone, rt, 2 h	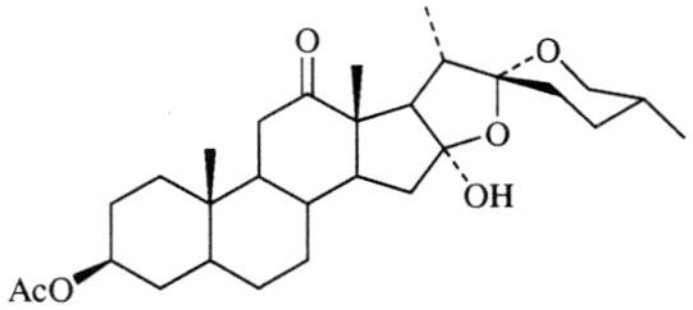 AcO, O, O, O, OH	(88)	470, 471
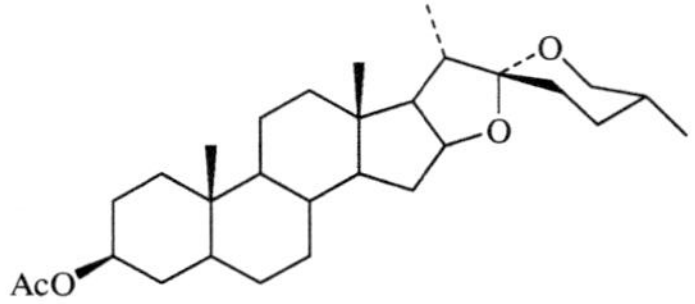 AcO, O, O	DMD, acetone, rt, 2 h	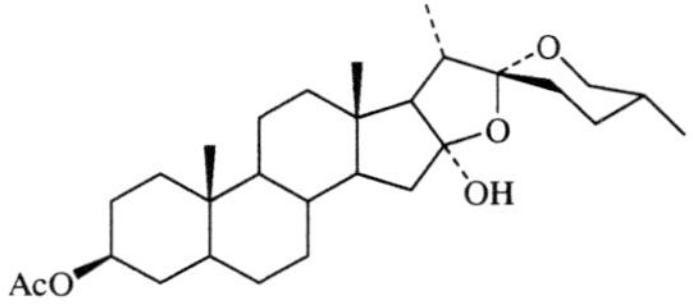 AcO, O, O, OH	(95)	470

TABLE 5A. C–H OXIDATION BY ISOLATED DIOXIRANES (*Continued*)

Substrate	Conditions	Product(s) and Yield(s) (%)	Refs.
C_{29}			
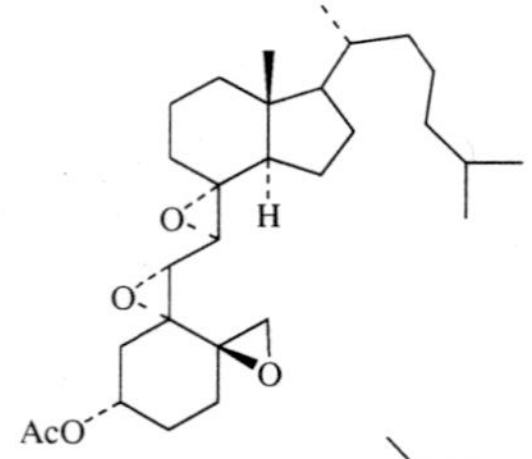 AcO, O, O, H, O	TFD (3 eq), CH_2Cl_2, TFP (9:1), 0°, 1.5 h	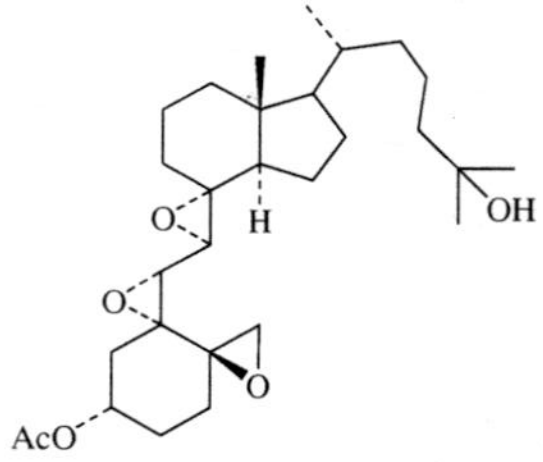 AcO, O, O, H, O, OH (82)	469
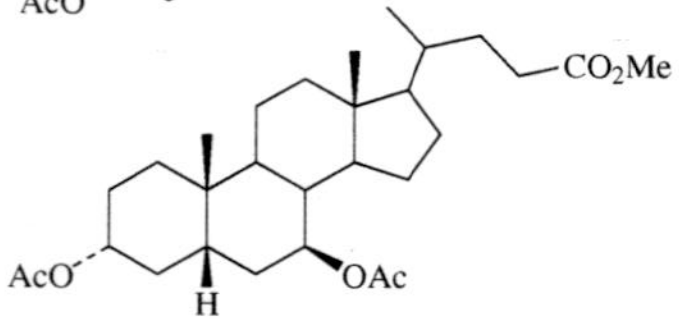 AcO, H, OAc, CO_2Me	DMD, $CHCl_3$, acetone, rt, 24 h	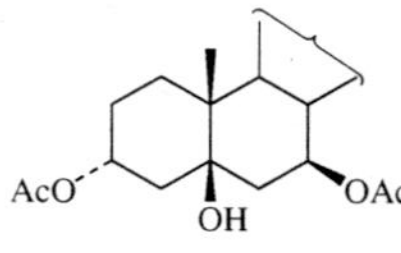 AcO, OH, OAc (12) 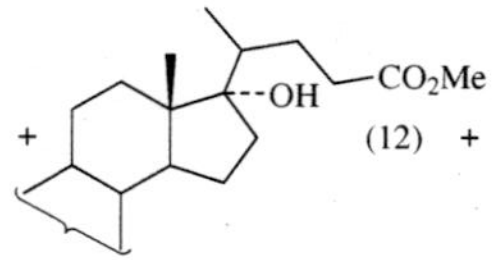 + OH, CO_2Me (12) 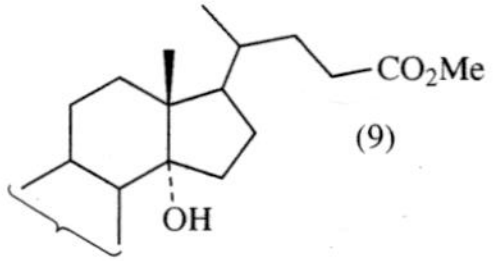 + OH, CO_2Me (9)	467
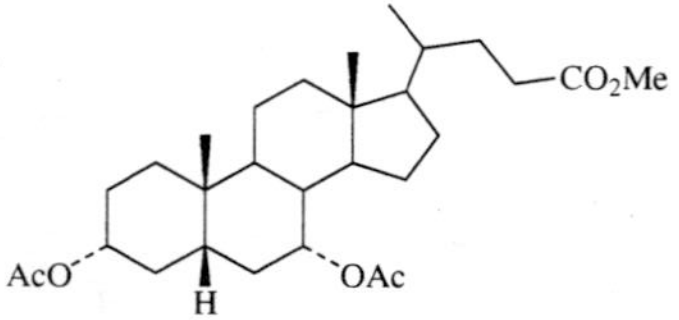 AcO, H, OAc, CO_2Me	DMD, $CHCl_3$, acetone, rt, 24 h	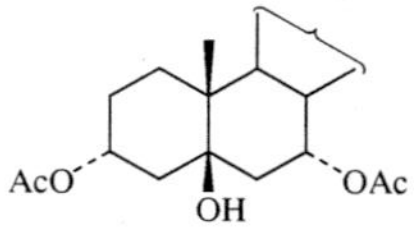 AcO, OH, OAc (22) 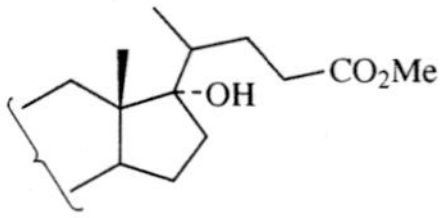 + OH, CO_2Me (12)	467
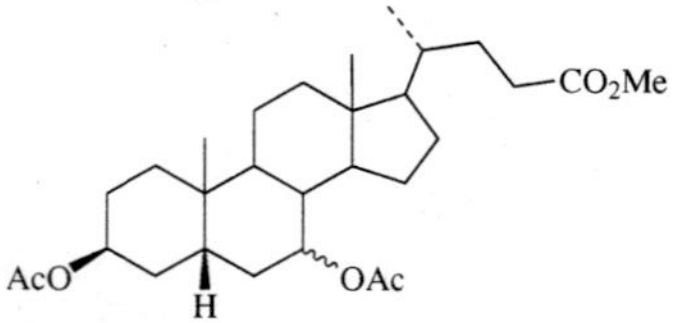 AcO, H, OAc, CO_2Me	DMD, CH_2Cl_2, $CHCl_3$, rt	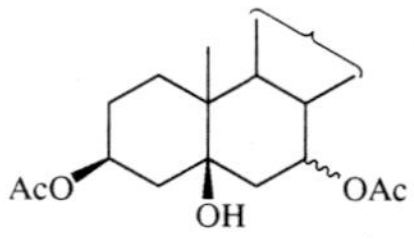 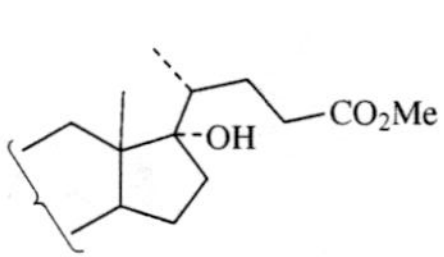 AcO, OH, OAc **I** + OH, CO_2Me **II**	465

DMD, CH_2Cl_2, $CHCl_3$, rt

I + II + III + IV + V (465)

Time	I + II	III	IV	V
12 h	(trace)	(6)	(6)	(4)
24 h	(4)	(10)	(11)	(6)
36 h	(7)	(15)	(16)	(11)
48 h	(14)	(20)	(22)	(15)

Time	I	II	III
12 h	(21)	(10)	(2)
24 h	(35)	(14)	(7)
36 h	(39)	(18)	(14)
48 h	(45)	(16)	(18)

TABLE 5A. C–H OXIDATION BY ISOLATED DIOXIRANES (*Continued*)

Substrate	Conditions	Product(s) and Yield(s) (%)	Refs.
C_{29} AcO, H	DMD, CH_2Cl_2, $CHCl_3$, rt	**I** (OH) + **II** (OH) + **III** (OH, OH) + **IV** (OH, OH)	465
		Time: 12 h — I (37), II (3), III + IV (5)	
		Time: 24 h — I (45), II (8), III + IV (12)	
		Time: 36 h — I (39), II (19), III + IV (29)	
AcO, O, O, O, OH, H	DMD (2.2 eq), CH_2Cl_2, rt, 7 d	AcO, O, O, O, OH, OH (82)	471
AcO, O, O, O	DMD, CH_2Cl_2, –50°, 30 min	AcO, O, O, O, OH, OH (99)	471

C_{30}

Substrate	Conditions	Product(s) and Yield(s) (%)	Refs.
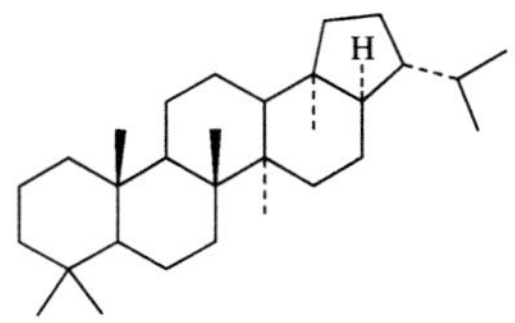	DMD, acetone, CH_2Cl_2, rt, 21 h	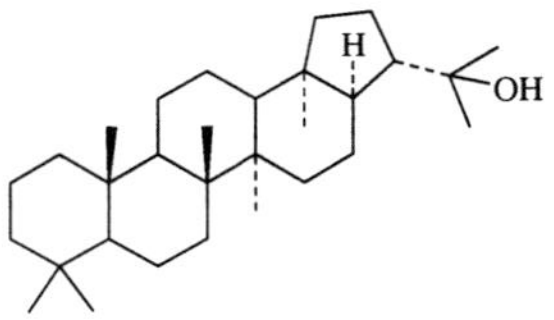 OH (10)	472
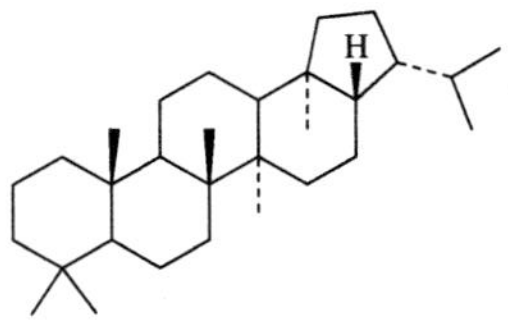	DMD, acetone, CH_2Cl_2, rt, 12 h	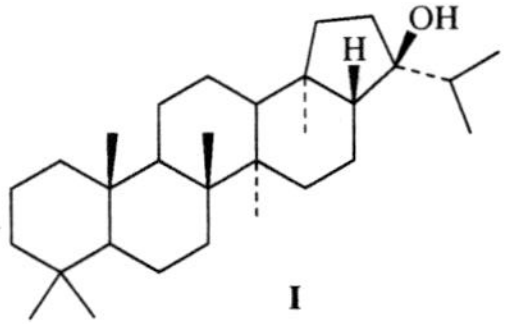 OH **I** (30) 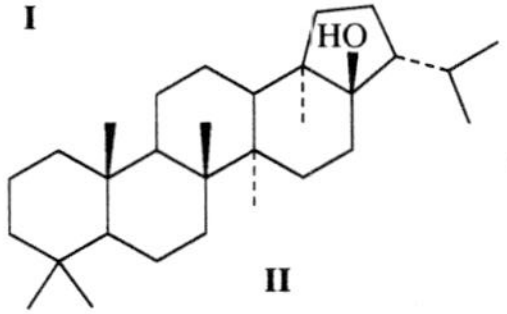 + HO **II** (20)	472
	DMD, acetone, CH_2Cl_2, rt, 21 h	**I** (50) + **II** (30)	472
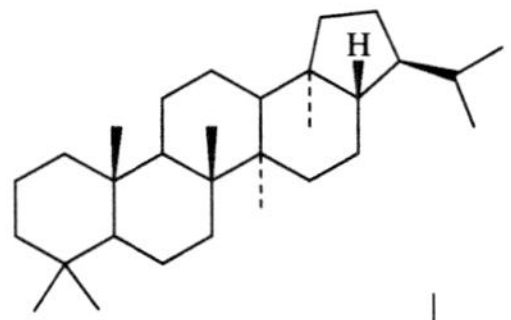	DMD, acetone, CH_2Cl_2, rt, 21 h	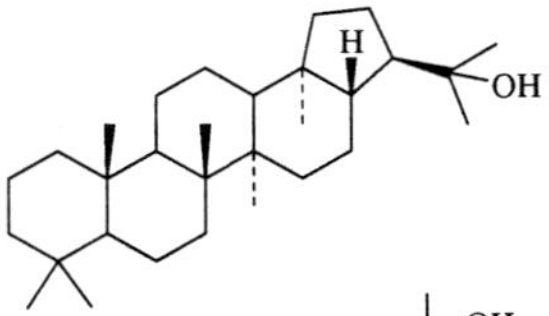 OH (15)	472
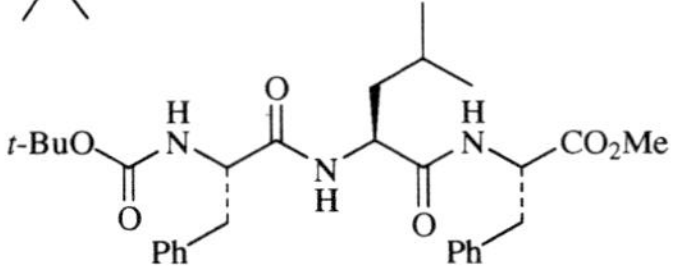 *t*-BuO, H N, O, N H, O, H N, CO_2Me, Ph, Ph	DMD, acetone, rt, 3 d	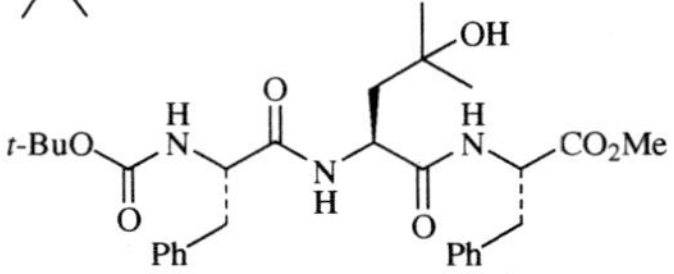 OH, *t*-BuO, H N, O, N H, O, H N, CO_2Me, Ph, Ph (49)	93

TABLE 5A. C–H OXIDATION BY ISOLATED DIOXIRANES (*Continued*)

Substrate	Conditions	Product(s) and Yield(s) (%)	Refs.
C_{30} 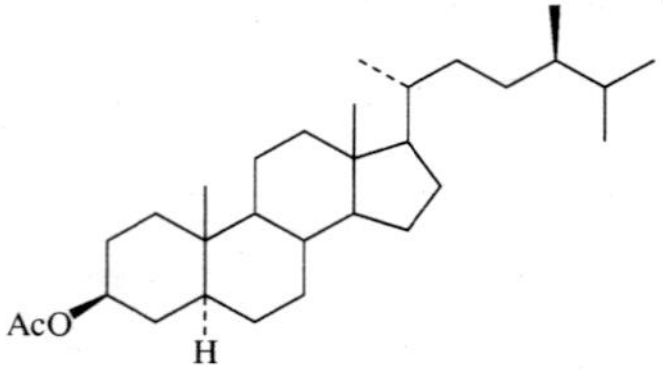 AcO, H	DMD, CH_2Cl_2, $CHCl_3$, rt, 24 h	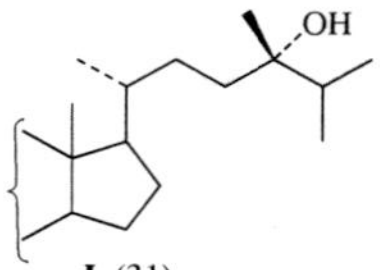 OH **I** (31) 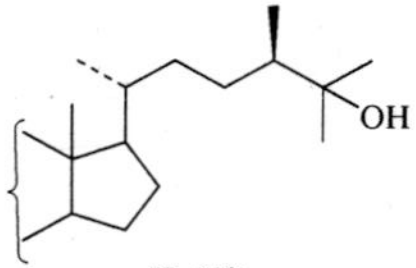 + OH **II** (40)	465
	DMD, CH_2Cl_2, $CHCl_3$, rt, 36 h	**I** (33) + **II** (30)	465
C_{31} 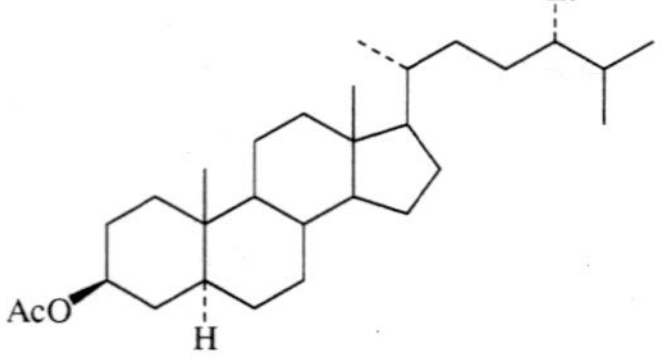 Et, AcO, H	DMD, CH_2Cl_2, $CHCl_3$, rt, 12 h	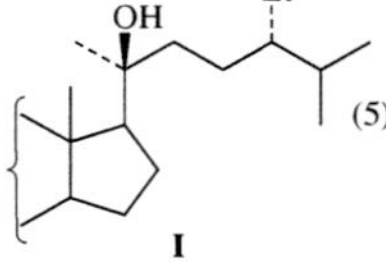 OH, Et **I** (5) 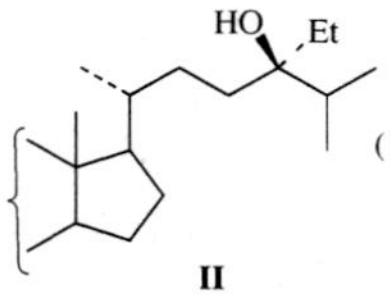 + HO, Et **II** (10) 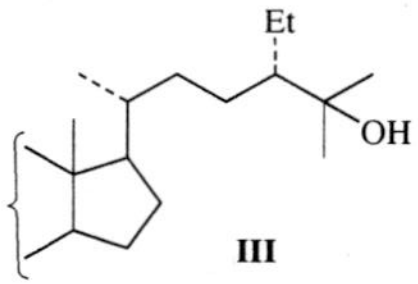 + Et, OH **III** (22)	465
	DMD, CH_2Cl_2, $CHCl_3$, rt, 24 h	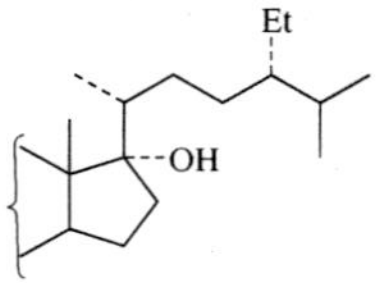 Et, OH (5) + **I** (8) + **II** (18) + **III** (22)	465

Conditions	Product(s) and Yield(s) (%)	Refs.
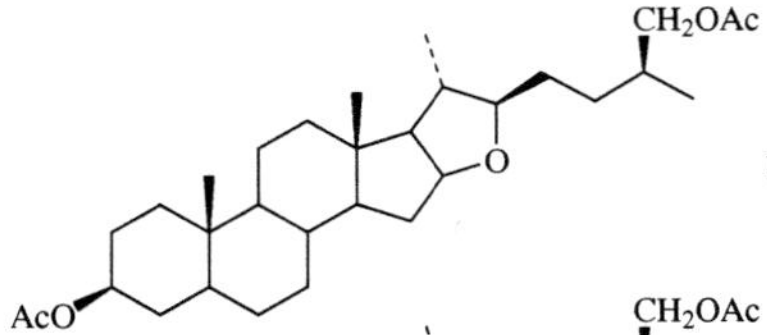 DMD, acetone, 0°, 4 h	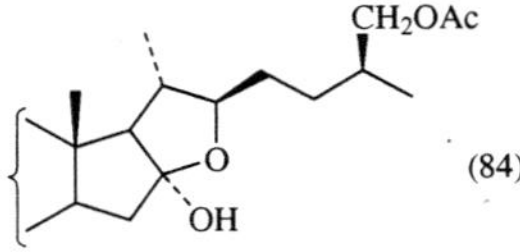 (84)	470
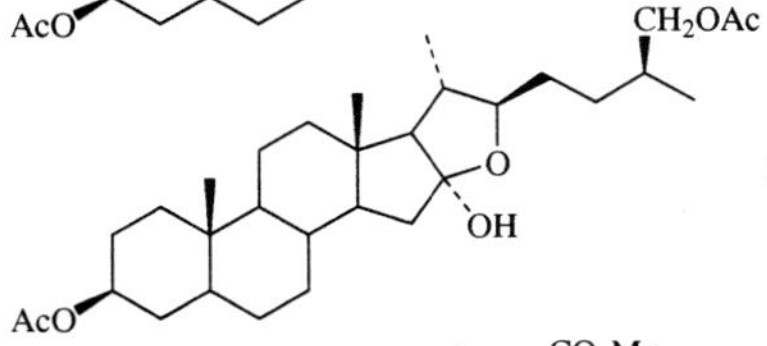 DMD, acetone	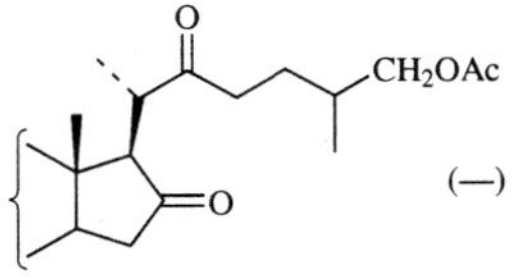 (—)	470
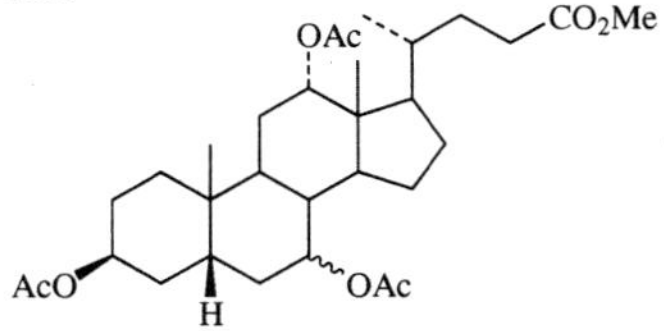 DMD, CH_2Cl_2, $CHCl_3$, rt, 12 h	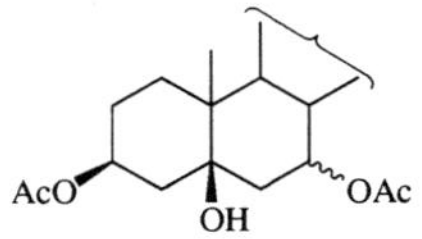 **I** (11)	465
DMD, CH_2Cl_2, $CHCl_3$, rt, 24 h	**I** (20) 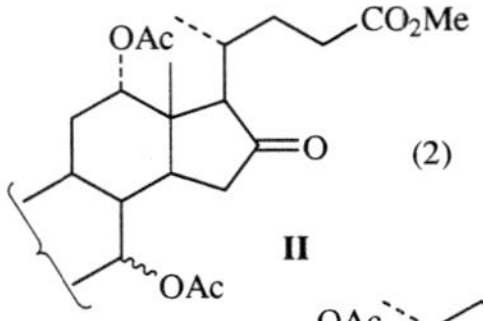 + **II** (2)	465
DMD, CH_2Cl_2, $CHCl_3$, rt, 36 h	**I** (33) + **II** (5) 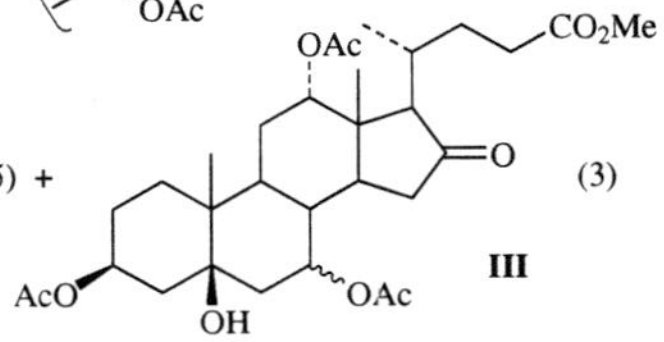 + **III** (3)	465
DMD, CH_2Cl_2, $CHCl_3$, rt	**I** + **II** + **III**	465

Time	**I**	**II**	**III**
48 h	(40)	(2)	(4)
60 h	(44)	(4)	(6)

TABLE 5A. C–H OXIDATION BY ISOLATED DIOXIRANES (*Continued*)

Substrate	Conditions	Product(s) and Yield(s) (%)	Refs.
C_{34} 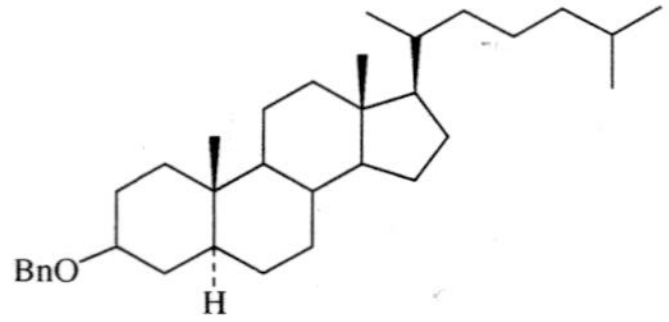 BnO, H	DMD (2 eq), CH_2Cl_2, acetone, dark, 22°, 15 h	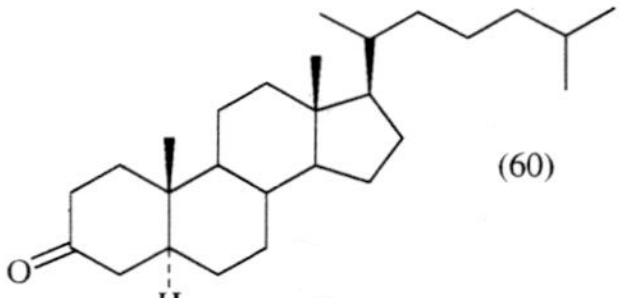 O, H **I** (60)	189
	DMD (1 eq), CH_2Cl_2, acetone, dark, 22°, 14 h	**I** 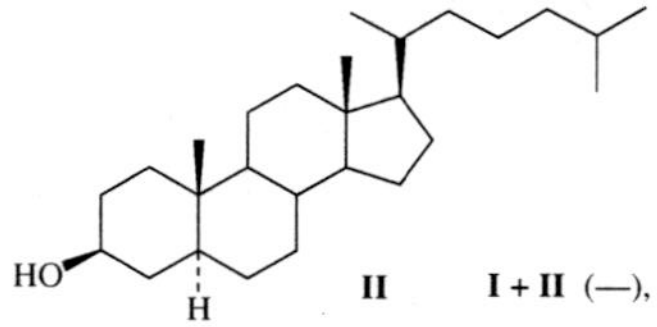 + HO, H **II** **I + II** (—), **I:II** = 3:1	189
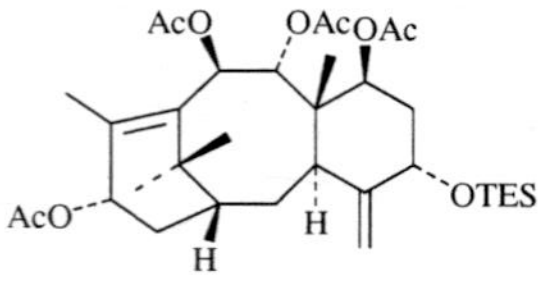 AcO, OAc, OAc, AcO, OTES, H, H	DMD (15 eq), acetone, rt, 48 h	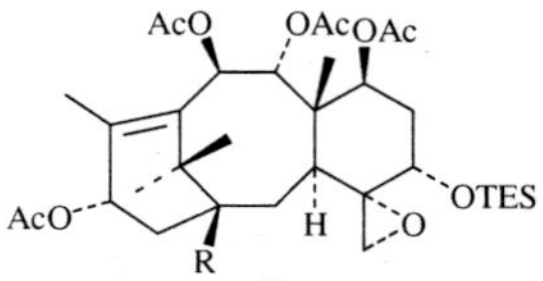 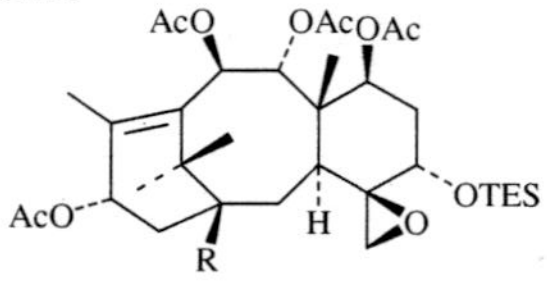 AcO, OAc, OAc, AcO, OTES, R, H, O **I** + AcO, OAc, OAc, AcO, OTES, R, H, O **II** R / **I** / **II** OH (13) (25) H (32) (28)	473
	DMD (30 eq), acetone, rt, 48 h	**I + II** R = OH (90)	473
I R = H	DMD (25-30 eq), acetone, rt, 48 h	**I** R = OH (—)	473
II R = H	DMD (25-30 eq), acetone, rt, 48 h	**II** R = OH (—)	473

	Conditions	Product(s) and Yield(s) (%)	Refs.
C_{35} 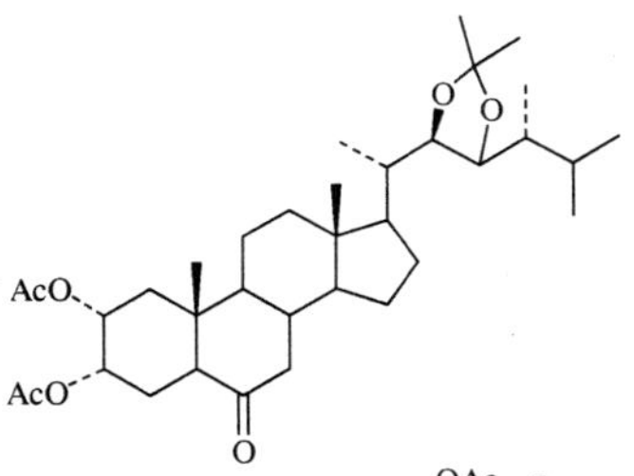	1. TFD, TFP, CH_2Cl_2, N_2, 0° 2. Dark, rt, 40 h	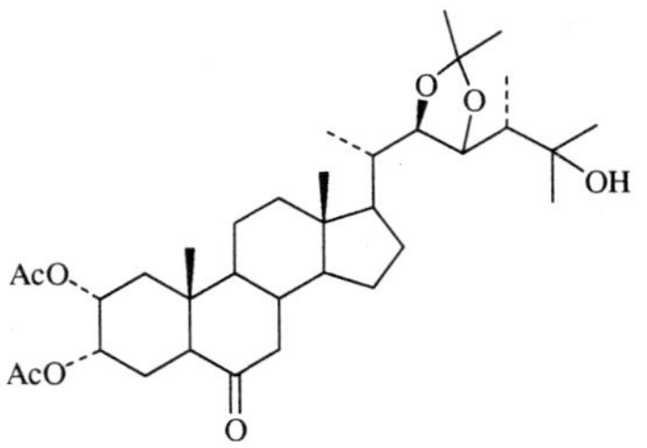 (62)	474
C_{36} 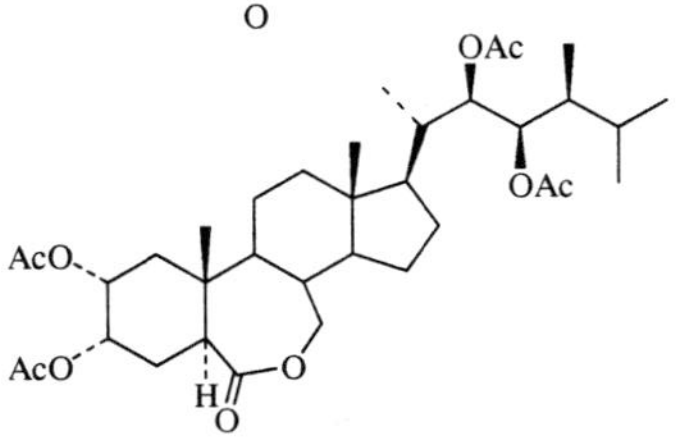	TFD, TFP, CH_2Cl_2, 0°, 3 h	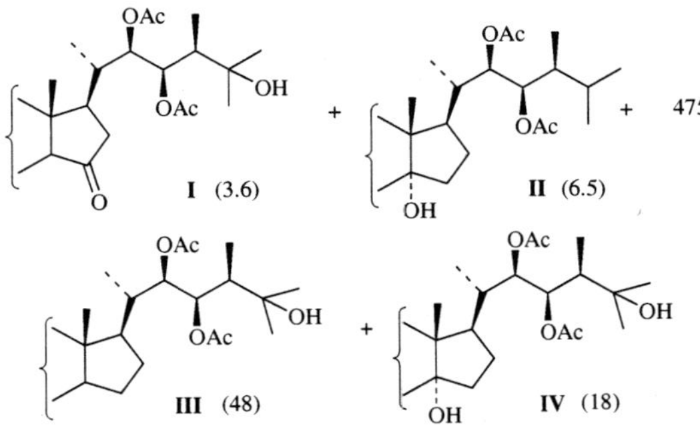 **I** (3.6) + **II** (6.5) + **III** (48) + **IV** (18)	475
	TFD, TFP, –30°, 5 h	**III** (61)	475
	TFD, TFP, CH_2Cl_2, rt, 24 h	**IV** (78)	475
	1. TFD, TFP, CH_2Cl_2, N_2, 0° 2. Dark, rt, 40 h	**III** (71)	474

TABLE 5A. C–H OXIDATION BY ISOLATED DIOXIRANES (*Continued*)

Substrate	Conditions	Product(s) and Yield(s) (%)	Refs.
C_{37} 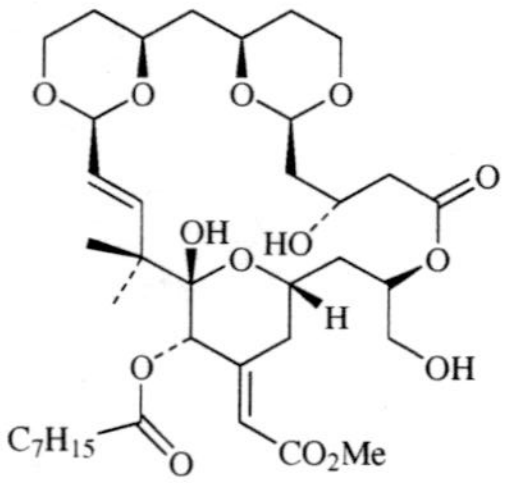	DMD, acetone, rt, 48 h	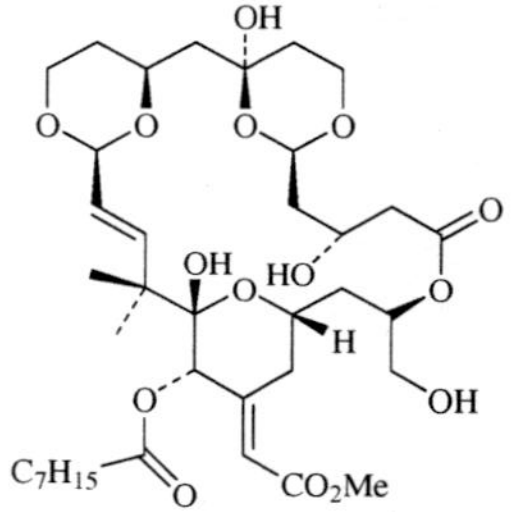 (70)	476
C_{38} 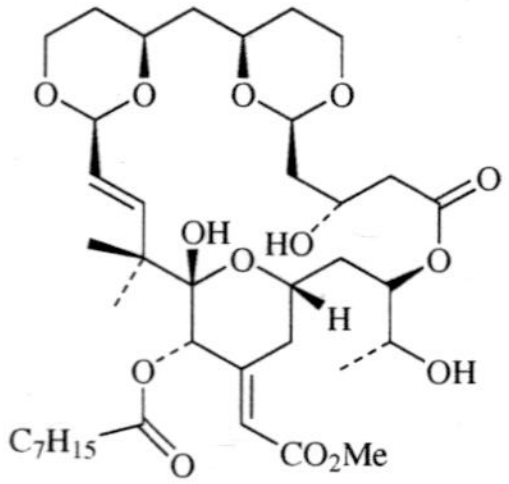	DMD, acetone, rt, 48 h	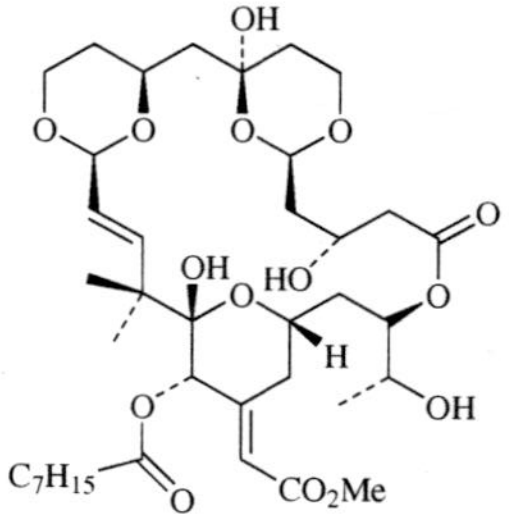 (9) + 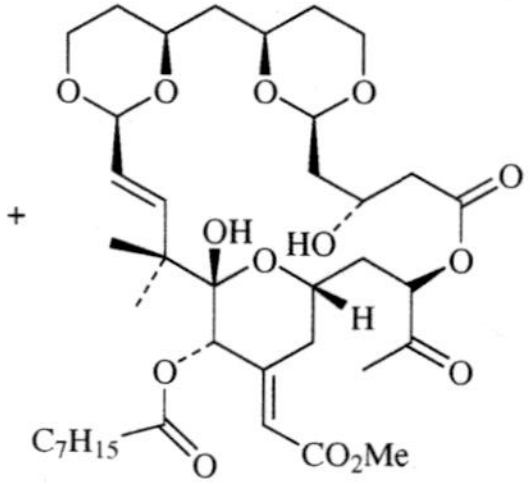 (26)	476

+ 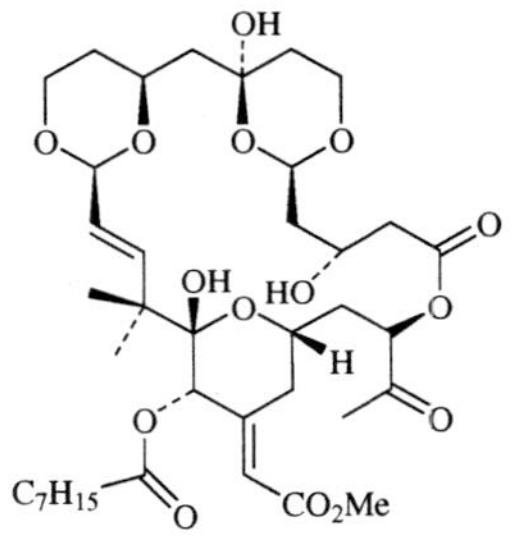 (46)

---

[a] Reaction was run using solutions purged with pure, oxygen-free nitrogen gas.

[b] The reaction was carried out under N_2.

[c] The diastereomer ratio was 10:1.

[d] The value includes 14% of an oxo aldehyde byproduct.

TABLE 5B. REGIOSELECTIVE C–H OXIDATION BY ISOLATED DIOXIRANES

Substrate	Conditions	Product(s) and Yield(s) (%)	Refs.
C_4			
1,3-butanediol (OH, OH)	DMD, acetone, rt, 18-22 h	4-hydroxy-2-butanone (90)	183
C_5			
1,2-pentanediol (OH, OH)	DMD, acetone, rt, 18-22 h	1-hydroxy-2-pentanone (100)	183
1,4-pentanediol (OH, OH)	DMD, acetone, rt, 18-22 h	5-hydroxy-2-pentanone (60)	183
C_{5-11}			
$R^2CH(R^1)CH(OH)CH(OH)R^3$	DMD, acetone, CH_2Cl_2, rt	$R^2CH(R^1)CH(OH)C(O)R^3$ (I) + $R^2CH(R^1)C(O)CH(OH)R^3$ (II)	428

R^1	R^2	R^3	Time	I	II
OH	Me	Me	—	(—)	(—)
MeO	Me	Me	—	(—)	(—)
AcO	Me	Me	24 h	(90)	(—)
Br	*n*-Pr	Me	48 h	(95)	(—)
MeO	H	*n*-Pr	48 h	(76)	(14)
AcO	H	*n*-Pr	36 h	(83)	(—)
N_3	H	*n*-C_5H_{11}	48 h	(95)	(—)
OH	H	*n*-C_5H_{11}	24 h	(59)	(—)
CH_2OH	H	*n*-Bu	48 h	(38)	(38)
OH	Me	*n*-C_5H_{11}	—	(—)	(—)
CH_2OMe	H	*n*-Bu	24 h	(30)	(15)
CH_2OAc	H	*n*-Bu	24 h	(60)	(—)
OMe	Me	*n*-C_5H_{11}	36 h	(60)	(—)
OAc	Me	*n*-C_5H_{11}	48 h	(70)	(—)

Substrate	Conditions	Product(s) and Yield(s) (%)	Refs.
$C_{5\text{-}13}$			
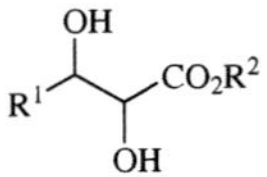 $R^1CH(OH)CH(OH)CO_2R^2$	DMD, acetone, CH_2Cl_2, rt, 72 h	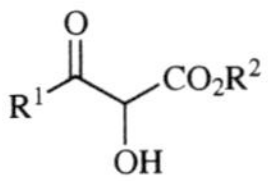 $R^1C(O)CH(OH)CO_2R^2$ R^1 / R^2: Me / Me (80); c-C_6H_{11} / Et (57); n-C_9H_{19} / Me (< 10)	449
C_6			
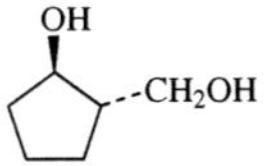 OH, CH$_2$OH (cyclopentane)	DMD, acetone, rt, 18-22 h	mixture	183
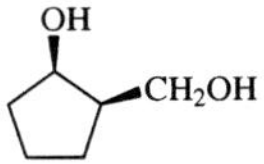 OH, CH$_2$OH (cyclopentane)	DMD, acetone, rt, 18-22 h	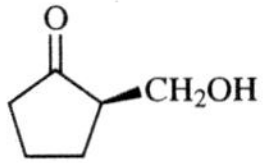 O, CH$_2$OH (cyclopentanone) (85)	183
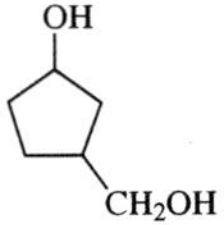 OH, CH$_2$OH (cyclopentane)	DMD, acetone, rt, 18-22 h	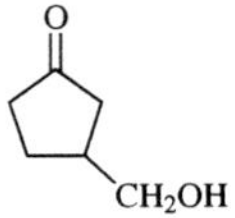 O, CH$_2$OH (cyclopentanone) (82)	434
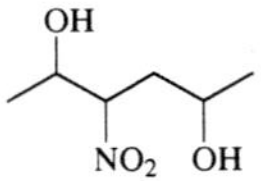 OH, NO$_2$, OH	DMD, acetone, rt, overnight	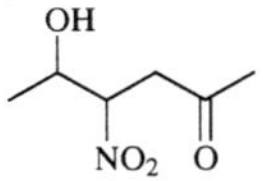 OH, NO$_2$, O (—)	434
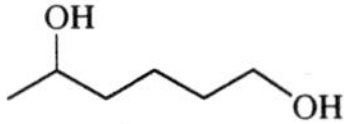 OH, OH	DMD, acetone, rt, 18-22 h	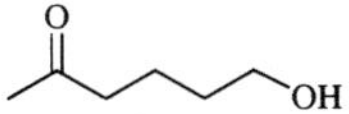 O, OH (60)	183

TABLE 5B. REGIOSELECTIVE C–H OXIDATION BY ISOLATED DIOXIRANES (*Continued*)

Substrate	Conditions	Product(s) and Yield(s) (%)	Refs.
$C_{6\text{-}8}$ 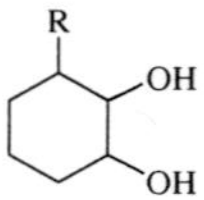	DMD, acetone, rt	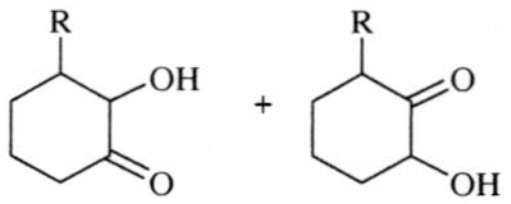 I + II	428
C_7 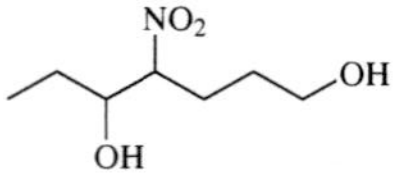	DMD, acetone, rt, overnight	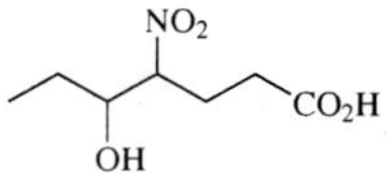 (80)	434
C_8 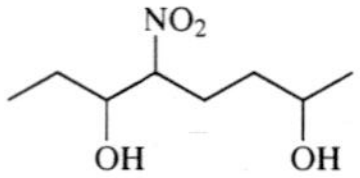	DMD, acetone, rt, overnight	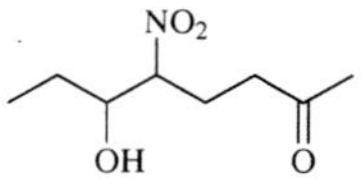 (55)	434
C_9 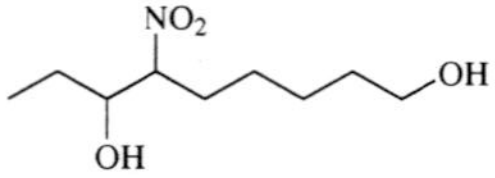	DMD, acetone, rt, overnight	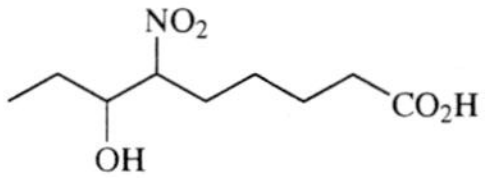 (> 96)	434
$C_{9\text{-}10}$ 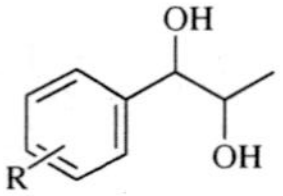	DMD, CH_2Cl_2, acetone, rt	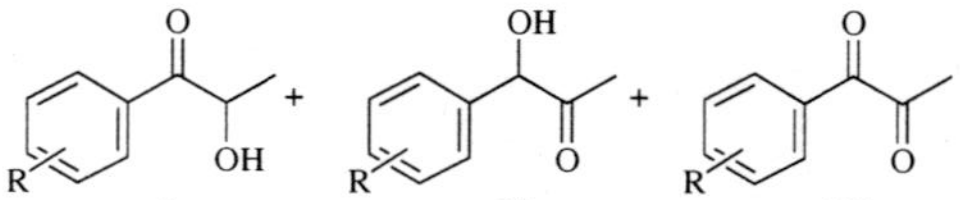 I + II + III	449

R	Time	I	II
Br	24 h	(> 95)	(—)
N_3	48 h	(> 95)	(—)
OH	48 h	(> 95)	(—)
OMe	48 h	(74)	(16)
AcO (syn, syn)	24 h	(> 95)	(—)
AcO (anti, syn)	24 h	(> 95)	(—)

R	Time	% Conv.	**I:II:III**
4-NO_2	12 h	> 96	52:32:16
2-NO_2	24 h	> 96	15:44:41
H	24 h	> 96	0:82:18
4-MeO	12 h	> 96	66:14:20

C_{10}

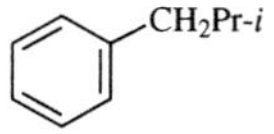

CH_2Pr-*i*

DMD, acetone, dark, rt, 3 d

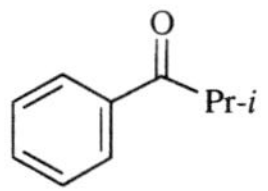

Pr-*i* (16) + OH (4)

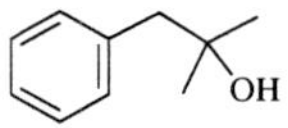

442

$C_{10\text{-}11}$

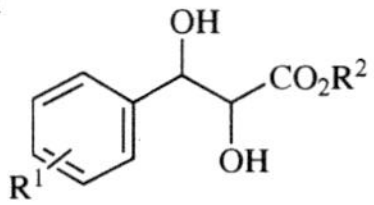

OH, CO_2R^2, OH, R^1

DMD, CH_2Cl_2, acetone, rt

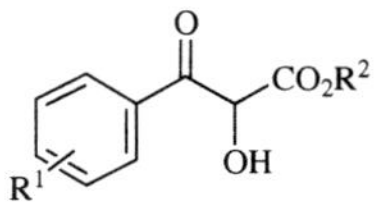

O, CO_2R^2, OH, R^1

R^1	R^2	Time	
H	Me	24 h	(13)
2-OH	Me	24-72 h	(—)
4-NO_2	Et	24 h	(4)
2-NO_2	Et	24-72 h	(—)
4-MeO	Me	24 h	(> 96)
2-MeO	Me	24-72 h	(—)

449

C_{11}

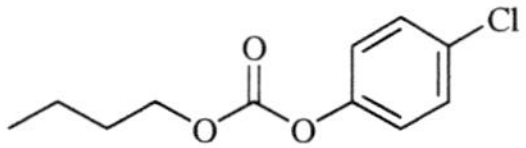

TFD, CH_2Cl_2, –20°, 48 h

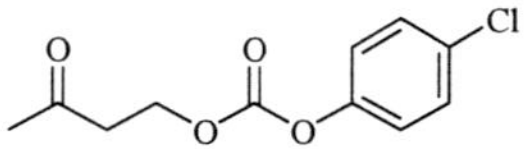

(> 98)

477

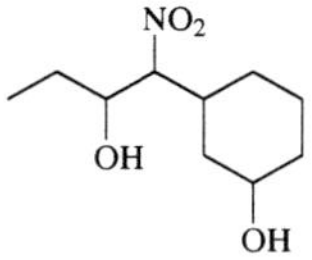

NO_2, OH, OH

DMD, acetone, rt, overnight

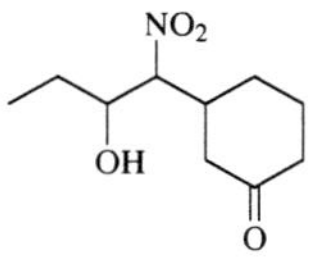

NO_2, OH, O (85)

434

C_{12}

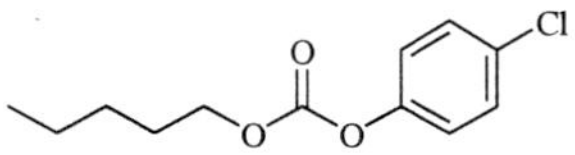

TFD, CH_2Cl_2, –20°, 48 h

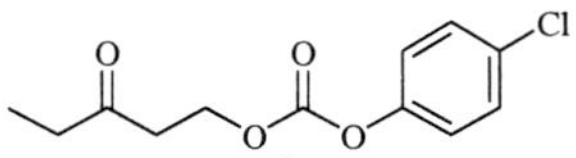

I 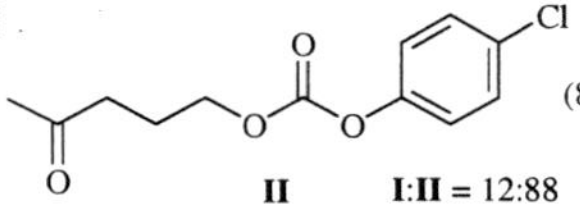 (—) + **II** (80)

I:**II** = 12:88

477

TABLE 5B. REGIOSELECTIVE C–H OXIDATION BY ISOLATED DIOXIRANES (*Continued*)

Substrate	Conditions	Product(s) and Yield(s) (%)	Refs.
C_{12}  OAc	TFD, CH_2Cl_2, –20°, 48 h	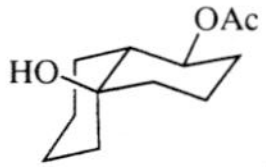 HO OAc (97)	477
C_{12-14} 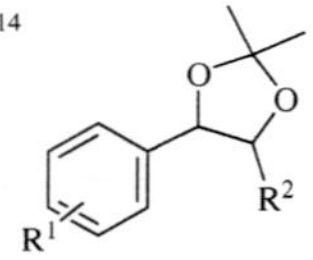 O O R^2 R^1	DMD, acetone, rt	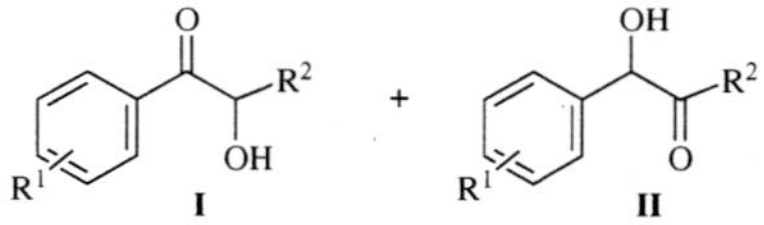 O R^2 OH R^1 **I** + OH R^2 O R^1 **II**	449
C_{13} O O O Cl	TFD, CH_2Cl_2, –20°, 48 h	O O O O Cl **I** (—) + O O O O Cl **II** (56) **I**:**II** = 40:60	477
O O O Cl	TFD, CH_2Cl_2, –20°, 48 h	O O O O Cl **I** (—) + O O **II** (25) + O O **III** (55) **I**:**II**:**III** = 14:29:57	477

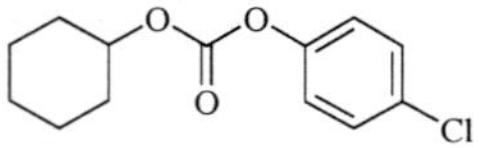

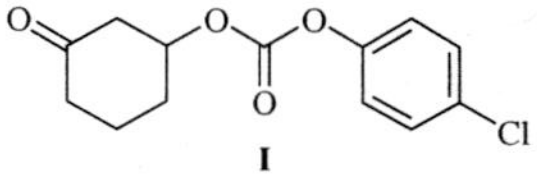

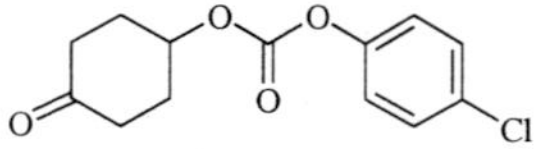

R^1	R^2	Time	**I**	**II**
4-NO_2	Me	24 h	(70)	(15)
H	Me	12 h	(85)	(8)
H	CO_2Me	–	(—)	(—)
4-MeO	Me	12 h	(> 96)	(—)
4-MeO	CO_2Me	48 h	(72)	(—)

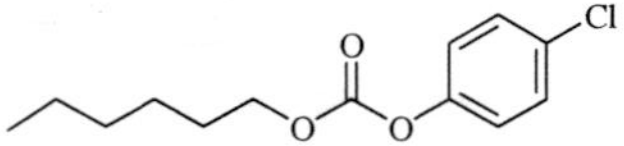

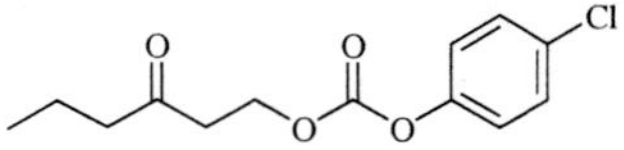

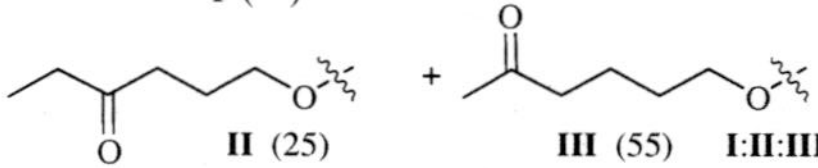

Substrate	Conditions	Product(s) and Yield(s) (%)	Refs.
	TFD, CH_2Cl_2, –20°, 48 h	Cl, O, O, O, O; I (—) + Cl, O, O, O, O; II (90) I:II = 8:92	477
Ph	DMD, acetone, dark, rt, 3 d	Ph OH (33) + OH, Ph (20)	442
Ph Ph + Ph, Ph Ph; 1 : 1	DMD, acetone, dark, rt, 3 d	O, Ph Ph (0.9) + Ph, OH, Ph Ph (8.7)	442
C_{14}			
O, O, O, Cl	TFD, CH_2Cl_2, –20°, 48 h	O, O, O, O, Cl; I (—) + O, O, O, O, Cl; II (95) I:II = 2:98	477
O, O, O, Cl	TFD, CH_2Cl_2, –20°, 48 h	O, O, O, O, Cl; I (5) + O, O, O, O, Cl; II (90); I:II = 9:91	477

TABLE 5B. REGIOSELECTIVE C–H OXIDATION BY ISOLATED DIOXIRANES (*Continued*)

Substrate	Conditions	Product(s) and Yield(s) (%)	Refs.
C_{14} 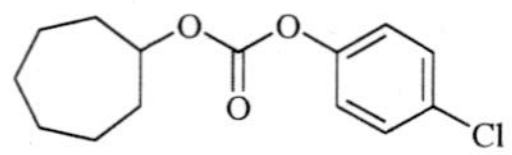	TFD, CH_2Cl_2, –20°, 48 h	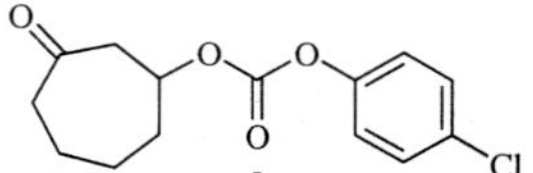 I (9) 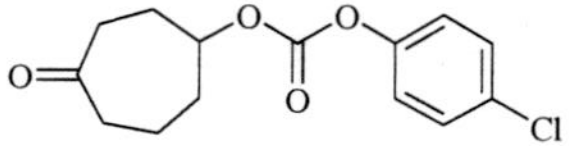 + II (83) I:II = 15:85	477
C_{19} 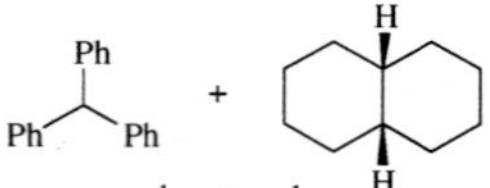 Ph, Ph, Ph + H, H 1 : 1	DMD, acetone, dark, rt, 3 d	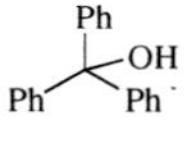 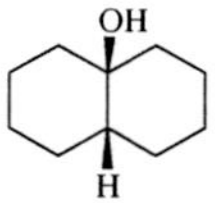 Ph, OH, Ph, Ph (0.6) + OH, H (46)	442

TABLE 5C. C–H OXIDATION BY IN SITU GENERATED DIOXIRANES

Substrate	Conditions	Product(s) and Yield(s) (%)	Refs.
C_2			
CH_3CHO	Oxone®, acetone, rt, 5 h	CH_3CO_2H 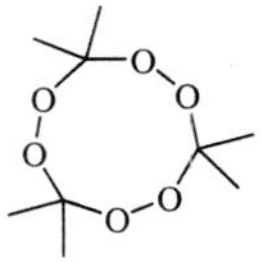 (—) + [structure] (2)	121
C_6			
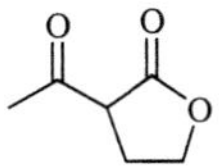 [structure]	Oxone®, acetone, phosphate buffer (pH 7.3-7.5), CH_2Cl_2, KF, 20°, 1 h	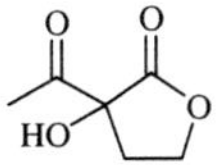 [structure] (85)	193
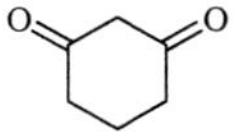 [structure]	Oxone®, acetone, H_2O, $NaHCO_3$	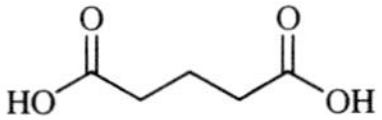 [structure] (14)	478
C_7			
[structure]	Oxone®, ketone **1**, $NaHCO_3$, MeCN, H_2O, EDTA, rt, 3.5 h Ketone = [structure] **1**	[structure] (91)	52
C_8			
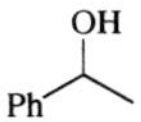 [structure]	Oxone®, ketone **1**, $NaHCO_3$, MeCN, H_2O, EDTA, rt, 3.5 h	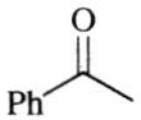 [structure] (80)	52
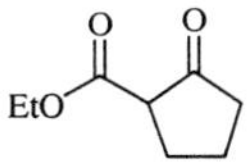 [structure]	Oxone®, acetone, phosphate buffer (pH 7.3-7.5), CH_2Cl_2, KF, 20°, 1 h	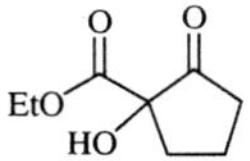 [structure] (85)	193

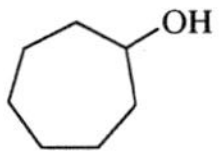

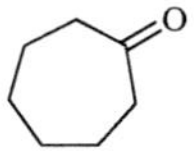

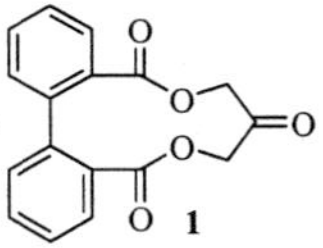

TABLE 5C. C–H OXIDATION BY IN SITU GENERATED DIOXIRANES (*Continued*)

Substrate	Conditions	Product(s) and Yield(s) (%)	Refs.
C_8			
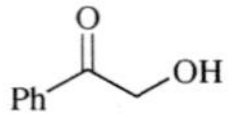 Ph, O, OH	Oxone®, acetone, H_2O, $NaHCO_3$	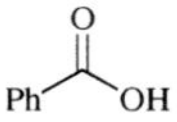 Ph, O, OH (96)	478
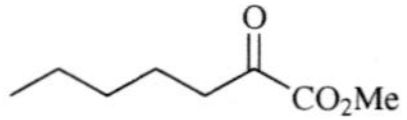 O, CO_2Me	Oxone®, MeCN, H_2O, $NaHCO_3$, rt, 24 h	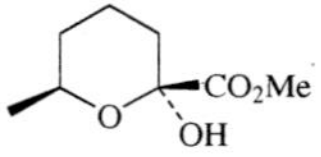 O, CO_2Me, OH (86)	195
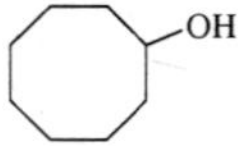 OH	Oxone®, ketone **1**, $NaHCO_3$, MeCN, H_2O, EDTA, rt, 4 h	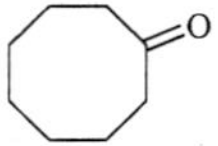 O (95)	52
C_9			
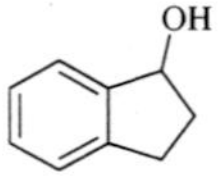 OH	Oxone®, ketone **1**, $NaHCO_3$, MeCN, H_2O, EDTA, rt, 5 h	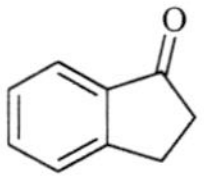 O (94)	52
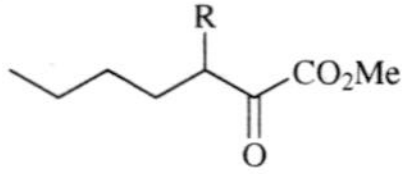 R, CO_2Me, O	Oxone®, MeCN, H_2O, $NaHCO_3$, Na_2EDTA, rt	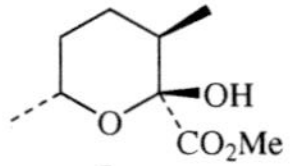 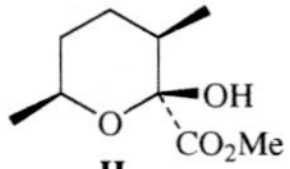 O, OH, CO_2Me **I** + O, OH, CO_2Me **II**	479
O, CO_2Me	Oxone®, MeCN, H_2O, $NaHCO_3$, Na_2EDTA, rt, 24 h	Et, O, OH, CO_2Me (70)	195
CF_3, O	Oxone®, MeCN, H_2O, $NaHCO_3$, Na_2EDTA, rt, 24 h	H, O, OH, CF_3, H (74)	480

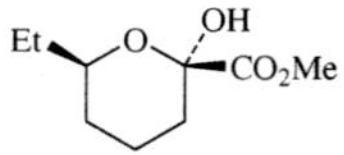

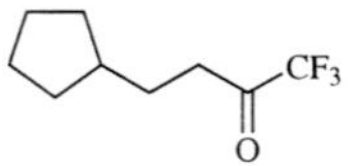

R	Time	**I + II**	**I:II**
Me	7 h	(80)	3.4:1
OMe	24 h	(9)	1:1

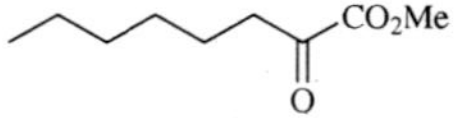

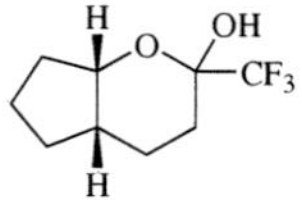

C$_{10}$

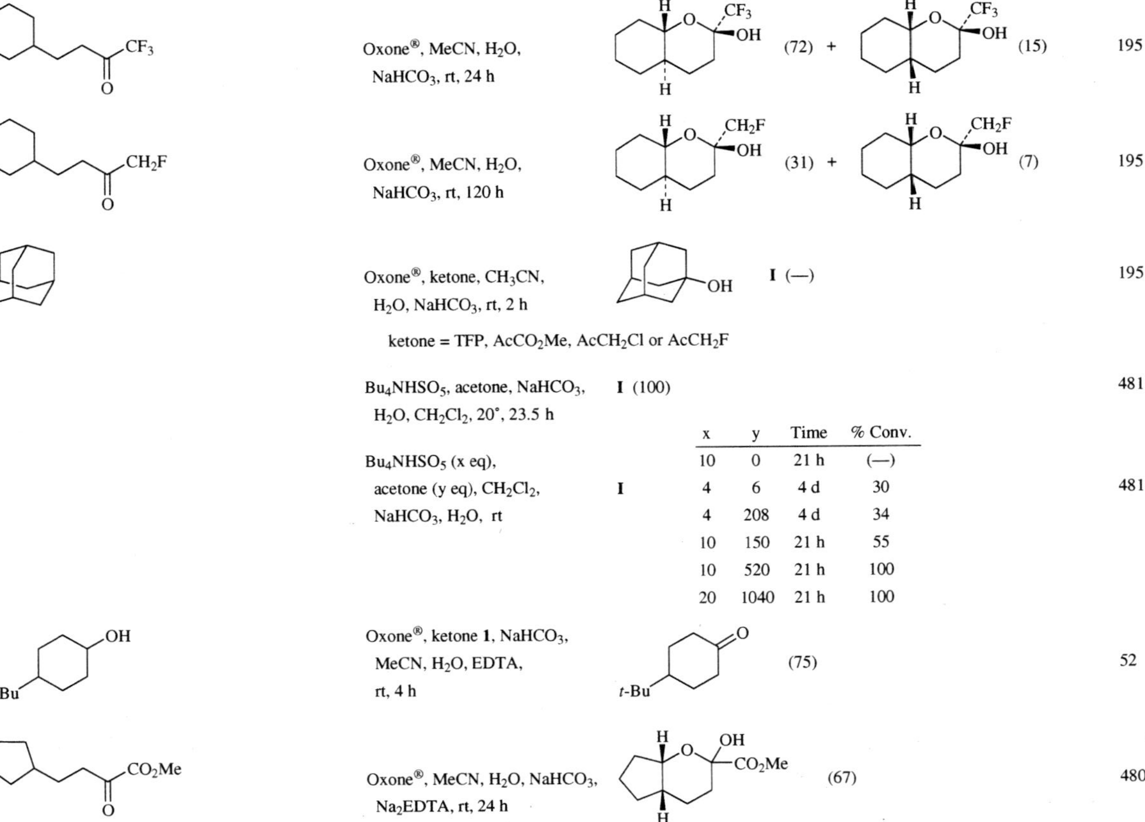

Substrate	Conditions	Product(s) and Yield(s) (%)	Refs.
[structure] (CF_3)	Oxone®, MeCN, H_2O, $NaHCO_3$, rt, 24 h	[structure] (72) + [structure] (15)	195
[structure] (CH_2F)	Oxone®, MeCN, H_2O, $NaHCO_3$, rt, 120 h	[structure] (31) + [structure] (7)	195
[structure]	Oxone®, ketone, CH_3CN, H_2O, $NaHCO_3$, rt, 2 h ketone = TFP, $AcCO_2Me$, $AcCH_2Cl$ or $AcCH_2F$	[structure] **I** (—)	195
	Bu_4NHSO_5, acetone, $NaHCO_3$, H_2O, CH_2Cl_2, 20°, 23.5 h	**I** (100)	481
	Bu_4NHSO_5 (x eq), acetone (y eq), CH_2Cl_2, $NaHCO_3$, H_2O, rt	**I**	481

x	y	Time	% Conv.
10	0	21 h	(—)
4	6	4 d	30
4	208	4 d	34
10	150	21 h	55
10	520	21 h	100
20	1040	21 h	100

Substrate	Conditions	Product(s) and Yield(s) (%)	Refs.
[structure] (t-Bu, OH)	Oxone®, ketone **1**, $NaHCO_3$, MeCN, H_2O, EDTA, rt, 4 h	[structure] (75)	52
[structure] (CO_2Me)	Oxone®, MeCN, H_2O, $NaHCO_3$, Na_2EDTA, rt, 24 h	[structure] (67)	480

TABLE 5C. C–H OXIDATION BY IN SITU GENERATED DIOXIRANES (*Continued*)

Substrate	Conditions	Product(s) and Yield(s) (%)	Refs.
C_{11}			
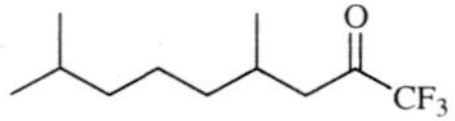 O, CF_3	Oxone®, MeCN, H_2O, $NaHCO_3$, Na_2EDTA, rt, 24 h	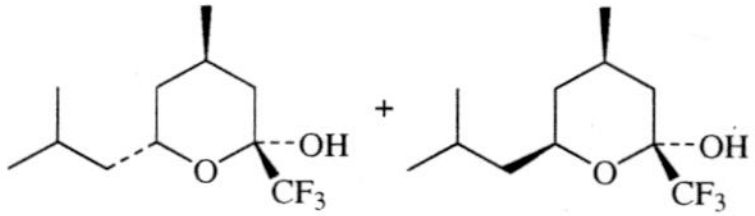 OH, CF_3, O (I) + OH, CF_3, O (II) (85) I:II = 1:15	479
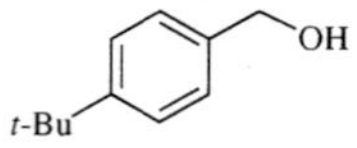 OH, *t*-Bu	Oxone®, ketone **1**, $NaHCO_3$, MeCN, H_2O, EDTA, rt, 12 h	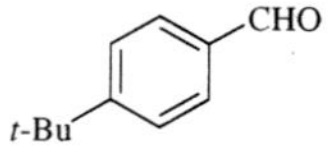 CHO, *t*-Bu (77)	52
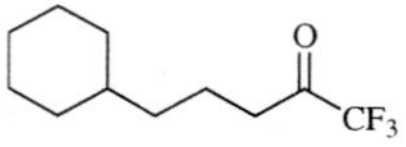 O, CF_3	Oxone®, MeCN, H_2O, $NaHCO_3$, rt, 24 h	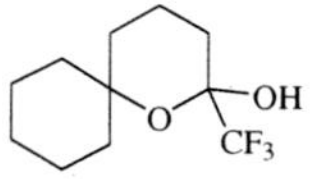 O, OH, CF_3 (78)	195
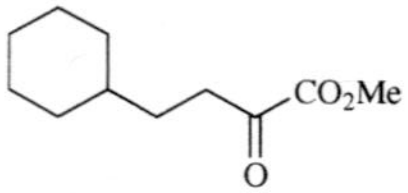 CO_2Me, O	Oxone®, MeCN, H_2O, $NaHCO_3$, rt, 24 h	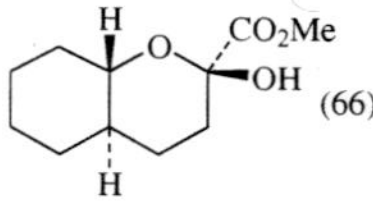 H, O, CO_2Me, OH, H (66) + H, O, CO_2Me, OH, H (17)	195
O, CH_2X	Oxone®, MeCN, H_2O, $NaHCO_3$, rt, 72 h	O, OH, CH_2X (I) + OH, O, CH_2X (II)	195
		X / I / II: Cl (56) (21); F (41) (14)	
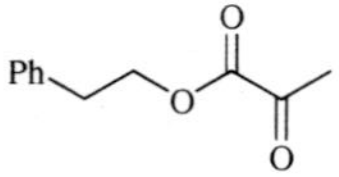 Ph, O, O, O	Oxone®, MeCN, H_2O, $NaHCO_3$, Na_2EDTA, rt, 120 h	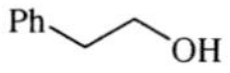 Ph, OH (93)	480

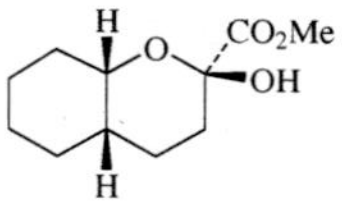

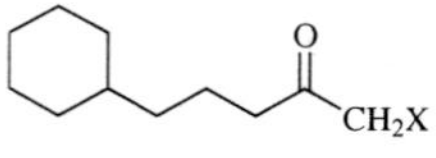

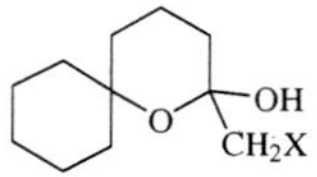

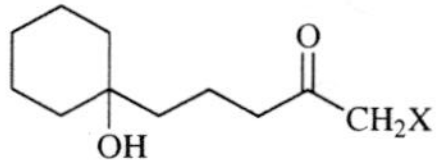

Substrate	Conditions	Product(s) and Yield(s) (%)	Refs.
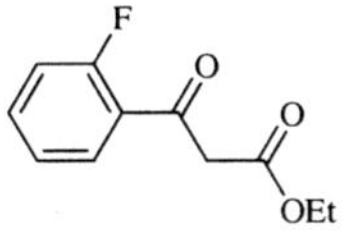	Oxone®, acetone, H_2O, $NaHCO_3$	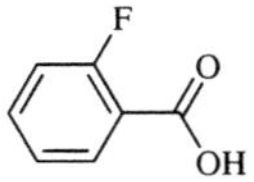 (90)	478
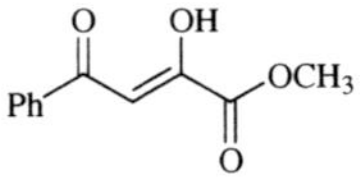	Oxone®, acetone, H_2O, $NaHCO_3$	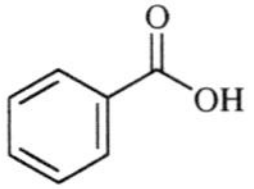 (96)	478
C_{12}			
	Oxone®, MeCN, H_2O, $NaHCO_3$, Na_2EDTA, rt, 4.5 h	**I** + **II** (62), **I**:**II** = 1:10	479
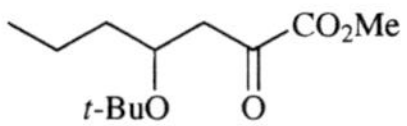	Oxone®, MeCN, H_2O, $NaHCO_3$, Na_2EDTA, rt, 5 h	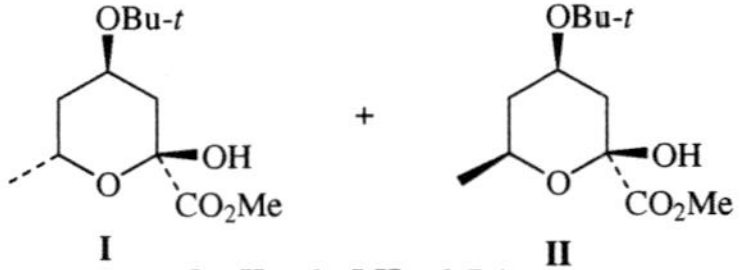 **I** + **II** (76), **I**:**II** = 2.7:1	479
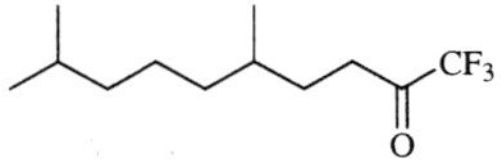	Oxone®, MeCN, H_2O, $NaHCO_3$, Na_2EDTA, rt, 24 h	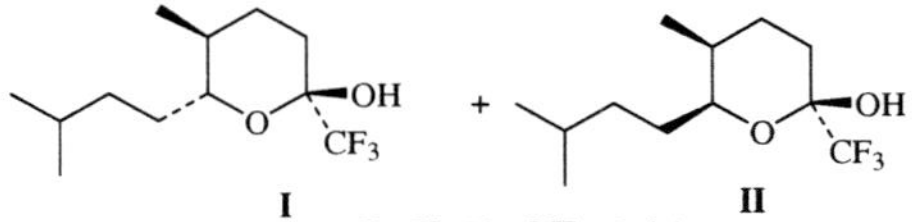 **I** + **II** (78), **I**:**II** = 3.6:1	479
C_{13}			
	Oxone®, acetone, phosphate buffer (pH 7.3-7.5), CH_2Cl_2, Bu_4NF, 20°, 1 h	(75)	193

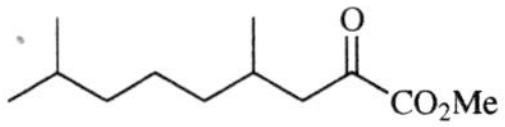

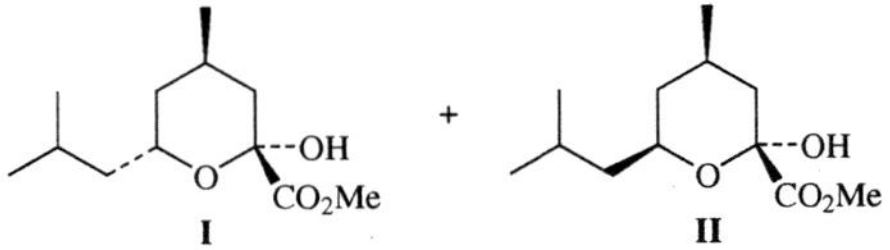

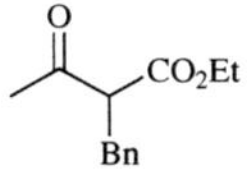

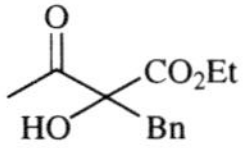

TABLE 5C. C–H OXIDATION BY IN SITU GENERATED DIOXIRANES (*Continued*)

Substrate	Conditions	Product(s) and Yield(s) (%)	Refs.
C_{14}			
	1. Oxone®, acetone, CH_2Cl_2, H_2O, $NaHCO_3$, 0°, 2 h 2. rt, 12 h	OH (38) + OAc (1)	450
OTBS O, CO_2Me	Oxone®, MeCN, H_2O, $NaHCO_3$, Na_2EDTA, rt, 27 h	OTBS, OH, CO_2Me **I** + OTBS, OH, CO_2Me **II** **I** + **II** (58), **I**:**II** = 2.7:1	479
O, CO_2Me	Oxone®, MeCN, H_2O, $NaHCO_3$, Na_2EDTA, rt, 6 h	CO_2Me (78) 73% ee + O (8)	480
O, CO_2Me	Oxone®, MeCN, H_2O, $NaHCO_3$, Na_2EDTA, rt, 24 h	OH, CO_2Me (73)	480
C_{15}			
OTBS O, CO_2Me	Oxone®, MeCN, H_2O, $NaHCO_3$, Na_2EDTA, rt, 20 h	OTBS, OH, CO_2Me **I** + OTBS, OH, CO_2Me **II** **I** + **II** (43), **I**:**II** = 2.3:1	479

	Reactant	Conditions	Product(s) and Yield(s) (%)	Refs.
	OTBS O, CO_2Me	Oxone®, MeCN, H_2O, $NaHCO_3$, Na_2EDTA, rt, 20 h	OTBS, OH, CO_2Me (**I**) + OTBS, OH, CO_2Me (**II**); **I** + **II** (59), **I**:**II** = 3.1:1	479
C_{16}	Ph, O, O, Ph	Oxone®, acetone, H_2O, $NaHCO_3$	O, Ph, OH (97)	478
	PhthN, CO_2Me, O	Oxone®, MeCN, H_2O, $NaHCO_3$, Na_2EDTA, rt, 48 h	NPhth, OH, CO_2Me (45)	479
C_{20-21}	R^1, R^2, AcO	Oxone®, acetone, phosphate buffer (pH 7.5), CH_2Cl_2, Na_2EDTA, Bu_4NHSO_4, 2°, 7 h	R^1, R^2, OH, AcO; R^1 R^2: C=O (79); OAc H (75)	460
C_{22}	O, H, H, H, O, O, O	Oxone®, MeCN, H_2O, $NaHCO_3$, Na_2EDTA, rt, 120 h	O, H, H, H, HO (16) + O, H, H, H, O (39) + O, O, H, H, H, O (15)	480

TABLE 5C. C–H OXIDATION BY IN SITU GENERATED DIOXIRANES (*Continued*)

Substrate	Conditions	Product(s) and Yield(s) (%)	Refs.
C_{22} 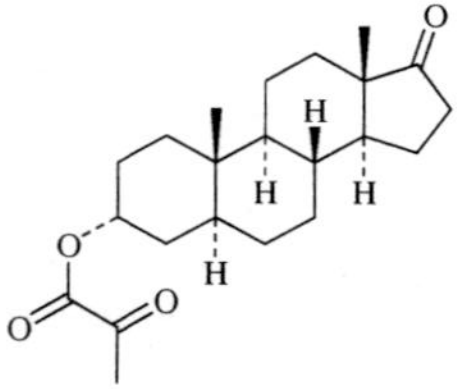	Oxone®, MeCN, H_2O, $NaHCO_3$, Na_2EDTA, rt, 120 h	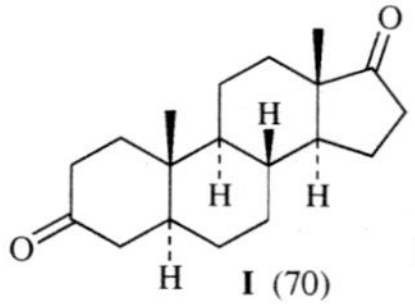 I (70) 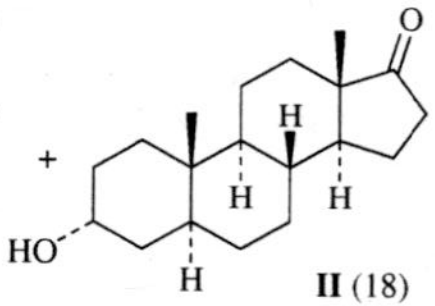 + II (18) (18)	480
C_{24} 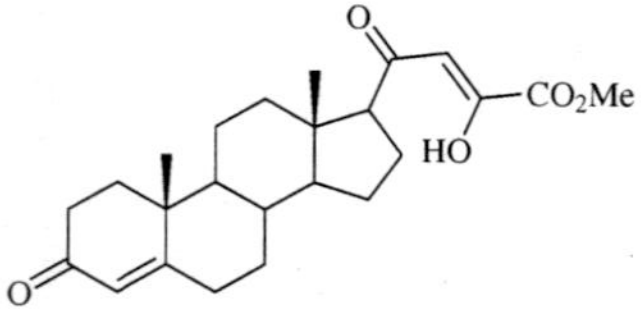 CO$_2$Me, HO	Oxone®, acetone, H_2O, $NaHCO_3$, 0° to rt, 15 min	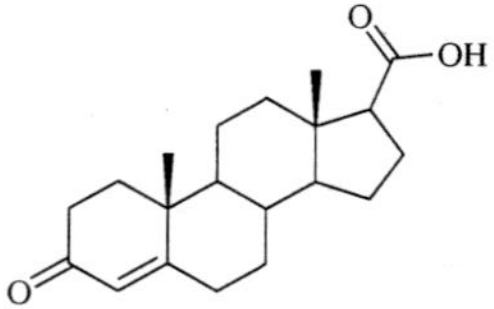 OH (87)	478
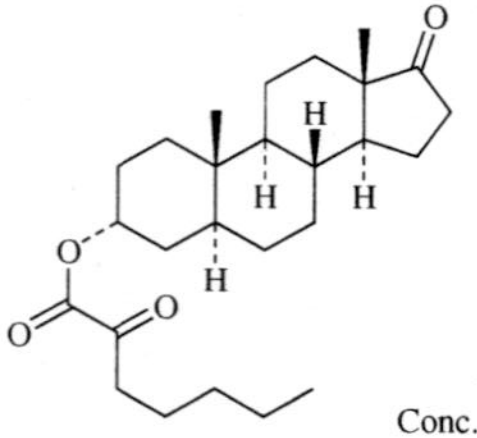 Conc. 1.5 mM; 10 mM	Oxone®, MeCN, H_2O, $NaHCO_3$, Na_2EDTA, rt, 120 h	II 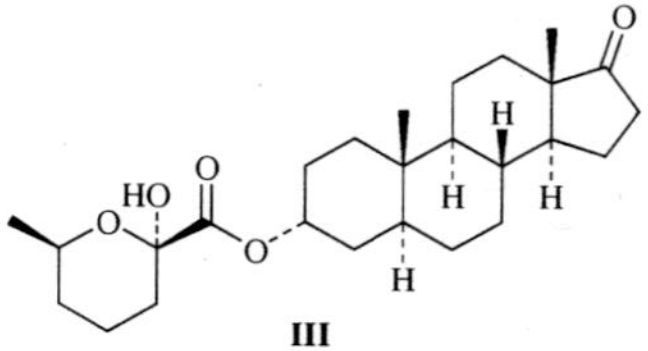 + III; II / III: (69) (34); (15) (42)	480

Substrate	Conditions	Product(s) and Yield(s) (%)	Refs.
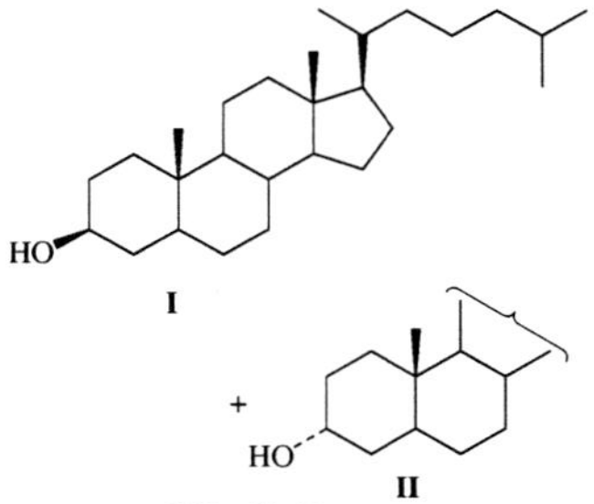 C_{27} HO, **I** + HO, **II**; **I**:**II** = 50:50	Oxone®, acetone, CH_2Cl_2, phosphate buffer (pH 7.5), Bu_4NHSO_4, 0-5°	**I** + **II** 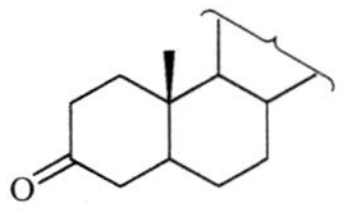 + (—) **I** + **II** (—), **I**:**II** = 35:65	482
C_{29-31} 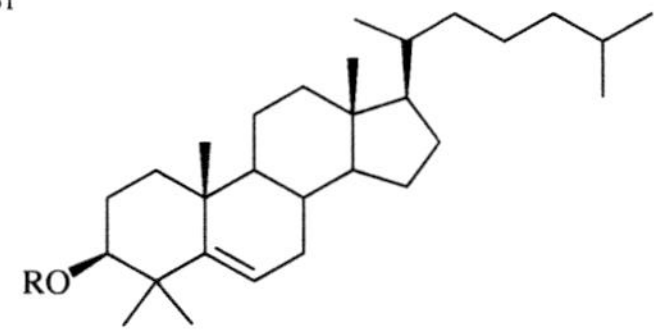 RO	Oxone®, acetone, CH_2Cl_2, phosphate buffer (pH 7.5), Bu_4NHSO_4, 0-5°	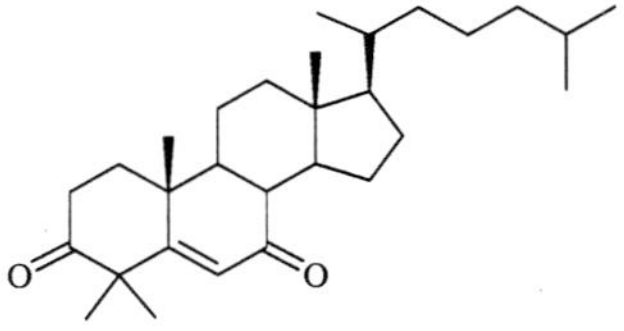 R: H (36); Ac (44)	482
C_{30} 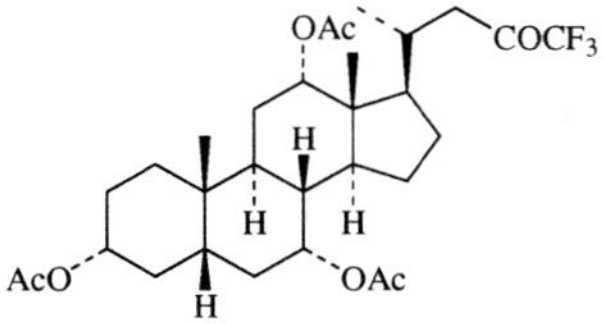 AcO, OAc, OAc, $COCF_3$	Oxone®, MeCN, H_2O, $NaHCO_3$, Na_2EDTA, rt, 24 h	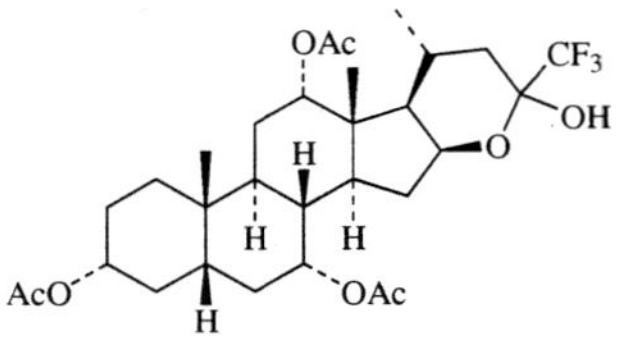 AcO, OAc, OAc, CF_3, OH, O (77)	480

TABLE 5C. C–H OXIDATION BY IN SITU GENERATED DIOXIRANES (*Continued*)

Substrate	Conditions	Product(s) and Yield(s) (%)	Refs.
C_{31} AcO, OAc, OAc, H, H, H, H, $COCF_3$	Oxone®, MeCN, H_2O, $NaHCO_3$, Na_2EDTA, rt, 24 h	AcO, OAc, OAc, H, H, H, H, O, OH, CF_3 (58)	195, 480
C_{35} CF_3, O, O, H	Oxone®, MeCN, H_2O, $NaHCO_3$, Na_2EDTA, rt, 41 d	HO, H (3) + O, H (4) + O, O, H (10) + O, O, H (17) + HO, OH (3) + O, OH (6)	480

TABLE 5D. ASYMMETRIC C–H OXIDATION BY IN SITU GENERATED OPTICALLY ACTIVE DIOXIRANES

Substrate	Conditions	Product(s) and Conversions(s) (%)	Refs.
C_9			
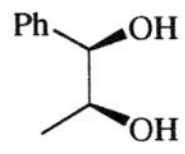 racemic	Oxone®, K_2CO_3, ketone, buffer (pH 10.5), MeCN, 0°, 2 h Ketone 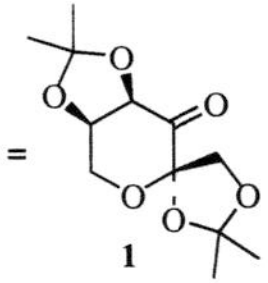 = 1	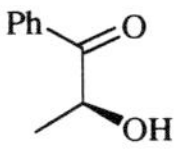 I 23% ee 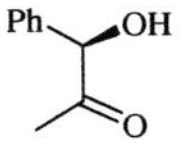 + II 8% ee **I** + **II** (34), **I**:**II** = 89:11	197
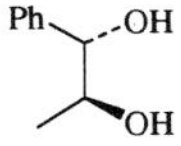 racemic	Oxone®, K_2CO_3, ketone **1**, buffer (pH 10.5), MeCN, 0°, 2 h	**I** + **II** (20), **I** 71% ee, **II** 11% ee, **I**:**II** = 84:16	197
C_{9-10}			
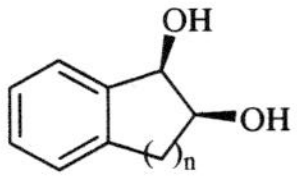 racemic	Oxone®, K_2CO_3, ketone **1**, buffer (pH 10.5), MeCN, 0°, 2 h	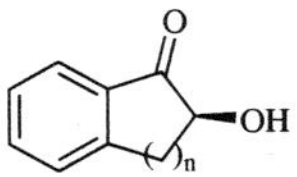 n / % ee 1 (30) 9 2 (18) 14	197
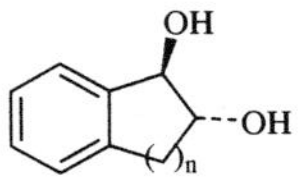 racemic	Oxone®, K_2CO_3, ketone **1**, buffer (pH 10.5), MeCN, 0°, 2 h	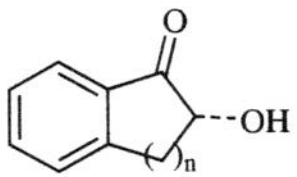 n / % ee 1 (26) 20 2 (8) 5	197
C_{12}			
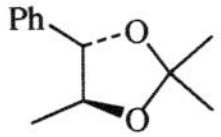	Oxone®, K_2CO_3, ketone **1**, buffer (pH 10.5), MeCN, DMM, 0°, 5 h	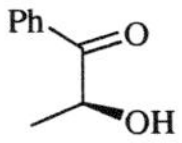 (< 5), 11% ee	196, 197
C_{12-19}			
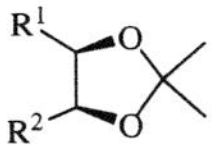 meso or rac	Oxone®, K_2CO_3, ketone **1**, buffer (pH 10.5), MeCN, DMM, 0°, 5 h	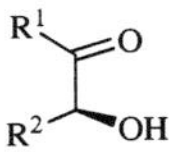 R^1 / R^2 / % ee Ph Me (6) 44 Ph Ph (10) 63 4-MeC_6H_4 4-MeC_6H_4 (10) 65	197

TABLE 5D. ASYMMETRIC C–H OXIDATION BY IN SITU GENERATED OPTICALLY ACTIVE DIOXIRANES (*Continued*)

Substrate	Conditions	Product(s) and Conversions(s) (%)	Refs.
$C_{14\text{-}16}$ 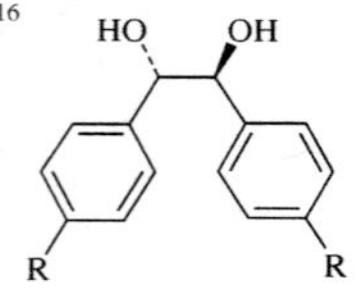 HO OH R R	Oxone®, ketone **1**, K_2CO_3, buffer (pH 10.5), MeCN, 0°	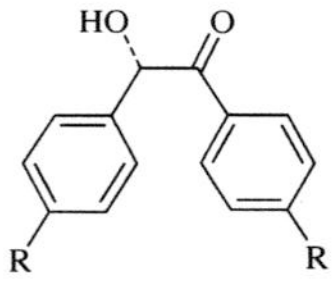 HO O R R R / Time / / % ee Br / 2 h / (10) / 74 Cl / 2 h / (11) / 70 H / 3 h / (51) / 65 CN / 2.5 h / (6) / 75 Me / 3 h / (12) / 61	197
C_{14} 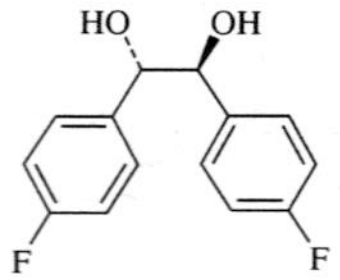 HO OH F F racemic	Oxone® (0.75 eq), ketone **1**, K_2CO_3, buffer (pH 10.5), MeCN, 0°, 1.5 h	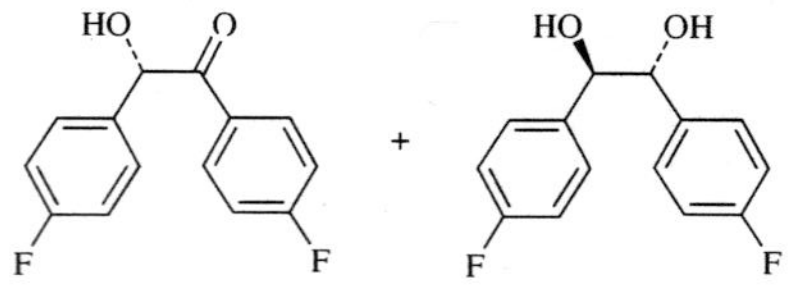 HO O F F + HO OH F F **I** + **II** (10), **I** 71% ee, **II** 11% ee	197
	Oxone® (1.5 eq), ketone **1**, K_2CO_3, buffer (pH 10.5), MeCN, 0°, 3 h	**I** + **II** (31), **I** 69% ee, **II** 28% ee	197
$C_{14\text{-}16}$ 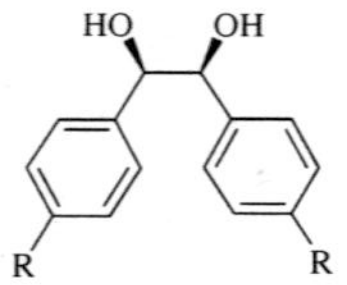 HO OH R R	Oxone®, ketone **1**, K_2CO_3, buffer (pH 10.5), MeCN, 0°	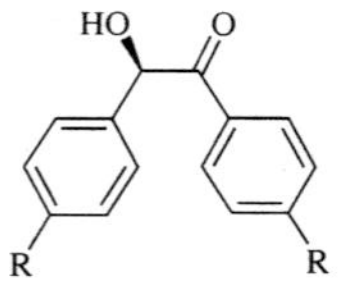 HO O R R R / Time / / % ee Br / 2 h / (61) / 58 Cl / 3 h / (56) / 54 F / 3 h / (89) / 58 H / 3 h / (95) / 45 CN / 3 h / (< 5) / 60 Me / 3 h / (92) / 30 MeO / 3 h / (95) / 24	197

TABLE 5E. Si–H OXIDATION BY ISOLATED DIOXIRANES

Substrate	Conditions	Product(s) and Yield(s) (%)	Refs.
C_{5-18} $R^2R^1R^3Si$–H	DMD, acetone, (CCl_4), rt	$R^2R^1R^3Si$–OH (> 99)	483
C_6 Et_3Si–H	TFD, TFP, CH_2Cl_2, –20°, < 1 min	Et_3Si–OH (> 98)	39
C_8 Ph–$SiMe_2$–H	TFD, TFP, CH_2Cl_2, –20°, < 1 min	Ph–$SiMe_2$–OH (> 98)	39
C_9 $(TMS)_3Si$–H	DMD, acetone, Ar	$(TMS)_3Si$–OH **I** + $(TMS)_3Si$–OAc **II**	483
	DMD, $(CF_3CO)_2O$, acetone, CCl_4, Ar, 20°, < 1 min	$(TMS)_3Si$–O–C(=O)CF_3 (90) + $(TMS)_3Si$–Cl (10)	483

R^1	R^2	R^3	Time
TMSO	Me	Me	15 min
Et	Et	Et	< 5 min
t-Bu	Me	Me	< 5 min
Ph	Me	H	30min
TMSO	TMSO	Me	30 min
Ph	Me	Me	< 5 min
TMSO	TMSO	TMSO	3 h
Ph	Me	TMS	< 1 min
Ph	Ph	Ph	10 min

Additive	Temp	Time	I	II
—	20°	< 1 min	(79)	(16)
O_2	20°	< 1 min	(> 99)	(—)
—	–70°	10 min	(> 99)	(—)
hν	–70°	10 min	(88)	(5)

TABLE 5E. Si–H OXIDATION BY ISOLATED DIOXIRANES (*Continued*)

Substrate	Conditions	Product(s) and Yield(s) (%)	Refs.
C_{12}			
	DMD, acetone, CCl_4, rt	(> 99)	483
C_{17}			
96.5% ee (+)	TFD, TFP, CH_2Cl_2, –20°, < 1 min; or, DMD, acetone, CH_2Cl_2, 0°, 18 min	(> 98), 97% ee	39
C_x			
	DMD, acetone, 0°	(—)	484

TABLE 6. OXIDATION OF ORGANOMETALLICS BY ISOLATED DIOXIRANES

Substrate	Conditions	Product(s) and Yield(s) (%)	Refs.
C_{7-12}			
R₅-Cp Fe(CO)₂ Si(H)(H)SiH₃	DMD, acetone, toluene, –78° to rt, 50 min	R₅-Cp Fe(CO)₂ Si(OH)(OH)SiH₃ R: H (89); Me (89)	227
C_{9-19}			
1-(OML_3)indene	1. DMD, THF, acetone, –78°, 1 min 2. NH_4F, H_2O, rt, 1-12 h	2-hydroxy-1-indanone M, L_3: Na, — (32); Ti, (*i*-PrO)$_3$ (50); Ti, Cp_2Cl (70)	212
C_{9-22}			
Ph-CH=C(OML_3)OR	1. DMD, THF, acetone, –78°, 1 min 2. NH_4F, H_2O, rt, 1-12 h	Ph-CH(OH)-C(=O)OR R, M, L_3: Me, Na, — (60); *t*-Bu, Ti, (*i*-PrO)$_3$ (12); *t*-Bu, Ti, Cp_2Cl (67)	212
C_{9-38}			
CpFe(CO)(L)$SiHR_2$	DMD, acetone; or acetone, CH_2Cl_2	CpFe(CO)(L)SiR_2OH L, R, Temp, Time: CO, Me, 0°, 6 h (46); CO, *t*-Bu, –78°, 0.3 h (98); CO, Ph, 0°, 2.5 h (86); CO, 2-MeC_6H_4, 0°, 0.5 h (85); PPh_3, 2-MeC_6H_4, –78°, 0.5 h (43)	220
C_{10}			
CpFe(CO)₂Si(H)(Me)Me	DMD, acetone, –78°, 1 h; rt, 1 h	CpFe(CO)₂Si(OH)(Me)Me (67)	485

TABLE 6. OXIDATION OF ORGANOMETALLICS BY ISOLATED DIOXIRANES (*Continued*)

Substrate	Conditions	Product(s) and Yield(s) (%)	Refs.
C_{10}			
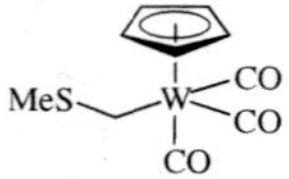 MeS–CH$_2$–W(CO)$_3$Cp	DMD (1.0 eq), acetone, CH_2Cl_2, –65°, 10 min	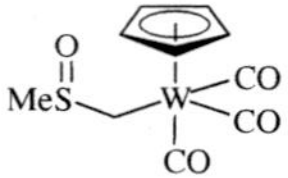 MeS(=O)–CH$_2$–W(CO)$_3$Cp (88)	486
	DMD (2.0 eq), acetone, CH_2Cl_2, –65°, 10 min	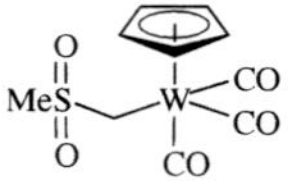 MeS(=O)$_2$–CH$_2$–W(CO)$_3$Cp (70)	486
C_{10-11}			
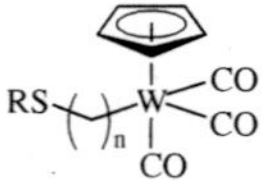 RS–(CH$_2$)$_n$–W(CO)$_3$Cp	DMD (1.0 eq), acetone, CH_2Cl_2, –65°, 10 min	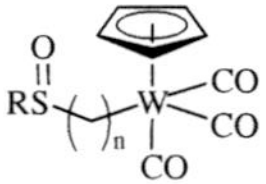 RS(=O)–(CH$_2$)$_n$–W(CO)$_3$Cp n R 1 Me (88) 2 Me (—) 1 Et (89)	486
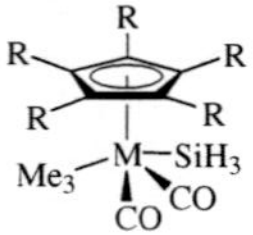 (C_5R_5)M(CO)$_2$(SiH$_3$)(Me$_3$)	DMD, acetone, toluene, –78°, 2 h	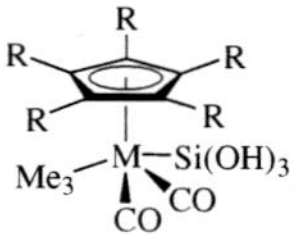 (C_5R_5)M(CO)$_2$(Si(OH)$_3$)(Me$_3$) (34-71) M R Mo H Mo Me W Me	221
C_{10-13}			
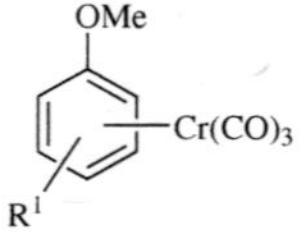 (R^1-anisole)Cr(CO)$_3$ R^1 = H, 2-CH$_2$OH, 3-CH$_2$OH, 4-CH$_2$OH, –OCH$_2$CH=CH–	DMD, acetone, 20°	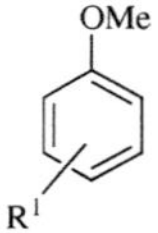 R^1-anisole (100) + Cr_2O_3 (—)	487
C_{10-14}			
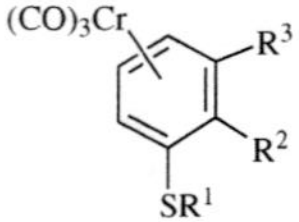 (CO)$_3$Cr(arene)–SR^1, R^2, R^3	1. DMD, acetone, N_2, –78°, 15 min 2. rt, 1 h	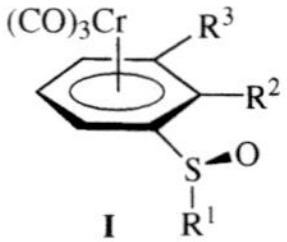 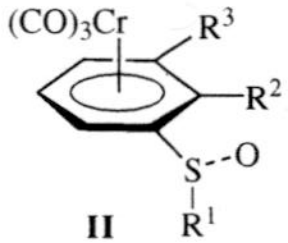 (CO)$_3$Cr(arene)–S(O)R^1, R^2, R^3 **I** + (CO)$_3$Cr(arene)–S(O)R^1, R^2, R^3 **II**	128, 129

R^1	R^2	R^3	**I**	**II**	**I:II**
Me	H	H	**I + II** (93)		1:1
Me	Me	H	(—)	(78)	10:90
Me	H	MeO	(33)	(33)	50:50
Me	MeO	H	(—)	(80)	7:93
Et	MeO	H	(—)	(39)	15:85
i-Pr	MeO	H	(56)	(—)	70:30
t-Bu	Me	H	(92)	(—)	98:2
Me	*t*-Bu	H	(—)	(45)	6:94
t-Bu	MeO	H	(77)	(—)	98:2

C_{11}

SMe, $Cr(CO)_3$, OMe — DMD, acetone, 20° — O=S–Me, $Cr(CO)_3$, OMe (—) — 487

C_{12}

O, O, $Cr(CO)_3$ — DMD, acetone, –28°, 1 min — O, OH, OH, O, $Cr(CO)_3$ (75) — 488

C_{13-25}

$Cr(CO)_3$, H, NMe_2, R — 1. DMD (0.95 eq), acetone, N_2, –78°, 15 min 2. –78° to rt, 2 h — $Cr(CO)_3$, R — 217

R	
I	(48)
Ac	(69)
CH_2=$CHCH_2$	(69)
Bu_3Sn	(20)

C_{13}

$Cr(CO)_3$, H, NMe_2 — 1. DMD, acetone, N_2, –78°, 15 min 2. –78° to rt, 2 h — $Cr(CO)_3$, H, $\overset{+}{N}Me_2$, ^{-}O (—) + $Cr(CO)_3$ (—) + other products — 217

TABLE 6. OXIDATION OF ORGANOMETALLICS BY ISOLATED DIOXIRANES (*Continued*)

Substrate	Conditions	Product(s) and Yield(s) (%)	Refs.
C_{13}			
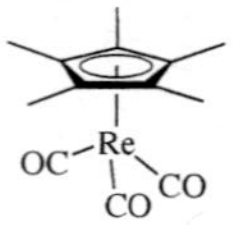 OC–Re(CO)(CO)	DMD, acetone, 0°	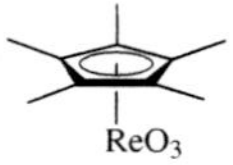 ReO_3 (74)	210
C_{13-14}			
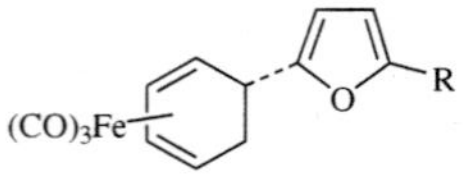 $(CO)_3Fe$ ... R	DMD, acetone, Ar, 1 min	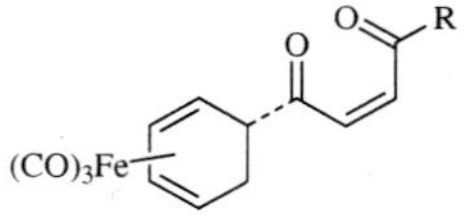 $(CO)_3Fe$ ... O, O, R R: H (—); Me (39)	214
C_{13-28}			
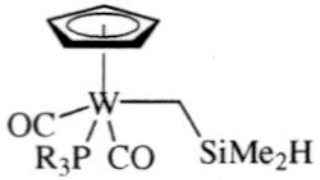 OC–W, R_3P, CO, $SiMe_2H$; R = Me, Ph	DMD, acetone	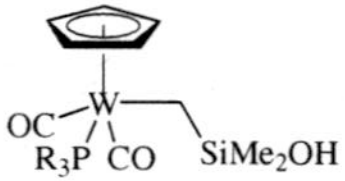 OC–W, R_3P, CO, $SiMe_2OH$ (—)	489
C_{14}			
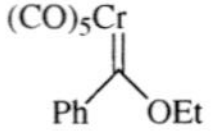 $(CO)_5Cr$, Ph, OEt	DMD, acetone, 20°, 3 h	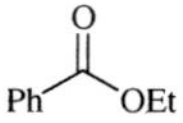 O, Ph, OEt (97)	204
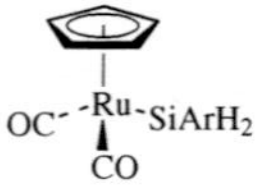 OC–Ru–$SiArH_2$, CO; Ar = 4-MeC_6H_4	DMD, acetone, toluene, –78° to rt, 60 min	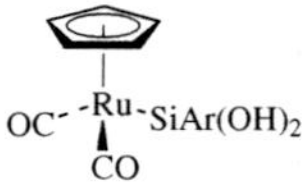 OC–Ru–$SiAr(OH)_2$, CO (68)	226
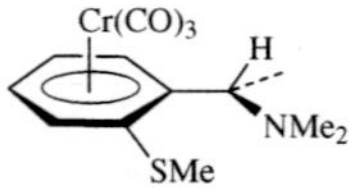 $Cr(CO)_3$, H, NMe_2, SMe	1. DMD (1.2 eq), acetone, N_2, –78°, 15 min 2. –78° to rt, 2 h	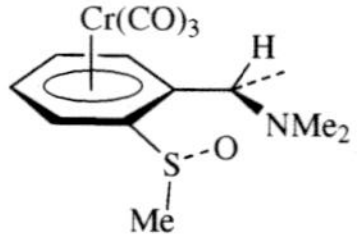 $Cr(CO)_3$, H, NMe_2, S–O, Me (63)	217
	1. DMD (2.4 eq), acetone, N_2, –78°, 15 min 2. –78° to rt, 2 h	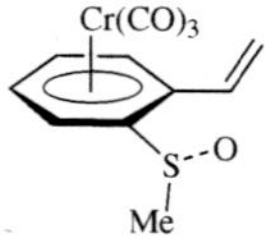 $Cr(CO)_3$, S–O, Me (80)	217

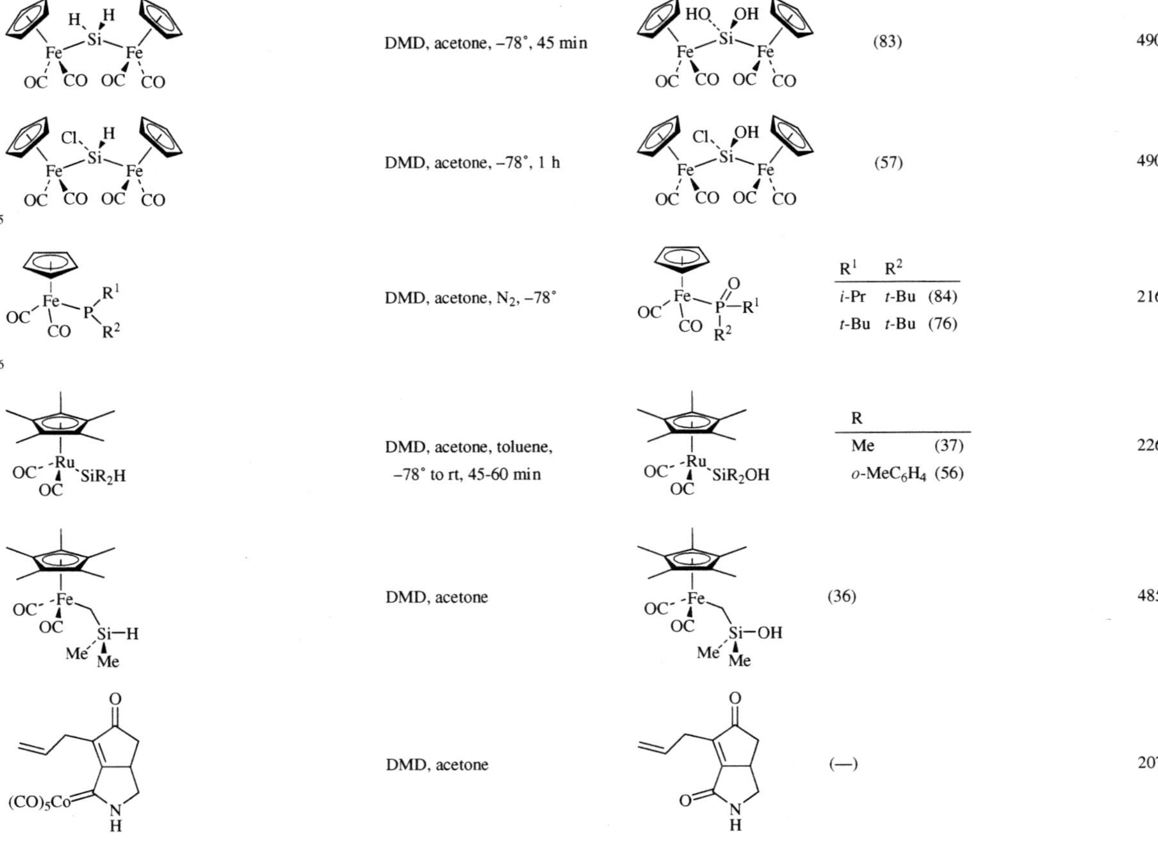

Substrate	Conditions	Product(s) and Yield(s) (%)	Refs.
	DMD, acetone, −78°, 45 min	(83)	490
C_{14-15}	DMD, acetone, −78°, 1 h	(57)	490
C_{14-26}	DMD, acetone, N_2, −78°	R^1 / R^2: *i*-Pr / *t*-Bu (84); *t*-Bu / *t*-Bu (76)	216
	DMD, acetone, toluene, −78° to rt, 45-60 min	R: Me (37); *o*-MeC_6H_4 (56)	226
C_{15}	DMD, acetone	(36)	485
	DMD, acetone	(—)	207

TABLE 6. OXIDATION OF ORGANOMETALLICS BY ISOLATED DIOXIRANES (*Continued*)

Substrate	Conditions	Product(s) and Yield(s) (%)	Refs.
C_{15-19}			
$Cp(\eta^5\text{-}C_5H_4)Fe$–Y–C($R^3$)=C($R^1$)$R^2$	DMD (x eq), acetone, CH_2Cl_2, gas	$Cp(\eta^5\text{-}C_5H_4)Fe$–Y–epoxide [C($R^3$)–O–C($R^1$)$R^2$]	213

Y	R^1	R^2	R^3	x	Time	Temp	Gas	Yield (%)
CO	Me	Me	H	6.0	5 h	20°	Ar	(24)
CO	Me	Me	H	9.0	9 h	20°	Ar	(44)
CO	Me	Me	H	12.0	14 h	20°	Ar	(58)
CO	H	Me	Me	6.0	6 h	20°	Ar	(59)
CO	H	Me	Me	6.0	24 h	20°	O_2	(38)
CH_2	Me	Me	H	3.0	40 min	0°	Ar	(66)
CH_2	Me	Me	H	3.0	10 min	20°	Ar	(75)
CH_2	Me	Me	H	3.0	1 min	56°	Ar	(63)
CH_2	Me	Me	H	4.0	1 min	56°	Ar	(76)
CH_2	Me	Me	H	3.0	25 min	20°	O_2	(28)
CH_2	H	Me	Me	3.0	15 min	20°	Ar	(67)
CH_2	H	Me	Me	3.0	30 min	20°	O_2	(53)
CO	H	Ph	H	6.0	8 h	20°	Ar	(30)
CH_2	H	Ph	H	6.0	45 min	20°	Ar	(38)

Substrate	Conditions	Product(s) and Yield(s) (%)	Refs.
C_{15-21}			
$Cp(CO)_2Fe$–Si(H)(R)–$Fe(CO)_2Cp$	DMD, acetone, –78° to rt, 25 min	$Cp(CO)_2Fe$–Si(HO)(R)–$Fe(CO)_2Cp$ R: Me (89); 4-MeC_6H_4 (95)	223
C_{16}			
$(CO)_5Cr$=C(OEt)–C≡C–Ph	DMD, O_2, acetone, –20°, 4 h	Ph–C≡C–CO_2Et **I** (90)	204, 205
	TFD, O_2, TFP, –20°, 4 h	**I** (91)	205

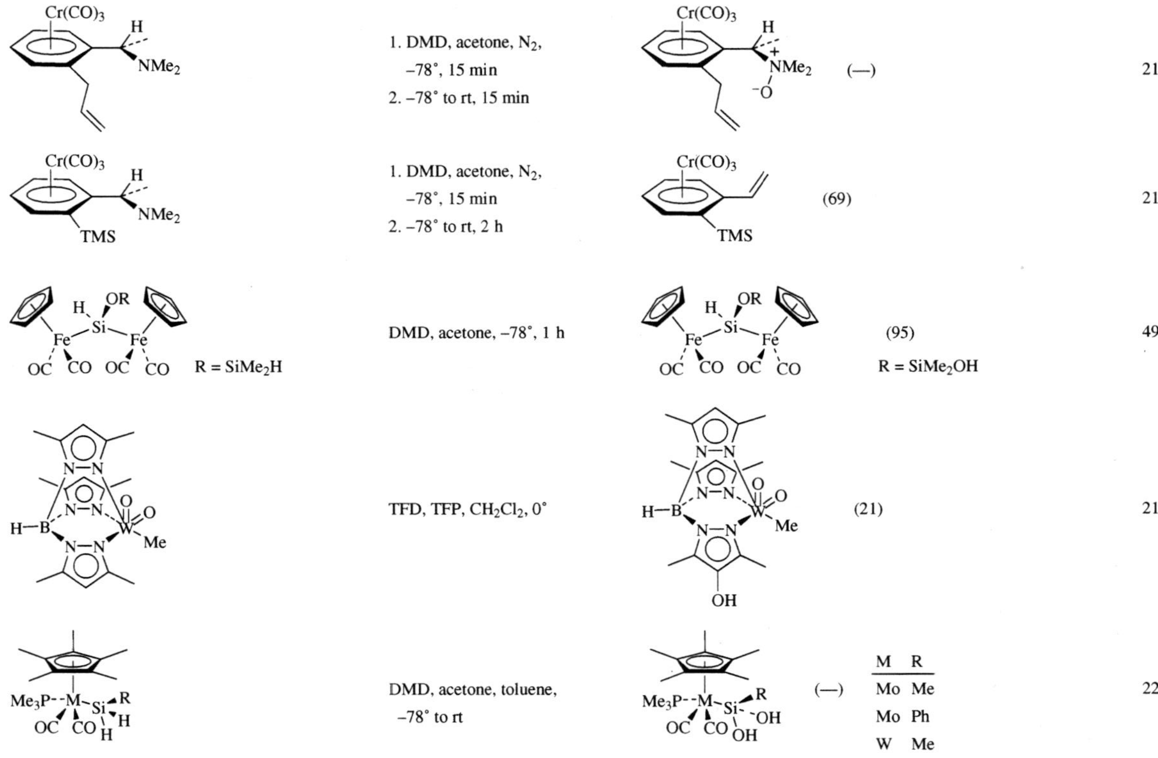

Reactant	Conditions	Product(s) and Yield(s) (%)	Refs.
	1. DMD, acetone, N_2, −78°, 15 min 2. −78° to rt, 15 min	(—)	217
	1. DMD, acetone, N_2, −78°, 15 min 2. −78° to rt, 2 h	(69)	217
R = $SiMe_2H$	DMD, acetone, −78°, 1 h	(95) R = $SiMe_2OH$	490
	TFD, TFP, CH_2Cl_2, 0°	(21)	211
	DMD, acetone, toluene, −78° to rt	(—) M R: Mo Me; Mo Ph; W Me	222

TABLE 6. OXIDATION OF ORGANOMETALLICS BY ISOLATED DIOXIRANES (*Continued*)

Substrate	Conditions	Product(s) and Yield(s) (%)	Refs.
C_{16-22}			
$CpW(L)(CO)_2PR^1R^2$	DMD, acetone, solvent, N_2, −78°	$CpW(L)(CO)_2P(O)R^1R^2$	216
L / R^1 / R^2	Solvent		
CO / *t*-Bu / *t*-Bu	toluene	(—)	
CO / H / 2,4,6-$Me_3C_6H_2$	—	(88)	
CO / Me / 2,4,6-$Me_3C_6H_2$	—	(80)	
CO / 2-MeC_6H_4 / 2-MeC_6H_4	—	(69)	
PMe_3 / Ph / Ph	—	(67)	
C_{17}			
$(CO)_5Cr{=}C(NHPr)C{\equiv}CPh$	DMD, O_2, acetone, −20°, 4 h	$PhC{\equiv}CCONHPr$ (94)	204
Et_4N^+ $TpMo(CO)_3^-$	DMD, acetone, Ar, rt, 15 min	Et_4N^+ $TpMoO_3^-$ (59)	210
Et_4N^+ $Tp^*Mo(CO)_3^-$	DMD, acetone, Ar, rt, 15 min	Et_4N^+ $Tp^*MoO_3^-$ (—)	210
$Cr(CO)_3$-complexed arene bearing $CH(NMe_2)$ (H dashed) and $CH{=}CHCO_2Me$	1. DMD, acetone, N_2, −78°, 15 min 2. −78° to rt, 2 h	$Cr(CO)_3$-complexed arene bearing $CH{=}CH_2$ and $CH{=}CHCO_2Me$ (44)	217

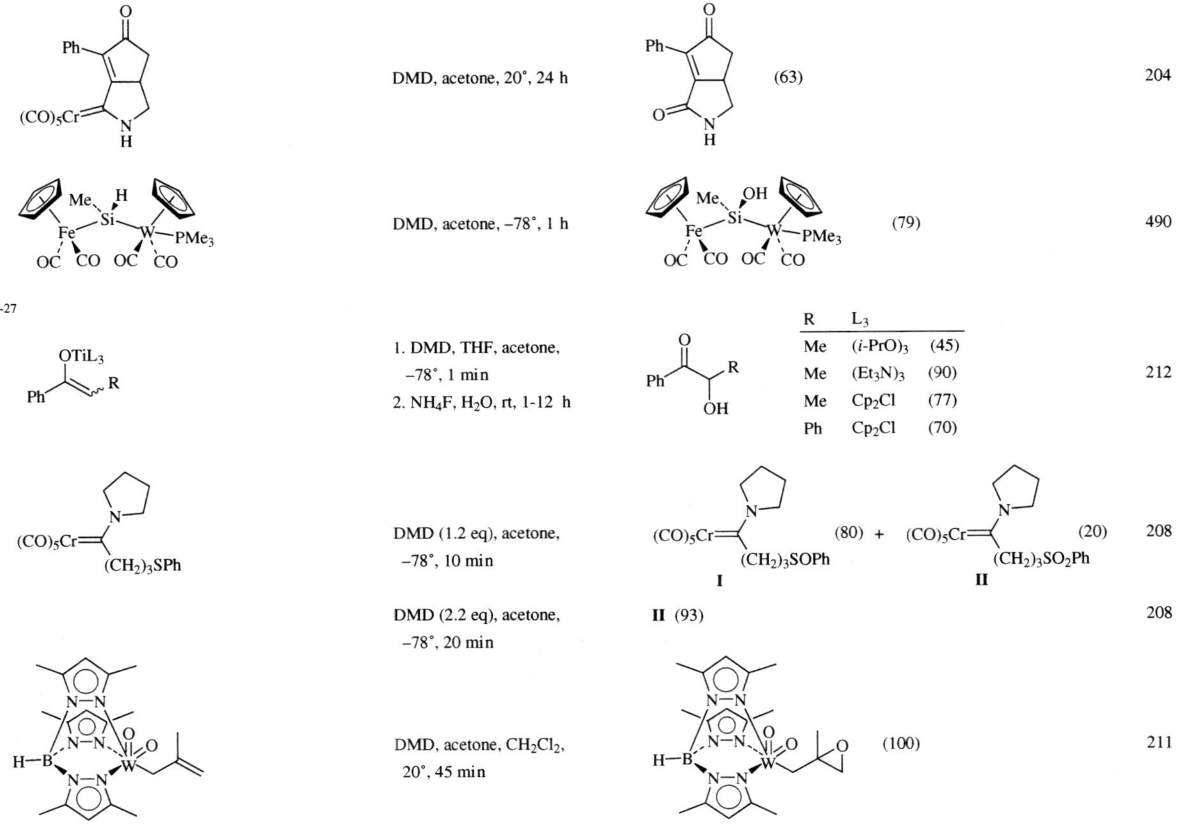

	Conditions	Product(s) and Yield(s) (%)	Refs.
C_{18}			
	DMD, acetone, 20°, 24 h	(63)	204
	DMD, acetone, –78°, 1 h	(79)	490
C_{18-27}			
	1. DMD, THF, acetone, –78°, 1 min 2. NH_4F, H_2O, rt, 1-12 h	R, L_3: Me, $(i\text{-}PrO)_3$ (45); Me, $(Et_3N)_3$ (90); Me, Cp_2Cl (77); Ph, Cp_2Cl (70)	212
C_{19}			
	DMD (1.2 eq), acetone, –78°, 10 min	**I** (80) + **II** (20)	208
	DMD (2.2 eq), acetone, –78°, 20 min	**II** (93)	208
	DMD, acetone, CH_2Cl_2, 20°, 45 min	(100)	211

TABLE 6. OXIDATION OF ORGANOMETALLICS BY ISOLATED DIOXIRANES (*Continued*)

Substrate	Conditions	Product(s) and Yield(s) (%)	Refs.
C_{19}			
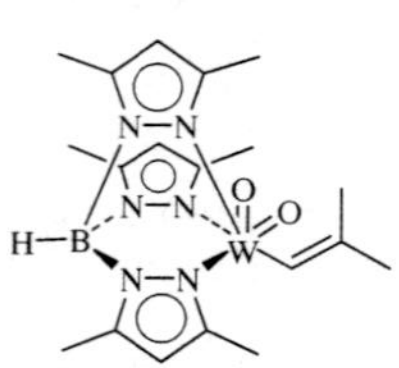	DMD, acetone, CH_2Cl_2, 20°, 45 min	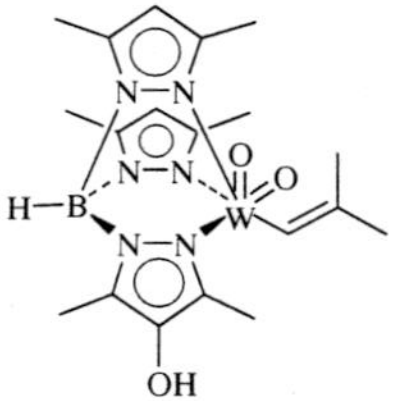 (39)	211
C_{19-20}			
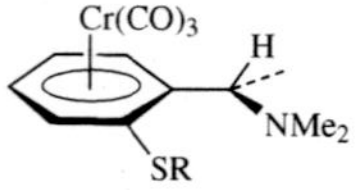 $Cr(CO)_3$, H, NMe_2, SR	1. DMD, acetone, N_2, –78°, 15 min 2. –78° to rt, 2 h	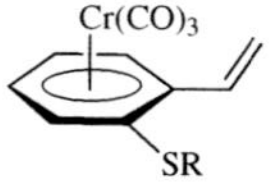 $Cr(CO)_3$, SR (63) R: Ph (84-85); 4-$MeOC_6H_4$ (81-95)	217
C_{20}			
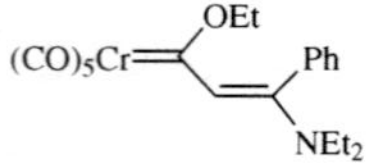 $(CO)_5Cr$, OEt, Ph, NEt_2	DMD, acetone, 20°	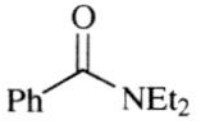 Ph, O, NEt_2 (52)	206
	1. DMD, CH_2Cl_2, acetone, –30° 2. Et_3OBF_4, CH_2Cl_2, acetone, –70°	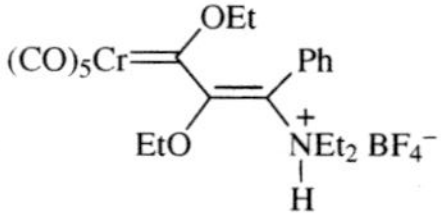 $(CO)_5Cr$, OEt, EtO, Ph, $\overset{+}{N}Et_2$ BF_4^-, H (—)	206
C_{21}			
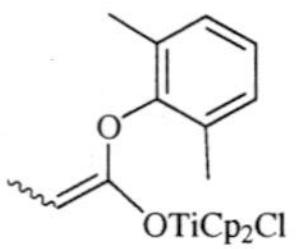 O, $OTiCp_2Cl$	1. DMD, THF, acetone, –78°, 1 min 2. NH_4F, H_2O, rt, 1-12 h	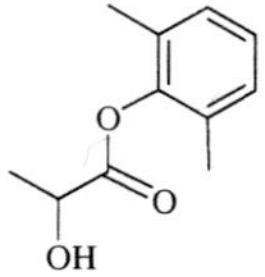 O, O, OH (65)	212

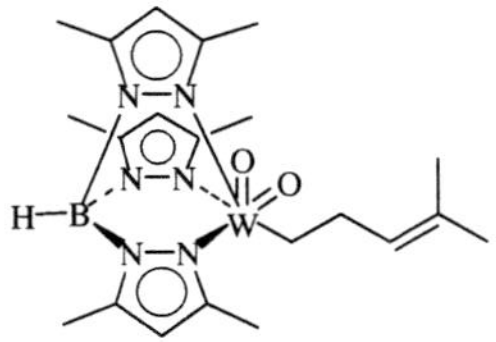

Substrate	Conditions	Product(s) and Yield(s) (%)	Refs.
	DMD, acetone, CH_2Cl_2, 20°, 0.5 h	I (100)	211
	TFD, TFP, CH_2Cl_2, 0°	I (83)	211
C_{21-24}	DMD, acetone, CH_2Cl_2, N_2, –20°, 20 h		215
C_{22}	TFD, TFP, Et_2O, rt	(—)	209
C_{22-24}	DMD, acetone, rt	I + II	491

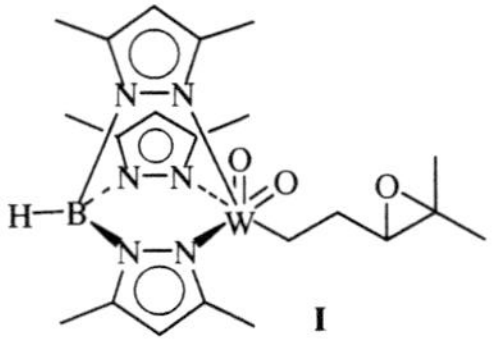

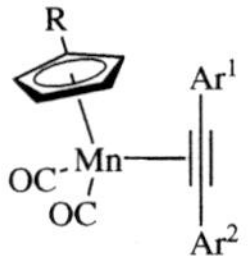

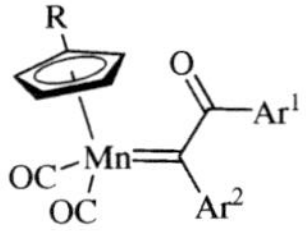

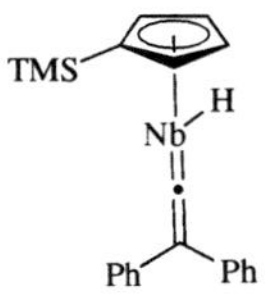

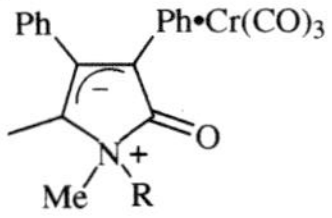

R	Ar_1	Ar^2	
H	Ph	Ph	(30)
Me	Ph	Ph	(25)
H	4-MeC_6H_4	4-MeC_6H_4	(42)
Me	Ph	4-$MeOC_6H_4$	(27)
Me	4-$MeOC_6H_4$	Ph	(—)
Me	4-MeC_6H_4	4-MeC_6H_4	(41)
Me	4-$MeOC_6H_4$	4-MeC_6H_4	(—)
Me	4-MeC_6H_4	4-$MeOC_6H_4$	(29)

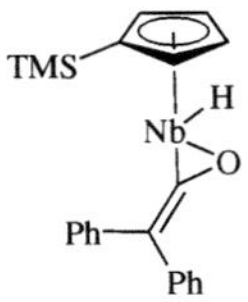

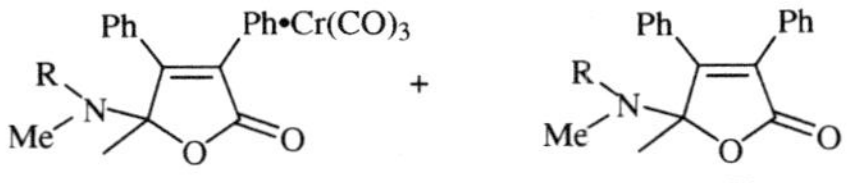

R	I	II
Me	(20)	(75)
c-C_3H_5	(47)	(—)

TABLE 6. OXIDATION OF ORGANOMETALLICS BY ISOLATED DIOXIRANES (*Continued*)

Substrate	Conditions	Product(s) and Yield(s) (%)	Refs.
C_{22-27}			
$(C_5R_5)M(CO)_2(PMe_3)SiPh_2H$	1. DMD, acetone, toluene, −78°, 30 min 2. rt, 1 h	$(C_5R_5)M(CO)_2(PMe_3)SiPh_2OH$ M, R: Cr, H (79); Mo, Me (64); W, Me (89)	224
C_{24}			
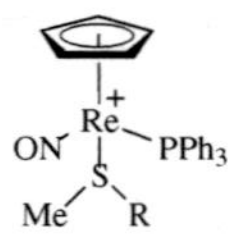 $[CpRe(NO)(PPh_3)(SMeR)]^+$	DMD, acetone, 0°, 45 min	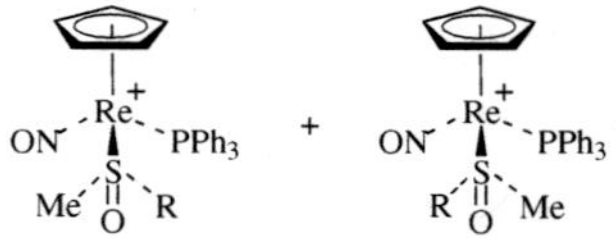 $[CpRe(NO)(PPh_3)(S(O)MeR)]^+$ **I** + **II** (diastereomeric sulfoxides)	492

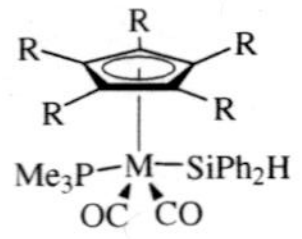

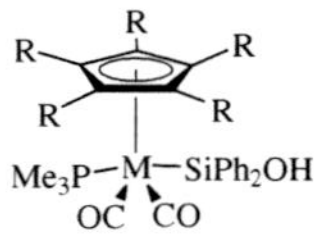

R	Time	I	II	% ee
BF_4^-	2.3 h	(67)	(—)	96
BF_4^-	2.0 h	(49)	(24)	100
BF_4^-	3.0 h	(84)	(10)	91
BF_4^-	5.6 h	(100)	(—)	85
TfO^-	2.6 h	(53)	(21)	39
TfO^-	3.0 h	(84)	(11)	38
TfO^-	2 d	(77)	(—)	79
TfO^-	4 d	(58)	(—)	100
TfO^-	—	(32)	(—)	—

Substrate	Conditions	Product(s) and Yield(s) (%)	Refs.
C_{25}			
$Cr(CO)_3$-arene with $CH(H)NMe_2$ and PPh_2	1. DMD (0.95 eq), acetone, N_2, −78°, 15 min 2. −78° to rt, 2 h	$Cr(CO)_3$-arene with $CH(H)NMe_2$ and $P(=O)Ph_2$ (85)	217

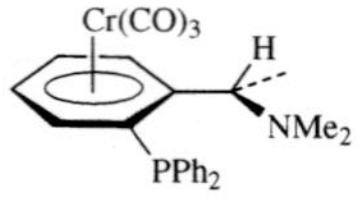

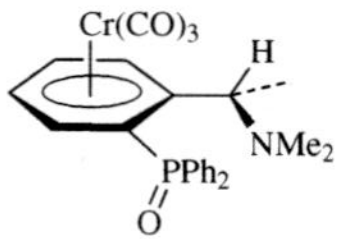

1. DMD (2.4 eq), acetone, N_2, −78°, 15 min
2. −78° to rt, 2 h

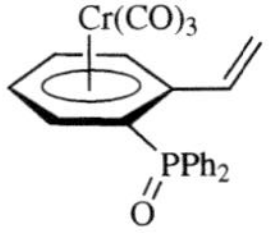

$Cr(CO)_3$, PPh₂=O vinyl arene product (73) — 217

C_{28-42}

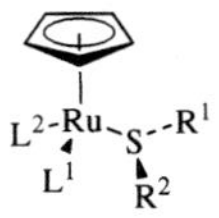

L^2-Ru(Cp)(L^1)–S(R^1)(R^2)

DMD, acetone, −40 to 0°

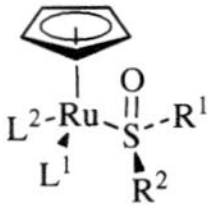

L^2-Ru(Cp)(L^1)–S(=O)(R^1)(R^2) — 127

L^1	L^2	R^1	R^2	Time		dr
$Me_2P(CH_2)_2PPh_2$		Me	*i*-Pr	45 min	(100)	80:20
Ph_3P	CO	Me	*i*-Pr	2 h	(45)	64:36
$Me_2P(CH_2)_2PPh_2$		Me	Ph	45 min	(100)	80:20
$Me_2P(CH_2)_2PPh_2$		Me	Bz	45 min	(100)	75:25
Ph_3P	CO	Me	Ph	2 h	(100)	54:46
$Me_2P(CH_2)_2PPh_2$		*i*-Pr	Bz	45 min	(10)	67:33
Ph_3P	CO	Me	Bz	2 h	(30)	62:38
(*S*,*S*)-$Ph_2PCH(Me)CH(Me)PPh_2$		Me	*i*-Pr	45 min	(100)	93:7
(*S*,*S*)-$Ph_2PCH(Me)CH(Me)PPh_2$		Me	Ph	45 min	(100)	73:27
(*S*,*S*)-$Ph_2PCH(Me)CH(Me)PPh_2$		Me	*c*-C_6H_{11}	45 min	(70)	92:8
(*S*,*S*)-$Ph_2PCH(Me)CH(Me)PPh_2$		Me	Bz	45 min	(100)	> 99:1
(*S*,*S*)-$Ph_2PCH(Me)CH(Me)PPh_2$		Et	Bz	45 min	(5)	95:5

C_{31}

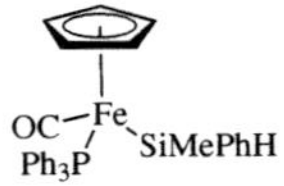

Cp(OC)(Ph_3P)Fe–SiMePhH

DMD, acetone, −78°, 5 h

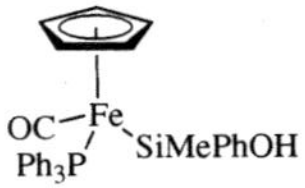

Cp(OC)(Ph_3P)Fe–SiMePhOH (80) — 225

TABLE 6. OXIDATION OF ORGANOMETALLICS BY ISOLATED DIOXIRANES *(Continued)*

Substrate	Conditions	Product(s) and Yield(s) (%)	Refs.
C_{31-48} L^1–Ru(Cp)(L^2)–SR	DMD, acetone, CH_2Cl_2, –40°, 10 min	L^1–Ru(Cp)(L^2)–SO_2R (ca. 90)	486
C_{32} [Co complex with Bu-t isocyanide ligands, thiolate S donors] $^-$ Et_4N^+	DMD (4.4 eq), acetone, –30°, 1 h	[Co complex with Bu-t isocyanide ligands, SO_2 donors] $^-$ Et_4N^+ (—)	493
C_{34-35} [Cp(Ph$_2$P(CH$_2$)$_n$PPh$_2$)Ru–S(Me)CH$_2$CH=CH$_2$]$^+$ PF_6^-	DMD, acetone, 20°, 2-20 d	[Cp(Ph$_2$P(CH$_2$)$_n$PPh$_2$)Ru–S(=O)(Me)CH$_2$CH=CH$_2$]$^+$ PF_6^- n = 1 (69) n = 2 (73)	219
C_{34-42} [Cp(Ph$_2$P(CH$_2$)$_n$PPh$_2$)Ru–S(R^1)CH(R^2)CH=C(R^3)R^4]$^+$ PF_6^-	DMD, acetone, 20°, 2-20 d	[Cp(Ph$_2$P(CH$_2$)$_n$PPh$_2$)Ru–S(=O)(R^1)CH(R^2)–(epoxide)C(R^3)R^4]$^+$ PF_6^-	219

L^1	L^2	R
CO	PPh_3	Bn
$Ph_2P(CH_2)_2PPh_2$		Me
$Ph_2P(CH_2)_2PPh_2$		Ph
$Ph_2P(CH_2)_2PPh_2$		Bn
PPh_3	PPh_3	Me
PPh_3	PPh_3	Ph
PPh_3	PPh_3	Bn

n	R^1	R^2	R^3	R^4		dr
1	Me	H	H	H	(35)	61:39
1	Et	H	H	H	(24)	55:45
2	Me	H	H	H	(33)	70:30
1	Me	H	Me	Me	(50)	72:28
2	Et	H	H	H	(27)	55:45
1	Me	—$(CH_2)_3$—		H	(37)	48:38:9:5
1	Et	H	Me	Me	(51)	74:26
2	Me	H	Me	Me	(46)	79:21
1	Et	—$(CH_2)_3$—		H	(36)	47:43:6:4
2	Me	—$(CH_2)_3$—		H	(41)	30:30:23:17
2	Et	H	Me	Me	(48)	71:29
1	Ph	H	H	H	(29)	72:28
2	Et	—$(CH_2)_3$—		H	(32)	84:6:6:4
2	Ph	H	H	H	(32)	78:22
1	Ph	H	Me	Me	(35)	71:29
1	Ph	—$(CH_2)_3$—		H	(32)	58:25:17
2	Ph	H	Me	Me	(43)	59:41

C_{44-51}

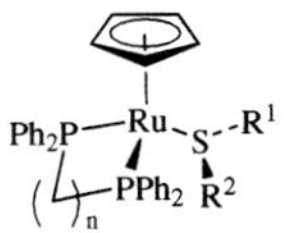

DMD, acetone, –40° to 0°, 45 min

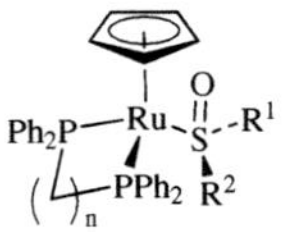

n	R^1	R^2	
1	Me	*i*-Pr	(100)
2	Me	*i*-Pr	(100)
1	Me	Ph	(100)
2	Me	Ph	(100)
1	Me	Bz	(100)
2	Me	Bz	(100)
1	*i*-Pr	Bz	(20)
2	*i*-Pr	Bz	(—)

127

TABLE 6. OXIDATION OF ORGANOMETALLICS BY ISOLATED DIOXIRANES (*Continued*)

Substrate	Conditions	Product(s) and Yield(s) (%)	Refs.
C_{36}			
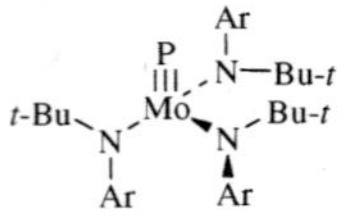 Ar = 3,5-$Me_3C_6H_3$	DMD, acetone, CH_2Cl_2, N_2, –78°	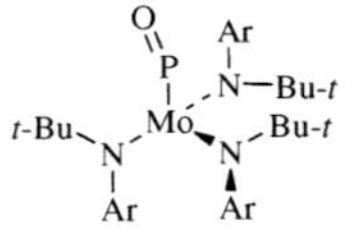 (74)	218
C_{37-38}			
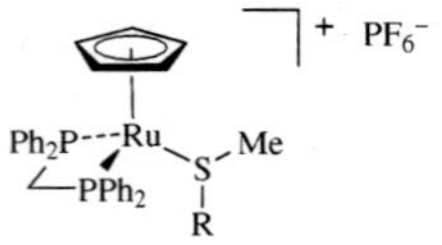 R = Ph, Bn	DMD, acetone, 0°, 45 min	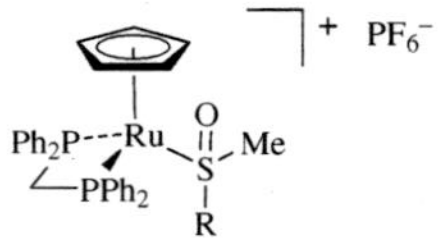 (> 90)	126
C_{37-41}			
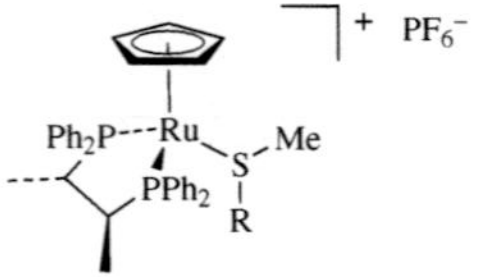	DMD, acetone, 0°, 45 min	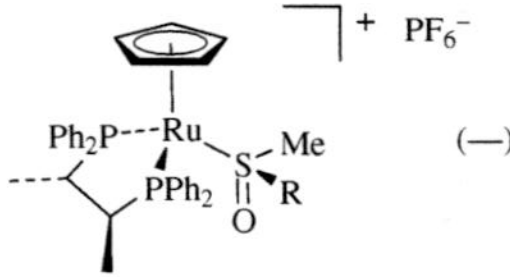 (—) R / de: *i*-Pr 86; Cy 84; Ph 46; Bz 98	492
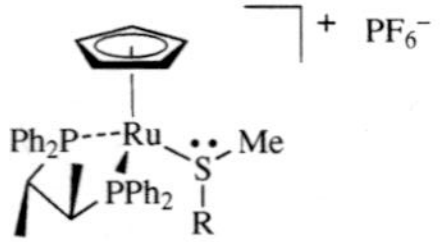 R: *i*-Pr; Ph; Bn	DMD, acetone, 0°, 45 min	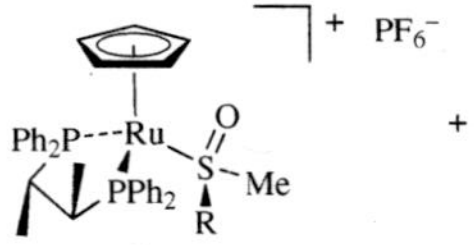 I + 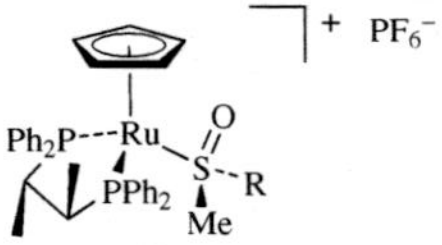 II I + II / I:II: (95) 7:93; (90) 73:27; (95) 99:1	126

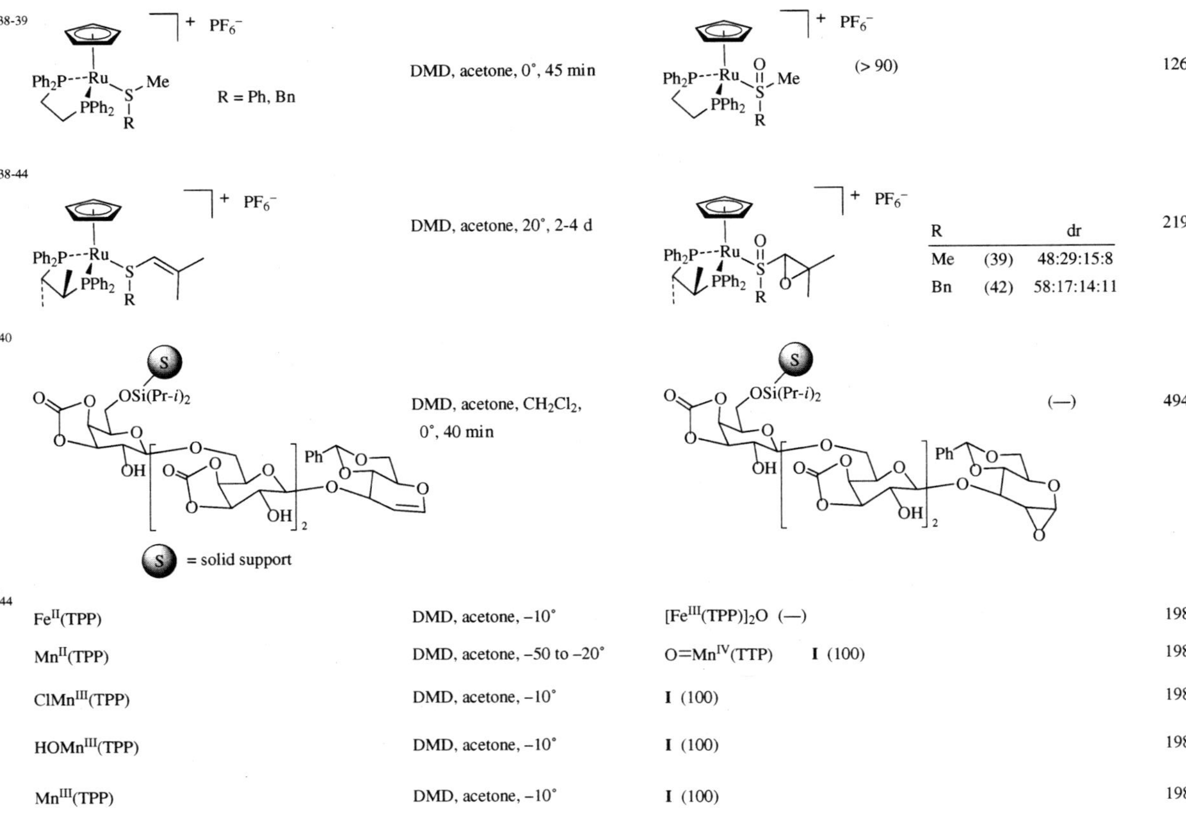

Reactant	Conditions	Product(s) and Yield(s) (%)	Refs.
C_{38-39} R = Ph, Bn	DMD, acetone, 0°, 45 min	(> 90)	126
C_{38-44}	DMD, acetone, 20°, 2-4 d	R: Me (39) dr 48:29:15:8; Bn (42) dr 58:17:14:11	219
C_{40} (S) = solid support	DMD, acetone, CH_2Cl_2, 0°, 40 min	(—)	494
C_{44}			
$Fe^{II}(TPP)$	DMD, acetone, –10°	$[Fe^{III}(TPP)]_2O$ (—)	198
$Mn^{II}(TPP)$	DMD, acetone, –50 to –20°	$O{=}Mn^{IV}(TTP)$ **I** (100)	198
$ClMn^{III}(TPP)$	DMD, acetone, –10°	**I** (100)	198
$HOMn^{III}(TPP)$	DMD, acetone, –10°	**I** (100)	198
$Mn^{III}(TPP)$	DMD, acetone, –10°	**I** (100)	198

TABLE 6. OXIDATION OF ORGANOMETALLICS BY ISOLATED DIOXIRANES (*Continued*)

Substrate	Conditions	Product(s) and Yield(s) (%)	Refs.
C_{44-46}			
[CpRu(R¹R²R³R⁴-diphosphine)(S(Me)(CH₂)₄NPhth)]⁺ PF_6^-	DMD, acetone, 0°, 2 h	[CpRu(diphosphine)(S(O)(Me)(CH₂)₄NPhth)]⁺ PF_6^-	495
R^1 = H, R^2 = H, R^3 = H, R^4 = H		(95)	
R^1 = H, R^2 = Me, R^3 = H, R^4 = Me		(90)	
R^1 = Me, R^2 = H, R^3 = Me, R^4 = H		(90)	
C_{56}			
Fe^{II}(TMP)	DMD, acetone, –10°	O=Fe^{IV}(TMP) (100)	198
C_{88}			
[Fe^{III}(TPP)]$_2$O	DMD, acetone, –10°	O=Fe^{IV}(TTP) (—)*	198
[Mn^{III}(TPP)]$_2$O	DMD, acetone, –10°	O=Mn^{IV}(TTP) (100)	198

TABLE 7. MISCELLANEOUS OXIDATIONS BY ISOLATED DIOXIRANES

Substrate	Conditions	Product(s) and Yield(s) (%)	Refs.
C_4 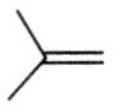	1. DMD, MeI, acetone, –70° 2. –40° to rt	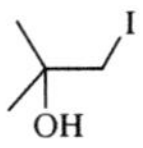 OH, I (78)	170
C_6 	1. DMD, MeI, acetone, –70° 2. –40° to rt	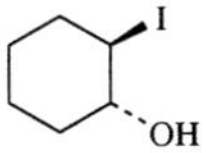 I, OH (85)	170
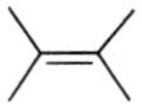	1. DMD, MeI, acetone, –70° 2. –40° to rt	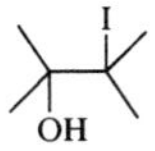 I, OH (—)	170
C_7 	DMD, acetone, CH_2Cl_2, 0° or 15°	 (—)	496
	1. DMD, MeI, acetone, –70° 2. –40° to rt	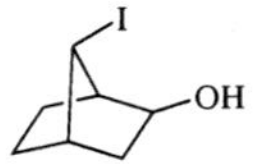 I, OH (72)	170
C_{8-9} 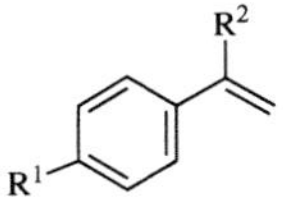 R[1], R[2]	1. DMD, MeI, acetone, –70° 2. –40° to rt	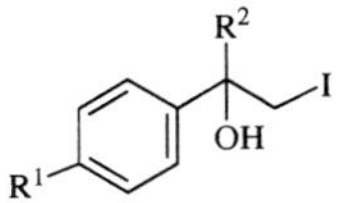 R[1], R[2], OH, I R[1] R[2] Cl H (82) H H (85) CF_3 H (44) H Me (72) Me H (75) MeO H (60)	170
C_9 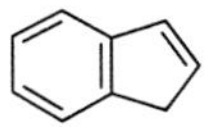	1. DMD, MeI, acetone, –70° 2. –40° to rt	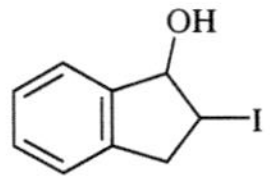 OH, I (82)	170

TABLE 7. MISCELLANEOUS OXIDATIONS BY ISOLATED DIOXIRANES (*Continued*)

Substrate	Conditions	Product(s) and Yield(s) (%)	Refs.
C_9			
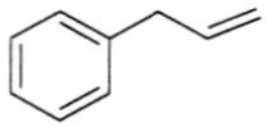	1. DMD, MeI, acetone, –70° 2. –40° to rt	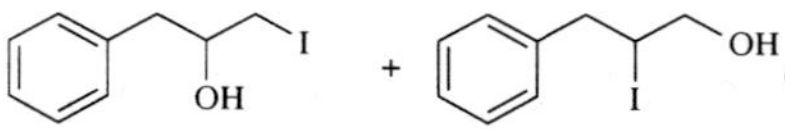 OH, I + I, OH (84)[a]	170
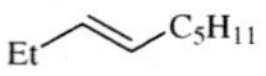 Et, C_5H_{11}	1. DMD, MeI, acetone, –70° 2. –40° to rt	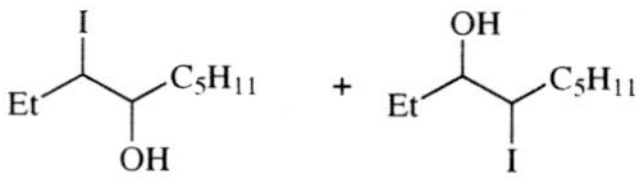 I, Et, C_5H_{11}, OH + OH, Et, C_5H_{11}, I (85)[b]	170
C_{13}			
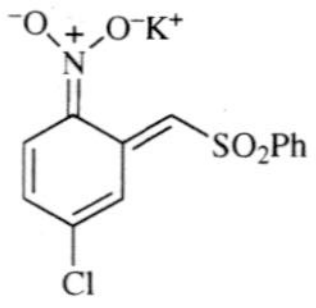 ^{-}O, +, $O^{-}K^{+}$, N, SO_2Ph, Cl	DMD, acetone, THF, Ar, 20°, 5 min	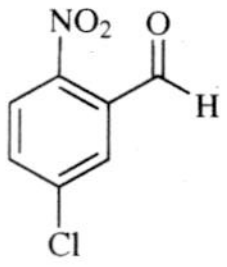 NO_2, O, H, Cl (43)	413
C_{14}			
 Ph, Ph	1. DMD, MeI, acetone, –70° 2. –40° to rt	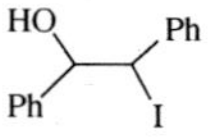 HO, Ph, Ph, I (—)	170
C_{18}			
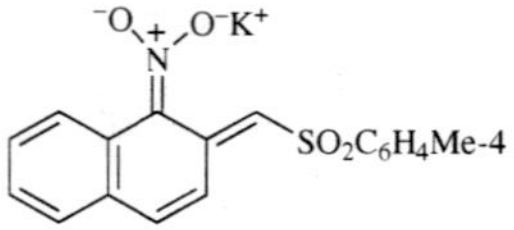 ^{-}O, +, $O^{-}K^{+}$, N, $SO_2C_6H_4Me$-4	DMD, acetone, THF, Ar, 20°, 5 min	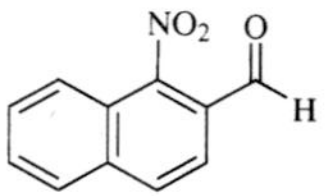 NO_2, O, H (30)	413

[a] These two products were obtained as a mixture in a ratio of 65:35.
[b] These two products were obtained as a mixture in a ratio of 50:50.

REFERENCES

[1] Adam, W.; Curci, R.; Edwards, J. O. *Acc. Chem. Res.* **1989**, *22*, 205.

[2] Murray, R. W. *Chem. Rev.* **1989**, *89*, 1187.

[3] Curci, R. In *Advances in Oxygenated Process*; Baumstark, A. L., Ed.; JAI: Greenwich, CT, 1990; Vol. 2, pp 1–59.

[4] Adam, W.; Hadjiarapoglou, L. P.; Curci, R.; Mello, R. In *Organic Peroxides*; Ando, W., Ed.; Wiley: New York, 1992; pp 195–219.

[5] Adam, W.; Hadjiarapoglou, L. P. *Top. Curr. Chem.* **1993**, *164*, 45.

[6] Curci, R.; Dinoi, A.; Rubino, M. F. *Pure Appl. Chem.* **1995**, *67*, 811.

[7] Lévai, A.; Adam, W.; Halász, J.; Nemes, C.; Patonay, T.; Tóth, G. *J. Heterocycl. Compounds* **1995**, *10*, 1345.

[8] Clennan, E. L. *Trends Org. Chem.* **1995**, *5*, 231.

[9] Adam, W.; Smerz, A. K. *Bull. Soc. Chim. Belg.* **1996**, *105*, 581.

[10] Murray, R. W.; Singh, M. In *Comprehensive Heterocyclic Chemistry II*; Padwa, A., Ed.; Elsevier: Oxford, 1996; Vol. 1A, pp 429–456.

[11] Adam, W.; Smerz, A. K.; Zhao, C.-G. *J. Prakt. Chem.* **1997**, *339*, 298.

[12] Denmark, S. E.; Wu, Z. *Synlett* **1999**, 847.

[13] Kazakov, V. P.; Voloshin, A. I.; Kazakov, D. V. *Russ. Chem. Rev.* **1999**, *68*, 253.

[14] Adam, W.; Degen, H.-G.; Pastor, A.; Saha-Möller, C. R.; Schambony, S. B.; Zhao, C.-G. In *Peroxide Chemistry: Mechanistic and Preparative Aspects of Oxygen Transfer*; Adam, W., Ed.; Wiley-VCH: Weinheim, 2000; pp 78–112.

[15] Adam, W.; Saha-Möller, C. R.; Zhao, C.-G. *Org. React.* **2002**, *61*, 219.

[16] Crandall, J. K.; Batal, D. J.; Sebesta, D. P.; Lin, F. *J. Org. Chem.* **1991**, *56*, 1153.

[17] Crandall, J. K.; Zucco, M.; Kirsch, R. S.; Coppert, D. M. *Tetrahedron Lett.* **1991**, *32*, 5441.

[18] Adam, W.; Golsch, D. *Chem. Ber.* **1994**, *127*, 1111.

[19] Deubel, D. V. *J. Org. Chem.* **2001**, *66*, 2686.

[20] Deubel, D. V. *J. Org. Chem.* **2001**, *66*, 3790.

[21] Adam, W.; Golsch, D. *Angew. Chem., Int. Ed. Engl.* **1993**, *32*, 737.

[22] Murray, R. W.; Singh, M.; Rath, N. *Tetrahedron: Asymmetry* **1996**, *7*, 1611.

[23] Adam, W.; Bottle, S. E.; Mello, R. *J. Chem. Soc., Chem. Commun.* **1991**, 771.

[24] Nelsen, S. F.; Scamehorn, R. G.; De Felippis, J.; Wang, Y. *J. Org. Chem.* **1993**, *58*, 1657.

[25] Buxton, P. C.; Ennis, J. N.; Marples, B. A.; Waddington, V. L.; Boehlow, T. R. *J. Chem. Soc., Perkin Trans. 2* **1998**, 265.

[26] Adam, W.; Briviba, K.; Duschek, F.; Golsch, D.; Kiefer, W.; Sies, H. *J. Chem. Soc., Chem. Commun.* **1995**, 1831.

[27] Lange, A.; Bauer, D. H. *J. Chem. Soc., Perkin Trans. 2* **1996**, 805.

[28] Miaskiewicz, K; Teich, N.; Smith, A. *J. Org. Chem.* **1997**, *62*, 6493.

[29] Mello, R.; Cassidei, L.; Fiorentino, M.; Fusco, C.; Hümmer, W.; Jäger, V.; Curci, R. *J. Am. Chem. Soc.* **1991**, *113*, 2205.

[30] Mello, R.; Fiorentino, M.; Fusco, C.; Curci, R. *J. Am. Chem. Soc.* **1989**, *111*, 6749.

[31] Adam, W.; Asensio, G.; Curci,, R.; González-Nuñez, M. E.; Mello, R. *J. Org. Chem.* **1992**, *57*, 953.

[32] Minisci, F.; Zhao, L.; Fontana, F.; Bravo, A. *Tetrahedron Lett.* **1995**, *36*, 1697.

[33] Minisci, F.; Zhao, L.; Fontana, F.; Bravo, A. *Tetrahedron Lett.* **1995**, *36*, 1895.

[34] Bravo, A.; Fontana, F.; Fronza, G.; Miele, A.; Minisci, F. *J. Chem. Soc., Chem. Commun.* **1995**, 1573.

[35] Bravo, A.; Fontana, F.; Fronza, G.; Minisci, F.; Zhao, L. *J. Org. Chem.* **1998**, *63*, 254.

[36] Curci, R.; Dinoi, A.; Fusco, C.; Lillo, M. A. *Tetrahedron Lett.* **1996**, *37*, 249.

[37] Adam, W.; Curci, R.; D'Accolti, L.; Dinoi, A.; Fusco, C.; Gasparrini, F.; Kluge, R.; Paredes, R.; Schulz, M.; Smerz, A. K.; Veloza, L. A.; Weinkötz, S.; Winde, R. *Chem. Eur. J.* **1997**, *3*, 105.

[38] Simakov, P. A.; Choi, S.-Y.; Newcomb, M. *Tetrahedron Lett.* **1998**, *39*, 8187.

[39] Adam, W.; Mello, R.; Curci, R. *Angew. Chem., Int. Ed. Engl.* **1990**, *29*, 890.

[40] Adam, W.; Mitchell, C. M.; Saha-Möller, C. R.; Weichold, O. In *Structure and Bonding, Metal-Oxo and Metal-Peroxo Species in Catalytic Oxidations*; Meunier, B., Ed.; Springer Verlag: Berlin Heidelberg, 2000; Vol. 97, pp 237–285.
[41] Shustov, G. V.; Rauk, A. *J. Org. Chem.* **1998**, *63*, 5413.
[42] Glukhovtsev, M. N.; Canepa, C.; Bach, R. D. *J. Am. Chem. Soc.* **1998**, *120*, 10528.
[43] Du, X.; Houk, K. N. *J. Org. Chem.* **1998**, *63*, 6480.
[44] Edwards, J. O.; Pater, R. H.; Curci, R.; Di Furia, F. *Photochem. Photobiol.* **1979**, *30*, 63.
[45] Curci, R.; Fiorentino, M.; Troisi, L.; Edward, J. O.; Pater, R. H. *J. Org. Chem.* **1980**, *45*, 4758.
[46] Jeyaraman, R.; Murray, R. W. *J. Am. Chem. Soc.* **1984**, *106*, 2462.
[47] Adam, W.; Hadjiarapoglou, L.; Smerz, A. *Chem. Ber.* **1991**, *124*, 227.
[48] Yang, D.; Wong, M.-K.; Yip, Y.-C. *J. Org. Chem.* **1995**, *60*, 3887.
[49] Frohn, M.; Wang, Z.-X.; Shi, Y. *J. Org. Chem.* **1998**, *63*, 6425.
[50] Denmark, S. E.; Wu, Z. *J. Org. Chem.* **1998**, *63*, 2810.
[51] Yang, D.; Yip, Y.-C.; Jiao, G.-S.; Wong, M.-K. *J. Org. Chem.* **1998**, *63*, 8952.
[52] Yang, D.; Yip, Y.-C.; Tang, M.-W.; Wong, M.-K; Cheung, K.-K. *J. Org. Chem.* **1998**, *63*, 9888.
[53] Carnell, A. J.; Johnstone, R. A. W.; Parsy, C. C.; Sanderson, W. R. *Tetrahedron Lett.* **1999**, *40*, 8029.
[54] Adam, W.; Saha-Möller, C. R.; Ganeshpure, P. A. *Chem. Rev.* **2001**, *101*, 3499.
[55] Crandall, J. K.; Batal, D. J. *J. Org. Chem.* **1988**, *53*, 1338.
[56] Crandall, J. K.; Batal, D. J. *Tetrahedron Lett.* **1988**, *29*, 4791.
[57] Crandall, J. K.; Batal, D. J.; Lin, F.; Reix, T.; Nadol, G. S.; Ng, R. A. *Tetrahedron* **1992**, *48*, 1427.
[58] Crandall, J. K.; Reix, T. *Tetrahedron Lett.* **1994**, *35*, 2513.
[59] Crandall, J. K.; Rambo, E. *Tetrahedron Lett.* **1994**, *35*, 1489.
[60] Crandall, J. K.; Rambo, E. *J. Org. Chem.* **1990**, *55*, 5929.
[61] Pasto, D. J.; Yang, S.-H.; Muellerleihe, J. A. *J. Org. Chem.* **1992**, *57*, 2976.
[62] Crandall, J. K.; Coppert, D. M.; Schuster, T.; Lin, F. *J. Am. Chem. Soc.* **1992**, *114*, 5998.
[63] Murray, R. W.; Singh, M. *J. Org. Chem.* **1993**, *58*, 5076.
[64] Curci, R.; Fiorentino, M.; Fusco, C.; Mello, R.; Ballistreri, F. P.; Failla, S.; Tomaselli, G. A. *Tetrahedron Lett.* **1992**, *33*, 7929.
[65] Mello, R.; Ciminale, F.; Fiorentino, M.; Fusco, C.; Prencipe, T.; Curci, R. *Tetrahedron Lett.* **1990**, *31*, 6097.
[66] Adam, W.; Schönberger, A. *Tetrahedron Lett.* **1992**, *33*, 53.
[67] Mohler, D. L.; Vollhardt, K. P. C.; Wolff, S. *Angew. Chem., Int. Ed. Engl.* **1995**, *34*, 563.
[68] Bernini, R.; Mincione, E.; Sanetti, A.; Mezzetti, M.; Bovicelli, P. *Tetrahedron Lett.* **2000**, *41*, 1087.
[69] Agarwal, S. K.; Boyd, D. R.; Jennings, W. B.; McGuckin, R. M.; O'Kane, G. A. *Tetrahedron Lett.* **1989**, *30*, 123.
[70] Murray, R. W.; Singh, M.; Rath, N. *Tetrahedron Lett.* **1996**, *37*, 8671.
[71] Elemes, Y.; Silverman, S. K.; Sheu, C.; Kao, M.; Foote, C. S.; Alvarez, M. M.; Whetten, R. L. *Angew. Chem., Int. Ed. Engl.* **1992**, *31*, 351.
[72] Fusco, C.; Seraglia, R.; Curci, R.; Lucchini, V. *J. Org. Chem.* **1999**, *64*, 8363.
[73] Adger, B. M.; Barrett, C.; Brennan, J.; McKervey, M. A.; Murray, R. W. *J. Chem. Soc., Chem. Commun.* **1991**, 1553.
[74] Adam, W.; Ahrweiler, M.; Sauter, M. *Angew. Chem., Int. Ed. Engl.* **1993**, *32*, 80.
[75] Adger, B. J.; Barrett, C.; Brennan, J.; McGuigan, P.; McKervey, M. A.; Tarbit, B. *J. Chem. Soc., Chem. Commun.* **1993**, 1220.
[76] Adam, W.; Sauter, M., unpublished results.
[77] Adam, W.; Hadjiarapoglou, L.; Mosandl, T.; Saha-Möller, C. R.; Wild, D. *J. Am. Chem. Soc.* **1991**, *113*, 8005.
[78] Adam, W.; Hadjiarapoglou, L.; Wang, X. *Tetrahedron Lett.* **1991**, *32*, 1295.
[79] Adam, W.; Bialas, J.; Hadjiarapoglou, L.; Sauter, M. *Chem. Ber.* **1992**, *125*, 231.
[80] Adam, W.; Hadjiarapoglou, L.; Peters, K.; Sauter, M. *Angew. Chem., Int. Ed. Engl.* **1993**, *32*, 735.
[81] Adam, W.; Hadjiarapoglou, L.; Peters, K.; Sauter, M. *J. Am. Chem. Soc.* **1993**, *115*, 8603.
[82] Adam, W.; Sauter, M.; Zünkler, C. *Chem. Ber.* **1994**, *127*, 1115.

[83] Adam, W.; Käb, G.; Sauter, M. *Chem. Ber.* **1994**, *127*, 433.
[84] Adam, W.; Sauter, M. *Liebigs Ann. Chem.* **1994**, 689.
[85] Adam, W.; Sauter, M. *Liebigs Ann. Chem.* **1992**, 1095.
[86] Adam, W.; Peters, K.; Sauter, M. *Synthesis* **1994**, 111.
[87] Adam, W.; Hadjiarapoglou, L.; Mosandl, T.; Saha-Möller, C.; Wild, D. *Angew. Chem., Int. Ed. Engl.* **1991**, *30*, 200.
[88] Adam, W.; Sauter, M. *Acc. Chem. Res.* **1995**, *28*, 289.
[89] Adam, W.; Sauter, M. *Tetrahedron* **1994**, *50*, 11441.
[90] Adam, W.; Ahrweiler, M.; Peters, K.; Schmiedeskamp, B. *J. Org. Chem.* **1994**, *59*, 2733.
[91] Adam, W.; Ahrweiler, M.; Paulini, K.; Reißig, H.-U.; Voerckel, V. *Chem. Ber.* **1992**, *125*, 2719.
[92] Zhang, X.; Foote, C. S. *J. Am. Chem. Soc.* **1993**, *115*, 8867.
[93] Saladino, R.; Mezzetti, M.; Mincione, E.; Torrini, I.; Paradisi, M. P.; Masteropietro, G. *J. Org. Chem.* **1999**, *64*, 8468.
[94] Kazakov, D. V.; Maistrenko, G. Y.; Polyakova, N. P.; Kazakov, V. P.; Adam, W.; Trofimov, A.; Zhao, C.-G.; Kiefer, W.; Schlücker, S. *Luminescence* **2002**, *17*, 293.
[95] Ferrer, M.; Sánchez-Baeza, F.; Messeguer, A. *Tetrahedron* **1997**, *53*, 15877.
[96] Murray, R. W.; Singh, M. *Synth. Commun.* **1989**, *19*, 3509.
[97] Murray, R. W.; Singh, M. *Tetrahedron Lett.* **1988**, *29*, 4677.
[98] Murray, R. W.; Singh, M.; Jeyaraman, R. *J. Am. Chem. Soc.* **1992**, *114*, 1346.
[99] Neset, S. M.; Benneche, T.; Undheim, K. *Acta. Chem. Scand.* **1993**, *47*, 1141.
[100] Wittman, M. D.; Halcomb, R. L.; Danishefsky, S. J. *J. Org. Chem.* **1990**, *55*, 1981.
[101] Murray, R. W.; Rajadhyaksha, S. N.; Mohan, L. *J. Org. Chem.* **1989**, *54*, 5783.
[102] Boyd, D. R.; Coulter, P. B.; McGuckin, M. R.; Sharma, N. D.; Jennings, W. B.; Wilson, V. E. *J. Chem. Soc., Perkin Trans. 1* **1990**, 301.
[103] Ihmels, H.; Maggini, M.; Prato, M.; Scorrano, G. *Tetrahedron Lett.* **1991**, *32*, 6215.
[104] Darkins, P.; McCarthy, N.; McKervey, M. A.; Ye, T. *J. Chem. Soc., Chem. Commun.* **1993**, 1222.
[105] Saba, A. *Synth. Commun.* **1994**, *24*, 695.
[106] Adam, W.; Hadjiarapoglou, L.; Mielke, K.; Treiber, A. *Tetrahedron Lett.* **1994**, *35*, 5625.
[107] Adam, W.; Makosza, M.; Saha-Möller, C. R.; Zhao, C.-G. *Synlett* **1998**, 1335.
[108] Pinnick, H. W. *Org. React.* **1990**, *38*, 655.
[109] Williams, D. R.; Brugel, T. A. *Org. Lett.* **2000**, *2*, 1023.
[110] Adam, W.; Makosza, M.; Stalinski, K.; Zhao, C.-G. *J. Org. Chem.* **1998**, *63*, 4390.
[111] Adam, W.; Makosza, M.; Zhao, C.-G.; Surowiec, M. *J. Org. Chem.* **2000**, *65*, 1099.
[112] Paradkar, V. M.; Latham, T. B.; Demko, D. M. *Synlett* **1995**, 1059.
[113] Altamura, A.; D'Accolti, L.; Detomaso, A.; Dinoi, A.; Firentino, M.; Fusco, C.; Curci, R. *Tetrahedron Lett.* **1998**, *39*, 2009.
[114] Crandall, J. K.; Reix, T. *J. Org. Chem.* **1992**, *57*, 6759.
[115] Gu, D.; Harpp, D. N. *Tetrahedron Lett.* **1993**, *34*, 67.
[116] Lupattelli, P.; Saladino, R.; Mincione, E. *Tetrahedron Lett.* **1993**, *34*, 6313.
[117] Saladino, R.; Crestini, C.; Bernini, R.; Mincione, E. *Tetrahedron Lett.* **1993**, *34*, 7785.
[118] Saladino, R.; Crestini, C.; Bernini, R.; Frachey, G.; Mincione, E. *J. Chem. Soc., Perkin Trans 1* **1994**, 3053.
[119] Saladino, R.; Crestini, C.; Bernini, R.; Mincione, E.; Ciafrino, R. *Tetrahedron Lett.* **1995**, *36*, 2665.
[120] Saladino, R.; Bernini, R.; Crestini, L.; Mincione, E.; Bergamini, A.; Marini, S.; Palamara, A. T. *Tetrahedron* **1995**, *51*, 7561.
[121] Murray, R. W.; Jeyaraman, R. *J. Org. Chem.* **1985**, *50*, 2847.
[122] Quallich, G. J.; Lackey, J. W. *Tetrahedron Lett.* **1990**, *31*, 3685.
[123] Danelon, G. O.; Mata, E. G.; Mascaretti, O. A. *Tetrahedron Lett.* **1993**, *34*, 7877.
[124] Gunda, T. E.; Tamás, L.; Sályi, S.; Nemes, C.; Sztaricskai, F. *Tetrahedron Lett.* **1995**, *36*, 7111.
[125] Berkowitz, D. B.; Danishefsky, S. J.; Schulte, G. K. *J. Am. Chem. Soc.* **1992**, *114*, 4518.
[126] Schenk, W. A.; Frisch, J.; Adam, W.; Prechtl, F. *Angew. Chem., Int. Ed. Engl.* **1994**, *33*, 1609.
[127] Schenk, W. A.; Frisch, J.; Dürr, M.; Burzlaff, N.; Stalke, D.; Fleischer, R.; Adam, W.; Prechtl, F.; Smerz, A. K. *Inorg. Chem.* **1997**, *36*, 2372.

[128] Pérez-Encabo, A.; Perrio, S.; Slawin, A. M. Z.; Thomas, S. E.; Wierzchleyski, A. T.; Williams, D. J. *J. Chem. Soc., Chem. Commun.* **1993**, 1059.
[129] Pérez-Encabo, A.; Perrio, S.; Slawin, A. M. Z.; Thomas, S. E.; Wierzchleyski, A. T.; Williams, D. J. *J. Chem. Soc., Perkin Trans. 1*, **1994**, 629.
[130] Colonna, S.; Gaggero, N. *Tetrahedron Lett.* **1989**, *30*, 6233.
[131] Colonna, S.; Gaggero, N.; Leone, M. *Tetrahedron* **1991**, *47*, 8385.
[132] Frohn, M.; Shi, Y. *Synthesis* **2000**, 1979.
[133] Pasquato, L.; De Lucchi, O.; Krotz, L. *Tetrahedron Lett.* **1991**, *32*, 2177.
[134] Ishii, A.; Tsuchiya, C.; Shimada, T.; Furusawa, K.; Omata, T.; Nakayama, J. *J. Org. Chem.* **2000**, *65*, 1799.
[135] Clennan, E. L.; Stensaas, K. L. *J. Org. Chem.* **1996**, *61*, 7911.
[136] Jin, Y.-N.; Ishii, A.; Sugihara, Y.; Nakayama, J. *Tetrahedron Lett.* **1998**, *39*, 3525.
[137] Glass, R. S.; Liu, Y. *Tetrahedron Lett.* **1994**, *35*, 3887.
[138] Miyahara, Y.; Inazu, T. *Tetrahedron Lett.* **1990**, *31*, 5955.
[139] Tsirk, A.; Gronowitz, S.; Hörnfeldt, A.-B. *Tetrahedron* **1995**, *51*, 7035.
[140] Pouzet, P.; Erdelmeier, I.; Ginderow, D.; Mornon, J.-P.; Dansette, P. M.; Mansuy, D. *J. Heterocycl. Chem.* **1997**, *34*, 1567.
[141] Nakayama, J.; Nagasawa, H.; Sugihara, Y.; Ishii, A. *J. Am. Chem. Soc* **1997**, *119*, 9077.
[142] Frachey, G.; Crestini, C.; Bernini, R.; Saladino, R.; Mincione, E. *Heterocycles* **1994**, *38*, 2621.
[143] Crestini, C.; Mincione, E.; Saladino, R.; Nicoletti, R. *Tetrahedron* **1994**, *50*, 3259.
[144] Saladino, R.; Mincione, E.; Crestini, C.; Mezzetti, M. *Tetrahedron* **1996**, *52*, 6759.
[145] Tabuchi, T.; Nojima, M.; Kusabayashi, S. *J. Chem. Soc., Chem. Commun.* **1990**, 625.
[146] Tabuchi, T.; Nojima, M.; Kusabayashi, S. *J. Chem. Soc., Perkin Trans. 1* **1991**, 3043.
[147] Watanabe, S.; Yamamoto, T.; Kawashima, T.; Inamoto, N.; Okazaki, R. *Bull. Chem. Soc. Jpn.* **1996**, *69*, 719.
[148] Gaggero, N.; D'Accolti, L.; Colonna, S.; Curci, R. *Tetrahedron Lett.* **1997**, *38*, 5559.
[149] Coburn, M. D. *J. Heterocycl. Chem.* **1989**, *26*, 1883.
[150] Sánchez-Baeza, F.; Durand, G.; Barceló, D.; Messeguer, A. *Tetrahedron Lett.* **1990**, *31*, 3359.
[151] Piettre, S. R. *Tetrahedron Lett.* **1996**, *37*, 4707.
[152] Ishii, A.; Matsubayashi, S.; Takahashi, T.; Nakayama, J. *J. Org. Chem.* **1999**, *64*, 1084.
[153] Nakayama, J.; Matsui, T.; Sigihara, Y.; Ishii, A.; Kumakura, S. *Chem. Lett.* **1996**, 269.
[154] Matsui, T.; Nakayama, J.; Sato, N.; Sugihara, Y.; Ishii, A.; Kumakura, S. *Phosphorus, Sulfur, Silicon, Relat. Elem.* **1996**, *118*, 227.
[155] Umezawa, T.; Sugihara, Y.; Ishii, A.; Nakayama, J. *J. Am. Chem. Soc.* **1998**, *120*, 12351.
[156] Kirschfeld, A.; Muthusamy, S.; Sander, W. *Angew. Chem., Int. Ed. Engl.* **1994**, *33*, 2212.
[157] Sander, W.; Schröder, K.; Muthusamy, S.; Kirschfeld, A.; Kappert, W.; Böse, R.; Kraka, E.; Sosa, C.; Cremer, D. *J. Am. Chem. Soc.* **1997**, *119*, 7265.
[158] Chappell, M. D.; Halcomb, R. L. *Tetrahedron Lett.* **1999**, *40*, 1.
[159] Wasserman, H. H.; Baldino, C.; Coates, S. J. *J. Org. Chem.* **1995**, *60*, 8231.
[160] Cassidei, L.; Fiorentino, M.; Mello, R.; Sciacovelli, O.; Curci, R. *J. Org. Chem.* **1987**, *52*, 699.
[161] Montgomery, R. E. *J. Am. Chem. Soc.* **1974**, *96*, 7820.
[162] Adam, W.; Kazakov, D. V.; Kazakov, V. P.; Kiefer, W.; Latypova, R. R.; Schlücker, S. *Photochem. Photobiol. Sci.* **2004**, *3*, 182.
[163] Yang, D.; Tang, Y.-C.; Chen, J.; Wang, X.-C. Bartberger, M. D.; Houk, K. N.; Olson, L. *J. Am. Chem. Soc.* **1999**, *121*, 11976.
[164] Adam, W.; Asensio, G.; Curci, R.; González-Núñez, M. E.; Mello, R. *J. Am. Chem. Soc.* **1992**, *114*, 8345.
[165] Ferrer, M.; Sánchez-Baeza, F.; Messeguer, A.; Adam, W.; Golsch, D.; Görth, F.; Kiefer, W.; Nagel, V. *Eur. J. Org. Chem.* **1998**, 2527.
[166] Dinoi, A.; Curci, R.; Carloni, P.; Damiani, E.; Stipa, P.; Greci, L. *Eur. J. Org. Chem.* **1998**, 871.
[167] Adam, W.; Bialas, J.; Hadjiarapoglou, L. *Chem. Ber.* **1991**, *124*, 2377.
[168] Bravo, A.; Fontana, F.; Fronza, G.; Minisci, F.; Serri, A. *Tetrahedron Lett.* **1995**, *36*, 6945.
[169] Mahadevan, A.; Fuchs, P. L. *J. Am. Chem. Soc.* **1995**, *117*, 3272.

[170] Asensio, G.; Andreu, C.; Boix-Bernardini, C.; Mello, R.; González- Nuñez, M. E. *Org. Lett.* **1999**, *1*, 2125.
[171] Studley, A.; Zhao, C.-G., unpublished results, University of Texas at San Antonio.
[172] Asensio, G.; Mello, R.; González-Núñez, M. E.; Castellano, G.; Corral, J. *Angew. Chem., Int. Ed. Engl.* **1996**, *35*, 217.
[173] Mello, R.; Fiorentino, M.; Sciacovelli, O.; Curci, R. *J. Org. Chem.* **1988**, *53*, 3890.
[174] Murray, R. W.; Jeyaraman, R.; Mohan, L. *J. Am. Chem. Soc.* **1986**, *108*, 2470.
[175] Kovall, F.; Baumstark, A. L. *Tetrahedron Lett.* **1994**, *35*, 8751.
[176] Bovicelli, P.; Sanetti, A.; Bernini, R.; Lupattelli, P. *Tetrahedron* **1997**, *53*, 9755.
[177] Teager, D. S.; Murray, R. K., Jr. *J. Org. Chem.* **1993**, *58*, 5548.
[178] Mello, R.; Cassidei, L.; Fiorentino, M.; Fusco, C.; Curci, R. *Tetrahedron Lett.* **1990**, *31*, 3067.
[179] Bovicelli, P.; Lupattelli, P.; Mincione, E.; Prencipe, T.; Curci, R. *J. Org. Chem.***1992**, *57*, 5052.
[180] Bovicelli, P.; Gambacorta, A.; Lupattelli, P.; Mincione, E. *Tetrahedron Lett.* **1992**, *33*, 7411.
[181] Adam, W.; Pastor, A.; Zhao, C.-G., unpublished results.
[182] Curci, R.; D'Accolti, L.; Detomaso, A.; Fusco, C.; Takeuchi, K.; Ohga, Y.; Eaton, P. E.; Yip, Y. C. *Tetrahedron Lett.* **1993**, *34*, 4559.
[183] Bovicelli, P.; Lupattelli, P.; Sanetti, A.; Mincione, E. *Tetrahedron Lett.* **1994**, *35*, 8477.
[184] D'Accolti, L.; Detomaso, A.; Fusco, C.; Rosa, A.; Curci, R. *J. Org. Chem.* **1993**, *58*, 3600.
[185] Adam, W.; Prechtl, F.; Richter, M. J.; Smerz, A. K. *Tetrahedron Lett.* **1995**, *36*, 4991.
[186] Baumstark, A. L.; Kovac, F.; Vasquez, P. C. *Can. J. Chem.* **1999**, *77*, 308.
[187] Ferrer, M.; Sánchez-Baeza, F.; Casas, J.; Messeguer, A. *Tetrahedron Lett.* **1994**, *35*, 2981.
[188] Csuk, R.; Dörr, P. *Tetrahedron* **1994**, *50*, 9983.
[189] Marples, B. A.; Muxworthy, J. P.; Baggaley, K. H. *Synlett* **1992**, 646.
[190] Curci, R.; D'Accolti, L.; Dinoi, A.; Fusco, C.; Rosa, A. *Tetrahedron Lett.* **1996**, *37*, 115.
[191] Curci, R.; D'Accolti, L.; Fiorentino, M.; Fusco, C.; Adam, W.; González-Nuñez, M. E.; Mello, R. *Tetrahedron Lett.* **1992**, *33*, 4225.
[192] Asensio, G.; González-Núñez, M. E.; Biox Bernardini, C.; Mello, R.; Adam, W. *J. Am. Chem. Soc.* **1993**, *115*, 7250.
[193] Adam, W.; Prechtl, F. *Chem. Ber.* **1991**, *124*, 2369.
[194] Adam, W.; Smerz, A. K. *Tetrahedron* **1996**, *52*, 5799.
[195] Yang, D.; Wong, M.-K.; Wang, X.-C.; Tang, Y.-C. *J. Am. Chem. Soc.* **1998**, *120*, 6611.
[196] Adam, W.; Saha-Möller, C. R.; Zhao, C.-G. *Tetrahedron: Asymmetry* **1998**, *9*, 4117.
[197] Adam, W.; Saha-Möller, C. R.; Zhao, C.-G. *J. Org. Chem.* **1999**, *64*, 7492.
[198] Wolowiec, S.; Kochi, J. K. *J. Chem. Soc., Chem. Commun.* **1990**, 1782.
[199] Adam, W.; Mitchell, C. M.; Saha-Möller, C. R.; Weichold, O. *J. Am. Chem. Soc.* **1999**, *121*, 2097.
[200] Adam, W.; Jekö, J.; Lévai, A.; Nemes, C.; Patonay, T.; Sebök, P. *Tetrahedron Lett.* **1995**, *36*, 3669.
[201] Lévai, A.; Adam, W.; Fell, R. T.; Gessner, R.; Patonay, T.; Simon, A.; Tóth, G. *Tetrahedron* **1998**, *54*, 13105.
[202] Adam, W.; Jekö, J.; Lévai, A.; Majer, Z.; Nemes, C.; Patonay, T.; Párkányi, L.; Sebök, P. *Tetrahedron: Asymmetry* **1996**, *7*, 2437.
[203] Adam, W.; Fell, R. T.; Lévai, A.; Patonay, T.; Perters, K.; Simon, A.; Tóth, G. *Tetrahedron: Asymmetry* **1998**, *9*, 1121.
[204] Lluch, A.-M.; Jordi, L.; Sánchez-Baeza, F.; Ricart, S.; Camps, F.; Messeguer, A.; Moretó, J. M. *Tetrahedron Lett.* **1992**, *33*, 3021.
[205] Gibert, M.; Ferrer, M.; Lluch, A.-M.; Sánchez-Baeza, F.; Messeguer, A. *J. Org. Chem.* **1999**, *64*, 1591.
[206] Lluch, A.-M.; Gibert, M.; Sánchez-Baeza, F.; Messeguer, A. *Tetrahedron* **1996**, *52*, 3973.
[207] Jordi, L.; Ricart, S.; Viñas, J. M.; Moretó, J. M. *Organometallics* **1997**, *16*, 2808.
[208] Baldoli, C.; Del Buttero, P.; Licandro, E.; Maiorana, S.; Papagni, A. *Synlett* **1995**, 666.
[209] Fermin, M. C.; Bruno, J. W. *J. Am. Chem. Soc.* **1993**, *115*, 7511.
[210] Wolowiec, S.; Kochi, J. K. *Inorg. Chem.* **1991**, *30*, 1215.
[211] Adam, W.; Putterlik, J.; Schuhmann, R. M.; Sundermeyer, J. *Organometallics* **1996**, *15*, 4586.
[212] Adam, W.; Müller, M.; Prechtl, F. *J. Org. Chem.* **1994**, *59*, 2358.

[213] Adam, W.; Schuhmann, R. M. *J. Organomet. Chem.* **1995**, *487*, 273.
[214] Adam, W.; Schuhmann, R. M. *Liebigs Ann. Chem.* **1996**, 635.
[215] Sun, S.; Edwards, J. O.; Sweigart, D. A.; D'Accolti, L.; Curci, R. *Organometallics* **1995**, *14*, 1545.
[216] Malisch, W.; Hindahl, K.; Grün, K.; Adam, W.; Prechtl, F.; Sheldrick, W. S. *J. Organomet. Chem.* **1996**, *509*, 209.
[217] Christian, P. W. N.; Gibson, S. E.; Gil, R.; Jones, P. C. V.; Marcos, C. F.; Prechtl, F.; Wierzchleyski, A. T. *Recl. Trav. Chim. Pays-Bas* **1995**, *114*, 195.
[218] Johnson, M. J. A.; Odom, A. L.; Cummins, C. C. *Chem. Commun.* **1997**, 1523.
[219] Schenk, W. A.; Steinmetz, B.; Hagel, M.; Adam, W.; Saha-Möller, C. R. *Z. Naturforsch.* **1997**, 1359.
[220] Adam, W.; Azzena, U.; Prechtl, F.; Hindahl, K.; Malisch, W. *Chem. Ber.* **1992**, *125*, 1409.
[221] Malisch, W.; Lankat, R.; Schmitzer, S.; Reising, J. *Inorg. Chem.* **1995**, *23*, 5701.
[222] Malisch, W.; Lankat, R.; Fey, O.; Reising, J.; Schmitzer, S. *J. Chem. Soc., Chem. Commun.* **1995**, 1917.
[223] Malisch, W.; Hindahl, K.; Käb, H.; Reising, J.; Adam, W.; Prechtl, F. *Chem. Ber.* **1995**, *128*, 963.
[224] Malisch, W.; Schmitzer, S.; Lankat, R.; Neumayer, M.; Prechtl, F.; Adam, W. *Chem. Ber.* **1995**, *128*, 1251.
[225] Malisch, W.; Neumayer, M.; Fey, O.; Adam, W.; Schuhmann, R. *Chem. Ber.* **1995**, *128*, 1257.
[226] Möller, S.; Fey, O.; Malisch, W.; Seelbach, W. *J. Organomet. Chem.* **1996**, *507*, 239.
[227] Malisch, W.; Jehle, H.; Möller, S.; Saha-Möller, C.; Adam, W. *Eur. J. Inorg. Chem.* **1998**, 1585.
[228] Ganeshpure, P. A.; Adam, W. *Synthesis* **1996**, 179.
[229] Adam, W.; Ganeshpure, P. A. *Synthesis* **1993**, 280.
[230] Davis, F. A.; Sheppard, A. C. *Tetrahedron* **1989**, *45*, 5703.
[231] Davis, F. A.; Chen, B. C. *Chem. Rev.* **1992**, *92*, 919.
[232] Davis, F. A.; Reddy, R. T.; Han, W.; Reddy, R. E. *Pure Appl. Chem.* **1993**, *65*, 633.
[233] Payne, G. B.; Deming, P. H.; Williams, P. H. *J. Org. Chem.* **1961**, *26*, 659.
[234] Payne, G. B. *Tetrahedron* **1962**, *18*, 763.
[235] Payne, G. B.; Williams, P. H. *J. Org. Chem.* **1961**, *26*, 651.
[236] Shi, Y. *Acc. Chem. Res.* **2004**, *37*, 488.
[237] Lewis, S. N. In *Oxidations*, Augustine, R. L., Ed.; Marcel Dekker: New York, 1969; Vol. 1, pp 213–58.
[238] Groves, J. T.; Viski, P. *J. Org. Chem.* **1990**, *55*, 3628.
[239] Pitchen, P.; Dunach, E.; Deshmukh, M. N.; Kagan, H. B. *J. Am. Chem. Soc.* **1984**, *106*, 8188.
[240] May, S. W.; Phillips, R. S. *J. Am. Chem. Soc.* **1980**, *102*, 5981.
[241] Auret, B. J.; Boyd, D. R.; Henbest, H. B.; Koss, S. *J. Chem. Soc. C* **1968**, 2371.
[242] Abushanab, E.; Reed, D.; Suzuki, F.; Shih, C. J. *Tetrahedron Lett.* **1977**, 3415.
[243] Page, P. C. B.; Heer, J. P.; Bethel, D.; Lund, B. A. *Phosphorus, Sulfur, Silicon, Relat. Elem.* **1999**, *153–154*, 247.
[244] Bethel, D.; Page, P. C. B.; Vahedi, H. *J. Org. Chem.* **2000**, *65*, 6756.
[245] Bohe, L.; Lusinchi, X. *Tetrahedron* **1999**, *55*, 155.
[246] Crabtree, R. H. *Chem. Rev.* **1985**, *85*, 245.
[247] Channa Reddy, C.; Hamilton, G. A.; Madyastha, K. M. *In Biological Oxidation Systems*; Academic Press: San Diego, CA, 1990; Vol. 1, p 534.
[248] Yang, J.; Breslow, R. *Angew. Chem., Int. Ed. Engl.* **2000**, *39*, 2692.
[249] Breslow, R.; Zhang, X.; Huang, Y. *J. Am. Chem. Soc.* **1997**, *119*, 4535.
[250] Fang, Z.; Breslow, R. *Org. Lett.* **2006**, *8*, 251.
[251] Barton, D. H. R.; Ozbalik, N. In *Activation and Functionalization of Alkanes*; Hill, C. L., Ed.; Wiley: New York, 1989; pp 281–301.
[252] Groves, J. T.; Nemo, T. E. *J. Am. Chem. Soc.* **1983**, *103*, 6243.
[253] Rychnovsky, S. D.; Vaidyanathan, R. *J. Org. Chem.* **1999**, *64*, 310.
[254] Bolm, C.; Magnus, A. S.; Hildebrand, P. P. *Org. Lett.* **2000**, *2*, 1173.
[255] Melvin, F.; McNeill, A.; Henderson, P. J. F.; Herbert, R. B. *Tetrahedron Lett.* **1999**, *40*, 1201.

[256] Zhao, M.; Li, J.; Mano, E.; Song, Z.; Tschaen, D. M.; Grabowski, E. J. J.; Reider, P. J. *J. Org. Chem.* **1999**, *64*, 2564.
[257] Song, Z. J.; Zhao, M.; Desmond, R.; Devine, P.; Tschaen, D. M.; Tillyer, R.; Frey, L.; Heid, R.; Xu, F.; Foster, B.; Li, J.; Reamer, R.; Volante, R.; Grabowski, E. J. J.; Dolling, U. H.; Reider, P. J.; Okada, S.; Kato, Y.; Mano, E. *J. Org. Chem.* **1999**, *64*, 9658.
[258] Einhorn, J.; Einhorn, C.; Ratajczak, F.; Pierre, J.-L. *J. Org. Chem.* **1996**, *61*, 7452.
[259] De Mico, A.; Margarita, R.; Parlanti, L.; Vescovi, A.; Piancatelli, G. *J. Org. Chem.* **1997**, *62*, 6974.
[260] Kochkar, H.; Lassalle, L.; Morawietz, M.; Hoelderich, W. F. *J. Catal.* **2000**, *194*, 343.
[261] Herrmann, W. A.; Zoller, J. P.; Fischer, R. W. *J. Organomet. Chem.* **1999**, *579*, 404.
[262] Dijksman, A.; Arends, I. W. C. E.; Sheldon, R. A. *Chem. Commun.* **1999**, 1591.
[263] Betzemeier, B.; Cavazzini, M.; Quici, S.; Knochel, P. *Tetrahedron Lett.* **2000**, *41*, 4343.
[264] Rychnovsky, S. D.; Malernon, T. L.; Rajapakse, H. *J. Org. Chem.* **1996**, *61*, 1194.
[265] Murray, R. W.; Singh, M. *Org. Synth.* **1997**, *74*, 91.
[266] Murray, R. W.; Jeyaraman, R.; Mohan, L. *Tetrahedron Lett.* **1986**, *27*, 2335.
[267] Zabrowski, D. L.; Moorman, A. E.; Beck, K. R., Jr. *Tetrahedron Lett.* **1988**, *29*, 4501.
[268] Moscher, H. S.; Turner, L.; Carlsmith, A. *Org. Synth. Coll. Vol. 4*, **1963**, 828.
[269] Bovicelli, P.; Lupattelli, P.; Sanetti, A.; Mincione, E. *Tetrahedron Lett.* **1995**, *36*, 3031.
[270] Seto, H.; Fujioka, S.; Koshino, H.; Yoshida, S.; Tsubuki, M.; Honda, T. *Tetrahedron* **1999**, *55*, 8341.
[271] Huang, J.; Hsung, R. P. *J. Am. Chem. Soc.* **2005**, *127*, 50.
[272] Crandall, J. K.; Rambo, E. *Tetrahedron* **2002**, *58*, 7027.
[273] Xiong, H.; Hsung, R. P.; Berry, C. R.; Rameshkumar, C. *J. Am. Chem. Soc.* **2001**, *123*, 7174.
[274] Xiong, H.; Huang, J.; Ghosh, S. K.; Hsung, R. P. *J. Am. Chem. Soc.* **2003**, *125*, 12694.
[275] Rameshkumar, C.; Xiong, H.; Tracey, M. R.; Berry, C. R.; Yao, L. J.; Hsung, R. P. *J. Org. Chem.* **2002**, *67*, 1339.
[276] Rameshkumar, C.; Hsung, R. P. *Angew. Chem., Int. Ed. Engl.* **2004**, *43*, 615.
[277] Altmura, A.; Fusco, C.; D'Accolti, L.; Mello, R.; Prencipe, T.; Curci, R. *Tetrahedron Lett.* **1991**, *32*, 5445.
[278] Bovicelli, P.; Mincione, E.; Antonioletti, R.; Bernini, R.; Colombari, M. *Synth. Commun.* **2001**, *31*, 2955.
[279] Adam, W.; Hadjiarapoglou, L. P.; Meffert, A. *Tetrahedron Lett.* **1991**, *32*, 6697.
[280] Chu, H.-W.; Wu, H.-T.; Lee, Y.-J. *Tetrahedron* **2004,** *60*, 2647.
[281] Lovely, C. J.; Du, H.; He, Y.; Dias, H. V. R. *Org. Lett.* **2004**, 6, 735.
[282] Adam, W.; Shimizu, M. *Synthesis* **1994**, 560.
[283] Oishi, S.; Nelson, S. D. *J. Org. Chem.* **1992**, *57*, 2744.
[284] Adam, W.; Balci, M.; Kilic, H. *J. Org. Chem.* **1998**, *63*, 8544.
[285] Mithani, S.; Drew, D. M.; Rydberg, E. H.; Taylor, N. J.; Mooibroek, S.; Dmitrienko, G. I. *J. Am. Chem. Soc.* **1997**, *119*, 1159.
[286] Murray, R. W.; Singh, M.; Rath, N. P. *J. Org. Chem.* **1997**, *62*, 8794.
[287] Murray, R. W.; Singh, M.; Rath, N. P. *J. Org. Chem.* **1996**, *61*, 7660.
[288] Adam, W.; Ahrweiler, M.; Sauter, M.; Schmiedeskamp, B. *Tetrahedron Lett.* **1993**, *34*, 5247.
[289] Colandrea, V. J.; Rajaraman, S.; Jimenez, L. *Org. Lett.* **2003**, *5*, 785.
[290] Hanna, I. *Tetrahedron Lett.* **1999**, *40*, 2521.
[291] Green, M. P.; Pichlmair, S.; Marques, M. M. B.; Martin, H. J.; Diwald, O.; Berger, T.; Mulzer, J. *Org. Lett.* **2004**, *6*, 3131.
[292] Adam, W.; Sauter, M. *Tetrahedron* **1994**, *50*, 8393.
[293] Adam, W.; Sauter, M. *Chem. Ber.* **1993**, *126*, 2697.
[294] Adam, W.; Reinhardt, D. *J. Chem. Soc., Perkin Trans. 2* **1994**, 1503.
[295] Lévai, A.; Koevar, M.; Tóth, G.; Simon, A.; Vraniar, L.; Adam, W. *Eur. J. Org. Chem.* **2002**, 1830.
[296] Schkeryantz, J. M.; Woo, J. C. G.; Siliphaivanh, P.; Depew, K. M.; Danishefsky, S. J. *J. Am. Chem. Soc.* **1999**, *121*, 11964.
[297] Kumaraswamy, S.; Jalisatgi, S. S.; Matzer, A. J.; Miljanic, O. S.; Volhardt, K. P. C. *Angew. Chem., Int. Ed. Engl.* **2004**, *43*, 3711.
[298] Kraehenbuel, K.; Picasso, S.; Vogel, P. *Helv. Chim. Acta* **1998**, *81*, 1439.

[299] Rodríguez, G.; Castedo, L.; Domínguez, D.; Sáa, C.; Adam, W. *J. Org. Chem.* **1999**, *64*, 4830.
[300] Rodríguez, G.; Castedo, L.; Domínguez, D.; Sáa, C.; Adam, W.; Saha-Möller, C. R. *J. Org. Chem.* **1999**, *64*, 877.
[301] Boyer, F.-D.; Es-Safi, N.-E.; Beauhaire, J.; Guerneve, C. L.; Ducrot, P.-H. *Bioorg. Med. Chem. Lett.* **2005**, *15*, 563.
[302] Vazquez, E.; Payack, J. F. *Tetrahedron Lett.* **2004**, *45*, 6549.
[303] Bovicelli, P.; Bernini, R.; Antonioletti, R.; Minicione, E. *Tetrahedron Lett.* **2002**, *43*, 5563.
[304] Sooter, J. A.; Marshall, T. P.; McKay, S. E. *Heterocycl. Commun.* **2003**, *9*, 221.
[305] Closa, M.; de March, P.; Figueredo, M.; Font, J. *Tetrahedron: Asymmetry* **1997**, *8*, 1031.
[306] Winkeljohn, W. R.; Vasquez, P. C.; Strekowski, L.; Baumstark, A. L. *Tetrahedron Lett.* **2004**, *45*, 8295.
[307] Dyker, G.; Hölzer, B. *Tetrahedron* **1999**, *55*, 12557.
[308] Gagnon, J. L.; Zajac Jr., W. W. *Tetrahedron Lett.* **1995**, *36*, 1803.
[309] Murray, R. W.; Singh, M. *J. Org. Chem.* **1990**, *55*, 2954.
[310] Katritzky, A. R.; Maimait, R.; Denisenko, S. N.; Steel, P. J.; Akhmedov, N. G. *J. Org. Chem.* **2001**, *66*, 5585.
[311] Adam, W.; Van Barneveld, C.; Golsch, D. *Tetrahedron* **1996**, *52*, 2377.
[312] Saladino, R.; Neri, V.; Crestini, C.; Tagliatesta, P. *J. Mol. Catalysis A: Chemical* **2004**, *214*, 219.
[313] Detomaso, A.; Curci, R. *Tetrahedron Lett.* **2001**, *42*, 755.
[314] Boyd, D. R.; Davies, R. J. H.; Hamilton, L.; McCullough, J. J.; Porter, H. P. *J. Chem. Soc., Perkin Trans. 1* **1991**, 2189.
[315] Camps, P.; Muñoz-Torrero, D.; Muñoz-Torrero, V. *Tetrahedron Lett.* **1995**, *36*, 1917.
[316] Eaton, P. E.; Wicks, G. E. *J. Org. Chem.* **1998**, *53*, 5353.
[317] Adcock, W.; Trout, N. A. *Magn. Reson. Chem.* **1998**, *36*, 181.
[318] Dave, P. R.; Axenrod, T.; Qi, Le; Bracuti, A. *J. Org. Chem.* **1995**, *60*, 1895.
[319] Busqué, F.; de March, P.; Figueredo, M.; Font, J.; Gallagher, T.; Milán, S. *Tetrahedron: Asymmetry* **2002**, *13*, 437.
[320] Davies, R. J. H.; Stevenson, C.; Kumar, S.; Lyle, J.; Cosby, L.; Malone, J. F.; Boyd, D. R.; Sharma, N. D.; Hunter, A. P.; Stein, B. K. *Chem. Commun.* **2002**, *13*, 1378.
[321] Bonvalet, C.; Bourelle, F.; Scholler, D.; Feigenbaum, A. *J. Chem. Res. (S)* **1991**, 348.
[322] Palmer, B. D.; van Zijl, P.; Denny, W. A.; Wilson, W. R. *J. Med. Chem.* **1995**, *38*, 1229.
[323] Boehlow, T. R.; Harburn, J. J.; Spilling, C. D. *J. Org. Chem.* **2001**, *66*, 3111.
[324] Murray, R. W.; Singh, M. *Magn. Reson. Chem.* **1991**, *29*, 962.
[325] Eaton, P. E.; Xiong, Y.; Gilardi, R. *J. Am. Chem. Soc.* **1993**, *115*, 10195.
[326] Hu, J.; Miller, M. J. *J. Am. Chem. Soc.* **1997**, *119*, 3462.
[327] Hu, J.; Miller, M. J. *J. Org. Chem.* **1994**, *59*, 4858.
[328] Hu, J.; Miller, M. J. *Tetrahedron Lett.* **1995**, *36*, 6379.
[329] Exner, K.; Hochstrate, D.; Keller, M.; Klärner, F.-G.; Prinzbach, H. *Angew. Chem., Int. Ed. Engl.* **1996**, *35*, 2256.
[330] Eles, J.; Kalaus, G.; Lévai, A.; Greiner, I.; Kajtar-Peredy, M.; Szabo, P.; Szabo, L.; Szantay, C. *J. Heterocycl. Chem.* **2002**, *39*, 767.
[331] Lim, H.-J.; Sulikowski, G. A. *Tetrahedron Lett.* **1996**, *37*, 5243.
[332] Golik, J.; Wong, H.; Krishnan, B.; Vyas, D. M.; Doyle, T. W. *Tetrahedron Lett.* **1991**, *32*, 1851.
[333] Templeton, J. F.; Ling, Y.; Zeglam, T. H.; Marat, K.; LaBella, F. S. *J. Chem. Soc., Perkin Trans. 1* **1992**, 2503.
[334] Yu, K.-L.; Ostrowski, J.; Chen, S. *Synth. Commun.* **1995**, *25*, 2819.
[335] Ronald, R.-K.; Manfred, Z.; Burkhard, K. *J. Chem. Soc., Dalton Trans.* **2003,** *1*, 141.
[336] Templeton, J. F.; Ling, Y.; Zeglam, T. H.; LaBella, F. S. *J. Med. Chem.* **1993**, *36*, 42.
[337] Judd, T.C.; Williams, R. M. *J. Org. Chem.* **2004**, *69*, 2825.
[338] Zinurova, E. G.; Kabal'nova, N. N.; Shereshovets, V. V.; Ivanova, E. V.; Shults, E. E.; Tolstikov, G. A.; Yunusov, M. S. *Russ. Chem. Bull.* **2001**, *50*, 720.
[339] Noecker, L.; Duarte, F.; Bolton, S. A.; McMahon, W. G.; Diaz, M. T.; Giuliano, R. M. *J. Org. Chem.* **1999**, *64*, 6275.

[340] Jasys, V. J.; Kelbaugh, P. R.; Nason, D. M.; Phillips, D.; Ronsnack, K. J.; Saccomano, N. A.; Stroh, J. G.; Volkmann, R. A. *J. Am. Chem. Soc.* **1990**, *112*, 6696.

[341] Bisseret, P.; Seeman, M.; Rohmer, M. *Tetrahedron Lett.* **1994**, *35*, 2687.

[342] Ishii, A.; Yamashita, R.; Saito, M.; Nakayama, J. *J. Org. Chem.* **2003**, *68*, 1555.

[343] Ishii, A.; Kawai, T.; Tekura, K.; Oshida, H.; Nakayama, J. *Angew. Chem., Int. Ed. Engl.* **2001**, *40*, 1924.

[344] Rozen, S.; Bareket, Y. *J. Org. Chem.* **1997**, *62*, 1457.

[345] Nagasawa, H.; Sugihara, Y.; Ishii, A.; Nakayama, J. *Phosphorus, Sulfur, Silicon, Relat. Elem.* **1999**, *153–154*, 395.

[346] Nagasawa, H.; Sugihara, Y.; Ishii, A.; Nakayama, J. *Bull. Chem. Soc. Jpn.* **1999**, *72*, 1919.

[347] Marchán, V.; Gibert, M.; Messeguer, A.; Pedroso, E.; Grandas, A. *Synthesis* **1999**, 43.

[348] Asensio, G; Mello, R.; González-Núñez, M. E. *Tetrahedron Lett.* **1996**, *37*, 2299.

[349] Murray, R. W.; Jeyaraman, R.; Pillay, M. K. *J. Org. Chem.* **1987**, *52*, 746.

[350] Derbsey, G.; Harpp, D. N. *J. Org. Chem.* **1995**, *60*, 1044.

[351] Ishii, A.; Kashiura, S.; Oshida, H.; Nakayama, J. *Org. Lett.* **2004**, *6*, 2623.

[352] González-Nuñez, M. E.; Mello, R.; Royo, J.; Rios, J. V.; Asensio, G. *J. Am. Chem. Soc.* **2002**, *124*, 9156.

[353] Boyd, D. R.; Sharma, N. D.; Haughey, S. A.; Malone, J. F.; King, A. W. T.; McMurray, B. T.; Alves-Areias, A.; Allen, C. C. R.; Holt, R.; Dalton, H. *J. Chem. Soc., Perkin Trans. 1* **2001**, *24*, 3288.

[354] Patonay, T.; Adam, W.; Lévai, A.; Kövér, P.; Németh, M.; Peters, E.-M.; Peters, K. *J. Org. Chem.* **2001**, *66*, 2275.

[355] Ishii, A.; Furusawa, K.; Omata, T.; Nakayama, J. *Heteroat. Chem.* **2002**, *13*, 351.

[356] Adam, W.; Hadjiarapoglou, L.; Lévai, A., unpublished results.

[357] Patonay, T.; Adam, W.; Jekö, J.; Kövér, K. E.; Lévai, A.; Németh, M.; Peters, K. *Heterocycles* **1999**, *51*, 85.

[358] Ho, M. T.; Treiber, A.; Dansette, P. M. *Tetrahedron Lett.* **1998**, *39*, 5049.

[359] Ishii, A.; Oshida, H.; Nakayama, J. *Bull. Chem. Soc. Jpn.* **2002**, *75*, 319.

[360] Adam, W.; Golsch, D.; Görth, F. C. *Chem. Eur. J.* **1996**, *2*, 255.

[361] Adam, W.; Haas, W.; Lohray, B. B. *J. Am. Chem. Soc.* **1991**, *113*, 6202.

[362] Nojima, T.; Hirano, Y.; Ishiguro, K.; Sawaki, Y. *J. Org. Chem.* **1997**, *62*, 2387.

[363] Hanaki, H.; Fukatsu, Y.; Harada, M.; Sawaki, Y. *Tetrahedron Lett.* **2004**, *45*, 5791.

[364] Nakayama, J.; Aoki, S.; Takayama, J.; Sakamoto, A.; Sugihara, Y.; Ishii, A. *J. Am. Chem. Soc.* **2004**, *126*, 9085.

[365] Szilagyi, A.; Pelyvas, I. F.; Majercsik, O.; Herczegh, P. *Tetrahedron Lett.* **2004**, *45*, 4307.

[366] Ivanova, N. A.; Shangiraeva, F. G.; Miftakhov, M. S. *Russ. J. Org. Chem.* **2003**, *39*, 1652.

[367] Adam, W.; Hadjiarapoglou, L. *Tetrahedron Lett.* **1992**, *33*, 469.

[368] Adam, W.; Hadjiarapoglou, L.; Peseke, K., unpublished results.

[369] Clennan, E. L.; Yang, K. *J. Org. Chem.* **1993**, *58*, 4504.

[370] Ishii, A.; Nakabayashi, M.; Jin, Y.-N.; Nakayama, J. *J. Organomet. Chem.* **2000**, *611*, 127.

[371] Diguarher, T. L.; Chollet, A.-M.; Bertrand, M.; Hennig, P.; Raimbaud, E.; Sabatini, M.; Guilbaud, N.; Pierre, A.; Tucker, G. C.; Casara, P. *J. Med. Chem.* **2003**, *46*, 3840.

[372] Kiss-Szikszai, A.; Patonay, T.; Jeko, J. *ARKIVOC* **2001**, *3*, 40.

[373] Perales, J. B.; Makino, N. F.; Van Vranken, D. L. *J. Org. Chem.* **2002**, *67*, 6711.

[374] Donnelley, D. M. X.; Fitzpatrick, B. M.; O'Reilly, B. A.; Finet, J.-P. *Tetrahedron* **1993**, *49*, 7967.

[375] Oshida, H.; Ishii, A.; Nakayama, J. *Tetrahedron Lett.* **2002**, *43*, 5033.

[376] Oshida, H.; Ishii, A.; Nakayama, J. *J. Org. Chem.* **2004**, *69*, 1695.

[377] Matloubi Moghaddam, F.; Khakshoor, O. *Molecules* **2001**, *6*, M229.

[378] Lévai, A.; Jeko, J. *ARKIVOC* **2003**, *5*, 19.

[379] Adam, W.; Hadjiarapoglou, L.; Saalfrank, R., unpublished results.

[380] Takayama, J.; Sugihara, Y.; Ishii, A.; Nakayama, J. *Tetrahedron Lett.* **2003**, *44*, 7893.

[381] Chan, T.-H.; Fei, C.-P. *J. Chem. Soc., Chem. Commun.* **1993**, 825.

[382] Whalen, L. J.; McEvoy, K. A.; Halcomb, R. L. *Bioorg. Med. Chem. Lett.* **2003**, *13*, 301.

[383] Mugnaini, C.; Botta, M.; Coletta, M.; Corelli, F.; Focher, F.; Marini, S.; Renzulli, M. L.; Verri, A. *Bioorg. Med. Chem.* **2003**, *11*, 357.
[384] Cohen, S. B.; Halcomb, R. L. *J. Org. Chem.* **2000**, *65*, 6145.
[385] Huang, Q.; DesMarteau, D. D. *Chem. Commun.* **1999**, *17*, 1671.
[386] Camporeale, M.; Fiorani, T.; Troisi, L.; Adam, W.; Curci, R.; Edwards, J. O. *J. Org. Chem.* **1990**, *55*, 93.
[387] Zhdankin, V. V.; Goncharenko, R. N.; Litvinov, D. N.; Koposov, A. Y. *ARKIVOC* **2005**, *4*, 8.
[388] Zhdankin, V. V; D. N.; Koposov.; Smart, J. T. *J. Am. Chem. Soc.* **2001**, *123*, 4095.
[389] Koposov, A. Y.; Litvinov, D. N.; Zhdankin, V. V. *Tetrahedron Lett.* **2004**, *45*, 2719.
[390] Zhdankin, V. V.; Koposov, A. Y.; Netzel, B. C.; Yashin, N. V.; Rempel, B. P.; Ferguson, M. J.; Tykwinski, R. R. *Angew. Chem., Int. Ed. Engl.* **2003**, *42*, 2194.
[391] Gallopo, A. R.; Edwards, J. O. *J. Org. Chem.* **1981**, *46*, 1684.
[392] Webb, K. S.; Seneviratne, V. *Tetrahedron Lett.* **1995**, *36*, 2377.
[393] Priewisch, B.; Ruck-Braun, K. *J. Org. Chem.* **2005**, *70*, 2350.
[394] Brik, M. E. *Tetrahedron Lett.* **1995**, *36*, 5519.
[395] Yang, D.; Yip, Y.-C.; Wang, X.-C. *Tetrahedron Lett.* **1997**, *40*, 7083.
[396] Mizufune, H.; Irie, H.; Katsube, S.; Okada, T.; Mizuno, Y.; Arita, M. *Tetrahedron* **2001**, *57*, 7501.
[397] Volkmann, R. A.; Kelbaugh, P. R.; Nason, D. M.; Jasys, V. J. *J. Org. Chem.* **1992**, *57*, 4352.
[398] Webb, K. S. *Tetrahedron Lett.* **1994**, *35*, 3457.
[399] Adam, W.; Chan, Y.-Y.; Cremer, D.; Gauss, J.; Scheutzow, D.; Schindler, M. *J. Org. Chem.* **1987**, *52*, 2800.
[400] Adam, W.; Haas, W.; Sieker, G. *J. Am. Chem. Soc.* **1984**, *106*, 5020.
[401] Rozwdowska, M. D.; Sulima, A.; Gzella, A. *Tetrahedron: Asymmetry* **2002**, *13*, 2329.
[402] Shu, L.; Wang, P.; Gan, Y.; Shi, Y. *Org. Lett.* **2003**, *5*, 293.
[403] Zhang, W. *Org. Lett.* **2003**, *5*, 1011.
[404] Enders, D.; Signore, G. D. *Heterocycles* **2004**, *64*,101.
[405] Baker, R. W.; Wallace, B. J. *Chem. Commun.* **1999**, 1405.
[406] Finke, P.E.; Meurer, L. C.; Oates, B.; Mills, S. G.; MacCoss, M.; Daugherty, B. L.; Gould, S. L.; DeMartino, J. A.; Siciliano, S. J.; Carella, A.; Carver, G.; Holmes, K.; Danzeisen, R.; Hazuda, D.; Kessler, J.; Lineberger, J.; Miller, M.; Schleif, W. A.; Emini, E. A. *Bioorg. Med. Chem. Lett.* **2001**, *11*, 265.
[407] Zhao, C.-G. Dissertation, University of Würzburg, 1999.
[408] Hamilton, R.; McKervey, M. A.; Rafferty, M. D.; Walker, B. J. *J. Chem. Soc., Chem. Commun.* **1994**, 37.
[409] Davidson, N. E.; Botting, N. P. *J. Chem. Res. (S)* **1997**, 410.
[410] Surowiec, M.; Makosza, M. *Tetrahedron* **2004**, *60*, 5019.
[411] Olah, G. A.; Liao, Q.; Lee, C.-S.; Prakash, G. K. S. *Synlett* **1993**, 427.
[412] Liao, M.; Yao, N.; Wang, J. *Synthesis* **2004**, 2633.
[413] Makosza, M.; Adam, W.; Zhao, C.-G.; Surowiec, M. *J. Org. Chem.* **2001**, *66*, 5022.
[414] Altamura, A.; Curci, R.; Edwards, J. O. *J. Org. Chem.* **1993**, *58*, 7289.
[415] Makosza, M.; Surowiec, M.; Peszewski, M. *ARKIVOC* **2004**, *2*, 172.
[416] Lukin, K. A.; Li, J.; Eaton, P. E.; Kanomata, N.; Hain, J.; Punzalan, E.; Gilardi, R. *J. Am. Chem. Soc.* **1997**, *119*, 9591.
[417] Makosza, M.; Surowiec, M. *Tetrahedron* **2003**, *59*, 6261.
[418] Zhao, Y.; Jiang, N.; Wang, J. *Tetrahedron Lett.* **2003**, *44*, 8339.
[419] Trost, B. M.; Patterson, D. E.; Hembre, E. J. *Chem. Eur. J.* **2001**, *7*, 3768.
[420] Wong, M.-K.; Yu, C.-W.; Yuen, W.-H.; Yang, D. *J. Org. Chem.* **2001**, *66*, 3606.
[421] Bigdeli, M. A.; Nikje, M. M. A.; Heravi, M. M. *Phosphorus, Sulfur Silicon Relat. Elem.* **2002**, *177*, 15.
[422] Petricci, E.; Renzulli, M.; Corelli, F.; Radi, M.; Botta, M. *Tetrahedron Lett.* **2002**, *43*, 9667.
[423] Blot, V.; Reboul, V.; Metzner, P. *J. Org. Chem.* **2004**, *69*, 1196.
[424] Akbalina, Z. F.; Abushakhmina, G. M.; Kabal'nova, N.N.; Zlotskii, S. S.; Shereshovets, V. V. *Russ. J. Gen. Chem.* **2002**, *72*, 1406.
[425] Curci, R.; D'Accolti, L.; Fusco, C. *Tetrahedron Lett.* **2001**, *42*, 7087.

[426] D'Accolti, L.; Fusco, C.; Annese, C.; Rella, M. R.; Turteltaub, J. S.; Williard, P. G.; Curci, R. *J. Org. Chem.* **2004**, *69*, 8510.
[427] Murray, R. W.; Gu, H. *J. Phys. Org. Chem.* **1996**, *9*, 751.
[428] Bovicelli, P.; Truppa, D.; Sanetti, A.; Bernini R.; Lupattelli, P. *Tetrahedron* **1998**, *54*, 14301.
[429] Vanni, R.; Garden, S. J.; Banks, J. T.; Ingold, K.U. *Tetrahedron Lett.* **1995**, *36*, 7999.
[430] Abou-Elzahab, M.; Adam, W.; Saha-Möller, C. R. *Liebigs Ann. Chem.* **1991**, 445.
[431] D'Accolti, L.; Fiorentino, M.; Fusco, C.; Crupi, P.; Curci, R. *Tetrahedron Lett.* **2004**, *45*, 8575.
[432] Angelis, Y. S.; Hatzakis, N. S.; Smonou, I.; Orfanopoulos, M. *Tetrahedron Lett.* **2001**, *42*, 3753.
[433] D'Accolti, L.; Dinoi, A.; Fusco, C.; Russo, A.; Curci, R. *J. Org. Chem.* **2003**, *68*, 7806.
[434] Ballini, R.; Papa, F.; Bovicelli, P. *Tetrahedron Lett.* **1996**, *37*, 3507.
[435] Dehmlow, E. V.; Heiligenstädt, N. *Tetrahedron Lett.* **1996**, *37*, 5363.
[436] Prechtl, F. Diploma Dissertation 1990, University of Würzburg.
[437] de Macedo Puyau, P.; Perie, J. J. *Synth. Commun.* **1998**, *28*, 2679.
[438] Baumstrak, A. L.; Beeson, M.; Vasquez, P. C. *Tetrahedron Lett.* **1989**, *30*, 5567.
[439] González-Nuñez, M. E.; Castellano, G.; Andreu, C.; Royo, J.; Baguena, M.; Mello, R.; Asensio, G. *J. Am. Chem. Soc.* **2001**, *123*, 7487.
[440] Adam, W.; Fröhling, B.; Peters, K.; Weinkötz, S. *J. Am. Chem. Soc.* **1998**, *120*, 8914.
[441] Murray, R. W.; Gu, D. *J. Chem. Soc., Perkin Trans. 2* **1994**, 451.
[442] Kuck, D.; Schuster, A. *Z. Naturforsch.* **1991**, *46b*, 1223.
[443] Fusco, C.; Fiorentino, M.; Dinoi, A.; Curci, R. *J. Org. Chem.* **1996**, *61*, 8681.
[444] Murray, R. W.; Gu, H. *J. Org. Chem.* **1995**, *60*, 5673.
[445] González-Nuñez, M. E.; Royo, J.; Castellano, G.; Andreu, C.; Boix, C.; Mello, R.; Asensio, G. *Org. Lett.* **2000**, *2*, 831.
[446] Lin, H.-C.; Wu, H.-J. *Tetrahedron* **2000**, *56*, 341.
[447] D'Accolti, L.; Kang, P.; Khan, S.; Curci, R.; Foote, C. S. *Tetrahedron Lett.* **2002**, *43*, 4649.
[448] Mezzetti, M.; Mincione, E.; Saladino, R. *Chem. Commun.* **1997**, 1063.
[449] Bovicelli, P.; Sanetti, A.; Lupattelli, P. *Tetrahedron* **1996**, *52*, 10969.
[450] Pramod, K.; Eaton, P. E.; Gilardi, R.; Flippen-Anderson, J. L. *J. Org. Chem.* **1990**, *55*, 6105.
[451] D'Accolti, L.; Fusco, C.; Lucchini, V.; Carpenter, G. B.; Curci, R. *J. Org. Chem.* **2001**, *66*, 9063.
[452] Bernini, R.; Mincione, E.; Sanetti, A.; Bovicelli, P.; Lupattelli, P. *Tetrahedron Lett.* **1997**, *38*, 4651.
[453] Adam, W.; Prechtl, F.; Richter, M. J.; Smerz, A. K. *Tetrahedron Lett.* **1993**, *34*, 8427.
[454] Lee, C.-S.; Audelo, M. Q.; Reibenpies, J.; Sulikowski, G. A. *Tetrahedron* **2002**, *58*, 4403.
[455] Boyer, F.-D.; Descoins, C. L.; Thanh, G. V.; Descoins, C.; Prange, T.; Ducrot, P.-H. *Eur. J. Org. Chem.* **2003**, *7*, 1172.
[456] Mincione, E.; Sanetti, A.; Bernini, R.; Felici, M.; Bovicelli, P. *Tetrahedron Lett.* **1998**, *39*, 8699.
[457] Bovicelli, P.; Lupattelli, P.; Mincione, E.; Prencipe, T.; Curci, R. *J. Org. Chem.* **1992**, *57*, 2182.
[458] Dixon, J. T.; Holzapfel, C. W.; van Heerden, F. R. *Synth. Commun.* **1993**, *23*, 135.
[459] Bovicelli, P.; Lupattelli, P.; Fiorini, V.; Mincione, E. *Tetrahedron Lett.* **1993,** *34*, 6103.
[460] Brown, D. S.; Marples, S. A.; Muxworthy, J. P.; Baggaley, K. H. *J. Chem. Res. (S)* **1992**, 28.
[461] van Heerden, F. R.; Dixon, J. T.; Holzapfel, C. W. *Tetrahedron Lett.* **1992**, *33*, 7399.
[462] Kuck, D.; Schuster, A.; Fusco, C.; Fiorentino, M.; Curci, R. *J. Am. Chem. Soc.* **1994**, *116*, 2375.
[463] Kolb, H. C.; Ley, S. V.; Slawin, A. M. Z.; Williams, D. J. *J. Chem. Soc., Perkin Trans. 1* **1992**, 2735.
[464] Sasaki, T.; Nakamori, R.; Yamaguchi, T.; Kasuga, Y.; Iida, T.; Nambara, T. *Chem. Phys. Lipids* **2001**, *109*, 135.
[465] Iida, T.; Yamaguchi, T.; Nakamori, R.; Hikosaka, M.; Mano, N.; Goto, J.; Nambara, T. *J. Chem. Soc., Perkin Trans. 1* **2001**, *18*, 2229.
[466] Iida, T.; Ogawa, S.; Shiraishi, K.; Kakiyama, G.; Goto, T.; Mano, N.; Goto, J. *ARKIVOC* **2003**, *8*, 170.
[467] Cerrè, C.; Hofmann, A. F.; Schteingart, C. D.; Jia, W.; Maltby, D. *Tetrahedron* **1997**, *53*, 435.
[468] Iida, T.; Hikosaka, M.; Kakiyama, G.; Shiraishi, K.; Schteingart, C. D.; Hagey, L. R.; Ton-Nu, H.-T.; Hofmann, A. F.; Mano, N.; Goto, J.; Nambara, T. *Chem. Pharm. Bull.* **2002**, *50*, 1327.
[469] Curci, R.; Detomaso, A.; Prencipe, T.; Carpenter, G. B. *J. Am. Chem. Soc.* **1994**, *116*, 8112.

[470] Bovicelli, P.; Lupattelli, P.; Fracassi, D.; Mincione, E. *Tetrahedron Lett.* **1994**, *35*, 935.
[471] Lee, J. S.; Fuchs, P. L. *Org. Lett.* **2003**, *5*, 2247.
[472] Bisseret, P.; Rohmer, M. *Tetrahedron Lett.* **1993**, *34*, 1131.
[473] Horiguchi, T.; Cheng, Q.; Oritani, T. *Tetrahedron Lett.* **2000**, *41*, 3907.
[474] Voigt, B.; Porzel, A.; Golsch, D.; Adam, W.; Adam, G. *Tetrahedron* **1996**, *52*, 10653.
[475] Seto, H.; Fujioka, S.; Koshino, H.; Yoshida, S.; Watanabe, T.; Takatsuto, S. *Tetrahedron Lett.* **1998**, *39*, 7525.
[476] Wender, P. A.; Hilinski, M. K.; Mayweg, A. V. W. *Org. Lett.* **2005**, *7*, 79.
[477] Asensio, G.; Castellano, G.; Mello, R.; González-Nuñez, M. E. *J. Org. Chem.* **1996**, *61*, 5564.
[478] Ashford, S. W.; Grega, K. C. *J. Org. Chem.* **2001**, *66*, 1523.
[479] Wong, M.-K.; Chung, N.-W.; He, L.; Yang, D. *J. Am. Chem. Soc.* **2003**, *125*, 158.
[480] Wong, M.-K.; Chung, N.-W.; He, L.; Wang, X.-C.; Yan, Z.; Tang, Y.-C.; Yang, D. *J. Org. Chem.* **2003**, *68*, 6321.
[481] Kumarathasan, R.; Hunter, N. R. *Org. Prep. Proced. Int.* **1991**, *23*, 651.
[482] Marples, B. A.; Muxworthy, J. P.; Baggaley, K. H. *Tetrahedron Lett.* **1991**, *32*, 533.
[483] Grabovskii, S. A.; Kabal'nova, N. N.; Shereshovets, V. V.; Chatgilialoglu, C. *Organometallics* **2002**, *21*, 3506.
[484] Han, Y.-K.; Pearce, E. M.; Kwei, T. K. *Macromolecules* **2000**, *33*, 1321.
[485] Malisch, W.; Hofmann, M.; Nieger, M.; Schöller, W. W.; Sundermann, A. *Eur. J. Inorg. Chem.* **2002**, 3242.
[486] Schenk, W. A.; Frisch, J.; Adam, W.; Prechtl, F. *Inorg. Chem.* **1992**, *31*, 3329.
[487] Lluch, A.-M.; Sánchez-Baeza, F.; Camps, F.; Messeguer, A. *Tetrahedron Lett.* **1991**, *32*, 5629.
[488] Leimweber, D.; Wartchow, R.; Butenschön, H. *Eur. J. Org. Chem.* **1999**, 167.
[489] Hofmann, M.; Malisch, W.; Hupfer, H.; Nieger, M. *Z. Naturforsch.* **2003**, *58*, 36.
[490] Malisch, W.; Vögler, M.; Schumacher, D.; Nieger, M. *Organometallics* **2002**, *21*, 2891.
[491] Chelain, E.; Goumont, R.; Hamon, L.; Palier, A.; Rudler, M.; Rudler, H.; Daran, J.-C.; Vaissermann, J. *J. Am. Chem. Soc.* **1992**, *114*, 8088.
[492] Otto, M.; Boone, B. J.; Arif, A. M.; Gladysz, J. A. *J. Chem. Soc., Dalton Trans.* **2001**, *8*, 1218.
[493] Rat, M.; de Sousa, R. A.; Vaissermann, J.; Leduc, P.; Mansuy, D.; Artaud, I. *J. Inorg. Biochem.* **2001**, *84*, 207.
[494] Randolph, J. T.; McClure, K. F.; Danishefsky, S. J. *J. Am. Chem. Soc.* **1995**, *117*, 5712.
[495] Schenk, W. A.; Dürr, M. *Chem. Eur. J.* **1997**, *3*, 713.
[496] Murray, R. W.; Pillay, M. K. *Tetrahedron Lett.* **1988**, *29*, 15.

Supplemental References for Table 1A. Oxidation of Allenes and Alkynes by Isolated Dioxiranes

[497] Al-Rashid, Z. F.; Hsung, R. P. *Org. Lett.* **2008**, *10*, 661.
[498] Ghosh, P.; Lotesta, S. D.; Williams, L. J. *J. Am. Chem. Soc.* **2007**, *129*, 2438.
[499] Lotesta, S. D.; Kiren, S.; Sauers, R.R.; Williams, L. J. *Angew. Chem. Int. Ed.* **2007**, *46*, 7108.
[500] Ramirez-Galicia, G.; Rubio, M. F. *Theochem* **2005**, *723*, 9.
[501] Zeller, K.-P.; K., Meike; Haiss, P. *Org. Biomol. Chem.* **2005**, *3*, 2310.

Supplemental References for Table 1B. Oxidation of Allenes and Alkynes by In Situ Generated Dioxiranes

[502] Zeller, K.-P.; K., Meike; Haiss, P. *Org. Biomol. Chem.* **2005**, *3*, 2310.

Supplemental References for Table 2A. Oxidation of Arenes and Heteroarenes by Isolated Dioxiranes

[503] Antipin, A. V.; Shishlov, N. M.; Grabovskii, S. A.; Kabal'nova, N. N. *Russ. Chem. Bull.* **2004**, *53*, 800.
[504] Berndt, T; Boege, O. *Phys. Chem. Chem. Phys.* **2001**, *3*, 4946.
[505] Boukouvalas, J.; Wang, J.-X.; Marion, O.; Ndzi, B. *J. Org. Chem.* **2006**, *71*, 6670.
[506] Boukouvalas, J.; Xiao, Y.; Cheng, Y.-X.; Loach, R. P. *Synlett* **2007**, 3198.

[507] Chen, L.-J.; Burka, L. T. *Chem. Res. Toxicol.* **2007**, *20*, 1741.
[508] Chen, L.-J.; DeRose, E. F.; Burka, L. T. *Chem. Res. Toxicol.* **2006**, *19*, 1320.
[509] Davies, R. J. H.; Stevenson, C.; Kumar, S.; Lyle, J.; Cosby, L.; Malone, J. F.; Boyd, D. R.; Sharma, N. D.; Hunter, A. P.; Stein, B. K. *Nucleosides Nucleotides Nucl. Acids* **2003**, *22*, 1355.
[510] Druckova, A.; Marnett, L. J. *Chem. Res. Toxicol.* **2006**, *19*, 1330.
[511] Lu, X.; Yuan, Q.; Zhang, Q. *Org. Lett.* **2003**, *5*, 3527.
[512] Manoharan, M. *J. Org. Chem.* **2000**, *65*, 1093.
[513] Murray, R. W.; Singh, M. *Enantiomer* **2000**, *5*, 245.
[514] Nauduri, D.; Greenberg, A. *Tetrahedron Lett.* **2004**, *45*, 4789.
[515] Rege, P. D.; Tian, Y.; Corey, E. J. *Org. Lett.* **2006**, *8*, 3117.
[516] Snider, B. B.; Zeng, H. *J. Org. Chem.* **2003**, *68*, 545.
[517] Suarez-Castillo, O. R.; Sanchez-Zavala, M.; Melendez-Rodriguez, M.; Castelan-Duarte, L. E.; Morales-Rios, M. S.; Joseph-Nathan, P. *Tetrahedron* **2006**, *62*, 3040.
[518] Treiber, A. *J. Org. Chem.* **2002**, *67*, 7261.
[519] Winkeljohn, W. R.; Leggett-Robinson, P.; Peets, M. R.; Strekowski, L.; Vasquez, P. C.; Baumstark, A. L. *Heterocycl. Commun.* **2007**, *13*, 25.
[520] Xu, Y.-J.; Zhang, Y.-F.; Li, J.-Q. *J. Phys. Chem. B* **2006**, *110*, 6148.
[521] Yang, X. Y.; Haug, C.; Yang, Y. P.; He, Z. S.; Ye, Y. *Chin. Chem. Lett.* **2003**, *14*, 130.
[522] Zhao, C.; Whalen, D. L. *Chem. Res. Toxicol.* **2006**, *19*, 217.

Supplemental References for Table 2B. Oxidation of Arenes and Heteroarenes by In Situ Generated Dioxiranes

[523] Boukouvalas, J.; Xiao, Y.; Cheng, Y.-X.; Loach, R. P. *Synlett* **2007**, 3198.
[524] Ogrin, D.; Chattopadhyay, J.; Sadana, A. K.; Billups, W. E.; Barron, A. R. *J. Am. Chem. Soc.* **2006**, *128*, 11322.
[525] Yang, D.; Wong, M.-K.; Yan, Z. *J. Org. Chem.* **2000**, *65*, 4179.

Supplemental References for Table 3A. Nitrogen Oxidation by Isolated Dioxiranes

[526] Breton, G. W.; Oliver, L. H.; Nickerson, J. E. *J. Org. Chem.* **2007**, *72*, 1412.
[527] Chapman, R. D.; Wilson, W. S.; Fronabarger, J. W.; Merwin, L. H.; Ostrom, G. S. *Thermochimica Acta* **2002**, *384*, 229.
[528] Ivanova, E. V.; Grabovskii, S. A.; Kabal'nikova, N. N.; Shereashovets, V. V. *Zh. Prikladnoi Khim.* **2000**, *73*, 2014.
[529] Judd, T. C.; Williams, R. M. *Angew. Chem. Int. Ed.* **2002**, *41*, 4683.
[530] Koseki, Y.; Katsura, S.; Kusano, S.; Sakata, H.; Sato, H.; Monzene, Y.; Nagasaka, T. *Heterocycles* **2003**, *59*, 527.
[531] McKay, S. E.; Sooter, J. A.; Bodige, S. G.; Blackstock, S. C. *Heterocycl. Commun.* **2001**, *7*, 307.
[532] Merckle, L.; de Andres-Gomez, A.; Dick, B.; Cox, R. J.; Godfrey, C. R. A. *ChemBioChem* **2005**, *6*, 1866.
[533] Prezhdo, V. V.; Bykova, A. S.; Daszkiewicz, Z.; Halas, M.; Iwaszkiewicz-Kostka, I.; Prezhdo, O. V.; Kyziol, J.; Blaszczak, Z. *Russ. J. Gen. Chem.* **2001**, *71*, 907.
[534] Rajca, A.; Vale, M.; Rajca, S. *J. Am. Chem. Soc.* **2008**, *130*, 90999105.
[535] Rege, P. D.; Tian, Y.; Corey, E. J. *Org. Lett.* **2006**, *8*, 3117.
[536] Rella, M. R.; Williard, P. G. *J. Org. Chem.* **2007**, *72*, 525.
[537] Thenmozhiyal, J. C.; Venkatraj, M.; Jeyaraman, R.; Ponnuswamy, S. *Ind. J. Chem. B* **2006**, *45B*, 1887.
[538] Winkeljohn, W. R.; Leggett-Robinson, P.; Peets, M. R.; Strekowski, L.; Vasquez, P. C.; Baumstark, A. L. *Heterocycl. Commun.* **2007**, *13*, 25.

Supplemental References for Table 3B. Sulfur and Selenium Oxidation by Isolated Dioxiranes

[539] Abdrakhmanova, A. R.; Kabal'nova, N. N.; Rol'nik, L. Z.; Yagafarova, G. G.; Shereshovets, V. V. *Bashkirskii Khim. Zh.* **2000**, *7*, 27.
[540] Betancor, C.; Dorta, R. L.; Freire, R.; Prange, T.; Suarez, E. *J. Org. Chem.* **2004**, *69*, 9323.

[541] Birsa, M.; Hopf, H. *Phosphorus Sulfur Silicon* **2005**, *180*, 1453.
[542] Bourles, E.; Alves de Sousa, R.; Galardon, E.; Selkti, M.; Tomas, A.; Artaud, I. *Tetrahedron* **2007**, *63*, 2466.
[543] Cermola, F.; Iesce, M. R. *J. Org. Chem.* **2002**, *67*, 4937.
[544] Gibson, C. L.; La Rosa, S.; Suckling, C. J. *Tetrahedron Lett.* **2003**, *44*, 1267.
[545] Gonzalez-Nunez, M. E.; Mello, R.; Royo, J.; Asensio, G.; Monzo, I.; Tomas, F.; Lopez, J. G.; Ortiz, F. L. *J. Org. Chem.* **2004**, *69*, 9090.
[546] Gonzalez-Nunez, M. E.; Mello, R.; Royo, J.; Rios, J. V.; Asensio, G. *J. Am. Chem. Soc.* **2002**, *124*, 9154.
[547] Guiney, D.; Gibson, C. L.; Suckling, C. J. *Org. Biomol. Chem.* **2003**, *1*, 664.
[548] Hanson, P.; Hendrickx, R. A. A. J.; Smith, J. R. L. *Org. Biomol. Chem.* **2008**, *6*, 745.
[549] Hanson, P.; Hendrickx, R. A. A. J.; Smith, J. R. L. *Org. Biomol. Chem.* **2008**, *6*, 762.
[550] Ishii, A.; Ohishi, M.; Matsumoto, K.; Takayanagi, T. *Org. Lett.* **2006**, *8*, 91.
[551] Ishii, A.; Ohishi, M.; Nakata, N. *Eur. J. Inorg. Chem.* **2007**, 5199.
[552] Ishii, A.; Suzuki, M.; Yamashita, R. *Tetrahedron* **2006**, *62*, 5441.
[553] Ishii, A.; Yamashita, R. *J. Sulfur Chem.* **2008**, *29*, 303.
[554] Kovacs, J.; Tóth, G.; Simon, A.; Lévai, A.; Koch, A.; Kleinpeter, E. *Magn. Reson. Chem.* **2003**, *41*, 193.
[555] Lévai, A. *ARKIVOC* **2003**, (*14*), 14.
[556] Lévai, A.; Patonay, T.; Tóth, G.; Kovacs, J.; Jeko, J. *J. Heterocycl. Chem.* **2002**, *39*, 817.
[557] Nakayama, J.; Nagasawa, H.; Sugihara, Y.; Ishii, A. *Heterocycles* **2000**, *52*, 365.
[558] Nakayama, J.; Yoshida, S.; Sugihara, Y.; Sakamoto, A. *Helvetica Chim. Acta* **2005**, *88*, 1451.
[559] Okuma, K.; Koda, M.; Maekawa, S.; Shioji, K.; Inoue, T.; Kurisaki, T.; Wakita, H.; Yokomori, Y. *Org. Biomol. Chem.* **2006**, *4*, 2745.
[560] Tardif, S. L.; Harpp, D. N. *Sulfur Lett.* **2000**, *23*, 169.

Supplemental References for Table 3C. Phosphorus Oxidation by Isolated Dioxiranes

N/A

Supplemental References for Table 3D. Oxygen Oxidation by Isolated Dioxiranes

[561] Adam, W.; K. W.; Schluecker, S.; Saha-Möller, C.; Kazakov, D. V.; Kazakov, V. P.; Latypova, R. R. *Bioluminesc. Chemiluminesc. : Prog. Curr. Appl.* **2002**, 129.
[562] Kazakov, D. V.; Kazakov, V. P.; Maistrenko, G. Y.; Mal'zev, D. V.; Schmidt, R. *J. Phys. Chem. A* **2007**, *111*, 4267.

Supplemental References for Table 3E. Halogen Oxidation by Isolated Dioxiranes

[563] Adam, W.; K. W.; Schluecker, S.; Saha-Möller, C.; Kazakov, D. V.; Kazakov, V. P.; Latypova, R. R. *Bioluminesc. Chemiluminesc. : Prog. Curr. Appl.* **2002**, 129.
[564] Arnone, A.; Nasini, G.; Panzeri, W.; Vajna de Pava, O.; Zucca, C. *J. Chem. Res.* **2003**, 683.
[565] Koposov, A. Y.; Karimov, R. R.; Geraskin, I. M.; Nemykin, V. N.; Zhdankin, V. V. *J. Org. Chem.* **2006**, *71*, 8452.
[566] Kuhakarn, C.; Kittigowittana, K.; Pohmakotr, M.; Reutrakul, V. *ARKIVOC* **2005**, (*1*), 143.
[567] Ogawa, S.; Hosoi, K.; Ikeda, N.; Makino, M.; Fujimoto, Y.; Iida, T. *Chem. Pharm. Bull.* **2007**, *55*, 247.
[568] Richardson, R. D.; Desaize, M.; Wirth, T. *Chem. Eur. J.* **2007**, *13*, 6745.
[569] Zhdankin, V. V.; Koposov, A. Y.; Litvinov, D. N.; Ferguson, M. J.; McDonald, R.; Luu, T.; Tykwinski, R. R. *J. Org. Chem.* **2005**, *70*, 6484.
[570] Zhdankin, V.; Goncharenko, R. N.; Litvinov, D. N.; Koposov, A. Y. *ARKIVOC* **2005**, (*4*), 8.

Supplemental References for Table 3F. Nitrogen Oxidation by In Situ Generated Dioxiranes

N/A

Supplemental References for Table 3G. Sulfur Oxidation by In Situ Generated Dioxiranes

[571] Colonna, S.; Pironti, V.; Drabowicz, J.; Brebion, F.; Fensterbank, L.; Malacria, M. *Eur. J. Org. Chem.* **2005**, 1727.
[572] Dieva, S. A.; Eliseenkova, R. M.; Efremov, Y. Y.; Sharafutdinova, D. R.; Bredikhin, A. A. *Russ. J. Org. Chem.* **2006**, *42*, 12.
[573] Khiar, N.; Mallouk, S.; Valdivia, V.; Bougrin, K.; Soufiaoui, M.; Fernandez, I. *Org. Lett.* **2007**, *9*, 1255.
[574] Sawwan, N.; Greer, A. *J. Org. Chem.* **2006**, *71*, 5796.
[575] Sisu, I.; Dinca, N.; Sisu, E. *Revista Chim.* **2006**, *57*, 615.

Supplemental References for Table 3H. Oxidation of Other Heteroatoms by In Situ Generated Dioxiranes

[576] Huang, Q.; DesMarteau, D. D. *Inorg. Chem.* **2000**, *39*, 4670.
[577] Kazakov, D. V.; Kazakov, V. P.; Maistrenko, G. Y.; Mal'zev, D. V.; Schmidt, R. *J. Phys. Chem. A* **2007**, *111*, 4267.
[578] Selvararani, S.; Medona, B.; Ramachandran, M. S. *Int. J. Chem. Kinet.* **2005**, *37*, 483.

Supplemental References for Table 4A. C=Y Oxidation by Isolated Dioxiranes

[579] Darkins, P.; Groarke, M.; McKervey, M. A.; Moncrieff, H. M.; McCarthy, N.; Nieuwenhuyzen, M. *J. Chem. Soc., Perkin Trans. 1*, **2000**, 381.
[580] Dodda, R.; Zhao, C.-G. *Org. Lett.* **2006**, *8*, 4911.
[581] Florio, S.; Makosza, M.; Lorusso, P.; Troisi, L. *ARKIVOC* **2006**, (*6*), 59.
[582] Groarke, M.; McKervey, M. A.; Moncrieff, H.; Nieuwenhuyzen, M. *Tetrahedron Lett.* **2000**, *41*, 1279.
[583] Groarke, M.; McKervey, M. A.; Nieuwenhuyzen, M. *Tetrahedron Lett.* **2000**, *41*, 1275.
[584] Makosza, M.; Kamienska-Trela, K.; Paszewski, M.; Bechcicka, M. *Tetrahedron* **2005**, *61*, 11952.
[585] Nakayama, J.; Yoshida, S.; Sugihara, Y.; Sakamoto, A. *Helvetica Chim. Acta* **2005**, *88*, 1451.
[586] Nose, Z.; Kovac, F. *Int. J. Chem. Kinet.* **2007**, *39*, 492.

Supplemental References for Table 4B. C=Y Oxidation by In Situ Generated Dioxiranes

N/A

Supplemental References for Table 5A. C−H Oxidation by Isolated Dioxiranes

[587] Akbalina, Z. F.; Zlotskii, S. S.; Kabal'nova, N. N.; Grigor'ev, I. A.; Kotlyar, S. A.; Shereshovets, V. V. *Russ. J. Appl. Chem.* **2002**, *75*, 1120.
[588] Boyer, F.-D.; Es-Safi, N.-E.; Beauhaire, J.; Le Guerneve, C.; Ducrot, P.-H. *Bioorg. Med. Chem. Lett.* **2005**, *15*, 563.
[589] D'Accolti, L.; Fiorentino, M.; Fusco, C.; Capitelli, F.; Curci, R. *Tetrahedron Lett.* **2007**, *48*, 3575.
[590] D'Accolti, L.; Fusco, C.; Lampignano, G.; Capitelli, F.; Curci, R. *Tetrahedron Lett.* **2008**, *49*, 5614.
[591] Fokin, A. A.; Tkachenko, B. A.; Gunchenko, P. A.; Gusev, D. V.; Schreiner, P. R. *Chem. Eur. J.* **2005**, *11*, 7091.
[592] Fokin, A. A.; Tkachenko, B. A.; Korshunov, O. I.; Gunchenko, P. A.; Schreiner, P. R. *J. Am. Chem. Soc.* **2001**, *123*, 11248.
[593] Freccero, M.; Gandolfi, R.; Sarzi-Amade, M.; Rastelli, A. *J. Org. Chem.* **2003**, *68*, 811.
[594] Freccero, M.; Gandolfi, R.; Sarzi-Amade, M.; Rastelli, A. *Tetrahedron Lett.* **2001**, *42*, 2739.
[595] González-Núñez, M. E.; Royo, J.; Mello, R.; Báguena, M.; Ferrer, J. M.; Ramírez de Arellano, C.; Asensio, G.; Prakash, G. K. S. *J. Org. Chem.* **2005**, *70*, 7919.
[596] Grabovskii, S. A.; Suvorkina, E. S.; Kabal'nova, N. N.; Khursan, S. L.; Shereshovets, V. V. *Russ. Chem. Bull.* **2000**, *49*, 1332.

[597] Grabovskiy, S. A.; Antipin, A. V.; Ivanova, E. V.; Dokichev, V. A.; Tomilov, Y. V.; Kabal'nova, N. N. *Org. Biomol. Chem.* **2007**, *5*, 2302.
[598] Grabovskiy, S. A.; Antipin, A. V.; Kabal'nova, N. N. *Kinet. Catal.* **2004**, *45*, 809.
[599] Grabovskiy, S. A.; Markov, E. A.; Ryzhkov, A. B.; Kabal'nova, N. N. *Russ. Chem. Bull.* **2006**, *55*, 1780.
[600] Grabovskiy, S. A.; Timerghazin, Q. K.; Kabal'nova, N. N. *Russ. Chem. Bull.* **2005**, *54*, 2384.
[601] Grabovsky, S. A.; Markov, E. A.; Maksutov, R. U.; Kabal'nova, N. N. *Bashkirskii Khim. Zh.* **2006**, *13*, 55.
[602] Hayes, C. J.; Sherlock, A. E.; Green, M. P.; Wilson, C.; Blake, A. J.; Selby, M. D.; Prodger, J. C. *J. Org. Chem.* **2008**, *73*, 2041.
[603] Hayes, C. J.; Sherlock, A. E.; Selby, M. D. *Org. Biomol. Chem.* **2006**, *4*, 193.
[604] Horiguchi, T.; Kiyota, H.; Cheng, Q.; Oritani, T. *Tennen Yuki Kagobutsu Toronkai Koen Yoshishu* **2000**, 745.
[605] Iida, T.i; Shiraishi, K.; Ogawa, S.; Goto, T.i; Mano, N. Goto, J.; Nambara, T. *Lipids* **2003**, *38*, 281.
[606] Kakiyama, G.; Iida, T.; Goto, T.; Mano, N.; Goto, J.; Nambara, T. *Chem. Pharm. Bull.* **2004**, *52*, 371.
[607] Kakiyama, G.; Nakamori, R.; Iida, T. *Kenkyu Kiyo - Nihon Daigaku Bunrigakubu Shizen Kagaku Kenkyusho* **2003**, *38*, 283.
[608] Kazakov, D. V.; Barzilova, A. B.; Kazakov, V. P. *Chem. Commun.* **2001**, 191.
[609] Li, L.; Teng, G.-F.; Li, Z.-H. *Gaodeng Xuexiao Huaxue Xuebao* **2007**, *28*, 2179.
[610] Mello, R.; González-Núñez, M. E.; Asensio, G. *Synlett* **2007**, 47.
[611] Mello, R.; Royo, J.; Andreu, C.; Báguena-Ano, M.; Asensio, G.; González-Núñez, M. E. *Eur. J. Org. Chem.* **2008**, 455.
[612] Nagy, M.; Keki, S.; Orosz, L.; Deak, G.; Herczegh, P.; Lévai, Al.; Zsuga, M. *Macromolecules* **2005**, *38*, 4043.
[613] Narumi, Y.; Watanabe, T.; Takatsuto, S. *Nihon Yukagakkaishi* **2000**, *49*, 173.
[614] Ogawa, S.; Hosoi, K.; Ikeda, N.; Makino, M.; Fujimoto, Y.; Iida, T. *Chem. Pharm. Bull.* **2007**, *55*, 247.
[615] Paquette, L. A.; Kreilein, M. M.; Bedore, M. W.; Friedrich, D. *Org. Lett.* **2005**, *7*, 4665.
[616] Rella, M. R.; Williard, P. G. *J. Org. Chem.* **2007**, *72*, 525.
[617] Schroeder, K.; Sander, W. *Eur. J. Org. Chem.* **2005**, 496.
[618] Suarez-Castillo, O. R.; Sanchez-Zavala, M.; Melendez-Rodriguez, M.; Castelan-Duarte, L. E.; Morales-Rios, M. S.; Joseph-Nathan, P. *Tetrahedron* **2006**, *62*, 3040.
[619] Wahl, F.; Weiler, A.; Landenberger, P.; Sackers, E.; Voss, T.; Haas, A.; Lieb, M.; Hunkler, D.; Wörth, J.; Knothe, L.; Prinzbach, H. *Chem. Eur. J.* **2006**, *12*, 6255.
[620] Wen, X.; Norling, H.; Hegedus, L. S. *J. Org. Chem.* **2000**, *65*, 2096.
[621] Wender, P. A.; Hilinski, M. K.; Mayweg, A. V. W. *Org. Lett.* **2005**, *7*, 79.
[622] Zhang, H. B.; Zhang, Y. S.; Zhong, H. M.; Liu, J. P. *Chin. Chem. Lett.* **2005**, *16*, 143.

Supplemental References for Table 5B. Regioselective C–H Oxidation by Isolated Dioxiranes

[623] Baumstark, A. L.; Kovac, Franci; V., Pedro C. *Heterocycl. Commun.* **2002**, *8*, 9.
[624] Boyer, F.-D.; Es-Safi, N.-E.; Beauhaire, J.; Le Guerneve, C.; Ducrot, P.-H. *Bioorg. Med. Chem. Lett.* **2005**, *15*, 563.
[625] D'Accolti, L.; Fiorentino, M.; Fusco, C.; Capitelli, F.; Curci, R. *Tetrahedron Lett.* **2007**, *48*, 3575.
[626] D'Accolti, L.; Fusco, C.; Lampignano, G.; Capitelli, F.; Curci, R. *Tetrahedron Lett.* **2008**, *49*, 5614.
[627] Fokin, A. A.; Tkachenko, B. A.; Gunchenko, P. A.; Gusev, D. V.; Schreiner, P. R. *Chem. Eur. J.* **2005**, *11*, 7091.
[628] Iida, T.i; Shiraishi, K.; Ogawa, S.; Goto, T.i; Mano, N. Goto, J.; Nambara, T. *Lipids* **2003**, *38*, 281.
[629] Kakiyama, G.; Iida, T.; Goto, T.; Mano, N.; Goto, J.; Nambara, T. *Chem. Pharma. Bull.* **2004**, *52*, 371.
[630] Mello, R.; González-Núñez, M. E.; Asensio, G. *Synlett* **2007**, 47.

[631] Rella, M. R.; Williard, P. G. *J. Org. Chem.* **2007**, *72*, 525.
[632] Wahl, F.; Weiler, A.; Landenberger, P.; Sackers, E.; Voss, T.; Haas, A.; Lieb, M.; Hunkler, D.; Wörth, J.; Knothe, L.; Prinzbach, H. *Chem. Eur. J.* **2006**, *12*, 6255.
[633] Wender, P. A.; Hilinski, M. K.; Mayweg, A. V. W. *Org. Lett.* **2005**, *7*, 79.

Supplemental References for Table 5C. C–H Oxidation by In Situ Generated Dioxiranes

[634] Akbalina, Z. F.; Zlotskii, S. S.; Kabal'nova, N. N.; Grigor'ev, I. A.; Kotlyar, S. A.; Shereshovets, V. V. *Russ. J. Appl. Chem.* **2002**, *75*, 1120.
[635] Zeller, K.-P.; K., Meike; Haiss, P. *Org. Biomol. Chem.* **2005**, *3*, 2310.

Supplemental References for Table 5D. Asymmetric C–H Oxidation by In Situ Generated Optically Active Dioxiranes

[636] Jakka, K.; Zhao, C.-G. *Org. Lett.* **2006**, *8*, 3013.

Supplemental References for Table 5E. Si–H Oxidation by Isolated Dioxiranes

[637] Malisch, W.; Hofmann, M.; Kaupp, G.; Kab, H.; Reising, J. *Eur. J. Inorg. Chem.* **2002**, 3235.
[638] Zaborovskiy, A. B.; Lutsyk, D. S.; Prystansky, R. E.; Kopylets, V. I.; Timokhin, V. I.; Chatgilialoglu, C. *J. Organomet. Chem.* **2004**, *689*, 2912.

Supplemental References for Table 6. Oxidation of Organometallics by Isolated Dioxiranes

[639] Alves de Sousa, R.; Galardon, E.; Rat, M.; Giorgi, M.; Artaud, I. *J. Inorg. Biochem.* **2005**, *99*, 690.
[640] Caskey, S. R.; Stewart, M. H.; Kivela, J. E.; Sootsman, J. R.; Johnson, M. J. A.; Kampf, J. W. *J. Am. Chem. Soc.* **2005**, *127*, 16750.
[641] Cubillos, J.; Holderich, W. *Revista Facult. Ingenieria, Univ. Antioquia* **2007**, *41*, 31.
[642] Hoffmann-Roeder, A.; Krause, N. *Synthesis* **2006**, 2143.
[643] Ishii, A.; Kashiura, S.; Hayashi, Y.; Weigand, W. *Chem. Eur. J.* **2007**, *13*, 4326.
[644] Ishii, A.; Ohishi, M.; Nakata, N. *Eur. J. Inorg. Chem.* **2007**, 5199.
[645] Malisch, W.; Hofmann, M.; Kaupp, G.; Kab, H.; Reising, J. *Eur. J. Inorg. Chem.* **2002**, 3235.
[646] Patonay, T.; Jeko, J.; Kiss-Szikszai, A.; Lévai, A. *Monatsh. Chem.* **2004**, *135*, 743.
[647] Wang, P.-D.; Yang, N.-F.; Ling, Y.; Li, J.-C.; Cao, J. *Youji Huaxue* **2007**, *27*, 885.
[648] Zareba, M.; Legiec, M.; Sanecka, B.; Sobczak, J.; Hojniak, M.; Wolowiec, S. *J. Mol. Catal. A: Chem.* **2006**, *248*, 144.

Supplemental References for Table 7. Miscellaneous Oxidations by Dioxiranes

[649] Abou-Yousef, H. *Polpu Chongi Gisul* **2001**, *33*, 25.
[650] Bach, R. D.; Dmitrenko, O. *J. Org. Chem.* **2002**, *67*, 2588.
[651] Call, H.-P.; Call, S. *Pulp Paper Can.* **2005**, *106*, 45.
[652] Dewil, R.; Appels, L.; Baeyens, J.; Degreve, J. *J. Hazard. Mater.* **2007**, *146*, 577.
[653] Economou, A. M. *Appita J.* **2000**, *53*, 312.
[654] Jaaskelainen, A.-S.; Poppius-Levlin, K.; Sundquist, J. *Paperi ja Puu* **2000**, *82*, 257.
[655] Kazakov, D.V.; Maistrenko, G. Y.; Kotchneva, O. A.; Latypova, R. R.; Kazakov, V. P. *Mendeleev Commun.* **2001**, 188.
[656] Kopania, E.; Wandelt, P. *Przemysl Chem.* **2003**, *82*, 1135.
[657] Minisci, F.; Gambarotti, C.; Pierini, M.; Porta, O.; Punta, C.; Recupero, F.; Lucarini, M.; Mugnaini, V. *Tetrahedron Lett.* **2006**, *47*, 1421.
[658] Nikje, M. A.; Mozaffari, Z.; Rfiee, A. *Designed Monomers Polymers* **2007**, *10*, 119.
[659] Pan, G. X.; Thomson, C. I.; Leary, G. J. *Holzforsch.* **2003**, *57*, 282.
[660] Paul, M.; Atluri, S.; Setlow, B.; Setlow, P. *J. Appl. Microbol.* **2006**, *101*, 1161.
[661] Wallace, W. H.; Bushway, K. E.; Miller, S. D.; Delcomyn, C. A.; Renard, J. J.; Henley, M. V. *Environm. Sci. Technol.* **2005**, *39*, 6288.

[662] Wang, H. H.; Hunt, K.; Wearing, J. T. *J. Pulp Paper Sci.* **2000**, *26*, 76.
[663] Wong, M.-K.; Chan, T.-C.; Chan, W.-Y.; Chan, W.-K.; Vrijmoed, L. L. P.; O'Toole, D. K.; Che, C.-M. *Environ. Sci. Technol.* **2006**, *40*, 625.
[664] Xiong, J.; Ye, J.; Wan, W.-J.; Xie, G.-H. *Huanan Ligong Daxue Xuebao, Ziran Kexueban* **2001**, *29*, 7.
[665] Zareba, M.; Legiec, M.; Sanecka, B.; Sobczak, J.; Hojniak, M.; Wolowiec, S. *J. Mol. Catal. A: Chem.* **2006**, *248*, 144.
[666] Zeller, K.-P.; Kowallik, M. Schuler, P. *Eur. J. Org. Chem.* **2005**, 5151.
[667] Zhang, W.; Yu, J.; Wu, L. *Fangzhi Xuebao* **2005**, *26*, 100.

INDEX

Oxidation of Organic Compounds by Dioxiranes, by Waldemar Adam, Cong-Gui Zhao, Chantu R. Saha-Möller, and Kavitha Jakka

ne,